Advanced Lectures in Mathematics
Volume 43

Proceedings of the Seventh International Congress of Chinese Mathematicians

Volume I

Companion to the volume
Proceedings of the Seventh International Congress of Chinese Mathematicians,
Volume II

edited by

Lizhen Ji · Lo Yang · Shing-Tung Yau

International Press
www.intlpress.com

高等教育出版社
HIGHER EDUCATION PRESS

Advanced Lectures in Mathematics, Volume 43

Proceedings of the Seventh International Congress of Chinese Mathematicians, Volume I

Companion to the volume
Proceedings of the Seventh International Congress of Chinese Mathematicians, Volume II

Editors:
Lizhen Ji (Department of Mathematics, University of Michigan)
Lo Yang (Institute of Mathematics, Chinese Academy of Sciences)
Shing-Tung Yau (Department of Mathematics, Harvard University)

ISBN: 978-1-57146-369-2

Printed in the United States of America.

22 21 20 19 9 8 7 6 5 4 3 2 1

Advanced Lectures in Mathematics

Honored guests of opening ceremony

The scene of opening ceremony

Prof. Shing-Tung Yau gave opening speech

Left to right: Prof. Shing-Tung Yau, Prof. Björn Engquist (the awardee of the International Cooperation Award), and Prof. Zhong-Ci Shi

Left to right: Prof. Shing-Tung Yau, Prof. Xi-Ping Zhu (the awardee of Chern Prize), and Prof. Lo Yang

Mr. Ronnie Chan (the awardee of Chern Prize) and Prof. Shing-Tung Yau

Left to right: Prof. Shing-Tung Yau, Prof. Wei Zhang (the awardee of Morningside Gold Medal of Mathematics), and Prof. Chunli Bai (President of CAS)

Prof. Edward Witten and Prof. Shing-Tung Yau

Left to right: Prof. Shing-Tung Yau, Prof. Yuefei Wang, Prof. Chunli Bai (President of CAS), and Prof. Yong Qiu (President of Tsinghua University)

Preface

Every three years, mathematicians of Chinese descent meet for five days to present their results on mathematics and related subjects. Traditionally, the first day of the meeting is organized by the Morningside Foundation, where the prestigious Morningside Awards, the Chern Award, and the International Cooperation Awards are given. In 2016, the International Congress of Chinese mathematicians (ICCM) was organized by the Morningside Center of Mathematics in cooperation with the Academy of Mathematics and Systems Science at Chinese Academy of Sciences. The ceremony was held in the Great Hall of the People in China. Nearly two thousand Chinese mathematicians from all over the world came to Beijing for this ceremony. Many outstanding mathematicians of non-Chinese descent also came and gave spectacular lectures. We are glad that a large number of speakers have been willing to write up their lectures, and we have finally collected these manuscriptsthrough the hard work of Guangqiang Tie, and others. These papers provide good insight into the current status of many subjects in mathematics. We are proud to present them collected here. We should like to thank all participants, and the Morningside Center of Mathematics, for making such an effort to make the ICCM meeting and the publication of its proceedings successful.

Lizhen Ji, Lo Yang and Shing-Tung Yau

Contents

Morningside Lectures

Proceedings of the 7th International
Congress of Chinese Mathematicians
ALM 43, pp. 3–15

Noether's Problem and Descent

Fedor Bogomolov and Yuri Tschinkel

Abstract

We study Noether's problem from the perspective of torsors under linear algebraic groups and descent.

Keywords and Phrases: Rationality, fibrations.

Introduction

This note is inspired by the theory of universal torsors developed by Colliot-Thélène and Sansuc in connection with arithmetic problems on del Pezzo surfaces [CTS87b]. This theory associates to a geometrically rational surface X over a field k, with $X(k) \neq \emptyset$, a torsor

$$\pi : \mathcal{T} \xrightarrow{T} X, \tag{1}$$

where T is the Néron-Severi torus of X, i.e., the character group of $T_{\bar{k}}$ is isomorphic, as a Galois module, to the geometric Néron-Severi lattice $\mathrm{NS}(X_{\bar{k}})$ of X. The torsor is viewed as a *descent* variety. Basic arithmetic problems on X, such as Weak Approximation, are reduced to the following geometric hypothesis: $\mathcal{T}$ is *rational* over k. The gist of this hypothesis is that on $\mathcal{T}$, the arithmetic complexity of X is *untwisted*: while X may have nontrivial (algebraic) Brauer group, it is eliminated via passage to a universal torsor. A positive answer to this would imply in particular that X is unirational over k, which is an open problem for del Pezzo surfaces of degree 1.

In higher dimensions, unirational varieties may have nontrivial transcendental Brauer group, or more generally, nontrivial higher unramified cohomology. Examples are conic bundles over rational surfaces considered by Artin–Mumford [AM72], quadric bundles [CTO89], or Brauer–Severi bundles. Our motivation was to understand whether or not these obstructions to stable rationality can be untwisted via passage to fibrations as in (1).

Definition 1. A variety X over a field k admits a *rational tower* if there exists a sequence of dominant rational maps

$$X_n \xrightarrow{\pi_n} X_{n-1} \to \cdots \to X_1 \xrightarrow{\pi_1} X_0 := X, \tag{2}$$

over k, such that

(1) the *source* of the tower, X_n, is rational over k,

(2) the generic fiber of π_i is geometrically rational and irreducible, over the function field $k(X_{i-1})$, for all i.

We say that X admits a *toric rational tower*, if in addition

(2') the generic fiber of π_i is birational to a torsor under an algebraic torus, over the function field $k(X_{i-1})$, for all i.

Conjecture 2. *Let X be a unirational variety over k. Then X admits a rational tower.*

This conjecture is motivated by topology: every continuous map is homotopy equivalent to a fibration with constant fiber. In the algebraic-geometric context, the conjecture implies in particular that *unramified cohomology* of the source of the tower, X_n, in Definition 1, is trivial. Of course, one can trivialize a given unramified cohomology class (with finite coefficients) via passage to a tower with geometrically rational fibers, as in (2). Indeed, let $K = k(X)$ be the function field of an algebraic variety over a field k, which contains all roots of unity. By the Bloch-Kato conjecture, proved by Voevodsky, the subring

$$\oplus_{i\geq 2} H^i(K) \subset H^*(K)$$

of Galois cohomology of K, with finite coefficients, equals the ideal generated by $H^2(K)$. Since elements of H^2 are trivialized on the corresponding Brauer-Severi variety, we can trivialize an arbitrary finite set of Galois cohomology classes via passage to a tower as in (2):

$$Y_m \to Y_{m-1} \to \cdots \to Y_1 \to Y_0 = X,$$

where the generic fibers are forms of projective spaces. However, a direct application of this construction does not imply Conjecture 2, as we cannot assert the rationality of Y_m. In fact, new unramified classes may appear in the process, and we cannot even ensure that Y_m has trivial unramified cohomology.

To motivate Conjecture 2 from the perspective of algebra, consider the following important class of unirational varieties: quotients V/G, where G is a finite group and V a (finite-dimensional) faithful linear representation of G over k. In [BT17] we showed that, over $k = \bar{\mathbb{F}}_p$, these varieties are *universal* for unramified cohomology: every unramified cohomology class of an algebraic variety over k is induced from V/G^c, where G^c is a central extension of an abelian group (see Section 1 for more details). Thus we expect that such quotients are universal from the birational point of view as well. For groups of this type, and a wide class of solvable groups G, we prove:

- V/G admits a toric rational tower.

Its source X_n is a k-rational variety, untwisting all unramified cohomology of V/G.

Acknowledgments. We are thankful to B. Bloom-Smith for useful comments. This work was partially supported by Laboratory of Mirror Symmetry NRU HSE, RF grant 14.641.31.0001. The first author was funded by the Russian Academic Excellence Project '5-100'. The first author was also supported by a Simons Fellowship and by the EPSRC program grant EP/M024830. The second author was partially supported by the NSF grant 1601912.

1 Background

Let X be a smooth projective geometrically rational variety over a field k. Among obstructions to k-rationality are the absence of k-rational points or the nontriviality of the Brauer group

$$\mathrm{Br}(X)/\mathrm{Br}(k).$$

More generally, we may consider *unramified cohomology*

$$H^i_{nr}(X, \mu_m^{\otimes i-1}), \quad H^i_{nr}(X, \mathbb{Q}/\mathbb{Z}(i-1)), \quad m \in \mathbb{N},$$

(see [Bog87a], [CTO89] for definitions and basic properties). These groups are birational invariants which vanish when $i > \dim(X)$ or when X is k-rational, when k is algebraically closed. For X smooth and projective, we have

$$\mathrm{Br}(X)[m] = H^2_{nr}(X, \mu_m),$$

where μ_m stands for m-th roots of 1. Below, we omit the coefficients, when they are clear from the context.

Conjecture 3. *Let X be a unirational variety over an algebraically closed field of characteristic zero. Then its unramified cohomology is finite.*

Finiteness of $H^i_{nr}(X, \mathbb{Q}/\mathbb{Z}(i-1))$, for unirational X, is known for:

- $i = 2$, classically,
- $i = 3$ [CTK13, Proposition 3.2].

When $X(k) \neq \emptyset$, the theory of Colliot-Thélène–Sansuc provides a fibration

$$\pi : \mathcal{T} \xrightarrow{T} X$$

as in (1), such that

- the generic fiber of π is a principal homogeneous space under an algebraic torus T and
- if $\mathcal{T}(k) \neq \emptyset$ then $\mathcal{T}$ is, *hypothetically*, a rational variety [CTS87b, Hypothesis H1, Section 2.8].

Special cases of this are known, e.g., when X admits a conic bundle over k, with 4 degenerate fibers [CTS87a]. As further evidence for the rationality hypothesis, one has the proof that, over k of characteristic zero, $H^2_{nr}(\mathcal{T})$ is trivial [CTS87b, Theorem 2.1.2]. A more recent consistency check is [Cao16a] and [Cao16b], where it is shown that

- when X is a geometrically rational surface, the nontrivial part of $H^3_{nr}(\mathcal{T})$ is finite and 2-primary.

An example of a geometrically rational variety of dimension 3 where Hypothesis H1 fails is given in [CTS87b, Remarque 2.8.18].

We now turn to unirational varieties which are not necessarily geometrically rational, which may have nontrivial *transcendental* Brauer group. Examples appeared in the context of *Noether's problem*:

Problem 4 (Noether). *Let V be a faithful representation of a finite group G over k. Is V/G (stably) rational?*

Noether's problem has a negative solution: counterexamples are based on explicit computations of unramified cohomology, which is essentially a combinatorial problem, in terms of the structure of Sylow subgroups of G. Groups G with nontrivial $H^2_{nr}(V/G)$ were constructed in [Sal84], [Bog87b], and [Mor12]; with trivial $H^2_{nr}(V/G)$ but nontrivial $H^3_{nr}(V/G)$ in [Pey08] and [HKY16].

On the other hand, unramified cohomology is trivial for actions of many classical finite groups. In some, but not all, of these cases, we have proofs of (stable) rationality (see, e.g., [BK85], [BBGVB11]).

Our goal in the following sections will be to *untwist* unramified cohomology of V/G by constructing rational towers as in Definition 1. Of particular importance are solvable groups G. Indeed, in [BT17] we proved a *universality* result:

Theorem 5. *Let X be an algebraic variety of dimension ≥ 2 over $k = \bar{\mathbb{F}}_p$, $\ell \neq p$ a prime, and*

$$\alpha \in H^i_{nr}(X, \mathbb{Z}/\ell^n)$$

an unramified class. Then there exist finite-dimensional k-vector spaces $V_j, j \in J$, depending on α, such that α is induced, via a rational map, from an unramified class in the cohomology of the quotient of

$$\mathbb{P} := \prod_{j \in J} \mathbb{P}(V_j)$$

by a finite abelian ℓ-group G^a, acting projectively on each factor.

In other words, central extensions of abelian groups capture all unramified cohomology invariants.

2 First properties

Throughout, G is a linear algebraic group over a field k and V a finite-dimensional linear faithful representation of G over k.

Let $\mathrm{Bir}(k)$ be the set algebraic varieties over k, modulo k-birational equivalence, which we denote by $\sim_k$. Let

$$\mathrm{Rat}(k) \subset \mathcal{LQ}(k) \subset \mathcal{GQ}(k) \subset \mathrm{Unirat}(k),$$

be the classes of algebraic varieties over k which are

- k-rational,
- k-birational to V/G (Linear Quotients),
- k-birational to X/G, where X is a k-rational algebraic variety and G is a subgroup of the group $\mathrm{BirAut}(X)$ of k-birational automorphisms of X (General Quotients),
- k-unirational,

respectively. Our goal is to connect these classes, via passage to fibrations with geometrically rational generic fibers as in Definition 1. We start with simple examples:

Example 6. Let G be a finite abelian group, or an extension of a finite cyclic group by a finite abelian group. If k is of characteristic coprime to $n := |G|$ and contains n-th roots of 1 then V/G is rational (see [Swa83, Theorem 6.1]).

This can fail, in general, even when G is cyclic. Over $\mathbb{Q}$, there are counterexamples, e.g., for $G = \mathbb{Z}/8\mathbb{Z}$ [Sal82, Theorem 5.11], or for $G = \mathbb{Z}/47\mathbb{Z}$ [Swa69], [Len74], [Pla17].

Example 7. Let V be a faithful linear representation of a finite group G over k, i.e., $G \hookrightarrow \mathrm{GL}(V)$. Then

$$V/G \sim_k \mathbb{P}(V)/G \times \mathbb{P}^1.$$

Example 8. Let G be a *connected* linear algebraic group and V a faithful linear representation of G over k. Then $Y := V/G$ admits a rational tower. Indeed, the total space of the corresponding (rational) fibration

$$V \to V/G = Y$$

is clearly k-rational, and the generic fiber is geometrically rational.

Example 9. Toric varieties: a universal torsor of a toric variety X_Σ over k is given by

$$\mathcal{T}_\Sigma = \mathbb{A}^n \setminus Z_\Sigma,$$

where Z_Σ is a locally closed subvariety; we have

$$X_\Sigma = \mathcal{T}_\Sigma /\!\!/ T_{NS},$$

where T_{NS} is the Néron-Severi torus of X. Thus X_Σ admits a toric rational tower.

Lemma 10. *Let G be a finite group and $Y = V/G \in \mathcal{LQ}(k)$. Assume that there exists an $X \in \mathrm{Rat}(k)$ with a generically free G-action, such that $X/G \in \mathrm{Rat}(k)$. Then there exists a fibration $Y_1 \to Y$ with geometrically rational generic fiber and k-rational Y_1.*

Proof. We have a fibration

$$(X \times V)/G \to V/G = Y,$$

with generic fiber geometrically isomorphic to X. On the other hand, we have a *vector bundle*

$$(X \times V)/G \to X/G$$

with fiber V, since the G-action on V is linear. Thus $(X \times V)/G$ is k-rational. □

Corollary 11. *Let X be a rational surface over an algebraically closed field k and G a linear algebraic group contained in* $\mathrm{BirAut}(X)$. *Then $Y = V/G$ admits a rational tower.*

Remark 12. Let $C \to \mathbb{P}^1$ be a Galois cover with Galois group G. Then there exists a fibration

$$\pi : (C \times V)/G \to V/G$$

with rational total space. However, the generic fiber of π is not necessarily geometrically rational; it is rational iff the genus $\mathsf{g}(C) = 0$.

3 Central extensions and wreath products

In this section, we investigate the existence of rational towers for $\mathcal{LQ}(k)$, over algebraically closed fields k of characteristic zero. We prove Conjecture 2 for V/G, for special groups G.

It is well-known that for finite abelian groups A and their faithful linear representations V, the quotient V/A is rational, provided the ground field k contains n-th roots of 1, where $n = |A|$. We turn to central extensions of abelian groups: among these, we have the free central extension

$$1 \to Z \to F^c(A) \to A \to 1,$$

with

$$Z := \wedge^2(A),$$

generated by commutators of lifts of elements of A, without nontrivial relations. The extension $F^c(A)$ unique, modulo isoclinism.

Lemma 13. *Consider a central extension of an abelian group*

$$1 \to Z \to G^c \to A \to 1.$$

Then there exists an exact sequence

$$1 \to \tilde{Z} \to \tilde{F}^c \to G^c \to 1, \tag{3}$$

where $\tilde{F}^c$ is isoclinic to $F^c(A)$, and $\tilde{Z}$ is abelian.

Proof. It suffices to add to G^c additional elements which kill the image of stable cohomology $H^2_{st}(A, \mathbb{Z}/\ell)$ in $H^2(\tilde{F}^c, \mathbb{Z}/\ell)$ (see [BPT10, Section 4] and [BT17, Section 2] for definitions and properties). □

Corollary 14. *There exist a faithful representation W of $F^c = F^c(A)$ and a fibration*

$$W/F^c \to V/G^c$$

whose generic fiber is birational to an algebraic torus over the function field of the base.

Proof. Let W' be a faithful representation of $\tilde{F}^c$ and put $W := W' \oplus V$. We have a natural surjective homomorphism $W \to V$ giving rise to a fibration

$$W/\tilde{F}^c \to V/G^c.$$

Its generic fiber is geometrically isomorphic to $W'/\tilde{Z}$, with $\tilde{Z}$ defined in (3). It suffices to recall that linear quotients of abelian groups are rational, when k contains roots of unity. □

Conjecture 15. *Let A be a finite abelian group, $F^c(A)$ its free central extension, and V a faithful linear representation of F^c, over an algebraically closed field k. Then V/F^c is stably rational.*

Note that stable rationality does not change within a fixed isoclinism class, over fields containing roots of unity.

Example 16. Conjecture 15 holds when A is cyclic. When $A \simeq \mathbb{Z}/\ell \oplus \mathbb{Z}/\ell$, $F^c(A)$ is the Heisenberg group. The problem reduces to a *monomial* action, and Conjecture 15 holds as well. Next, consider:

$$A := \mathbb{Z}/\ell \oplus \mathbb{Z}/\ell \oplus \mathbb{Z}/\ell, \qquad \text{for } \ell = 2.$$

Modulo isoclinism, we can represent

$$F^c = F^c(A) \subset Q_{8,1} \times Q_{8,2} \times Q_{8,3},$$

where each $Q_{8,i}$ is the group of quaternions over $\mathbb{Z}/2$: Choose a $\mathbb{Z}/2$-basis $\{e_1, e_2, e_3\}$ of A and a basis π_1, π_2, π_3 of surjective homomorphisms

$$\pi_i : A \to \mathbb{Z}/2 \oplus \mathbb{Z}/2,$$

each trivial on one of the generators e_1, e_2, or e_3. This induces diagrams

$$\begin{array}{ccccccccc} 1 & \longrightarrow & (\mathbb{Z}/2)^3 & \longrightarrow & F^c & \longrightarrow & (\mathbb{Z}/2)^3 & \longrightarrow & 1 \\ & & \downarrow & & \downarrow & & \downarrow & & \\ 1 & \longrightarrow & (\mathbb{Z}/2) & \longrightarrow & Q_{8,i} & \longrightarrow & (\mathbb{Z}/2)^2 & \longrightarrow & 1 \end{array}$$

Let V_i be the standard 2-dimensional representation of $Q_{8,i}$. Then

$$W := V_1 \oplus V_2 \oplus V_3$$

is a faithful representation of F^c. We have a $k^\times$-action on each component that commutes with the action of F^c. It follows that

$$W/F^c \sim_k \big((\mathbb{P}^1 \times \mathbb{P}^1 \times \mathbb{P}^1)/G\big) \times (k^\times)^3,$$

where $G = (\mathbb{Z}/2)^3$ acts monomially on $k(x_1, x_2, x_3)$, as follows:

$$\sigma(x_i) = c_{i,\sigma} x_i^{a_{i,\sigma}}, \quad c_{i,\sigma}, a_{i,\sigma} \in \{\pm 1\}, \quad \forall \sigma \in G. \tag{4}$$

By [Yam12, Theorem 10], Type (3,3,3,1) in the notation therein, the field of invariants $k(x_1, x_2, x_3)^G$ is rational over k, when k is algebraically closed. It follows that W/F^c is stably rational.

Lemma 17. *[BP11, Lemma 2.4] Let G, H be finite groups, acting faithfully on X and Y, respectively. Assume that $X/G, Y/H \in \mathrm{Rat}(k)$. Let $K := H \wr G$ be the wreath product, with its natural action on $W := X^{|H|} \times Y$. Then W/K is rational.*

Proof. Observe that the quotient $X^{|H|}/H^{|G|}$ is a product of rational varieties $(X/G)^{|H|}$. The group H acts on $(X/G)^{|H|}$ by permutations, and this action is equivalent to the linear (free permutation) action of H on $V^{\oplus|H|}$, where $V \sim_k X/G$. Thus the quotient W/K is k-birational to a vector bundle over Y/H, and hence rational. □

Definition 18. A finite solvable group G is called *special*, if there exists a filtration

$$G = G_0 \supsetneq G_1 \supsetneq \cdots \supsetneq G_r = 1$$

such that, for all $i = 0, \dots, r-1$, one has

- G_i is a normal subgroup in G,
- the kernel of the projection

$$p_i : G/G_{i+1} \to G/G_i$$

  is abelian and
- there exists a section $s_i : G/G_i \to G/G_{i+1}$

Proposition 19. *Let G be a special solvable group and V a faithful linear representation of G. Then there exists a toric rational tower*

$$X_n \to \cdots \to X_0 := V/G.$$

Proof. By induction. We assume that the claim holds for $G_{(i)} := G/G_i$. The group $G_{(i+1)}$ is uniquely defined by the action of $G_{(i)}$ on the abelian group $A_i := G_i/G_{i+1}$. Then there exists a surjection

$$\beta_i : B_i := \mathbb{Z}/m_i[G_i]^{r_i} \to A_i, \tag{5}$$

for some r_i and m_i. Its kernel $\mathrm{Ker}(\beta_i)$ is abelian.

This induces a surjection from the wreath product $G_{(i)} \wr (\mathbb{Z}/m_i)^{r_i}$ onto $G_{(i+1)}$. Let W_i be a faithful representation of B_i. Then there is a G-equivariant projection

$$W_i^{|G_i|} \to W,$$

onto some linear representation W of A_i, corresponding to the surjection of modules (5). Let V_i be a faithful representation of $G_{(i)}$ and consider the surjection

$$V_i \oplus W_i^{|G_{(i)}|} \to V_i \oplus W.$$

Its kernel Ker_i is a linear space with a faithful action of the abelian group $\mathrm{Ker}(\beta_i)$. The quotient $\mathrm{Ker}/\mathrm{Ker}(\beta_i)$ is toric, it is the fiber of the projection

$$(V_i \oplus W_i^{|G_{(i)}|})/W_i \to (V_i \oplus W)/G_{(i+1)}. \tag{6}$$

By induction hypothesis and Lemma 17, the left side of (6) admits a tower of toric fibrations with rational source. This implies the claim for $i+1$. □

Remark 20. Our argument is similar to [Sal82, Theorem 3.3] that was focused on the Inverse Galois Problem.

Corollary 21. *Assume that G is a special solvable group. Then there exists a rational G-variety X with $X/G \in \mathrm{Rat}(k)$.*

Proof. By induction, as in the proof of Proposition 19. It suffices to observe that X_{i+1} is a quotient of a vector bundle over X_i by an abelian linear action on the generic fiber. The corresponding quotient is rational, when X_i is rational. □

Thus we obtain many groups with nontrivial (equivalent to linear) embedding into Cremona groups, with rational quotients.

Corollary 22. *Let V be a faithful representation of an ℓ-group G^c, a finite central extension of an abelian ℓ-group A. Then V/G^c admits a tower of toric fibrations with rational source.*

Proof. We apply induction on the ℓ-rank of A. The claim is trivial when rank equals 1. Let $A' := A \oplus \mathbb{Z}/\ell^m$. Let $F^c := F^c(A')$ be the free central extension of A'. We have a surjection

$$F^c(A') \twoheadrightarrow F^a(A),$$

with a canonical section and abelian kernel. Now we apply Proposition 19. □

As mentioned in Section 1, Noether's problem, i.e., the rationality of V/G, has a negative solution. However, one can consider more general, nonlinear, generically free G-actions on rational varieties. A version of Conjecture 2, and of Noether's problem, is the following

Conjecture 23. *Let G be a finite group. Then there exists a k-rational G-variety X with generically free G-action such that X/G is k-rational.*

4 Group theory

In this section we consider finite simple groups G. Our main observation is that the Sylow subgroups $\mathrm{Syl}_\ell(G)$ of most simple groups G satisfy the assumptions of Proposition 19; the corresponding $V/\mathrm{Syl}_\ell(G)$ admit a rational tower. Below we sketch a proof in a special case.

Let $G = \mathrm{PGL}_n(\mathbb{F}_q)$, with $q = p^m$, for $m \in \mathbb{N}$.

1. Consider $\mathrm{Syl}_p(G)$. It is conjugate to a subgroup of upper-triangular matrices $U_n \subset G$. There is a natural projection

$$U_r \to U_{r-1}, \quad \forall r,$$

with a section s_{r-1} and abelian kernel, as in the assumptions of Proposition 19, which induces the corresponding structure on $\mathrm{Syl}_p(G)$.

2. Consider $\mathrm{Syl}_\ell(G)$, with $\ell \neq p$. Every such subgroup is a subgroup of the normalizer $N(T)$ of a (possibly nonsplit) maximal torus $T \subset G$. We have an extension
$$1 \to T \to N(T) \to W_T(G) \to 1.$$
Note that $\mathrm{Syl}_\ell(G)$ admits a projection onto the corresponding $\mathrm{Syl}_\ell(W_T(G))$ and the induced extension, by an abelian ℓ-group, splits (for $\ell \neq 2$). Since Proposition 19 holds for
$$\mathrm{Syl}_\ell(W_T(G)) = \mathrm{Syl}_\ell(\mathfrak{S}_{n-1})$$
(it is an iterated wreath product extension by cyclic ℓ-groups), it also holds for $\mathrm{Syl}_\ell(G)$.

Similar arguments apply to other finite groups of Lie type. Additional considerations are needed for some sporadic simple groups, and small primes.

5 Actions in small dimensions

In this section, we survey related results on rationality of quotients X/G, for low-dimensional rational varieties X over (possibly nonclosed) fields k of characteristic zero.

Let V be a faithful linear representation of a finite group G over k. When G is abelian, V/G is rational; however, $\mathbb{P}(V)/G$ need not be k-rational, even when $\dim(V) = 4$ [AHK00, Example 2.3].

When $\dim(V) \leq 3$, V/G is rational over algebraically closed fields k: indeed, by Example 7, it suffices to consider the unirational surface $\mathbb{P}(V)/G$ which is rational. However, the situation is different over nonclosed fields.

Dimension 2. There is a large literature on rational G-surfaces X over nonclosed fields, i.e., actions of finite groups $G \subset \mathrm{Aut}(X)$ (see [Man67]). We have $X/G \sim_k S/G$, where S is a G-minimal del Pezzo surface or a conic bundle. If S is a conic bundle then $G \subset \mathrm{PGL}_2(\bar{k})$ (see [Tre16a, Theorem 1.3]). We start with several rationality results: let X be a smooth del Pezzo surface with $X(k) \neq \emptyset$ and $G \subset \mathrm{Aut}(X)$.

- If $\deg(X) = K_X^2 \geq 5$ then $X/G \in \mathrm{Rat}(k)$ [Tre14, Corollary 1.4], [Tre16b, Theorem 1.1].
- If $\deg(X) = 3$ and $|G| \neq 3$ then $X/G \in \mathrm{Rat}(k)$ [Tre16b, Theorem 1.3].

Example 24. [Tre16b, Section 5] Let $X \subset \mathbb{P}^3$ be a smooth cubic surface given by
$$f_3(x, y) + zt(ux + vy) + z^3 + wt^3 = 0,$$
where f_3 is a form of degree 3, $(x : y : z : t)$ are coordinates in $\mathbb{P}^3$, and u, v, w are parameters. Assume that the Galois group of f_3 is $\mathbb{Z}/2$. Then X admits an action of $G = \mathbb{Z}/3$, and X/G is k-birational to a nonrational over k, minimal degree 4 del Pezzo surface S, admitting a conic bundle $S \to \mathbb{P}^1$ over k. By [CTS87a], X/G admits a rational tower: a universal torsor is k-rational.

On the other hand, over nonclosed k, the set of k-birational types of quotients of conic bundles over $\mathbb{P}^1$ may be infinite [Tre16a, Theorem 1.8]. E.g., this holds when $G = \mathfrak{A}_5$ and

- not every element of k is a square,
- $\sqrt{-1}, \sqrt{5} \in k$.

Problem 25. *Establish the existence of rational towers for Del Pezzo surfaces over nonclosed fields.*

Dimension 3. Over algebraically closed k, stable rationality of quotients $\mathbb{P}^3/G$ is unknown for

- central $\mathbb{Z}/2$-extensions of the following groups:
$$\tilde{\mathfrak{S}}_5, \tilde{\mathfrak{A}}_6, \tilde{\mathfrak{S}}_6, \tilde{\mathfrak{A}}_7, \mathrm{SL}_2(\mathbb{F}_7),$$
- extensions of $\mathfrak{A}_5$, respectively, $\mathfrak{A}_6$ by a group N of order 64.

For all other subgroups $G \subset \mathrm{GL}_4(k)$, $\mathbb{P}^3/G$ is stably rational [Pro10]. Rationality over nonclosed fields has not been addressed.

References

[AHK00] Hamza Ahmad, Mowaffaq Hajja, and Ming-chang Kang. Rationality of some projective linear actions. *J. Algebra*, 228(2):643–658, 2000.

[AM72] M. Artin and D. Mumford. Some elementary examples of unirational varieties which are not rational. *Proc. London Math. Soc. (3)*, 25:75–95, 1972.

[BBGVB11] Fedor Bogomolov, Christian Böhning, and Hans-Christian Graf Von Bothmer. Rationality of quotients by linear actions of affine groups. *Sci. China Math.*, 54(8):1521–1532, 2011.

[BK85] F. A. Bogomolov and P. I. Katsylo. Rationality of some quotient varieties. *Mat. Sb. (N.S.)*, 126(168)(4):584–589, 1985.

[Bog87a] F. A. Bogomolov. The Brauer group of quotient spaces of linear representations. *Izv. Akad. Nauk SSSR Ser. Mat.*, 51(3):485–516, 688, 1987.

[Bog87b] F. A. Bogomolov. The Brauer group of quotient spaces of linear representations. *Izv. Akad. Nauk SSSR Ser. Mat.*, 51(3):485–516, 688, 1987.

[BP11] Fedor Bogomolov and Tihomir Petrov. Unramified cohomology of alternating groups. *Cent. Eur. J. Math.*, 9(5):936–948, 2011.

[BPT10] Fedor Bogomolov, Tihomir Petrov, and Yuri Tschinkel. Unramified cohomology of finite groups of Lie type. In *Cohomological and geometric approaches to rationality problems*, volume 282 of *Progr. Math.*, pages 55–73. Birkhäuser Boston, Inc., Boston, MA, 2010.

[BT17] Fedor Bogomolov and Yuri Tschinkel. Universal spaces for unramified Galois cohomology. In *Brauer groups and obstruction problems*, volume 320 of *Progr. Math.*, pages 57–86. Birkhäuser/Springer, Cham, 2017.

[Cao16a] Yang Cao. Cohomologie non ramifiée de degré 3 : variétés cellulaires et surfaces de del Pezzo de degré au moins 5, 2016. `1607.01739`.

[Cao16b] Yang Cao. Troisième groupe de cohomologie non ramifié des torseurs universels sur les surfaces rationnelles, 2016. `arXiv:1609.05117`.

[CTK13] Jean-Louis Colliot-Thélène and Bruno Kahn. Cycles de codimension 2 et H^3 non ramifié pour les variétés sur les corps finis. *J. K-Theory*, 11(1):1–53, 2013.

[CTO89] J.-L. Colliot-Thélène and M. Ojanguren. Variétés unirationnelles non rationnelles: au-delà de l'exemple d'Artin et Mumford. *Invent. Math.*, 97(1):141–158, 1989.

[CTS87a] J.-L. Colliot-Thélène and A. N. Skorobogatov. R-equivalence on conic bundles of degree 4. *Duke Math. J.*, 54(2):671–677, 1987.

[CTS87b] Jean-Louis Colliot-Thélène and Jean-Jacques Sansuc. La descente sur les variétés rationnelles. II. *Duke Math. J.*, 54(2):375–492, 1987.

[HKY16] Akinari Hoshi, Ming-chang Kang, and Aiichi Yamasaki. Degree three unramified cohomology groups. *J. Algebra*, 458:120–133, 2016.

[Len74] H. W. Lenstra, Jr. Rational functions invariant under a finite abelian group. *Invent. Math.*, 25:299–325, 1974.

[Man67] Ju. I. Manin. Rational surfaces over perfect fields. II. *Mat. Sb. (N.S.)*, 72 (114):161–192, 1967.

[Mor12] Primož Morávec. Groups of order p^5 and their unramified Brauer groups. *J. Algebra*, 372:420–427, 2012.

[Pey08] Emmanuel Peyre. Unramified cohomology of degree 3 and Noether's problem. *Invent. Math.*, 171(1):191–225, 2008.

[Pla17] Bernat Plans. On Noether's rationality problem for cyclic groups over $\mathbb{Q}$. *Proc. Amer. Math. Soc.*, 145(6):2407–2409, 2017.

[Pro10] Yuri G. Prokhorov. Fields of invariants of finite linear groups. In *Cohomological and geometric approaches to rationality problems*, volume 282 of *Progr. Math.*, pages 245–273. Birkhäuser Boston, Inc., Boston, MA, 2010.

[Sal82] David J. Saltman. Generic Galois extensions and problems in field theory. *Adv. in Math.*, 43(3):250–283, 1982.

[Sal84] David J. Saltman. Noether's problem over an algebraically closed field. *Invent. Math.*, 77(1):71–84, 1984.

[Swa69] Richard G. Swan. Invariant rational functions and a problem of Steenrod. *Invent. Math.*, 7:148–158, 1969.

[Swa83] Richard G. Swan. Noether's problem in Galois theory. In *Emmy Noether in Bryn Mawr (Bryn Mawr, Pa., 1982)*, pages 21–40. Springer, New York-Berlin, 1983.

[Tre14] Andrey S. Trepalin. Rationality of the quotient of $\mathbb{P}^2$ by finite group of automorphisms over arbitrary field of characteristic zero. *Cent. Eur. J. Math.*, 12(2):229–239, 2014.

[Tre16a] Andrey Trepalin. Quotients of conic bundles. *Transform. Groups*, 21(1):275–295, 2016.

[Tre16b] Andrey Trepalin. Quotients of cubic surfaces. *Eur. J. Math.*, 2(1):333–359, 2016.

[Yam12] Aiichi Yamasaki. Negative solutions to three-dimensional monomial Noether problem. *J. Algebra*, 370:46–78, 2012.

Courant Institute of Mathematical Sciences, N.Y.U.
251 Mercer str., New York, NY 10012, USA
and National Research University Higher School of Economics
Russian Federation, AG Laboratory, HSE
7 Vavilova str., Moscow, Russia, 117312
E-mail address: bogomolo@cims.nyu.edu

Courant Institute of Mathematical Sciences, N.Y.U.
251 Mercer str., New York, NY 10012, USA
and Simons Foundation, 160 Fifth Av., New York, NY 10010, USA
E-mail address: tschinkel@cims.nyu.edu

Proceedings of the 7th International
Congress of Chinese Mathematicians
ALM 43, pp. 17–151

Gauge Theory and Integrability, I

Kevin Costello, Edward Witten and Masahito Yamazaki

Abstract

Several years ago, it was proposed that the usual solutions of the Yang-Baxter equation associated to Lie groups can be deduced in a systematic way from four-dimensional gauge theory. In the present paper, we extend this picture, fill in many details, and present the arguments in a concrete and down-to-earth way. Many interesting effects, including the leading nontrivial contributions to the R-matrix, the operator product expansion of line operators, the framing anomaly, and the quantum deformation that leads from $\mathfrak{g}[[z]]$ to the Yangian, are computed explicitly via Feynman diagrams. We explain how rational, trigonometric, and elliptic solutions of the Yang-Baxter equation arise in this framework, along with a generalization that is known as the dynamical Yang-Baxter equation.

Mathematics Subject Classification: 81R12, 70S15.

Keywords and Phrases: Gauge theory, integrable models.

Contents

1 Introduction

Integrable systems of 1 + 1-dimensional many-body physics and two-dimensional statistical mechanics first emerged in Bethe's discovery of the Bethe Ansatz [1] and Onsager's solution of the two-dimensional Ising model [2], respectively. Subsequent study led to remarkable generalizations and new discoveries, continuing to the present day.

Much of the wisdom about integrable models can be distilled into the Yang-Baxter equation [3, 4, 5, 6] and its interpretation via quantum groups [7, 8]. Many of the classic papers on this subject are reprinted in [9]. See also, for example, [10] for an introduction.

In the present paper, we will aim to explain, simplify, and further develop a new approach to the Yang-Baxter equation and related integrable systems of statistical mechanics that was proposed several years ago by one of us [11, 12]. In this approach, the solutions of the Yang-Baxter equation and their properties are deduced from a four-dimensional gauge theory that can be regarded as a T-dual version of three-dimensional Chern-Simons gauge theory. An informal introduction to this approach can be found in [13]. In spirit, this approach is in keeping with a vision that was proposed long ago by Atiyah [14]: the Yang-Baxter equation in two dimensions is deduced by starting with a theory in higher dimensions.

The purpose of the present paper is to further develop this approach, filling in many details and hopefully presenting the arguments in a concrete and down-to-earth way. In section 2, we review the basic facts about integrability and the Yang-Baxter equation that will be needed. In section 3, we introduce the relevant four-dimensional gauge theory. We explain why it leads automatically to solutions of the Yang-Baxter equation and more specifically why possible choices of compactification to two dimensions lead to rational, trigonometric, and elliptic solutions of that equation. We explain why in this theory, one can at the classical level introduce Wilson line operators associated not just to representations of the gauge group G but to representations of an infinite-dimensional algebra $\mathfrak{g}[[z]]$. This generalization turns out to be crucial in understanding the theory. Quantum mechanically, as we learn later, $\mathfrak{g}[[z]]$ will be promoted to the Yangian deformation of $\mathfrak{g}[[z]]$ (or its trigonometric or elliptic generalization). In section 3, we also review some simple examples of solutions of the Yang-Baxter equation and see how one can deduce from these simple examples that there must be a framing anomaly for Wilson line operators.

In sections 4 - 8, we study some increasingly subtle quantum effects in this theory. In section 4, we compute directly the first nontrivial term in the quantum R-matrix. General theorems [7, 8] actually determine the whole structure in terms of this lowest order term together with formal properties of the theory. However, our goal in the present paper is to see everything as explicitly as possible rather than relying on abstract arguments.

In section 5, we study the first nontrivial quantum correction to the operator product expansion (OPE) of Wilson line operators. We show that to get a closed OPE, one has to consider line operators associated to representations of $\mathfrak{g}[[z]]$, not just representations of the underlying finite-dimensional gauge group G. We also explain that this first quantum correction to the classical OPE implies that in yet higher orders, there will have to be further deformations, which will deform $\mathfrak{g}[[z]]$ to the Yangian (or one of its generalizations).

In section 6, we compute the framing anomaly for Wilson line operators in this theory, recovering from a Feynman diagram calculation the result that was predicted on more abstract grounds in section 3. The framing anomaly found here is somewhat analogous to the framing anomaly for Wilson operators in three-dimensional Chern-Simons theory, but its consequences are more far-reaching.

In section 7, we generalize the analysis from Wilson line operators to networks of Wilson lines – graphs that can be drawn in the plane in which the line segments are Wilson line operators and the "vertices" are invariant couplings that describe (for example) the "fusion" of two Wilson line operators to make a single one. In this context, there is a quantum anomaly that generalizes and can largely be deduced from the framing anomaly.

In section 8, we reconsider, following the elementary considerations in section 5, the deformation from $\mathfrak{g}[[z]]$ to the Yangian. At the two-loop level, that is, in order[1] $\hbar^2$, there is a potential anomaly in the coupling of two gauge bosons to a Wilson line operator. To avoid or cancel the anomaly, Wilson lines must be associated (in the rational case) to representations of a quantum deformation of $\mathfrak{g}[[z]]$ known as the Yangian. A surprising consequence of this is that an ordinary Wilson operator associated to a finite-dimensional representation of G may be anomalous and hence absent in the quantum theory. For example, for $G = SO(N)$ (or any simple Lie group other than $SU(N)$), there is no Wilson line operator associated to the adjoint representation.

In sections 9 and 10, we analyze the variants of the construction that lead to trigonometric and elliptic solutions of the Yang-Baxter equation, respectively. In section 11, we explain how a generalization of the Yang-Baxter equation known as the "dynamical Yang-Baxter equation" [21, 22, 23, 24] fits in this framework. In brief, one finds an ordinary Yang-Baxter equation when one expands around a classical gauge theory solution that has no moduli; moduli lead to a dynamical Yang-Baxter equation.

[1]There does not seem to be a standard terminology for counting loops in Feynman diagrams that contain Wilson operators. We refer to a contribution that is of order $\hbar^n$ relative to a leading order contribution as an n-loop effect.

All of our explicit computations in the present paper are in the lowest nontrivial order in $\hbar$ in which some quantum effect occurs. In a companion paper [25], we will explain how to construct the Yangian algebra, and its trigonometric and elliptic generalizations, "exactly" in the present framework, not just in lowest order of perturbation theory. We have put the word "exactly" in quotes because the theory, in the form in which it has been developed so far, is a perturbative theory, so "exactly" really means "to all orders in perturbation theory." It is anticipated that the D4-NS5 system of string theory would provide the framework for a nonperturbative description, along the lines of the study of the D3-NS5 system in [26], but this has not yet been developed.

Finally, we recall the existence of another and superficially quite different relationship between integrable spin systems and four-dimensional gauge theory [27, 28]. A connection between the two approaches is not yet known.

2 Review of Integrability

In this section, we review some standard facts about integrable models, aiming just to explain what is needed for the purposes of this paper. The goal of the rest of the paper will be to explain these facts from the standpoint of four-dimensional gauge theory.

2.1 The Yang-Baxter Equation

We consider a system of particles whose internal quantum numbers take values in some vector space V. It is often convenient to pick a basis $\{e_i\}_{i=1}^{\dim V}$ of V. A particle is also characterized by a complex parameter z that in the context of integrable systems is known as the spectral parameter. It will play a crucial role in what follows. The particles live in a two-dimensional spacetime and travel on (possibly curved) one-dimensional worldlines in this spacetime. When two worldlines cross (Fig. 1), their spectral parameters are assumed to be unchanged, but their internal state is transformed by a matrix that in general depends on the spectral parameters.

We write this matrix as $R(z_1, z_2) : V \otimes V \to V \otimes V$, or in more detail, in the chosen basis, as $R_{ij}^{kl}(z_1, z_2)$. However, although there are interesting R-matrices that lack this property,[2] we will be concerned in this paper with the case that the R-matrix depends only on the difference $z = z_1 - z_2$ of the two spectral parameters. (In some applications of the Yang-Baxter equation, the spectral parameter is interpreted as a particle momentum or rapidity, and the fact that the R-matrix depends only on the difference of spectral parameters is interpreted as a consequence of Galilean invariance or Lorentz invariance.)

The Yang-Baxter equation says that when three worldlines cross in a pairwise fashion, the arrangement in which they cross does not matter (Fig. 2). We denote

[2]The basic example is the chiral Potts model [15, 16], in which the spectral parameter takes values in a curve of genus greater than 1. Interestingly this model arises as a root-of-unity degeneration of the model of [17], which in turn arises from supersymmetric indices of four-dimensional $\mathcal{N} = 1$ quiver gauge theories [18, 19, 20].

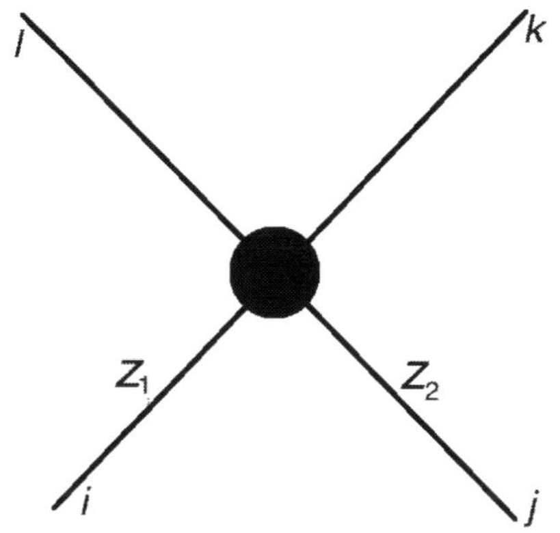

Figure 1: Crossing of two worldlines in a two-dimensional spacetime. The "blob" indicates a scattering process the amplitude for which will (in the context of the present paper) ultimately be computed via gauge theory. When two particles cross, their spectral parameters z_1 and z_2 are unchanged but their "internal" state is transformed.

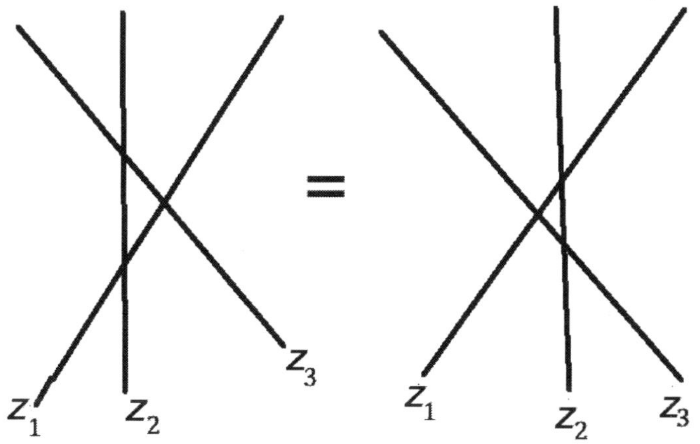

Figure 2: The Yang-Baxter equation asserts the equivalence between these two pictures.

the three particles as a, b, c, and write, for example, V_a for the vector space of internal states of particle a, z_a for its spectral parameter, and $R_{ab}(z_a - z_b) : V_a \otimes V_b \to V_a \otimes V_b$ for the corresponding R-matrix.[3] Then the Yang-Baxter equation reads[4]

$$R_{12}(z_1-z_2)R_{13}(z_1-z_3)R_{23}(z_2-z_3) = R_{23}(z_2-z_3)R_{13}(z_1-z_3)R_{12}(z_1-z_2) \ . \tag{2.1}$$

In terms of the basis $\{e_i\}$ of V, the equation takes the imposing form

$$\begin{aligned} &\sum_{o,p,q} R_{12}(z_1 - z_2)^{nm}_{qo} R_{13}(z_1 - z_3)^{ql}_{ip} R_{23}(z_2 - z_3)^{op}_{jk} \\ &= \sum_{r,s,t} R_{23}(z_2 - z_3)^{ml}_{rt} R_{13}(z_1 - z_3)^{nt}_{sk} R_{12}(z_1 - z_2)^{sr}_{ij} \ , \end{aligned} \tag{2.2}$$

where the meaning of the indices is more clear in a picture (Fig. 3).

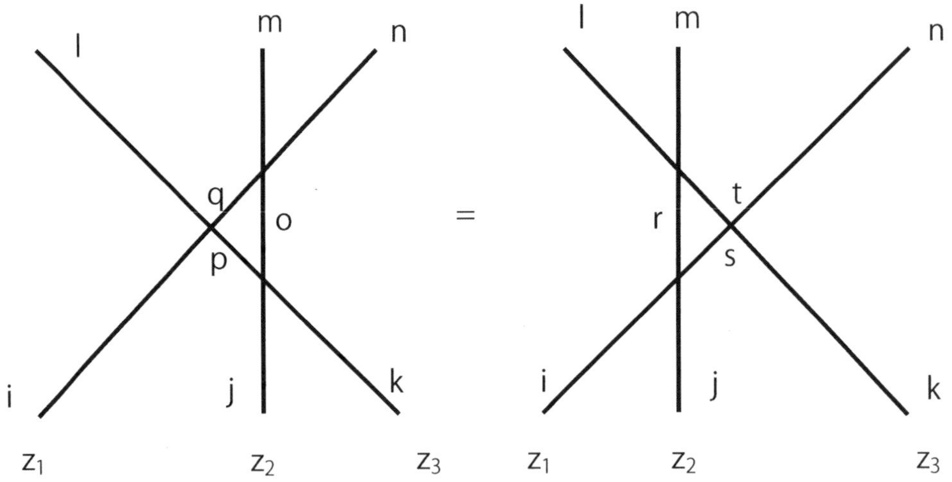

Figure 3: A graphical representation of the Yang-Baxter equation. On the left, one sums over labels p, q, o, and on the right one sums over r, s, t. An appropriate R-matrix element is attached to each vertex.

The Yang-Baxter equation can be naturally supplemented with a condition sometimes called "unitarity," which asserts that a picture in which two worldlines cross and then cross back is equivalent to one in which they do not cross at all (Fig. 4). In formulas, the relation is

$$R_{21}(z_2 - z_1)R_{12}(z_1 - z_2) = 1 \ . \tag{2.3}$$

[3]We also denote $R_{ab} \otimes 1 : V_a \otimes V_b \otimes V_c \to V_a \otimes V_b \otimes V_c$ simply as R_{ab}.

[4]The general form of this equation without assuming that the spectral parameter depends only on the difference of rapidities is simply

$$R_{12}(z_1, z_2)R_{13}(z_1, z_3)R_{23}(z_2, z_3) = R_{23}(z_2, z_3)R_{13}(z_1, z_3)R_{12}(z_1, z_2) \ .$$

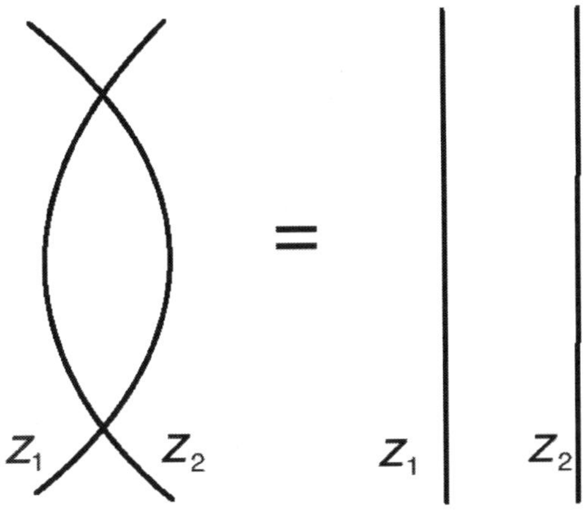

Figure 4: In the context of the Yang-Baxter equation, "unitarity" is the equivalence of these two pictures.

All solutions of the Yang-Baxter equation studied in this paper satisfy such a unitarity condition. There is also a crossing relation, which we will come to in sections 3.5 and 3.6.

The Yang-Baxter equation is highly over-constrained, especially if the dimension of V is large: the R-matrix has $\mathcal{O}((\dim V)^4)$ coefficients, while the Yang-Baxter equation has $\mathcal{O}((\dim V)^6)$ components. Moreover, the presence of the spectral parameter further constrains the possible solutions to the Yang-Baxter equation. Nevertheless, the Yang-Baxter equations does have solutions and these lead to a remarkably rich theory.

2.2 Quasi-Classical R-matrix

While no complete classification is known of the general solution of the Yang-Baxter equation, there are more complete results for the case of a so-called quasi-classical R-matrix,[5] a concept that we now explain.

A quasi-classical R-matrix is a solution $R_\hbar(z)$ of the Yang-Baxter equation that depends on another continuous parameter $\hbar$ as well as on the spectral parameter z, and that is holomorphic near $\hbar = 0$ with $R_0(z) = 1$. Thus $R_\hbar(z)$ has an expansion near $\hbar = 0$ that begins

$$R_\hbar(z) = I + \hbar\, r(z) + \mathcal{O}(\hbar^2)\,. \tag{2.4}$$

Here $r(z)$ is called the classical r-matrix.

By considering the $\mathcal{O}(\hbar^2)$ term of the Yang-Baxter equation (2.1), one learns that the classical r-matrix obeys an equation that is known as the classical Yang-

[5] There are known solutions of Yang-Baxter equations which are not quasi-classical. The chiral Potts model [15, 16] is again a basic example.

Baxter equation:

$$[r_{12}(z_1 - z_2), r_{13}(z_1 - z_3) + r_{23}(z_2 - z_3)] + [r_{13}(z_1 - z_3), r_{23}(z_2 - z_3)] = 0 \,. \quad (2.5)$$

Note that this equation is quadratic in the classical r-matrix r, whereas the original Yang-Baxter equation was cubic in the R-matrix R.

Now, Belavin and Drinfeld [29] classified solutions of the classical Yang-Baxter equation (2.5), modulo trivial equivalences,[6] under certain assumptions. The assumptions were motivated by the examples which were known at that time. The solution is assumed to be associated to the Lie algebra $\mathfrak{g}$ of a semi-simple[7] Lie group G. No reality condition will be important in the present paper, so we consider G to be a complex semi-simple Lie group with complex Lie algebra $\mathfrak{g}$. The classical r-matrix r is assumed to be an element of $\mathfrak{g} \otimes \mathfrak{g}$:

$$r(z) = \sum_{a,b} r_{ab}(z)(t^a \otimes t^b) \,, \quad (2.6)$$

where t^a are a basis of $\mathfrak{g}$. The classical r-matrix is assumed to be non-degenerate, namely $\det_{a,b}(r_{ab}(z)) \not\equiv 0$.

Then the result shows that the poles of $r(z)$ in the complex plane spans a lattice, which is either of rank 0, 1 or 2. Solutions of the classical Yang-Baxter equation of any of the three types are almost uniquely determined[8] by the choice of $\mathfrak{g}$ and the rank of the lattice. The solutions for rank 0, 1, and 2 can be written explicitly in terms of rational, trigonometric, and elliptic functions and are known as rational, trigonometric, and elliptic solutions of the Yang-Baxter equation.

In the language of representation theory, rational, trigonometric and elliptic solutions have their algebraic counterparts, namely the Yangian $Y_\hbar(\mathfrak{g})$, the quantum affine algebra $U_{q,\hbar}(\mathfrak{g})$ and the elliptic algebra $E_{q,\tau,\hbar}(\mathfrak{g})$. Solutions of the full (rather than classical) Yang-Baxter equation depend on the choice of a representation of one of these algebras and the R-matrix $R_\hbar(z) \in \mathrm{End}(V \otimes V)$ is then an intertwiner for the tensor products of these representations. In the algebraic approach, the spectral parameter enters as part of the data needed to specify a representation.

Elliptic solutions of the classical Yang-Baxter equation exist only for $\mathfrak{g} = \mathfrak{sl}_N$, whereas trigonometric and rational solutions exist for any semisimple $\mathfrak{g}$. Rational

[6]The classical Yang-Baxter equation is invariant under conjugation and under adding to r a multiple of the identity. The latter possibility reflects the fact that the Yang-Baxter equation is invariant under multiplying $R(z)$ by a function of z.

[7]In gauge theory, it is natural to consider the somewhat larger class of Lie groups consisting of those whose Lie algebra admits an invariant, non degenerate bilinear form. An important example is a reductive group, which is locally the product of a semi-simple group and a torus (an abelian group). We will find at least two reasons to consider reductive groups in this paper. One reason is that the simplest example for some purposes is actually the case $G = GL_N(\mathbb{C})$, which is reductive but not semi-simple. Another reason is that in the framework we will follow in this paper, trigonometric solutions of the Yang-Baxter equation are most easily understood starting with a gauge group that is reductive but not semi-simple.

[8]In the elliptic case, there is a discrete choice to be made. For $\mathfrak{g} = \mathfrak{sl}_N$, one is free to pick a generator of the finite group $\mathbb{Z}_n$. There also are some subtleties in the trigonometric case, involving the possibility of an "external field." We explain in sections 9 and 10 what these issues mean from the point of view of four-dimensional gauge theory.

solutions of the classical Yang-Baxter equation have G as a group of symmetries, while trigonometric solutions admit only the maximal torus of G as a symmetry group and elliptic solutions for $\mathfrak{sl}_N$ have only a finite group $\mathbb{Z}_N \times \mathbb{Z}_N$ of automorphisms.

In this paper, we will see how quasi-classical R-matrices with these properties can emerge from four-dimensional gauge theory.

3 Four-Dimensional Gauge Theory

3.1 The Starting Point

The four-dimensional gauge theory that is relevant to our subject [11, 12] may be described as follows.

The theory in question is only defined on a four-manifold with some additional structure. We start with the basic case, which is a product 4-manifold $\mathbb{R}^2 \times \mathbb{C}$ with real coordinates[9] x, y on $\mathbb{R}^2$ and a holomorphic coordinate z on $\mathbb{C}$.

The fundamental field of our theory is a 3-component partial connection

$$A = A_x \mathrm{d}x + A_y \mathrm{d}y + A_{\bar{z}} \mathrm{d}\bar{z} \; , \tag{3.1}$$

where we did not include the component $A_z \mathrm{d}z$ that is of type $(1,0)$ along $\mathbb{C}$. The fields A_x, A_y, and $A_{\bar{z}}$ all depend nontrivially on z and $\bar{z}$ as well as x or y; that is, they are not constrained to vary holomorphically or antiholomorphically on $\mathbb{C}$. Since A_z is missing, it would not be possible to place a reasonable reality condition on this space of fields. Instead we take the gauge group to be a complex Lie group G with complex Lie algebra $\mathfrak{g}$, and view A_x, A_y, and $A_{\bar{z}}$ as independent complex fields. The construction will make use of an invariant and nondegenerate bilinear form on $\mathfrak{g}$, which we will denote as Tr. The notation is motivated by the fact that if $\mathfrak{g}$ is semisimple or more generally if it is reductive (the direct sum of a semi simple Lie algebra with an abelian one), then an invariant quadratic form can be defined as the trace in a suitable representation. However, our discussion in this paper applies whether Tr actually has this interpretation or not. For a simple summand of $\mathfrak{g}$, the Killing form of $\mathfrak{g}$ gives an invariant nondegenerate quadratic form. We choose an orthonormal basis t_a of $\mathfrak{g}$ with respect to this Killing form and normalize Tr by

$$\mathrm{Tr}(t_a t_b) = \delta_{ab} \; . \tag{3.2}$$

The action of our theory is given by

$$S = \frac{1}{2\pi} \int_{\mathbb{R}^2 \times \mathbb{C}} \mathrm{d}z \wedge \mathrm{CS}(A) \; , \tag{3.3}$$

where $\mathrm{CS}(A)$ is the Chern-Simons three-form

$$\mathrm{CS}(A) := \mathrm{Tr}\left(A \wedge \mathrm{d}A + \frac{2}{3} A \wedge A \wedge A \right) = \varepsilon^{ijk} \mathrm{Tr}\left(A_i \partial_j A_k + \frac{2}{3} A_i A_j A_k \right) \; . \tag{3.4}$$

[9] When convenient, we will denote x and y as x^1 and x^2.

Here and afterwards the indices $i, j, \ldots$ run over x, y and $\bar{z}$ (ε is a totally antisymmetric tensor with $\varepsilon^{xy\bar{z}} = 1$). The VEV (vacuum expectation value) of an observable $\mathcal{O}$ is given by the path-integral

$$\langle \mathcal{O} \rangle = \frac{\displaystyle\int \mathcal{D}A \, \mathcal{O} \exp\left(\frac{\mathrm{i}S}{\hbar}\right)}{\displaystyle\int \mathcal{D}A \, \exp\left(\frac{\mathrm{i}S}{\hbar}\right)} . \tag{3.5}$$

The action S is obviously not invariant under four-dimensional diffeomorphisms, because the use of the 1-form $\mathrm{d}z$ spoils the four-dimensional symmetry. Nor does it have the three-dimensional diffeomorphism symmetry of three-dimensional Chern-Simons theory; this is the symmetry that enables one to define quantum invariants of knots. But we still have two-dimensional diffeomorphism symmetry – invariance under orientation-preserving diffeomorphisms of $\mathbb{R}^2$ (or of its generalization Σ that will be introduced later). This will ultimately lead to the Yang-Baxter equation and the unitarity relation.

We understand the action S as a holomorphic function of complex variables A_x, A_y, $A_{\bar{z}}$, and this implies that the construction that we will be describing is somewhat formal. There is no difficulty in formally carrying out perturbation theory in such a holomorphic theory. That approach was taken in [11, 12] and it is the approach that we will follow here. (We expect that a nonperturbative definition of the theory can be given by considering the D4-NS5 system of string theory, along the lines of the study of the D3-NS5 system in [26], but we will not pursue this in the present paper.) The parameter $\hbar$ that appears in the action is, at the quantum level, the loop-counting parameter. In the semi-classical limit $\hbar \to 0$, this parameter will be identified with the parameter of the same name that appears in the quasi-classical R-matrix (2.4). The parameter $\hbar$ has dimensions of length, in the sense that for $C = \mathbb{C}$, the theory is invariant under a common rescaling of z and $\hbar$. The factor of $1/(2\pi)$ in the action is included here to match with the literature on integrable models.

A reflection of the fact that the construction is formal and leads (in the form we present here) only to a perturbative theory is the following. There is no quantization condition for $\hbar$ that will ensure that the action is gauge-invariant mod $2\pi\mathbb{Z}$. This contrasts with three-dimensional Chern-Simons theory, which is defined with such a condition.

The action is invariant, modulo surface terms that are irrelevant in perturbation theory, under gauge transformations acting in the usual way.

$$A_i \mapsto g^{-1} A_i g + g^{-1} \partial_i g \, , \quad (i = x, y, \bar{z}) \, . \tag{3.6}$$

This is true because the Chern-Simons three-form is gauge-invariant modulo an exact form. Alternatively, we can integrate by parts to put the action in a manifestly gauge-invariant form, after discarding surface terms that are irrelevant in perturbation theory:

$$S = -\frac{1}{2\pi} \int_{\mathbb{R}^2 \times \mathbb{C}} z \, \mathrm{Tr} \, F \wedge F \, . \tag{3.7}$$

This is the standard topological term of the Yang-Mills theory, where the θ-angle now depends linearly on z.

Some readers might be more comfortable starting with a standard 4-component connection

$$A = A_x \mathrm{d}x + A_y \mathrm{d}y + A_{\bar{z}} \mathrm{d}\bar{z} + A_z \mathrm{d}z \ , \tag{3.8}$$

with gauge transformations acting in the usual way on all four components, and again with the action (3.3). In this case, one finds that due to the presence of the differential form $\mathrm{d}z$ in the action (3.3), the A_z component drops out from the action, and hence we have an extra gauge symmetry

$$A \mapsto A + \chi \, \mathrm{d}z \ . \tag{3.9}$$

We can then fix this extra gauge symmetry by choosing a gauge $A_z = 0$. The 4-component gauge transformation for the 4-component gauge field (3.9) is not consistent with this gauge since it will in general generate a non-trivial A_z component. However, a combination of the 4-component gauge transformation, with the extra gauge symmetry (3.9) with $\chi = -A_z$, remains as a residual gauge symmetry. This is the 3-component gauge transformation (3.6).

In the following, we will always choose $A_z = 0$, so that A is the 3-component connection and the only remaining gauge symmetry is the conventional gauge transformation (3.6).

A possibly more familiar theory that is defined in a similar way with a partial connection is holomorphic Chern-Simons theory. This theory is defined on a Calabi-Yau threefold X with holomorphic 3-form Ω. The dynamical variable is a $(0,1)$ connection $A = \sum_{i=1}^{3} A_i \mathrm{d}\bar{z}^i$ and the action is the integral of the Chern-Simons $(0,3)$-form, wedged with Ω:

$$S = \int_X \Omega \wedge \mathrm{CS}(A^{(0,1)}) \ . \tag{3.10}$$

The definition of this action depends only on the complex structure and holomorphic volume form Ω of the 3-fold X.

In this light, the four-dimensional theory of (3.3) is intermediate between ordinary Chern-Simons theory in three dimensions and holomorphic Chern-Simons theory on a Calabi-Yau threefold. These theories arise as effective theories of branes in the topological A-model and B-model respectively [30] and are related by mirror symmetry. The four-dimensional theory that we will be studying here is intermediate between the two cases and on an appropriate four-manifold can be related by T-duality – mirror symmetry in some but not all dimensions of spacetime – to either one of them.

The classical equations of motion of the theory read

$$F_{xy} = 0 \ , \quad F_{x\bar{z}} = F_{y\bar{z}} = 0 \ . \tag{3.11}$$

This means that the gauge field defines a flat bundle on $\mathbb{R}^2$, which then varies holomorphically as we move along $\mathbb{C}$.

The equations (3.11) imply that all local gauge-invariant quantities that can be constructed from the field A actually vanish. This is the reason that the theory works at the quantum level. Because the loop-counting parameter $\hbar$ has dimensions of length or inverse mass, the theory is unrenormalizable by power-counting. But this does not cause difficulty because all conceivable counterterms actually vanish by the equations of motion. The theory thus can be quantized in perturbation theory [11, 12]. However, it is affected by framing anomalies somewhat similar to those of three-dimensional Chern-Simons theory, but more subtle.

The fact that the theory is unrenormalizable by power counting actually leads to a very important simplification. After gauge-fixing, when one concretely constructs the theory in perturbation theory, it is infrared-free. The fact that the theory is infrared-free makes it straightforward, once one introduces Wilson line operators, to deduce a local procedure to compute their expectation values. From this local procedure, one then can immediately recover the Yang-Baxter equation of an integrable system. (This will be explained in detail in section 3.4.) By contrast, three-dimensional Chern-Simons theory is renormalizable by power counting and does not lead as directly to a local picture.

3.2 Generalization

We comment next on replacing $\mathbb{R}^2 \times \mathbb{C}$ by a more general 4-manifold.

In this paper, we will exclusively study the special case that the 4-manifold is a product of two Riemann surfaces,

$$M = \Sigma \times C\ , \tag{3.12}$$

where Σ is a smooth oriented 2-manifold, and C is a complex manifold endowed with a holomorphic (or sometimes meromorphic, as discussed shortly) 1-form ω, which plays a role similar to Ω in the holomorphic Chern-Simons theory of eqn. (3.10). We will sometimes refer to Σ as the "topological plane" and C as the "holomorphic plane," though in general neither one of them is really a plane.

We can then define a natural generalization of the action (3.3) by

$$I = \frac{1}{2\pi}\int_M \omega \wedge \mathrm{CS}(A)\ . \tag{3.13}$$

As long as ω is closed, this action is gauge-invariant modulo total derivatives that do not affect perturbation theory.

Now we should discuss the possible role of zeroes and poles of ω. Naively, since the action involves only the ratio $\omega/\hbar$, a zero of ω corresponds to a point at which $\hbar \to \infty$. Thus, in a theory that one only knows how to define perturbatively, it should not be straightforward to make sense of the behavior near a zero of ω. We expect that essentially new ingredients are needed to make sense of that behavior. We will not explore this issue in the present paper.

Conversely, near a pole of ω, $\hbar$ is effectively going to zero and perturbation theory should be within reach. However, poles of ω are still subtle for the following reason. If ω has a pole at a point $p \in C$, then $\mathrm{d}\omega$ does not vanish near p but is a distribution supported at p. Accordingly, gauge-invariance will fail unless we place some suitable conditions near p on the gauge field A and the gauge parameter g. If ω has a double pole at p, one can restore gauge invariance by asking that $A = 0$ at p and $g = 1$ at p. What one has do if ω has a simple pole at p is more subtle and will be described in section 9.

Only simple and double poles are relevant, as one sees if one considers the possibilities for a complex Riemann surface C with a holomorphic one-form ω that is allowed to have poles but not zeroes. By the Riemann-Roch theorem, the number of zeroes of any meromorphic differential ω minus the number of its poles is $2g - 2$, where g is the genus of C. Thus if ω has no zeroes, C must have genus 0 or 1. Moreover, for $g = 0$ we have either (1) a single pole with multiplicity 2, which corresponds to $C = \mathbb{C}$ with differential $\mathrm{d}z$ (which has a double pole at ∞) or (2) two simple poles, in which case we can take $C = \mathbb{C}^\times = \mathbb{C}/\mathbb{Z}$ with differential $\omega = \frac{\mathrm{d}z}{z}$, which has simple poles at 0 and ∞. For $g = 1$, there are no poles at all; C is a complex torus or elliptic curve $\mathbb{C}/(\mathbb{Z} + \tau\mathbb{Z})$ (with modulus τ) with the holomorphic differential $\mathrm{d}z$. In each case, the choice of ω is unique up to a normalization constant that can be absorbed in rescaling $\hbar$.

Summarizing, we have the following three possibilities for C:

$$\begin{aligned}
&C = \mathbb{C}\,, && \omega = \mathrm{d}z\,, && \text{double pole at } \{\infty\}\,, && \text{(rational)}\,,\\
&C = \mathbb{C}^\times\,, && \omega = \frac{\mathrm{d}z}{z}\,, && \text{poles at } \{0,\infty\}\,, && \text{(trigonometric)}\,,\\
&C = E = \mathbb{C}/(\mathbb{Z}+\tau\mathbb{Z})\,, && \omega = \mathrm{d}z\,, && \text{no poles}\,, && \text{(elliptic)}\,.
\end{aligned} \tag{3.14}$$

As indicated, the three choices of C match the three broad classes of quasi-classical R-matrices that were summarized in section 2.2, if we assume that C parametrizes the spectral parameter of the classical r-matrix. Developing this relationship is the purpose of the present paper.

A notable fact is that the three examples are all abelian groups. This is no coincidence, of course. Since the holomorphic differential ω on C has no zeroes, its inverse is a holomorphic vector field $\zeta = \omega^{-1}$ that generates an abelian group symmetry. For the three cases, in the coordinates used in eqn. (3.14), the group action is $z \to z + a$ in the case that C is the complex plane or an elliptic curve, or $z \to \lambda z$ in the case of $\mathbb{C}^\times$. It is because of this group action, which is a symmetry of the action (3.3) and the theory constructed from it, that the R-matrix $R(z_1, z_2)$ that we eventually construct is a function only of the difference $z_1 - z_2$ or the ratio z_1/z_2, as the case may be.

Though this will not be developed in the rest of the paper, we will briefly describe a more general possible choice of 4-manifold. Suppose that the 4-manifold M admits a complex-valued closed 1-form ω. We require that $\mathrm{Re}\,\omega$ and $\mathrm{Im}\,\omega$ are everywhere linearly independent. This means that locally $\omega = \mathrm{d}f + \mathrm{i}\,\mathrm{d}g$, where f and g are real-valued functions and $\mathrm{d}f$ and $\mathrm{d}g$ are linear independent. M can then locally be foliated by the smooth two-manifolds that are defined by setting f and g to constants. Thus M has a two-dimensional integrable foliation. The gauge field A is a 3-component partial $\mathfrak{g}$-valued connection, or alternatively it is an ordinary connection with the extra gauge symmetry

$$A \mapsto A + \chi\,\omega\,. \tag{3.15}$$

The action is still given by (3.13).

The considerations to this point have been purely classical, but there are important quantum corrections. As we discuss briefly in section 3.6 (see [11, 12] for a detailed account), at the quantum level there is a framing anomaly which means that we can only define the theory on some, but not all, 4-manifolds M of the type mentioned above. For the product manifold of (3.12), the framing anomaly implies that Σ must be equipped with a framing. In particular, if compact, Σ must be a two-torus. For integrable lattice models associated to solutions of the Yang-Baxter equation, the most important examples are that Σ is $\mathbb{R}^2$ or a two-torus.

3.3 Wilson Lines

Now let us consider the gauge-invariant operators of the theory. There are no local ones because they all vanish by the equations of motion. The simplest gauge-invariant operators – and the only ones that we will study in the present paper – are Wilson line operators.

In ordinary gauge theory, a natural gauge-invariant quantity is the trace, in some representation ρ of the gauge group, of the holonomy of the connection around a closed loop K. Quantum field theorists usually write this quantity as

$$W_\rho(K) = \mathrm{Tr}_\rho P \exp\left(\oint_K A_i(x^1, x^2, z, \bar{z})\mathrm{d}x^i\right), \tag{3.16}$$

where P denotes path-ordering along the loop K, and Tr_ρ the trace in the representation ρ. In the present context, K cannot be an arbitrary loop in the four-manifold $\Sigma \times C$. On the contrary, because we only have a partial connection with no $\mathrm{d}z$ term, there is no notion of parallel transport in the C direction.[10] Accordingly, we are restricted to the case that K is a loop in the topological plane Σ, at a specified point[11] $z = z_0$ in C. This already makes contact in a preliminary way with some aspects of the standard Yang-Baxter picture that we reviewed in section 2.1. A Wilson operator is supported on a 1-manifold K in the two-manifold Σ (which one can think of as the worldline of a particle in a two-dimensional spacetime), and it is a labeled by a spectral parameter, that is, by a point in C, and by a choice of a representation ρ of G. Here ρ will play the role of the vector space V of internal states of a particle, introduced at the beginning of section 2.1.

We can also introduce more general Wilson operators, which do not have analogs in standard gauge theories. The existence of these operators is related to the fact that the loop K is highly restricted, as described in the last paragraph. At the classical level, these Wilson loops are labeled by a representation $\widehat{\rho}$ of

[10] Either there is no A_z and no way to define parallel transport along a path on which z is not constant, or there is an A_z but also an extended gauge invariance (3.15), and parallel transport in the z direction is not gauge-invariant.

[11] This classical statement will later be subject to some revision because of the framing anomaly.

the infinite-dimensional Lie algebra $\mathfrak{g}[[z]] = \prod_{n\geq 0}(\mathfrak{g}\otimes z^n)$ of series in z whose coefficients are in the finite-dimensional Lie algebra $\mathfrak{g}$. (The same algebra $\mathfrak{g}[[z]]$ will appear regardless of the choice of C because the considerations will be local along C.) To be more exact, Wilson operators supported at $z=0$ will be associated to representations of $\mathfrak{g}[[z]]$. Wilson operators supported at $z=z_0$ are similarly associated to representations of $\mathfrak{g}[[z-z_0]]$ (which is obtained from $\mathfrak{g}[[z]]$ by $z\to z-z_0$).

Since the relevant concepts may be unfamiliar, we pause for an explanation. Roughly speaking, an element of $\mathfrak{g}[[z]]$ is a $\mathfrak{g}$-valued function of z. If $\mathfrak{g}$ has a basis t_a, $a=1,\dots,\dim\mathfrak{g}$, then a basis, in the relevant sense, of the space of $\mathfrak{g}$-valued functions of z is provided by

$$\mathsf{t}_{a,n}(z) = t_a z^n, \quad n\geq 0\,. \tag{3.17}$$

So to define a representation of $\mathfrak{g}[[z]]$, we need to give, for every a and n, a matrix (or operator) $t_{a,n}$ that represents the action of $\mathsf{t}_{a,n}(z)$. Concretely, if $\mathfrak{g}$ has a basis t_a with $[t_a,t_b]=f_{ab}{}^c t_c$, then the natural commutation relation for $\mathfrak{g}$-valued functions of z is $[t_a z^n, t_b z^m] = f_{ab}{}^c t_c z^{n+m}$. Therefore, the corresponding representation matrices should obey

$$[t_{a,n},t_{b,m}] = f_{ab}{}^c t_{c,n+m}\,. \tag{3.18}$$

It is important that there is no central extension here and that this algebra has finite-dimensional representations. We will be primarily interested in finite-dimensional representations, and more specifically representations with the property that there is some n_0 such that $t_{a,n}=0$ for $n\geq n_0$. To orient the reader, we consider the first nontrivial example, which arises for $n_0=2$. The nonzero generators are just $t_{a,0}$, which generates the finite-dimensional algebra $\mathfrak{g}$, and $t_{a,1}$, which commutes with itself and transforms in the adjoint representation of $\mathfrak{g}$. To construct a representation $\widehat{\rho}$ of this algebra, we can take a direct sum $\rho\oplus\rho$ of two copies of any representation ρ of $\mathfrak{g}$. If t_a are the representation matrices of ρ, then a representation of $\mathfrak{g}[[z]]$ is given by

$$t_{a,0} = \begin{pmatrix} t_a & 0\\ 0 & t_a\end{pmatrix}\,, \quad t_{a,1} = \begin{pmatrix} 0 & t_a\\ 0 & 0\end{pmatrix}\,. \tag{3.19}$$

This representation is indecomposable as a representation of $\mathfrak{g}[[z]]$, though it is decomposable as a representation of $\mathfrak{g}$. There are many elaborations on this theme with $n_0\geq 2$.

Because the line K in (3.16) is supported at a point in C, its definition depends only on the components $A_i(x^1,x^2,z,\bar z)$ of the gauge field, with $i=1,2$. We will preserve this fact in defining (at the classical level) more general Wilson operators. For fixed x^1 and x^2, $A_i(x^1,x^2,z,\bar z)$ is a $\mathfrak{g}$-valued function of z and $\bar z$. Formally setting $\bar z=0$, we get (for each point in the topological plane and each i) a $\mathfrak{g}$-valued function of z, namely $\widehat{A}_i(x^1,x^2,z)=A_i(x^1,x^2,z,0)$. A precise definition of what we mean by setting $\bar z$ to 0 with z fixed is that we define

$$\widehat{A}_i(x^1,x^2,z) := \sum_{k\geq 0}\frac{z^k}{k!}\,\frac{\partial^k}{\partial z^k}A_i(x^1,x^2,z,\bar z)\bigg|_{z=\bar z=0}\,. \tag{3.20}$$

We do not need to worry about convergence of this series, because we consider representations that are annihilated by a sufficiently high power of z. Thus for any given representation, we can terminate the series after finitely many terms and consider $\widehat{A}_i$ to have a polynomial dependence on z.

Next we consider gauge transformations. The generator of such a gauge transformation is a $\mathfrak{g}$-valued function $u(x^1, x^2, z, \bar{z})$. We can restrict such a function to $\bar{z} = 0$, in the same sense described in the last paragraph, and extract a $\mathfrak{g}$-valued function $\widehat{u}(x^1, x^2, z) = u(x^1, x^2, z, 0)$, which we can interpret as a $\mathfrak{g}[[z]]$-valued function of x^1, x^2. The theory we are studying is invariant under a gauge transformation of A_i generated by $u(x^1, x^2, z, \bar{z})$. When we restrict to $\bar{z} = 0$, the action of $u(x^1, x^2, z, \bar{z})$ on A_i becomes an action of $\widehat{u}(x^1, x^2, z)$ on $\widehat{A}_i(x^1, x^2, z)$. Here we can view $\widehat{u}(x^1, x^2, z)$ as a $\mathfrak{g}[[z]]$-valued function on Σ, and its action of $\widehat{A}_i$ is the natural action of such a function on $\widehat{A}_i$, viewed as a $\mathfrak{g}[[z]]$-valued gauge field on Σ.

So finally if $\widehat{\rho}$ is a representation of $\mathfrak{g}[[z]]$ of the allowed class, we can define a corresponding Wilson operator by modifying eqn. (3.16) in an almost trivial way:

$$W_{\widehat{\rho}}(K) = \mathrm{Tr}_{\widehat{\rho}}\, P \exp\left(\oint_K A_i(x^1, x^2, z, 0) \mathrm{d}x^i \right). \tag{3.21}$$

Here $A_i(x^1, x^2, z, 0)$ is expanded around $z = 0$ with fixed $\bar{z} = 0$. Wilson lines supported at some other point $z = z_0$ in C are similarly defined by expanding around $z = z_0$ with fixed $\bar{z} = \bar{z}_0$. (In this case, one considers representations of $\mathfrak{g}[[z - z_0]]$.)

However, here we should point out a crucial subtlety that is important for applications of the extension to $\mathfrak{g}[[z]]$. It is very undesirable to take a trace in eqn. (3.21) because this causes much of the interesting structure to disappear. We can see that by going back to eqn. (3.19). This representation is indecomposable as a representation of $\mathfrak{g}[[z]]$, and it does not just come from a representation of the finite-dimensional algebra $\mathfrak{g}$. The holonomy operator along a given path in this representation is different from what it would be if one sets $t_{a,1} = 0$, which would give a decomposable representation of $\mathfrak{g}[[z]]$. But if we take the *trace* of the holonomy, then $t_{a,1}$ will play no role because it is strictly upper triangular, and we cannot distinguish the given representation from its decomposable cousin.

Because the theory is infrared-free, as explained at the end of section 3.1, the holonomy itself rather than its trace is a meaningful observable. To see this, we take $\Sigma = \mathbb{R}^2$ and we consider a 1-manifold $K \subset \Sigma$ (supported at a point in C) that is not compact but has its ends at infinity along Σ. Because the theory is infrared-free, the gauge field can be considered to vanish at infinity along Σ and then the holonomy along K is a gauge-invariant observable, with no need to take its trace. It is really this that gives power to the fact that the theory has Wilson operators associated to a large class of representations of $\mathfrak{g}[[z]]$. The importance of the extension to $\mathfrak{g}[[z]]$ will not be fully clear until we analyze the operator product expansion of Wilson operators in section 5.

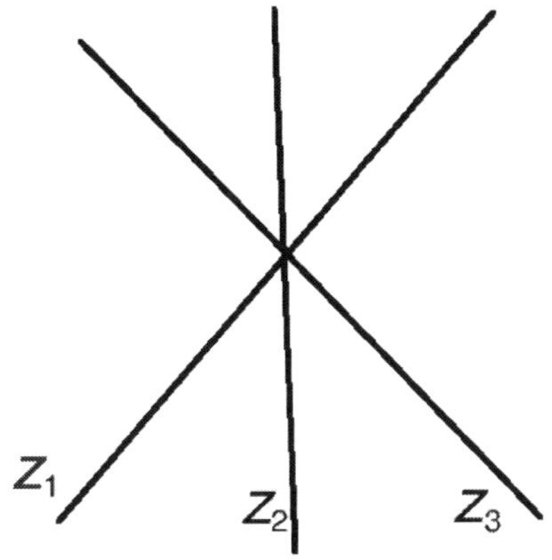

Figure 5: Three lines meeting at a point in Σ, but with distinct values of the spectral parameters z_i.

3.4 The Yang-Baxter Equation and Unitarity

Having defined Wilson operators, we can now return to Figs. 2 and 4 in which the Yang-Baxter equation and the associated unitarity relation are illustrated. We now interpret these figures as representing configurations of Wilson lines $K_i \in \Sigma$ (supported at the indicated points $z_i \in C$). It is now not hard to argue for the equivalence of the left and right hand sides of these pictures. Two-dimensional diffeomorphism symmetry means that we are free to move the K_i around as long as we do not change the topology of the situation. However, this alone is not quite enough to prove the equivalences suggested in the pictures. For example, in Fig. 2, we are free to move the "middle" Wilson line to the left or right, as long as we do not try to pass through a configuration (Fig. 5) in which the three lines all meet at a point; such a configuration is not equivalent by a diffeomorphism of Σ to a configuration without a triple intersection. In trying to prove the equivalences between the left and right of Fig. 2 by moving the middle Wilson line from left to right, we have to ask whether there is a discontinuity in the path integral at the moment that a triple intersection occurs. However, as long as the spectral parameters z_1, z_2, z_3 are all distinct, none of the lines are meeting in four dimensions and it is manifest that the configuration that has a triple intersection when projected to Σ is not associated to any singularity.[12] In particular there is no discontinuity and the pictures on the left and right of Fig. 2 are equivalent as long as the z_i are distinct (in fact, it is enough that they are not all equal). Likewise the pictures on the left and right of Fig. 4 are equivalent as long as $z_1 \neq z_2$.

It takes more than this to argue that the theory has an R-matrix that satisfies the Yang-Baxter equation and the unitarity relation. The usual R-matrix formalism, as summarized in section 2.1, involves a much more specific interpretation of

[12]An interpretation of the Yang-Baxter equation and the spectral parameter somewhat along these lines was conjectured by M. F. Atiyah in the 1980's [14].

the pictures. Each line in Fig. 1 is supposed have associated to it a space V of "internal states" accessible to a particle. In the present framework, the meaning of this is clear: a line is a Wilson line associated to some representation ρ of $\mathfrak{g}$ (or more generally of $\mathfrak{g}[[z]]$), and ρ corresponds to V. But the usual R-matrix picture is much simpler than one would expect in quantum field theory in general. In the usual R-matrix picture, each line segment between two crossings is labeled by a basis vector e_i of V, and to a crossing one associates a local factor, the R-matrix element $R_{ij}^{kl}(z_1 - z_2)$, which depends only on the data at a particular crossing and not on any other details in which the local picture is embedded. Moreover, this R-matrix element depends only on the difference $z_1 - z_2$. In quantum field theory in general, one would not expect a local picture like this. Finally, though in standard presentations of R-matrix theory one might take this for granted and skip it over, it is noteworthy that in R-matrix theory, the two-dimensional regions bounded by the lines do not carry any labels. This is a nontrivial point and in fact there is a generalization of the Yang-Baxter equation (the dynamical Yang-Baxter equation [21, 22, 23, 24]) in which the bulk regions do carry labels, above and beyond the labels carried by the line segments.

In trying to explain these facts in the present context, the most basic question is why there is a local picture of any sort. The reason for this is that the theory is infrared-free, as was noted at the end of section 3.1. Concretely, in constructing perturbation theory, as we will do starting in section 4, one picks a Riemannian metric on $\Sigma \times C$. If one scales up the metric on Σ by a large factor, so that different crossings are very far apart (compared to the distances between the points in C at which a given set of Wilson line operators are supported), then the infrared-free nature of the theory guarantees that some kind of local picture will be possible.

To explain more, let us first ask what would happen in the absence of any line operators. The theory under study is topological in the Σ direction, so (ignoring further subtleties that arise because we are dealing with a theory whose action is a holomorphic function of complex variables) in general we would expect the theory to have a space $\mathcal{H}$ of quantum states. These would roughly correspond to vacua of a standard quantum field theory. In general, one would expect to label the regions between the lines – that is, any region of Σ that is not near one of the Wilson operators – by a basis vector of $\mathcal{H}$. Accordingly, if $\mathcal{H}$ has dimension bigger than 1, we would get something like the dynamical Yang-Baxter equation [21, 22, 23, 24], with labels for regions as well as line segments, rather than the standard Yang-Baxter equation in which regions between the lines are unlabeled. We discuss this situation in section 11.

To get something as simple as the standard Yang-Baxter equation, we want $\mathcal{H}$ to be one-dimensional, which will happen if the space of classical solutions of the theory, modulo gauge transformations, is a point. This is also the condition that eliminates the subtleties associated with having a holomorphic action; perturbation theory is straightforward in principle if there is only one classical solution to expand around, and it has only a finite group of automorphisms.[13] The simplest

[13]If there is a unique classical solution up to gauge transformation, but it has a nontrivial automorphism group H, then in developing perturbation theory one wants to divide by the volume of H. If Hf is not a finite group, this volume might be hard to interpret. However, this

case is that $C = \mathbb{C}$. In quantizing the theory on $\Sigma \times \mathbb{C}$ for any Σ, we require that the gauge field A and the generator of a gauge transformation both vanish at infinity. With this choice, the only (stable[14]) classical solution, up to a gauge transformation, is $A = 0$. So we are in the situation in which the bulk regions do not carry labels and perturbation theory is straightforward in principle.

It is likewise possible when C is $\mathbb{C}^{\times}$ or an elliptic curve to ensure that the classical phase space is a point, leading to straightforward perturbation theory and (as we argue shortly) a conventional Yang-Baxter equation. The details are more involved and we defer a discussion to sections 9 and 10.

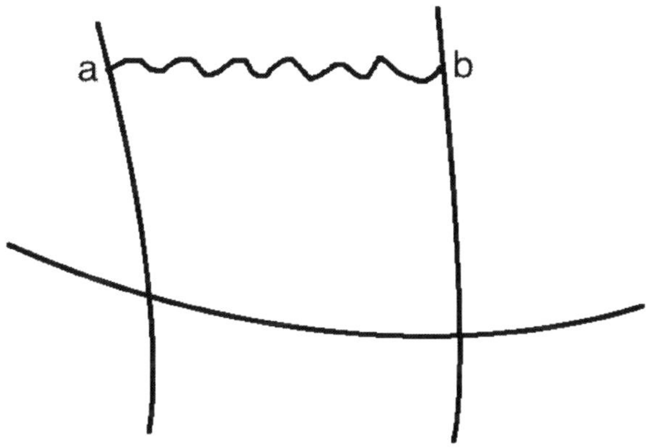

Figure 6: The infrared-free nature of the theory under study means that when we scale up the metric of Σ after gauge-fixing, gluon exchange between Wilson lines that are not crossing becomes irrelevant.

Let us now imagine doing perturbation theory in an infrared-free theory in the presence of a configuration of Wilson lines. What sort of perturbative corrections are significant? A typical example of an effect that is not significant is gauge boson exchange between two Wilson lines that are not crossing (Fig. 6). By scaling up the metric of Σ, the points a and b in the figure can be made arbitrarily far apart, regardless of where they lie on the Wilson lines in question, and the contribution of gluon exchange between them goes to zero. So this can be ignored.

A typical contribution that cannot be ignored is a gauge boson exchange between two lines that are crossing (Fig. 7). In this case, the points a and b can be near the crossing point, so they cannot be assumed to be far away in Σ. We

issue involves only an overall constant factor in the path integral, independent of what collection of Wilson lines one considers.

[14]There are many classical solutions on $\Sigma \times \mathbb{C}$ that correspond to bundles on $\mathbb{C}$ (trivialized at infinity) that are unstable in the sense of algebraic geometry. This likely makes them unsuitable as a starting point for perturbation theory. At any rate, the fact that is really important for us is that the trivial connection on $\Sigma \times \mathbb{C}$ is a classical solution that has no infinitesimal deformations or gauge automorphisms. In perturbation theory around this solution, we do not meet unstable bundles.

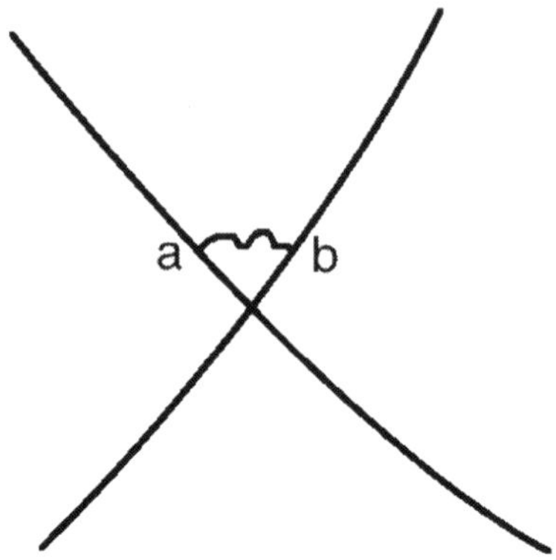

Figure 7: The infrared-free nature of the theory under study means that when we scale up the metric of Σ after gauge-fixing, gluon exchange between Wilson lines that are not crossing becomes irrelevant.

can still scale up the metric in the picture in order to exploit the infrared-free nature of the theory. But all that happens when we do this is that the lines that are crossing turn into straight lines near the point in Σ where they cross, and they become very widely separated from any other crossing points. So the diagram of Fig. 7 may be nontrivial – and it is nontrivial, as we will calculate in section 4 – but it will be local: it will not depend on the details of a larger picture in which the crossing of Fig. 7 might be embedded.

Now we can put the pieces together and explain why something along the lines of standard Yang-Baxter theory will emerge. We assume a situation with a unique classical solution[15] $A = 0$. Moreover, we know that quantum effects are negligible except near crossings. Thus away from crossings, we can assume that $A = 0$ everywhere. This means that a line segment between crossings just describes a free particle in the relevant representation of $\mathfrak{g}$ (or of $\mathfrak{g}[[z]]$) and can be labeled by a basis vector in that representation. And crucially, the amplitude associated to a given crossing can only depend on the local data at that crossing – the representations and labels of the lines that are crossing. Thus the equivalence of the two pictures of Fig. 2, which follows from rather general arguments that were given above, turns into the more precise numerical equivalence of Fig. 3, with a local R-matrix at each crossing. The same reasoning applies to the unitarity relation of Fig. 4. Because of the infrared-free nature of the theory, it turns into the concrete unitarity relation $R_{21}R_{12} = 1$ of Yang-Baxter theory.

For the case that C is $\mathbb{C}$ or an elliptic curve, the local R-matrix $R_{ij}^{kl}(z_1, z_2)$ is actually a function only of the difference $z_1 - z_2$, because the classical action

[15]In the trigonometric and elliptic cases, this argument needs to be stated a little more carefully. There is a unique classical solution, and although it is not gauge-equivalent globally to $A = 0$, this is true locally. Given this, the argument proceeds as in the text.

(3.3) is invariant under shifting z by a constant. For the case of $C = \mathbb{C}^\times$ with differential $\mathrm{d}z/z$, the equivalent statement is that the R-matrix (written in these multiplicative coordinates) is a function of the ratio z_1/z_2.

For the case that $C = \mathbb{C}$, related to the Yangian, a few further nice things happen which make this case particularly simple and elegant. First of all, the action is invariant under a common rescaling of z and $\hbar$, so actually the R-matrix is a function only of a single variable $(z_1 - z_2)/\hbar$. Second, in quantizing the theory with $C = \mathbb{C}$, we divide only by gauge transformations that are 1 at infinity along C. But we are left with gauge transformations that are constant at infinity along C, and these behave as global symmetries. Thus the R-matrix for $C = \mathbb{C}$ has G as an automorphism group. (This is not true for the other choices of C, as we will see in sections 9 and 10.) The properties stated in this paragraph make it straightforward to understand some simple examples. We present some of these elementary examples in the next section. We present them because they are fun – though probably well-known to many readers – and also because they enable one to see in a completely direct and elementary way why the theory must have a framing anomaly.

3.5 Elementary Examples

For some elementary examples, we take $G = GL_N$ (or SL_N, which would be equivalent for the purposes of this analysis), and we will consider the case that ρ is the fundamental representation of G or its dual. We denote these representations as V and V^*, respectively. In all cases, we will take $C = \mathbb{C}$, so that the R-matrix has G symmetry.

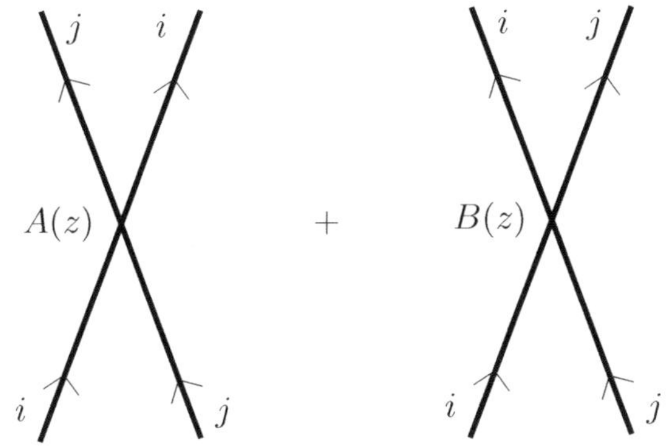

Figure 8: The R-matrix for crossing of two Wilson lines in the fundamental representation V. The non-zero matrix elements of R correspond to processes in which the final states equal the initial states, possibly modulo a permutation.

First we consider the R-matrix for crossing of two copies of V. It will be a G-invariant linear map $R(z_1 - z_2) : V \otimes V \to V \otimes V$. Such an operator is a linear combination of the identity and the operator $P : V \otimes V \to V \otimes V$ that exchanges the two factors: $R(z) = A(z) + PB(z)$, with $z = z_1 - z_2$. In this particular case, $R(z)$ can also be written fairly conveniently as a matrix with all its indices:

$$R_{ij}^{i'j'}(z) = \delta_i^{i'} \delta_j^{j'} A(z) + \delta_i^{j'} \delta_j^{i'} B(z) \,. \tag{3.22}$$

Here i and j refer to "incoming" lines and i' and j' to outgoing ones. The $A(z)$ term describes two lines crossing without "charge exchange," while $B(z)$ describes crossing with charge exchange. The non-zero matrix elements of R are depicted in Fig. 8.

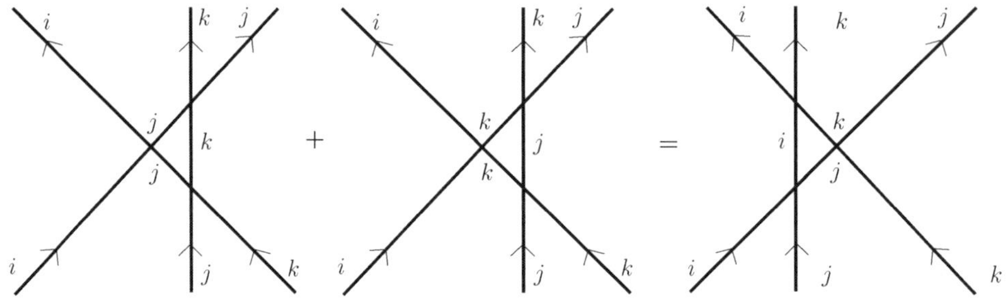

Figure 9: The Yang-Baxter equation for crossing of three Wilson lines in the fundamental representation V.

The Yang-Baxter equation is in general invariant under multiplying the R-matrix by a scalar function – a z-dependent multiple of the identity. So it is only sensitive to the ratio $U(z) = B(z)/A(z)$. It is not difficult to work out the Yang-Baxter equation in this case (Fig. 9) and to learn that it is equivalent to

$$U(z_1 - z_3)U(z_2 - z_3) + U(z_1 - z_2)U(z_1 - z_3) = U(z_1 - z_2)U(z_2 - z_3) \,. \tag{3.23}$$

After dividing by the product $U(z_1-z_2)U(z_1-z_3)U(z_2-z_3)$, we learn that $1/U(z)$ is a multiple of z, so (remembering that the R-matrix for $C = \mathbb{C}$ is a function of $\hbar/z$) U must be a constant multiple of $\hbar/z$. Determining the constant from the $\mathcal{O}(\hbar)$ contribution to the R-matrix (see section 4), we find

$$U(z) = \frac{\hbar}{z}. \tag{3.24}$$

Now we consider the R-matrix for crossing of a copy of V with a copy of the dual representation V^*. The R-matrix is now a linear map $R(z) : V \otimes V^* \to V \otimes V^*$. Again $R(z)$ is determined by the G symmetry in terms of two functions: $R(z) = C(z) + QD(z)$. Here Q is the G-invariant projection operator from $V \otimes V^*$ to its G-invariant subspace. A picture is rather clear (Fig. 10), but now a formula analogous to eqn. (3.22) is less transparent:

$$R_{i\,j'}^{j\,i'}(z) = \delta_i^{i'} \delta_{j'}^{j} C(z) + \delta_i^{j} \delta_{j'}^{i} D(z) \,. \tag{3.25}$$

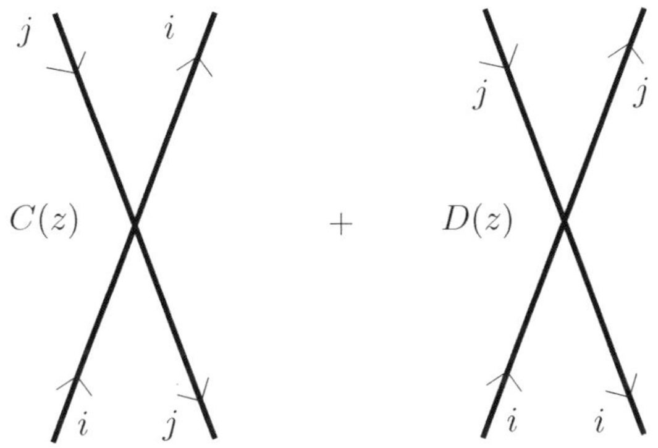

Figure 10: The R-matrix for crossing of two Wilson lines in the dual representations V and V^*.

Figure 11: The Yang-Baxter equation for three lines labeled by representations V, V, and V^*.

(As before, lower indices refer to incoming lines and upper indices to outgoing ones.) The Yang-Baxter equation will involve only the ratio $W(z) = D(z)/C(z)$.

It is again not difficult to write down the Yang-Baxter equation. We learn (Fig. 11) that

$$U(z_1 - z_2)W(z_2 - z_3) + W(z_1 - z_3)W(z_2 - z_3) = U(z_1 - z_2)W(z_1 - z_3)\,. \quad (3.26)$$

This is equivalent to $1/W(z_2 - z_3) - 1/W(z_1 - z_3) = 1/U(z_1 - z_2)$, with the general solution

$$W(z) = -\frac{\hbar}{z - \hbar b}\,, \quad (3.27)$$

with a constant b. This constant b can actually be determined by the unitarity

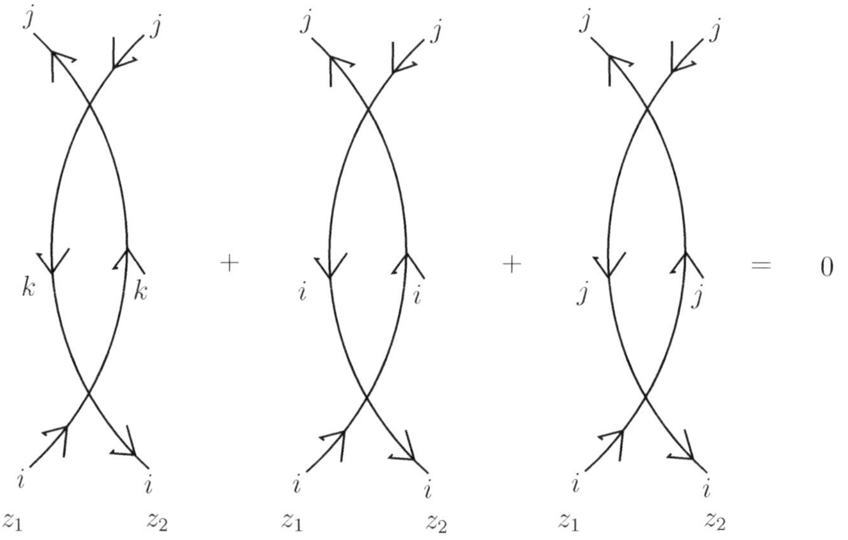

Figure 12: The "unitarity" relation for lines labeled by V and V^* implies that the given sum vanishes for $i \neq j$.

relation $R_{21}(-z)R_{12}(z) = 1$. A specific matrix element of this relation, for the case of crossing of V and V^* (see Fig. 12) gives

$$NW(z)W(-z) + W(z) + W(-z) = 0\,, \quad (3.28)$$

leading to

$$W(z) = -\frac{\hbar}{z + \frac{\hbar N}{2}}\,. \quad (3.29)$$

We need not separately consider the R-matrix for crossing of two copies of V^*, because it simply equals the R-matrix for crossing of two copies of V. The reason is that the outer automorphism of GL_N or SL_N that exchanges V and V^* is a symmetry of the action (3.3), and of the theory derived from it.

Not determined by these arguments are overall scalar functions in the R-matrices for $V \otimes V \to V \otimes V$ and $V \otimes V^* \to V \otimes V^*$. These can be partly but not entirely determined by the unitarity relation; to some extent these overall scalar functions depend on arbitrary choices in quantizing the theory.

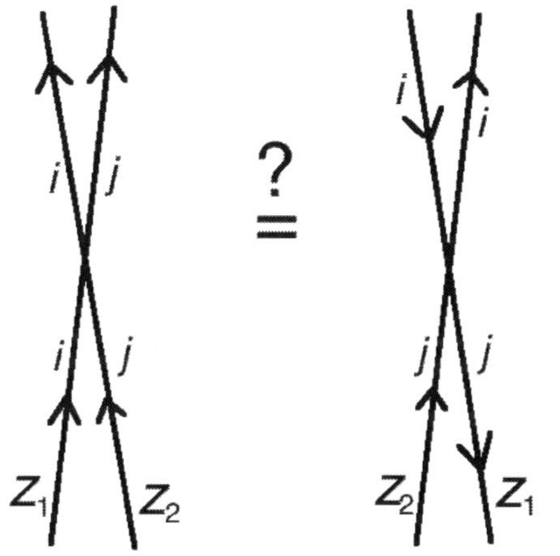

Figure 13: On the left are two Wilson lines in the fundamental representation of GL_N or SL_N; they differ from the vertical by small angles $\pm\alpha$. The one on the left has spectral parameter z_1 and the one on the right has spectral parameter z_2. To get the picture on the right, the z_2 Wilson line has been rotated clockwise by the small angle 2α, and the z_1 Wilson line has been rotated clockwise by the much larger angle $\pi - 2\alpha$. Naively the two pictures should be equivalent, implying that $U(z_1 - z_2) = W(z_2 - z_1)$, a claim that turns out to be false.

In the theory as we have developed it so far, the angles at which two Wilson lines cross are of no consequence. We can use the above formulas to test this expectation in an interesting way. We consider (see the left of Fig. 13) two Wilson lines in the representation V, with spectral parameters z_1 and z_2, and differing from the vertical by small angles $\pm\alpha$. Rotating one line clockwise by a small angle 2α and the other one clockwise by a larger angle $\pi - 2\alpha$, we arrive at the right hand side of the figure, which depicts the crossing of a pair of nearly vertical Wilson lines in the representations V and V^*. The rotation converts a charge exchange for VV crossing to an "annihilation" process for VV^*. Naively, the two parts of Fig. 13 should be equivalent. This would imply

$$U(z) \stackrel{?}{=} W(-z)\,. \tag{3.30}$$

A look back at our previous formulas shows, however, that this is false. What is true instead is that

$$U\left(z - \frac{\hbar N}{2}\right) = W(-z)\,. \tag{3.31}$$

3.6 First Look at the Framing Anomaly

What accounts for this discrepancy? The answer is that the theory has a framing anomaly for Wilson operators, which will be explored from another point of view in section 6. The framing anomaly is analogous to the perhaps familiar framing anomaly for Wilson operators in Chern-Simons theory, but more subtle. It can be formulated in different but topologically equivalent ways. However, in an approach natural in perturbation theory, the framing anomaly can be formulated as follows. To begin with we take the topological plane to really be a plane $\Sigma = \mathbb{R}^2$, and we quantize with a gauge choice that uses a flat metric on $\mathbb{R}^2$. (See section 4.) We consider a Wilson operator supported on a general curve $K \subset \Sigma$, and we let $\varphi(p)$ be the angle between the tangent vector to K at a given point $p \in K$ and some chosen direction in $\mathbb{R}^2$ (e.g. the vertical). (We define this angle to increase if K bends in a clockwise direction.) Thus φ is not quite well-defined as a function on K, but it is well-defined up to an additive constant, and its differential $\mathrm{d}\varphi$ is well-defined.

The framing anomaly means that what is constant along K is not the spectral parameter z, as one would expect from a classical analysis, but $z - \hbar \mathsf{h}^\vee \varphi/(2\pi)$, where $\mathsf{h}^\vee$ is the dual Coxeter number of the gauge group. As a perhaps surprising example of the implications of this statement, a Wilson operator whose support is a simple closed curve in Σ is anomalous and does not exist in the quantum theory, because in going around a simple closed loop, φ increases by 2π.

Now in our problem, we can formulate the comparison between the two parts of Fig. 13 in a slightly different way. To go from the left to the right of the figure, we make the z_1 Wilson line bend in the plane by an angle $\Delta\varphi = \pi$ before crossing the z_2 Wilson line, as in Fig. 14. But because of the framing anomaly, when we do this z_1 is shifted to $z_1 - \hbar \mathsf{h}^\vee \Delta\varphi/(2\pi) = z_1 - \hbar N/2$, where N is the dual Coxeter number of SL_N (or GL_N). Thus what should coincide with $W(-z)$ is not $U(z)$ but $U(z - \hbar N/2)$, and this is precisely what we found in eqn. (3.31).

In light of the framing anomaly, one might ask the following question. In the usual formulation of the Yang-Baxter equation for crossing of three lines, what are the angles at which the lines cross? The answer is clear if one considers the case that the three representations involved are all the same. It is usually then assumed that the three R-matrices R_{12}, R_{23}, and R_{13} are given by the same matrix-valued function of z. For this to be true, the relative angles must be the same at all three crossings. Since the actual relation is that the 13 crossing angle is the sum of the 12 and 23 crossing angles, the three angles are equal only in the limit that they all go to zero. So the usual formulation of the Yang-Baxter equation refers to the case of nearly parallel lines in the limit that the crossing angles vanish. That is actually why Fig. 13 has been drawn with lines at small angles $\pm\alpha$ to the vertical, and it was implicitly assumed in our discussion of Fig. 14.

The reader might be slightly perplexed that we began our explanation of the framing anomaly by restricting to the special case $\Sigma = \mathbb{R}^2$. To understand this point properly, one has to analyze the framing anomaly for four-manifolds as well as the framing anomaly for Wilson operators. This is somewhat beyond the scope of the present paper. However, the upshot is that to avoid an anomaly, the two-

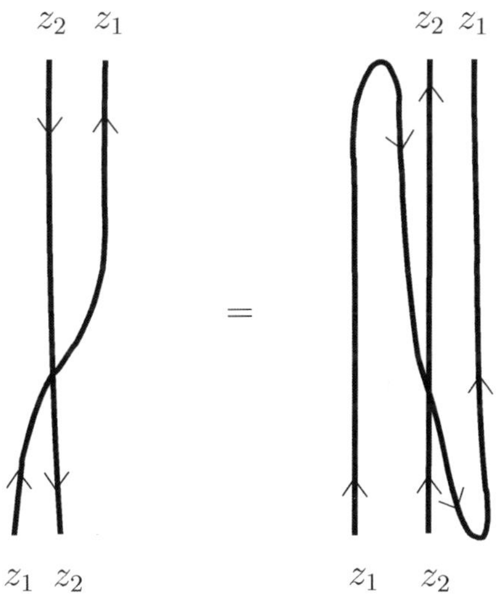

Figure 14: An improved view of the unitarity relation that was explored in Fig. 13, formulated to take the framing anomaly into account. Two lines both labeled by V enter at the bottom and emerge at the top at nearly vertical angles. If they cross at nearly vertical angles, we get an R-matrix element for a VV crossing, while if one of them turns by an angle very close to π before the crossing, then we get a VV^* crossing, but with z_1 replaced by $z_1 - N\hbar/2$ because of the framing anomaly. Thus the naive relation $U(z) = W(-z)$ suggested by Fig. 13 is corrected to $U(z - N\hbar/2) = W(-z)$.

manifold Σ must be "framed," meaning that its tangent bundle must be trivialized. (This is actually analogous to the framing anomaly of three-dimensional Chern-Simons theory, which is defined on a framed three-manifold.) On a framed two-manifold Σ, one can define along any embedded oriented one-manifold $K \subset \Sigma$ a function with the properties of the φ used above. For this, recall that a "framing" is a pair of everywhere linearly independent vector fields v_1 and v_2 on Σ. Given a framing, one can pick a metric such that v_1 and v_2 are everywhere orthonormal, and then one can define φ as the angle between the tangent vector to K and the v_1 direction. Note that the condition that Σ should be framed is very restrictive; for example, a compact framed two-manifold must have zero Euler characteristic and hence must have genus 1. Thus – as we also saw from the anomaly for a closed loop in the plane – the framing anomaly in the four-dimensional theory discussed here is much more restrictive than its three-dimensional cousin.

If we think of the vertical direction in the above figures as the Euclidean "time," then the reader will note that we have interpreted a V^*-valued particle moving forward in time as a V-valued particle moving backward in time. This is reminiscent of the relation between particles and antiparticles in relativistic quantum field theory, and indeed in the application of R-matrix theory to integrable models of relativistic quantum field theory, this operation becomes crossing symmetry whereby an S-matrix element with a particle in the initial state, after analytic continuation to negative energy, is interpreted as an S-matrix element with an antiparticle in the final state. That is why in R-matrix theory, the relation between R-matrix elements associated to a pair of dual representations ρ and ρ^* is often called "crossing."

4 R-Matrix from Crossing Wilson Lines

Here and in the next two sections, we will perform concrete Feynman diagram calculations to compute (1) the $\mathcal{O}(\hbar)$ term in the R-matrix; (2) the $\mathcal{O}(\hbar)$ quantum correction to the operator product expansion (OPE) for Wilson line operators; (3) the framing anomaly. In fact, in the case of the framing anomaly, the $\mathcal{O}(\hbar)$ term that we compute gives the complete answer; in the other cases, there are higher order contributions to the effects that we calculate, although they can be determined by general principles (such as the Yang-Baxter equation and associativity of the operator product expansion) once the lowest order terms are known. We will perform independent Feynman diagram calculations of the three effects, but actually the three effects can be deduced from each other to a large extent. We already deduced the framing anomaly from the $\mathcal{O}(\hbar)$ term in the R-matrix in section 3.6, and we will explain in section 5 why the quantum correction to the OPE is inevitable given the quantum correction to the R-matrix.

In all cases, we take $C = \mathbb{C}$, corresponding to a rational solution of the Yang-Baxter equation. In the case of the framing anomaly, the considerations are manifestly local, so the choice of C does not matter. For the R-matrix and the OPE, once the result is known for $\mathbb{C}$, it can be deduced from global considerations for the other choices of C. We leave this for sections 9 and 10.

We will compute in a way that involves a choice of metric on $\Sigma \times C$. As explained in section 3.4, for the output of Feynman diagrams to have a straightforward interpretation in the usual language of R-matrix theory, we have to scale up the metric on Σ by a large factor. In the limit, Σ becomes $\mathbb{R}^2$ near the crossing and the supports of the two Wilson lines that are crossing become straight lines in $\mathbb{R}^2$. In $\mathcal{O}(\hbar)$, the angle at which the lines cross does not matter (the reader can verify this by a slight generalization of the calculation that we will describe), so we can take the two lines to be the x-axis and the y-axis in the xy plane. In higher orders, the crossing angle would matter via the framing anomaly.

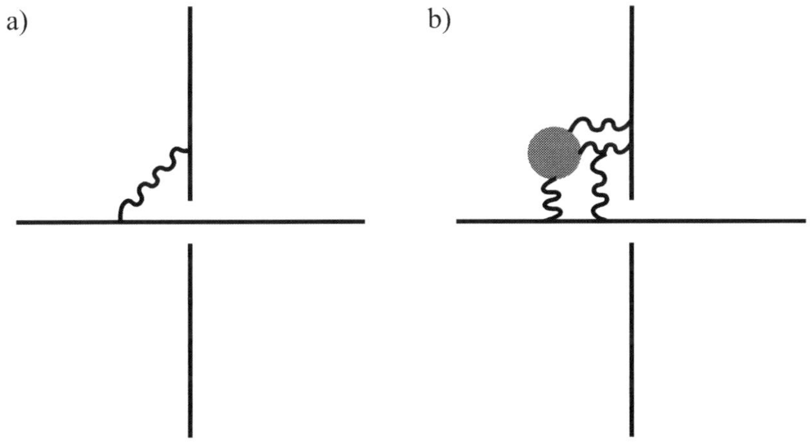

Figure 15: (a) The leading order Feynman diagram for gluon exchange between crossing Wilson lines. (b) A higher order diagram, with further complications hidden in the "blob."

The $\mathcal{O}(\hbar)$ contribution to the R-matrix involves one gluon exchange between the two lines, as sketched in Fig. 15. We evaluate the contribution of this diagram for the case that the Wilson lines are associated to representations ρ and ρ' of $\mathfrak{g}$ supported respectively at $z = z_1$ and $z = z_2$.

The metric that we will use on $\mathbb{R}^2 \times C$ is $\mathrm{d}s^2 = \mathrm{d}x^2 + \mathrm{d}y^2 + \mathrm{d}z\mathrm{d}\bar{z} = g_{\mu\nu}\mathrm{d}x^\mu \mathrm{d}x^\nu$, or in another language

$$g_{xx} = g_{yy} = 1\,, \quad g_{z\bar{z}} = g_{\bar{z}z} = \frac{1}{2}\,, \quad \text{other components zero}\,. \tag{4.1}$$

The corresponding inverse metric is

$$g^{xx} = g^{yy} = 1\,, \quad g^{z\bar{z}} = g^{\bar{z}z} = 2\,, \quad \text{other components zero}\,. \tag{4.2}$$

For a gauge-fixing condition, we pick

$$0 = \frac{\partial}{\partial x}A_x + \frac{\partial}{\partial y}A_y + 4\frac{\partial}{\partial z}A_{\bar{z}}\,. \tag{4.3}$$

This is the closest analog of the usual Lorentz gauge for this theory with a partial gauge connection. The factor of four is explained by noting that if the gauge field A satisfies this equation and also the linearized equations of motion $\mathrm{d}z \wedge \mathrm{d}A = 0$, then each component of A is harmonic for the metric we have chosen.

In this gauge, the four-dimensional propagator for the gauge field is then given by

$$\begin{aligned}
\langle A_x^a(x,y,z,\overline{z}) A_y^b(x',y',z',\overline{z}')\rangle &= -\delta^{ab}\frac{4}{4\pi}\frac{\partial}{\partial z}\frac{1}{(x-x')^2+(y-y')^2+|z-z'|^2} \\
&= \delta^{ab}\frac{1}{2\pi}\frac{2(\overline{z}-\overline{z}')}{\left((x-x')^2+(y-y')^2+|z-z'|^2\right)^2}\,, \\
\langle A_{\overline{z}}^a(x,y,z,\overline{z}) A_x^b(x',y',z',\overline{z}')\rangle &= -\delta^{ab}\frac{1}{4\pi}\frac{\partial}{\partial y}\frac{1}{(x-x')^2+(y-y')^2+|z-z'|^2} \\
&= \delta^{ab}\frac{1}{2\pi}\frac{y-y'}{\left((x-x')^2+(y-y')^2+|z-z'|^2\right)}\,, \\
\langle A_y^a(x,y,z,\overline{z}) A_{\overline{z}}^b(x',y',z',\overline{z}')\rangle &= -\delta^{ab}\frac{1}{4\pi}\frac{\partial}{\partial x}\frac{1}{(x-x')^2+(y-y')^2+|z-z'|^2} \\
&= \delta^{ab}\frac{1}{2\pi}\frac{x-x'}{\left((x-x')^2+(y-y')^2+|z-z'|^2\right)^2}\,.
\end{aligned} \tag{4.4}$$

For later purposes, we can reinterpret the propagator as a two-form on two copies of $\mathbb{R}^2 \times C$. Setting $x = x' - x''$, $y = y' - y''$, $z = z' - z''$, $\overline{z} = \overline{z}' - \overline{z}''$, we define

$$\begin{aligned}
P^{ab}(x,y,z,\overline{z}) &:= \frac{1}{2}\sum_{i,j=x,y,\overline{z}} \langle A_i^a(x',y',z',\overline{z}') A_j^b(x'',y'',z'',\overline{z}'')\rangle \mathrm{d}x^i \wedge \mathrm{d}x^j \\
&= -\frac{\delta^{ab}}{4\pi}\left(\mathrm{d}y\wedge\mathrm{d}\overline{z}\frac{\partial}{\partial x} + \mathrm{d}\overline{z}\wedge\mathrm{d}x\frac{\partial}{\partial y} + 4\mathrm{d}x\wedge\mathrm{d}y\frac{\partial}{\partial z}\right)\frac{1}{(x^2+y^2+z\overline{z})} \\
&= \frac{\delta^{ab}}{2\pi}\left(x\mathrm{d}y\wedge\mathrm{d}\overline{z} + y\mathrm{d}\overline{z}\wedge\mathrm{d}x + 2\overline{z}\mathrm{d}x\wedge\mathrm{d}y\right)\frac{1}{(x^2+y^2+z\overline{z})^2}\,.
\end{aligned} \tag{4.5}$$

In the following, we often use the propagator two-form with adjoint indices stripped off:

$$P^{ab}(x,y,z,\overline{z}) = \delta^{ab} P(x,y,z,\overline{z})\,. \tag{4.6}$$

The defining equations of the 2-form $P(x,y,z,\overline{z})$ are

$$\frac{\mathrm{i}}{2\pi}\mathrm{d}z\wedge\mathrm{d}P(x,y,z,\overline{z}) = \delta_{x,y,z,\overline{z}=0}\,, \tag{4.7}$$

$$\left(\partial_x \iota_{\partial_x} + \partial_y \iota_{\partial_y} + 4\partial_z \iota_{\partial_{\overline{z}}}\right) P(x,y,z,\overline{z}) = 0\,. \tag{4.8}$$

where $\delta_{x,y,z,\overline{z}=0}$ is a delta-function distribution localized at $x = y = z = \overline{z} = 0$, and ι_V indicates contraction with a vector field V.

To verify the normalizations, we can check these equations explicitly. It is easy to verify that $\mathrm{d}z \wedge \mathrm{d}P = 0$ away from the origin $x = y = z = \overline{z} = 0$, and that eqn. (4.8) is satisfied.

Note that, when restricted to the unit three-sphere,

$$P(x, y, z, \overline{z}) = \frac{1}{2\pi}\left(x\mathrm{d}y \wedge \mathrm{d}\overline{z} - y\mathrm{d}x \wedge \mathrm{d}\overline{z} + 2\overline{z}\mathrm{d}x \wedge \mathrm{d}y\right) . \tag{4.9}$$

Since the forms on the two sides of this equation are the same on the unit three-sphere, Stokes' theorem tell us that applying the operator $\mathrm{d}z \wedge \mathrm{d}$ to both sides and integrating over the ball of radius 1 will give the same answer. Therefore we have the identity (using the coordinates u, v with $z = u + \mathrm{i}v$)

$$\begin{aligned} \frac{\mathrm{i}}{2\pi}\int_{x^2+y^2+z\overline{z}\leq 1} \mathrm{d}z \wedge \mathrm{d}P(x, y, z, \overline{z}) &= \frac{\mathrm{i}}{2\pi}\frac{1}{2\pi}\int_{x^2+y^2+z\overline{z}\leq 1} 4\mathrm{d}x\mathrm{d}y\mathrm{d}z\mathrm{d}\overline{z} \\ &= \frac{\mathrm{i}}{2\pi}\frac{1}{2\pi}\int_{x^2+y^2+z\overline{z}\leq 1} (-8\mathrm{i})\mathrm{d}x\mathrm{d}y\mathrm{d}u\mathrm{d}v \\ &= \frac{\mathrm{i}}{2\pi}\frac{1}{2\pi}(-8\mathrm{i})\frac{\pi^2}{2} = 1 . \end{aligned} \tag{4.10}$$

The four-dimensional bulk gauge field A^a couples to a Wilson line in the representation ρ by a factor $t_{a,\rho}$, which is the matrix by which the Lie algebra element t_a acts in the representation ρ. For the case of a single gluon coupling to a Wilson line, this factor does not depend on where on the line the gluon is inserted. For gluon exchange between two Wilson lines, as in Fig. 15, we simply get such a factor on each line. The propagator between a Wilson line supported on the x axis at $z = z_1$ and one supported on the y axis at $z = z_2$ then evaluates to

$$\begin{aligned} I_1 &= \hbar\left((t_{a,\rho} \otimes t_{b,\rho'}\right) \int \mathrm{d}x\mathrm{d}y'\, P^{ab}(x - x', y - y', z_1 - z_2, \overline{z}_1 - \overline{z}_2) \\ &= \hbar\, c_{\rho,\rho'} \int \mathrm{d}x\mathrm{d}y'\, P(x - x', y - y', z_1 - z_2, \overline{z}_1 - \overline{z}_2) , \end{aligned} \tag{4.11}$$

where the color factor reads

$$c_{\rho,\rho'} = \sum_a t_{a,\rho} \otimes t_{a,\rho'} . \tag{4.12}$$

Here $c_{\rho,\rho'}$ can be viewed as the image of an element $c = \sum_a t_a \otimes t_a$ of $\mathfrak{g} \otimes \mathfrak{g}$ in the representation $\rho \otimes \rho'$ of $\mathfrak{g} \otimes \mathfrak{g}$. The factor of $\hbar$ is the loop counting parameter.

It is straightforward to evaluate the integral in (4.11), with the result

$$I_1 = \hbar\, c_{\rho,\rho'}\frac{1}{2\pi}\int \mathrm{d}x\mathrm{d}y \frac{2(\overline{z}_1 - \overline{z}_2)}{(x^2 + y^2 + |z_1 - z_2|^2)^2} = \frac{\hbar\, c_{\rho,\rho'}}{z_1 - z_2} . \tag{4.13}$$

This reproduces the standard semi-classical expansion of the rational R-matrix

$$R = I + \hbar r + \mathcal{O}(\hbar^2) = I + \frac{\hbar\, c_{\rho,\rho'}}{z_1 - z_2} + \mathcal{O}(\hbar^2) . \tag{4.14}$$

General theorems (see [7, p. 814] or [8, p. 418]) imply that the full rational R-matrix is determined up to a prefactor by the general conditions that it obeys together with the leading order term that we have just computed. Some interesting special cases of this statement are rather easy, as we have reviewed in section 3.5.

5 OPE of Parallel Wilson Lines

5.1 Overview

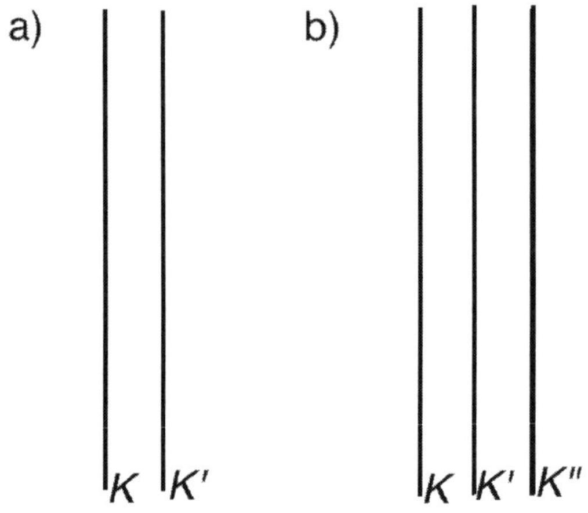

Figure 16: (a) In a diffeomorphism-invariant theory, two parallel line operators can be considered to be arbitrarily close, so they behave as a single line operator. This is the operator product expansion for line operators. (b) In two dimensions, this operator product expansion is not necessarily commutative, but it is always associative, because given three parallel line operators, there is no natural sense of one pair being "closer" than the other. The triple product $KK'K''$ can be identified as either $(KK')K''$ or $K(K'K'')$ by moving together one pair of adjacent line operators or the other.

The next topic that we will consider is the operator product expansion (OPE) of Wilson lines. We begin with generalities about line operators in diffeomorphism-invariant theories.

Consider parallel line operators K and K' in a theory with diffeomorphism invariance in any dimension ≥ 2 (Fig. 16(a)). Diffeomorphism invariance means that there is no natural notion of whether K and K' are "near" or "far" and therefore that we can think of them as being arbitrarily near. This implies that it must be possible in any diffeomorphism invariant theory to interpret a product KK' of two parallel line operators as a single line operator K''. This is the operator product expansion for line operators.

Although there is in a diffeomorphism invariant theory no natural notion of K and K' being "near" or "far," something special happens in two dimensions: there can be a natural notion of whether K is to the left or right of K'. The product KK' with K to the left of K' may be different from the product $K'K$ with K to the right. Thus, in two dimensions, the OPE of line operators is not necessarily commutative.[16]

Although not necessarily commutative in two dimensions, the OPE of line operators is always associative. That is because (Fig. 16(b)) given three parallel line operators K, K', and K'', there is no natural notion of K and K' being closer or farther than K' and K''. There is just one product of line operators that depends only on how they are arranged from left to right.

Our problem has a few special features. The line operators that we will be studying are indeed supported on a line K in the smooth two-manifold Σ which has diffeomorphism symmetry,[17] but they are also supported at a point in the complex Riemann surface C. The OPE for line operators supported at distinct points in C is trivial – and in particular commutative – as they can pass through each other in Σ without any singularity. They interesting case is the OPE for line operators that are supported at the same point in C. We may as well take this point to be $z = 0$.

Although abstractly the product KK' of line operators K and K' will always be another line operator, if we try to consider too small a class of line operators, we might find that the product KK' is not in the class that we started with. That is actually what happens in the theory described in the present paper if we consider only the most obvious class of line operators: Wilson line operators associated to representations of the finite-dimensional Lie algebra $\mathfrak{g}$. We will see in section 5.2 that this class of line operators is not closed under operator products. To get an OPE for Wilson line operators, we have to consider the more general class of Wilson operators associated to representations of $\mathfrak{g}[[z]]$.

Given two parallel Wilson line operators, on general grounds their product in a diffeomorphism-invariant theory is another line operator. But it is nontrivial to exhibit this new line operator as another Wilson operator for some representation of $\mathfrak{g}[[z]]$. To exhibit this, we have to do a calculation, starting with two parallel line operators, taking the limit, in a concrete quantization scheme, as they approach each other, and searching for a single Wilson operator that will reproduce the effects of the two Wilson operators that we started with.

At the classical level, the OPE for Wilson operators is trivial. Given Wilson operators associated to representations ρ and ρ' of $\mathfrak{g}[[z]]$, their product is the Wilson line associated to the tensor product representation $\rho \otimes \rho'$. There is a quantum correction to this and we will compute it to lowest order in $\hbar$.

We will perform the computation assuming that ρ and ρ' are representations of the finite-dimensional algebra $\mathfrak{g}$, or equivalently that they are representations of $\mathfrak{g}[[z]]$ in which the generators $t_{a,n}$ vanish for $n > 0$. What we will show is that in order $\hbar$, the tensor product representation acquires a nonzero $t_{a,1}$ (there is no correction to $t_{a,n}$ for any $n \neq 1$). It is fundamentally because of this fact that it is

[16]In three dimensions, there is a more subtle analog of this: KK' and $K'K$ are always isomorphic, but there can be different isomorphisms between them, depending on the direction in which K and K' are moved around each other, leading to a notion of braiding of line operators.

[17]This diffeomorphism symmetry is mildly broken by a framing anomaly, but not in a way that affects the present discussion.

important to consider Wilson operators associated to representations of $\mathfrak{g}[[z]]$ that do not come from representations of $\mathfrak{g}$.

5.2 Lowest Order Computation

We start with a Wilson line in the representation ρ' of $\mathfrak{g}$ at $y = 0$, and one in the representation ρ'' at $y = \epsilon$. As just explained, we assume that these are "ordinary" Wilson lines, associated to representations of $\mathfrak{g}$.

The leading order Feynman diagram is given in Fig. 17. This diagram represents the coupling of an external gauge field to the two Wilson lines. We will calculate what happens, to leading order in $\hbar$, when we put these lines beside each other and send $\epsilon \to 0$ from above. Modulo $\hbar$, the result will simply be the Wilson line associated to the tensor product representation $\rho = \rho' \otimes \rho''$ of $\mathfrak{g}$. There is an order $\hbar$ correction, in which the z-derivative of the gauge field A is coupled to the Wilson line in a non-trivial way.

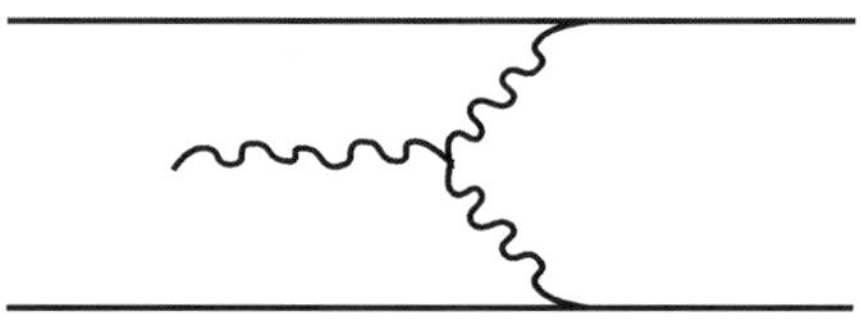

Figure 17: The Feynman diagram representing the $\mathcal{O}(\hbar)$ correction to the OPE of Wilson lines. The two lines are at $y = 0$ and $y = \epsilon$, respectively. We consider the limit $\epsilon \to 0$.

For the evaluation of this diagram we need an extra Feynman rule not needed so far, namely the bulk interaction vertex, away from any Wilson lines. This can be read off from the action (3.3), and is

$$\frac{\mathsf{i}}{2\pi} f^{abc} \mathrm{d}z \; . \tag{5.1}$$

Note that this vertex contains a 1-form $\mathrm{d}z$, since this is present in the action. To evaluate the diagram of Fig. 17, we have to integrate one interaction vertex over the line $y = 0$, one over the line $y = \epsilon$, and one over $\mathbb{R}^4 = \mathbb{R}^2 \times \mathbb{C}$. Connecting the vertices by propagators, we have to integrate

$$\begin{aligned} I_2 = & \frac{\mathsf{i}}{2\pi} \hbar (t^a \otimes t^b f_{abc}) \\ & \times \int_{x_1, x_2 = -\infty}^{\infty} \int_{x,y,z,\bar{z}} P(x_1 - x, y, z) \wedge \mathrm{d}z \wedge A^c(x,y,z) \wedge P(x_2 - x, y - \epsilon, z) \; , \end{aligned} \tag{5.2}$$

where P is the four-dimensional propagator two-form introduced previously in (4.5). Since we have two propagators and one bulk vertex, we have a factor of $\hbar^{2-1} = \hbar$ in front, as in the discussion of the R-matrix. (The meaning of this is that the diagram is of order $\hbar$ compared to a contribution in which the gauge boson couples directly to one of the Wilson operators, with the bulk interaction playing no role.)

Integrating first over x_1 and x_2, we obtain

$$I_2 = \frac{\mathsf{i}}{2\pi}\hbar(t^a \otimes t^b f_{abc}) \int_{x,y,z,\bar{z}} \mathrm{d}z \wedge A^c(x,y,z,\bar{z}) \wedge P'(y,z,\bar{z}) \wedge P'(y-\epsilon,z,\bar{z}) \,, \tag{5.3}$$

where we defined a three-dimensional one-form propagator with color indices stripped off on the plane $\mathbb{R}^3$ parametrized by $(y,z,\bar{z})$:

$$P'(y,z,\bar{z}) := \int \mathrm{d}x\, P(x,y,z,\bar{z}) = -\frac{1}{4}\left(-\mathrm{d}\bar{z}\frac{\partial}{\partial y} + 4\mathrm{d}y\frac{\partial}{\partial z}\right)\frac{1}{(y^2+z\bar{z})^{\frac{1}{2}}} \,. \tag{5.4}$$

Note that we obtained a numerical factor of π from the x integral.

The most fundamental thing to explain about eqn. (5.3) is why, for $\epsilon \to 0$, it produces a local coupling of A to a line operator at $y = z = 0$. The reason for this is simply that for $\epsilon \to 0$, the integrand in eqn. (5.3) vanishes as long as y and z are not both 0. That is true simply because the integrand for $\epsilon = 0$ is proportional to $P'(y,z,\bar{z}) \wedge P'(y,z,\bar{z})$, and this trivially vanishes because P' is a 1-form.

Accordingly, the small ϵ limit of $\mathrm{d}z \wedge P'(y,z,\bar{z}) \wedge P'(y-\epsilon,z,\bar{z})$ is a distribution supported at $y = z = 0$. By dimensional analysis and rotation symmetry in the z plane, this distribution must be a multiple of $\partial_z \delta^3(y,z,\bar{z})$, where $\delta^3(y,z,\bar{z})$ is a three-form delta function satisfying $\int_{\mathbb{R}\times\mathbb{C}} \delta^3(y,z,\bar{z}) = 1$. We will now show that

$$\lim_{\epsilon\to 0} \mathrm{d}z \wedge P'(y,z,\bar{z}) \wedge P'(y-\epsilon,z,\bar{z}) = \frac{\pi}{\mathsf{i}}\frac{\partial}{\partial z}\delta^3(y,z,\bar{z}) \,. \tag{5.5}$$

This will imply that

$$\begin{aligned} I_2 &= \frac{\mathsf{i}}{2\pi}\hbar(t^a \otimes t^b f_{abc}) \int_{x,y,\bar{z}} A^c(x,y,z,\bar{z}) \left(\frac{\pi}{\mathsf{i}}\frac{\partial}{\partial z}\delta^3(y,z,\bar{z})\right) \\ &= -\hbar\frac{1}{2}(t^a \otimes t^b f_{abc}) \int \mathrm{d}x \frac{\partial}{\partial z} A^c(x,y,z,\bar{z}) \,. \end{aligned} \tag{5.6}$$

The coupling to $\partial_z A$ shows that the composite Wilson line operator obtained by bringing two such operators together has $t_{a,1} \neq 0$, that is it is associated to a representation of $\mathfrak{g}[[z]]$ but not to a representation of $\mathfrak{g}$. Moreover, it is easy to see from this formula that although we will take $\epsilon > 0$ in our calculation, the result is actually proportional to the sign of ϵ. Changing the sign of ϵ would lead, after replacing y with $y + \epsilon$, to the same calculation but with the roles of the two line operators exchanged, replacing $t^a \otimes t^b$ in eqn. (5.6) with $t^b \otimes t^a$. Since this changes the sign of $f_{abc}t^a \otimes t^b$, eqn. (5.6) implies that the quantum OPE for Wilson operators is noncommutative: it gives a result that depends on which of the two Wilson operators is on the "left" of the other.

To show (5.5), let us express the three-dimensional one-form propagator (5.4) as

$$P'(y,z,\bar{z}) = -\mathrm{d}\bar{z}\,\partial_y Q(y,z,\bar{z}) + 4\mathrm{d}y\,\partial_z Q(y,z,\bar{z}) \tag{5.7}$$

with

$$Q(y,z,\bar{z}) := -\frac{1}{4}\frac{1}{(y^2+z\bar{z})^{\frac{1}{2}}} \,. \tag{5.8}$$

Then the left hand side of (5.5) reads (we here drop $z, \bar{z}$ from the arguments of Q, to simplify the expressions)

$$\begin{aligned}
&-4\mathrm{d}y\wedge\mathrm{d}z\wedge\mathrm{d}\bar{z}\,\left(\partial_y Q(y)\,\partial_z Q(y-\epsilon)-\partial_z Q(y)\,\partial_y Q(y-\epsilon)\right)\\
&=\mathrm{d}y\wedge\mathrm{d}z\wedge\mathrm{d}\bar{z}\left[\frac{\partial}{\partial z}(a)+\frac{\partial}{\partial y}(b)\right],
\end{aligned} \tag{5.9}$$

with

$$(a) := -2\partial_y Q(y)\,Q(y-\epsilon)+2Q(y)\,\partial_y Q(y-\epsilon)=\frac{2}{4^2}\frac{\epsilon\left(y(\epsilon-y)+|z|^2\right)}{(y^2+|z|^2)^{\frac{3}{2}}\left((y-\epsilon)^2+|z|^2\right)^{\frac{3}{2}}}\,, \tag{5.10}$$

$$(b) := -2Q(y)\,\partial_z Q(y-\epsilon)+2\partial_z Q(y)\,Q(y-\epsilon)=\frac{1}{4^2}\frac{-\epsilon\bar{z}(\epsilon-2y)}{(y^2+|z|^2)^{\frac{3}{2}}\left((y-\epsilon)^2+|z|^2\right)^{\frac{3}{2}}}\,.$$

These are all explicitly proportional to ϵ, and thus, as claimed earlier, everything vanishes for $\epsilon\to 0$ as long as y, z are not both zero. We expect to extract from (5.9) in the limit $\epsilon\to 0$ a multiple of $\partial_z\delta^3(y,z)$. Since the (a) term appears in eqn. (5.9) in the form $\partial_z(a)$, we expect that (a) by itself without this derivative would simply produce a multiple of $\delta^3(y,z)$.

This means that the contribution of the (a) term to the coefficient of $\partial_z\delta^3(y,z)$ can be obtained by evaluating the integral of (a):

$$C := \int \mathrm{d}y\mathrm{d}z\mathrm{d}\bar{z}\,(a)\,. \tag{5.11}$$

To demonstrate eqn. (5.5), it suffices to show that $C=\frac{\pi}{\mathrm{i}}$. This integral is absolutely convergent. A simple scaling argument shows that it is independent of ϵ, so we may as well set $\epsilon=1$.

In polar coordinates ($\mathrm{d}z\mathrm{d}\bar{z}=-2\mathrm{i}r\mathrm{d}r\mathrm{d}\theta$), after integration over θ, we get

$$C=\frac{-8\mathrm{i}\pi}{4^2}\int_{-\infty}^{\infty}\mathrm{d}y\int_0^{\infty}\mathrm{d}r\,\frac{r\left(y(1-y)+r^2\right)}{(y^2+r^2)^{\frac{3}{2}}\left((y-1)^2+r^2\right)^{\frac{3}{2}}}\,. \tag{5.12}$$

We can evaluate this integral with a formula familiar from the evaluation of Feynman diagrams:

$$\frac{1}{A^\alpha B^\beta}=\frac{\Gamma(\alpha+\beta)}{\Gamma(\alpha)\Gamma(\beta)}\int_0^1\mathrm{d}t\frac{t^{\alpha-1}(1-t)^{\beta-1}}{(tA+(1-t)B)^{\alpha+\beta}}\,. \tag{5.13}$$

Choosing $\alpha=\beta=3/2$ and remembering $\Gamma(3)/\Gamma(3/2)^2=8/\pi$, we have

$$C=\frac{-64\mathrm{i}}{4^2}\int_0^1\mathrm{d}t\,\sqrt{t(1-t)}\int_{-\infty}^{\infty}\mathrm{d}y\int_0^{\infty}\mathrm{d}r\frac{r\left(y(1-y)+r^2\right)}{\left((1-t)y^2+t(y-1)^2+r^2\right)^3}\,. \tag{5.14}$$

Using

$$\int_0^\infty \mathrm{d}r \frac{r(a+r^2)}{(b+r^2)^3} = \frac{a+b}{4b^2}\,, \tag{5.15}$$

we obtain

$$\begin{aligned} C &= \frac{-16\mathrm{i}}{4^2}\int_0^1 \mathrm{d}t\,\sqrt{t(1-t)}\int \mathrm{d}y\,\frac{y(1-y)+(1-t)y^2+t(y-1)^2}{((1-t)y^2+t(y-1)^2)^2} \\ &= \frac{-16\mathrm{i}}{4^2}\int_0^1 \mathrm{d}t\,\sqrt{t(1-t)}\frac{\pi}{\sqrt{t(1-t)}} = \frac{\pi}{\mathrm{i}}\,. \end{aligned} \tag{5.16}$$

We can do a similar analysis for (b), but (b) takes the form

$$\bar{z}F(y,|z|) \tag{5.17}$$

and hence vanishes when integrated over the angle θ. This means that there is no singular contribution from (b). This concludes the verification of (5.5).

5.3 Relation to the R-Matrix

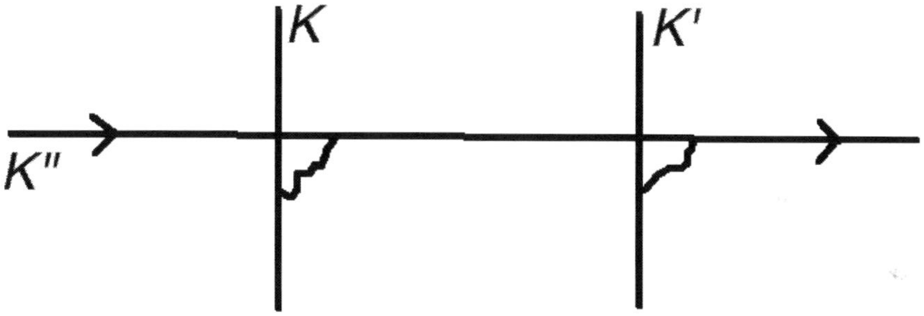

Figure 18: A Wilson operator K'' crossing two parallel Wilson operators K and K'. K and K' are supported at $z=0$ and K'' at some other point $z=z_0$. Drawn is a Feynman diagram with a gluon exchanged between K and K'' as well as one between K' and K''.

Though we have performed an independent computation, the result can actually be deduced from a knowledge of the $\mathcal{O}(\hbar)$ contribution to the R-matrix, computed in section 4. Consider (Fig. 18) two parallel Wilson lines K and K' both supported at $z=0$ and a third Wilson line K'' at some other point $z=z_0$ that is crossing them. Assuming that K and K' are far apart (compared to $|z_0|$) there is a unique lowest order Feynman diagram in which K'' interacts nontrivially with both K and K'. This is the diagram sketched in the figure with a gluon exchanged from K'' to K and another from K'' to K'. The group theory factors associated to the two gluon exchanges are respectively $t_a \otimes 1 \otimes t_a$ and $1 \otimes t_b \otimes t_b$ (where the three factors refer respectively to K, K', and K''). The product of these is $t_a \otimes t_b \otimes t_b t_a$, where the factors acting on K'' are ordered as $t_b t_a$ because K' is to the right of K (so that path ordering along K'' puts t_b to the left of t_a).

Now suppose that K and K' are brought closer together. Two-dimensional diffeomorphism invariance means that we have to get the same result after summing over all diagrams, but the contributions of individual diagrams can depend on the distance from K to K'. In particular, the diagram that we considered before still contributes when K and K' are close compared to $|z_0|$, but their contribution is not the same as before. That is because (Fig. 19(a)) even though K' is to the right of K, the gluon emitted from K' might be absorbed on K'' to the left of the gluon emitted from K. So the contribution of the diagram that we considered previously is now modified by a term proportional to $t_a \otimes t_b \otimes [t_a, t_b] = f_{abc} t_a \otimes t_b \otimes t_c$. So there must be another diagram that is significant when K and K' are nearby and that gives a group theory factor of this form. That diagram is shown in Fig. 19(b).

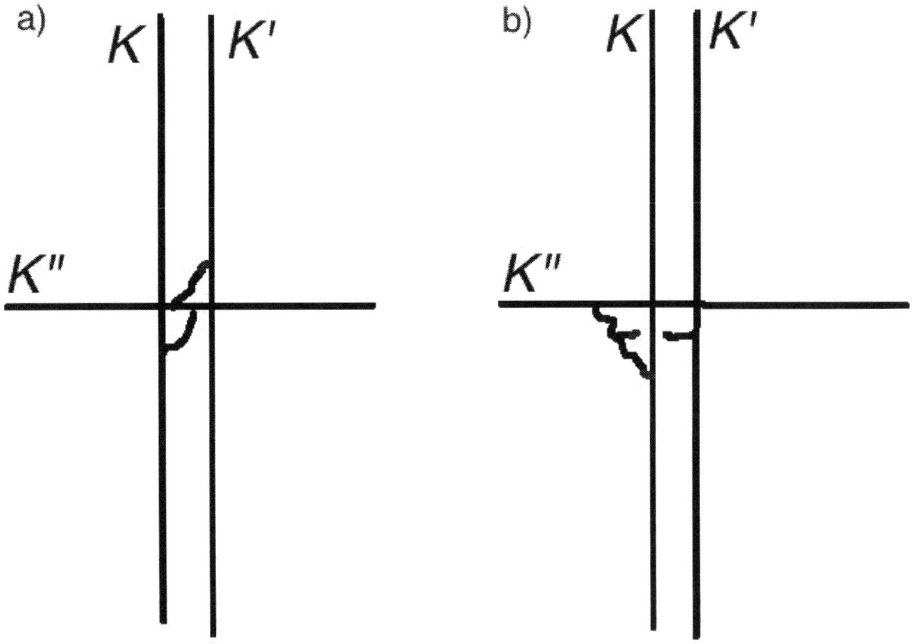

Figure 19: When K and K' are brought nearby, the contribution of the diagram in (a) is modified relative to Fig. 18. To maintain two-dimensional topological invariance, this is balanced by a new contribution in (b), which is related to the existence of a quantum correction to the OPE.

But this last diagram is just the one (Fig. 17) that we studied to find the quantum correction to the OPE, except that now instead of considering an arbitrary external A field, as in the previous discussion, we have provided a third Wilson operator that is the source of this A field. The analysis of Fig. 19(b) for

$K \to K'$ is essentially the same as the analysis that we have already performed of Fig. 17.

The upshot is that the quantum correction to the OPE is an inevitable consequence of the quantum correction to the R-matrix, and vice-versa.

5.4 Interpretation of the Result

In the analysis of Fig. 17, we started with two representations ρ' and ρ'' of $\mathfrak{g}$, that is representations of $\mathfrak{g}[[z]]$ with generators $t'_{a,n} = t''_{a,n} = 0$ for $n > 0$. Let us write just t'_a and t''_a for $t'_{a,0}$ and $t''_{a,0}$. The above computation showed that the Wilson operator obtained by fusing the two we started with couples to $\partial_z A$ and therefore has a nonzero $t_{a,1}$. In fact $t_{a,1}$ of the fused Wilson line can be read off from eqn. (5.6):

$$t_{a,1} = -\frac{\hbar}{2} \sum_{b,c} f_{abc} t'_b \otimes t''_c \; . \tag{5.18}$$

On the other hand, the computation gave no contributions to generators $t_{a,n}$ of the composite Wilson operator for $n \neq 1$. So to this order, $t_{a,0}$ is given by the classical formula

$$t_{a,0} = t'_a \otimes 1 + 1 \otimes t''_a \; , \tag{5.19}$$

and the higher generators vanish, $t_{a,n} = 0$, $n > 1$. Here $t_{a,0}$ are the generators of $\mathfrak{g}$ in the classical tensor product $\rho = \rho' \otimes \rho''$.

This result certainly shows that to get a closed OPE, we must consider Wilson operators derived from representations of $\mathfrak{g}[[z]]$, not just $\mathfrak{g}$. But its consequences go far beyond that. The formula (5.18) actually implies that quantum corrections to the theory actually deform $\mathfrak{g}[[z]]$ itself – or to be more precise that at the quantum level, Wilson operators of the theory correspond to representations not of $\mathfrak{g}[[z]]$ itself but of a quantum deformation of this algebra (or more accurately, of its universal enveloping algebra, as we will see).

The basic reason for this is that the formula (5.18) is not consistent with the commutation relations of $\mathfrak{g}[[z]]$, so it implies further deformations. In $\mathfrak{g}[[z]]$, one has the commutation relation

$$[t_{a,1}, t_{b,1}] = f_{abc} t_{c,2} \; . \tag{5.20}$$

Recalling the Jacobi identity

$$f_{uva} f_{abc} + f_{vba} f_{auc} + f_{bua} f_{avc} = 0 \; , \tag{5.21}$$

we deduce from this that

$$f_{uva} [t_{a,1}, t_{b,1}] + (\dots) = 0 \; , \tag{5.22}$$

where the omitted terms are obtained by cyclic permutations of the indices uvb.

This is an identity in $\mathfrak{g}[[z]]$, but a short calculation will show that $t_{a,1}$ as defined in eqn. (5.18) does not obey this identity. Instead it satisfies a deformed version of this identity that we will describe presently.

Because $\mathfrak{g}[[z]]$ is being deformed, what did we mean in claiming that to get a closed OPE for line operators, we should start with representations of $\mathfrak{g}[[z]]$? The precise statement is not that line operators in the quantum theory correspond to representations of $\mathfrak{g}[[z]]$, but that had we started at the classical level with arbitrary representations of the $\mathfrak{g}[[z]]$, then consideration of products of line operators would not have forced us to consider new objects. By contrast, if we start at the classical level with representations of $\mathfrak{g}$ only, then getting a closed OPE does require introducing many more line operators.

Now let us go back to the question of how to interpret the failure of the quantum-induced $t_{a,1}$ to obey the commutation relations of $\mathfrak{g}[[z]]$. Since $t_{a,1}$ in eqn. (5.18) is bilinear in the generators t'_a and t''_a, the left hand side of eqn. (5.21), after evaluating the commutator, is cubic in these generators; more specifically it is a sum of terms of bidegree $(2,1)$ and $(1,2)$ in t'_a and t''_a.

The upshot is that to account for the failure of eqn. (5.22), we have to deform this commutation relation by adding, in order $\hbar^2$, a certain cubic polynomial in $t_{a,0}$. The generators $t_{a,0}$ and $t_{a,1}$ of the deformed algebra satisfy

$$f_{uva}[t_{a,1}, t_{b,1}] + (\dots) = \hbar^2 Q_{uvb}(t_{a,0}) \,, \tag{5.23}$$

where $Q_{uvb}(t_{a,0})$, which is completely antisymmetric in its indices uvb, is for each uvb a homogeneous symmetric cubic polynomial in the $t_{a,0}$.

What polynomial is needed can be deduced by trying to make sure that eqn. (5.23) is consistent with eqns. (5.18) and (5.19). However, when one tries to do this, one runs into a snag. No matter what Q_{uvb} may be, if $t_{a,0} = t'_a \otimes 1 + 1 \otimes t''_a$, then $Q_{uvb}(t_{a,0})$ is a sum of terms of bidegree $(3,0)$, $(2,1)$, $(1,2)$, and $(0,3)$ in t'_a and t''_a. We can pick Q_{uvb} so that the terms in eqn. (5.23) of bidegree (2,1) and (1,2) work out correctly. But there is nothing we can do about the terms of bidegree $(3,0)$ and $(0,3)$, unless they vanish by themselves.

The interpretation of this is as follows. The quantum deformed algebra satisfies (5.23) (with analogous deformations of the commutation relations involving other generators). In contrast to $\mathfrak{g}[[z]]$, a representation of $\mathfrak{g}$ does not automatically lift or extend to a representation of the deformed algebra. If we start with matrices t_a that represent $\mathfrak{g}$, we can always get a representation of $\mathfrak{g}[[z]]$ by setting $t_{a,0} = t_a$ and $t_{a,n} = 0$ for $n > 0$. But this only works in the deformed algebra if $Q_{uvb}(t_a) = 0$. Otherwise, the deformed commutation relations force $t_{a,1}$ to be nonzero. So quantum mechanically, the Wilson operators whose product we are trying to determine are anomalous, and need to be modified[18] with a contribution to $t_{a,1}$ of $\mathcal{O}(\hbar)$, unless $Q_{uvb}(t'_a) = Q_{uvb}(t''_a) = 0$. Once we impose this restriction, we do not need to worry about the terms in eqn. (5.23) of bidegree $(3,0)$ or $(0,3)$. We need consider only the $(2,1)$ and $(1,2)$ terms in that equation.

The explicit polynomial $Q_{uvb}(t_{a,0})$, though not very illuminating, can be worked out by analyzing those terms.[19] A notable fact is that because this polynomial is of degree greater than 1, the deformation from $\mathfrak{g}[[z]]$ by including $\hbar^2 Q_{uvb}$

[18] If they cannot be so modified, they are simply anomalous and do not have counterparts in the quantum theory.

[19] We will also derive it by a direct Feynman diagram analysis in section 8. Section 8.6.1 contains a fairly thorough analysis of this polynomial. See also [8], p. 376.

on the right hand side of eqn. (5.23) cannot be understood as a deformation of the Lie algebra $\mathfrak{g}[[z]]$. It can instead be understood as part of an associative algebra deformation of the universal enveloping algebra of $\mathfrak{g}[[z]]$, denoted $U(\mathfrak{g}[[z]])$. This makes sense because a polynomial in elements of $U(\mathfrak{g}[[z]])$ is itself an element of $U(\mathfrak{g}[[z]])$, so $Q_{uvb}(t_{a,0})$ is an element of $U(\mathfrak{g}[[z]])$. Eqn. (5.23) (and its analogs for other components that we have not calculated) can be understood as giving an associative algebra deformation (which in particular entails a Lie algebra deformation) of $U(\mathfrak{g}[[z]])$.

5.5 The Yangian

The deformation of $U(\mathfrak{g}[[z]])$ that was uncovered in section 5.4 is known as the Yangian. In fact, eqn. (5.18) is a standard formula describing the difference in lowest order between the tensor product of representations of the Yangian and the tensor product of representations of $\mathfrak{g}[[z]]$.

How can we know that the relevant deformation of $U(\mathfrak{g}[[z]])$ is the Yangian, given that we have only performed some simple computations in lowest nontrivial order? The answer to this question is that according to general theorems, the Yangian is the only deformation of the OPE that agrees with the lowest order deformation that we have found in eqn. (5.18) and possesses certain general properties. Thus (as already stated in [11, 12]), the associativity of the OPE, together with the fact that the OPE in $\mathcal{O}(\hbar)$ receives a non-trivial correction, determines the OPE uniquely to all orders in perturbation theory, up to changes of variables. This is guaranteed by a theorem of Drinfeld, which also says that the resulting 1-parameter deformation of the $U(\mathfrak{g}[[z]])$ is the Yangian $Y_\hbar(\mathfrak{g})$, the algebra underlying the rational solution of the Yang-Baxter equation associated to $\mathfrak{g}$.

To make these statements more precise and closer in spirit to those in [11, 12], we can use the language of category theory to say that the "category of Wilson lines" $\mathcal{C}_\hbar$ is given by the category of representations of $Y_\hbar(\mathfrak{g})$.

This category $\mathcal{C}_\hbar$ has the following four properties.

First, $\mathcal{C}_\hbar$ is a non-trivial one-parameter deformation of the category of representations[20] of $U(\mathfrak{g}[[z]])$.

Second, the OPE of the Wilson lines defines a functor, by fusing two objects to produce a third object

$$\mathcal{C}_\hbar \times \mathcal{C}_\hbar \to \mathcal{C}_\hbar \,. \tag{5.24}$$

Moreover, the OPE is associative, though not commutative. These two conditions define a *monoidal category*, and hence $\mathcal{C}_\hbar$ is monoidal.[21]

Third, the category $\mathcal{C}_\hbar$ allows for a kind of braiding[22], because it has certain

[20] At the classical level, representations of $\mathfrak{g}[[z]]$ are the same as representations of $U(\mathfrak{g}[[z]])$, so Wilson operators can be identified with representations of either of these algebras. But the statement in the text must be formulated for $U(\mathfrak{g}[[z]])$, because the Yangian is a deformation of $U(\mathfrak{g}[[z]])$, not of $\mathfrak{g}[[z]]$.

[21] By the same argument, the category of line operators in a two-dimensional topological quantum field theory (TQFT) is always a monoidal category, as is the category of boundary line operators in a 3-dimensional TQFT with a topological boundary condition.

[22] This is different from the braiding familiar from line operators in three-dimensional topological field theories.

additional structure. Here we again consider two Wilson lines parallel in the topological plane, but now at different points z and $z+\lambda$ of the complex plane. Since $z \neq z+\lambda$, the Wilson lines never coincide with each other, even when they coincide in the topological plane. Moreover, since we have two directions at our disposal in the holomorphic plane C, we can move the two Wilson lines around in C, to replace z with $z+\lambda$, without encountering any singularities. In the language of [11, 12], the translation $z \mapsto z+\lambda$ of the Wilson line by a parameter λ defines a functor $T_\lambda : \mathcal{C}_\hbar \to \mathcal{C}_\hbar$. If W, V are two Wilson lines supported at 0, then the explanations above means that we have a natural isomorphism

$$R_{V,W} : T_\lambda W \otimes V \cong V \otimes T_\lambda W \ , \tag{5.25}$$

where $\otimes$ indicates the monoidal structure on $\mathcal{C}_\hbar$ coming from the OPE of parallel Wilson lines.

A theorem of Drinfeld [7] now implies that $\mathcal{C}_\hbar$ is uniquely fixed by these properties to be the category of representations of the Yangian $Y_\hbar(\mathfrak{g})$. The monoidal structure on $\mathcal{C}_\hbar$ comes from the coproduct on the Yangian, and the map $R_{V,W}$ comes from the R-matrix of the Yangian $Y_\hbar(\mathfrak{g})$.

These arguments are admittedly somewhat abstract. In a companion paper [25], we will explain, in a concrete and down to earth way, how to extract a representation of the Yangian – exactly, not just to order $\hbar$ – from any Wilson line operator of this theory. The ability to do this is somewhat analogous to the ability in section 3.5 to determine in an elementary way some solutions of the Yang-Baxter equation for GL_N. Hopefully this will help convince the reader that the conclusions from the abstract arguments really apply to the quantum field theory.

6 The Framing Anomaly and Curved Wilson Lines

In section 3.6, we deduced from some explicit elementary examples that there must be a framing anomaly for Wilson lines. The framing anomaly says that what is constant along a Wilson line is not z but $z - \hbar \mathsf{h}^\vee \varphi/\pi$, where $\mathsf{h}^\vee$ is the dual Coxeter number of the gauge group, and φ is the angle between the tangent vector to the Wilson line and some chosen direction in the topological plane. The purpose of the present section is to perform a direct Feynman diagram computation exhibiting this effect.

In contrast to the calculations that we performed for the R-matrix and for the OPE of Wilson operators, for the framing anomaly, a lowest order computation gives the exact result. For the case of the Yangian – that is for $C = \mathbb{C}$ – this is clear from the fact that this theory is invariant under adding a constant to z and under a common rescaling of z and $\hbar$. The assertion that $z - \hbar \mathsf{h}^\vee \varphi/\pi$ is constant along a Wilson operator is consistent with those symmetries, but a modified statement with additional terms of higher order in $\hbar$ would not be. The fact that the lowest order framing anomaly gives the full answer is also evident in the examples considered in section 3.6.

For other choices of C related to trigonometric or elliptic solutions of the Yang-Baxter equation, there is no scaling symmetry. However, because the analysis of the framing anomaly is local along C, the form of the anomaly is the same for the other cases as long as one picks variables so that locally along C the action takes the same form as for $C = \mathbb{C}$. (In the coordinates of eqn. (3.14), this is automatically true if C is an elliptic curve, and it is true for $C = \mathbb{C}^\times$ after replacing z with $\log z$.)

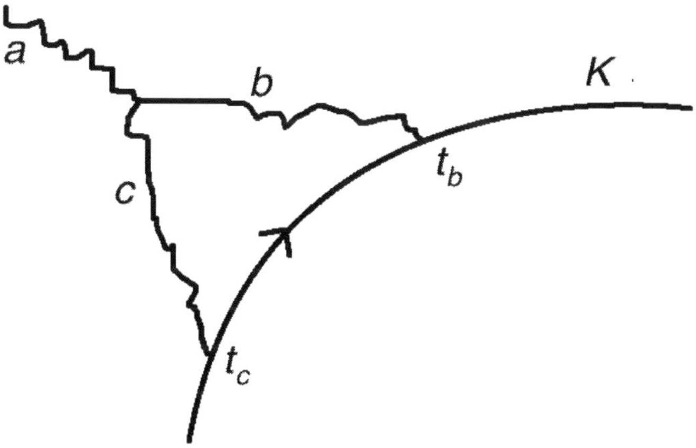

Figure 20: The one-loop diagram that leads to a framing anomaly for Wilson operators.

The relevant diagram that will give the anomaly is shown in Fig. 20. This diagram makes a convergent and well-defined contribution to the effective action of the theory. However, this contribution is not gauge-invariant. A failure of gauge-invariance will arise from the behavior when all three interaction vertices coincide at a common point, which will inevitably be along the support K of the Wilson operator.

Before describing what form the anomaly will take, we first do a small exercise in group theory. We write the gauge field A explicitly as $A = \sum_a A_a t_a$, and recall that the trilinear vertex involving a coupling of three gauge fields is proportional to f_{abc}. A gauge boson of type b or c couples to the Wilson line via a factor of t_b or t_c (understood here as a matrix acting in the appropriate representation). So as far as group theory is concerned, the indicated diagram generates a coupling of an external gauge field A_a to the Wilson operator via a matrix

$$\sum_{a,b,c} A_a f_{abc} t_b t_c \, . \tag{6.1}$$

Here because of antisymmetry of f_{abc}, we can replace $t_b t_c$ by $\frac{1}{2}[t_b, t_c] = \frac{1}{2} f_{bcd} t_d$. Then as $\sum_{b,c} f_{abc} f_{bcd} = 2\mathsf{h}^\vee \delta_{ad}$, where $\mathsf{h}^\vee$ is known as the dual Coxeter number[23] of $\mathfrak{g}$. Thus just from the point of view of group theory, we get a coupling $\mathsf{h}^\vee \sum_a A_a t_a$ of the external gauge field to the Wilson line. Thus the dependence on $\mathfrak{g}$ is only an overall factor of $\mathsf{h}^\vee$.

To calculate the anomaly, it suffices to consider the case of a Wilson line supported on a curve $K \subset \mathbb{R}^2$ that is nearly a straight line, and at a point $z = z_0$ in $\mathbb{C}$. In fact, since a simple description in terms of Yangians and integrable systems only arises in a limit in which the metric of $\mathbb{R}^2$ is scaled up, we are really supposed to consider only the case that K has a very large radius of curvature and therefore can everywhere be locally well-approximated as a straight line. We can pick coordinates so that K, in the region of interest, is very close to the x-axis in $\mathbb{R}^2$, and is described by a curve $y = y(x)$, with y everywhere small. We will find that the amplitude I_1 that comes from the diagram of Fig. 20 is not gauge-invariant. Under a gauge transformation $A \to A + D\varepsilon$, with $\varepsilon = \sum_a t_a \varepsilon_a(x, y, z)$ a gauge parameter, the variation of I_1 will be

$$\delta I_1 = -\frac{\hbar\,\mathsf{h}^\vee}{2\pi} \int_K \mathrm{d}x \left(\frac{\mathrm{d}^2 y}{\mathrm{d}x^2} \partial_z \varepsilon(x, y(x), z_0) \right) . \tag{6.2}$$

Here as above $\mathsf{h}^\vee$ is the dual Coxeter number. (A factor of $\hbar$ is present because this is the anomaly in a diagram with two propagators, each giving a factor of $\hbar$, and one bulk vertex, proportional to $1/\hbar$.) This formula for δI_1 as an integral over K is written in a way that is only valid in a portion of K that is close to the x-axis. Since we will assume $y(x)$ small, we can replace $\varepsilon(x, y(x), z_0)$ by $\varepsilon(x, 0, z_0)$, which we will write simply as $\varepsilon(x, z_0)$. Now let us discuss how to cancel this anomaly by a correction to the classical definition of the Wilson operator. We recall that, in linear order, an external gauge field A couples to the Wilson operator via

$$I_{\mathrm{cl}} = \int_K \mathrm{d}x A_x(x, z_0) . \tag{6.3}$$

This is gauge-invariant, as long as z_0 is constant. To cancel the anomaly δI_1, we must modify the classical coupling I_{cl} so that it is no longer gauge-invariant. We do this by replacing z_0 by $z_0 - \hbar\,\mathsf{h}^\vee \frac{1}{2\pi} \mathrm{d}y/\mathrm{d}x$. Thus we replace I_{cl} with

$$I'_{\mathrm{cl}} = \int_K \mathrm{d}x A_x \left(x, z_0 - \frac{\hbar\,\mathsf{h}^\vee}{2\pi} \frac{\mathrm{d}y}{\mathrm{d}x} \right) . \tag{6.4}$$

Under a gauge transformation $\delta A_x(x, y, z) = \partial_x \varepsilon(x, y, z)$, the variation of I'_{cl} is[24]

$$\delta I'_{\mathrm{cl}} = -\frac{\hbar\,\mathsf{h}^\vee}{2\pi} \int_K \mathrm{d}x \left(\frac{\mathrm{d}^2 y}{\mathrm{d}x^2} \partial_z \varepsilon(x, z_0) \right) . \tag{6.5}$$

This cancels the anomaly δI_1.

We learn, then, that including the anomaly, we must set not $z = z_0$ but $z = z_0 - \frac{1}{2\pi} \hbar\,\mathsf{h}^\vee \mathrm{d}y/\mathrm{d}x$. The formulas as written, however, are only valid under

[23] Recall our normalization of the Killing form in eqn. (3.2). The factor of 2 here is inserted to match with the standard definition of $\mathsf{h}^\vee$ in the literature.

[24] We use $\mathrm{d}\varepsilon/\mathrm{d}x = \partial_x \varepsilon + (\partial_x z) \partial_z \varepsilon$. We work to first order in y and replace $\partial_z \varepsilon(x, z_0 - \hbar\,\mathsf{h}^\vee \mathrm{d}y/\mathrm{d}x)$ with $\partial_z \varepsilon(x, z_0)$.

the assumption that $\mathrm{d}y/\mathrm{d}x$ is small. To get a more general formula, we should re-express $\mathrm{d}y/\mathrm{d}x$ in terms of the angle φ between the tangent vector to K and some chosen direction in the xy plane, for instance the direction of increasing x. If we define the sign of φ to increase when K rotates in a clockwise direction, then $\mathrm{d}y/\mathrm{d}x$ can be identified (when it is small) with $-\varphi$. So a more general form of the relation between z and the slope of K is $z = z_0 + \hbar\,\mathsf{h}^\vee\varphi/(2\pi)$. A different way to express this result is to say that along K, what is constant is not z but $z - \hbar\,\mathsf{h}^\vee\varphi/(2\pi)$. This is the way the anomaly was stated in section 3.6.

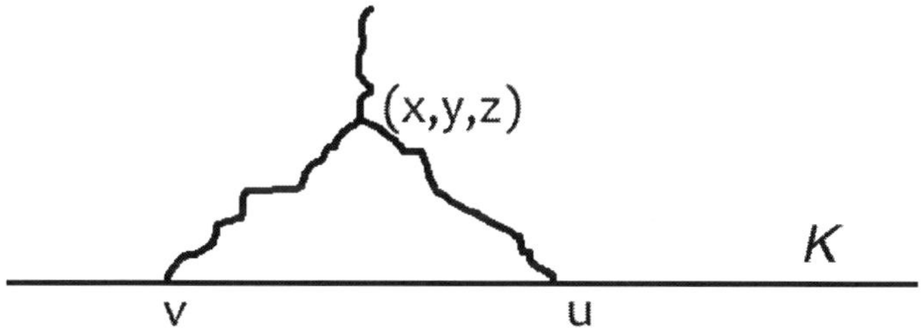

Figure 21: The one-loop diagram with K taken to be a straight line – the x axis.

Before trying to explain how this anomaly comes from the diagram of Fig. 20, let us first explain how the amplitude coming from that diagram is defined. The only possible problem in defining the integral associated to this diagram comes from the region of integration in which all three interaction vertices – one in bulk and two along the curve K – coincide in $\mathbb{R}^4$. In this region, K can be approximated as a straight line. Let us take the line to be the x-axis, that is the curve $y = 0$, and discuss the integral in this case. For simplicity, we choose $z_0 = 0$. As in the figure, we take the vertices along K to be at $x = u, v$ with $v \leq u$, and we place the bulk vertex at a generic point (x, y, z). Making use of the explicit form of the propagators, the necessary integral turns out to be (up to a constant factor, about which we will be more precise later)

$$\begin{aligned}&\int_{-\infty<v\leq u<\infty} \mathrm{d}u\,\mathrm{d}v \int_{\mathbb{R}^4} \mathrm{d}x\mathrm{d}y\mathrm{d}z\mathrm{d}\bar{z}\\ &\cdot A_x(x,y,z)\frac{2y\bar{z}}{((x-v)^2+y^2+|z|^2)^2((x-u)^2+y^2+|z|^2)^2}\,.\end{aligned} \tag{6.6}$$

Here $A = A_x\mathrm{d}x + A_y\mathrm{d}y + A_{\bar{z}}d\bar{z}$ is a background field. However, for the case of a straight line in a flat $\mathbb{R}^2$ (and only for this case), only the A_x component contributes, as is indicated in eqn. (6.6). This depends on the detailed form of the propagator.

Any possible divergence in the integral will be a local effective action along K that will be represented as an integral along K. Thus in assessing the convergence of the integral, we should leave one variable unintegrated, say x. (Alternatively, we could ensure that there is no divergence in the integral over x by taking A

to have compact support in the x direction.) So we set, for example, $x = 0$. If $A_x(0, y, z)$ is independent of y and z, then the integral (6.6) is linearly divergent by power counting. (We have to integrate over five variables $u, v, y, z, \bar{z}$, and the integrand scales as $(\text{length})^{-4}$.) However, the integrand is odd under $y \to -y$ and separately under $z \to -z$. Accordingly, the integral is well-defined by a sort of principal value prescription. We take $\mathbb{R}^4_\eta$ to be the subspace of $\mathbb{R}^4$ defined by $(y^2 + |z|^2)^{1/2} \geq \eta$, and we replace the integral by

$$\lim_{\eta \to 0} \int_{-\infty < v \leq u < \infty} \mathrm{d}u \, \mathrm{d}v \int_{\mathbb{R}^4_\eta} \mathrm{d}x \mathrm{d}y \mathrm{d}z \mathrm{d}\bar{z} \\ \cdot A_x(x, y, z) \frac{2y\bar{z}}{((x-v)^2 + y^2 + |z|^2)^2((x-u)^2 + y^2 + |z|^2)^2} \, . \tag{6.7}$$

This integral is well-defined. Concretely, if we expand $A_x(0, y, z)$ in powers of y, z and $\bar{z}$ near $y = z = 0$, the first term that can contribute is proportional to yz and leads to a convergent integral. (Other nonvanishing contributions come from terms with additional factors of y^2 or $z\bar{z}$.)

We have described this for the case that K is a straight line, but since any curve can be approximated locally as a straight line, the general case is similar. The one-loop integral can always be defined by a principal value procedure in which one constrains the bulk vertex to be a distance at least η from K and then takes the limit $\eta \to 0$. The only change from the discussion in the last paragraph is that instead of breaking the symmetry under $y \to -y$ by picking out a term in $A(0, y, z)$ that is linear in y, we could make use of the curvature of K to break the symmetry. This again leads to a convergent integral.

Now that we have defined the 1-loop amplitude that we want to study, we can assess its gauge-invariance. For this, it is convenient to perform the u and v integrals in (6.7), leading to an expression of the general form

$$\int_{\mathbb{R}^4} A \wedge \Theta_0 \, , \tag{6.8}$$

where Θ_0 is 3-form on $\mathbb{R}^4$. To be more precise, Θ_0 is a distributional 3-form on $\mathbb{R}^4$, that is, it is a distribution on smooth 1-forms A. We note that although we have described a specific procedure to define Θ_0, there was nothing really distinguished about this procedure and we could have used a different one. The effect of using a different procedure would be to add to Θ_0 a distribution Θ_0' supported on K. But

$$\int_{\mathbb{R}^4} A \wedge \Theta_0' = \int_K A\beta \, , \tag{6.9}$$

where β is some 0-form on K, which depends on Θ_0'. A shift of this type in the effective action can be interpreted as a change in the classical line operator whose quantum properties we are trying to study. We are only interested in anomalies modulo those that can be removed by such a redefinition of the underlying classical line operator.

We want to assess gauge-invariance of eqn. (6.7) under $A \to A + D\varepsilon$, where $D = \mathrm{d}x\partial_x + \mathrm{d}y\partial_y + \mathrm{d}\bar{z}\partial_{\bar{z}}$. It suffices to consider the case $\varepsilon = z\varepsilon'$ where ε' is regular at $z = 0$. In fact, any ε can be written $\varepsilon = z\varepsilon'(x, y, z, \bar{z}) + \varepsilon''(x, y, \bar{z})$, where ε'' is holomorphic in $\bar{z}$. In the following computation, any contribution from ε'' will disappear after integration over the phase or argument of z. (It will also become clear that we can assume ε' to be independent of z and $\bar{z}$ since terms proportional to z or $\bar{z}$ are not singular enough to be relevant, and that likewise only the restriction of ε' to K matters.) Since D commutes with z, it is equivalent to replace Θ_0 with $\Theta = z\Theta_0$ and to assess the invariance of

$$\int_{\mathbb{R}^4} A \wedge \Theta \tag{6.10}$$

under $A \to A + D\varepsilon'$. After integration by parts, this means that we need to study the distributional four-form $D\Theta$. This form vanishes away from K because of the classical gauge-invariance of the theory. So we expect $D\Theta$ to be a distribution with support on K. Finally, as Θ is proportional to an explicit factor of $\mathrm{d}z$ which comes from the bulk interaction vertex, we can replace $D\Theta$ by $\mathrm{d}\Theta$, where d is the ordinary exterior derivative $\mathrm{d} = D + \mathrm{d}z\partial_z$.

Thus, we are reduced to studying $\mathrm{d}\Theta$, which should equal a distribution supported along K. In fact, we claim

$$\mathrm{d}\Theta = \alpha\delta_K \, , \tag{6.11}$$

where δ_K is a three-form delta function that is Poincaré dual to K, and α is a one-form supported on K. The anomaly is then given by α. (*A priori*, instead of δ_K, we might have gotten an expression involving normal derivatives of δ_K. Indeed, this would have happened had we not included an explicit factor of z in the definition of Θ. With that term extracted, we will see that the most singular contribution is a delta function rather than a derivative of one.) We will use the following simple procedure to study $\mathrm{d}\Theta$. We assume that K is close to the x-axis, so that it can be parametrized by x. Then we expand the three-form Θ as the sum of two terms, one proportional to $\mathrm{d}x$ and one not:

$$\Theta = \mathrm{d}x\Lambda + \Lambda' \, . \tag{6.12}$$

Here Λ' cannot contribute to the anomaly. The reason is that as Λ', by definition, is a 3-form that vanishes if contracted with ∂_x, and is smooth away from K, $\mathrm{d}\Lambda'$ cannot generate a delta function supported on K unless Λ' already has a delta function supported on K. But such a term is irrelevant; it could be eliminated by redefining Θ along the lines of eqn. (6.9).

So in studying $\mathrm{d}\Theta$, we can replace Θ with $\mathrm{d}x\Lambda$, in other words with the part of Θ that is proportional to $\mathrm{d}x$. We restrict Λ to, say, $x = 0$, and look for a delta function contribution in $\mathrm{d}\Lambda$ at $y = z = 0$. We can assume that K is tangent to the x-axis at $x = 0$ and is described near $x = 0$ by $y = \frac{1}{2}fx^2$, where $f = \mathrm{d}^2y/\mathrm{d}x^2$. Note that $\Lambda = 0$ for the case that K is the x-axis. (This is equivalent to the statement that the only component of A that appears in eqn. (6.6) above is A_x.) So we expand Λ in powers of f near $f = 0$. We will see momentarily that the term in Λ linear in f does indeed lead to a delta function in $\mathrm{d}\Lambda$. Higher order terms in f are not singular enough to make such a contribution.

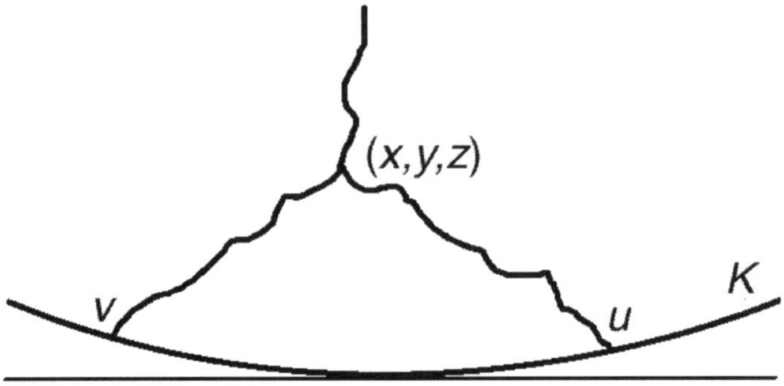

Figure 22: Here we take K to be tangent to the x-axis (the horizontal line in the figure) at, say, $x = 0$.

Explicit calculation (see Appendix D) shows that the contribution to Λ that is linear in f is at $x = 0$,

$$\Lambda = -\frac{3f\mathrm{i}}{32\pi^2}\hbar\,\mathsf{h}^\vee \frac{-yz\bar{z}(\mathrm{d}z \wedge \mathrm{d}\bar{z}) + 2z\bar{z}^2(\mathrm{d}y \wedge \mathrm{d}z)}{(y^2 + |z|^2)^{\frac{5}{2}}} . \tag{6.13}$$

By Stokes's theorem, the coefficient of the delta function can be extracted as

$$\int_{S_\eta} \Lambda , \tag{6.14}$$

where S_η is the two-sphere $\left(y^2 + |z|^2\right)^{1/2} = \eta$ (the integral is independent of η, so we can choose $\eta = 1$). We then need to evaluate

$$\int_{S_{\eta=1}} \left(-yz\bar{z}(\mathrm{d}z \wedge \mathrm{d}\bar{z}) + 2z\bar{z}^2(\mathrm{d}y \wedge \mathrm{d}z)\right) . \tag{6.15}$$

By choosing $y = \cos\theta, z = \sin\theta e^{\mathrm{i}\phi}$, this becomes[25]

$$2\pi \int_0^\pi \mathrm{d}\theta \left(-2\mathrm{i}\sin^3\theta\cos^2\theta - 2\mathrm{i}\sin^5\theta\right) = (2\pi)(-2\mathrm{i})\left(\frac{4}{15} + \frac{16}{15}\right) = -\frac{16\pi\mathrm{i}}{3} . \tag{6.16}$$

Therefore the coefficient in front of the delta function is $-f\hbar\,\mathsf{h}^\vee/(2\pi)$:

$$\mathrm{d}\Lambda' = -\frac{1}{2\pi} f\hbar\,\mathsf{h}^\vee \delta^3(y, z) , \tag{6.17}$$

where $\delta^3(y, z)$ is the Poincaré dual to the point $x = y = z = 0$ in the hypersurface $x = 0$.

Since $f = \mathrm{d}^2y/\mathrm{d}x^2$, this result when inserted back in (6.10) corresponds to an anomaly

$$-\frac{\hbar\,\mathsf{h}^\vee}{2\pi} \int_K \mathrm{d}x \frac{\mathrm{d}^2y}{\mathrm{d}x^2} \varepsilon' . \tag{6.18}$$

[25]We use $\mathrm{d}z \wedge d\bar{z} = -2\mathrm{i}\sin\theta\cos\theta d\theta \wedge d\phi$, $dy \wedge \mathrm{d}z = -\mathrm{i}\sin^2\theta e^{-\mathrm{i}\phi} d\theta \wedge d\phi$.

But ε' was defined by removing a factor of z from ε. So an equivalent and more illuminating way to describe the result is that the anomaly in the 1-loop amplitude under a gauge transformation generated by ε is

$$-\frac{1}{2\pi}\hbar\,\mathsf{h}^\vee \int_K \mathrm{d}x \frac{\mathrm{d}^2 y}{\mathrm{d}x^2}\partial_z \varepsilon\,. \tag{6.19}$$

This is the form of the anomaly that was promised in eqn. (6.2).

7 Networks of Wilson Lines

7.1 Overview

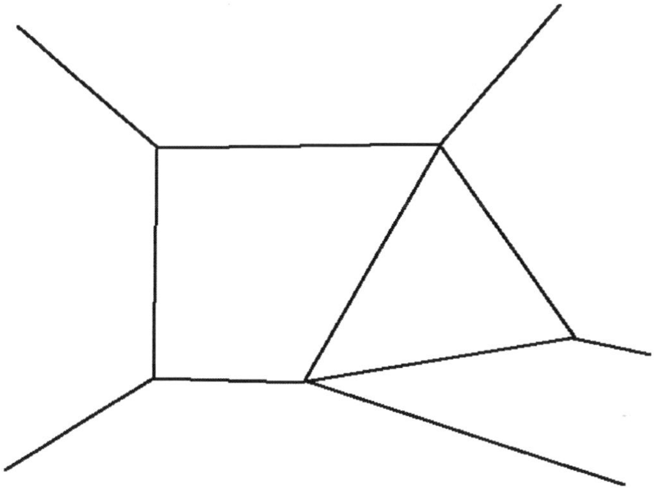

Figure 23: A "network" of Wilson lines is a graph in which the line segments are Wilson lines, labeled by representations of G, and a vertex represents a gauge-invariant coupling among the representations that label the lines that meet at that vertex.

By a "network" of Wilson lines, we mean simply a graph (Fig. 23) made of Wilson lines. Each line segment in the graph is labeled by a representation of the gauge group G (in general a different representation for each segment), and a vertex in the graph represents, in physics language, a gauge-invariant coupling among the representations that meet at that vertex. At the classical level – modulo the framing anomaly and its generalization for networks – the whole network will be at a fixed value[26] of the spectral parameter z.

[26]The analog of this in purely three-dimensional Chern-Simons theory is to consider a not necessarily planar graph made from Wilson lines and embedded in spacetime in an arbitrary fashion. For example, the quantum 6j symbol is the expectation value of a tetrahedral graph [32].

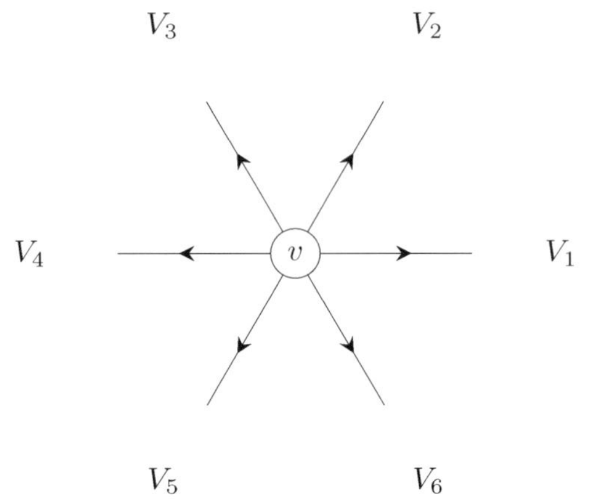

Figure 24: A vertex labeled by an invariant tensor $v \in V_1 \otimes \dots V_6$. Here and later, lines are labeled by the corresponding representations.

Since we are already familiar with Wilson lines, the new ingredient in building such a network is the vertex. So let us discuss this in more detail. In general, if we are given a collection of representations $V_1, \dots, V_n$, with a G-invariant element

$$v \in V_1 \otimes \cdots \otimes V_n \,, \tag{7.1}$$

then we can at the classical level form a Wilson line vertex in which n Wilson lines labeled by these n representations meet, as depicted in Fig. 24.

This construction is possible, at least classically, in any gauge theory. Classically, the vertex is defined as follows. For each Wilson line, we define the holonomy operator as usual, starting at the vertex where all the Wilson lines meet. The G-invariant tensor $v \in V_1 \otimes \cdots \otimes V_n$ provides an initial state to which we apply the holonomy operator on each Wilson line. The resulting operator is invariant under gauge transformations which vanish at the other end of the Wilson line. If each Wilson line has its second end either at another gauge-invariant vertex or at infinity (where we assume fields and gauge parameters to be trivial), then we get a gauge-invariant network of Wilson lines, as in Fig. 23.

In the theory under discussion in the present paper, not every classical Wilson line corresponds to a line operator of the quantum theory. There is an obstruction that we first encountered in section 5.4. Likewise, it turns out that not every classical vertex corresponds to a vertex in the quantum theory.

As we will see, the condition that a classical vertex should correspond to a vertex in the quantum theory is that a certain $\mathcal{O}(\hbar)$ anomaly should vanish. This anomaly comes from a lowest order Feynman diagram with one gluon exchange.

In this section, we will establish the result just stated; moreover, we will compute the $\mathcal{O}(\hbar)$ obstruction to "quantizing" a vertex and find useful conditions under which it vanishes.

Finally, we will give concrete, interesting, and (as we will see in a companion paper) useful examples of quantum vertices.

Of course, to give examples of vertices, we first need examples of Wilson lines that we are allowed to work with. We will use the following sufficient condition that a Wilson line associated to a representation of G (as opposed to a more general Wilson line associated to a representation of $\mathfrak{g}[[z]]$) can be quantized. Suppose the representation V satisfies the following algebraic condition:[27]

(†) Every G-invariant map from $\wedge^2\mathfrak{g}$ to $\mathrm{End}(V)$ factors through a copy of the adjoint representation.

Then a Wilson line associated to the representation V can be "quantized," that is, it exists in the quantum theory. Unfortunately, the proof of this statement is rather technical, and we have relegated it to Appendix C. We should point out that the condition (†) is only a sufficient criterion for a representation to be quantizable in this sense, but is far from being necessary. For example, for $G = SL_n$, it is known from other arguments that all representations of G are quantizable, but in general the condition (†) is not satisfied.

We should perhaps remark that what we call vertices correspond, in the theory of integrable relativistic scattering, to couplings of external particles to bound states or poles of the S-matrix. However, we will not try to make contact with the insight that comes from that point of view.

7.2 Vanishing of Higher Order Anomalies

Our first task is to show that an anomaly obstructing quantization of a vertex can arise only in lowest nontrivial order, from one-gluon exchange.

The anomaly that describes the failure of the configuration in Fig. 24 to be gauge-invariant at the quantum level will be a local operator of ghost number 1, made from the gauge field and the ghost field c, supported at the vertex. It will be valued in the vector space $\widehat{V} = \otimes V_i$ which lives at the vertex. As in our discussion of the framing anomaly, the anomaly is determined by local considerations, so the choice of the complex Riemann surface C will not matter. We may as well take $C = \mathbb{C}$ and take the vertex to be supported at $z = 0$. Likewise we can take the topological two-manifold Σ to be the xy plane with the vertex at the origin.

A key constraint is that the anomaly will be invariant under any symmetry of the theory that is also a symmetry of the classical vertex. A special case of diffeomorphism invariance in the Σ directions is invariance under scaling of the xy plane. The configuration in Fig. 24 with n Wilson lines emerging radially from a common vertex is invariant under this scaling, which therefore must be a symmetry of the anomaly. This tells us that the anomaly cannot depend on the x and y components of the gauge field A, and it cannot involve any x and y derivatives.

The classical theory is also invariant under the symmetry that simultaneously scales $z, \overline{z}$ and $\hbar$ by a real number, and under the symmetry in which z is rotated through θ, $\overline{z}$ through $-\theta$, and $\hbar$ through θ. It follows that the anomaly is also

[27] An equivalent statement is that the only irreducible representation of G that appears both in $\wedge^2\mathfrak{g}$ and in $\mathrm{End}(V) = V \otimes V^*$ is the adjoint representation.

preserved by these symmetries. From this it follows that the anomaly cannot have any $\overline{z}$ derivatives, nor can it depend on $A_{\overline{z}}$. Thus the anomaly must be constructed from the ghost field c only, and it is linear in c since it has ghost number 1. Moreover, in order $\hbar^k$, the anomaly must have k z-derivatives, that is, it must be proportional to $\partial_z^k \mathsf{c}(0)$.

Finally, the anomaly must be invariant under constant gauge transformations. Since $\partial_z^k \mathsf{c}$ transforms in the adjoint representation, it must be combined with a copy of the adjoint representation in $\widehat{V} = \otimes_i V_i$. Thus, an order $\hbar^k$ contribution to the anomaly must be an operator of the form

$$\sum_a \partial_z^k \mathsf{c}^a(0)\alpha_a \tag{7.2}$$

for some collection of elements $\alpha_a \in \widehat{V}$ that transform in the adjoint representation of $\mathfrak{g}$. Here c^a are the components of the ghost field c relative to a basis t_a, $a = 1, \ldots, \dim \mathfrak{g}$ of $\mathfrak{g}$.

It remains to impose the condition that the anomaly must be BRST closed. We will see that this condition is satisfied if and only if $k = 1$. In implementing the condition of BRST invariance, we have to remember that the local operator that represents the anomaly lives at the endpoint of n Wilson lines. In general, the BRST transformation of a Wilson operator $W(p, q)$ with ends at p and q is

$$\{Q, W(p, q)\} = \mathsf{c}(p)W - W\mathsf{c}(q) \, . \tag{7.3}$$

Here $\mathsf{c}(p)$ and $\mathsf{c}(q)$ are operators acting in the representation carried by the Wilson line. We are interested in the case that q is the location $x = y = z = 0$ of the vertex under study, and we are only interested in the $\mathsf{c}(q)$ term in (7.3) (the other term will participate in a similar cancellation at the other end of the Wilson line in question). For the ith Wilson line that ends at the vertex, we can write $\mathsf{c}(q)$ in more detail as

$$\mathsf{c}(q) = \sum_a \mathsf{c}^a(0) t_{a;i} \, , \tag{7.4}$$

where $t_{a;i}$ is the operator by which the Lie algebra generator t_a acts in the representation V_i.

The quantity that must vanish for BRST invariance of the anomaly (7.2) is therefore

$$-\sum_{i=1}^{n} \mathsf{c}^a(0) t_{a;i} \partial_z^k \mathsf{c}^b(0)\alpha_b + \{Q, \partial_z^k \mathsf{c}^b(0)\}\alpha_b \, . \tag{7.5}$$

The statement that the α_b transform in the adjoint representation means that

$$\sum_{i=1}^{n} t_{a;i}\alpha_b = f_{ab}^c \alpha_c \, . \tag{7.6}$$

Using the standard commutator relation

$$\{Q, \mathsf{c}^a(0)\} = \frac{1}{2} f^a{}_{bc} \mathsf{c}^b(0)\mathsf{c}^c(0) \, , \tag{7.7}$$

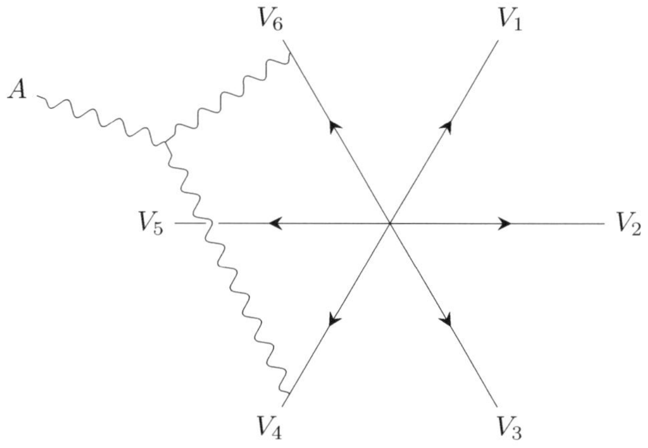

Figure 25: One-loop anomaly to the Wilson network in Fig. 24. Gluons are attached to two of the outgoing Wilson lines, in this case the ones labeled by V_4 and V_6. The full anomaly comes from a sum of such diagrams, with gluons attached to any two distinct Wilson lines.

from which it follows that

$$\{Q, \partial_z^k \mathsf{c}^a(0)\} = \frac{1}{2}\sum_{r=0}^{k} f^a_{bc}\partial_z^r \mathsf{c}^b(0)\partial_z^{k-r}\mathsf{c}^c(0) \; , \tag{7.8}$$

we see that eqn. (7.5) is satisfied if and only if $k = 1$.

We conclude that anomalies can occur only in order $\hbar$, that is in the lowest nontrivial order, due to one-gluon exchange.

7.3 Calculating the Anomaly

Let us now turn to calculating the anomaly. Any $\mathcal{O}(\hbar)$ anomaly must come from the Feynman diagrams depicted in Fig. 25.

This diagram involves only two of the Wilson lines in the network, and the anomaly will receive a contribution from each pair of Wilson lines.

For a fixed pair of Wilson lines, we have already done essentially this computation in section 6: it is the same as the calculation leading to the framing anomaly. There are a few small differences. First, in the calculation we are now considering, both Wilson lines are labelled with arrows pointing away from the vertex. At the classical level, this configuration is the same as one where one arrow is incoming, and the other is outgoing, but we use the representations V_i and V_j^* instead of V_i and V_j.

Also, in our calculation of the framing anomaly, we labelled both segments of the line by a single representation of the group, but now the two representations may be different. This affects only the group theory factor, and even that factor can still be written similarly to what we had before.

Finally, in our previous analysis, the Wilson line was allowed to bend in an arbitrary fashion, but now the tangent direction is changing only by a delta function jump at the origin.

Taking these details into account, we can readily write down the anomaly associated to Fig. 25, where the two gluons attach to Wilson lines i and j. To get the full anomaly, we will have to sum over pairs i, j. Let

$$v \in V_1 \otimes \cdots \otimes V_n \tag{7.9}$$

denote the invariant tensor that we place at the vertex classically. The group theory factor associated to the diagram in which the gluons are attached to lines i and j is

$$\sum f_a{}^{bc} t_{b;i} t_{c;j} v \,. \tag{7.10}$$

(This reduces to eqn. (6.1) if v is an invariant bilinear form that establishes an isomorphism between V_i and V_j^*.)

Let θ_i denote the angle between the ith Wilson line and the vertical, measured in the clockwise direction. Let us assume that $\theta_i < \theta_j$ for $i < j$. This means that V_1 is the first line segment we encounter if we start at the vertical and move clockwise. Then, the anomaly associated to the i and j line segments with $i < j$ is proportional to $\theta_j - \theta_i - \pi$. This follows from our calculation of the framing anomaly, which was the special case of this calculation when the representations V_j was V_i^*. In that case, a straight line does not have an anomaly, and the anomaly is proportional to the deviation of the line from being straight. This deviation for the geometry we are considering now is $\theta_j - \theta_i - \pi$.

Thus, the total anomaly to the existence of the vertex at the quantum level, summing all pairs of Wilson lines and otherwise borrowing our previous result, is

$$\frac{\hbar}{2\pi} \partial_z \mathsf{c}^a \left(\sum_{1 \leq i < j \leq n} (\theta_j - \theta_i - \pi) f_a{}^{bc} t_{b;i} t_{c;j} \right) v \,. \tag{7.11}$$

This can be viewed as a G-invariant linear map from $\mathfrak{g}$ to $V_1 \otimes \cdots \otimes V_n$.

A possibly counter-intuitive feature of the anomaly is that it is discontinuous in the angles θ_i. If θ_i increases so that it crosses θ_{i+1}, there is a discontinuity in the anomaly of the form $-\hbar f_a{}^{bc} t_{b;i} t_{c;i+1} v$ (an exchange of i and $i+1$, . This can be explained as follows. When $\theta_i = \theta_{i+1}$, what we are computing is the anomaly for a system with $n-1$ Wilson lines, where the lines V_i and V_{i+1} have been fused to a single Wilson line. In section 5, we have calculated the $\mathcal{O}(\hbar)$ correction to the action $\mathfrak{g}[[z]]$ on the fusion of two Wilson lines. The level 1 generator acting on the fused Wilson line turned out to be

$$\mp \frac{1}{2} \hbar f_a{}^{bc} t_{b;i} t_{c;i+1} \,, \tag{7.12}$$

where the sign depends on whether V_i is brought to V_{i+1} from the left or the right. Thus the configurations with V_i approaching V_{i+1} from left or right are different, and therefore they can have different anomalies.

The above formula for the fusion of V_i with V_{i+1} shows that the anomaly for n Wilson lines with θ_i tending to θ_{i+1} from below coincides with the anomaly for $n-1$ Wilson lines where V_i and V_{i+1} have been fused, with V_i merging with V_{i+1} from the right. (The fused Wilson line has a level one Yangian generator. A vertex in which it couples to ordinary Wilson lines that lack such a generator is not gauge-invariant at the classical level, and this contributes to the anomaly for such a vertex.)

Similarly, the anomaly for n Wilson lines where θ_i tends to θ_{i+1} from above coincides with the anomaly for $n-1$ Wilson lines where V_i has been fused with V_{i+1} from the left.

Thus the fact that the anomaly has a discontinuity when θ_i crosses θ_{i+1} reflects the fact that the algebra of Wilson line operators is non-commutative: the result of fusing two parallel Wilson lines depends on whether they are brought together form the left or the right.

7.4 Cancelling the Anomaly

There are two ways to try to cancel this anomaly. We can shift the relative positions of the Wilson lines in the z-plane by an amount of order $\hbar$. Suppose we shift the ith Wilson line from $z_i = 0$ to to $z_i = b_i\hbar$, for some constants b_i. This introduces a new anomaly $\left(\sum b_i \hbar t_{a;i}\right) v$, and one can hope that with a judicious choice of b_i this will cancel the anomaly.

The other thing we can try to do is to change the angles of the Wilson lines. However, it turns out that moving the angle of the ith Wilson line has the same effect as shifting it in the z plane. This is clear from the framing anomaly, but we can derive it from eqn. (7.11). Suppose that we change θ_i to $\theta_i + w_i$ for some fixed value of i, with w_i a constant, leaving the other Wilson lines unchanged. Then, the anomaly shifts by

$$\frac{\hbar}{2\pi}\left(+ \sum_{j<i} w_i f_a{}^{bc} t_{b;j} t_{c;i} - \sum_{i<j} w_i f_a{}^{bc} t_{b;i} t_{c;j} \right) v \,. \tag{7.13}$$

Using the fact that $\sum_j t_{j;a} v = 0$, we can rewrite this as

$$\frac{\hbar}{2\pi}\left(w_i f_a{}^{bc} t_{b;i} t_{c;i} \right) v = \frac{\hbar\, \mathsf{h}^\vee}{\pi} w_i t_{a;i} v \,. \tag{7.14}$$

This is the same shift we find if we move the ith Wilson line by w_i in the z plane, up to a factor of the dual Coxeter number $\mathsf{h}^\vee$ which arises from the framing anomaly.[28]

Comparing the two formulas makes it clear that if it is possible to cancel the anomaly by adjusting the angles, then it is also possible to do so by shifting the relative positions in the z-plane. The converse is not quite true. In our

[28] If $\mathsf{h}^\vee = 0$, then shifting the angles has no effect and we have to try to cancel the anomaly by shifting the values of z. For simple Lie groups, $\mathsf{h}^\vee$ is always positive, but there actually are interesting solutions of the Yang-Baxter equation for supergroups with $\mathsf{h}^\vee = 0$ [33].

computation of the anomaly, we have made an assumption about the cyclic order of the Wilson lines in the xy plane. This assumption only allows us to vary the angles while preserving certain inequalities. In some cases, the anomaly can be canceled by shifting the z_i but not by varying the angles, because the relevant inequalities would be violated.

7.5 Indecomposable Vertices

Before discussing specific examples, let us discuss a useful condition that can be satisfied by a classical vertex.

For each i, the states $t_{a;i}v$ transform in the adjoint representation of G, since v is G-invariant. This gives n copies of the adjoint representation, but G-invariance of v is equivalent to one relation between them:

$$\sum_{i=1}^{n} t_{a;i}v = 0 . \tag{7.15}$$

Thus, these states form at most $n-1$ linearly independent copies of the adjoint representation. We will call a vertex *indecomposable* if there are no further relations between the states $t_{a;i}v$, so that the number of linearly independent copies of the adjoint representation made in this way is precisely $n-1$. The motivation for the terminology is that if there is an additional relation, one can show by general considerations of group theory that we can decompose the set of representations into disjoint subsets R and S with the feature that the operators $\sum_{r\in R} t_{a;r}$ and $\sum_{s\in S} t_{a;s}$ each annihilate the element v. This means that v is the tensor product (or a sum of tensor products) of invariant elements in $\otimes_{r\in R}V_r$ and in $\otimes_{s\in S}V_s$. Thus, our vertex decomposes as a tensor product of two vertices, which can be shifted to different values of z and analyzed independently.

Accordingly, it is reasonable to restrict our attention to indecomposable vertices. An indecomposable vertex has at least $n-1$ possible anomalies (since the $t_{a;i}v$ provide $n-1$ copies of the adjoint representation in $\widehat{V} = V_1 \otimes V_2 \otimes \cdots \otimes V_n$). If there are only these $n-1$ linearly independent copies of the adjoint in $\widehat{V}$, then by shifting the $n-1$ relative positions of the n Wilson lines in the z-plane, there is a unique way to cancel the anomaly. Indeed, anomalies of the form $t_{a;i}V$ are precisely the ones that can be eliminated by shifting the values of z. If certain inequalities are obeyed, the anomalies can also be cancelled by shifting the relative angles rather than the values of z.

If the number of copies of the adjoint representation in $\widehat{V}$ is actually greater than $n-1$, one would expect that generically anomaly cancellation is not possible.

7.6 Vertices Constrained by Symmetries

We will describe various concrete examples of anomaly-free quantum vertices.[29] The simplest examples, which do not require any computations, arise when there

[29] These detailed examples will not be needed in the rest of the present paper, though some of them play an important role in the companion paper [25].

are enough symmetry constraints to determine the angles between the Wilson lines and ensure that the anomaly vanishes.

For a simple and also useful case, suppose that $V_1, \dots, V_n$ are all the same representation V, and consider a vertex associated to a G-invariant vector $v \in V^{\otimes n}$ that is either cyclically invariant or cyclically anti-invariant (in other words, assume that it either is invariant or changes sign under a cyclic permutation of the n copies of V). Such a vertex is anomaly-free, assuming that the angles between successive Wilson lines are equal, so as to respect the symmetry. (Fig. 24 has been drawn with equal angles, so it is invariant under a $2\pi/n$ rotation and potentially represents a vertex with cyclic symmetry or antisymmetry.) The proof is simple. Since the anomaly depends linearly on v, it is cyclically invariant or anti-invariant if v is. By assumption, there are no cyclically invariant or anti-invariant copies of the adjoint representation in $\widehat{V}$, so the anomaly is zero.

In most cases, a vertex with cyclic symmetry or antisymmetry has a further symmetry. This happens because a cyclically symmetric configuration of n lines meeting at equal angles in the plane, as in Fig. 24, is actually invariant under suitable reflections of the plane. In the gauge theory under study in the present paper, a reflection of the xy plane is a symmetry if accompanied by $z \to -z$. The latter is also a symmetry of ordinary Wilson lines (associated to representations of $\mathfrak{g}$, not $\mathfrak{g}[[z]]$) that are supported at $z = 0$. So as long as the vector $v \in V^{\otimes n}$ used in constructing the vertex is either even or odd under the reflection symmetry (all examples we will consider will have this property), the corresponding cyclically symmetric classical vertex actually has dihedral symmetry, generated by $2\pi/n$ rotations and also certain reflections. If the vertex is anomaly-free because of the conditions stated in the last paragraph, it will automatically possess the dihedral symmetry.

We will now describe explicit examples of anomaly-free vertices that are cyclically invariant or anti-invariant (and thus also dihedrally invariant). In each case, we have to first make sure that the representation V that we want to use is itself anomaly-free, in other words that there is a quantum Wilson line in this representation. For this, we will use the criterion (†) that was stated at the end of section 7.1: any G-invariant map from $\wedge^2\mathfrak{g}$ to $\mathrm{End}(V) = V \otimes V^*$ factors through the adjoint representation, or equivalently the only irreducible representation of $\mathfrak{g}$ that appears both in $\wedge^2\mathfrak{g}$ and in $V \otimes V^*$ is the adjoint representation.

Here are some examples:

1. V is the fundamental representation of SL_n, which satisfies condition (†), and $v \in V^{\otimes n}$ is the essentially unique[30] invariant tensor. There are $n-1$ copies of the adjoint in $V^{\otimes n}$, so this vertex can be quantized. Since v is cyclically invariant or anti-invariant (depending on whether n is odd or even), the vertex can be quantized so the Wilson lines all have the same value of z and the angle between them is $2\pi/n$.
2. Take V to be the $\mathbf{7}$ of G_2, which satisfies condition (†), and take

$$v \in \wedge^3\mathbf{7} \subset \mathbf{7}^{\otimes 3}$$

[30]In such a statement, we always mean unique up to a constant multiple.

to be the essentially unique invariant tensor. Using the tables on p. 298 of [34], one finds that every map from $\wedge^2 \mathbf{14}$ to $\mathbf{7} \otimes \mathbf{7}$ factors through $\mathbf{14}$. This implies that the Wilson line associated to $\mathbf{7}$ can be quantized. Further, the adjoint appears precisely twice in $\mathbf{7}^{\otimes 3}$. Therefore, the vertex associated to the tensor v can be quantized with angles $2\pi/3$ between the Wilson lines.

3. Consider the three 8 dimensional representations $\mathbf{8}_v$, $\mathbf{8}_c$, $\mathbf{8}_s$ of Spin(8), which are permuted by triality. For each representation, criterion (†) holds, so that there are no anomalies to quantizing the corresponding Wilson lines. Let

$$v \in \mathbf{8}_v \otimes \mathbf{8}_c \otimes \mathbf{8}_s$$

be the essentially unique invariant tensor. Note that v is also invariant under triality, together with a cyclic permutation of the representations. There are only two copies of the adjoint in $\mathbf{8}_v \otimes \mathbf{8}_c \otimes \mathbf{8}_s$, so this vertex can be quantized. If the Wilson lines have relative angles $2\pi/3$, then a rotation through $2\pi/3$ together with an application of the triality symmetry of Spin(8) is a symmetry of the configuration. This tells us that the only consistent quantization is the one where the Wilson lines have relative angles $2\pi/3$.

4. Take V to be the $\mathbf{26}$ of F_4. Using table 45 of [35], we find that this representation satisfies condition (†) and so the corresponding Wilson line can be quantized. We consider the vertex associated to the essentially unique invariant tensor $v \in \mathrm{Sym}^3\, \mathbf{26}$. Again using table 45 of [35], one can check that there are only two copies of the adjoint in $\mathbf{26}^{\otimes 3}$. Therefore this vertex quantizes, with angles $2\pi/3$ between adjacent lines.

5. Take V to be the $\mathbf{27}$ of E_6. Table 48 of [35] implies that this representation satisfies condition (†). (See section 7.10.) We take $v \in \mathrm{Sym}^3\, \mathbf{27}$ to be the unique invariant element. Again using table 48 of [35], we find that there are only two copies of the adjoint in $\mathbf{27}^{\otimes 3}$. Therefore this vertex quantizes, with angles $2\pi/3$ between adjacent lines.

7.7 A General Formula for the Angles at a Trivalent Vertex

Now we turn our attention to general trivalent vertices. Assuming that only two copies of the adjoint occur in $\widehat{V} = V_1 \otimes V_2 \otimes V_3$, a classical trivalent vertex can be quantized. Moreover, it is possible to find a simple general formula for the relative angles that are required. (We will express our results in terms of angles rather than in terms of shifts in z because that corresponds to a simpler classical picture, but when some relative angles come out to be negative and thus inconsistent with an assumed cyclic ordering of the vertices, one can use the alternative approach in terms of shifting the relative values of the z_i.)

Suppose that V_{ρ_1}, V_{ρ_2}, V_{ρ_3} are three irreducible highest weight representations of a group G of highest weights ρ_1, ρ_2, ρ_3, and suppose that the corresponding Wilson lines can be quantized. Consider a classical vertex associated to an invariant tensor $v \in V_{\rho_1} \otimes V_{\rho_2} \otimes V_{\rho_3}$. As we have seen, if there are exactly two copies of the adjoint representation in $V_{\rho_1} \otimes V_{\rho_2} \otimes V_{\rho_3}$, then this vertex can be quantized.

Suppose the Wilson lines are arranged in the plane with cyclic order $V_{\rho_1}, V_{\rho_2}, V_{\rho_3}$. Let θ_{12}, θ_{23}, θ_{31} be the angles between the Wilson lines. We will derive a formula for the angles θ_{ij}.

Let $c(\rho_i)$ denote the action of the quadratic Casimir of $\mathfrak{g}$ on V_{ρ_i}. We define

$$\begin{aligned}\beta_1 &= c(\rho_1) - c(\rho_2) - c(\rho_3)\ ,\\ \beta_2 &= c(\rho_2) - c(\rho_3) - c(\rho_1)\ ,\\ \beta_3 &= c(\rho_3) - c(\rho_1) - c(\rho_2)\ .\end{aligned} \tag{7.16}$$

We will show that the angles between the three Wilson lines are given by the formula

$$\begin{aligned}\theta_{12} &= \pi - \pi\frac{\beta_1\beta_2}{\beta_1\beta_2+\beta_1\beta_3+\beta_2\beta_3}\ ,\\ \theta_{23} &= \pi - \pi\frac{\beta_2\beta_3}{\beta_1\beta_2+\beta_1\beta_3+\beta_2\beta_3}\ ,\\ \theta_{31} &= \pi - \pi\frac{\beta_1\beta_3}{\beta_1\beta_2+\beta_1\beta_3+\beta_2\beta_3}\ .\end{aligned} \tag{7.17}$$

The derivation of this formula is as follows. The anomaly to quantizing the vertex vanishes if

$$((\theta_{12}-\pi)f_a{}^{bc}t_{b;1}t_{c;2} + (\pi-\theta_{31})f_a{}^{bc}t_{b;1}t_{c;3} + (\theta_{23}-\pi)f_a{}^{bc}t_{b;2}t_{c;3})v = 0\ . \tag{7.18}$$

This equation should hold for every value of a.

Let us choose our basis t_a to be orthonormal with respect to the chosen invariant pairing on the Lie algebra $\mathfrak{g}$. Applying the operator $t_{a;1}$ to eqn. (7.18) and summing over a we find

$$((\theta_{12}-\pi)f^a_{bc}t_{a;1}t_{b;1}t_{c;2} + (\pi-\theta_{31})f^a_{bc}t_{a;1}t_{b;1}t_{c;3} + (\theta_{23}-\pi)f^a_{bc}t_{a;1}t_{b;2}t_{c;3})\, v = 0\ . \tag{7.19}$$

Since

$$\sum_a f^a_{bc}t_{a;1}t_{b;1} = 2\mathsf{h}^\vee t_{c;1}\ , \tag{7.20}$$

we can rewrite this equation as

$$((\theta_{12}-\pi)\mathsf{h}^\vee t_{c;1}t_{c;2} + (\pi-\theta_{31})\mathsf{h}^\vee t_{c;1}t_{c;3} + (\theta_{23}-\pi)f^a_{bc}t_{a;1}t_{b;2}t_{c;3})\, v = 0\ . \tag{7.21}$$

Next, since

$$t_{c;3}v = -t_{c;1}v - t_{c;2}v\ , \tag{7.22}$$

we have

$$\begin{aligned}f_{abc}t_{a;1}t_{b;2}t_{c;3}v &= -f_{abc}t_{a;1}t_{b;2}(t_{c;1}+t_{c;2})\\ &= f_{acb}t_{a;1}t_{c;1}t_{b;2} - f_{bca}t_{a;1}t_{b;2}t_{c;2}\\ &= t_{b;1}t_{b;2}\mathsf{h}^\vee - t_{a;1}t_{a;2}\mathsf{h}^\vee\\ &= 0\ .\end{aligned} \tag{7.23}$$

Thus, our equation becomes

$$((\theta_{12}-\pi)\mathsf{h}^\vee t_{c;1}t_{c;2} + (\pi-\theta_{31})\mathsf{h}^\vee t_{c;1}t_{c;3})\, v = 0\ . \tag{7.24}$$

Now,

$$\sum_c t_{c;1}t_{c;2} = \sum_c \tfrac{1}{2}(t_{c;1}+t_{c;2})^2 - \tfrac{1}{2}t_{c;1}^2 - \tfrac{1}{2}t_{c;2}^2 \,. \tag{7.25}$$

Let $c(\rho_i)$ denote the eigenvalue of the quadratic Casimir $\sum t_a^2$ on the representation V_{ρ_i}. The operator $\sum_a t_{a;i}^2$ acts on $V_{\rho_1}\otimes V_{\rho_2}\otimes V_{\rho_3}$ by $c(\rho_i)$.

Acting on an element v, we have

$$\begin{aligned}\sum_c t_{c;1}t_{c;2}v &= \sum_c \tfrac{1}{2}(t_{c;1}+t_{c;2})^2 v - \tfrac{1}{2}t_{c;1}^2 v - \tfrac{1}{2}t_{c;2}^2 v \\ &= \tfrac{1}{2}t_{c;3}^2 v - \tfrac{1}{2}t_{c;1}^2 v - \tfrac{1}{2}t_{c;2}^2 v \\ &= \tfrac{1}{2}c(\rho_3)v - \tfrac{1}{2}c(\rho_1)v - \tfrac{1}{2}c(\rho_i)v \,.\end{aligned} \tag{7.26}$$

Thus (after dividing by $\mathsf{h}^\vee/2$), our equation becomes

$$(\theta_{12}-\pi)(c(\rho_3)-c(\rho_1)-c(\rho_2)) = (\theta_{31}-\pi)(c(\rho_2)-c(\rho_1)-c(\rho_3)) \,. \tag{7.27}$$

Similar arguments give us two additional equations, which are the cyclic permutations of the equation we have just derived:

$$\begin{aligned}(\theta_{12}-\pi)(c(\rho_3)-c(\rho_1)-c(\rho_2)) &= (\theta_{23}-\pi)(c(\rho_1)-c(\rho_2)-c(\rho_3)) \,, \\ (\theta_{12}-\pi)(c(\rho_3)-c(\rho_1)-c(\rho_2)) &= (\theta_{31}-\pi)(c(\rho_2)-c(\rho_1)-c(\rho_3)) \,.\end{aligned} \tag{7.28}$$

Evidently, one of these equations is redundant.

In terms of β_i defined in eqn. (7.16), our equations become

$$(\theta_{12}-\pi)\beta_3 = (\theta_{23}-\pi)\beta_1 \tag{7.29}$$

plus its cyclic permutations. Their unique solution is given by (7.17).

As a consistency check, note that

$$\theta_{12}+\theta_{23}+\theta_{31} = 2\pi \,.$$

Note also that if the three representations are the same, then the angle between any two Wilson lines is $2\pi/3$, as expected for symmetry reasons.

Let us specialize the formula to the case that the two representations V_{ρ_1}, V_{ρ_2} are the same (or at least have the same eigenvalue of the quadratic Casimir). In that case, letting $c(\rho)=c(\rho_1)=c(\rho_2)$, we have

$$\begin{aligned}\beta_1 &= \beta_2 = -c(\rho_3) \,, \\ \beta_3 &= c(\rho_3) - 2c(\rho) \,.\end{aligned} \tag{7.30}$$

The angles between the Wilson lines become

$$\begin{aligned}\theta_{12} &= \pi - \pi\frac{c(\rho_3)}{4c(\rho)-c(\rho_3)} \,, \\ \theta_{31} &= \pi\frac{2c(\rho)}{4c(\rho)-c(\rho_3)} \,, \\ \theta_{23} &= \pi\frac{2c(\rho)}{4c(\rho)-c(\rho_3)} \,.\end{aligned} \tag{7.31}$$

As an example, let analyze the vertex connecting two copies of the fundamental representation of $\mathfrak{sl}_n$ with the dual of the exterior square of the fundamental representation. Let us normalize the quadratic Casimir so that its value on the fundamental representation V is 1 (the normalization plays no role in our formula). Then, its value on $\wedge^2 V^*$ is $\frac{2(n-2)}{n-1}$. If we take V_{ρ_1}, V_{ρ_2} to be the two copies of the fundamental representation in the above calculation, then we have

$$\begin{aligned} \theta_{12} &= \pi - \pi \frac{\frac{2(n-2)}{n-1}}{4 - \frac{2(n-2)}{n-1}} = \pi \frac{2}{n} \,, \\ \theta_{31} &= \pi \frac{2}{4 - \frac{2(n-2)}{n-1}} = \pi \frac{n-1}{n} \,, \\ \theta_{23} &= \pi \frac{n-1}{n} \,. \end{aligned} \tag{7.32}$$

This is a special case of a more general formula that we compute next.

7.8 Trivalent Vertices Linking Fundamental Representations of $\mathfrak{sl}_n$

We will describe trivalent vertices involving three of the fundamental representations of $\mathfrak{sl}_n$, which are the kth rank antisymmetric tensors $\wedge^k V$, where V is the fundamental representation and $1 \leq k \leq n-1$. These representations are all quantizable (since in fact all representations of $\mathfrak{sl}_n$ are quantizable), though criterion (†) generally does not hold.

In what follows, it is useful to recall that, in an appropriate normalization, the eigenvalue $c(V)$ of the quadratic Casimir on a representation $V = V_\rho$ satisfies

$$c(V) = \frac{l(v)}{\dim(V)} \,, \tag{7.33}$$

where $l(v)$ is the Dynkin index of the representation. Dynkin indices of various representations can be found in tables such as those in [35].

Let V be the vector representation of $\mathfrak{sl}_n$ so that $\wedge^k V$, $k = 1, \ldots, n-1$ are the fundamental representations. The Dynkin index of $\wedge^k V$ is $\binom{n-2}{k-1}$, so that

$$c(\wedge^k V) = \frac{\binom{n-2}{k-1}}{\binom{n}{k}} = \frac{k(n-k)}{n(n-1)} \,. \tag{7.34}$$

Consider three fundamental representations $\wedge^{k_i} V$ where $k_1 + k_2 + k_3 = n$. There is an invariant element of the tensor product of these representations coming from the map

$$\wedge^{k_1} V \otimes \wedge^{k_2} V \otimes \wedge^{k_3} V \to \wedge^n V = \mathbb{C} \,. \tag{7.35}$$

The adjoint representation only appears twice in the tensor product of the three representations $\wedge^{k_i} V$. It follows that the vertex corresponding to the invariant tensor can be quantized. From formulae (7.17), we can determine the angles.

We set

$$\beta_1 = c(\wedge^{k_1} V) - c(\wedge^{k_2} V) - c(\wedge^{k_2} V) \tag{7.36}$$

plus cyclic permutation. Then

$$\beta_1 = \frac{1}{n(n-1)} \left(k_1(n-k_1) - k_2(n-k_2) - k_3(n-k_3)\right) = \frac{-2k_2k_3}{n(n-1)}$$

plus cyclic permutations. Bearing in mind that the a scaling of all the β_i will not affect the angles, we find

$$\begin{aligned} \theta_{12} &= \pi \frac{k_1+k_2}{n}, \\ \theta_{23} &= \pi \frac{k_2+k_3}{n}, \\ \theta_{31} &= \pi \frac{k_1+k_3}{n}. \end{aligned} \tag{7.37}$$

7.9 Vertices Related to Γ-Matrices

Consider the group Spin(n) where n is even. Let V denote the vector representation, and S_+, S_- the two irreducible (complex) spin representations of opposite chirality. These are both of dimension $2^{\frac{n}{2}-1}$. We will use criterion (†) to show that these representations can be quantized.

If $n = 2 \mod 4$, then the vector representation appears in $S_+ \otimes S_+$ and in $S_- \otimes S_-$. If $n = 0 \mod 4$, then the vector representation appears in $S_+ \otimes S_-$. In each case, we can try to quantize the vertex linking the vector representation to two spin representations.

In the calculation that follows, we will assume $n \geq 8$ to avoid any low-dimensional coincidences.

Note that the endomorphisms of the direct sum $S_+ \oplus S_-$ make up the Clifford algebra Cl_n built from the vector representation V. As a representation of $\mathfrak{so}(n)$, the Clifford algebra is isomorphic to direct sum of the exterior powers of V.

To use condition (†) to show that the spinor representations can be quantized, we need to classify maps

$$\wedge^2 \mathfrak{so}(n) \to \wedge^k V \tag{7.38}$$

for all values of k. An exercise in classical invariant theory tells us that the only such maps that can possibly exist are when $k = 2, 4, n-4, n-2$. If $k = 2, n-2$ then $\wedge^k V$ is the adjoint representation, so such maps will not contribute anomalies. If $k = 4, n-4$ then the only map $\mathfrak{so}(n)^{\otimes 2} \to \wedge^k V$ comes from the wedge product map

$$\mathfrak{so}(n) \otimes \mathfrak{so}(n) = \wedge^2 V \otimes \wedge^2 V \to \wedge^4 V \tag{7.39}$$

(noting that $\wedge^4 V = \wedge^{n-4} V$). This map is symmetric, not antisymmetric, and so is not relevant to condition (†). We conclude that in all cases, the Wilson lines attached to the spin representations exist in the quantum theory.

Next, let us analyze whether the vertex linking the vector representation V with two copies of a spin representation can be quantized. We will start with the case when $n = 2 \mod 4$, in which case the vector representation appears in $S_+ \otimes S_+$. We need to show that there are exactly two copies of the adjoint representation in $V \otimes S_+ \otimes S_+$.

To do this, we need to recall how to describe $S_+ \otimes S_+$ in terms of exterior powers of V (this computation is familiar from the study of central extensions of supersymmetry algebras). Since $n = 2 \mod 4$, S_+ and S_- are dual representations. The endomorphisms of $S_+ \oplus S_-$ which reverse the parity of a spinor are the odd elements of the Clifford algebra built from V. This space of endomorphisms is $S_+ \otimes S_+ \oplus S_- \otimes S_-$. Thus,

$$S_+ \otimes S_+ \oplus S_- \otimes S_- = \oplus_{k \text{ odd}} \wedge^k V \,. \tag{7.40}$$

We can further decompose the right hand side of this equation to find

$$S_+ \otimes S_+ = V \oplus \wedge^3 V \oplus \cdots \oplus \wedge_+^{n/2} V \,, \tag{7.41}$$

where $\wedge_+^{n/2} V$ indicates those elements which are self-dual under the Hodge star operator.

In $V \otimes S_+ \otimes S_+ = V \otimes (V \oplus \wedge^3 V \oplus \cdots \oplus \wedge_+^{n/2} V)$, the only copies of the adjoint representation are those in $V \otimes V$ and in $V \otimes \wedge^3 V$, each of which contains one copy of the adjoint. So overall there are precisely two copies, and the vertex connecting V, S_+ and S_+ can be quantized.

Next let us check whether the vertex can be quantized in the case that $n = 0 \mod 4$. In this case V appears in $S_+ \otimes S_-$ and the representations S_+, S_- are self-dual. We again find that $S_+ \otimes S_- \oplus S_- \otimes S_+$ is the space of odd elements in the Clifford algebra, so that

$$S_+ \otimes S_- = V \oplus \wedge^3 V \oplus \cdots \oplus \wedge^{n/2-1} V \,. \tag{7.42}$$

The only copies of the adjoint that appear in $V \otimes S_+ \otimes S_-$ are those in $V \otimes V$ and $V \otimes \wedge^3 V$, so again the vertex can be quantized.

Now we know that the spinor representations and the vertices connecting spinor and vector representations can be quantized. The final step is to calculate the angles between the lines at the vertex. To do this, we need to know the Dynkin indices of the representations $V, S_\pm$. The Dynkin index of V is 2. We can calculate the Dynkin indices of $S_\pm$ as follows.

Recall that the Dynkin index is additive under direct sums of representations, and under tensor product there is the following formula:

$$l(R_1 \otimes R_2) = l(R_1) \dim(R_2) + \dim(R_1) l(R_2) \,, \tag{7.43}$$

where R_1, R_2 are representations, $l(R_i)$ is the Dynkin index and $\dim(R_i)$ is the dimension. Further, if R is any representation, then

$$l(\wedge^k R) = \binom{\dim(R) - 2}{k - 1} l(R) \,. \tag{7.44}$$

The representations S_+ and S_- are related by a diagram automorphism of the Dynkin diagram $D_{n/2}$ of the Lie algebra $\mathfrak{so}(n)$. Therefore they have the same

Dynkin index. Using the fact that $(S_+ \oplus S_-)^{\otimes 2}$ is the sum of all the exterior powers of V, we find

$$2^{n/2+2} l(S_\pm) = 2 \sum_{k=1}^{n-1} \binom{n-2}{k-1} = 2^{n-1} . \tag{7.45}$$

Thus,

$$l(S_\pm) = 2^{\frac{n}{2}-3} . \tag{7.46}$$

Now let us compute the angles between the three Wilson lines. We label the lines where $V_1 = S_+$, $V_2 = S_-$ and $V_3 = V$ is the vector representation. The quadratic Casimirs in each representation are the ratios of the Dynkin index to the dimension. They are

$$\begin{aligned} c(V) &= \frac{2}{n} , \\ c(S_\pm) &= \frac{2^{\frac{n}{2}-3}}{2^{\frac{n}{2}-1}} = 2^{-2} . \end{aligned} \tag{7.47}$$

According to formula (7.31), we have

$$\begin{aligned} \theta_{12} &= \pi \frac{n-4}{n-2} , \\ \theta_{31} &= \pi \frac{n}{2n-4} , \\ \theta_{23} &= \pi \frac{n}{2n-4} . \end{aligned} \tag{7.48}$$

Note that when $n = 8$, the angle between any two lines is $2\pi/3$, which is consistent with what we determined earlier using the triality symmetry of the vertex in this case. When $n = 6$, the vertex we are considering is that relating two copies of the vector representation of $\mathfrak{sl}(4)$ with the exterior square of the fundamental. The formula for the angle for a vertex connecting three fundamental representations of $\mathfrak{sl}_n$ agrees, in this case, with the formula given here.

7.10 A Vertex Connecting Representations of E_6

Just for fun, let us use our formula to calculate the angles in a vertex associated to representations of the exceptional group E_6. The fundamental representation of E_6 will be denoted by $\mathbf{27}$, and its dual by $\overline{\mathbf{27}}$. There are four representations of dimension 351, which come in dual pairs. We will use the conventions of [35] and denote them by $\mathbf{351}$, $\overline{\mathbf{351}}$, $\mathbf{351}'$, $\overline{\mathbf{351}}'$. The $\mathbf{27}^3$ vertex was already considered in section 7.6, so here we primarily consider a more elaborate example. (We also will complete the discussion of the $\mathbf{27}^3$ by showing that the $\mathbf{27}$ can be quantized.)

According to table 48 of [35], $\overline{\mathbf{351}}'$ appears once in $\mathbf{27} \otimes \mathbf{27}$. Thus there is an invariant tensor in $\mathbf{27} \otimes \mathbf{27} \otimes \mathbf{351}'$.

We would like to quantize this to a vertex connecting three Wilson lines. To do this, we first need to show that the Wilson lines themselves quantize. It is sufficient, according to condition (†), to show that any map from the exterior square of the adjoint representation to the endomorphisms of the **27** or **351**′ factors through the adjoint representation.

According to table 48 of [35], the exterior square of the adjoint representation decomposes as

$$\wedge^2 \mathbf{78} = \mathbf{78} \oplus \mathbf{2925} \,. \tag{7.49}$$

To show that the **27** and **351**′ quantize, we need to show that **2925** does not appear in $\overline{\mathbf{27}} \otimes \mathbf{27}$ or in $\overline{\mathbf{351}}' \otimes \mathbf{351}'$. Table 48 of [35] shows that it does not, so these representations quantize.

Next, to show that the vertex in $\mathbf{27} \otimes \mathbf{27} \otimes \mathbf{351}'$ quantizes, we need to show that the adjoint representation appears precisely twice in this tensor product. Table 48 of [35] tells us that

$$\mathbf{27} \otimes \mathbf{27} = \overline{\mathbf{27}} \oplus \overline{\mathbf{351}} \oplus \overline{\mathbf{351}}' \,. \tag{7.50}$$

If we tensor this with **351**′, table 48 of [35] tells us that the adjoint appears once in $\overline{\mathbf{351}}' \otimes \mathbf{351}'$, once in $\overline{\mathbf{351}} \otimes \mathbf{351}'$, and not at all in $\overline{\mathbf{27}} \otimes \mathbf{351}$. Therefore the vertex quantizes.

Next, let us compute the angles. Let us label the representations as $V_1 = \mathbf{27}$, $V_2 = \mathbf{27}$, $V_3 = \mathbf{351}$. Table 47 of [35] tells us that the Dynkin index of **27** is 6 and that of **351**′ is 6×28. The values of the quadratic Casimirs are $6/27$ and $6 \times 28/351$. We can change the normalization so that the values of the quadratic Casimirs are 1 and $28/13$. The angles are

$$\begin{aligned} \theta_{12} &= \pi - \pi \frac{\frac{28}{13}}{4 - \frac{28}{13}} = -\pi \frac{1}{6} \,, \\ \theta_{23} &= \theta_{31} = \pi \frac{2}{4 - \frac{28}{13}} = \pi \tfrac{13}{12} \,. \end{aligned} \tag{7.51}$$

Since, in this example, $\theta_{12} < 0$, we have to shift the Wilson lines in the z-plane instead of just placing them at angles in the topological plane.

Many more examples can be analyzed in a similar way.

7.11 A 4-Valent Vertex for the 56 of E_7

The smallest representation of E_7 is the **56**. It is a pseudoreal or symplectic representation, so there is an invariant antisymmetric form $\omega \in \wedge^2\mathbf{56}$. In addition, there is a completely symmetric quartic invariant $\psi \in \mathrm{Sym}^4\mathbf{56}$. It is natural to ask whether this vertex can be quantized, like the $\mathbf{27}^3$ of E_6, which was one of our examples in section 7.6. The answer is that it can, though the proof is not as simple as for the $\mathbf{27}^3$ vertex. Both of these examples will be useful in [25].

A configuration of four Wilson lines cannot have full S_4 permutation symmetry. The maximum possible symmetry is a dihedral subgroup D_4. This is simply the symmetry group of four Wilson lines with equal relative angles $\pi/2$, say running along the $\pm x$ and $\pm y$ axes. Dihedral symmetry was discussed in section 7.6. In this case, the group D_4 is generated by rotations of the xy plane by an angle

$\pi/2$, along with a reflection that preserves the given configuration of Wilson lines. Thus in all D_4 has eight elements.

We would like to understand the possible vertices that are D_4 invariant, and the possible anomalies that are compatible with the D_4 symmetry. Let us first enumerate the E_7-invariant elements in $\mathbf{56}^{\otimes 4}$ that are also D_4-invariant. Note that we can identify $\mathbf{56}$ with its dual, using the E_7-invariant symplectic form ω. We can therefore identify E_7-invariant elements of $\mathbf{56}^{\otimes 4}$ with maps of E_7 representations

$$\mathbf{56}^{\otimes 2} \to \mathbf{56}^{\otimes 2} . \tag{7.52}$$

According to the tables of [35] or [36], $\mathbf{56}^{\otimes 2}$ decomposes as a sum of 4 distinct irreducible representations. Therefore, there are 4 E_7 invariant linear operators on $\mathbf{56}^{\otimes 2}$, given by the projectors onto these 4 irreducible subrepresentations. Correspondingly, there are 4 invariant tensors in $\mathbf{56}^{\otimes 4}$.

We can enumerate these tensors as follows. One of them is the completely symmetric invariant ψ with which we began. The other three are more elementary. Let e_i be a basis of $\mathbf{56}$; in this basis, the antisymmetric form ω corresponds to a matrix ω^{ij}. The other three E_7 invariants in $\mathbf{56}^{\otimes 4}$ are given by the formulas

$$\begin{aligned} &\omega^{ij}\omega^{kl} e_i \otimes e_j \otimes e_k \otimes e_l \,, \\ &\omega^{ik}\omega^{lj} e_i \otimes e_j \otimes e_k \otimes e_l \,, \\ &\omega^{il}\omega^{jk} e_i \otimes e_j \otimes e_k \otimes e_l \,. \end{aligned} \tag{7.53}$$

Among these three tensors, there is a single linear combination which is invariant under D_4, namely

$$(\omega^{ij}\omega^{kl} + \omega^{jk}\omega^{li}) e_i \otimes e_j \otimes e_k \otimes e_l \,. \tag{7.54}$$

We conclude that there are a total of two dihedrally invariant tensors in $\mathbf{56}^{\otimes 4}$.

Given the explicit form of the invariant (7.54), an elementary computation using the general formula (7.11) for the anomaly shows that a vertex constructed using this invariant has an anomaly. This means that it will be possible to use this invariant as a counterterm to help in canceling an anomaly.

We will prove that there is a unique linear combination of these two dihedrally invariant tensors which quantizes to a vertex linking the Wilson lines. To show this, we have to show that there is precisely one possible anomaly.

7.11.1 Anomalies

Anomalies are proportional to $\partial_z \mathsf{c}$. Since an element of D_4 that acts as a reflection of the xy plane also acts as $z \to -z$, changing the sign of $\partial_z \mathsf{c}$, the group theory invariant that multiplies $\partial_z \mathsf{c}$ in an anomaly is not D_4-invariant. Rather, it is D_4 anti-invariant, that is, invariant under rotations in D_4 but odd under reflections. We will show that the adjoint representation of E_7 (which is the $\mathbf{133}$) occurs precisely once in the D_4 anti-invariant part of $\mathbf{56}^{\otimes 4}$. Since the invariant (7.54) does have a nonzero anomaly, this implies that by adding a multiple of (7.54) to the invariant ψ, one can construct an anomaly-free $\mathbf{56}^{\otimes 4}$ vertex.

As a first step, let us compute the part of $\mathbf{56}^{\otimes 4}$ that is anti-invariant under a dihedral subgroup $D_2 \subset D_4$. There actually are two possible embeddings of

D_2 in D_4. We make the following choice. If the four Wilson lines run along the $\pm x$ and $\pm y$ axes, we consider a subgroup $D_2 \cong \mathbb{Z}_2 \times \mathbb{Z}_2$ that is generated by a reflection that acts by $(x, y) \to (x, -y)$ and one that acts as $(x, y) \to (-x, y)$. Thus if we label the four Wilson lines in cyclic order as 1,2,3, and 4, one reflection acts by exchanging 1 and 3, keeping fixed 2 and 4, and the other exchanges 2 and 4, keeping fixed 1 and 3. Thus, the part of $\mathbf{56}^{\otimes 4}$ that is odd under each reflection is $\wedge^2\mathbf{56} \otimes \wedge^2\mathbf{56}$. The two factors are associated to the pairs 13 and 24.

From the tables in [35] or [36], one has $\wedge^2\mathbf{56} \cong \mathbf{1} \oplus \mathbf{1539}$. So the D_2 anti-invariant part of $\mathbf{56}^{\otimes 4}$ is $(\mathbf{1} \oplus \mathbf{1539}) \otimes (\mathbf{1} \oplus \mathbf{1539})$.

Now we want to identify the part of this that is anti-invariant under D_4, not just under D_2. So we have to consider the action of a $\pi/2$ rotation. The D_4 anti-invariants are simply the D_2 anti-invariants that are invariant under a $\pi/2$ rotation. However, there is a small surprise when we try to impose invariance under a $\pi/2$ rotation on the above description of the D_2 anti-invariants.

A $\pi/2$ rotation exchanges the two factors of $\wedge^2\mathbf{56}$ that we used in the above analysis, but with an important minus sign. This happens as follows. We recall that the two factors of $\wedge^2\mathbf{56}$ are associated respectively to the pair of Wilson lines 13 and 24. A $\pi/2$ rotation maps 13 to 24, but it maps 24 to 31; replacing 31 with 13 acts as -1 on one of the two copies of $\wedge^2\mathbf{56}$. Thus the D_4 anti-invariant part of $\mathbf{56}^{\otimes 4}$ is the antisymmetric part of $(\mathbf{1} \oplus \mathbf{1539}) \otimes (\mathbf{1} \oplus \mathbf{1539})$, or more explicitly it is[31] $\mathbf{1539} \oplus \wedge^2\mathbf{1539}$.

From the tables of [36], one learns that the adjoint or $\mathbf{133}$ of E_7 occurs precisely once in $\mathbf{1539}^{\otimes 2}$. This one occurrence actually is in $\wedge^2\mathbf{1539}$, not in $\mathrm{Sym}^2\mathbf{1539}$, because, as $\mathbf{1539}$ is a real representation of E_7, the adjoint must occur at least once in $\wedge^2\mathbf{1539}$. So as claimed above, there is precisely one possible anomaly.

8 Two-Loop Correction To Gauge Invariance

8.1 Preliminaries

Consider a Wilson line in our four-dimensional theory in a general representation V. If we choose a basis t_a of the Lie algebra $\mathfrak{g}$, the Wilson line is characterized classically by matrices

$$t_{a,k} : V \to V \ , \tag{8.1}$$

where, as before, the matrices $t_{a,k}$ tell us how $\partial_z^k A$ is coupled to the Wilson line. If V is a representation of G and not of $\mathfrak{g}[[z]]$, then $t_{a,k} = 0$ for $k > 0$.

At the classical level, gauge-invariance requires that these matrices must satisfy the commutation relations

$$[t_{a,k}, t_{b,l}] = f_{ab}{}^{c} t_{c,k+l} \ . \tag{8.2}$$

[31] As a check on this, the dimension of $\mathbf{1539} \oplus \wedge^2\mathbf{1539}$ is 1185030. This is the right dimension for the D_4 anti-invariant part of $\mathbf{56}^{\otimes 4}$. For example, an exercise using the character table of D_4 (or based on elementary considerations) tells us that the D_4 anti-invariants in $(\mathbb{C}^d)^{\otimes 4}$ are of dimension $\frac{1}{8}\left(d^4 - 2d^3 - d^2 + 2d\right)$.

These relations receive quantum corrections; the condition for a Wilson operator to be anomaly-free at the quantum level is different from eqn. (8.2). We have seen in section 5.4 that a correction must occur at order $\hbar^2$. We gave one derivation of this statement, which relied on an analysis of the fusion of parallel Wilson lines. In a companion paper [25], we will give another derivation based on the RTT presentation of the Yangian algebra.

Both derivations are a little indirect, and one might wish for a more direct one. In this section we will provide a direct derivation via Feynman diagrams.

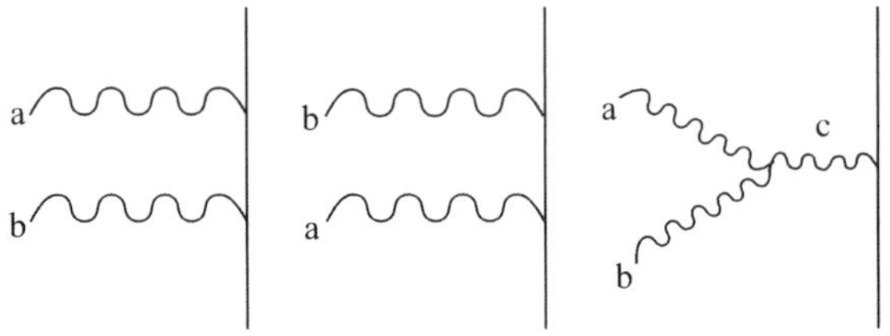

Figure 26: The three tree-level diagrams describing coupling of a pair of gauge bosons to a charged particle or a Wilson line operator.

Some elementary remarks may help one anticipate what sort of diagrams will be relevant. In an ordinary gauge theory with a finite dimensional Lie algebra $\mathfrak{g}$, how does one usually see in perturbation theory that the matrices t_a by which gauge fields A^a couple to a charged particle or a Wilson line must satisfy the expected commutation relations? At tree level, there are three diagrams (F ig. 26). They are proportional respectively to $t_a t_b$, $t_b t_a$, and $f_{ab}^c t_c$, where f_{abc} are the structure constants that appear in the bulk A^3 vertex. One tests these diagrams for gauge-invariance or BRST invariance by making a gauge transformation $A^a \to A^a + \mathrm{d}\mathsf{c}^a$ for one of the external gauge bosons, with c^a the ghost field. A standard calculation shows that the sum of the three diagrams is gauge-invariant and BRST-invariant if and only if $[t_a, t_b] = f_{ab}^c t_c$. The violation of BRST-invariance if this condition is not satisfied is bilinear in A and c, because the diagrams have two external bosons, one of which is replaced by $\mathrm{d}\mathsf{c}^a$ when one tests for BRST-invariance. Likewise, in the analysis that follows, the anomalies come from diagrams with two external gauge bosons, and are bilinear in A and c.

After verifying gauge invariance at tree level, one should go on to determine whether quantum corrections to the coupling of two gauge bosons to a charged particle or to a Wilson line preserve gauge invariance. In conventional four-dimensional quantum field theory, the answer in general is that there is a problem: in certain theories, one-loop triangle diagrams have an anomaly that irreparably spoils gauge invariance. In the model that we will study, there is no problem in order $\hbar$, but we will find an anomaly in order $\hbar^2$ – roughly speaking, in two-loop order. This anomaly, however, will not represent a complete breakdown of gauge invariance. Rather, it will represent a deformation of the gauge symmetry algebra – a quantum correction to the classical commutation relations (8.2).

Such a quantum deformation as the outcome of an analysis of anomalies is perhaps unfamiliar, for the following reason. It does not occur in the case of a semi-simple gauge group, because the classification of semi-simple groups is discrete. The infinite-dimensional gauge algebra $\mathfrak{g}[[z]]$ is, however, susceptible to continuous deformation. (Actually, because the anomalous diagrams have more than one gauge boson attached to the Wilson line, what we will get is a deformation of the universal enveloping algebra of $\mathfrak{g}[[z]]$, not a deformation of $\mathfrak{g}[[z]]$ as a Lie algebra.)

In our theory, because $\mathfrak{g}[[z]]$ will be deformed, the gauge invariance of a Wilson line depends on modifying the matrices $t_{a,n}$ to provide a representation of the deformed algebra rather than of $\mathfrak{g}[[z]]$. If this is possible for a given Wilson line, we say that this Wilson line (or the associated representation of $\mathfrak{g}[[z]]$) "quantizes." A general Wilson line does not quantize in this sense. In section 8.7, we describe an explicit counterexample.

8.2 Anomalies on Wilson Lines

We will first provide a general analysis of what anomalies can occur in the coupling of gauge fields to a given Wilson line operator. We will derive a cohomological interpretation of the anomaly. Combined with a cohomology calculation given in appendix C, this allows us to prove the following results.

Theorem 8.1. *Let V be a representation of a simple Lie algebra $\mathfrak{g}$ which has no Abelian factors and which is not $\mathfrak{sl}_2$.*

1. *The first possible anomaly to quantizing the Wilson line arises at 2 loops, that is in order $\hbar^2$.*
2. *Let $\wedge_0^2\mathfrak{g}$ be the kernel of the bracket map $\wedge^2\mathfrak{g} \to \mathfrak{g}$. If there are no non-trivial G-invariant maps $\wedge_0^2\mathfrak{g} \to \mathrm{End}(V)$, then there are no anomalies.*
3. *At two loops, the number of possible anomalies is exactly the dimension of the space of G-invariant maps $\wedge_0^2\mathfrak{g} \to \mathrm{End}(V)$.*

These results imply, for instance, that the vector representations of the classical groups SL_n, SO_n and Sp_{2n} all quantize.

Let us analyze the possible forms of the anomalies to quantizing a Wilson line. By the anomaly of a Wilson line, we mean its BRST variation. This will always be the integral over the Wilson line of a local operator of ghost number 1. Thus, this operator will be linear in the ghost field c (and its derivatives) with *a priori* an arbitrary polynomial dependence on the gauge field A (and its derivatives).

As usual, we work on $\mathbb{R}^2 \times C$, with the usual real coordinates x, y and complex coordinate z. We take a Wilson line to be parametrized by x, at $y = z = 0$. Such a Wilson line is invariant under the classical symmetries that independently rescale x and y. The anomaly will have the same invariances. This implies that the anomaly must be linear in A_x and independent of A_y, and cannot have any x or y derivatives.

Because the classical symmetry which scales the x and y directions is unbroken by the coupling of a Wilson line, any anomaly[32] must respect this symmetry.

[32] In general, given a quantum system defined up to order k in $\hbar$, the anomaly to quantizing

Since the anomaly must be an integral over the Wilson line, it must involve only the x-component A_x of the connection, and can not involve the y-component or any x and y derivatives. Otherwise, it would not respect the scaling symmetries in the xy plane. For the same reason, the anomaly must be at most a linear function of A_x.

This argument does not exclude the possibility that the anomaly can include some polynomial in $A_{\overline{z}}$, or its $\overline{z}$ derivatives. Let us now see why these can not occur (in section 7.2 we have performed a similar analysis for the anomaly to the existence of a vertex). The classical action functional of the theory is invariant under the symmetry which scales z $\overline{z}$, and $\hbar$ by a real parameter c. It is also invariant under the symmetry by which z and $\hbar$ are rotated through an angle $e^{2\pi i\theta}$, and where $\overline{z}$ is therefore rotated by $e^{-2\pi i\theta}$. Any anomaly must be invariant under these symmetries (where we include the factor of $\hbar^k$ naturally present in a k-loop anomaly).

The most general anomaly at k loops will involve n_1 copies of $A_{\overline{z}}$, n_2 $\overline{z}$ derivatives, and n_3 z-derivatives. A copy of $A_{\overline{z}}$ in the anomaly behaves the same with respect to these symmetries as a $\overline{z}$-derivative, as it involves contracting the connection A with the vector field $\partial_{\overline{z}}$. The symmetries under scaling and rotation in the z-plane tell us that

$$\begin{aligned} n_1 + n_2 + n_3 - k &= 0 \,, \\ -n_1 - n_2 + n_3 - k &= 0 \,. \end{aligned} \tag{8.3}$$

From this we see that $n_1 = n_2 = 0$ and $n_3 = k$. Therefore, the anomaly can not involve the $\overline{z}$-component of the connection or any $\overline{z}$ derivatives, and the number of z derivatives in a k-loop anomaly must be precisely k.

As a further constraint, note that if the anomaly does not depend at all on A_x, and so depends only on the ghost field c, then it must be given by an integral of $\partial_x \mathsf{c}$ (in view of the scaling symmetry in the x direction). We can write $\partial_x \mathsf{c} = D_x \mathsf{c} - [A_x, \mathsf{c}]$, where $D_x \mathsf{c}$ is the covariant derivative of c. Since insertion of $D_x \mathsf{c}$ in a Wilson operator gives a total derivative that would not contribute, we can replace $\partial_x c$ with $-[A_x, \mathsf{c}]$, and thus it is not necessary to consider terms that are independent of A_x.

The constraints we have considered so far tell us that the most general k-loop anomaly is of the form

$$\sum_{k_1+k_2=k} \Theta_{a,k_1,b,k_2} \int_{y=z=0} \partial_z^{k_1} A_x^a \partial_z^{k_2} \mathsf{c}^b \mathrm{d}x \,, \tag{8.4}$$

where for each value of k_1, k_2, Θ^{a,k_1,b,k_2} is a G-invariant linear map

$$\mathfrak{g} \otimes \mathfrak{g} \to \mathrm{End}(V) \,,$$

where V is the representation of G from which we build the Wilson line.

the system to order $k+1$ in $\hbar$ must respect any symmetries present in the system at order k. To see this, we note that if we change the order $k+1$ counter-terms, the anomaly changes by a BRST-exact term. Therefore the cohomology class of the anomaly only depends on the theory up to order k, and is preserved by any symmetries present in the system up to order k.

There is one more constraint. Since the anomaly is the BRST variation of the Wilson line, it is itself BRST-invariant. Applying this constraint leads us to the equation

$$\begin{aligned}&\sum_{k_1+k_2=k} \Theta_{a,k_1,b,k_2} \int_{y=z=0} \partial_z^{k_1}(\mathrm{d}\mathsf{c}^a + f^a{}_{cd}\mathsf{c}^c A_x^d)\left(\partial_z^{k_2}\mathsf{c}^b\right)\mathrm{d}x \\ &+ \sum_{k_1+k_2=k} \Theta_{d,k_1,a,k_2} \int_{y=z=0} \partial_z^{k_1} A_x^d \partial_z^{k_2}\left(\frac{1}{2}f^a{}_{cb}\mathsf{c}^c\mathsf{c}^b\right)\mathrm{d}x \\ &+ \sum_{k_1+k_2=k} [\rho_c, \Theta_{d,k_1,b,k_2}] \int_{y=z=0} \mathsf{c}^c\left(\partial_z^{k_1}A_x^d\right)\left(\partial_z^{k_2}\mathsf{c}^b\right)\mathrm{d}x = 0\end{aligned} \tag{8.5}$$

In the third term in this equation, the operator $\rho_a : V \to V$ indicates the classical action of the operator t_a on V. The appearance of this term was be explained in detail in a similar context in section 7.2.

This equation can be separated into a term which is linear in A and one which has no A-dependence. Both of these terms must vanish. Let us first analyze the term which is independent of A. If we integrate by parts, and use the fact that the ghost field is an anti-commuting variable (i.e. a fermionic field of spin 0), we find that this equation tells us

$$\Theta_{a,k_1,b,k_2} + \Theta_{b,k_2,a,k_1} = 0\,. \tag{8.6}$$

Next, let us analyze the term in eqn. (8.5) which is linear in A. We obtain equations by setting the coefficients of $\int \left(\partial_z^{k_0}\mathsf{c}^c\right)\left(\partial_z^{k_1}A^d\right)\left(\partial_z^{k_2}\mathsf{c}^b\right)$ to zero. These equations are

$$\begin{aligned}&\Theta_{a,k_0+k_1,b,k_2} f^a{}_{cd}\binom{k_0+k_1}{k_0} - \Theta_{a,k_2+k_1,c,k_0} f^a{}_{bd}\binom{k_1+k_2}{k_1} \\ &+ \Theta_{d,k_1,a,k_0+k_2} f^a{}_{cb}\binom{k_0+k_2}{k_0} + \delta_{k_0=0}[\rho_c, \Theta_{d,k_1,b,k_2}] - \delta_{k_2=0}[\rho_b, \Theta_{d,k_1,c,k_0}] = 0\,.\end{aligned} \tag{8.7}$$

Note that this quantity is totally antisymmetric under permutation of (k_0, c), (k_1, d) and (k_2, b), as follows from eqn. (8.6).

We can summarize this result as follows. Let us view the collection Θ_{a,k_1,b,k_2} as a linear map

$$\begin{aligned}\Theta : \quad \wedge^2\mathfrak{g}[[z]] &\to \mathrm{End}(V)\,, \\ \Psi \qquad\quad & \qquad \Psi \\ (t_a z^{k_1}) \wedge (t_b z^{k_2}) &\mapsto k_1! k_2! \Theta_{a,k_1,b,k_2}\,.\end{aligned} \tag{8.8}$$

Let $C^*(\mathfrak{g}[[z]], \mathrm{End}(V))$ denote the Chevalley-Eilenberg cochain complex, which in degree n consists of linear maps $\wedge^n\mathfrak{g}[[z]] \to \mathrm{End}(V)$.

Then, eqn. (8.7) is precisely the condition that Θ is a closed element of this Chevalley-Eilenberg cochain complex.

8.2.1 Cancellation by Counter-Terms.

Let us now turn to analyzing when such an anomaly can be cancelled by a counter-term. Suppose that the first anomaly arises at k loops and is given by the expression in eqn. (8.4) involving some Θ_{a,k_1,b,k_2} satisfying 8.6.

To try to cancel the anomaly, we change the coupling of the gauge field A so that $\partial_z^k A^a$ is coupled by some operator $\rho_{a,k} \in \mathrm{End}(V)$. We will assume that $\rho_{a,k}$ transforms under the adjoint representation of G on $\mathrm{End}(V)$. Simple Lie algebras and their representations have a discrete classification and cannot be deformed, so we will assume that $\rho_{a,k}$ vanishes for $k = 0$.

The term $\rho_{a,k} \int \partial_z^k A^a$ that we add to the action may not be gauge invariant at the classical level. The failure to be gauge invariant is given by the expression

$$\rho_{a,k} \int \partial_z^k \mathrm{d}\mathsf{c}^a + \rho_{a,k} f_{abc} \int \partial_z^k \left(\mathsf{c}^b A^c\right) + [\rho_b, \rho_{a,k}] \int \mathsf{c}^b \partial_z^k A^a . \tag{8.9}$$

In order to cancel the anomaly at k loops, we need

$$\sum_{k_1+k_2=k} \Theta_{a,k_1,b,k_2} \int_{y=z=0} \partial_z^{k_1} A_x^a \partial_z^{k_2} \mathsf{c}^b \mathrm{d}x = \rho_{a,k} f_{abc} \int \partial_z^k \left(\mathsf{c}^b A^c\right) + [\rho_b, \rho_{a,k}] \int \mathsf{c}^b \partial_z^k A^a . \tag{8.10}$$

This can happen if and only if

$$\Theta_{a,k_1,b,k_2} = \rho_{c,k} f_{ba}^c \binom{k}{k_1} + [\rho_a, \rho_{b,k}] . \tag{8.11}$$

This is precisely the condition that the cocycle in $C^*(\mathfrak{g}[[z]], \mathrm{End}(V))$ associated to the anomaly (as in eqn. (8.8)) is exact.

In sum, we have found that possible anomalies to coupling to a Wilson line, modulo counter-terms which can cancel these anomalies, are given by the Chevalley-Eilenberg cohomology group

$$H^2(C^*(\mathfrak{g}[[z]], \mathrm{End}(V)) . \tag{8.12}$$

We will repeatedly use the fact that for G a simple Lie group, the Chevalley-Eilenberg cohomology (of any degree) with values in any representation can always be represented by a G-invariant cocycle, roughly by averaging over a maximal compact subgroup of G. (Of course, the expressions that we generate from Feynman diagrams in expanding around the trivial flat connection will always be G-invariant.) Note that Chevalley-Eilenberg cohomology is also called Lie algebra cohomology, and under that name often appears in analysis of BRST cohomology in quantum field theory.

One nice consequence of the description of anomalies by Chevalley-Eilenberg cohomology is that we can understand what anomalies can exist by applying standard arguments from homological algebra. When translated into field-theory language, the arguments would be quite complicated.

For instance, the following is a relatively easy result:

Lemma 8.2. *For a Wilson line in a representation V of a simple group G (as opposed to a representation of $\mathfrak{g}[[z]]$), there are no one-loop anomalies which cannot be cancelled by the introduction of a counter-term.*

Further, all two-loop anomalies are equivalent to ones given by expressions of the form

$$\sum \Theta_{a,1,b,1} \int_{y=z=0} \partial_z A_x^a \partial_z \mathsf{c}^b \mathrm{d}x \;, \tag{8.13}$$

where, as before, $\Theta_{a,1,b,1}$ is a linear operator on the vector space V and is antisymmetric under the exchange of the a, b indices. Viewed as a linear map $\wedge^2\mathfrak{g} \to \mathrm{End}(V)$, Θ can be assumed to be G-invariant.

Two-loop anomalies which factor through a map $\wedge^2\mathfrak{g} \to \mathfrak{g}$ can be cancelled by a counter-term.

Proof. The proof of the lemma uses the notion of relative Lie algebra cochains of $\mathfrak{g}[[z]]$ with respect to the subalgebra $\mathfrak{g}$. Relative Lie algebra cochains are G-invariant elements of $C^*(z\mathfrak{g}[[z]], \mathrm{End}(V))$, that is, G-invariant cochains which do not involve $\mathfrak{g}$. The relative cochains are equipped with the same Chevalley-Eilenberg differential as before. We let $H^*((\mathfrak{g}[[z]] \mid \mathfrak{g}), \mathrm{End}(V))$ denote the relative cohomology. General results on the cohomology of simple Lie algebras imply that there is an isomorphism

$$\oplus_{i+j=k} H^i(\mathfrak{g}) \otimes H^j((\mathfrak{g}[[z]] \mid \mathfrak{g}), \mathrm{End}(V)) \cong H^k(\mathfrak{g}[[z]], \mathrm{End}(V)) \;. \tag{8.14}$$

Since $\mathfrak{g}$ is simple, $H^*(\mathfrak{g})$ is zero in degrees 1 and 2, and $H^0(\mathfrak{g}) = \mathbb{C}$. Therefore relative and absolute Lie algebra cohomology coincide in degrees 1 and 2.

At weight 1 under the $\mathbb{C}^\times$ action that scales z, the relative cohomology is zero in degrees 2 and higher. This is simply because we cannot built a bilinear of weight 1 from $z\mathfrak{g}[[z]]$. Therefore there are no anomalies of weight 1.

The cohomology of weight 2 under this $\mathbb{C}^\times$ action is the part that contributes to the two-loop anomaly. Relative 2-cocycles of weight 2 must be given by a G-invariant bilinear and antisymmetric map $z\mathfrak{g} \otimes z\mathfrak{g} \to \mathrm{End}(V)$. Because each tensor factor involves one z, the corresponding two-loop anomaly is of the form given in eqn. (8.13) (with one z derivative of A and one of c). So every two-loop anomaly is of this form.

Concerning the last statement in the lemma, we observe first that for simple G, any G-invariant map $\wedge^2\mathfrak{g} \to \mathfrak{g}$ is actually a multiple of the commutator map. We do not know a unified proof of this statement, but for classical groups it follows from invariant theory,[33] and for exceptional groups, it can be verified by consulting tables (such as those in [36]). Given this, we need to show that if a two-cocycle

$$\wedge^2 (z\mathfrak{g}) \to \mathrm{End}(V) \tag{8.15}$$

factors through the commutator map $\wedge^2\mathfrak{g} \to \mathfrak{g}$, then it is the coboundary of a G-invariant map $z^2\mathfrak{g} \to \mathrm{End}(V)$. But the coboundary of a G-invariant map $z^2\mathfrak{g} \to \mathrm{End}(V)$ is given by the composition

$$\wedge^2 (z\mathfrak{g}) \to z^2\mathfrak{g} \to \mathrm{End}(V), \tag{8.16}$$

[33] That is, one explicitly considers an element of $\mathfrak{g}$ as a two-index tensor, and one considers the possible contractions of indices to make a G-invariant map from $\wedge^2\mathfrak{g}$ to $\mathfrak{g}$.

where the first map is the commutator. Therefore any two-cocycle that factors through the commutator is a coboundary. □

This lemma tells us that find that the first possible anomalies can arise at two loops, from diagrams with two external gauge fields.

In appendix C, we prove the following result (relying heavily on the results of [31]). This result generalizes the previous lemma to all loops. The proof is much more difficult.

Proposition 8.3. *Let $\mathfrak{g}$ be a simple Lie algebra which is not*[34] $\mathfrak{sl}_2$. *Let $\wedge_0^2\mathfrak{g}$ be the kernel of the Lie bracket map from $\wedge^2\mathfrak{g}$ to $\mathfrak{g}$. (Note that there are no copies of the adjoint in the representation $\wedge_0^2\mathfrak{g}$). Then the inclusion map*

$$H^2(\mathfrak{g}[[z]], \wedge_0^2\mathfrak{g}) \to H^2(\mathfrak{g}[[z], \mathfrak{g}\otimes\mathfrak{g}) \tag{8.17}$$

is an isomorphism.

In other words, if we decompose $\mathfrak{g}\otimes\mathfrak{g}$ as $\mathrm{Sym}^2\,\mathfrak{g}\oplus\mathfrak{g}\oplus\wedge_0^2\mathfrak{g}$, then all of the cohomology we are interested in comes from $\wedge_0^2\mathfrak{g}$.

We are interested in $H^2(\mathfrak{g}[[z]], \mathrm{End}(V))$ for a representation V. In general, $\mathrm{End}(V)$ is a complicated, highly reducible representation of G. Chevalley-Eilenberg cohomology is defined for every representation, and if $\mathrm{End}(V)$ is a direct sum of irreducibles, then the desired cohomology is the direct sum of cohomology with values in these irreducibles.

However, the desired cohomology can be represented by G-invariant cocycles and such a cocycle, ignoring the z-dependence, is a G-invariant map from $\mathfrak{g}\otimes\mathfrak{g}$ to $\mathrm{End}(V)$. Hence, any two-cocycle valued in $\mathrm{End}(V)$ must come from a two-cocycle valued in $\mathfrak{g}\otimes\mathfrak{g}$ under some map of G-representations from $\mathfrak{g}\otimes\mathfrak{g}$ to $\mathrm{End}(V)$. The same is true for the corresponding second cohomology classes. The proposition gives an important refinement of this statement: any non-trivial second cohomology class must come from a map $\wedge_0^2\mathfrak{g}\to\mathrm{End}(V)$. This means that when we study Feynman diagrams, we can ignore terms in which two gauge bosons A^a and A^b are coupled symmetrically in a and b, and we can also ignore terms in which they are coupled via f_{ab}^c.

As a corollary, we find that if there are no maps of representations $\wedge_0^2\mathfrak{g}\to\mathrm{End}(V)$, then the Wilson line living in V can not have an anomaly. This is the criterion (†) that was introduced in section 7.1 and used in various applications.

8.3 Enumerating Feynman diagrams

So far, we have performed a cohomological analysis of possible anomalies that can appear when one tries to quantize a Wilson line. We have seen that the first such anomalies can appear at two loops. Next, we will enumerate all possible two-loop

[34]The case of $\mathfrak{sl}_2$ is slightly different. There we find that $H^2(\mathfrak{g}[[z]], \mathfrak{g}\otimes\mathfrak{g})$ is isomorphic to $H^2(\mathfrak{g}[[z]], \mathrm{Sym}^2\,\mathfrak{g})$, which is one-dimensional and in weight 3 under the $\mathbb{C}^\times$ action scaling z. Since it is known by other methods that every representation of $\mathfrak{sl}_2$ quantizes, we will avoid the special case of $\mathfrak{sl}_2$ in what follows.

diagrams that may contribute an anomaly. We will find that, for various reasons, all but one of the diagrams we enumerate can not produce an anomaly.

Throughout our analysis, we will assume our Wilson line is placed at $z = 0$ and is invariant under the symmetry which rescales z and $\hbar$. This means that $t_{a,k}$ is accompanied by $\hbar^k$, and in particular at the classical level $t_{a,k} = 0$ for $k > 0$.

Let us now analyze possible two-loop diagrams which can contribute anomalies of this form. The only diagrams that are relevant are two-loop diagrams which may be attached to the Wilson line at an arbitrary number of points, but which have precisely two external lines to which the gauge field is coupled.

We must also bear in mind an additional subtlety. In principle one can introduce one-loop counter-terms whereby $\partial_z A$ is coupled to the Wilson line in some way. It could happen that these are forced on us to cancel any one-loop anomalies, but even if this is not the case, such one-loop counter-terms do not violate any symmetries and so one is always free to introduce them. Therefore there are two possible vertices at which a gluon can meet a Wilson line. There is the usual classical interaction, but also the interaction coming from a one-loop counter-term. We will depict these interactions as an ordinary vertex, and a circle labelled 1, respectively. Counting of the loop parameter tells us that one-loop diagrams, one of whose vertices on the Wilson line is the one-loop vertex, are counted as two-loop diagrams.

One can cut down the number of diagrams we need to consider by observing that every diagram must have at least two vertices on the Wilson line. Diagrams with one vertex on the Wilson line can not contribute, because the corresponding anomaly will be given by a two-cocycle valued in some copy of $\mathfrak{g} \subset \mathrm{End}(V)$. Our cohomological analysis in proposition 8.3 tells us that any anomaly of this form can be cancelled by a counter-term.

In evaluating whether a diagram can contribute to the anomaly, note that the anomaly must have one z-derivative at each external line, and no other derivatives. We can thus calculate whether or not a diagram contributes to the anomaly by taking the two external fields to be the gauge field $A = z\delta_{x=0}t_a$, and the ghost field $\mathsf{c} = zt_b$. This tells us, in particular, that it is not possible to have an anomaly from a diagram in which the external lines are connected directly to the Wilson line by a tree-level vertex. This is because the Wilson line is at $z = 0$, and there are no z-derivatives in how the Wilson line is coupled classically.

A further constraint on the diagrams that can appear is provided by the observation that if we evaluate the propagator with both ends on a straight Wilson line we find zero. Therefore in any diagram which can contribute to the anomaly this configuration can not occur.

These observations cut down substantially the diagrams that can appear. We will further exclude a few simple diagrams which are topologically trees, but whose coupling to the Wilson line involves one and two-loop counter-terms. These diagrams are similar to those in Fig. 26. In section 8.6 we will analyze the effect of these diagrams, and see that they contribute to the two-loop correction to the algebra $\mathfrak{g}[[z]]$.

All remaining diagrams are depicted in Fig. 27. Each figure admits a number of variants where the order of the vertices on the Wilson line is permuted. These variants are not depicted.

We will find that the only diagram that can contribute is diagram (A1). All other diagrams will be excluded by a variety of arguments.

8.3.1 Excluding Diagrams (A2), (C1), (C2) and (D2)

Consider any two-loop diagram with the feature that the two vertices attached to the external lines are connected by a propagator. The only non-trivial anomalies are antisymmetric in the external lines. After antisymmetrizing and using the Jacobi identity, the anomaly from such a diagram is of the form

$$\int f^{cab} \partial_z \mathsf{c}_a \partial_z A_b \mu_c \tag{8.18}$$

for some matrices $\mu_c : V \to V$. Any anomaly of this form can be cancelled, because it comes from a map $\wedge^2 \mathfrak{g} \to \mathrm{End}(V)$ which factors through the adjoint representation.

This tells us that diagrams (A2), (C1), (C2) and (D2) can not contribute a non-trivial anomaly.

8.3.2 Excluding Diagrams of Type (C3)

Diagram of type (C3) can be excluded for Lie algebraic reasons. Any anomaly that can appear from a diagram of type (C3) is of the form:

$$\mathrm{Tr}_{\mathfrak{g}} \left(t_a t_c t_b t_d \right) \int \partial_z \mathsf{c}_a \partial_z A_b \rho(t_c) \rho(t_d) \ . \tag{8.19}$$

(Here the trace is taken in the adjoint representation, and the external lines have indices a, b).

This anomaly can be cancelled by a two-loop counter-term if the tensor $\mathrm{Tr}_{\mathfrak{g}} \left(t_a t_c t_b t_d \right)$ is symmetric in a and b, or if it can be written in terms of the commutator $f^e{}_{ab} t_e$ of t_a and t_b. Since the adjoint representation is equipped with a symmetric non-degenerate pairing, we have the identity

$$\mathrm{Tr}_{\mathfrak{g}}(t_a t_c t_b t_d) = \mathrm{Tr}_{\mathfrak{g}}(t_d t_b t_c t_a) \ . \tag{8.20}$$

Together with cyclic symmetry of the trace, this tells us that $\mathrm{Tr}(t_a t_c t_b t_d)$ is symmetric in a and b. This implies that any anomaly associated to the tensor $\mathrm{Tr}_{\mathfrak{g}}(t_a t_c t_b t_d)$ can be cancelled by a counter-term.

8.3.3 Excluding Diagrams (C4) and (C5)

Diagrams (C4) and (C5) contain an internal gauge particle loop. In (C4), this loop has two external lines, both of which are internal lines in the full diagram, while (C5) has an internal loop with three external lines, of which in the full diagram two are internal and one is external.

The group theory factor that comes from the internal loop in (C4) is an invariant in $\mathfrak{g} \otimes \mathfrak{g}$. Any such invariant is a multiple of δ_{cd}, that is, the Killing

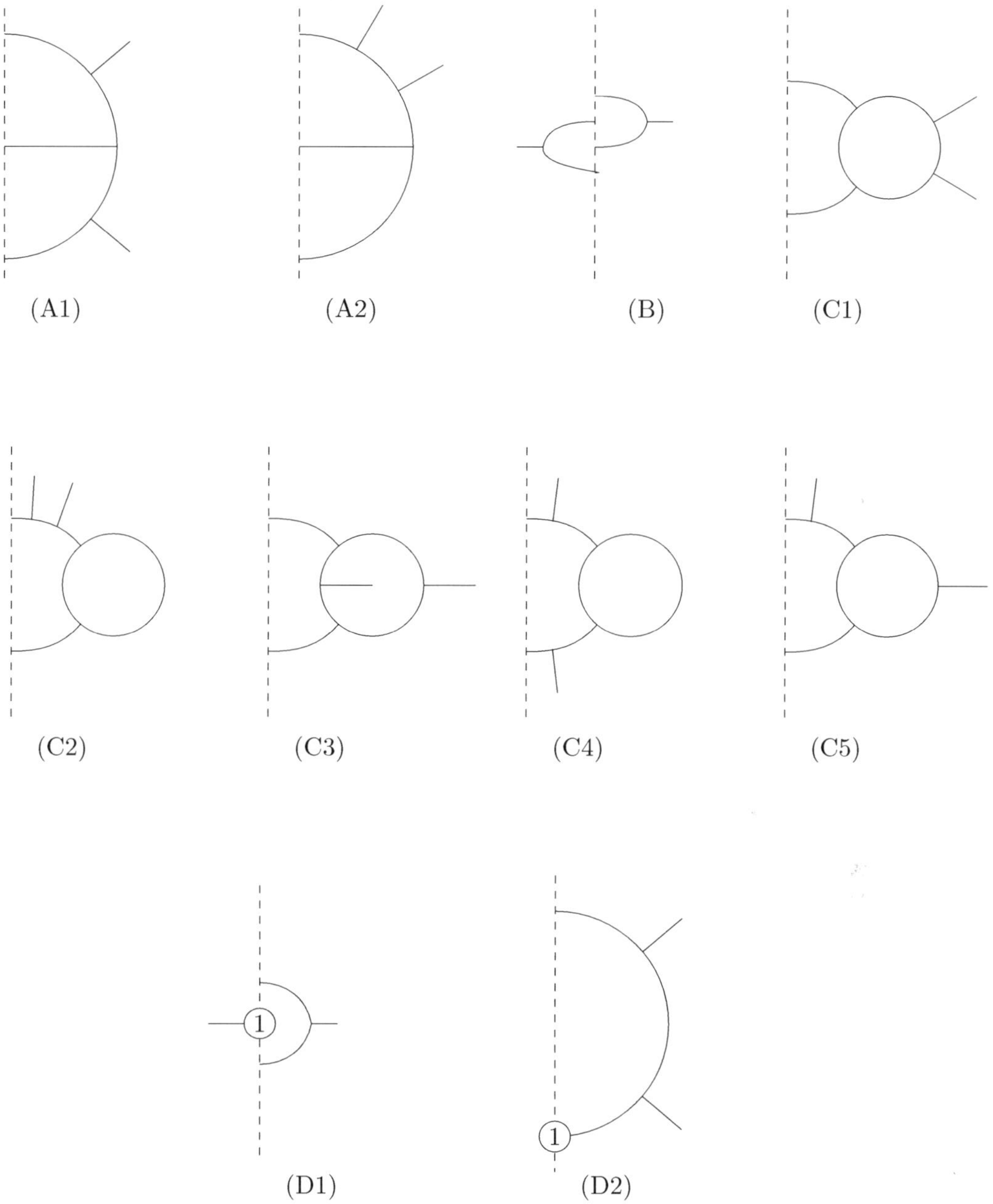

Figure 27: Two-loop Feynman diagrams with two external lines which can potentially contribute to the anomaly. The dashed vertical line represents a Wilson line, and the vertices on the Wilson line labelled 1 indicate a coupling of the gauge field and Wilson line by a one-loop counter-term. Each diagram admits a number of variants, not shown, in which the positions of the vertices on the Wilson line are permuted.

form, and accordingly, the overall group theory factor of the diagram in (C4) is a multiple of what it would be if the internal loop were collapsed to a point. Then in (C4) the two external gauge fields would be connected by a propagator. So this diagram can be excluded by the same argument as in section 8.3.1.

In (C5), since the internal loop has three external lines, it produces an invariant in $\mathfrak{g} \otimes \mathfrak{g} \otimes \mathfrak{g}$. For any simple Lie algebra other than SL_N, $N > 2$, such an invariant actually lies in[35] $\wedge^3 \mathfrak{g}$ and is a multiple of the invariant associated to the structure constants f_{abc}. When this is the case, the group theory factor of (C5) is the same as if the internal loop were collapsed to a point. When this is done, the two external gauge fields are connected by a propagator and this diagram can be excluded as before.

For SL_N, $N > 2$, there actually is another invariant in $\mathfrak{g} \otimes \mathfrak{g} \otimes \mathfrak{g}$; in fact, it lies in $\mathrm{Sym}^3 \mathfrak{g}$. However, this invariant is odd under the outer automorphism of SL_N, which is a symmetry of the theory under discussion, and therefore cannot appear.

8.3.4 Excluding Diagrams (B) and (D1)

Diagrams (B) and (D1) cannot be excluded by Lie algebraic arguments. Instead we will show by an explicit calculation that the amplitudes for such diagrams are zero, so they can not contribute an anomaly. The computation involves only the part of the diagram on the right in the figures; this is the part that (B) and (D1) have in common.

To evaluate the amplitudes for these diagrams, we will employ a point-splitting regularization in which any pair of vertices on the Wilson lines must be separated by a distance of at least ϵ. Once we employ such a point-splitting regularization, we can compute the regularized amplitude for each connected component of the diagrams (B) and (D1) separately. The two connected components only interact with each other by how they affect the domain of integration of the vertices on the Wilson lines. We will get a vanishing that does not depend on this domain.

We therefore need to compute the amplitude of the diagram

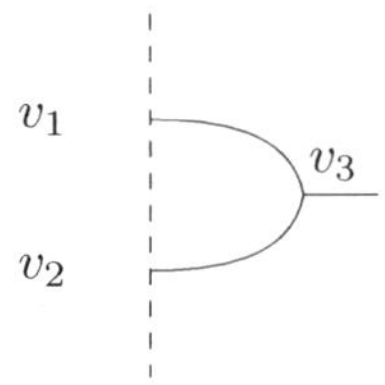

The coordinates for the vertex v_i will be (x_i, y_i, z_i). For vertices v_1, v_2 we have $y_i = z_i = 0$.

[35]Invariants in $\mathfrak{g}\otimes\mathfrak{g}\otimes\mathfrak{g}$ correspond to G-invariant maps $\mathfrak{g}\otimes\mathfrak{g} \to \mathfrak{g}$. That there is only one such invariant for most G and two for SL_N, $N > 2$, can be shown by invariant theory for classical G and by consulting tables such as those in [36] for exceptional G. In SL_N, $N > 2$, the two invariants are $\mathrm{Tr}\, A[B,C]$ and $\mathrm{Tr}\, A\{B,C\}$ and are respectively even and odd under the outer automorphism. The fact that an invariant in $\mathrm{Sym}^3\mathfrak{g}$ exists only for SL_N, $N > 2$ is important in particle physics in classifying possible anomalies in gauge theory.

Recall (see eqn. (4.5)) that the propagator two-form is

$$P = \frac{1}{2\pi} \frac{x\mathrm{d}y \wedge \mathrm{d}\overline{z} - y\mathrm{d}x \wedge \mathrm{d}\overline{z} + 2\overline{z}\mathrm{d}x \wedge \mathrm{d}y}{\left(x^2 + y^2 + |z|^2\right)^2} . \tag{8.21}$$

The amplitude for the diagram is then

$$\begin{aligned} c \int_{x_1,x_2,x_3,y_3,z_3} & A\mathrm{d}z_3 \wedge \frac{(x_3 - x_1)\mathrm{d}y_3\mathrm{d}\overline{z}_3 - y_3\mathrm{d}(x_3 - x_1)\mathrm{d}\overline{z}_3 + 2\overline{z}_3\mathrm{d}(x_3 - x_1)\mathrm{d}y_3}{d(v_3, v_1)^4} \\ & \wedge \frac{(x_3 - x_2)\mathrm{d}y_3\mathrm{d}\overline{z}_3 - y_3\mathrm{d}(x_3 - x_2)\mathrm{d}\overline{z}_3 + 2\overline{z}_3\mathrm{d}(x_3 - x_2)\mathrm{d}y_3}{d(v_3, v_2)^4} , \end{aligned} \tag{8.22}$$

where $d(v_i, v_j)$ indicates the distance between the vertices v_i and v_j, and A is the gauge field on the external line.

Since we are integrating over x_1, x_2, we need only keep terms which involve $\mathrm{d}x_1$ and $\mathrm{d}x_2$. If we retain only those terms, we find that the integrand is $\mathrm{d}x_1 \wedge dx_2$ times the square of the one-form

$$y_3\mathrm{d}\overline{z}_3 - 2\overline{z}_3\mathrm{d}y_3 \tag{8.23}$$

and is therefore zero.

8.4 The Anomaly Associated to Diagram (A1).

Let us now come to the only remaining diagram, namely (A1) in Fig. 27. We need to note that we first need to include two other diagrams, as depicted in Fig.28, where we permute the three vertices along the Wilson lines.

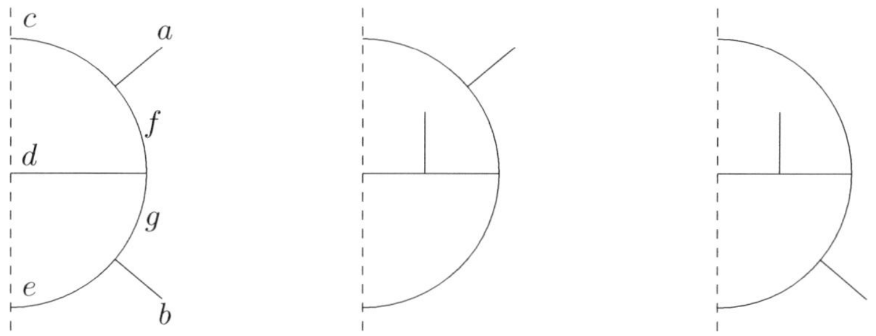

Figure 28: Permutations of diagram (A1).

8.4.1 Color Factor

It turns out that all these possibilities have the same color factor, up to those contributions which can be canceled by counterterms.

To see this, let us assign color indices a, b to external lines (ghost field and the gauge field) and c, d, e to points on Wilson lines. In the first diagram in Fig. 28 we have c, d, e along the Wilson line from top to bottom, but in the second we have c, e, d and third d, c, e.

Let us consider the difference of color factors between first and the second diagram. Since this involves the change in the relative position of d and e, the difference gives the commutator $[\rho(t^d), \rho(t^e)] = f^{de}{}_f \rho(t^f)$. Hence the difference in color factors can be represented graphically as the diagram on the left of Fig. 29. Similarly, the difference of the color factor between the second the third diagram gives the second diagram of Fig. 29.

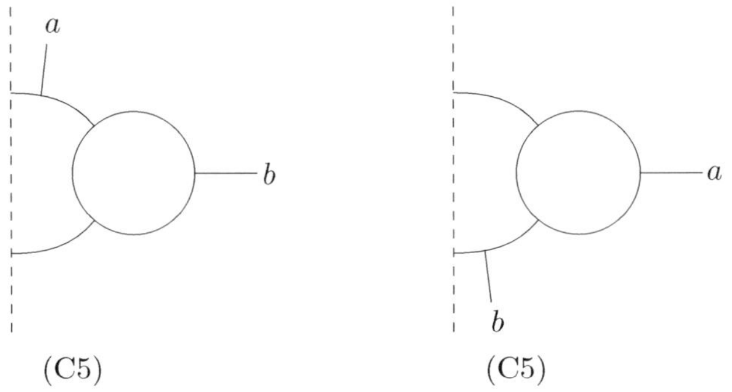

Figure 29: The differences of the color factors between first and second, and first and third diagrams in Fig. 28 have the color structures as shown in this figure.

We verified above (section 8.3.3) that the color factor for diagrams of this type is such that any anomaly can always be cancelled by a counter-term.

This tells us that for each of the three diagrams in Fig. 28, the anomaly is symmetric under permutation of the order of the vertices connected to the Wilson line, up to anomalies which can be cancelled by a counter-term. Since the three diagrams in Fig. 28 are exchanged by such a permutation, we conclude that they all have the same color factor (again, modulo terms which can be cancelled by a counter-term).

The non-trivial part of the color factor is (recall the normalization of the Killing form in eqn. (3.2))

$$f^{acf} f^{fgd} f^{geb} \{\rho(t^c), \rho(t^d), \rho(t^e)\} = ([[t^a, t^c], t^d], [t^b, t^e]) \{\rho(t^c), \rho(t^d), \rho(t^e)\} , \tag{8.24}$$

where a, b are indices for the external lines and

$$\{\rho(t^{a_1}), \rho(t^{a_2}), \rho(t^{a_3})\} = \frac{1}{3!} \sum_{\sigma \in S_3} \rho(t^{a_{\sigma(1)}}) \rho(t^{a_{\sigma(2)}}) \rho(t^{a_{\sigma(3)}}) , \tag{8.25}$$

where the sum is over the permutations of the indices $1, 2, 3$.

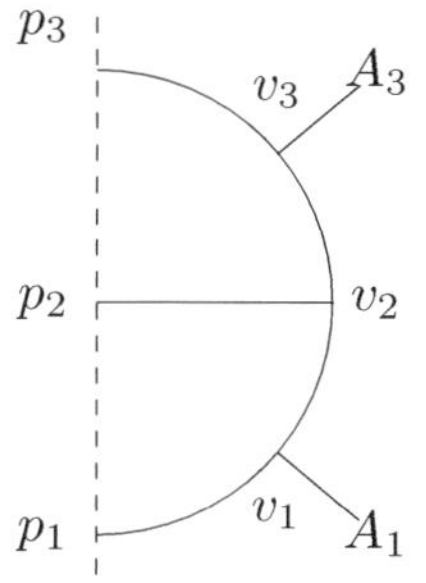

Figure 30: The Feynman diagram with vertices labelled. The A_i on the external lines indicate gauge fields.

8.5 Numerical Factor

Let us next initiate the evaluation of the numerical factor. We choose the Wilson line to be placed along a straight line at $y = z = 0$.

We let p_1, p_2, p_3 denote the position of the vertices on the Wilson line, and $v_1, v_2, v_3 \in \mathbb{R} \times \mathbb{C}$ be the positions of the internal vertices. The coordinates for vertex v_i are denoted x_i, y_i, z_i for $i = 1, 2, 3$. The labeling on the vertices for the first diagram in Fig. 28 is given in Fig. 30.

Let us write down the numerical factor of the amplitude, where we include all possible orderings of the vertices on the Wilson lines. We let P be the propagator viewed as a two-form on $\mathbb{R}^2 \times \mathbb{C}$ (with singularities at the origin). The space over which we integrate consists of a copy of $\mathbb{R}^2 \times \mathbb{C}$ for each internal vertex, and $\mathbb{R}$ for each vertex on the Wilson line. Every vertex gives a map from this space to $\mathbb{R}^2 \times \mathbb{C}$, where we project on the relevant factor for an internal vertex and project to $\mathbb{R}$ and then include on the x-axis for the vertices on the Wilson lines. For a pair of vertices (v_i, v_j) or (v_i, p_j) we define two-forms by

$$P(v_i, v_j) = (v_i - v_j)^* P\,, \quad P(v_i, p_j) = (v_i - p_j)^* P\,, \tag{8.26}$$

where we pull-back the two-form P on $\mathbb{R}^2 \times \mathbb{C}$ via the map $v_i - v_j$ or $v_i - p_j$.

The complete amplitude, including all the relevant diagrams and the color factor, is then

$$\begin{aligned}&\tfrac{1}{2}\left(\tfrac{\mathrm{i}}{2\pi}\right)^3 \int_{p_1,p_2,p_3 \in \mathbb{R}} \int_{v_1,v_2,v_3} (P(0,v_1))\mathrm{d}z_1 A_1^a P(v_1,v_2)\mathrm{d}z_2 P(v_2,p_2) P(v_2,v_3)\mathrm{d}z_3 A_3^b P(v_3,p_3)\\ &\times f^{afc} f^{fgd} f^{gbe} \{\rho(t^c), \rho(t^d)\,, \rho(t^e)\}\,,\end{aligned} \tag{8.27}$$

where A_2, A_3 indicate the external gauge fields. The prefactor of $\frac{1}{2}$ comes from the diagram automorphism.

8.5.1 A comment on signs

In the numerical factor, we have implicitly oriented each edge of the graph. If we reverse the orientation of the graph, the propagator changes sign, because

$P(v_i, v_j) = -P(v_j, v_i)$. Similarly, in defining the integral, we have implicitly ordered the set of 6 vertices (three on the Wilson line and three integrated over the four-dimensional space-time). If we permute the order of these vertices we change the integral by a sign. For the vertices on the Wilson line, the sign arises because a permutation can reverse the orientation. For the bulk vertices, since each is accompanied by the one-form $\mathrm{d}z$, permuting them will change the order in which we take the wedge product of these forms, and therefore introduce a sign.

These signs are also implicit in the definition of the color factor, so there is no overall sign ambiguity. One way to see this is to treat the elements of the Lie algebra as being anti-commuting variables. This is a natural thing to do, because the form degree of the forms we are integrating differs by one from the ghost number of the corresponding fields of the physical theory. For example, a one-form is of odd degree but is a field of ghost number zero. We can force Lie algebra valued forms to have the correct parity if we declare that the Lie algebra factor is fermionic.

If we treat the Lie algebra elements as fermionic, then the color factor $\delta_{ab} t^a_{v_i} \otimes t^b_{v_j}$ associated to the propagator $P(v_i, v_j)$ changes sign if we reverse the order of an edge, just like the $P(v_i, v_j)$ does. The color factor associated to a vertex is a graded symmetric function of the three Lie algebra elements. Because it is a fermionic function – since it depends on three fermionic Lie algebra elements – the overall color factor depends via a sign on the ordering of the set of vertices. Similarly, the color factor associated to a vertex on the Wilson line depends on a single Lie algebra element, together with an endomorphism of the bosonic vector space of states on the Wilson line. Therefore the color factor depends on the ordering of the set of vertices on the Wilson line, just as the numerical factor does.

To get the correct signs in the total amplitude, the rule is to choose an orientation of each edge on the graph and an ordering on the set of vertices. Then, compute both the color factor and the numerical factor using this choice. Their product is the amplitude. Because the color factor and the numerical factor change signs in the same way if we change the ordering of the vertices and the orientation of the edges, there is no ambiguity. The color factor in eqn. (8.27) was computed in this way.

8.5.2 Regularizing the integral

We will use a point-splitting regulator in which we restrict the domain of integration to the region where $p_3 - p_1 \geq \epsilon$, if we assume that the three points are ordered so that $p_1 < p_2 < p_3$. To verify that this is a good regulator, we need to show that the integral converges absolutely in this domain. We are only concerned with UV divergences, so we will further restrict the domain of integration to a region where all vertices are in some ball around the origin.

To verify convergence, we will use the following bound in absolute value of the propagator:

$$|P| \leq (x^2 + y^2 + z\overline{z})^{-3/2} . \tag{8.28}$$

This bound arises because the denominator in P is $(x^2 + y^2 + |z|^2)^{-2}$, while the numerator is a linear function of the variables $x, y, \overline{z}$. Any linear function on $\mathbb{R}^4$

is bounded in absolute value by some multiple of $(x^2 + y^2 + |z|^2)^{1/2}$. It suffices to verify that the integral converges when each propagator $P(v_i, v_j)$ or $P(p_i, v_j)$ is replaced by $d(v_i, v_j)^{-3}$ or $d(p_i, v_j)^{-3}$, where d is the Euclidean distance.

We can bound the external gauge fields by a constant, so we can drop them from the integral. We are thus left with the integral

$$\int_{p_2,p_3>\epsilon,v_1,v_2,v_3} \frac{1}{d(p_3,v_3)^3 d(v_2,v_3)^3 d(v_2,p_2)^3 d(v_2,v_1)^3 d(v_1,p_1=0)^3}, \tag{8.29}$$

where by overall translation invariance we set $p_1 = 0$. In our domain of integration, the most divergent region is when v_1, v_2, v_3, p_2 are all near 0 (or all near p_3). Focusing on this region is equivalent to taking p_3 to be far away from the other points, which allows us to drop the $d(p_3, v_3)^{-3}$ term since it is non-singular. We are then considering the integral

$$\int_{p_2,v_1,v_2,v_3} \frac{1}{d(0,v_1)^3 d(v_1,v_2)^3 d(v_2,p_1)^3 d(v_2,v_3)^3} \mathrm{d}^4 v_1 \mathrm{d}^4 v_2 \mathrm{d}^4 v_3 \mathrm{d} p_2 \,. \tag{8.30}$$

There are 13 integration variables, and the integrand has weight -12 under scaling all the variables. Therefore the integral converges absolutely on a domain where the integration variables v_i, p_2 are bounded from above.

8.5.3 The Failure of the Amplitude to be Gauge Invariant

If we include all three diagrams that contribute to the anomaly, we find that the regularized amplitude is

$$\frac{1}{2}\left(\frac{\mathrm{i}}{2\pi}\right)^3 \int_{\substack{p_i,v_j \\ p_{\min}<p_{\max}-\epsilon}} P(0,v_1)\mathrm{d}z_1 A_1^a P(v_1,v_2)\mathrm{d}z_2 P(v_2,p_2)P(v_2,v_3)\mathrm{d}z_3 A_3^b P(v_3,p_3)$$
$$\times f^{afc} f^{fgd} f^{gbe} \{\rho(t^c), \rho(t^d), \rho(t^e)\} \,, \tag{8.31}$$

where $p_{\min}, p_{\max}$ are the coordinates on the Wilson line with the minimum and maximum value. The factor of $\frac{1}{2}$ appears because there are 6 ways of ordering three points on the line, but only 3 diagrams we need to consider.

Let us now investigate the failure of the integral to be gauge invariant. Let us perform a linearized gauge transformation $A \mapsto A + \mathrm{d}\mathsf{c}$ to the external gauge field A. By integration by parts, the result can be written as a sum of terms where we apply the exterior derivative d to one of the propagators, or else we integrate over one of the boundary components of the domain of integration. We will show that all such terms vanish except the term where we integrate over the boundary component in which $p_{\max} = p_{\min} + \epsilon$.

We use the identity (4.7). Applying the exterior derivative to a propagator has the effect of yielding a contribution where we integrate over the region where the vertices v_i, v_j (or p_i, v_j) at either end of the propagator are identified.

If we contract the propagator connecting p_3 and v_3, the result must vanish. This is because we can assume that the external gauge field A and the ghost c

are both divisible by z. This assumption is justified because the anomaly always involves a z-derivative on each external field. If we integrate over the region where $v_3 = p_3$, then since $z_3 = 0$ on this domain, the integrand vanishes.

If we contract the propagator connecting p_2 and v_2, the result vanishes by the argument presented in section 8.3.4.

If we contract the propagator connecting v_1 and v_2, or v_3 or v_1, the results cancel by the Jacobi identity.

Next, let us consider the boundary components where two adjacent points on the Wilson line can meet. They can meet from above or below, with different signs. After using the Jacobi identity, the color factor associated to a boundary component like this is that given by the diagram

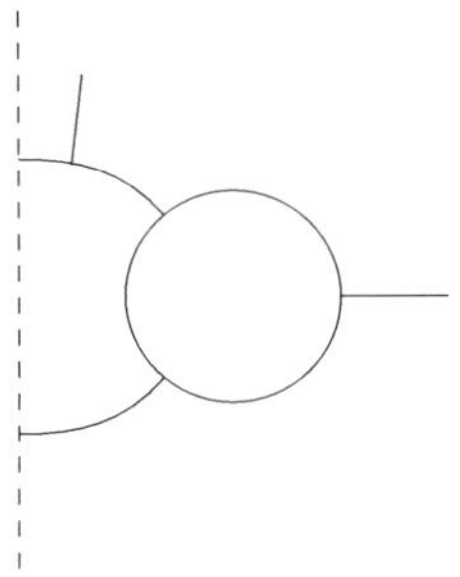

We have seen in section 8.3.3 that anomalies with a color factor of this form can be cancelled by a counter-term, and so are not relevant.

The remaining boundary component is the one with $p_{\max} = p_{\min} + \epsilon$. This is the one that will contribute to the anomaly.

The integral describing the anomaly is then

$$\left(\frac{\mathrm{i}}{2\pi}\right)^3 \int_{\substack{p_i, v_j \\ p_{\min}=p_{\max}-\epsilon}} P(p_1, v_1)\mathrm{d}z_1 \mathsf{c}_1^a P(v_1, v_2)\mathrm{d}z_2 P(v_2, p_2) P(v_2, v_3)\mathrm{d}z_3 A_3^b P(v_3, p_3) \\ \times f^{afc} f^{fgd} f^{gbe} \{\rho(t^c), \rho(t^d), \rho(t^e)\}\,, \tag{8.32}$$

where on vertex v_1 the external gauge field has been replaced by the ghost c. The prefactor of $\frac{1}{2}$ has been cancelled by the factor of 2 coming from the two possible external lines to the ghost field. The diagram automorphism interchanges these two, so we are left with the single integral above but without the prefactor of $\frac{1}{2}$.

This integral can be written as a sum of six terms, according to the six possible orderings of the points on the Wilson line. A reflection in a plane orthogonal to the Wilson line shows that there are only three independent integrals, which we can take to be the three where the points p_1, p_2, p_3 are cyclically ordered. We find that we need to compute

$$2\left(\tfrac{\mathrm{i}}{2\pi}\right)^3 \int_{p_1<p_2<p_3=p_1+\epsilon} \int_{v_1,v_2} P(p_1,v_1)\mathrm{d}z_1 \mathsf{c}_1^a P(v_1,v_2)\mathrm{d}z_2 P(v_2,p_2)P(v_2,v_3)\mathrm{d}z_3 A_3^b P(v_3,p_3)$$
$$+\,2\left(\tfrac{\mathrm{i}}{2\pi}\right)^3 \int_{p_2<p_3<p_1=p_2+\epsilon} \int_{v_1,v_2} P(p_1,v_1)\mathrm{d}z_1 \mathsf{c}_1^a P(v_1,v_2)\mathrm{d}z_2 P(v_2,p_2)P(v_2,v_3)\mathrm{d}z_3 A_3^b P(v_3,p_3)$$
$$+\,2\left(\tfrac{\mathrm{i}}{2\pi}\right)^3 \int_{p_3<p_1<p_2=p_3+\epsilon} \int_{v_1,v_2} P(p_1,v_1)\mathrm{d}z_1 \mathsf{c}_1^a P(v_1,v_2)\mathrm{d}z_2 P(v_2,p_2)P(v_2,v_3)\mathrm{d}z_3 A_3^b P(v_3,p_3)\,. \tag{8.33}$$

The factor of 2 here is because each integral represents the contribution from one of two possible orderings on the points p_i.

8.5.4 Calculating the Anomaly Integral

Because we know that the anomaly must involve a z-derivative for each external line and no other derivatives, we can detect the anomaly by assuming that the external ghost field c is z, and the external gauge field A is $z\delta_{x=0}$.

The choice of external gauge field fixes the location of the vertex v_3; by translation invariance we choose instead to fix the location of the vertex p_1 to be $p_1 = 0$.

We will first evaluate the integral of the diagram in Fig. 30, where $p_1 < p_2 < p_3$, and we integrate over the region where $p_3 = p_1 + \epsilon$, and we set $p_1 = 0$. We denote p_2 by p.

The integral for the first diagram of Fig. 28 is

$$2\left(\tfrac{\mathrm{i}}{2\pi}\right)^3 \int_{p=0}^{\epsilon} \int_{v_1,v_2,v_3} P(0,v_1)\wedge z_1\mathrm{d}z_1 \wedge P(v_1,v_2)\wedge \mathrm{d}z_2 \wedge P(v_2,p)\wedge P(v_2,v_3)\wedge z_3\mathrm{d}z_3 \wedge P(v_3,\epsilon)\,. \tag{8.34}$$

Now that the integrand has 5 propagator 2-forms, and 1 vertex 1-forms, making a 13-form, which is consistent since we wish to integrate over $1+4\times 3 = 13$ variables.

We evaluate this integral in Appendix E. Reintroducing the color factor and the factor of two present in eqn. (8.33), we find that the anomaly is given by

$$\frac{\hbar^2}{12} f^{afc} f^{fgd} f^{gbe} \{\rho(t^c), \rho(t^d), \rho(t^e)\}\,. \tag{8.35}$$

8.6 The Anomaly as a Correction to the Algebra

In this section we will show that for a Wilson line to be defined at the quantum level, there must be quantum corrections to the coupling of the gauge field which satisfy a certain algebraic relation. This relation forces the Wilson line to be built from a representation of the Yangian, not of the Lie algebra $\mathfrak{g}[[z]]$.

Suppose we have a Wilson line in a representation V of our gauge group. Suppose that the level-one generators $t_{a,1}$ of the Yangian act by some operators $\rho(t_{a,1})$. Consider the four two-loop diagrams in Figure 31.

As before, a vertex on the Wilson line labelled with the number 1 or 2 indicates a coupling of the first or second z-derivative of the gauge field. (We tacitly

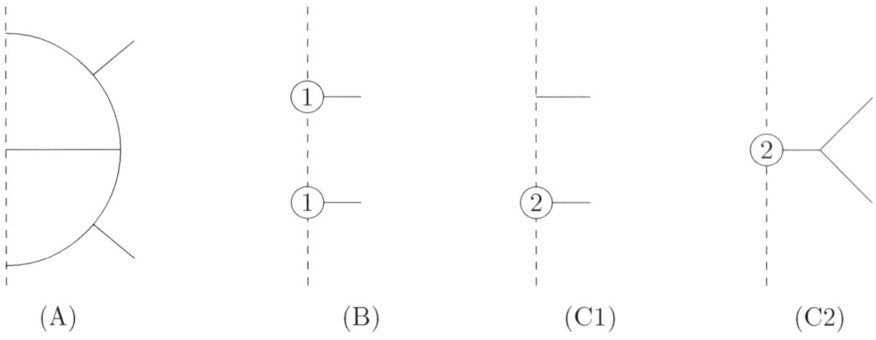

Figure 31: Four Feynman diagrams contributing to the anomaly.

include the variations on these diagrams where the ordering of the vertices on the Wilson lines has been permuted).

The amplitude of each of these diagrams can fail to be gauge invariant. For the first diagram (A), this involves the two-loop computation that we have been describing. For the other diagrams, the failure to be gauge invariant is much more straightforward, as we will now see.

Suppose that at the quantum level, the first and second derivative of the gauge field are coupled to the Wilson line by operators $\rho_{a,1} : V \to V$ and $\frac{1}{2}\rho_{a,2} : V \to V$, where a is an adjoint index.

Then, the amplitude for diagram (B) is given by

$$\int_{p_1 < p_2 \in \mathbb{R}} \partial_z A^a(p_1) \partial_z A^b(p_2) \rho_{a,1} \circ \rho_{b,1} \, . \tag{8.36}$$

If we change the gauge field by a gauge transformation $A^a \mapsto A^a + \mathrm{d}\mathsf{c}^a$ we find (by integration by parts and Stokes' theorem) that the integral becomes

$$\int_{p \in \mathbb{R}} \partial_z \mathsf{c}^a(p) \partial_z A^b(p) [\rho_{a,1}, \rho_{b,1}] \, . \tag{8.37}$$

Similarly, for diagram (C1), the amplitude is

$$\frac{1}{2} \int_{p_1 < p_2 \in \mathbb{R}} (\partial_z^2 A^a(p_1)) A^b(p_2) \rho_{a,2} \circ \rho_{b,0} + \frac{1}{2} \int_{p_1 < p_2 \in \mathbb{R}} (A^a(p_1)) \partial_z^2 A^b(p_2) \rho_{a,0} \circ \rho_{b,2} \, . \tag{8.38}$$

The failure of this to be gauge invariant is given by

$$\frac{1}{2} \int_{p \in \mathbb{R}} (\partial_z^2 \mathsf{c}^a(p)) A^b(p) [\rho_{a,2}, \rho_{b,0}] + \frac{1}{2} \int_{p \in \mathbb{R}} \mathsf{c}^a(p) \partial_z^2 A^b(p) [\rho_{a,0}, \rho_{b,2}] \, . \tag{8.39}$$

Next, diagram (C2) has amplitude

$$\frac{1}{4} \int_{p \in \mathbb{R}, v = (x,y,z)} \mathrm{d}z \partial_{z_p}^2 P(p, v) A^a(v) A^b(v) f_{ab}{}^c \rho_{c,2} \, . \tag{8.40}$$

Here v indicates the position of the interior vertex, and ∂_{z_p} indicates we apply a z-derivative to the p-coordinate of the propagator. The failure of this to be gauge invariant is, using the fact that $\mathrm{d}z_v \mathrm{d}P(p,v) = -\delta_{p=v}$ and imposing the equations of motion for A,

$$-\tfrac{1}{2}\int_{p\in\mathbb{R}} \partial_z^2\left(\mathsf{c}^a(p)A^b(b)\right) f_{ab}{}^c\rho_{c,2}\,. \tag{8.41}$$

Finally, the anomaly for diagram (A) is, as we have seen,

$$\tfrac{\hbar^2}{12}\int_{p\in\mathbb{R}} \partial_z \mathsf{c}^a(p)\partial_z A^b(p) f^{afc} f^{fgd} f^{gbe}\{\rho(t^c),\rho(t^d),\rho(t^e)\}\,. \tag{8.42}$$

The sum of the anomalies for the diagrams (A), (B) and (C) must vanish. This gives rise to the equations

$$[\rho_{a,2},\rho_{b,0}] = f_{ab}{}^c\rho_{c,2}\,, \tag{8.43}$$

$$[\rho_{a,1},\rho_{b,1}] = f_{ab}{}^c\rho_{c,2} - \tfrac{\hbar^2}{12}Q_{ab}(\rho_{\bullet,0})\,, \tag{8.44}$$

where the first equation is from the cancellation of the coefficients of $(\partial_z^2\mathsf{c}^a(p))A^b(p)$ and $\mathsf{c}^a(p)\partial_z^2 A^b(p)$, and the second from those of $(\partial_z\mathsf{c}^a(p))\partial_z A^b(p)$. In the equations above we used the short-hand notation $\rho_{a,n} = \rho(t_{a,n})$ and we defined

$$Q_{ab}(\rho_{\bullet,0}) = f^{afc}f^{fgd}f^{gbe}\{\rho(t^c),\rho(t^d),\rho(t^e)\}\,. \tag{8.45}$$

The first equation is the commutation relation we found classically. The second equation tells us that the operators $\rho_{a,1}$ do not commute to give $\rho_{c,2}$, as we would find classically, but a linear combination of $\rho_{c,2}$ and a certain cubic polynomial in the level 0 generators $\rho_{d,0}$.

We can decompose the exterior square $\wedge^2\mathfrak{g}$ into $\mathfrak{g}\oplus\wedge_0^2\mathfrak{g}$, where $\wedge_0^2\mathfrak{g}$ is the kernel of the Lie bracket map from $\wedge^2\mathfrak{g}\to\mathfrak{g}$. By using this decomposition, we can write the second relation as a sum of two independent relations:

$$f^{ab}{}_c[\rho_{a,1},\rho_{b,1}] = \mathsf{h}^\vee\rho_{c,2} - \tfrac{\hbar^2}{12}f^{ab}{}_cQ_{ab}(\rho_{\bullet,0})\,, \tag{8.46}$$

$$\Lambda^{ab}[\rho_{a,1},\rho_{b,1}] = -\tfrac{\hbar^2}{12}\Lambda^{ab}Q_{ab}(\rho_{\bullet,0})\,, \tag{8.47}$$

where $\Lambda^{ab}\in\wedge^2\mathfrak{g}$. The first relation can always be satisfied by redefining the operator $\rho_{c,2}$ to

$$\rho'_{c,2} = \rho_{c,2} - \frac{\hbar^2}{12\mathsf{h}^\vee}f^{ab}{}_cQ_{ab}(\rho_{\bullet,0})\,. \tag{8.48}$$

The second relation, however, can not satisfied in such a trivial way.

What we have found from this analysis is that, in order for a classical Wilson line to quantize modulo $\hbar^3$, we must be able to define the operators $\rho_{a,1}$ so that they satisfy the relation (8.47). This relation is one of the relations in the Yangian algebra which makes it into a non-trivial deformation of the universal enveloping algebra of $\mathfrak{g}[[z]]$.

The reader will note that the diagrams in Fig. 31 are simply the classical diagrams of Fig. 26, with quantum corrections to the vertices, plus the anomaly diagram (A). Note that we did not add to diagrams (C2) in the figure additional contributions involving a quantum correction to the bulk vertex. Such contributions can be omitted, using Proposition 8.3, as they involve a cohomology class that factors through a map to $\mathfrak{g}$.

8.6.1 Matching with Standard Relations in the Yangian

Let us explain how to match the relation we have found with one of the known descriptions of the Yangian.

For any simple Lie algebra $\mathfrak{g}$, the sequence

$$\wedge^3 \mathfrak{g} \to \wedge^2 \mathfrak{g} \to \mathfrak{g} \tag{8.49}$$

is exact in the middle, where each map in the sequence is obtained by applying the Lie bracket to two entries in the exterior power. This sequence is part of the Chevalley-Eilenberg homology complex of $\mathfrak{g}$. For any simple Lie algebra, $H_2(\mathfrak{g}, \mathbb{C}) = 0$, which is why this sequence is exact in the middle.

Therefore, the kernel of the map from $\wedge^2 \mathfrak{g} \to \mathfrak{g}$ is the image of the map from $\wedge^3 \mathfrak{g} \to \wedge^2 \mathfrak{g}$. The antisymmetric tensor $\Lambda^{ab} \in \wedge^2 \mathfrak{g}$ appearing in eqn. (8.47) is in the kernel of this map. Therefore, we lose no information by considering the relation in eqn. (8.47) when Λ^{ab} is taken to be in the image of the map from $\wedge^3 \mathfrak{g}$, that is, when Λ^{ab} is of the form

$$\Lambda^{ab} = \Gamma^{ade} f_{de}{}^{b} \tag{8.50}$$

for some antisymmetric tensor $\Gamma^{ade} \in \wedge^3 \mathfrak{g}$.

Relation (8.47) takes the form, in this notation,

$$\Gamma^{ade} f_{de}{}^{b} [\rho_{a,1}, \rho_{b,1}] = -\tfrac{\hbar^2}{12} \Gamma^{ade} f_{de}{}^{b} Q_{ab}(\rho_{\bullet,0}) \,. \tag{8.51}$$

For an element $t \in \mathfrak{g}$, we let $\rho(t)$ denote the action of t on our representation, and $\rho(J(t))$ denote the action of the corresponding level one generator on our representation. Then our relation is

$$\Gamma^{abc}[\rho(J(t_a)), \rho(J([t_b, t_c]))] = -\tfrac{\hbar^2}{12} \Gamma^{abc}([t_e,[t_a,t_d]],[t_f,[t_b,t_c]])\{\rho(t_d),\rho(t_e),\rho(t_f)\} \,, \tag{8.52}$$

where as before $\{\rho(t_d), \rho(t_e), \rho(t_f)\}$ is 1/6 times the sum of the products of these operators in each of the six possible orders (see eqn. (8.25)). The Jacobi identity and the $\mathfrak{g}$-invariance of the Killing form allows us to rewrite

$$\begin{aligned}([t_e,[t_a,t_d]],[t_f,[t_b,t_c]]) &= ([t_e,[t_a,t_d]],[[t_f,t_b],t_c]) + ([t_e,[t_a,t_d]],[t_b,[t_f,t_c]]) \\ &= ([t_c,[t_e,[t_a,t_d]]],[t_f,t_b]) - ([t_b,[t_e,[t_a,t_d]]],[t_f,t_c]) \,,\end{aligned}$$

Thus anti-symmetrizing in a, b, c and again using the Jacobi identity, we obtain

$$\begin{aligned}&\Gamma^{abc}([t_e,[t_a,t_d]],[t_f,[t_b,t_c]]) \\ &\quad = 2\Gamma^{abc}([t_c,[t_e,[t_a,t_d]]],[t_f,t_b]) \\ &\quad = 2\Gamma^{abc}([[t_c,t_e],[t_a,t_d]],[t_f,t_b]) + 2\Gamma^{abc}([t_e,[t_c,[t_a,t_d]],[t_f,t_b]) \\ &\quad = -2\Gamma^{abc}([[t_c,t_e],[t_a,t_d]],[t_b,t_f]) + \Gamma^{abc}([t_e,[[t_c,t_a],t_d]],[t_f,t_b]) \,,\end{aligned}$$

where on the last line we have again used anti-symmetry of Γ^{abc} to conclude that we can replace $[t_c,[t_a,t_d]]$ by $\frac{1}{2}[[t_c,t_a],t_d]$. The second factor in the final line, after combining with the factor $\{\rho(t_d),\rho(t_e),\rho(t_f)\}$, gives back the expression we started with, with an overall minus sign:

$$\begin{aligned}&\Gamma^{abc}([t_e,[[t_c,t_a],t_d]],[t_f,t_b])\{\rho(t_d),\rho(t_e),\rho(t_f)\}\\ &\quad=-\Gamma^{abc}([t_e,[t_f,t_b]],[[t_c,t_a],t_d]])\{\rho(t_d),\rho(t_e),\rho(t_f)\}\\ &\quad=-\Gamma^{abc}([t_e,[t_a,t_d]],[t_f,[t_b,t_c]])\{\rho(t_d),\rho(t_e),\rho(t_f)\}\ ,\end{aligned}$$

where we used $\mathfrak{g}$-invariance of the Killing form, anti-symmetry in a,b,c and symmetry in d,e,f. This allows us to conclude that

$$\begin{aligned}&\Gamma^{abc}([t_e,[t_a,t_d]],[t_f,[t_b,t_c]])\{\rho(t_d),\rho(t_e),\rho(t_f)\}\\ &\quad=-\Gamma^{abc}([[t_c,t_e],[t_a,t_d]],[t_b,t_f])\{\rho(t_d),\rho(t_e),\rho(t_f)\}\ .\end{aligned}\tag{8.53}$$

Thus the quantum-corrected commutation relation (8.52) becomes

$$\Gamma^{abc}[\rho(J(t_a)),\rho(J([t_b,t_c]))]=\frac{\hbar^2}{12}\Gamma^{abc}([[t_c,t_e],[t_a,t_d]],[t_b,t_f])\{\rho(t_d),\rho(t_e),\rho(t_f)\}\ .\tag{8.54}$$

Note that by anti-symmetry of Γ^{abc} and symmetry of $\{\rho(t_d),\rho(t_e),\rho(t_f)\}$ we have

$$\begin{aligned}&\Gamma^{abc}[\rho(J(t_a)),\rho(J[t_b,t_c])]\\ &=\tfrac{1}{3}\Gamma^{abc}\left([\rho(J(t_a)),\rho(J([t_b,t_c]))]+[\rho(J(t_b)),\rho(J([t_c,t_a]))]+[\rho(J(t_c)),\rho(J([t_a,t_b]))]\right)\ .\end{aligned}\tag{8.55}$$

Also, $([[t_c,t_e],[t_a,t_d]],[t_f,t_b])\{\rho(t_d),\rho(t_e),\rho(t_f)\}$ is antisymmetric in a,b,c. Therefore our relation is equivalent to

$$\begin{aligned}&[\rho(J(t_a)),\rho(J([t_b,t_c]))]+[\rho(J(t_b)),\rho(J([t_c,t_a]))]+[\rho(J(t_c)),\rho(J([t_a,t_b]))]\\ &\quad=\tfrac{\hbar^2}{4}([[t_c,t_e],[t_a,t_d]],[t_b,t_f])\{\rho(t_d),\rho(t_e),\rho(t_f)\}\\ &\quad=\tfrac{\hbar^2}{4}([t_a,t_d],[[t_b,t_e]],[t_c,t_f]])\{\rho(t_d),\rho(t_e),\rho(t_f)\}\ .\end{aligned}\tag{8.56}$$

This is the same as the known relation for the Yangian, see e.g. eqn. (4) in [8, Theorem 12.1.1].

8.7 An Example of an Anomalous Wilson Line

It is known that the adjoint representation of any simple Lie algebra not of type A does not lift to a representation of the Yangian. This indicates that a Wilson line in the adjoint representation should be anomalous. In this section we will see explicitly that our two-loop anomaly cannot be cancelled for the adjoint representation of SO_N.

Let us start with some group-theory background. If $M_1,\ldots,M_k\in\mathfrak{so}_N$ (for N sufficiently large), the ring of invariant function of these k elements of $\mathfrak{so}_N$ is generated by the invariant functions given by traces, that is, the functions

$\mathrm{Tr}(M_{i_1}\dots M_{i_r})$ where $i_1,\dots,i_r$ are in the set $\{1,\dots,k\}$. Repeated indices are allowed and the trace is taken in the fundamental representation.

For finite N, there are trace relations among these functions. For N large compared to the degree of the polynomial in the variables M_i, the only relation that survives is the relation that says

$$\mathrm{Tr}(M_{i_1}\dots M_{i_r}) = (-1)^r\,\mathrm{Tr}(M_{i_r}\dots M_{i_1})\,. \tag{8.57}$$

We will need to understand the decomposition of $\mathfrak{so}_N\otimes\mathfrak{so}_N$ into irreducible representations. The number of irreducible representations is the same as the number of invariant tensors in $\mathfrak{so}_N^{\otimes 4}$. Let M_1, M_2, N_1, N_2 denote elements of $\mathfrak{so}_N$. The analysis above allows us to enumerate the invariant functions of these matrices which are linear in each variable. We find that there are 6 invariant tensors, three with a single trace and three with two traces. Of these, 4 of them are symmetric when we exchange M_1 and M_2, or when we exchange N_1 and N_2. These correspond to projectors onto irreducible representations inside $\mathrm{Sym}^2\,\mathfrak{so}_N$. Two of them are antisymmetric under the same permutations, and correspond to projectors onto irreducible representations in $\wedge^2\mathfrak{so}_N$. The two antisymmetric tensors are

$$\begin{aligned}&\mathrm{Tr}([M_1,M_2][N_1,N_2])\,,\\ &\mathrm{Tr}(M_1N_1)\,\mathrm{Tr}(M_2N_2)-\mathrm{Tr}(M_1N_2)\,\mathrm{Tr}(M_2N_1)\,.\end{aligned} \tag{8.58}$$

The first antisymmetric tensor gives rise to the projection from $\wedge^2\mathfrak{so}_N$ onto the adjoint representation, and the second to the projection onto the irreducible representation $\wedge^2_0\mathfrak{so}_N$.

Note also that there is only one copy of the adjoint representation in $\mathfrak{so}_N\otimes\mathfrak{so}_N$. Any copy of the adjoint representation is associated to an invariant element in $\mathfrak{so}_N^{\otimes 3}$, and so an invariant function of M_1, M_2, M_3 which is linear in each M_i. The discussion above tells us that the only such invariant function is $\mathrm{Tr}(M_1M_2M_3)$.

If we have a Wilson line in the adjoint representation of SO_N, it follows from this that the level one generators must be coupled by a multiple of the level 0 generators. Therefore, the commutator of the level 1 generators must be proportional to the level 0 generators. Relation 8.44 can never hold unless

$$Q_{ab}(\rho_{\bullet,0}) = f^{afc}f^{fgd}f^{gbe}\{\rho(t^c),\rho(t^d),\rho(t^e)\} \tag{8.59}$$

is a multiple of some $\rho(t^c)$. We will show the projection of the tensor Q_{ab} onto $\wedge^2_0(\mathfrak{so}_N)$ is non-zero, so the anomaly is non-zero.

Let us introduce incoming and outgoing states t_{in}, t_{out} on the adjoint Wilson line. The three diagrams which give us an anomaly all, as we have seen, have the same color factor. Since ρ is in the adjoint representation, we can compute the color factor using a diagram in which the Wilson line is placed on the same footing as the other lines in the Feynman diagram. If we do this, the diagram looks like

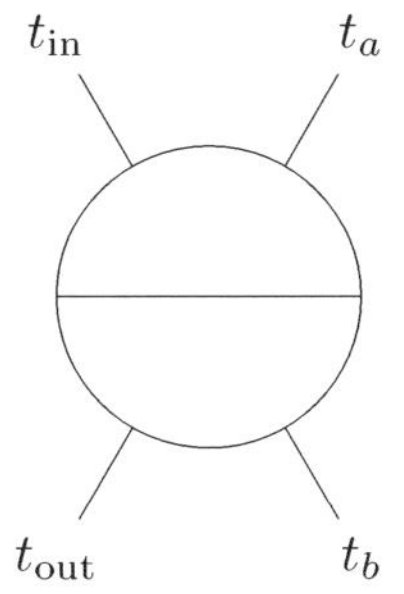

where the segment connecting t_{in}, t_{out} is what was the Wilson line. Here we are implicitly anti-symmetrizing in a and b, since the anomaly is always antisymmetric in the external lines.

This makes it clear that the color factor of the anomaly in the adjoint representation is

$$(t_{\text{out}}, Q_{ab}(\rho_{\bullet,0})t_{\text{in}}) = \operatorname{Tr}(t_{\text{in}}t_a t_f t_b t_{\text{out}} t_f) - \operatorname{Tr}(t_{\text{in}} t_b t_f t_a t_{\text{out}} t_f) \,, \tag{8.60}$$

where the trace is taken in the adjoint representation, and we sum over f.

Let us view the anomaly (8.60) as a linear operator from $\mathfrak{so}_N^{\otimes 2} \to \mathfrak{so}_N^{\otimes 2}$, where the first two copies of $\mathfrak{so}_N$ are given by t_a, t_b and the second two by $t_{\text{in}}, t_{\text{out}}$. Since this is an $SO(N)$ invariant operator, it is a linear combination of the projection onto the 6 irreducible representations in $\mathfrak{so}_N^{\otimes 2}$. The anomaly is non-zero as long as the coefficient of the projection onto $\wedge_0^2 \mathfrak{so}_N$ is non-zero.

Equivalently, we can expand the expression (8.60) as a linear combination of the 6 invariant tensors in $\mathfrak{so}_N^{\otimes 4}$. To show that the anomaly is non-zero it suffices to show that, in this expansion, the coefficient of

$$\operatorname{Tr}(t_{\text{in}}t_a)\operatorname{Tr}(t_{\text{out}}t_b) - \operatorname{Tr}(t_{\text{in}}t_b)\operatorname{Tr}(t_{\text{out}}t_a) \tag{8.61}$$

is non-zero. This is because this invariant tensor corresponds to projection onto $\wedge_0^2 \mathfrak{so}_N$. Since the anomaly is antisymmetric in a and b we need to verify that the coefficient of $\operatorname{Tr}(t_{\text{in}}t_a)\operatorname{Tr}(t_{\text{out}}t_b)$ is non-zero.

We can evaluate the expression in equation (8.60) for $\mathfrak{so}_N$ using the double-line notation familiar from evaluation of the color factors for $\mathfrak{gl}_N$ gauge theories. Since we are using $\mathfrak{so}_N$ instead of $\mathfrak{gl}_N$, the double-line technique works a little differently. There are two types of double-line edges, which we can think of as a flat ribbon and a ribbon with a half-twist. This is because the Casimir for $\mathfrak{so}_N$ is represented as

$$\sum E_{ij} \otimes E_{ji} - \sum E_{ij} \otimes E_{ij} \,, \tag{8.62}$$

where E_{ij} is the elementary matrix. Only the first term appears in the Casimir of $\mathfrak{gl}_N$, which is why only the first type of double-line edge appears when we study $\mathfrak{gl}_N$ color factors.

Using the double line notation, we find that the color factor we are computing is given by a sum over connected unoriented surfaces of Euler characteristic -1, with 4 marked points on the boundary. Since we are interested in the double-trace terms in the anomaly, we need to consider surfaces where two of the marked points (labelled by t_{in}, t_a) are on one boundary component, and the other two or on a different component.

The order N two-trace term in the anomaly is given by the planar diagram

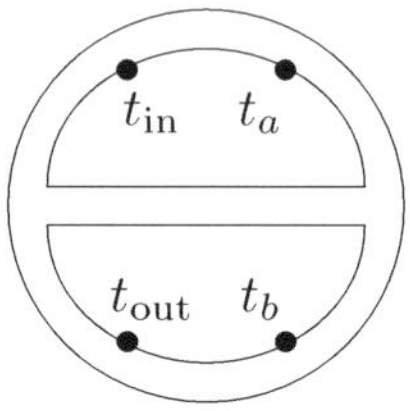

which contributes

$$N\,\mathrm{Tr}(t_{\rm in}t_a)\,\mathrm{Tr}(t_b t_{\rm out})\ . \tag{8.63}$$

(We have not drawn the similar diagram which contributes the same expression with a and b exchanged and a different sign).

There is also an order 1 term, given by a double-line diagram which has the topology of a Möbius band with a circle removed. For N sufficiently large ($N = 10$ certainly suffices) the coefficient of this term is much smaller than that of the order N term, so the anomaly is non-zero.

9 Trigonometric Solutions of the Yang-Baxter Equations

9.1 Preliminaries

Our goal in this section will be to understand from the present perspective the trigonometric solutions of the Yang-Baxter equation. For this we have to consider the case that $C = \mathbb{C}^\times$, or equivalently $C = \mathbb{CP}^1$ with a differential $\omega = \mathrm{d}z/z$ that has two simple poles at 0 and ∞.

The basic consequence of a pole in ω is the following. One has

$$\bar\partial \frac{\mathrm{d}z}{z} = 2\pi\mathrm{i}\delta^2(z)\ , \tag{9.1}$$

where $\delta^2(z)$ is a delta function normalized by $\int |\mathrm{d}^2z|\delta^2(z) = 1$. Accordingly, when we vary the action (3.3) to derive the equations of motion, we pick up a "boundary term" supported at $z = 0$:

$$\delta S = \cdots - \frac{\mathrm{i}}{\hbar}\int_{\Sigma\times\{0\}} \mathrm{Tr}\, A\wedge\delta A\ , \tag{9.2}$$

where bulk terms have been omitted. The equations of motion of the classical theory tell us to set to zero the bulk terms and also the boundary term $\mathrm{Tr}\, A\wedge\delta A|_{\Sigma\times\{0\}}$. We will abbreviate the latter as $\mathrm{Tr}\, A\wedge\delta A|$ (and we use a similar notation in general for restriction to a singularity or boundary).

Actually, the boundary terms in the equations of motion play a distinguished role. Quantum field theory is constructed by integrating over fields that do not necessarily satisfy the equations of motion in bulk. The bulk equations of motion are only satisfied in the classical limit. But boundary terms in the equations of motion have to vanish exactly. This is needed in order to get in bulk a symmetric propagator, as expected for bosons (or an antisymmetric one in an analogous problem with fermions). In general, in quantum field theory in the presence of boundaries or defects, one typically runs into "boundary terms" analogous to $\mathrm{Tr}\, A \wedge \delta A$. As a starting point in quantization, one always needs to impose a minimal condition that sets the boundary terms to 0. For example, if one quantizes on a manifold X an ordinary scalar field with the usual Lagrangian $\frac{1}{2}|\mathrm{d}\phi|^2$, the boundary term is $\int_{\partial X} \delta\phi\partial_n\phi$ (where ∂_n is the normal derivative). The simplest way to dispose of this boundary term is to set either $\phi| = 0$ (Dirichlet boundary conditions) or $\partial_n\phi| = 0$ (Neumann boundary conditions). With either of these conditions, the theory can be quantized. By contrast, a stronger condition such as $\phi = \partial_n\phi = 0$ is too strong and does not lead to a quantum theory of a scalar field.

What condition on A will we use to set $\mathrm{Tr}\, A \wedge \delta A| = 0$? The most obvious condition might seem to be $A| = 0$. However, this condition is too strong, analogous to $\phi| = \partial_n\phi| = 0$ for the scalar field. If we impose it, a suitable propagator will not exist and we will not be able to do perturbation theory.[36]

We can gain some intuition by considering the case of a double pole. Of course, we have already studied double poles in the context of rational solutions of the Yang-Baxter equation. In that analysis, we considered the case that $\omega = \mathrm{d}z$ with a double pole at infinity. For our present purposes, it is more convenient to place the pole at a finite point, so we take $\omega = \mathrm{d}z/z^2$, with a double pole at the origin. Since

$$\bar{\partial}\frac{\mathrm{d}z}{z^2} = -2\pi\mathrm{i}\partial_z\delta^2(z)\ , \tag{9.3}$$

the vanishing of the surface term now requires

$$0 = (\partial_z \mathrm{Tr}\, A \wedge \delta A)|\ . \tag{9.4}$$

A completely natural way to satisfy this condition is to set $A| = 0$. Varying A with the constraint $A| = 0$, we have also $\delta A| = 0$, so that $\mathrm{Tr}\, A \wedge \delta A$ has a double zero at $z = 0$ and eqn. (9.4) is obeyed. The condition $A| = 0$ is gauge-invariant if we likewise constrain the generator ε of a gauge transformation to obey $\varepsilon| = 0$.

Not only is $A| = 0$ a natural way to satisfy eqn. (9.4), it is what we have actually done in studying rational solutions of the Yang-Baxter equation. Starting in section 4, we used a propagator which vanishes at infinity, which amounts to taking A to vanish at infinity (where we placed the double pole). As explained in section 3.4, this choice has the advantage of ensuring that up to gauge transformation, the only classical solution is the trivial one $A = 0$, a condition that is needed

[36] Later, we will impose a weaker condition and find a unique solution for the r-matrix. A similar analysis assuming that $A| = 0$ will show that no possible r-matrix exists. This is actually clear from the fact that the unique r-matrix we get with a weaker condition does not obey $A| = 0$.

if we want to obtain something as simple as the usual Yang-Baxter equation. As we will explain below, there are other reasonable ways to satisfy $(\partial_z \mathrm{Tr}\, A \wedge \delta A)| = 0$, but they lead to something more complicated.

If $A| = 0$ is a satisfactory condition in the presence of a double pole, it can hardly be the right condition for a simple pole. If we start with a differential $\omega = \mathrm{d}z/z^2$ with a double pole, and perturb it to, say, $\omega' = \mathrm{d}z/(z - z_1)(z - z_2)$, with small z_1, z_2, the condition $A| = 0$ at a double pole must somehow be split between the two simple poles at z_1 and z_2.

As remarked above, when one studies field theory in the presence of a boundary or defect, the right procedure is always to impose a minimally restrictive condition that sets this term to zero. But the condition $A| = 0$ is not a minimal condition to ensure that $\mathrm{Tr}\, A \wedge \delta A| = 0$. For this, it suffices to pick a middle-dimensional complex subspace $\mathfrak{l}_0 \subset \mathfrak{g}$ that is "isotropic" (or "Lagrangian") for the quadratic form Tr, in the sense that for $a, b \in \mathfrak{l}_0$, $\mathrm{Tr}\, ab = 0$. (We take $\mathfrak{l}_0$ to be middle-dimensional because this is the maximum possible dimension for an isotropic subspace, leading to the weakest possible condition on A.) Then we require that $A|$ is $\mathfrak{l}_0$-valued, that is, that it is an $\mathfrak{l}_0$-valued 1-form along Σ. Having imposed this condition on A, we impose it also on δA, and then we see that with $A|$ and $\delta A|$ being $\mathfrak{l}_0$-valued 1-forms, $\mathrm{Tr}\, A \wedge \delta A| = 0$.

One may object that the condition for $A|$ to be $\mathfrak{l}_0$-valued is not gauge-invariant. However, what we actually need is not to maintain the full gauge symmetry along the locus of the pole, but only to maintain enough gauge symmetry so that the usual "longitudinal" part of $A|$ can be gauged away. For this, we simply ask that $\mathfrak{l}_0$ should be a subalgebra of $\mathfrak{g}$ (and not just a subspace) and we ask that the generator ε of a gauge transformation should satisfy the condition that $\varepsilon|$ is $\mathfrak{l}_0$-valued. These conditions on A and ε are compatible in the sense that, with both $A|$ and $\varepsilon|$ constrained to be $\mathfrak{l}_0$-valued, the usual gauge transformation law $\delta A = D\varepsilon$ makes sense, and moreover the action (3.3) is gauge-invariant. In effect, what has happened is the following. Along the locus of the pole, we have set to zero some components of A, and the other components have their usual gauge-invariance.

We are really interested in a situation in which ω has two simple poles, say $\omega = \mathrm{d}z/z$ with poles at 0 and at ∞. Treating each simple pole as above, we pick two middle-dimensional isotropic subalgebras of $\mathfrak{g}$, say $\mathfrak{l}_0$ and $\mathfrak{l}_\infty$, in general with no relation between them. We require A and ε to be $\mathfrak{l}_0$-valued when restricted to $\Sigma \times \{0\}$, and $\mathfrak{l}_\infty$-valued when restricted to $\Sigma \times \{\infty\}$.

To get from this construction a solution of the usual Yang-Baxter equation (as opposed to the "dynamical Yang-Baxter equation," which we study in section 11), we need a further condition that is familiar from section 3.4: the trivial solution $A = 0$ should have no deformations and no continuous unbroken gauge symmetries. The two conditions are equivalent for the following reason. Let $\mathfrak{g}_{0,\infty}$ be the sheaf of holomorphic $\mathfrak{g}$-valued functions on $\mathbb{CP}^1$ that are $\mathfrak{l}_0$-valued at 0 and $\mathfrak{l}_\infty$-valued at ∞. The Lie algebra of the group of gauge symmetries of the trivial solution $A = 0$ is $H^0(\mathbb{CP}^1, \mathfrak{g}_{0,\infty})$, and the tangent space to $A = 0$ in the moduli space of classical solutions of the theory is $H^1(\mathbb{CP}^1, \mathfrak{g}_{0,\infty})$. Thus the condition that $A = 0$ has no deformations is $H^1(\mathbb{CP}^1, \mathfrak{g}_{0,\infty}) = 0$, and the condition that it has no continuous gauge symmetries is $H^0(\mathbb{CP}^1, \mathfrak{g}_{0,\infty}) = 0$. These two conditions are equivalent because of the Riemann-Roch theorem, which in the present situation implies that

$$\dim H^0(\mathbb{CP}^1, \mathfrak{g}_{0,\infty}) - \dim H^1(\mathbb{CP}^1, \mathfrak{g}_{0,\infty}) = 0 \,. \tag{9.5}$$

Concretely, $H^0(\mathbb{CP}^1, \mathfrak{g}_{0,\infty})$ is simply[37] $\mathfrak{l}_0 \cap \mathfrak{l}_\infty$. Thus, our two conditions are equivalent to

$$\mathfrak{l}_0 \cap \mathfrak{l}_\infty = 0 \,. \tag{9.6}$$

On dimensional grounds, this is equivalent to

$$\mathfrak{l}_0 + \mathfrak{l}_\infty = \mathfrak{g} \,. \tag{9.7}$$

We can understand directly the role of this last condition. If $\mathfrak{g}' = \mathfrak{l}_0 + \mathfrak{l}_\infty$ and $\mathfrak{g} = \mathfrak{g}' \oplus \mathfrak{g}''$, then the $\mathfrak{g}''$-valued part of A cannot be gauged away[38] and therefore the trivial solution $A = 0$ has deformations. So absence of deformations means that $\mathfrak{g}'' = 0$ and $\mathfrak{g} = \mathfrak{l}_0 + \mathfrak{l}_\infty$ (since $\mathfrak{l}_0 \cap \mathfrak{l}_\infty = 0$, this is equivalent to $\mathfrak{g} = \mathfrak{l}_0 \oplus \mathfrak{l}_\infty$).

Accordingly, to get a trigonometric solution of the Yang-Baxter equation, rather than its "dynamical" generalization, we need $\mathfrak{l}_0 + \mathfrak{l}_\infty = \mathfrak{g}$. We have arrived precisely at the notion of a "Manin triple" (see for example [8, p. 26]). A Manin triple is a complex Lie algebra $\mathfrak{g}$ with an invariant, nondegenerate quadratic form Tr, and a decomposition $\mathfrak{g} = \mathfrak{l}_0 + \mathfrak{l}_\infty$, where $\mathfrak{l}_0$ and $\mathfrak{l}_\infty$ are middle-dimensional isotropic subalgebras. So trigonometric solutions of the Yang-Baxter equation, or at least the ones that we will study, are associated to Manin triples.[39]

At this stage, it is perhaps also clear that in contrast to the rational solutions of Yang-Baxter, the trigonometric ones do not have G as a group of global symmetries. In the rational case, given the differential $\omega = \mathrm{d}z/z^2$, the condition on the generator of a gauge transformation at $z = 0$ was $\varepsilon| = 0$. This condition leaves constant gauge transformations at $z = 0$ as global symmetries. In the trigonometric case, a constant unbroken gauge symmetry, to be compatible with the conditions at 0 and at ∞, has to be an element of G that conjugates $\mathfrak{l}_0$ to itself, and also conjugates $\mathfrak{l}_\infty$ to itself. In practice, in the main example that will be introduced in section 9.2, this means that the group of global symmetries is the maximal torus of G. (This symmetry accounts for the block diagonal form of eqn. (9.14).)

Before leaving this subject, let us note that we stated the condition $\mathfrak{l}_0+\mathfrak{l}_\infty = \mathfrak{g}$ in a somewhat naive way. In gauge theory, any comparison between $\mathfrak{l}_0$ and $\mathfrak{l}_\infty$ involves parallel transport from 0 to ∞. What the condition $\mathfrak{l}_0 + \mathfrak{l}_\infty = \mathfrak{g}$ really means is that, after conjugating $\mathfrak{l}_0$ and $\mathfrak{l}_\infty$ into general position, $\mathfrak{l}_0 + \mathfrak{l}_\infty = \mathfrak{g}$. The

---

[37] A global holomorphic section of $\mathfrak{g}_{0,\infty}$ is a $\mathfrak{g}$-valued constant (or it would have singularities somewhere) that must be valued in $\mathfrak{l}_0 \cap \mathfrak{l}_\infty$ because of the conditions at 0 and at ∞.

[38] In trying to do so, one runs into the fact that $H^1(\mathbb{CP}^1, \mathcal{O}(-p-q)) \neq 0$, where $\mathcal{O}(-p-q)$ is the sheaf of holomorphic functions that vanish at two points p and q (here those two points are $z = 0$ and ∞).

[39] The construction as we have described it involves a Manin triple of $\mathfrak{g}$, but one could actually in a somewhat similar way use a Manin triple of $\mathfrak{g}[z, z^{-1}]$ (endowed with the nondegenerate quadratic form $(a, b) = \oint \frac{\mathrm{d}z}{z} \mathrm{Tr}\, a(z)b(z)$). This would involve a construction somewhat like that above, with a more complicated set of conditions on A and ε. The relation between Manin triples of $\mathfrak{g}$ and of $\mathfrak{g}[z, z^{-1}]$ is that a Manin triple of $\mathfrak{g}$ determines a Manin triple of $\mathfrak{g}[z, z^{-1}]$, namely $\mathfrak{g}[z, z^{-1}] = \widehat{\mathfrak{l}}_0 \oplus \widehat{\mathfrak{l}}_\infty$, with $\widehat{\mathfrak{l}}_0 = z\mathfrak{g}[z] \oplus \mathfrak{l}_0$, $\widehat{\mathfrak{l}}_\infty = z^{-1}\mathfrak{g}[z^{-1}] \oplus \mathfrak{l}_\infty$.

case that would lead to a solution of the dynamical Yang-Baxter equation rather than the ordinary one is that $\mathfrak{l}_0$ and $\mathfrak{l}_\infty$ are such that even after conjugating them into general position, $\mathfrak{l}_0 + \mathfrak{l}_\infty \neq \mathfrak{g}$.

9.2 Example

Given $\mathfrak{g}$, can we pick $\mathfrak{l}_0$ and $\mathfrak{l}_\infty$ to make a Manin triple? In general, the answer is certainly "no," since if $\mathfrak{g}$ is of odd dimension – for example $\mathfrak{sl}_2$ – it has no middle-dimensional isotropic subspaces. However, the following is a useful construction of examples that are related to the usual trigonometric solutions of the Yang-Baxter equation.

For any simple Lie algebra $\mathfrak{g}$, add to $\mathfrak{g}$ another copy $\widetilde{\mathfrak{h}}$ of the Cartan subalgebra $\mathfrak{h}$ of $\mathfrak{g}$ to make a Lie algebra $\widetilde{\mathfrak{g}} = \mathfrak{g} \oplus \widetilde{\mathfrak{h}}$. Equip $\widetilde{\mathfrak{h}}$ with a quadratic form given by the restriction of the Killing form on $\mathfrak{g}$. In this way $\widetilde{\mathfrak{g}}$ acquires an invariant nondegenerate pairing. We will construct a Manin triple for $\widetilde{\mathfrak{g}}$.

Choose a decomposition $\mathfrak{g} = \mathfrak{n}_- \oplus \mathfrak{h} \oplus \mathfrak{n}_+$ into nilpotent and Cartan subalgebras. Set

$$\begin{aligned} \mathfrak{h}_+ &= \{(X, \mathrm{i}X) \mid X \in \mathfrak{h}\} \subset \mathfrak{h} \oplus \widetilde{\mathfrak{h}}\,, \\ \mathfrak{h}_- &= \{(X, -\mathrm{i}X) \mid X \in \mathfrak{h}\} \subset \mathfrak{h} \oplus \widetilde{\mathfrak{h}}\,. \end{aligned} \tag{9.8}$$

Then we can choose a pair of complementary Lagrangian subalgebras of $\widetilde{\mathfrak{g}}$ by

$$\begin{aligned} \mathfrak{l}_0 &= \mathfrak{n}_+ \oplus \mathfrak{h}_+\,, \\ \mathfrak{l}_\infty &= \mathfrak{n}_- \oplus \mathfrak{h}_-\,. \end{aligned} \tag{9.9}$$

This gives a Manin triple for $\widetilde{\mathfrak{g}}$.

9.3 The r-Matrix

Now we consider our four-dimensional theory on $\mathbb{R}^2 \times \mathbb{P}^1$, where we use the 1-form dz/z on $\mathbb{P}^1$ and the Manin triple just described. Let us place Wilson lines at $z_1, z_2 \in \mathbb{C}^\times$, and have them cross in the topological plane as in Fig. 15. Suppose the Wilson lines are in representations V, W of $\mathfrak{g}$. We will show that the result of this crossing will be the trigonometric R-matrix.

Let us first calculate explicitly what happens to leading order in $\hbar$. To leading order in $\hbar$, the result of the crossing Wilson lines is described by $R = 1 + \hbar r$, where r is a $\widetilde{\mathfrak{g}} \otimes \widetilde{\mathfrak{g}}$-valued meromorphic function. To obtain $r(z_1, z_2)$, we can pick a gauge and then repeat the Feynman diagram computations of the previous section. Instead let us here note that $r(z_1, z_2)$ must satisfy the following properties:

1. $r(z_1, z_2)$ has a first-order pole at $z_1 = z_2$, and is regular elsewhere. The residue of this pole is $\hbar c$ where $c \in \widetilde{\mathfrak{g}} \otimes \widetilde{\mathfrak{g}}$ is the Casimir element, dual to the chosen invariant pairing. This is the same singular behavior as in the rational case, since OPE singularities are local and are not affected by global topology.
2. At $z_1 = 0$, $r(z_1, z_2)$ is in $\mathfrak{l}_0 \otimes \mathfrak{l}_\infty$, and at $z_1 = \infty$ it is in $\mathfrak{l}_\infty \otimes \mathfrak{l}_0$. This just reflects the corresponding conditions on A.

3. $r(z_1, z_2)$ is sent to $-r(z_1, z_2)$ if we simultaneously swap z_1 and z_2 and exchange the tensor factors in $\widetilde{\mathfrak{g}} \otimes \widetilde{\mathfrak{g}}$. (This reflects the fact that the action of the theory – and likewise the propagator and the Feynman diagram used to compute $r(z_1, z_2)$ – are all odd under an orientation-reversing symmetry of the topological plane that exchanges the two oriented Wilson lines that are crossing. Note that the rational r-matrix $r = t_a \otimes t_a/(z_1 - z_2)$ has the same property.)
4. Finally, $r(z_1, z_2)$ is a function only of the ratio z_1/z_2, since the differential $\omega = \mathrm{d}z/z$ the conditions we placed at 0 and ∞ are all invariant under the $\mathbb{C}^\times$ action on $C = \mathbb{C}^\times$.

There is a unique function satisfying these properties. To write it down, we observe that the Casimir $c = \sum_a t_a \otimes t_a \in \widetilde{\mathfrak{g}} \otimes \widetilde{\mathfrak{g}}$ can be written in a unique way as a sum

$$c = c(\mathfrak{l}_0, \mathfrak{l}_\infty) + c(\mathfrak{l}_\infty, \mathfrak{l}_0)\,, \tag{9.10}$$

where $c(\mathfrak{l}_0, \mathfrak{l}_\infty) \in \mathfrak{l}_0 \otimes \mathfrak{l}_\infty$, and similarly $c(\mathfrak{l}_\infty, \mathfrak{l}_0) \in \mathfrak{l}_\infty \otimes \mathfrak{l}_0$.

Then the unique function $r(z_1, z_2)$ satisfying the desired conditions is

$$2\pi\mathrm{i}\, r(z_1, z_2) = \frac{c(\mathfrak{l}_0, \mathfrak{l}_\infty)}{1 - \frac{z_1}{z_2}} - \frac{c(\mathfrak{l}_\infty, \mathfrak{l}_0)}{1 - \frac{z_2}{z_1}}. \tag{9.11}$$

Let us write this out more explicitly. Let us choose a basis $X_\alpha^\pm, H_i$ of $\mathfrak{g}$, so that $X_\alpha^\pm \in \mathfrak{n}_\pm$ and H_i form a basis of $\mathfrak{h}$. We assume that in this basis the chosen pairing on $\mathfrak{g}$ is such that $\langle X_{\alpha^+}, X_{\beta^-} \rangle = \delta_{\alpha\beta}$, and $\langle H_i, H_j \rangle = \delta_{ij}$. Let $\widetilde{H}_i$ be the corresponding basis of the other copy $\widetilde{h}$ of the Cartan. Then, in this basis, we have

$$\begin{aligned} 2\pi\mathrm{i}\, r(z_1, z_2) &= \frac{1}{1 - \frac{z_1}{z_2}} \sum_\alpha X_\alpha^+ \otimes X_\alpha^- + \frac{1}{1 - \frac{z_1}{z_2}} \frac{1}{2} \sum (H_j + \mathrm{i}\widetilde{H}_j) \otimes (H_j - \mathrm{i}\widetilde{H}_j) \\ &\quad - \frac{1}{1 - \frac{z_2}{z_1}} \sum_\alpha X_\alpha^- \otimes X_\alpha^+ - \frac{1}{1 - \frac{z_2}{z_1}} \frac{1}{2} \sum (H_j - \mathrm{i}\widetilde{H}_j) \otimes (H_j + \mathrm{i}\widetilde{H}_j) \\ &= \frac{1}{1 - \frac{z_1}{z_2}} \sum_\alpha X_\alpha^+ \otimes X_\alpha^- - \frac{1}{1 - \frac{z_2}{z_1}} \sum_\alpha X_\alpha^- \otimes X_\alpha^+ \\ &\quad + \frac{1}{2} \frac{z_2 + z_1}{z_2 - z_1} \left(\sum H_j \otimes H_j + \sum \widetilde{H}_j \otimes \widetilde{H}_j \right) + \frac{i}{2} \sum \left(\widetilde{H}_j \otimes H_j - H_j \otimes \widetilde{H}_j \right). \end{aligned}$$

To evaluate what happens when Wilson lines in particular representations V, W of $\widetilde{\mathfrak{g}}$ cross, one applies the homomorphism

$$\rho_V \otimes \rho_W : \mathfrak{g} \otimes \mathfrak{g} \to \mathrm{End}(V) \otimes \mathrm{End}(W) \tag{9.12}$$

to the function $r(z_1, z_2)$ (where ρ_V, ρ_W indicate the maps coming from the $\mathfrak{g}$-action on V and W).

9.4 Specializing to $\mathfrak{sl}_2$

As an example, let us consider the case that $\mathfrak{g} = \mathfrak{sl}_2$. Let us use the standard basis e, f, h of $\mathfrak{sl}_2$ where $\langle e, f\rangle = 1$ and $\langle h, h\rangle = 2$. In this basis,

$$\begin{aligned} 2\pi\mathrm{i}\, r(z_1, z_2) =& \frac{1}{1-\frac{z_1}{z_2}} e \otimes f - \frac{1}{1-\frac{z_2}{z_1}} f \otimes e \\ &+ \frac{1}{4}\frac{z_2+z_1}{z_2-z_1}\left(\sum h\otimes h + \sum \widetilde{h}\otimes\widetilde{h}\right) + \frac{\mathrm{i}}{4}\sum\left(\widetilde{h}\otimes h - h\otimes\widetilde{h}\right). \end{aligned} \tag{9.13}$$

Let us see what this looks like if our representations V, W are both the spin 1/2 representation of $\mathfrak{sl}_2$, and the basis element $\widetilde{H}$ of the second copy of the Cartan of $\mathfrak{sl}_2$ acts with constants s_1, s_2 in the two representations. Let us choose a basis e_+, e_- of the spin 1/2 representation of $\mathfrak{sl}_2$, and a corresponding basis $e_+ \otimes e_+$, etc., of the tensor product $V \otimes W$. In this basis, the matrix $r(z_1, z_2)$ looks like

$$\frac{1}{2\pi\mathrm{i}}\begin{pmatrix} r^{++}_{++} & & & \\ & r^{+-}_{+-} & r^{-+}_{+-} & \\ & r^{+-}_{-+} & r^{-+}_{-+} & \\ & & & r^{--}_{--} \end{pmatrix}, \tag{9.14}$$

where

$$\begin{aligned} r^{++}_{++} &= (1+s_1 s_2)\frac{z_2+z_1}{z_2-z_1} + \mathrm{i}(s_1 - s_2)\,, \\ r^{+-}_{+-} &= (-1+s_1 s_2)\frac{z_2+z_1}{z_2-z_1} + \ i(-s_1 - s_2)\,, \\ r^{-+}_{+-} &= \frac{4}{1-\frac{z_1}{z_2}}\,, \\ r^{+-}_{-+} &= \frac{-4}{1-\frac{z_2}{z_1}}\,, \\ r^{-+}_{-+} &= (-1+s_1 s_2)\frac{z_2+z_1}{z_2-z_1} + \mathrm{i}(s_1 + s_2)\,, \\ r^{--}_{--} &= (1+s_1 s_2)\frac{z_2+z_1}{z_2-z_1} + \mathrm{i}(s_2 - s_1)\,. \end{aligned} \tag{9.15}$$

If we set $s_1 = s_2 = 0$, we get the usual trigonometric r-matrix for $\mathfrak{sl}_2$, associated to the six-vertex model of statistical mechanics. For comparison with the literature, one might want to symmetrize our r-matrix by conjugation, replacing the off-diagonal components by

$$r^{-+}_{+-} = \frac{4\left(\frac{z_1}{z_2}\right)^{\frac{1}{2}}}{1-\frac{z_1}{z_2}} = r^{+-}_{-+}\,. \tag{9.16}$$

without changing the diagonal entries.

In the above basis, the s-dependent terms only contribute to the diagonal matrix elements of r. They come in two groups. First, we have $s_1 s_2 \frac{z_2+z_1}{z_2-z_1}$ times the identity operator. This we can absorb into the definition of the overall constant normalization factor of the R-matrix. The remaining terms are nontrivial. They reproduce the known generalization of the six-vertex model to include horizontal and vertical fields given by $\mathrm{i}s_1$ and $-\mathrm{i}s_2$ (see eqn. (B.5) in Appendix).

10 Elliptic Solutions Of the Yang-Baxter Equation

10.1 Preliminaries

In this section we discuss the elliptic case, where the holomorphic curve C is an elliptic curve E. Since the differential $\mathrm{d}z$ does not have any poles we do not need to consider boundary terms like those encountered in section 9. However, there is an important topological subtlety to consider.

In formulating the theory with gauge group G, we begin by considering a topological G-bundle over $\Sigma \times E$. Our considerations leading to a solution of the Yang-Baxter equation are local along Σ, so we can take $\Sigma = \mathbb{R}^2$, in which case the choice of a topological G-bundle $\mathcal{V} \to \Sigma \times E$ amounts to the choice of topological G-bundle $\mathcal{V} \to E$. However, in general there is a choice to be made, because, for a connected group G, a G-bundle over E is classified topologically by an invariant[40] $\zeta \in \pi_1(G)$.

Therefore, in setting up the theory, we have the freedom to make an arbitrary choice of the element ζ. Once we make this choice, a classical solution will be defined by a $\bar{\partial}$ operator $\partial_{\bar{z}} + [A_{\bar{z}}, \dots]$ on $\mathcal{V}$, modulo gauge transformations. Such an operator gives a holomorphic structure to the bundle $\mathcal{V}$ and the gauge-invariant data is precisely the holomorphic equivalence class of this bundle.

The moduli space of classical solutions is therefore the moduli space of holomorphic G-bundles over E with topological class ζ. The tangent space to the moduli space is $H^1(E, \mathrm{ad}(\mathcal{V}))$ (where $\mathrm{ad}(\mathcal{V})$ is the adjoint bundle associated to $\mathcal{V}$). Thus the condition of the theory having a unique classical solution – and therefore leading to straightforward perturbation theory and a solution of the classical Yang-Baxter equation – is equivalent to $H^1(E, \mathrm{ad}(\mathcal{V})) = 0$.

By Serre duality, $H^1(E, \mathrm{ad}(\mathcal{V}))$ is dual to $H^0(E, \mathrm{ad}(\mathcal{V}))$ for an elliptic curve E. Here $H^0(E, \mathrm{ad}(\mathcal{V}))$ is the Lie algebra of the automorphism group of a holomorphic $G_{\mathbb{C}}$ bundle $\mathcal{V}$. Thus if and only if $\mathcal{V}$ has no infinitesimal deformations, the Lie algebra of its automorphism group will be trivial and $\mathcal{V}$ has only a finite group of automorphisms. (Instead of Serre duality, we could have invoked here the Riemann-Roch theorem, as we did at a similar point in section 9.1.)

[40]For example, for $G = SO(3)$, this invariant is the second Stiefel-Whitney class $w_2(\mathcal{V})$. To define it in general, we observe that for G connected and p a point in E, the restriction of $\mathcal{V}$ to $E\backslash p$ is trivial, so $\mathcal{V}$ can be constructed by gluing a trivial bundle over $E\backslash p$ to a trivial bundle over a small disc D containing p. Since $D\backslash p$ is homotopic to a circle, the gluing function that is used here is classified up to homotopy by a class $\zeta \in \pi_1(G)$.

To find, therefore, an elliptic solution of the Yang-Baxter equation (as opposed to its "dynamical" generalization), we need to find a complex Lie group G and a holomorphic G-bundle $\mathcal{V} \to E$ such that $H^0(E, \mathrm{ad}(\mathcal{V})) = H^1(E, \mathrm{ad}(\mathcal{V})) = 0$. However, the options for such a G and $\mathcal{V}$ are very limited. The only cases are $G = PGL_N = GL_N/GL_1$ with $N \geq 2$, with $\mathcal{V}$ chosen so that ζ is a generator of the finite group $\pi_1(G) \cong \mathbb{Z}_N$. For such G and ζ, there is a unique stable holomorphic vector bundle $\mathcal{V}$, and it does obey $H^0(E, \mathrm{ad}(\mathcal{V})) = H^1(E, \mathrm{ad}(\mathcal{V})) = 0$.

Thus elliptic solutions of the Yang-Baxter equation (or at least those that we can construct) are classified by a choice of $N \geq 2$ – determining the group PGL_N – and a generator ζ of $\mathbb{Z}_N$, or equivalently a primitive N^{th} root of 1. The automorphism group of such a solution is the automorphism group of $\mathcal{V}$, which in all cases is $\mathbb{Z}_N \times \mathbb{Z}_N$.

10.2 Rigid Holomorphic Bundles

We will now describe concretely the rigid holomorphic bundles $\mathcal{V} \to E$ associated a choice of N and ζ.

It is well-known that there exist pairs of $N \times N$ matrices A, B obeying

$$AB = BA\, e^{2\pi \mathrm{i}/N} \,. \tag{10.1}$$

Moreover, A and B are unique up to conjugation and multiplication by scalars. For example, we can pick $A = \mathrm{diag}(1, e^{2\pi \mathrm{i}/N}, e^{4\pi \mathrm{i}/N}, \ldots, e^{2\pi \mathrm{i}(N-1)/N})$, and B a matrix that cyclically permutes the eigenspaces of A. In the opposite direction, the equation shows that (after possibly multiplying A by a scalar) the eigenvalues of A are the Nth roots of 1, each with multiplicity 1. So A can be put in the claimed form, and then it is not hard to see that (up to conjugation and multiplication by a scalar) B must be as claimed.

Now given $\zeta \in \mathbb{Z}_N$, the matrices A^ζ and B do not commute, but obey $A^\zeta B = BA^\zeta \exp(2\pi \mathrm{i}\zeta/N)$. But this means that A^ζ and B commute if projected to PGL_N. So we can define a flat PGL_N bundle $\mathcal{V} \to E$ whose monodromies around a pair of generators of $\pi_1(E) = \mathbb{Z} \oplus \mathbb{Z}$ are A^ζ and B. Being flat, this bundle is automatically stable and holomorphic.

The automorphism group of this bundle always includes a subgroup $\mathbb{Z}_N \times \mathbb{Z}_N$ generated by A and B. This is the full automorphism group if ζ is a generator of $\mathbb{Z}_N$. (One may prove this starting with the fact that if ζ is a generator of $\mathbb{Z}_N$, then the eigenvalues of the monodromy A^ζ are nondegenerate.) Accordingly, if ζ is a generator, then $H^0(E, \mathcal{V}) = H^1(E, \mathcal{V}) = 0$, and we are in the favorable situation that will lead to a solution of the Yang-Baxter equation.

If ζ is not a generator, then this automorphism group has a strictly positive dimension and likewise the bundle $\mathcal{V}$ can be deformed (as a holomorphic bundle and even as a flat one).

To conclude this section, we will describe $\mathrm{ad}(\mathcal{V})$ as a holomorphic bundle over E. We consider A^ζ and B as matrices acting by conjugation on $\mathfrak{g}$, the Lie algebra of PGL_N. The eigenvalues are pairs μ, μ' of Nth roots of 1, with each pair occurring exactly once except the trivial pair $\mu = \mu' = 1$. Each joint eigenspace of A^ζ and B corresponds to a holomorphic line bundle over E, and these lines bundles are all of order N because the eigenvalues of A^{ζ} and B are of order N. There are N^2 equivalence classes of holomorphic line bundle over E of order N. The trivial line bundle is of order N, but there are N^2-1 nontrivial ones, and $\mathrm{ad}(\mathcal{V}) \to E$ is the direct sum of all N^2-1 nontrivial line bundles of order N. (Thus as a holomorphic bundle, $\mathrm{ad}(\mathcal{V})$ does not depend on the generator ζ of $\mathbb{Z}_N$, though its Lie algebra structure does depend on ζ.)

10.3 Specializing to $N = 2$

To achieve some minor simplifications, we will specialize to $N = 2$. There is now only one choice of ζ. Accordingly, $\mathcal{V}$ is unique up to isomorphism. As a holomorphic vector bundle over E, $\mathrm{ad}(\mathcal{V})$ is the direct sum of the three non-trivial line bundles over E. However, we would like to describe the Lie algebra structure on $\mathrm{ad}(V)$.

We identify E with the complex z-plane modulo $z \cong z+1$ and $z \cong z+\tau$, where the complex number τ is constrained to have $\mathrm{Im}\,\tau > 0$ and is the modulus of E. A flat line bundle can be described by its monodromies α and β under $z \to z+1$ and $z \to z+\tau$. A flat line bundle is of order 2 if $\alpha^2 = \beta^2 = 1$. To describe the 3 non-trivial line bundles of order 2, we take

$$(\alpha_i, \beta_i) = \begin{cases} (1,-1) & \text{for } \mathcal{L}_1 \ , \\ (-1,1) & \text{for } \mathcal{L}_2 \ , \\ (-1,-1) & \text{for } \mathcal{L}_3 \ . \end{cases} \tag{10.2}$$

Since $(\alpha_1\alpha_2, \beta_1\beta_2) = (\alpha_3, \beta_3)$, there is a natural isomorphism $\phi_{12} : \mathcal{L}_1 \otimes \mathcal{L}_2 \cong \mathcal{L}_3$. The same holds with any cyclic permutations of the labels 123. The Lie algebra structure on $\mathcal{V}$ can be defined as follows. If s_1 and s_2 are local sections of $\mathcal{L}_1$ and $\mathcal{L}_2$, then $[s_1, s_2]$ is the local section of $\mathcal{L}_3$ defined by

$$[s_1, s_2] = \phi_{12}(s_1 \otimes s_2) \ . \tag{10.3}$$

This statement and its cyclic permutations define the Lie algebra structure.

However, it may be useful to describe this structure in a slightly more explicit way. The matrices A and B of section 10.2 commute as matrices acting on the Lie algebra $\mathfrak{sl}_2$. Their joint eigenfunctions, up to a choice of basis, are the standard generators t_1, t_2, t_3 of $\mathfrak{sl}_2$, obeying $[t_1, t_2] = t_3$ and cyclic permutations. Thus one might prefer to think of a section of $\mathrm{ad}(\mathcal{V})$ more explicitly as $\sum_{i=1}^3 s_i t_i$, where s_i is a section of $\mathcal{L}_i$ and t_i is an element of a standard basis of $\mathfrak{sl}(2)$. In this description, the commutator of $s = \sum_{i=1}^3 s_i t_i$ with $s' = \sum_{i=1}^3 s'_i t_i$ is $s'' = \sum_{i=1}^3 s''_i t_i$ where

$$s''_3 = \phi_{12}(s_1 \otimes s'_2 - s_2 \otimes s'_1) \ , \tag{10.4}$$

and cyclic permutations. Thus, one can think of eqn. (10.3) as a formula for the commutator of $s_1 t_1$ with $s_2 t_2$.

Now let us determine the lowest order nontrivial contribution to the r-matrix. It will be a (meromorphic) section of[41] $\mathrm{ad}(\mathcal{V}) \boxtimes \mathrm{ad}(\mathcal{V})$ over $\Sigma \times \Sigma$. This section

[41] Here $\mathrm{ad}(\mathcal{V}) \boxtimes \mathrm{ad}(\mathcal{V})$ is simply the tensor product of the $\mathrm{ad}(\mathcal{V})$ bundle over the first copy of Σ with the $\mathrm{ad}(\mathcal{V})$ bundle over the second copy; if $\pi_i : \Sigma \times \Sigma \to \Sigma$, $i = 1, 2$ are the two projections, then $\mathrm{ad}(\mathcal{V}) \boxtimes \mathrm{ad}(\mathcal{V}) = \pi_1^*(\mathrm{ad}(\mathcal{V})) \otimes \pi_2^*(\mathrm{ad}(\mathcal{V}))$.

can be determined by reasoning similar to what we used in section 9.3, though the details are simpler.

A priori, a general form of r would be $r(z_1, z_2) = \sum_{i,j=1}^{3} w_{i,j}(z_1, z_2) t_i \otimes t_j$, where $w_{i,j}(z_1, z_2)$ is a section of $\mathcal{L}_i \boxtimes \mathcal{L}_j$. However, the $\mathbb{Z}_2 \times \mathbb{Z}_2$ automorphism group ensures that $w_{i,j} = 0$ for $i \neq j$. Thus we reduce to $r(z_1, z_2) = \sum_{i=1}^{3} w_i(z_1, z_2) t_i \otimes t_i$. Here the functions $w_i(z_1, z_2)$ are determined by the following (somewhat redundant) set of conditions:

1. $r(z_1, z_2)$ has a first-order pole at $z_1 = z_2$, with residue $\hbar c = \hbar \sum_k t_k \otimes t_k$ as in the rational case, since OPE singularities are local.
2. $r(z_1, z_2) = -r(z_2, z_1)$ (we need not combine this with an exchange of the two tensor factors of $\mathfrak{sl}_2 \otimes \mathfrak{sl}_2$, since $r = \sum_i w_i t_i \otimes t_i$ is invariant under this switch).
3. Finally, $r(z_1, z_2)$ is a function only of $z_1 - z_2$, because of the translation symmetry of the elliptic curve E and the differential $\omega = \mathrm{d}z$.

The functions $w_k(z_1, z_2)$ that satisfy these conditions actually have a simple interpretation. Since the canonical bundle K of E is trivial, its possible square roots $K^{1/2}$ are line bundles of order 2, and thus we can think of $\mathcal{L}_k$, for $k = 1, 2$ or 3, as one of the three even spin bundles of Σ, that is, as representing a possible $K^{1/2}$. Let ψ_k be a (holomorphic) fermion field on Σ valued in this spin bundle, with action

$$I = \frac{1}{2\pi} \int_E \psi_k \bar{\partial} \psi_k \,. \tag{10.5}$$

Then the two-point function $\langle \psi_k(z_1) \psi_k(z_2) \rangle$ satisfies precisely the conditions of the functions $w_k(z_1, z_2)$.

With or without this interpretation, it is straightforward to write a formula for w_k:

$$w_k(z_1, z_2) = \sum_{n,m \in \mathbb{Z} \setminus \{(0,0)\}} \frac{\alpha_k^n \beta_k^m}{z_1 - z_2 - n - m\tau} \,. \tag{10.6}$$

The factors of $\alpha_k^n \beta_k^m$ ensure that $w_k(z_1, z_2)$ is a section of $\mathcal{L}_k \boxtimes \mathcal{L}_k$, and the desired properties are all obvious. We have defined w_k so that for $z_1 \to z_2$ it behaves as $1/(z_1 - z_2)$. Thus, taking the normalization from the rational case, the r-matrix is

$$r = \hbar \sum_{k=1}^{3} w_k(z_1 - z_2) t_k \,. \tag{10.7}$$

This coincides with the known expression in the literature, see e.g. [37, p. 539].

11 The Dynamical Yang-Baxter Equation

11.1 Reduction to an Abelian Subgroup

Our goal in this section is to take a first look at the case that G and C and the other relevant choices are such as to lead to a moduli space of classical solutions, not just a single isolated classical solution. This makes matters fundamentally more difficult and our analysis here will be preliminary.

A basic case to bear in mind is that C is an elliptic curve E, G is simple, and the topological invariant ζ introduced in section 10.1 vanishes, so that the G-bundle $\mathcal{V} \to E$ is topologically trivial. Under this condition, the structure group of a stable holomorphic G-bundle $\mathcal{V} \to E$ reduces always to the (complex) maximal torus $T \subset G$. The moduli space $\mathcal{M}$ of stable holomorphic G-bundles over E is then simply the corresponding moduli space $\mathcal{M}'$ for T, divided by the Weyl group $\mathcal{W}$: $\mathcal{M} = \mathcal{M}'/\mathcal{W}$. The automorphism group of a generic stable holomorphic G-bundle is simply T, though the symmetry is enhanced at points in $\mathcal{M}'$ at which the Weyl action is not free. For our purposes, we can just identify the moduli space of classical solutions with $\mathcal{M}'$ and ignore the Weyl group action and possible symmetry enhancement. (That is because we will eventually impose a constraint that keeps us away from special points.)

For C an elliptic curve, the story is similar[42] for any ζ, with T replaced by another torus $T_\zeta \subset T$, $\mathcal{M}'$ replaced with the moduli space $\mathcal{M}'_\zeta$ of T_ζ-bundles over E, and the Weyl group replaced by a subgroup. Rather than explain these details, we will simply continue with the case that $\zeta = 0$.

A complex line bundle over E can be represented by a very simple gauge field $A_{\bar{z}} = b$, with b a complex constant. Here b is subject to some equivalences, but they will not be important in what follows. Similarly, for gauge group G, a classical solution whose structure group reduces to T can be represented by $A_{\bar{z}} = b$, where now b is a $\mathfrak{t}$-valued constant. For example, if $G = SL_N$, this means that $A_{\bar{z}} = \mathrm{diag}(b_1, b_2, \ldots, b_N)$, with $\sum_{i=1}^N b_i = 0$. In particular, this classical solution is invariant under the translation symmetries of E.

In general in nonabelian gauge theory with gauge group G, when we expand around a classical solution whose structure group reduces to a proper subgroup H of G, we should look for a low energy description in the form of an H gauge theory. In the present case, we have the further fact that to study the Yang-Baxter equation or its analog, we can take the topological plane Σ to be simply a plane $\mathbb{R}^2$, and thus we work on $\mathbb{R}^2 \times E$ where E is compact and $\mathbb{R}^2$ is not. In this situation, in quantum field theory in general, we can look for a low energy effective field theory on $\mathbb{R}^2$. Our present problem is actually diffeomorphism-invariant in the $\mathbb{R}^2$ direction, so the low energy description is really a topological field theory.

Combining these facts, we can aim to find an effective description in the form of a two-dimensional topological gauge theory with structure group T. To find this description at least at a formal level, we make use of the translation symmetry of the classical solutions and throw away "massive modes" that lack this symmetry. Likewise, we keep only the $\mathfrak{t}$-valued part of the four-dimensional gauge field A. This means that $A_{\bar{z}}$ is reduced to the $\mathfrak{t}$-valued field b, which now depends on coordinates x and y of $\mathbb{R}^2$ but not on z and $\bar{z}$, and likewise the rest of the gauge field, namely $A_x \mathrm{d}x + A_y \mathrm{d}y$, becomes a purely two-dimensional gauge field that we will denote simply as A.

Specializing the underlying action (3.3) to this situation, we get an effective

[42]More exotic examples with moduli, such as the case $\mathfrak{l}_0 + \mathfrak{l}_\infty \neq \mathfrak{g}$ mentioned in section 10.1, might be more complicated.

two-dimensional abelian action

$$S = -\kappa \int_\Sigma \mathrm{Tr}\, bF\,, \tag{11.1}$$

where Tr is the restriction to $\mathfrak{t}$ of the quadratic form of the same name on $\mathfrak{g}$, $F = \mathrm{d}A$ is the two-dimensional gauge curvature, and $\kappa = \mathrm{i}\,\mathrm{Im}\,\tau/\pi\hbar$. This is the action that we will use somewhat formally in analyzing the present problem.

The reader may object that in quantum field theory in general, it is not correct to simply set the massive modes to zero, as we have done. Instead, one has to integrate them out, producing in general corrections to the effective action of the fields that are retained in the effective description. Moreover, if we wish to include Wilson loops and their crossings, we have to integrate out the massive fields in the presence of those operators. All this is true, but the implications for our problem are limited because two-dimensional diffeomorphism symmetry severely constrains what couplings can be generated by integrating out massive modes. If Σ is not $\mathbb{R}^2$ but is a curved two-manifold with scalar curvature R, then it is possible to generate a coupling $\int_\Sigma \mathrm{d}^2x\sqrt{g}Rf(b)$, for some function $f(b)$. One should expect such a term, but it will not be important for our purposes because in analyzing the crossing of line operators, one can assume that Σ is flat. In the presence of line operators, integrating out the massive modes will generate the framing anomaly of $\mathfrak{g}$. That is an important effect, but we are already familiar with it and will not discuss it further in the present section. Finally, integrating out the massive modes certainly affects the R-matrix that governs crossing of two line operators. But here we will just discuss the formal properties of this R-matrix, and will not try to calculate it. So we will not have to explicitly discuss the contribution of the massive modes to the R-matrix.

If A is regarded as a two-dimensional gauge field with structure group a *compact* torus T, and similarly b is real-valued, taking values in the corresponding Lie algebra $\mathfrak{t}_c$, then eqn. (11.1) becomes the action of a simple but much-studied two-dimensional topological field theory, often called BF theory.[43] We are not quite in that situation, since in eqn. (11.1), b, A_x, and A_y are all complex-valued fields and the action is a holomorphic function of those fields. Moreover, we are in a situation in which there is a nontrivial moduli space of classical solutions (parameterized locally by $b \in \mathfrak{t}$), so if we want to define correlation functions, we cannot just do perturbation theory in a formal way; we need some recipe for what to do with the moduli. One point of view that may seem somewhat formal but that seems to provide a satisfactory framework for our considerations below (for instance, the derivation of the dynamical Yang-Baxter equation makes sense in this framework) is to view the theory as a machine that generates a differential form on the moduli space of classical solutions, without worrying about how to integrate it. Alternatively, it might be possible to define an operator that fixes the values of the moduli. This idea is explained at the end of section 11.3. Finally, in principle one may use the D4-NS5 system to define a nonperturbative integration cycle for the underlying four-dimensional theory (3.3). Specializing this to the

[43]In that context, κ is real, assuming we want a unitary theory. In the context of reduction from the four-dimensional theory, κ is a complex number, since $\hbar$ is complex.

present situation, one would then learn in principle what to do with the moduli in the effective theory (11.1).

In practice, we will simply draw inspiration from conventional two-dimensional BF theory.[44] Based on this, we will suggest a simple procedure to make contact with the dynamical Yang-Baxter equation [21, 22, 23, 24]. Given that we assume $\zeta = 0$, the solution of the dynamical Baxter equation that arises in this way will have T as a group of automorphisms, since a generic classical solution with $\zeta = 0$ has automorphism group T, and we will be avoiding the exceptional cases.

11.2 BF Theory and the Dynamical Yang-Baxter Equation

The classical equation of motion of the gauge field A in BF theory, in the absence of line operators, simply says that $\mathrm{d}b = 0$. Thus b is simply constant.

What happens in the presence of a Wilson line operator? To start with, we take the gauge group to be $G = U(1)$. An irreducible representation of $U(1)$ is determined by the choice of an integer n, the "charge." The Wilson operator supported on a curve K for the representation of charge n is $\exp\left(\mathrm{i}n \int_K A\right)$. Including this factor, the argument of the path integral becomes $\exp(\mathrm{i}S) \exp\left(\mathrm{i}n \int_K A\right) = \exp(\mathrm{i}S')$, with

$$S' = -\kappa \int_\Sigma \mathrm{Tr}\, bF + n \int_K A\,. \tag{11.2}$$

In other words, the Wilson operator for $U(1)$ effectively contributes an additional term to the classical action. The classical equation of motion now becomes

$$\kappa \mathrm{d}b + n\delta_K = 0\,, \tag{11.3}$$

where δ_K is a one-form delta function Poincaré dual to K. The import of this is that $b' = \kappa b$ jumps by n in crossing K from right to left (Fig. 32(a)).

Now we can consider the case of two Wilson operators crossing. But in doing so, we may as well generalize[45] beyond a simple crossing of two line operators of charges n_1 and n_2. Since we are discussing a purely two-dimensional theory, any crossing really does involve a physical intersection of the two line operators, and charge exchange is possible. In general, we can consider a case with charges n_1 and n_2 coming in and charges n_3 and n_4 going out, the only constraint being that $n_1 + n_2 = n_3 + n_4$. The behavior of $b' = \kappa b$ is then as sketched in Fig. 32(b).

[44] The purely two-dimensional analysis that follows is similar to section 2.5 of [38].

[45] The purpose of discussing this generalization here is to give the simplest possible motivation for the picture of Fig. 32(b). A similar generalization is not possible in the context of the four-dimensional theory, assuming that the lines that are crossing have different values of the spectral parameter. Such a generalization is possible in two-dimensional BF theory with a nonabelian gauge group, though we will not need that case.

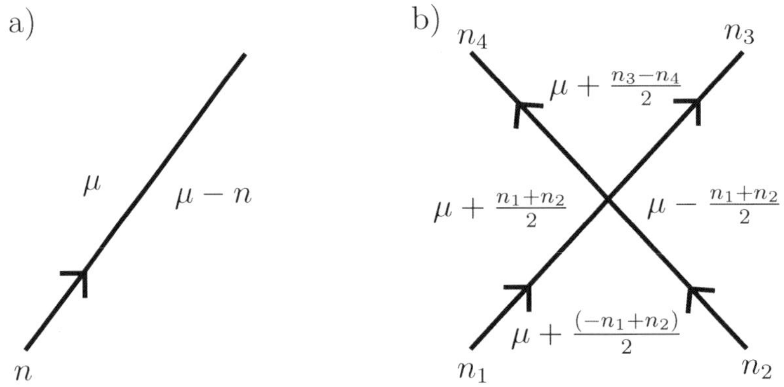

Figure 32: (a) The field $b' = \kappa b$ is constant away from Wilson operators and jumps by n in crossing a charge n Wilson operator from right to left. In the example shown, b' jumps from $\mu - n$ to μ. (b) Two Wilson operators of charges n_1 and n_2 come in to this intersection and two of charges n_3 and n_4 go out (here $n_1 + n_2 = n_3 + n_4$). The field b' jumps as shown.

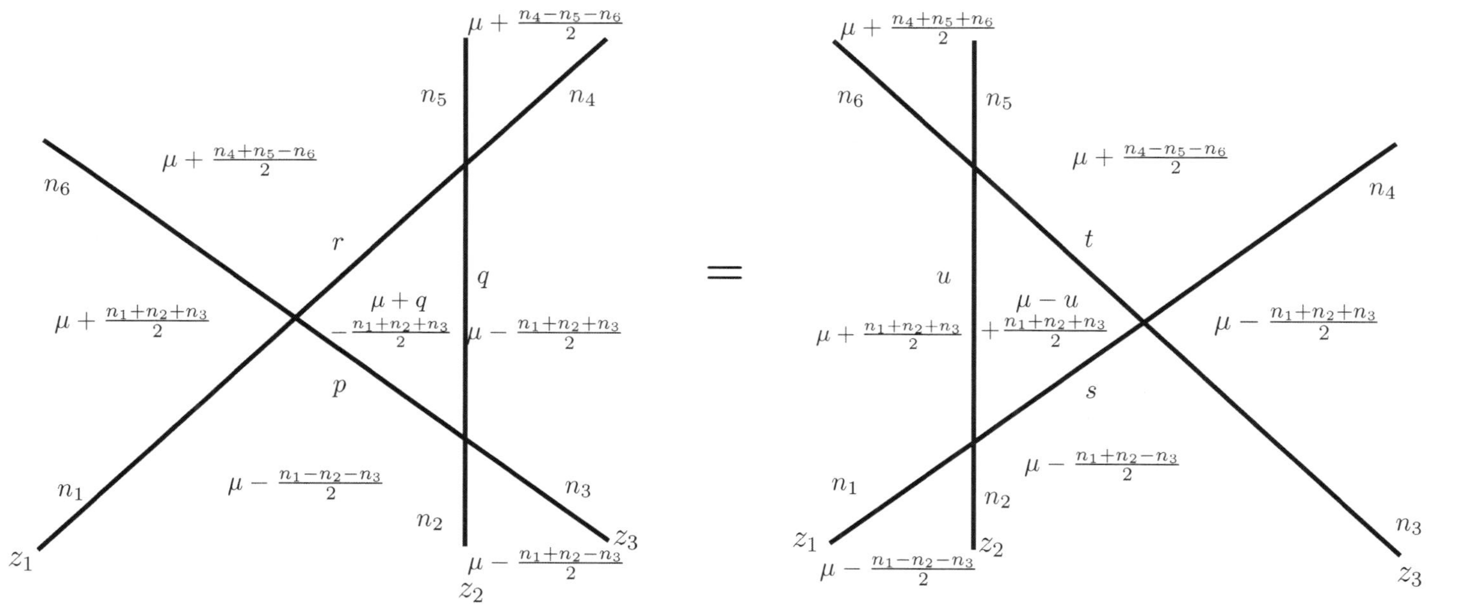

Figure 33: The dynamical Yang-Baxter equation. Both line segments and bulk regions are labeled. A line segment is labeled by a basis vector in a representation of G attached to that line, and a bulk region is labeled by a weight of G. The picture is drawn here for $G = SU(2)$, so that a basis vector is determined by its integer weight. Two-dimensional BF theory has all the ingredients to draw such a picture (without spectral parameters) but there is no reason to expect this equivalence to hold. When we go to four dimensions, the lines are labeled by spectral parameters in the familiar way (as shown here) and the usual arguments show an equivalence between the two pictures.

Now we consider the case of two-dimensional BF theory with a nonabelian gauge group. First let us take the gauge group to be $SU(2)$. As long as the field b is nonzero, the gauge symmetry is spontaneously broken from $SU(2)$ to $U(1)$. As a result, apart from possible subtleties when b vanishes, we can aim for a description of BF theory of $SU(2)$ in terms of BF theory of $U(1)$. It is actually possible (see section 2.3 of [39]) to compute rather precisely the effective $U(1)$ action that arises from integrating out the charged modes.[46] We omit this, as the results are not very essential for us, as explained in section 11.1. (However, we describe some aspects of a precise quantum treatment in section 11.3.) What is important for us is that a line operator of the $SU(2)$ theory can be written as a sum of line operators of the effective $U(1)$ theory. For Wilson line operators, this decomposition is fairly evident. A Wilson operator associated to an irreducible representation ρ of $SU(2)$ decomposes in the $U(1)$ description as a sum of line operators corresponding to the weight spaces of ρ. Let us denote a Wilson operator for the representation ρ of $SU(2)$ as W_ρ and one for the charge n representation of $U(1)$ as W'_n. Then, for example, if ρ is the two-dimensional representation of $SU(2)$, it decomposes under $U(1)$ as the sum of one-dimensional weight spaces of charges 1 and -1, so the corresponding formula for Wilson operators is $W_\rho = W'_1 + W'_{-1}$.

Now we can reconsider Fig. 32(b), with a slightly different interpretation. We consider the crossing of two Wilson operators of $SU(2)$ in some representations ρ and ρ'. On either of the two Wilson operators, on either side of the crossing, we can replace W_ρ or $W_{\rho'}$ by a sum of $U(1)$ Wilson operators, labeled by the weight spaces in ρ or ρ'. Thus the integers $n_1, \dots, n_4$ in Fig. 32(b) are now not arbitrary but denote weights of ρ or ρ'. In passing through the point at which the two Wilson operators cross, the charges obey $n_1 + n_2 = n_3 + n_4$ as before, because the effective description has $U(1)$ symmetry.

The case of a compact nonabelian gauge group G of any rank r is similar. Away from special values of b, the G gauge symmetry is broken to the maximal torus T. The charges of an irreducible representation of T are now an r-plet of integers $\vec{n}$. Likewise $b' = \kappa b$ becomes an r-plet $\vec{b}'$. Its jumping in crossing a Wilson operator is $\vec{b}' \to \vec{b}' + \vec{n}$, just as before. One detail is slightly different. For $r > 1$, the weight spaces of an irreducible representation may have multiplicity greater than 1. (For example, the adjoint representation of $SU(3)$ has six weight spaces each of multiplicity 1 and one of multiplicity 2.) Accordingly, when we write W_ρ as a sum of Wilson operators of an effective T gauge theory, some charges may appear more than once. So in general away from crossings, an effective Wilson operator of the low energy theory carries some additional labels in addition to its "charges" $\vec{n}$.

Finally we come to the main point. In actually calculating a path integral appropriate to the situation of Fig. 32(b), we would run into a factor associated to the crossing. This factor would depend on the charges that label the incoming and outgoing lines in the effective abelian theory and on the parameter μ that determines the labels of the bulk regions in the figure. We can denote this factor as a generalized R-matrix element $R_{n_1 n_2}^{n_3 n_4}(\mu)$. (For simplicity, in writing the R-

[46]The relevant procedure also has an analog in three-dimensional Chern-Simons theory on a Seifert manifold [40].

matrix in detail as a matrix, we take $G = SU(2)$ so that each line is simply labeled by an integer.)

Having defined the generalized R-matrix of two-dimensional BF theory, one can ask if it obeys a generalized Yang-Baxter equation, now with bulk labels as well as labels for line segments between crossings (Fig. 33). However, there is no reason to expect this. In a purely two-dimensional theory, one has to pass through a singularity to interpolate between the left and right hand sides of the figure. As far as we know, the generalized Yang-Baxter equation is not satisfied in this situation.

Hopefully the reader can anticipate what comes next. We consider not a purely two-dimensional BF theory, but an effective abelian BF theory in two dimensions that arises as in the discussion of eqn. (11.2) from a four-dimensional theory on $\mathbb{R}^2 \times E$ with complex gauge group G and with $\zeta = 0$. A Wilson operator is now labeled by a representation ρ of G (or more generally by a representation of a quantum deformation of $\mathfrak{g}[[z]]$) and also by a spectral parameter $z \in E$. Between crossings, a Wilson operator is further labeled by a basis vector of ρ. The bulk parameter $b' = \kappa b$ jumps in crossing a Wilson operator of the low energy theory. Now, we associate to a crossing a generalized R-matrix element $R^{n_3 n_4}_{n_1 n_2}(z_1 - z_2; \mu)$. The difference from before as that as long as $z_1 \neq z_2$, there is no singularity associated to a crossing. Therefore the two sides of Fig. 33 (with spectral parameters now included) are equivalent. The equivalence is known as the dynamical Yang-Baxter equation.

11.3 Quantum Treatment

We have treated the parameter b' classically, starting with purely two-dimensional BF theory. Actually, the locally constant value of b' has a natural meaning in the quantum theory.

Let us ask what are the quantum states when BF theory – to begin with for $G = U(1)$ – is quantized on a circle S. A physical state is a gauge-invariant function of the connection A restricted to S. Such a function is $\Psi_n = \exp(\mathrm{i}n \oint_S A)$ for some integer n. On the other hand, b' is the momentum conjugate to A, so (for p a point in S), $b'(p)$ can be identified with $-\mathrm{i}\delta/\delta A(p)$. Explicitly acting with this on Ψ_n, we find that $b'(p)\Psi_n = n\Psi_n$, for any p, independent of p.

Quantum mechanically, the constant value of $b'(p)$ should be interpreted as an eigenvalue of this operator, and (for $U(1)$) the eigenvalues are integers, as we have just seen. We can also understand in this language why crossing a Wilson line operator has the effect of shifting b' by an integer. For example, we can regard the Wilson line operator $W_m = \exp(\mathrm{i}m \oint_S A)$ as an operator that acts on physical states on S. Since $W_m \Psi_n = \Psi_{n+m}$, acting with W_m shifts the value of b' by m.

Now let us repeat this analysis for a compact but possibly nonabelian gauge group G. Of course, the analog of Ψ_n is what we might call Ψ_ρ, the trace of the holonomy around S in the representation ρ:

$$\Psi_\rho = \mathrm{Tr}_\rho U, \quad U = P\exp\left(\oint_S A\right) . \tag{11.4}$$

However, we would prefer to express this in a language that is better suited for the reduction to an effective abelian description. For this, we observe that a gauge-invariant function of A is precisely a function of the holonomy U that is invariant under conjugation. In other words, gauge-invariant functions of A are functions on G/G, where G acts on itself by conjugation.

The quotient G/G is the same as $T/\mathcal{W}$, the quotient of the maximal torus T by the Weyl group $\mathcal{W}$. This might lead one to expect that gauge-invariants functions of A would correspond to Weyl-invariant functions on T, but actually they correspond in a natural way to Weyl *anti*-invariant functions. We say that a function on T is Weyl anti-invariant if it is odd under each of the elementary reflections that generate $\mathcal{W}$. The association of a representation ρ of G with a Weyl anti-invariant function on T is given by the Weyl character formula; the Weyl anti-invariant function corresponding to ρ is the numerator of the usual Weyl character formula for ρ. Let us just explain what this means for $SU(2)$. A maximal torus of $SU(2)$ is the $U(1)$ subgroup $\mathrm{diag}(e^{\mathrm{i}\theta}, e^{-\mathrm{i}\theta})$. The character of the n-dimensional representation ρ_n of $SU(2)$ is

$$e^{\mathrm{i}(n-1)\theta} + e^{\mathrm{i}(n-3)\theta} + \cdots + e^{-\mathrm{i}(n-1)\theta} = \frac{\sin(n\theta)}{\sin\theta} . \tag{11.5}$$

The Weyl anti-invariant function corresponding to ρ_n is the numerator, or

$$\sin(n\theta) = \frac{1}{2\mathrm{i}}\left(e^{\mathrm{i}n\theta} - e^{-\mathrm{i}n\theta}\right) . \tag{11.6}$$

The functions $\sin(n\theta)$, $n = 1, 2, 3, \ldots$ are a basis for the Hilbert space of Weyl anti-invariant functions on T. This is the Hilbert space of BF theory of $SU(2)$, quantized on a circle. In this description, we can conveniently see the effective $U(1)$ description of $SU(2)$ BF theory. In this effective $U(1)$ description, the holonomy is $e^{\mathrm{i}\theta}$ and b' is the canonical momentum $b' = -\mathrm{i}\partial/\partial\theta$. We see that $\sin(n\theta)$ is not an eigenstate of b' but rather is a linear combination of eigenstates $e^{\mathrm{i}n\theta}$ and $e^{-\mathrm{i}n\theta}$ with eigenvalues $\pm n$. These values are Weyl conjugate, since the Weyl group of $SU(2)$ is $\mathbb{Z}_2$, acting as -1 on the Lie algebra $\mathfrak{t}$ of T. The implication is that b' should not really be regarded as an integer (or a real number, as in the classical description) but as a Weyl orbit of nonzero integers. It is fairly natural to pick from each Weyl orbit $n, -n$ the positive representative, and if we do this then the values of b' in the effective abelian description are positive integers. For any compact simple G, the analog is that the values of b' are dominant weights in the interior of a positive Weyl chamber.

Now let us see what happens to the value of b' in crossing a Wilson operator, say the operator W_m associated to the m-dimensional representation ρ_m. The character of ρ_m is $F_m(\theta) = e^{\mathrm{i}(m-1)\theta} + \cdots + e^{-\mathrm{i}(m-1)\theta}$. Crossing W_m has the effect of multiplying the quantum state by $F_m(\theta)$. We have

$$F_m(\theta)\sin(n\theta) = \sum_{j=-m+1,-m+3,\ldots,m-1} \sin((n+j)\theta) , \tag{11.7}$$

and if $m \leq n$, then $n + j$ is always positive. This means that in crossing W_m, b' can jump by j for any $j = -m+1, -m+3, \ldots, m-1$, that is, any weight of the representation ρ_m. This is the result that was claimed in section 11.2. For $m > n$, it is possible for $n + j$ to be nonpositive, and some terms on the right hand side of eqn. (11.7) vanish or cancel. This leads to some modification of the formalism when b' is not large.

These subtleties do not really affect the discussion in section 11.2 very much. In that discussion, b' was treated as a generic real number, but it would not have been much different to regard b' as a generic positive integer, where here "generic" is equivalent to "sufficiently large." Thus, in studying any concrete collection of Wilson operators, associated with representations of dimensions $m_1, \ldots, m_s$, the reasoning in section 11.2 is valid if b' is sufficiently large compared to those dimensions.

This quantum treatment of purely two-dimensional BF theory, however, highlights what is missing in our understanding of the four-dimensional theory. We certainly do not have available a quantum treatment that would identify definite allowed values of b' in the theory obtained by compactification from four dimensions. In fact, since the four-dimensional theory is a somewhat formal construction with a holomorphic action, it is not clear to what extent one should expect to have such a quantum treatment.

Neither – at least at first sight – do we wish to integrate over the possible values of b', as a natural cycle for such an integral does not present itself. In section 11.1, we already mentioned several possible ways to deal with this issue. Here we just elaborate on one possibility. In purely two-dimensional BF theory on a two-manifold Σ, it is fairly natural to pick a point $p \in \Sigma$, not in the support of any Wilson operator, and specify the value of b' there. The values of b' elsewhere would then be determined by the crossing rules of section 11.2. It is tempting to believe that a similar constraint should be imposed in the four-dimensional theory, to spare ourselves from having to sum or integrate over b'. Moreover, a constraint setting b' to a generic value at some particular point p, when supplemented with the crossing rules, will ensure that b' never takes a value at which the automorphism group is enhanced and the effective abelian description breaks down. However, we have not seriously tried to study a quantum field theory operator that would impose this constraint.

Acknowledgments

The authors would like to thank Davide Gaiotto, Robbert Dijkgraaf, Michio Jimbo, Cumrun Vafa, and especially Jacques H. H. Perk for discussions. K. C. is supported by the NSERC Discovery Grant program and by the Perimeter Institute for Theoretical Physics. Research at Perimeter Institute is supported by the Government of Canada through Industry Canada and by the Province of Ontario through the Ministry of Research and Innovation. E. W. is partially supported by National Science Foundation grant NSF Grant PHY-1606531. M. Y. is partially supported by WPI program (MEXT, Japan), by JSPS Program for Advancing Strategic International Networks to Accelerate the Circulation of Talented Researchers, by JSPS KAKENHI Grant No. 15K17634, and by JSPS-NRF Joint Research Project.

A Rational R-matrix for SO_N

In this Appendix we discuss the rational R-matrix for the fundamental representation of $G = SO_N$, extending the similar discussion for fundamental and anti-fundamental representations for $G = GL_N$ (or SL_N) in section 3.5. While the result in itself is known since the old work of [6], in the framework of this paper it is an illuminating exercise to check the consistency between unitarity, crossing and the framing anomaly.

Let us consider the R-matrix $R(z_1 - z_2) : V \otimes V \to V \otimes V$ for the fundamental representation V of $G = SO_N$. Since this representation is equivalent to its own conjugate, a Wilson line in this representation carries no orientation.

As we discussed in the main text, the rational R-matrix has G as a symmetry. Consequently, the R-matrix is constrained to be of the form

$$R_{ij}^{i'j'}(z) = \delta_i^{i'} \delta_j^{j'} E(z) + \delta_i^{j'} \delta_j^{i'} F(z) + \delta_{ij} \delta^{i'j'} G(z) \, , \tag{A.1}$$

with three unknown functions $E(z), F(z)$ and $G(z)$. In this literature this is also written as

$$R_{ij}^{i'j'}(z) = E(z) I + F(z) P + G(z) Q \, , \tag{A.2}$$

where P is a permutation, and $Q = {}^t P$ its transposition. Compared to the case of GL_N in eqn. (3.22), we here have one extra structure $\delta_{ij}\delta^{i'j'}$ consistent with G-symmetry.

The Yang-Baxter equation constrains only the ratios of the three functions: $X(z) = F(z)/E(z), Y(z) = G(z)/E(z)$. As we can see from Fig. 34, the resulting constraints takes precisely the same form as in (3.23) and (3.26), where the role of $U(z)$ and $W(z)$ are played by $X(z)$ and $Y(z)$ respectively:

$$\begin{aligned} X(z_1 - z_3) X(z_2 - z_3) + X(z_1 - z_2) X(z_1 - z_3) &= X(z_1 - z_2) X(z_2 - z_3) \, , \\ Y(z_1 - z_3) X(z_2 - z_3) = Y(z_1 - z_2) Y(z_1 - z_3) &+ Y(z_1 - z_2) X(z_2 - z_3) \, . \end{aligned} \tag{A.3}$$

By the similar reasoning as before we obtain

$$X(z) = \frac{\hbar}{z} \, , \quad Y(z) = -\frac{\hbar}{z - c\hbar} \, , \tag{A.4}$$

where c is some undetermined constant.

The undetermined constant c is constrained by unitarity (recall eqn. (2.3)), which gives the relation (see Fig. 34; note the difference from eqn. (3.28)):

$$N Y(z) Y(-z) + Y(z) + Y(-z) + X(z) Y(-z) + X(-z) Y(z) = 0. \tag{A.5}$$

This leads to

$$X(z) = \frac{\hbar}{z} \, , \quad Y(z) = -\frac{\hbar}{z + \frac{N-2}{2}\hbar} \, . \tag{A.6}$$

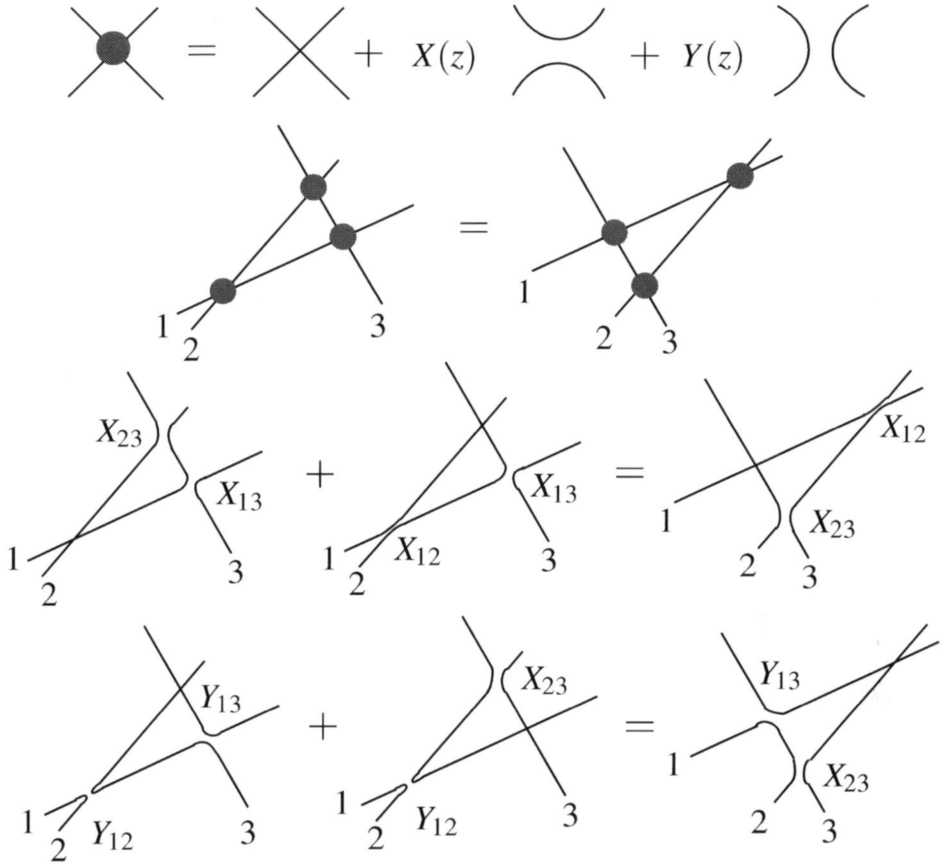

Figure 34: Graphical representation for the Yang-Baxter equation for the SO_N R-matrix given in eqn. (A.1). Here we used the shorthand notation $X_{ij} = X(z_i - z_j)$ and $Y_{ij} = Y(z_i - z_j)$, with z_i associated with the i-th line. The two equations here give eqns. (A.3).

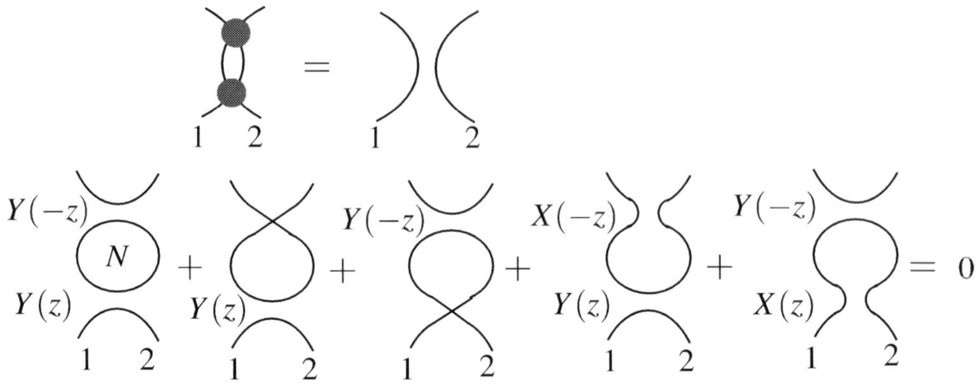

Figure 35: Graphical representation for the unitarity relation for the SO_N R-matrix given in eqn. (A.1), with $z = z_1 - z_2$. The factor of N comes from the color factor for the bubble in the first term of the equation.

These two functions satisfy

$$X\left(z - \hbar\frac{N-2}{2}\right) = Y(-z) , \tag{A.7}$$

which is consistent with the crossing and framing anomaly, since $N-2$ is the dual Coxeter number of SO_N.

We can also try to determine the overall factor $E(z)$. The diagonal component of the unitarity relation gives

$$E(z)E(-z) + F(z)F(-z) = E(z)E(-z)(1 + X(z)X(-z)) = 1 , \tag{A.8}$$

or equivalently

$$E(z)E(-z) = \frac{z^2}{z^2 - \hbar^2} . \tag{A.9}$$

This solution can easily be solved, for example, by

$$E(z) = \frac{z}{z \pm \hbar} . \tag{A.10}$$

However, this is actually not the solution we want, since we also want to impose the condition that this factor does not spoil the unitarity relation:

$$E\left(z - \hbar\frac{N-2}{2}\right) = E(-z) , \quad \text{or} \quad E\left(-\hbar\frac{N-2}{2} - z\right) = E(z) . \tag{A.11}$$

The minimal solutions for (A.9) and (A.11) are

$$E(z) = \frac{Q_{\pm\hbar, -\hbar(N-2)/2}(z)}{Q_{0, -\hbar(N-2)/2}(z)} , \tag{A.12}$$

where the function $Q_{\alpha,\beta}(z)$, satisfying the functional identities

$$Q_{\alpha,\beta}(z)Q_{\alpha,\beta}(-z) = \frac{1}{\frac{(z+\alpha)}{2\beta}\frac{(-z+\alpha)}{2\beta}}\,, \quad Q_{\alpha,\beta}(z) = Q_{\alpha,\beta}(\beta - z)\,, \tag{A.13}$$

is written as

$$Q_{\alpha,\beta}(z) = \frac{\Gamma\left(\frac{z+\alpha}{2\beta}\right)\Gamma\left(\frac{-z+\alpha+\beta}{2\beta}\right)}{\Gamma\left(\frac{z+\alpha+\beta}{2\beta}\right)\Gamma\left(\frac{-z+\alpha+2\beta}{2\beta}\right)}\,. \tag{A.14}$$

Note that both expressions in (A.12), with either plus or minus sign, are minimal. In the literature, one is associated with the S-matrix for the $O(N)$ sigma model, while another to that for the Gross-Neveu model [6].

To study the behavior at $\hbar \sim 0$, let us first note that

$$Q_{\gamma\delta\hbar,\delta\hbar}(z) = \frac{\Gamma\left(\frac{z}{2\hbar\delta} + \frac{1}{2}\right)\Gamma\left(-\frac{z}{2\hbar\delta} + \frac{1+\gamma}{2}\right)}{\Gamma\left(\frac{z}{2\hbar\delta} + \frac{1+\gamma}{2}\right)\Gamma\left(-\frac{z}{2\hbar\delta} + \frac{2+\gamma}{2}\right)}\,. \tag{A.15}$$

From the Stirling's formula

$$\log\Gamma(z+c) \sim z\log z - z + c\log z - \frac{1}{2}\log z + \frac{1}{2}\log 2\pi + \frac{c^2 - c}{z} + \mathcal{O}\left(\frac{1}{z^2}\right) \tag{A.16}$$

(c being constant), we obtain

$$Q_{\gamma\delta\hbar,\delta\hbar}(z) \sim e^{-\log\frac{z}{2\hbar\delta} - \hbar\frac{(\gamma^2-2\gamma-1)\delta}{2z} + \mathcal{O}\left(\left(\frac{\hbar}{z}\right)^2\right)}\,, \tag{A.17}$$

and hence we learn that the overall factor has a perturbative expansion in powers of $\hbar/z$, starting with identity:

$$E(z) = \frac{Q_{\pm\hbar,-\hbar(N-2)/2}(z)}{Q_{0,-\hbar(N-2)/2}(z)} = \begin{cases} 1 - \frac{N^2-8}{4(N-2)}\frac{\hbar}{z} + \mathcal{O}\left(\left(\frac{\hbar}{z}\right)^2\right) & \text{(plus sign)}\,, \\ 1 - \frac{N^2-8N+8}{4(N-2)}\frac{\hbar}{z} + \mathcal{O}\left(\left(\frac{\hbar}{z}\right)^2\right) & \text{(minus sign)}\,. \end{cases} \tag{A.18}$$

In perturbation theory, these formulas appear satisfactory, but nonperturbatively, one would like to understand the meaning of the poles of the function $E(z)$. As remarked in the introduction, the D4-NS5 system is likely to provide a nonperturbative framework in which such questions could be addressed, but we will not pursue that in the present paper.

We can repeat a similar exercise for the gauge group $G = Sp_{2N}$. We again find that the constraint from the Yang-Baxter equation and unitarity are consistent with crossing symmetry and the framing anomaly, with the dual Coxeter number given by $N + 1$.

B Examples of Trigonometric and Elliptic R-matrix

In section (3.5) and appendix A, we discussed several examples of rational R-matrices. In this appendix, we present examples of trigonometric and elliptic R-matrices known in the literature, for the simplest case of the fundamental representation of $G = SL_2$. We verify that their classical limit reproduces the classical r-matrices discussed in the sections 9.4 and 10.3. The R-matrices below are quasi-classical, and are normalized to be $R_{\hbar=0} = I$, which is the canonical normalization in gauge theory. If expanded in perturbation theory, they lead to r-matrices that we computed in sections 9 and 10.

As discussed in appendix A the complete understanding of the overall factor of the R-matrix goes beyond the perturbative analysis of this paper. Expressions below should be understood modulo an overall scalar factor ambiguities for the R-matrix, and hence modulo the shift by identity matrices for the classical r-matrix.

2.1 Rational

As discussed in the main text, there are three types of quasi-classical R-matrix known in the literature: rational, trigonometric and elliptic. Let us begin with the rational R-matrix, which we already discussed in section 3.5:

$$R_\hbar^{\text{rational}}(z) = \left(z + \frac{\hbar}{2}\right) I + \hbar\, c = zI + \hbar P\,, \tag{B.1}$$

where $z \in \mathbb{C}$, $c = (\vec{\sigma} \cdot \vec{\sigma})/2$ is the quadratic Casimir (4.12) for $\mathfrak{g} = \mathfrak{sl}_2$, and $P = c + I/2$ is the permutation operator of the two spins.

This R-matrix has manifest SL_2 symmetry. Choosing a basis, the R-matrix reads

$$R_\hbar^{\text{rational}}(z) = \begin{array}{c} \\ |\uparrow\uparrow\rangle \\ |\uparrow\downarrow\rangle \\ |\downarrow\uparrow\rangle \\ |\downarrow\downarrow\rangle \end{array} \begin{array}{c} \begin{array}{cccc} |\uparrow\uparrow\rangle & |\uparrow\downarrow\rangle & |\downarrow\uparrow\rangle & |\downarrow\downarrow\rangle \end{array} \\ \begin{pmatrix} z+\hbar & & & \\ & z & \hbar & \\ & \hbar & z & \\ & & & z+\hbar \end{pmatrix} \end{array} . \tag{B.2}$$

Note that up to normalization only the ratio of z and $\hbar$ enters into this solution, as expected on dimensional grounds.

2.2 Trigonometric Solution

The trigonometric solution is

$$R_\hbar^{\text{trig}}(z) = \begin{pmatrix} a(z,\hbar) & & & \\ & b(z,\hbar) & c(z,\hbar) & \\ & c(z,\hbar) & b(z,\hbar) & \\ & & & a(z,\hbar) \end{pmatrix}, \tag{B.3}$$

where

$$\begin{aligned} a(z,\hbar) &= \sinh(w+\hbar) \ , \\ b(z,\hbar) &= \sinh(w) \ , \\ c(z,\hbar) &= \sinh(\hbar) \ , \end{aligned} \tag{B.4}$$

with $z = e^w \in \mathbb{C}^\times$ and $q := e^\hbar$.

In terms of the quasi-classical parameter $\hbar$ and the cylindrical coordinate w, this R-matrix is written in terms of sine functions, as the name "trigonometric" suggests. When we impose unitarity, this R-matrix should be multiplied by a suitable overall normalization factor.

The six-vertex model allows for an integrable deformation, corresponding to an inclusion of the "electric field" h, v, in vertical and horizontal directions:

$$R^{\rm trig}_{\hbar,\theta}(z) = \begin{pmatrix} a(z,\hbar)e^{h+v} & & & \\ & b(z,\hbar)e^{-h+v} & c(z,\hbar) & \\ & c(z,\hbar) & b(z,\hbar)e^{h-v} & \\ & & & a(z,\hbar)e^{-h-v} \end{pmatrix} . \tag{B.5}$$

The parameters h, v correspond to s_1, s_2 in section 9.4: $h_1 = \mathrm{i}s_1, v = -\mathrm{i}s_2$.

When this R-matrix is used to construct a model in lattice statistical mechanics in the usual way, its six non-zero matrix elements become the Boltzmann weights for six allowed configurations at a vertex. The model accordingly is generally called the six-vertex model. Among the six matrix elements of the R-matrix, two combinations are inessential; one parameter is the overall normalization of the R-matrix and one can be removed by conjugation. This leaves four parameters, which we have here denoted z, $\hbar$ and h, v. The one-dimensional quantum spin chain associated with this model is the XXZ spin chain.

2.3 Elliptic Solution

The elliptic R-matrix has a new parameter, namely the modulus of the torus τ:

$$R^{\rm elliptic}_{\hbar,\tau}(z) = e^{i\pi\hbar F(z,\tau)} \begin{pmatrix} a(z,\tau,\hbar) & & & d(z,\tau,\hbar) \\ & b(z,\tau,\hbar) & c(z,\tau,\hbar) & \\ & c(z,\tau,\hbar) & b(z,\tau,\hbar) & \\ d(z,\tau,\hbar) & & & a(z,\tau,\hbar) \end{pmatrix} , \tag{B.6}$$

where we defined

$$\begin{aligned} a(z,\tau,\hbar) &= \frac{\vartheta_0(\pi\hbar|2\tau)}{\vartheta_0(0|2\tau)} \frac{\vartheta_1(\pi(z+\hbar)|2\tau)}{\vartheta_1(\pi z|2\tau)} , \\ b(z,\tau,\hbar) &= \frac{\vartheta_0(\pi\hbar|2\tau)}{\vartheta_0(0|2\tau)} \frac{\vartheta_0(\pi(z+\hbar)|2\tau)}{\vartheta_0(\pi z|2\tau)} , \\ c(z,\tau,\hbar) &= \frac{\vartheta_1(\pi\hbar|2\tau)}{\vartheta_0(0|2\tau)} \frac{\vartheta_0(\pi(z+\hbar)|2\tau)}{\vartheta_1(\pi z|2\tau)} , \\ d(z,\tau,\hbar) &= \frac{\vartheta_1(\pi\hbar|2\tau)}{\vartheta_0(0|2\tau)} \frac{\vartheta_1(\pi(z+\hbar)|2\tau)}{\vartheta_0(\pi z|2\tau)} . \end{aligned} \tag{B.7}$$

Here $\vartheta_0(z|\tau)$ and $\vartheta_1(z|\tau)$ are Jacobi theta functions

$$\begin{aligned}\vartheta_0(z|\tau) &= \prod_{n=1}^{\infty}(1-2e^{(2n-1)\pi i\tau}\cos 2z + e^{(4n-2)\pi i\tau})(1-e^{2n\pi i\tau})\,,\\ \vartheta_1(z|\tau) &= 2e^{\pi i\tau/2}\sin u\prod_{n=1}^{\infty}(1-2e^{2n\pi i\tau}\cos 2z + e^{4n\pi i\tau})(1-e^{2n\pi i\tau})\,,\end{aligned} \tag{B.8}$$

and $F(z,\tau)$ is a normalization factor on which we will comment later. Note also that, for a better match with the main text, we have used the Jacobi theta function with elliptic modulus 2τ (not τ); care is need for comparison with the literature.

This R-matrix has eight non-zero matrix elements and represents the eight-vertex model solved by Baxter [5]. The associated one-dimensional quantum spin chain is the XYZ spin chain.

The periodicity of the Jacobi theta functions are given by

$$\begin{aligned}&\vartheta_1(\pi(z+1)|2\tau) = -\vartheta_1(\pi z|2\tau)\,, \quad \vartheta_1\left(\pi\left(z+\tau\right)|2\tau\right) = ie^{-\frac{i\pi\tau}{2}}e^{-i\pi z}\vartheta_0(\pi z|2\tau)\,,\\ &\vartheta_0(\pi(z+1)|2\tau) = \vartheta_0(\pi z|2\tau)\,, \quad \vartheta_0\left(\pi\left(z+\tau\right)|2\tau\right) = ie^{-\frac{i\pi\tau}{2}}e^{-i\pi z}\vartheta_1(\pi z|2\tau).\end{aligned} \tag{B.9}$$

This means if we choose the normalization factor $F(z,\hbar)$ to be a function on the two-torus satisfying

$$F(z+1,\tau) = F(z,\tau)\,, \quad F(z+\tau,\tau) = F(z,\tau)+1\,, \tag{B.10}$$

then all the entries of the R-matrix are periodic under the shift $z \to z+1, z \to z+\tau$, up to a possible overall sign factor; for such F we can for example choose

$$F(z,\tau) = \frac{i}{\pi}\log\vartheta_0\left(\pi\left(z-\frac{\tau-1}{2}\right)|2\tau\right)\,. \tag{B.11}$$

Note also that the normalization of the R-matrix here is chosen so that $a(z,\tau) = b(z,\tau) = 1$ when $\hbar = 0$.

The classical r-matrix is then (note $\vartheta_1(0) = 0$)

$$r_{\tau}^{\text{elliptic}}(z,\tau) = \sum_{k=1}^{3} w_k(z)\,\sigma_k\otimes\sigma_k \tag{B.12}$$

with

$$\begin{aligned}w_1(z,\tau) &= \pi\frac{\vartheta_1'(0|2\tau)}{\vartheta_0(0|2\tau)}\left[\frac{\vartheta_0(\pi z|2\tau)}{\vartheta_1(\pi z|2\tau)}+\frac{\vartheta_1(\pi z|2\tau)}{\vartheta_0(\pi z|2\tau)}\right]\,,\\ w_2(z,\tau) &= \pi\frac{\vartheta_1'(0|2\tau)}{\vartheta_0(0|2\tau)}\left[\frac{\vartheta_0(\pi z|2\tau)}{\vartheta_1(\pi z|2\tau)}-\frac{\vartheta_1(\pi z|2\tau)}{\vartheta_0(\pi z|2\tau)}\right]\,,\\ w_3(z,\tau) &= \partial_z\log\frac{\vartheta_1(\pi z|2\tau)}{\vartheta_0(\pi z|2\tau)}\,.\end{aligned} \tag{B.13}$$

From (B.9) we find that[47]

$$\begin{aligned} w_1(z+1) &= -w_1(z)\,, \quad w_1(z+\tau) = w_1(z)\,, \\ w_2(z+1) &= -w_2(z)\,, \quad w_2(z+\tau) = -w_2(z)\,, \\ w_3(z+1) &= w_3(z)\,, \quad w_3(z+\tau) = -w_3(z)\,. \end{aligned} \tag{B.15}$$

This is consistent with eqn. (10.7).

2.4 Relations

Among the three solutions for $G = SL_2$, the elliptic solution is the most general, and the other two can be obtained by limits (together with suitable overall rescaling):

$$R^{\text{elliptic}}_{\hbar,\tau}(z) \xrightarrow{\tau\to i\infty} R^{\text{trig}}_{\hbar}(z) \xrightarrow{\hbar\to 0} R^{\text{rational}}(z/\hbar)\,. \tag{B.16}$$

However, the trigonometric R-matrix has a generalization with a non-zero "field" turned on; this generalization cannot be obtained from a reduction of the eight-vertex model. In section 10.3, we explained this fact from an analysis of the four-dimensional gauge theory on $\mathbb{R}^2 \times \mathbb{C}^\vee$.

C Anomalies to Quantizing Wilson lines

In this section we will perform the cohomology calculation referenced in section 8.

The main result we will show is the following.

Theorem C.1. *Let $\mathfrak{g}$ be a simple Lie algebra with no Abelian factors which is not $\mathfrak{sl}_2$. Let $\wedge^2_0\mathfrak{g}$ denote the kernel of the Lie bracket map from $\wedge^2\mathfrak{g} \to \mathfrak{g}$. Let V be a representation of $\mathfrak{g}$. Then*

1. *If there are no G-invariant maps from $\wedge^2_0\mathfrak{g} \to \mathrm{End}(V)$ then*

$$H^2(\mathfrak{g}[[z]], \mathrm{End}(V)) = 0\,. \tag{C.1}$$

2. *Let $H^2_{(k)}(\mathfrak{g}[[z]], \mathrm{End}(V))$ denote the part of the cohomology of weight k under the scaling of z (this group appears in the study of k-loop anomalies). Then, there is an isomorphism*

$$H^2_{(2)}(\mathfrak{g}[[z]], \mathrm{End}(V)) \cong \mathrm{Hom}_G(\wedge^2_0\mathfrak{g}, \mathrm{End}(V))\,, \tag{C.2}$$

where on the right hand side we have the space of G-invariant linear operators from $\wedge^2_0\mathfrak{g}$ to $\mathrm{End}(V)$.

---

[47] In particular, we have from (B.9)

$$\frac{\vartheta_1(\pi(z+1)|2\tau)}{\vartheta_0(\pi(z+1)|2\tau)} = -\frac{\vartheta_1(\pi z|2\tau)}{\vartheta_0(\pi z|2\tau)}, \quad \frac{\vartheta_1(\pi(z+\tau|2\tau))}{\vartheta_0(\pi(z+\tau|2\tau))} = \frac{\vartheta_0(\pi(z|2\tau))}{\vartheta_1(\pi(z|2\tau))}. \tag{B.14}$$

If $\mathfrak{g} = \mathfrak{sl}_2$*, then* $H^2_{(k)}(\mathfrak{g}[[z]], \mathrm{End}(V)) = 0$ *unless* $k = 3$*, and the dimension of* $H^2_{(3)}(\mathfrak{g}[[z]], \mathrm{End}(V))$ *is the number of copies of the* 5*-dimensional irreducible representation of* $\mathfrak{sl}_2$ *in* $\mathrm{End}(V)$*.*

The proof will take a number of steps, which we will write as separate propositions. The first step is the following.

Proposition C.2. *Let* $\mathfrak{g}$ *be a simple Lie algebra with no Abelian factors. Let* V *be a representation of* $\mathfrak{g}$ *with the feature that the image of any* G*-invariant linear map* $\mathfrak{g}^{\otimes 2} \to \mathrm{End}(V)$ *consists of copies of the trivial and adjoint representations of* $\mathfrak{g}$*. Then,*

$$H^2(\mathfrak{g}[[z]], \mathrm{End}(V)) = 0 \ . \tag{C.3}$$

The vanishing of this cohomology group implies that there are no anomalies to quantizing the Wilson line in the representation V.

Proof. In this calculation, we will use the Lie algebra $\mathfrak{g}[z]$ of polynomials with coefficients in $\mathfrak{g}$, instead of the Lie algebra $\mathfrak{g}[[z]]$ of series with coefficients in $\mathfrak{g}$. As long as we focus on cohomology classes of a fixed weight under the $\mathbb{C}^\times$ action which scales z, the cohomology groups are the same in each case. This is because a cochain of weight k under this $\mathbb{C}^\times$ action involves z^m for $m \leq k$.

Let $R \subset \mathrm{End}(V)$ be the largest sub-representation which contains all the representations occurring in the decomposition of $\mathfrak{g}\otimes\mathfrak{g}$ into irreducible components. There is a natural injective map

$$C^*(\mathfrak{g}[z], R) \to C^*(\mathfrak{g}[z], \mathrm{End}(V)) \ . \tag{C.4}$$

It is a standard fact of Lie algebra cohomology that we find the same cohomology groups if we use G-invariant cocycles. It follows from the definition of R that, on G-invariant cocycles, this map is an isomorphism in degrees ≤ 2. From this we see that this map must be an isomorphism on degree 2.

This reduces the problem to showing that $H^2(\mathfrak{g}[z], R) = 0$. The assumptions we state in the proposition are that R is a direct sum of copies of the adjoint and trivial representations. We therefore need to show that $H^2(\mathfrak{g}[z], \mathbb{C}) = 0$ and that $H^2(\mathfrak{g}[z], \mathfrak{g}) = 0$.

Since we are working in low cohomological degrees, both of these statements can be proved by hand with a direct computation. However, it is more convenient to invoke a general result of Fishel, Grojnowski, and Teleman [31], who calculate the Lie algebra cohomology of $\mathfrak{g}[z]$ with various coefficients.

Their results (Theorem B, p. 6 of [31]) tell us that $H^2(\mathfrak{g}[z], \mathbb{C}) = 0$. We need to show that $H^2(\mathfrak{g}[z], \mathfrak{g}) = 0$ as well. This follows a little indirectly from their analysis, as we will now explain.

We write $\mathfrak{g}[z^{-1}]$ for the space of $\mathfrak{g}$-valued functions of z of the form

$$\alpha_{a_0} t_{a_0} + z^{-1} \alpha_{a_1} t_{a_1} + \dots .$$

These are meromorphic functions on $\mathbb{C}$, which have a finite order pole at 0 and no other poles, and which are regular at infinity. There is a natural residue pairing

$$\langle t_a f(z), t_b g(z^{-1}) \rangle = 2\delta_{ab} \frac{1}{2\pi i} \oint f(z) g(z^{-1}) z^{-1} \mathrm{d}z \tag{C.5}$$

between $\mathfrak{g}[z^{-1}]$ and $\mathfrak{g}[z]$.

The space $\mathfrak{g}[z^{-1}]$ has the structure of module for $\mathfrak{g}[z]$ whereby

$$(t_a z^k) \cdot (t_b z^{-l}) = f_{ab}^c t_c z^{k-l} \delta_{k-l \leq 0} \,. \tag{C.6}$$

This identifies $\mathfrak{g}[z^{-1}]$ with the dual representation to the adjoint representation of $\mathfrak{g}[z]$, that is, with the coadjoint representation. (Because these are infinite dimensional modules, one has to take care with the meaning of "dual": we mean the restricted dual, where we take the direct sum over the weight spaces of the $\mathbb{C}^\times$ action which scales z. This use of restricted duals and the corresponding restricted cohomology elides the difference between $\mathfrak{g}[z]$ and $\mathfrak{g}[[z]]$, so we will use the algebra $\mathfrak{g}[z]$ in the course of the proof).

Fishel, Grojnowski and Teleman show (Theorem B, p. 6 of [31]) that the cohomology of $\mathfrak{g}[z]$ with coefficients in the coadjoint representation $\mathfrak{g}[z^{-1}]$ is a graded tensor product of the cohomology of the compact group G with a graded vector space consists of $\mathbb{C}[z^{-1}]$ in degree 0 and $z^{-1}\mathbb{C}[z^{-1}]$ in degree 1. We have written these cohomology groups in a way compatible with the $\mathbb{C}^\times$ which scales z. In particular, since the cohomology of the group G lives in degree 0 and degrees ≥ 3, we find that there is no second cohomology of $\mathfrak{g}[z]$ with coefficients in $\mathfrak{g}[z^{-1}]$.

To describe the cocycles for these cohomology classes, it is more convenient to use cohomology relative to the subalgebra $\mathfrak{g} \subset \mathfrak{g}[z]$. A relative k-cocycle is given by a G-invariant linear map

$$\wedge^k (z\mathfrak{g}[z]) \to \mathfrak{g}[z^{-1}] \,. \tag{C.7}$$

The absolute cohomology groups are the tensor product of the relative cohomology groups with the cohomology of the group G. Therefore they coincide in degrees less than 3.

The relative 1-cocycle associated to an element $z^{-k} \in z^{-1}\mathbb{C}[z^{-1}]$ is a linear map

$$\begin{array}{ccc} z\mathfrak{g}[z] & \to & \mathfrak{g}[z^{-1}] \\ \Psi & & \Psi \\ t_a f(z) & \mapsto & t_a \pi_-\left(z^{-k} f(z)\right) \,, \end{array} \tag{C.8}$$

where π_- indicates the operation of projection of an element of $\mathfrak{g}[z, z^{-1}]$ onto the non-positive powers of z.

There is a short exact sequence of $\mathfrak{g}[z]$-representations

$$0 \to \mathfrak{g} \to \mathfrak{g}[z^{-1}] \xrightarrow{z} \mathfrak{g}[z^{-1}] \to 0 \,, \tag{C.9}$$

where the second map is given by multiplying by z, using the convention that we only retain negative powers of z.

This short exact sequence leads to a long exact sequence on cohomology

$$\cdots \to H^1(\mathfrak{g}[z], \mathfrak{g}) \to H^1(\mathfrak{g}[z], \mathfrak{g}[z^{-1}]) \to H^1(\mathfrak{g}[z], \mathfrak{g}[z^{-1}]) \to H^2(\mathfrak{g}[z], \mathfrak{g}) \to 0 \,, \tag{C.10}$$

where we use the fact that $H^2(\mathfrak{g}[z], \mathfrak{g}[z^{-1}]) = 0$.

This leads to an exact sequence

$$\cdots \to H^1(\mathfrak{g}[z], \mathfrak{g}) \to z^{-1}\mathbb{C}[z^{-1}] \xrightarrow{z} z^{-1}\mathbb{C}[z^{-1}] \to H^2(\mathfrak{g}[z], \mathfrak{g}) \to 0\,. \tag{C.11}$$

From this we conclude that $H^2(\mathfrak{g}[z], \mathfrak{g}) = 0$, as desired. □

For any representation R of $\mathfrak{g}$, let $H^i_{(k)}(\mathfrak{g}[z], R)$ denote the cohomology group of weight $-k$ under the $\mathbb{C}^\times$ action which scales z. This is the cohomology group that will play a role in quantization at k loops.

Next, we will prove a more difficult cohomology vanishing result, also based on the results of [31].

Proposition C.3. *For any simple Lie algebra with no Abelian factors which is not $\mathfrak{sl}_n$ with $n > 2$, we have*

$$H^2_{(k)}(\mathfrak{g}[z], \mathrm{Sym}^2\, \mathfrak{g}) = 0 \textit{ unless } k = 3\,. \tag{C.12}$$

Further, there is an exact sequence

$$0 \to H^2_{(3)}(\mathfrak{g}[z], \mathrm{Sym}^2\, \mathfrak{g}) \to \mathbb{C} \to H^2_{(2)}(\mathfrak{g}[z], \wedge^2 \mathfrak{g}) \to H^3_{(3)}(\mathfrak{g}[z], \mathrm{Sym}^2\, \mathfrak{g}) \to 0\,. \tag{C.13}$$

Note that this proposition applies to $\mathfrak{sl}_2$.

Proof. The proof again uses the results of Fishel, Grojnowski and Teleman [31]. Let us consider the $\mathfrak{g}[z]$-module $\mathrm{Sym}^2\, \mathfrak{g}[z^{-1}]$. The results of [31] show that

$$\begin{aligned} H^2(\mathfrak{g}[z], \mathrm{Sym}^2\, \mathfrak{g}[z^{-1}]) &= \wedge^2 z^{-1}\mathbb{C}[z^{-1}]\,, \\ H^1(\mathfrak{g}[z], \mathrm{Sym}^2\, \mathfrak{g}[z^{-1}]) &= (\mathrm{Sym}^3\, \mathfrak{g})^G \otimes z^{-1}\mathbb{C}[z^{-1}]\,, \\ H^0(\mathfrak{g}[z], \mathrm{Sym}^2\, \mathfrak{g}[z^{-1}] &= \mathbb{C}[z^{-1}]\,. \end{aligned} \tag{C.14}$$

We have written this isomorphism in a way compatible with the $\mathbb{C}^\times$-action which scales z. Note that, unless $\mathfrak{g}$ is $\mathfrak{sl}_n$ for $n > 2$, $(\mathrm{Sym}^3\, \mathfrak{g})^G = 0$. Since we restrict to Lie algebras which are not of this form, we have $H^1(\mathfrak{g}[z], \mathrm{Sym}^2\, \mathfrak{g}[z^{-1}]) = 0$.

Define operators

$$\begin{aligned} z^n &: \mathrm{Sym}^2\, \mathfrak{g}[z^{-1}] \mapsto \mathrm{Sym}^2\, \mathfrak{g}[z^{-1}] \\ z^n \left((z^{-k} t_a)(z^{-l} t_b) \right) &= (z^{n-k} t_a)(z^{-l} t_b)\delta_{k-n \geq 0} + (z^{-k} t_a)(z^{n-l} t_b)\delta_{l-n \geq 0}. \end{aligned} \tag{C.15}$$

for $n \geq 1$. Note that the operator z is surjective. We let M denote the kernel of z. The space M is spanned by the elements

$$\begin{aligned} &A_{ab}(t_a)(t_b z^{-2k}) - A_{ab}(t_a z^{-1})(t_b z^{1-2k}) \pm \ldots \\ &\quad + (-1)^{k+1} A_{ab}(t_a z^{1-k})(t_b z^{-k-1}) + \tfrac{1}{2}(-1)^k A_{ab}(t_a z^{-k})(t_b z^{-k})\,, \end{aligned} \tag{C.16}$$

where A_{ab} is a symmetric tensor, and the elements

$$\begin{aligned} &B_{ab}(t_a)(t_b z^{-2k-1}) - B_{ab}(t_a z^{-1})(t_b z^{-2k}) \pm \dots \\ &\quad + (-1)^{k+1} B_{ab}(t_a z^{1-k})(t_b z^{-k-2}) + (-1)^k B_{ab}(t_a z^{-k})(t_b z^{-k-1}) \,, \end{aligned} \tag{C.17}$$

where B_{ab} is antisymmetric.

Thus, as a representation of $\mathfrak{g}$, the module M decomposes as

$$M = \mathrm{Sym}^2\,\mathfrak{g} \oplus z^{-1} \wedge^2 \mathfrak{g} \oplus z^{-2}\,\mathrm{Sym}^2\,\mathfrak{g} \dots \,, \tag{C.18}$$

where the powers of z indicate the $\mathbb{C}^\times$-weights.

As a $\mathfrak{g}[z]$-module, the structure is more complicated: elements in $z\mathfrak{g}$ moves us between summands of different weight.

Now, let us identify the cohomology of $\mathfrak{g}[z]$ with coefficients in M. Because we have a short exact sequence

$$0 \to M \to \mathrm{Sym}^2(\mathfrak{g}[z^{-1}]) \xrightarrow{z} \mathrm{Sym}^2(\mathfrak{g}[z^{-1}]) \to 0 \,, \tag{C.19}$$

we get a long exact sequence in cohomology

$$\begin{aligned} 0 &\to H^0(\mathfrak{g}[z], M) \to \mathbb{C}[z^{-1}] \xrightarrow{z} \mathbb{C}[z^{-1}] \\ &\to H^1(\mathfrak{g}[z], M) \to 0 \to 0 \\ &\to H^2(\mathfrak{g}[z], M) \to \wedge^2(z^{-1}\mathbb{C}[z^{-1}]) \xrightarrow{z} \wedge^2(z^{-1}\mathbb{C}[z^{-1}]) \\ &\qquad \to H^3(\mathfrak{g}[z], M) \to 0 \,. \end{aligned} \tag{C.20}$$

The endomorphism z of $\wedge^2(z^{-1}\mathbb{C}[z^{-1}])$ is defined by

$$z(z^{-k} \wedge z^{-l}) = z^{1-k} \wedge z^{-l} \delta_{k-1 \geq 1} + z^{-k} \wedge z^{1-l} \delta_{l-1 \geq 1} \,. \tag{C.21}$$

The kernel of the operator z is spanned by the elements

$$z^{-1} \wedge z^{-2k} - z^{-2} \wedge z^{1-2k} \pm \dots + (-1)^k z^{-k} \wedge z^{-1-k} \tag{C.22}$$

for $k \geq 1$. Thus, $H^2(\mathfrak{g}[z], M)$ is spanned by these elements, in weights $3, 5, \dots$. To sum up, we find

$$\begin{aligned} H^0(\mathfrak{g}[z], M) &= \mathbb{C} \,, \\ H^1(\mathfrak{g}[z], M) &= 0 \,, \\ H^2(\mathfrak{g}[z], M) &= \mathbb{C} \cdot z^{-3} \oplus \mathbb{C} \cdot z^{-5} \oplus \dots \,, \\ H^3(\mathfrak{g}[z], M) &= 0 \,. \end{aligned} \tag{C.23}$$

where the powers of z indicate the weights under the $\mathbb{C}^\times$-action.

Let $M_0 \subset M$ be the kernel of the operator z^2. This operator is surjective on $H^2(M)$, with kernel $\mathbb{C}$ in weight -3, and surjective on M, but acts by zero on $H^0(M)$ and $H^1(M)$. We have a long exact sequence

$$\begin{aligned} 0 &\to H^0(\mathfrak{g}[z], M_0) \to \mathbb{C} \xrightarrow{0} \mathbb{C} \\ &\to H^1(\mathfrak{g}[z], M_0) \to 0 \to 0 \\ &\to H^2(\mathfrak{g}[z], M_0) \to z^{-3}\mathbb{C}[z^{-2}] \xrightarrow{z^2} z^{-3}\mathbb{C}[z^{-2}] \\ &\to H^3(\mathfrak{g}[z], M_0) \to 0 \,. \end{aligned} \tag{C.24}$$

This sequence allows us to calculate the cohomology groups of $H^*(\mathfrak{g}[z], M_0)$. Before we state the answer, we should note that the boundary maps in this long exact sequence shift the weight under the $\mathbb{C}^\times$-action by -2. This is because of the appearance of the operator z^2, of weight 2, in the short exact sequence of modules leading to this long exact sequence of cohomology groups.

From this exact sequence we see that

$$\begin{aligned} H^0(\mathfrak{g}[z], M_0) &= \mathbb{C} \,, \\ H^1(\mathfrak{g}[z], M_0) &= \mathbb{C} \cdot z^{-2} \,, \\ H^2(\mathfrak{g}[z], M_0) &= \mathbb{C} \cdot z^{-3} \,. \end{aligned} \tag{C.25}$$

Since M_0 is the intersection of the kernel of z and the kernel of z^2 in $\mathrm{Sym}^2 \,\mathfrak{g}[z^{-1}]$, one can show that there is an exact sequence

$$0 \to \mathrm{Sym}^2 \,\mathfrak{g} \to M_0 \to z^{-1} \wedge^2 \mathfrak{g} \to 0 \,, \tag{C.26}$$

from which we derive a long exact sequence

$$\begin{aligned} &0 \to H^0(\mathfrak{g}[z], \mathrm{Sym}^2 \,\mathfrak{g}) \to H^0(\mathfrak{g}[z], M_0) \to H^0(\mathfrak{g}[z], z^{-1} \wedge^2 \mathfrak{g}) \\ &\quad \to H^1(\mathfrak{g}[z], \mathrm{Sym}^2 \,\mathfrak{g}) \to H^1(\mathfrak{g}[z], M_0) \to H^1(\mathfrak{g}[z], z^{-1} \wedge^2 \mathfrak{g}) \\ &\quad \to H^2(\mathfrak{g}[z], \mathrm{Sym}^2 \,\mathfrak{g}) \to H^2(\mathfrak{g}[z], M_0) \to H^2(\mathfrak{g}[z], z^{-1} \wedge^2 \mathfrak{g}) \to H^3(\mathfrak{g}[z], \mathrm{Sym}^2 \,\mathfrak{g}) \to 0 \,. \end{aligned} \tag{C.27}$$

Now, $H^0(\mathfrak{g}[z], \wedge^2 \mathfrak{g}) = 0$ because the are no G-invariant elements in $\wedge^2 \mathfrak{g}$ for any simple Lie algebra $\mathfrak{g}$. Further $H^1(\mathfrak{g}[z], \mathrm{Sym}^2 \,\mathfrak{g})$ is zero because there are no copies of the adjoint representation in $\mathrm{Sym}^2 \,\mathfrak{g}$ (recall that we are assuming that $\mathfrak{g}$ is not $\mathfrak{sl}_n$ for $n > 2$).

We also have $H^1(\mathfrak{g}[z], z^{-1} \wedge^2 \mathfrak{g}) = z^{-2}\mathbb{C}$, because there is one copy of the adjoint in $\wedge^2 \mathfrak{g}$.

From this, we find that we have an exact sequence

$$\begin{aligned} &0 \to \mathbb{C} \cdot z^{-2} \to \mathbb{C} \cdot z^{-2} \xrightarrow{\delta} H^2(\mathfrak{g}[z], \mathrm{Sym}^2 \,\mathfrak{g}) \\ &\quad \to H^2(\mathfrak{g}[z], M_0) \to H^2(\mathfrak{g}[z], z^{-1} \wedge^2 \mathfrak{g}) \to H^3(\mathfrak{g}[z], \mathrm{Sym}^2 \,\mathfrak{g}) \to 0 \,. \end{aligned} \tag{C.28}$$

The boundary map

$$\delta : \mathbb{C} \cdot z^{-2} \to H^2(\mathfrak{g}[z], \mathrm{Sym}^2 \,\mathfrak{g}) \tag{C.29}$$

must be zero.

We conclude that $H^2(\mathfrak{g}[z], \mathrm{Sym}^2 \,\mathfrak{g})$ is a subspace of

$$H^2(\mathfrak{g}[z], M_0) = \mathbb{C} \cdot z^{-3}. \tag{C.30}$$

In particular, $H^2(\mathfrak{g}[z], \mathrm{Sym}^2 \,\mathfrak{g})$ is only non-zero in weight 3.

□

Next, let us perform the same analysis in the case that $\mathfrak{g}$ is $\mathfrak{sl}_n$.

Proposition C.4. *Suppose that $\mathfrak{g} = \mathfrak{sl}_n$ for $n > 2$. Then,*

$$H^2_{(k)}(\mathfrak{sl}_n[z], \mathrm{Sym}^2 \mathfrak{sl}_n) = 0 \text{ unless } k = 3 . \tag{C.31}$$

As before, there is an exact sequence

$$0 \to H^2_{(3)}(\mathfrak{sl}_n[z], \mathrm{Sym}^2 \mathfrak{sl}_n) \to \mathbb{C} \to H^2_{(2)}(\mathfrak{sl}_n[z], \wedge^2 \mathfrak{sl}_n) \to H^3_{(3)}(\mathfrak{sl}_n[z], \mathrm{Sym}^2 \mathfrak{sl}_n) \to 0 . \tag{C.32}$$

Proof. The cohomology groups $H^*_{(k)}(\mathfrak{sl}_n[z], \mathrm{Sym}^2 \mathfrak{sl}_n)$ are acted on by the outer automorphism group $\mathbb{Z}/2$ of $\mathfrak{sl}_n$ (which is the automorphism group of the Dynkin diagram). We let $H^*_{(k),\mathrm{even}}$ and $H^*_{(k),\mathrm{odd}}$ refer to the cohomology groups which are even and odd under this action. We will first prove that the cohomology groups which are even under the outer automorphism satisfy the statement of the proposition, and then show that the groups which are odd under the outer automorphism are zero.

Note that in the proof of proposition C.3 we used the fact that $\mathfrak{g}$ is not $\mathfrak{sl}_n$ with $n > 2$ in two places. Firstly, we used the fact that Lie algebras which are not of this type do not have an invariant element in $\mathrm{Sym}^3 \mathfrak{g}$. An invariant element in $\mathrm{Sym}^3 \mathfrak{g}$ contributes, according to the results of [31], an element of $H^1(\mathfrak{g}[z], \mathrm{Sym}^2(\mathfrak{g}[z^{-1}])$. However, because the invariant element in $\mathrm{Sym}^3 \mathfrak{sl}_n$ is odd under the outer automorphism, the corresponding element of $H^1(\mathfrak{sl}_n[z], \mathrm{Sym}^2(\mathfrak{sl}_n[z^{-1}]))$ is also odd, and so does not contribute when we analyze the cohomology which is even under the outer automorphism.

The other place where we used the assumption that $\mathfrak{g}$ is not $\mathfrak{sl}_n$ with $n > 2$ was when we asserted that $H^1(\mathfrak{g}[z], \mathrm{Sym}^2 \mathfrak{g})$ must be zero. This is essentially the same point, as any element of $H^1(\mathfrak{g}[z], \mathrm{Sym}^2 \mathfrak{g})$ must come from an invariant element in $\mathfrak{g} \otimes \mathrm{Sym}^2 \mathfrak{g}$, and the only invariant element (when $\mathfrak{g} = \mathfrak{sl}_n$, $n > 2$) is totally symmetric. Because this element is odd under the outer automorphism of $\mathfrak{sl}_n$, we find again that it can not contribute to the cohomology which is even under this automorphism.

Therefore, when we restrict to the cohomology which is even under the outer automorphims, proposition C.3 holds with the same proof.

We need to calculate the cohomology which is odd under the outer automorphism. We are interested in $H^2_{\mathrm{odd}}(\mathfrak{sl}_n[z], \mathrm{Sym}^2 \mathfrak{sl}_n)$. Any odd two-cocycle can be represented as a sum of $\mathfrak{sl}_n$-invariant linear operators

$$(z^k \mathfrak{sl}_n) \otimes (z^l \mathfrak{sl}_n) \to \mathrm{Sym}^2 \mathfrak{sl}_n , \tag{C.33}$$

Any such linear operator can be thought of as an invariant tensor in $\mathfrak{sl}_n^{\otimes 4}$, which is invariant under the permutation of the first two factors. If A_1, A_2, B, C denote elements of $\mathfrak{sl}_n$ then there are three such tensors, invariant under permutation of A_1 and A_2:

$$\begin{aligned} &\mathrm{Tr}(A_1 A_2 B C) + \mathrm{Tr}(A_2 A_1 B C) , \\ &\mathrm{Tr}(A_1 A_2 C B) + \mathrm{Tr}(A_2 A_1 C B) , \\ &\mathrm{Tr}(A_1 B A_2 C) + \mathrm{Tr}(A_2 B A_1 C) . \end{aligned}$$

The only linear combination of these three tensors which is odd under the outer automorphism is

$$\mathrm{Tr}(A_1 A_2[B, C]) + \mathrm{Tr}(A_2 A_1[B, C]) \qquad \text{(C.34)}$$

(recall that the outer automorphism sends $A \in \mathfrak{sl}_n$ to $-A^T$).

This invariant tensor can be viewed as a linear operator

$$(z^k \mathfrak{sl}_n) \otimes (z^l \mathfrak{sl}_n) \to \mathfrak{sl}_n \subset \mathrm{Sym}^2 \mathfrak{sl}_n\,. \qquad \text{(C.35)}$$

We have just shown that the natural map

$$H^2(\mathfrak{sl}_n[z], \mathfrak{sl}_n) \to H^2_{\mathrm{odd}}(\mathfrak{sl}_n[z], \mathrm{Sym}^2 \mathfrak{sl}_n) \qquad \text{(C.36)}$$

is surjective. However, we already know that $H^2(\mathfrak{sl}_n[z], \mathfrak{sl}_n) = 0$, so we conclude that $H^2_{\mathrm{odd}}(\mathfrak{sl}_n[z], \mathrm{Sym}^2 \mathfrak{sl}_n) = 0$ also. □

The next result we need is the following.

Proposition C.5. *Let $\wedge^2_0 \mathfrak{g} \subset \wedge^2 \mathfrak{g}$ denote the kernel of the Lie bracket map $\wedge^2 \mathfrak{g} \to \mathfrak{g}$. Then,*

$$H^2_{(2)}(\mathfrak{g}[z], \wedge^2 \mathfrak{g}) = \mathrm{Hom}^G(\wedge^2_0 \mathfrak{g}, \wedge^2_0 \mathfrak{g})\,, \qquad \text{(C.37)}$$

where on the right hand side we have the space of G-invariant maps from $\wedge^2_0 \mathfrak{g}$ to itself.

Further, the map

$$\mathbb{C} = H^2_{(3)}(\mathfrak{g}[z], M_0) \to H^2_{(2)}(\mathfrak{g}[z], \wedge^2 \mathfrak{g}) \qquad \text{(C.38)}$$

is surjective for $\mathfrak{g} \neq \mathfrak{sl}_2$ (where we are using the notation M_0 from the proof of proposition C.3).

Proof. We can calculate $H^2_{(2)}(\mathfrak{g}[z], \wedge^2 \mathfrak{g})$ using Lie algebra cohomology relative to $\mathfrak{g}$. A basis for the relative 2-cochains is the space of G-invariant maps $\wedge^2(z\mathfrak{g}) \to \wedge^2 \mathfrak{g}$. Every such relative two-cochain is closed, because there are no relative three-cochains of weight 2. Elements of the form

$$f^a{}_{bc}(zt_a \wedge zt_b) \otimes R^a_{ef}(t_e \wedge t_f) \qquad \text{(C.39)}$$

are exact, where R^a_{ef} defines some G-invariant map from $\mathfrak{g}$ to $\wedge^2 \mathfrak{g}$.

For all simple Lie algebras, there are no copies of the adjoint representation in $\wedge^2_0(\mathfrak{g})$. This argument tells us that the space of closed, but not exact, 2-cochains of weight -2 is the space of G-invariant maps $\wedge^2 \mathfrak{g} \to \wedge^2_0 \mathfrak{g}$.

For $G \neq SL_2$, this space is non-trivial. The final thing we need to check is that for $G \neq SL_2$, the map

$$\mathbb{C} = H^2_{(3)}(\mathfrak{g}[z], M_0) \to H^2_{(2)}(\mathfrak{g}[z], \wedge^2 \mathfrak{g}) \qquad \text{(C.40)}$$

is non-zero. This follows from the explicit description of the cocycles of $\mathfrak{g}[z]$ with values in $\mathrm{Sym}^* \mathfrak{g}[z^{-1}]$ given in [31]. □

The following corollary sums up what we have learned about the cohomology groups controlling the obstructions to quantizing a Wilson line.

Corollary C.6. *For a representation V of a simple Lie algebra $\mathfrak{g} \neq \mathfrak{sl}_2$, consider the group $H^2_{(k)}(\mathfrak{g}[z], \mathrm{End}(V))$ which contains possible anomalies to quantizing the Wilson line associated to V at k loops. Then, $H^2_{(k)}(\mathfrak{g}[z], \mathrm{End}(V)) = 0$ unless there exists a non-trivial G-invariant map from $\wedge^2_0 \mathfrak{g}$ to $\mathrm{End}(V)$.*

Proof. Two-cocycles representing classes in $H^2(\mathfrak{g}[z], \mathrm{End}(V))$ are given by G-invariant linear maps $\wedge^2 z\mathfrak{g}[z] \to \mathrm{End}(V)$. Any such G-invariant map must factor through some copies of $\mathrm{Sym}^2 \mathfrak{g}$, $\mathfrak{g}$, or $\wedge^2_0 \mathfrak{g}$. We have seen that the cohomology with coefficients in $\mathrm{Sym}^2 \mathfrak{g}$ and $\mathfrak{g}$ is zero. So, assuming there are no non-trivial maps from $\wedge^2_0 \mathfrak{g} \to \mathrm{End}(V)$, the cohomology groups are all zero. □

In fact, one can strengthen this result with some further work. For instance, one can show that $H^2_{(k)}(\mathfrak{g}[z], \wedge^2_0 \mathfrak{g}) = 0$ unless $k = 2, 4, 6$. We have already described the rank of the cohomology group when $k = 2$. Computing explicitly the rank when $k = 4, 6$ is a bit involved, and since we do not ultimately need to know the answer, we have not included these calculations.

D Derivation of Eqn. (6.13)

In the Feynman diagram of Fig. 22, we have three vertices, two vertices u, v on the Wilson line K

$$\left(u, \frac{fu^2}{2}, 0, 0\right), \quad \left(v, \frac{fv^2}{2}, 0, 0\right), \tag{D.1}$$

and one vertex w in the bulk, which we parametrize as $(x, y, z, \bar{z})$. At the vertex u of the Wilson line, we have the line element

$$\int_K A ds = (A_x(u) + A_y(u) f u)\, du\,. \tag{D.2}$$

In the following, to save space we sometimes write $A_x(u) := A_x(x(u), y(u), 0, 0)$. The similar expression applies to the vertex v.

The diagram has a symmetry factor of 1/2, which we can incorporate by restricting the integration region to be $u < v$.

Now we connect the vertices by a propagator, to the vertex w. We are interested in linear order in f.

One possibility is to take A_x from both, and then obtain a factor of f from the expansion of the propagator. However, this contributes an expression of the form

$$\pm \langle A_x(u) A_y(w) \rangle \langle A_x(v) A_{\bar{z}}(w) \rangle A_x(w) \tag{D.3}$$

(or with u and v exchanged), and since we have $A_x(w)$ this does not contribute to Λ.

Another possibility is to take one A_x and one A_y from u and v:

$$(fv)A_x(u)A_y(v) \quad \text{or} \quad (fu)A_y(u)A_x(v)\,. \tag{D.4}$$

When we consider the propagator, there are three different possibilities for (D.4):

$$\begin{aligned}
&(f(v-u))\langle A_x(u)A_y(w)\rangle\langle A_y(v)A_{\bar z}(w)\rangle A_x(w)\,,\\
&(f(v-u))\langle A_x(u)A_{\bar z}(w)\rangle\langle A_y(v)A_x(w)\rangle A_y(w)\,,\\
&(f(v-u))\langle A_x(u)A_y(w)\rangle\langle A_y(v)A_x(w)\rangle A_{\bar z}(w)\,.
\end{aligned} \tag{D.5}$$

Using the expression for the propagator (4.5), the expressions (D.5) can be simplified, and when combined with differential forms we have

$$\begin{aligned}
&\frac{1}{(2\pi)^2}(f(v-u))\frac{(2\bar z)(x-u)}{d(u,w)^4 d(v,w)^4}A_x(w)\cdot dz(dudy)(dvd\bar z)dx\,,\\
&\frac{1}{(2\pi)^2}(f(v-u))\frac{(y-\frac{fu^2}{2})(2\bar z)}{d(u,w)^4 d(v,w)^4}A_y(w)\cdot dz(dud\bar z)(dvdx)dy\,,\\
&\frac{1}{(2\pi)^2}(f(v-u))\frac{(-1)(2\bar z)(2\bar z)}{d(u,w)^4 d(v,w)^4}A_{\bar z}(w)\cdot dz(dudy)(dvdx)d\bar z\,,
\end{aligned} \tag{D.6}$$

with distance $d(u,w)$ defined by

$$d(u,w)^2 = (x-u)^2 + \left(y-\frac{fu^2}{2}\right)^2 + |z|^2\,. \tag{D.7}$$

We also need to supplement these expressions by a factor $\hbar\,\mathsf{h}^\vee$, where $\mathsf{h}^\vee$ is the color factor explained in the text and $\hbar$ is the loop counting parameter. In the leading order in f, we have $d(u,w)^2 \simeq (x-u)^2 + y^2 + |z|^2$, and hence we obtain

$$\begin{aligned}
-\frac{1}{(2\pi)^2}\frac{\mathsf{i}}{2\pi}\hbar\,\mathsf{h}^\vee \int_{u<v} \mathrm{d}u\mathrm{d}v \int_{\mathbb{R}^4} \mathrm{d}x\mathrm{d}y\mathrm{d}z\mathrm{d}\bar z\,(f(v-u))&\\
\times\frac{2\bar z(x-u)A_x(w) + 2y\bar z A_y(w) + 4\bar z^2 A_{\bar z}(w)}{((x-u)^2+y^2+|z|^2)^2\,((x-v)^2+y^2+|z|^2)^2}&\,,
\end{aligned} \tag{D.8}$$

The equation above can be expressed in the form (6.8) if we define

$$\begin{aligned}
\Theta_0 = -\frac{1}{(2\pi)^2}\frac{\mathsf{i}}{2\pi}\hbar\,\mathsf{h}^\vee \int_{u<v} \mathrm{d}u\mathrm{d}v\, f(v-u)&\\
\times\frac{2(\bar z)(x-u)(\mathrm{d}y\wedge\mathrm{d}z\wedge\mathrm{d}\bar z) - 2y\bar z(\mathrm{d}x\wedge\mathrm{d}z\wedge\mathrm{d}\bar z) + 4\bar z^2(\mathrm{d}x\wedge\mathrm{d}y\wedge\mathrm{d}z)}{((x-u)^2+y^2+|z|^2)^2\,((x-v)^2+y^2+|z|^2)^2}&\,.
\end{aligned} \tag{D.9}$$

In the notation of the main text, we have $\Theta = z\Theta_0$ and $\Theta = \mathrm{d}x\wedge\Lambda + \Lambda'$ to obtain

$$\begin{aligned}
\Lambda = -\frac{1}{(2\pi)^2}\frac{2f\mathsf{i}}{2\pi}\hbar\,\mathsf{h}^\vee&\\
\times\int_{u<v}\mathrm{d}u\mathrm{d}v\,(v-u)\frac{-yz\bar z(\mathrm{d}z\wedge\mathrm{d}\bar z) + 2z\bar z^2(\mathrm{d}y\wedge\mathrm{d}z)}{((x-u)^2+y^2+|z|^2)^2\,((x-v)^2+y^2+|z|^2)^2}&\,.
\end{aligned} \tag{D.10}$$

We obtain (6.13) after doing the u and v integrals:

$$\int_{-\infty\leq u<v\leq\infty} \mathrm{d}u\mathrm{d}v\,(v-u)\frac{1}{\left(u^2+a^2\right)^2\left(v^2+a^2\right)^2} = \frac{1}{a^5}\frac{3\pi}{8}\,. \tag{D.11}$$

E Evaluation of Eqn. (8.34)

In this appendix we present details on the evaluation of the integral (8.34).

Let us first evaluate the part relevant for the v_1 integral inside the expression (8.34):

$$\int_{v_1} P(v_0,v_1)\wedge \mathrm{d}z_1(z_1)\wedge P(v_1,v_2)\,, \tag{E.1}$$

where we have introduced, temporarily, a vertex labelled $v_0$with coordinates $v_0 = (x_0, y_0, z_0, \overline{z}_0)$. Later we will set these coordinates to zero.

From the explicit expression for the propagator in eqn. (4.5) we have

$$P(v_i,v_j) = \frac{1}{2\pi}\frac{x_{ij}\mathrm{d}y_{ij}\wedge \mathrm{d}\bar{z}_{ij} - y_{ij}\mathrm{d}x_{ij}\wedge \mathrm{d}\bar{z}_{ij} + 2\overline{z}_{ij}\mathrm{d}x_{ij}\wedge \mathrm{d}y_{ij}}{d(v_i,v_j)^4}\,, \tag{E.2}$$

where the distance $d(v_i, v_j)$ between two points v_i, v_j is given by $d(v_i,v_j)^2 = x_{ij}^2 + y_{ij}^2 + |z_{ij}|^2$, and $x_{ij} = x_i - x_j, z_{ij} = z_i - z_j$, etc.

We can calculate that

$$P(v_0,v_1)\wedge P(v_1,v_2) = \frac{1}{(2\pi)^2}\mathrm{d}x_1\mathrm{d}y_1\mathrm{d}\overline{z}_1$$
$$\frac{(x_{01}y_{12}\mathrm{d}\overline{z}_2 - 2x_{01}\overline{z}_{12}\mathrm{d}y_2 + 2y_{01}\overline{z}_{12}\mathrm{d}x_2 - y_{01}x_{12}\mathrm{d}\overline{z}_2 + 2\overline{z}_{01}x_{12}\mathrm{d}y_2 - 2\overline{z}_{01}y_{12}\mathrm{d}x_2)}{d(v_0,v_1)^4 d(v_1,v_2)^4}\,. \tag{E.3}$$

where we have dropped all terms involving $\mathrm{d}x_0$, $\mathrm{d}y_0$, $\mathrm{d}\overline{z}_0$ because we will not be integrating over the vertex v_0. From this we see that

$$P(v_0,v_1)\wedge z_1\mathrm{d}z_1 P(v_1,v_2) = \frac{1}{(2\pi)^2}\mathrm{d}x_1\mathrm{d}y_1\mathrm{d}z_1\mathrm{d}\overline{z}_1\left[\partial_{\overline{z}_0}\frac{1}{d(v_0,v_1)^2 d(v_1,v_2)^4}\right]$$
$$(x_{01}y_{12}\mathrm{d}\overline{z}_2 - 2x_{01}\overline{z}_{12}\mathrm{d}y_2 + 2y_{01}\overline{z}_{12}\mathrm{d}x_2 - y_{01}x_{12}\mathrm{d}\overline{z}_2 + 2\overline{z}_{01}x_{12}\mathrm{d}y_2 - 2\overline{z}_{01}y_{12}\mathrm{d}x_2)\,. \tag{E.4}$$

after we set the v_0 coordinates to zero.

Moving the position of the derivative in $\overline{z}_0$, we obtain

$$P(v_0,v_1)\wedge z_1\mathrm{d}z_1 P(v_1,v_2) = \frac{1}{(2\pi)^2}\mathrm{d}x_1\mathrm{d}y_1\mathrm{d}z_1\mathrm{d}\overline{z}_1$$
$$\times\partial_{\overline{z}_0}\left[\frac{(x_{01}y_{12}\mathrm{d}\overline{z}_2 - 2x_{01}\overline{z}_{12}\mathrm{d}y_2 + 2y_{01}\overline{z}_{12}\mathrm{d}x_2 - y_{01}x_{12}\mathrm{d}\overline{z}_2 + 2\overline{z}_{01}x_{12}\mathrm{d}y_2 - 2\overline{z}_{01}y_{12}\mathrm{d}x_2)}{d(v_0,v_1)^2 d(v_1,v_2)^4}\right]$$
$$\frac{1}{(2\pi^2)^2}\mathrm{d}x_1\mathrm{d}y_1\mathrm{d}z_1\mathrm{d}\overline{z}_1\frac{(2x_{12}\mathrm{d}y_2 - 2y_{12}\mathrm{d}x_2)}{d(v_0,v_1)^2 d(v_1,v_2)^4}\,. \tag{E.5}$$

By integration by parts, the integral over $x_1, y_1, z_1, \overline{z}_1$ of all the terms on the first two lines vanishes. For example,

$$\begin{aligned}
&\int_{v_1} (x_{01}y_{12} - y_{01}x_{12}) \frac{1}{d(v_0,v_1)^2} \frac{1}{d(v_1,v_2)^4} \\
&= \tfrac{1}{2} \int_{v_1} (-x_{01}) \frac{1}{d(v_0,v_1)^2} \left(\partial_{y_1} \frac{1}{d(v_1,v_2)^2} \right) + \tfrac{1}{2} \int_{v_1} y_{01} \frac{1}{d(v_0,v_1)^2} \left(\partial_{x_1} \frac{1}{d(v_1,v_2)^2} \right) \\
&= \tfrac{1}{2} \int_{v_1} x_{01} \left(\partial_{y_1} \frac{1}{d(v_0,v_1)^2} \right) \frac{1}{d(v_1,v_2)^2} - \tfrac{1}{2} \int_{v_1} y_{01} \left(\partial_{x_1} \frac{1}{d(v_0,v_1)^2} \right) \frac{1}{d(v_1,v_2)^2} \\
&= \int_{v_1} (x_{01}y_{01} - x_{01}y_{01}) \frac{1}{d(v_0,v_1)^4} \frac{1}{d(v_1,v_2)^2} = 0 \, .
\end{aligned}$$

We are left with the integral

$$-\frac{1}{(2\pi^2)^2} \int_{x_1,y_1,z_1} \mathrm{d}x_1 \mathrm{d}y_1 \mathrm{d}z_1 \mathrm{d}\overline{z}_1 \frac{(2x_{12}\mathrm{d}y_2 - 2y_{12}\mathrm{d}x_2)}{d(v_0,v_1)^2 d(v_1,v_2)^4} \, . \tag{E.6}$$

We can evaluate this integral with the help of eqn. (5.13). Choosing $\alpha = 2, \beta = 1$ and using $\Gamma(3)/(\Gamma(2)\Gamma(1)) = 2$, we obtain

$$-\frac{2}{(2\pi)^2} \int_0^1 \mathrm{d}t \, t \int \mathrm{d}x_1 \mathrm{d}y_1 \mathrm{d}z_1 \mathrm{d}\overline{z}_1 \frac{(2x_{12}\mathrm{d}y_2 - 2y_{12}\mathrm{d}x_2)}{(\star)^3} , \tag{E.7}$$

where

$$\begin{aligned}
(\star) &= (1-t) \left[x_1^2 + y_1^2 + |z_1|^2 \right] + t \left[x_{12}^2 + y_{12}^2 + |z_{12}|^2 \right] \\
&= (x_1 - tx_2)^2 + (y_1 - ty_2)^2 + |z_1 - tz_2|^2 + t(1-t) \left[x_2^2 + y_2^2 + |z_2|^2 \right] \, .
\end{aligned} \tag{E.8}$$

After shifting integration variables we obtain

$$\begin{aligned}
&-\frac{2}{(2\pi)^2} \int_0^1 \mathrm{d}t \, t \int \mathrm{d}x_1 \mathrm{d}y_1 \mathrm{d}z_1 \mathrm{d}\overline{z}_1 \frac{(2(x_1 - (1-t)x_2)\mathrm{d}y_2 - 2(y_1 - (1-t)y_2)\mathrm{d}x_2)}{\left(x_1^2 + y_1^2 + |z_1|^2 + t(1-t) \left[x_2^2 + y_2^2 + |z_2|^2 \right]\right)^3} \\
&= \frac{4}{(2\pi)^2} \int_0^1 \mathrm{d}t \, t(1-t) \int \mathrm{d}x_1 \mathrm{d}y_1 \mathrm{d}z_1 \mathrm{d}\overline{z}_1 \frac{(x_2\mathrm{d}y_2 - y_2\mathrm{d}x_2)}{\left(x_1^2 + y_1^2 + |z_1|^2 + t(1-t) \left[x_2^2 + y_2^2 + |z_2|^2 \right]\right)^3} ,
\end{aligned} \tag{E.9}$$

where we dropped the pieces odd in x_1 or y_1, and hence do not contribute to the integral. After evaluating the integral over x_1, y_1 and then over $z_1, \overline{z}_1$, the integral of t becomes trivial, and we obtain (note that $\mathrm{d}z\mathrm{d}\overline{z} = -2ir\mathrm{d}r\mathrm{d}\theta$ for the polar coordinate $z = re^{i\theta}$)

$$\begin{aligned}
&\frac{4}{(2\pi)^2} \frac{\pi}{2} (x_2\mathrm{d}y_2 - y_2\mathrm{d}x_2) \int_0^1 \mathrm{d}t \, t(1-t) \int \mathrm{d}z_1 \mathrm{d}\overline{z}_1 \frac{1}{\left(|z_1|^2 + t(1-t) \left[x_2^2 + y_2^2 + |z_2|^2 \right]\right)^2} \\
&= \frac{4}{(2\pi)^2} \frac{\pi}{2} (-2\mathrm{i})\pi \, (x_2\mathrm{d}y_2 - y_2\mathrm{d}x_2) \frac{1}{(x_2^2 + y_2^2 + |z_2|^2)} \, .
\end{aligned} \tag{E.10}$$

We therefore find

$$\int_{v_1} P(v_0, v_1) \wedge z_1 \mathrm{d}z_1 \wedge P(v_1, v_2) = \frac{1}{\mathsf{i}} \frac{(x_2 \mathrm{d}y_2 - y_2 \mathrm{d}x_2)}{d(v_0, v_2)^2}\,. \tag{E.11}$$

Similarly, the part of (8.34) relevant for the v_3 integral is

$$\int_{v_3} P(v_2, v_3) \wedge z_3 \mathrm{d}z_3 \wedge P(v_3, p_3) = -\frac{1}{\mathsf{i}} \frac{((x_2 - \epsilon)\mathrm{d}y_2 - y_2 \mathrm{d}x_2)}{d(v_0, v_2)^2}\,. \tag{E.12}$$

We can now come back to the evaluation of (8.34). Using the results (E.11) and (E.12), we obtain

$$2\left(\tfrac{\mathsf{i}}{2\pi}\right)^3 \frac{1}{2\pi(\mathsf{i})(-\mathsf{i})} \int_{p=0}^{\epsilon} \int_{x,y,z,\bar{z}} \frac{(x\mathrm{d}y - y\mathrm{d}x)(\mathrm{d}z)(y\mathrm{d}\bar{z}\mathrm{d}p)((x-\epsilon)\mathrm{d}y - y\mathrm{d}x)}{d(0,v)^2 d(p,v)^4 d(\epsilon,v)^2}\,, \tag{E.13}$$

where we dropped the index 2 from $v_2 = (x_2, y_2, z_2, \bar{z}_2)$ to simplify the expressions. This gives

$$2\frac{\mathsf{i}}{(2\pi)^4} \int_{p=0}^{\epsilon} \int_{x,y,z,\bar{z}} \frac{\epsilon y^2 \mathrm{d}x\mathrm{d}y\mathrm{d}z\mathrm{d}\bar{z}\mathrm{d}p}{(x^2+y^2+|z|^2)((x-p)^2+y^2+|z|^2)^2((x-\epsilon)^2+y^2+|z|^2)}\,. \tag{E.14}$$

We need to take into account two more diagrams in Fig. 28. It turns out that the evaluation is rather similar, with the only different exchange of the role of the points $0, p$ and ϵ on the Wilson line. This means in addition to (E.14) we have two extra contributions

$$\begin{aligned} &2\frac{\mathsf{i}}{(2\pi)^4} \int_{p=0}^{\epsilon} \int_{x,y,z,\bar{z}} \frac{\epsilon y^2 \mathrm{d}x\mathrm{d}y\mathrm{d}z\mathrm{d}\bar{z}\mathrm{d}p}{(x^2+y^2+|z|^2)^2((x-p)^2+y^2+|z|^2)((x-\epsilon)^2+y^2+|z|^2)}\,,\\ &2\frac{\mathsf{i}}{(2\pi)^4} \int_{p=0}^{\epsilon} \int_{x,y,z,\bar{z}} \frac{\epsilon y^2 \mathrm{d}x\mathrm{d}y\mathrm{d}z\mathrm{d}\bar{z}\mathrm{d}p}{(x^2+y^2+|z|^2)((x-p)^2+y^2+|z|^2)((x-\epsilon)^2+y^2+|z|^2)^2}\,. \end{aligned} \tag{E.15}$$

Summing all the three contributions, we obtain

$$-\frac{\mathsf{i}}{(2\pi)^4} \int_{p=0}^{\epsilon} \int_{x,y,z,\bar{z}} (\epsilon y)\frac{\partial}{\partial y} \frac{\mathrm{d}x\mathrm{d}y\mathrm{d}z\mathrm{d}\bar{z}\mathrm{d}p}{(x^2+y^2+|z|^2)((x-p)^2+y^2+|z|^2)((x-\epsilon)^2+y^2+|z|^2)}\,, \tag{E.16}$$

which after integrating by parts gives

$$\frac{\mathsf{i}}{(2\pi)^4} \int_{p=0}^{\epsilon} \int_{x,y,z,\bar{z}} \epsilon \frac{\mathrm{d}x\mathrm{d}y\mathrm{d}z\mathrm{d}\bar{z}\mathrm{d}p}{(x^2+y^2+|z|^2)((x-p)^2+y^2+|z|^2)((x-\epsilon)^2+y^2+|z|^2)}\,. \tag{E.17}$$

After scaling the integration variables by ϵ (assuming $\epsilon > 0$) ϵ depends drops out, as expected:

$$\frac{\mathrm{i}}{(2\pi)^4}\int_{p=0}^{1}\int_{x,y,z,\bar{z}}\frac{\mathrm{d}x\mathrm{d}y\mathrm{d}z\mathrm{d}\bar{z}\mathrm{d}p}{(x^2+y^2+|z|^2)((x-p)^2+y^2+|z|^2)((x-1)^2+y^2+|z|^2)}. \tag{E.18}$$

In angular coordinates in $(y, z, \bar{z})$-plane, the volume form $\mathrm{d}y\mathrm{d}z\mathrm{d}\bar{z}$ is $-8\pi\mathrm{i}r^2\mathrm{d}r\mathrm{d}\Omega_{S^2}$ where $\mathrm{d}\Omega_{S^2}$ is the volume form on the two-sphere of volume 1. Integrating over the two-sphere we get

$$\frac{8\pi}{(2\pi)^4}\int_{p=0}^{1}\mathrm{d}p\int_x \mathrm{d}x\int \mathrm{d}r r^2\frac{1}{(x^2+r^2)((x-p)^2+r^2)((x-1)^2+r^2)}. \tag{E.19}$$

After the r integral we obtain

$$\frac{8\pi}{(2\pi)^4}\frac{\pi}{2}\int_{p=0}^{1}\mathrm{d}p\int_x \mathrm{d}x\frac{1}{(|x|+|x-p|)(|x|+|x-1|)(|x-p|+|x-1|)}. \tag{E.20}$$

We can evaluate this integral by dividing into four cases $x < 0, 0 < x < p, p < x < 1, 1 < x$. These integrals are interchanged by the change of coordinates $x \mapsto 1-x$, $p \mapsto 1-p$, so that the only two independent integrals are the cases when $p < x < 1$ and $1 < x$. We have

$$\int_{0<p<x<1}\frac{1}{(2x-p)(1-p)} = \frac{\pi^2}{8}, \tag{E.21}$$

$$\int_{0<p<1<x}\frac{1}{(2x-p)(2x-1)(2x-1-p)} = \frac{\pi^2}{24}. \tag{E.22}$$

Therefore we obtain

$$\frac{8\pi}{(2\pi)^4}\frac{\pi}{2}2\left(\frac{\pi^2}{8}+\frac{\pi^2}{24}\right) = \frac{1}{12}. \tag{E.23}$$

Including the the factors of $\hbar^2$ need for the two-loop diagram, we reproduce the numerical factor of (8.35).

References

[1] H. Bethe, "Zur Theorie der Metalle. I. Eigenwerte und Eigenfunktionen der linearen Atomkette," Zeit. fur Physik **71** (1931) 205-226.

[2] L. Onsager, "Crystal Statistics I: A Two-Dimensional Model With An Order-Disorder Transition," Phys. Rev. **65** (1944) 117-49.

[3] J. B. McGuire, "Study of Exactly Soluble One-Dimensional N-Body Problems," J. Math. Phys. **5**, 622 (1964).

[4] C. N. Yang, "Some Exact Results for the Many Body Problems in One Dimension With Repulsive Delta Function Interaction," Phys. Rev. Lett. **19**, 1312 (1967).

[5] R. J. Baxter, “Eight-Vertex Model in Lattice Statistics,” Phys. Rev. Lett. **26**, 832 (1971).
[6] A. B. Zamolodchikov and A. B. Zamolodchikov, “Factorized s Matrices in Two-Dimensions as the Exact Solutions of Certain Relativistic Quantum Field Models,” Annals Phys. **120**, 253 (1979).
[7] V. G. Drinfeld, “Quantum Groups,” in *Proceedings of the International Congress of Mathematicians, Berkeley, 1986*, American Mathematical Society, 1987.
[8] V. Chari and A. Pressley, “A Guide to Quantum Groups,” Cambridge University Press, 1994.
[9] M. Jimbo, ed., *Yang-Baxter Equation In Integrable Systems* (World-Scientific, 1989).
[10] J. H. H. Perk and H. Au-Yang, “Yang-Baxter Equation,” in *Encyclopedia of Mathematical Physics*, eds. J.-P. Françoise, G.L. Naber and S. T. Tsou (Oxford: Elsevier, 2006), Vol. 5, pp. 465-73, with an expanded version in arXiv:math-ph/0606053.
[11] K. Costello, “Integrable Lattice Models From Four-Dimensional Field Theories,” Proc. Symp. Pure Math. **88** (2014) 3-24, arXiv:1303.2632.
[12] K. Costello, “Supersymmetric Gauge Theory and the Yangian,” arXiv:1303.2632 [hep-th].
[13] E. Witten, “Integrable Lattice Models From Gauge Theory,” arXiv:1611.00592 [hep-th].
[14] M. F. Atiyah, “New Invariants Of 3- and 4-Dimensional Manifolds,” in R. O. Wells, ed., *The Mathematical Heritage Of Hermann Weyl* (American Mathematical Society, 1988).
[15] H. Au-Yang, B. M. McCoy, J. H. H. Perk, S. Tang and M. L. Yan, “Commuting Transfer Matrices in the Chiral Potts Models: Solutions of Star Triangle Equations with Genus > 1,” Phys. Lett. A **123**, 219 (1987).
[16] R. J. Baxter, J. H. H. Perk and H. Au-Yang, “New Solutions of the Star Triangle Relations for the Chiral Potts Model,” Phys. Lett. A **128**, 138 (1988).
[17] V. V. Bazhanov and S. M. Sergeev, “A Master solution of the quantum Yang-Baxter equation and classical discrete integrable equations,” Adv. Theor. Math. Phys. **16**, no. 1, 65 (2012), arXiv:1006.0651 [math-ph].
[18] V. P. Spiridonov, “Elliptic beta integrals and solvable models of statistical mechanics,” Contemp. Math. **563**, 181 (2012) [arXiv:1011.3798 [hep-th]].
[19] M. Yamazaki, “Quivers, YBE and 3-manifolds,” JHEP **1205**, 147 (2012), arXiv:1203.5784 [hep-th].
[20] M. Yamazaki, “New Integrable Models from the Gauge/YBE Correspondence,” J. Statist. Phys. **154**, 895 (2014), arXiv:1307.1128 [hep-th].
[21] J.-L. Gervais and A. Neveu, “Novel Triangle Relation And Absence of Tachyons in Liouville String Field Theory,” Nucl. Phys. **B238** (1984) 125-41.
[22] G. Felder, “Elliptic Quantum Groups,” arXiv:hep-th/9412207.
[23] G. Felder, “Conformal Field Theory and Integrable Systems Associated to Elliptic Curves,” in *Proceedings of the International Congress of Mathematicians, Zürich 1994*, Birkhäuser, 1994; “Elliptic quantum groups”, hep-

th/9412207.
[24] P. Etinghof and O. Schiffman, "Lectures on the Dynamical Yang-Baxter Equations," arXiv:math/9908064.
[25] K. Costello, E. Witten and M. Yamazaki, "Gauge Theory and Integrability, II," *to appear.*
[26] E. Witten, "Fivebranes And Knots," Quantum Topol. **3** (2012) 1-137, arXiv:1101.3216 [hep-th].
[27] N. A. Nekrasov and S. L. Shatashvili, "Supersymmetric Vacua and Bethe Ansatz," Nucl. Phys. Proc. Suppl. **192-3** (2009) 91-112, arXiv:0901.4744.
[28] N. A. Nekrasov and S. L. Shatashvili, "Quantization Of Integrable Systems And Four Dimensional Gauge Theories," in P. Exner, ed. *Proceedings, 16th International Congress on Mathematical Physics*, arXiv:0901.4748.
[29] A. A. Belavin and V. G. Drinfeld, "Solutions of the Classical Yang-Baxter equation for Simple Lie Algebras," Funktsional. Anal. i Prilozhen. **16**, 1 (1982).
[30] E. Witten, "Chern-Simons Gauge Theory as a String Theory," Prog. Math. **133**, 637 (1995) [hep-th/9207094].
[31] S. Fishel, I. Grojnowski and C. Teleman, "The strong Macdonald conjecture and Hodge theory on the loop Grassmannian", Ann. of Math. **168**, 175 (2008).
[32] E. Witten, "Gauge Theories, Vertex Models, and Quantum Groups," Nucl.Phys. **B330** (1990) 285-346.
[33] N. Beisert, "The $SU(2|2)$ Dynamic R-Matrix," Adv. Theor. Math. Phys. **12** (2008) 945-79.
[34] P. Ramond, "Group Theory: A Physicist's Survey," Cambridge, UK: Univ. Pr. (2010) 310 p.
[35] R. Slansky, "Group Theory for Unified Model Building," Phys. Rept. **79**, 1 (1981).
[36] W. G. McKay, J. Patera, and D. W. Rand, *Tables Of Representations Of Simple Lie Algebras: Exceptional Simple Lie Algebras* (Centre de Recherches Mathematiques, Univ. of Montréal, 1990).
[37] L. D. Faddeev and L. A. Takhtajan, "Hamiltonian Methods In The Theory Of Solitons," Springer, 1987.
[38] E. Witten, "On Quantum Gauge Theories In Two Dimensions," Commun. Math. Phys. **141** (1991) 153-209.
[39] M. Blau and G. Thompson, "Lectures on $2-d$ Gauge Theories: Topological Aspects and Path Integral Techniques," in *1993 Summer School in High-Energy Physics and Cosmology,* ed. E. Gava et. al. (World Scientific, 1994), arXiv:hep-th/9402097.
[40] M. Blau and G. Thompson, "Chern-Simons Theory on S^1-Bundles: Abelianization and q-Deformed Yang-Mills Theory," arXiv:hep-th/0601068.

Perimeter Institute for Theoretical Physics
Waterloo, ON N2L 2Y5, Canada

School of Natural Sciences, Institute for Advanced Study

Einstein Drive, Princeton, NJ 08540, USA

Kavli Institute for the Physics and Mathematics of the Universe (WPI)
University of Tokyo, Kashiwa, Chiba 277-8583, Japan

Proceedings of the 7th International
Congress of Chinese Mathematicians
ALM 43, pp. 153–172

The Nash–Kuiper Theorem and the Onsager Conjecture

Camillo De Lellis

Abstract

We give an account of the analogies between the Nash–Kuiper C^1 solutions of the isometric embedding problem and the weak solutions of the incompressible Euler equations which violate the energy conservation. Such analogies have lead to the recent resolution of a well-known conjecture of Lars Onsager in the theory of fully developed turbulence.

Mathematics Subject Classification: 35Q31, (35A01,35D30,53A99,53C21,76F02).

Keywords and Phrases: Onsager conjecture, Euler equations, isometric embedding, h-principle, Nash–Kuiper theorem.

1 The Nash-Kuiper Theorem

Let (Σ^n, g) be a smooth n-dimensional Riemannian manifold. A map $u : \Sigma \to \mathbb{R}^N$ is *isometric* if it preserves the length of curves, i.e. if

$$\ell_g(\gamma) = \ell_e(u \circ \gamma) \qquad \text{for any } C^1 \text{ curve } \gamma \subset \Sigma, \tag{1}$$

where $\ell_g(\gamma)$ denotes the length of γ with respect to the metric g:

$$\ell_g(\gamma) = \int \sqrt{g(\gamma(t))[\dot\gamma(t), \dot\gamma(t)]}\, dt\,. \tag{2}$$

As customary, in local coordinates we can express the metric tensor g as[1] $g = g_{ij} dx_i \otimes dx_j$. For a C^1 map u, condition (1) is equivalent to the system of partial differential equations

$$\partial_i u \cdot \partial_j u = g_{ij}\,. \tag{3}$$

In the usual language of Riemannian geometry, (3) means that g is the pullback of the Euclidean metric through the map u.

The existence of isometric immersions (resp. embeddings) of Riemannian manifolds into some Euclidean space is a classical problem, explicitly formulated

The research of C.D.L. has been supported by the grant 200021_159403 of the Swiss National Foundation.

[1]Here and in the rest of this note we follow Einstein's summation convention.

for the first time by Schläfli, see [46]: in the latter Schläfli conjectured that the system is solvable *locally* if the dimension N of the target is at least $s_n := \frac{n(n+1)}{2}$. Such conjecture stands to reason because (3) consists precisely of s_n equations in N unknowns. In the first half of the twentieth century Janet [36], Cartan [14] and Burstin [13] proved Schläfli's conjecture for analytic metrics.

For the very particular case of 2-dimensional spheres endowed with metrics of positive Gauss curvature, Weyl in [51] raised the question of the existence of (global!) isometric embeddings in $\mathbb{R}^3$. The Weyl's problem was solved by Lewy in [39] for analytic metrics and Nirenberg settled the case of smooth metrics in his PhD thesis in 1949; a different proof was given independently by Pogorelov [43] around the same time, building upon the work of Alexandrov [1] (see also [44]).

An important aspect of the Weyl's problem is the rigidity of the solutions found by Lewy, Nirenberg and Pogorelov. Indeed, already before the work of Lewy, Cohn–Vossen and Herglotz proved independently that C^2 isometric immersions of positively curved spheres are uniquely determined up to rigid motions, cf. [17, 33] and see also [49] for a thorough discussion.

Before the appearance of Nash's celebrated works, it was natural to expect that the assumption of C^2 regularity in the works of Cohn–Vossen and Herglotz was just of technical nature. But in his 1954 note [41] Nash astonished the geometry world and proved that the only true obstructions to the existence of isometric immersions are topological and that as soon as $N \geq n+1$ and there are no such obstructions, then there are in fact plenty of such immersions. Nash's Theorem was therefore in stark contrast with the intuition that codimension 1 smooth isometric immersions are rather rigid for $n = 2$ and that for $n > 2$, given that the system (3) is heavily overdetermined, existence of solutions should occur rarely.

In order to state Nash's Theorem we need some terminology.

Definition 1.1. *Let (Σ, g) be a Riemannian manifold. An immersion $v : \Sigma \to \mathbb{R}^N$ is short if we have it "shrinks" the length of curves. For C^1 immersions and in local coordinates such condition is equivalent to the inequality $(\partial_i v \cdot \partial_j v) w^i w^j \leq g_{ij} w^i w^j$ for any tangent vector w.*

Theorem 1.2. *Let (Σ, g) be a smooth closed n-dimensional Riemannian manifold and $v : \Sigma \to \mathbb{R}^N$ a C^∞ short immersion with $N \geq n+1$. Then, for any $\varepsilon > 0$ there exists a C^1 isometric immersion $u : \Sigma \to \mathbb{R}^N$ such that $\|u - v\|_{C^0} \leq \varepsilon$. If v is, in addition, an embedding, then u can be assumed to be an embedding as well.*

Indeed Nash gave a proof of Theorem 1.2 for $N \geq n+2$ and just remarked that it could be proved for $N \geq n+1$ with some additional work; the details were then given in two subsequent notes by Kuiper, [38]. For this reason Theorem 1.2 is called nowadays the Nash–Kuiper Theorem on C^1 isometric embeddings.

2 Rigidity versus Flexibility

Isometries of Riemannian manifolds behave then in rather different ways depending on their smoothness: from the one hand we have the rigidity of C^2 isometries, witnessed in the classical result of Cohn-Vossen and Herglotz, and on the other hand we have the flexibility of C^1 isometries stated in the Nash–Kuiper Theorem. A natural question is whether there is a threshold regularity which distinguishes between the two behaviors.

The Hölder spaces give a classical way to measure intermediate smoothness between C^1 and C^2: a C^1 map v belongs to the Hölder space $C^{1,\alpha}$ (with $0 < \alpha \leq 1$) if

$$|Dv(x) - Dv(y)| \leq C|x - y|^\alpha$$

for some constant C independent of $x, y \in \Sigma$. It is thus natural to look at $C^{1,\alpha}$ isometries of Riemannian manifolds and ask whether there is an α_0 for which such isometries display flexibility phenomena à la Nash–Kuiper for $\alpha < \alpha_0$ and rigidity phenomena à la Cohn-Vossen–Herglotz for $\alpha > \alpha_0$. The first mathematician who tackled such problem is Borisov, who published a series of works on the topic in the late fifties (see below). Later such question is mentioned by Gromov in [31] and by Yau in [52] and in the recent work [32] Gromov advanced the conjecture that the threshold α_0 is in fact $\frac{1}{2}$.

In a series of papers in the 1950s, cf. [2, 3, 4, 5], Borisov showed that the rigidity of the Weyl problem can in fact be extended to $C^{1,\theta}$ immersions provided θ is sufficiently large.

Theorem 2.1. *Let $(\mathbb{S}^2, g)$ be a surface with C^2 metric and positive Gauss curvature, and let $u \in C^{1,\theta}(\mathbb{S}^2; \mathbb{R}^3)$ be an isometric immersion with $\theta > 2/3$. Then $u(\mathbb{S}^2)$ is the boundary of an open convex set.*

Borisov's Theorem is more general, but his statement needs the introduction of Pogorelov's concept of bounded extrinsic curvature, cf. [20]:

Theorem 2.2. *If (Σ, g) is a surface with C^2 metric and positive Gauss curvature and $u \in C^{1,\theta}(\mathbb{S}^2; \mathbb{R}^3)$ is an isometric immersion with $\theta > 2/3$, then $u(\Sigma)$ has bounded extrinsic curvature in the sense of Pogorelov.*

Building upon the work of Pogorelov, [44], Theorem 2.1 can be immediately derived from Theorem 2.2.

The concept of Pogorelov's bounded extrinsic curvature can be easily explained as follows. When u is a smooth immersion, Gauss' Theorema Egregium asserts that the determinant of the differential dN of the Gauss map N of $u(\Sigma)$ equals the Gauss curvature κ. In particular, by the area formula, for any measurable subset A of $u(\Sigma)$ the surface area $|N(A)|$ of $N(A) \subset \mathbb{S}^2$ can be computed with the area formula and it is bounded by a constant times the area of A: such constant is simply the maximum of the absolute value of κ. On the other hand $N(A)$ and A are well defined as soon as u is a C^1 map and thus if an inequality of the form $|N(A)| \leq C|A|$ holds for every measurable A, we can assert that $u(\Sigma)$ has bounded curvature in a generalized sense.

Ultimately Theorem 2.2 states that a (suitable form of) Gauss' Theorema Egregium holds for $C^{1,\theta}$ immersions as soon as $\theta > \frac{2}{3}$. In [20] we discovered a very short proof of Borisov's Theorem, which exploits the same key computation of Constantin-E-Titi's proof, see [18], of part (a) of Onsager's conjecture, cf. Conjecture 3.1 below.

For sufficiently small Hölder exponents, instead, the Nash-Kuiper construction remains valid:

Theorem 2.3. *Let (Σ, g) be a C^2 Riemannian manifold of dimension n. Any short immersion $u : \Sigma \to \mathbb{R}^{n+1}$ can be uniformly approximated with $C^{1,\theta}$ isometric immersions, where:*

(a) $\theta < \frac{1}{1+n(n+1)}$ when Σ is a closed ball;
(b) $\theta < \frac{1}{1+n(n+1)^2}$ when Σ is a general compact n-manifold;
(c) $\theta < \frac{1}{5}$ if Σ is a 2-dimensional disk.

The maps can be chosen to be embeddings if u is an embedding.

Case (a) of this theorem was announced in [6] by Borisov, based on his habilitation thesis, under the additional assumption that g is analytic. A proof with $n = 2$ appeared more than 40 years later, cf. [7]. The general statements (a) and (b) of Theorem 2.3 have been proved in [20], whereas the improved bound for 2-dimensional disks, namely statement (c), has been shown rather recently in [23].

As a byproduct of the above constructions, there is no way of making sense of Gauss' Theorema Egregium for $C^{1,\theta}$ immersions when θ is sufficiently small. A natural question is then to ask whether these two behaviors are distinguished by a sharp threshold. In what follows we will call such question "Borisov–Gromov problem".

3 Onsager's Conjecture

Around 10 years ago László Székelyhidi and I pointed out a striking analogy between the Borisov–Gromov problem and a well-known conjecture in the theory of mathematical fluid dynamics, proposed in 1949 by the celebrated Norwegian theoretical physicist Lars Onsager (we refer to the survey articles [26] and [29] for a throrough discussion of this and several other points mentioned below). The unveiling of such analogy was a consequence of our work [24]: in that paper we applied methods which are reminiscent of those used by Nash in [41] in order to explain the existence of weak solutions to the incompressible Euler equations which do not preserve the total kinetic energy.

The incompressible Euler equations describe the motion of a perfect incompressible fluid. Written down by L. Euler over 250 years ago, these are the continuum equations corresponding to the conservation of momentum and mass of arbitrary fluid regions. In Eulerian variables they can be written as

$$\begin{cases} \partial_t v + (v \cdot \nabla) v + \nabla p = 0 \\ \operatorname{div} v = 0, \end{cases} \tag{4}$$

where $v = v(x,t)$ is the velocity and $p = p(x,t)$ is the pressure. We will focus on the 3-dimensional case with periodic boundary conditions. In other words we take the spatial domain to be the flat 3-dimensional torus $\mathbb{T}^3 = \mathbb{R}^3/(2\pi\mathbb{Z})^3$. A classical solution on a given time interval $[0,T]$ is defined to be a pair $(v,p) \in C^1(\mathbb{T}^3 \times [0,T])$ which solves (4) pointwise.

As far as weak solutions are concerned, there are several notions (see for instance the survey article [26] and the lecture notes [50]). One commonly considered in the literature consists of pairs $(v,p) : \mathbb{T}^3 \times [0,1] \to \mathbb{R}^3 \times \mathbb{R}$ which solve (4) in the sense of distributions[2].

For classical solutions (i.e. if $v \in C^1$) the total energy

$$e(t) := \frac{1}{2}\int_{\mathbb{T}^3} |v(x,t)|^2\, dx$$

is conserved by the flow induced by (4), so that $e(t) = e(0)$. The proof is an easy computation. We can scalar multiply the first equation by v to derive:

$$\sum_j v_j \partial_t v_j + \sum_j v_j \sum_k v_k \partial_{x_k} v_j + \sum_j v_j \partial_{x_j} p = 0\,,$$

which we can rewrite as

$$\partial_t \frac{|v|^2}{2} + (v \cdot \nabla)\left(\frac{|v|^2}{2} + p\right) = 0\,.$$

Using the fact that v is divergence free we finally achieve

$$\partial_t \frac{|v|^2}{2} + \operatorname{div}\left(\left(\frac{|v|^2}{2} + p\right) v\right) = 0\,,$$

which integrated in the space variable implies

$$\frac{d}{dt}\int_{\mathbb{T}^3} \frac{|v|^2(x,t)}{2}\, dx = 0\,.$$

Onsager in [42] was the first to suggest the existence of weak solutions which might dissipate the energy. Based on calculations in Fourier space, he formulated the following conjecture (in fact he had a non-rigorous argument for part (a)).

Conjecture 3.1. *Consider periodic 3-dimensional weak solutions of (4), where the velocity v satisfies the uniform Hölder condition*

$$|v(x,t) - v(x',t)| \le C|x - x'|^\theta, \tag{5}$$

for constants C and θ independent of x, x' and t.

(a) If $\theta > \frac{1}{3}$, then the total kinetic energy of v is constant;

---

[2] Recall the classical computation that $(v \cdot \nabla)v = \operatorname{div}(v \otimes v)$ if $\operatorname{div} v = 0$, so that distributional solutions are defined for any $v \in L^2(\mathbb{T}^3 \times [0,1])$.

(b) *For any $\theta < \frac{1}{3}$ there are solutions v for which the total kinetic energy* is not *constant.*

As already mentioned, Onsager in [42] actually suggested the existence of solutions for which the energy is strictly decreasing: in order to distinguish them from "non-conservative" weak solutions we will call them *dissipative solutions.* Moreover the threshold $\frac{1}{3}$ has a deep physical meaning, since it is related to the Kolmogorov's energy spectrum in his theory of fully developed turbulence, that Onsager rediscovered independently.

4 Energy conservation

The "positive part" of the Conjecture, namely statement (a), was proved by Constantin, E and Titi in [18] (a previous work of Eyink, [30], reached the critical threshold $\frac{1}{3}$ in a different scale of spaces). The argument of Constantin, E and Titi is very elegant and surprisingly simple. Fix a weak solutions (v, p) and consider a standard family of mollifiers φ_ε *in space.* As usual we will denote by v_ε the mollification of v with the kernel φ_ε. Clearly

$$\operatorname{div} v_\varepsilon = 0\,.$$

On the other hand the momentum balance in the Euler equations has a nonlinear term and for this reason v_ε is not a solution. We can however regard it as an "approximate solution":

$$\partial_t v_\varepsilon + \operatorname{div} v_\varepsilon \otimes v_\varepsilon + \nabla p_\varepsilon = \operatorname{div} \underbrace{(v_\varepsilon \otimes v_\varepsilon - (v\otimes v)_\varepsilon)}_{=:T_\varepsilon}\,.$$

It is not difficult to show that v_ε has enough smoothness to carry on the computations of the previous section, which therefore show

$$\frac{d}{dt}\frac{1}{2}\int \frac{|v_\varepsilon|^2}{2}(x,t)\,dx = \int T_\varepsilon : Dv_\varepsilon\,. \tag{6}$$

Since the right hand side of such equation has a commutator structure, a clever, yet elementary, computation allows to show that it converges to 0 as $\varepsilon \downarrow 0$ as soon as $v \in C^{0,1/3+\varepsilon}$ (in fact one can allow for slightly less regularity, cf. [18] and [16]).

Theorem 2.2 is proved in [20] with a very similar strategy. The isometric immersion u is approximated with a standard mollification procedure by smooth immersions u_ε. After writing the area formula for the Gauss map N_ε of $u_\varepsilon(\Sigma)$, we can understand Gauss' Theorema Egregium as a suitable family of integral identities. Something similar can be done for the Euler equations as well: we can write a suitable local version of the energy conservation as an integral identity involving a smooth test function; the equation (6) corresponds to the particular case of choosing a test function identically equal to 1.

Theorem 2.2 is achieved in [20] by passing into the limit in such identities as $\varepsilon \downarrow 0$. The convergence follows from a commutator estimate which has a striking similarity with the one of [18].

5 Energy dissipation: L^2, L^∞ and C^0

Concerning part (b), the first construction ever of an L^2 solution that violates the energy conservation is due to Scheffer in [45]. A different argument was later given by Shnirelman in [47], who was also able, a few years later, to give the first construction of a solution which dissipates the energy, cf. [48]. In [24] we gave a rather simple proof of these results, constructing bounded weak solutions of the incompressible Euler equations which violate the usual conservation energy and the uniqueness of the Cauchy problem in several ways (see also [25]). The key was to regard solutions of the system (4) as divergence-free matrix fields satisfying a suitable algebraic constraint: in particular we realized that this point of view allowed to use well established techniques from the theory of differential inclusions, cf. [15, 8, 21, 40, 37].

In the latter field, the authors consider systems of partial differential equations which prescribe the values of the gradients of the solutions and thus clearly the system (3) is a differential inclusion. A couple of decades ago the groundbreaking paper [40] of Müller and Šverak established a fruitful connection between the techniques used in the theory of differential inclusions and Gromov's h-principle (more precisely his convex integration methods) of which the Nash–Kuiper theorem is a primary example, cf. [31].

Our intuition that a suitable approach à la Nash could provide a line of attack for part (b) of the Conjecture was confirmed by the following result, which we proved in [27], using a suitable "convex integration scheme".

Theorem 5.1. *Given any positive smooth function e on $[0,T]$ there is a pair $(v,p) : \mathbb{T}^3 \times [0,T] \to \mathbb{R}^3 \times \mathbb{R}$ of continuous functions which solves (4) in the distributional sense and satisfies $\frac{1}{2}\int_{\mathbb{T}^3} |v|^2(x,t)\,dx = e(t)$.*

The construction of continuous solutions of (4) follows the basic strategy of Nash in the sense that at each step of the iteration we add a highly oscillatory correction. Note that both (4) and the equation of isometries (3) are quadratic – the oscillatory perturbation is chosen in such a way as to minimize the linearization.

Indeed we construct a sequence of subsolutions (v_q, p_q, R_q), i.e. solutions of

$$\begin{cases} \partial_t v_q + \operatorname{div} v_q \otimes v_q + \nabla p_q = -\operatorname{div} R_q \\ \operatorname{div} v_q = 0 \end{cases} \tag{7}$$

and iteratively remove the error R_q, which is a symmetric 3×3 matrix field. As a first observation note that if one is only interested in measuring the "distance" of a smooth pair (v_q, p_q) from being a solution of (4), then only the traceless part of R_q is relevant: we can write

$$R_q = \rho_q \mathrm{Id} + \mathring{R}_q,$$

where $\mathring{R}_q$ is a traceless 3×3 symmetric matrix, since $\operatorname{div}(\rho_q \mathrm{Id}) = \nabla \rho_q$. Hence if $\mathring{R}_q = 0$ then v_q is a solution of the Euler equations (perhaps with a different pressure).

Recall that we also aim in Theorem 5.1 at satisfying a certain energy profile for the total kinetic energy. We choose therefore a sequence $e_q = e_q(t)$ with $e_q(t) \to e(t)$ and set

$$\rho_q(t) := \frac{1}{3(2\pi)^3}\left(e_{q+1}(t) - \frac{1}{2}\int_{\mathbb{T}^3} |v_q(x,t)|^2\,dx\right),$$

$$R_q(x,t) := \rho_q(t)\mathrm{Id} + \mathring{R}_q(x,t)\,.$$

In the next section we explain the key points of the iteration and also which kind of Hölder regularity one could expect for the final solution.

6 A Nash-type iteration

Our aim is to build a sequence of triples $(v_q, p_q, \mathring{R}_q)$ solving (7) which converge uniformly to a triple $(v, p, 0)$ (actually in what follows we will mostly focus on the velocity v). The sequence will be achieved iteratively by adding a suitable perturbation to v_q and p_q. We thus set

$$w_q = v_q - v_{q-1}.$$

The size of w_q will be controlled with two parameters. The *amplitude* δ_q bounds the C^0 norm:

$$\|w_q\|_0 \lesssim \delta_q^{1/2}\,. \tag{8}$$

Up to negligible errors the Fourier transform of the perturbation w_q will be localized in a shell centered around a given *frequency* λ_q. Hence

$$\|\nabla w_q\|_0 \lesssim \delta_q^{1/2}\lambda_q\,. \tag{9}$$

Along the iteration we will have $\delta_q \to 0$ and $\lambda_q \to \infty$ at a rate that is at least exponential. For the sake of definiteness we may think

$$\lambda_q := \lambda^q \quad \text{and } \delta_q := \lambda_q^{-2\theta_0} \tag{10}$$

for some $\lambda > 1$ (although in the actual proofs a slightly super-exponential growth is required). The positive number θ_0 is the threshold Hölder regularity which we are able to achieve through the iteration, since it can be easily shown by interpolation that $\|v_q - v_{q-1}\|_\alpha = \|w_q\|_\alpha \lesssim \delta_q^{1/2}\lambda_q^\alpha \lesssim \lambda_q^{\alpha-\theta_0}$ and thus $\{v_q\}_q$ is a Cauchy sequence in C^α whenever $\alpha < \theta_0$.

The perturbation w_{q+1} is added to "balance" the error R_q and indeed we will see that $R_q \sim w_{q+1} \otimes w_{q+1}$. For this reason we will have

$$\|\mathring{R}_q\|_0 \le c_0\delta_{q+1} \tag{11}$$

$$\|\nabla\mathring{R}_q\|_0 \lesssim \delta_{q+1}\lambda_q \tag{12}$$

The main part of the perturbation w_{q+1} satisfies (ideally, as we will see later) an Ansatz of the type

$$w_o(x,t) = W\Big(v_q(x,t), R_q(x,t), \lambda_{q+1}x, \lambda_{q+1}t\Big)\,, \tag{13}$$

where W is a function which we are going to specify next. The pressure p_{q+1} will be defined similarly as $p_{q+1} = p_q + P(v_q, R_q, \lambda_{q+1}x, \lambda_{q+1}t)$, but we will not enter into the details in our discussion, since its role is anyway secondary.

First of all, the oscillatory nature of the perturbation requires us to impose that W is periodic in the variable $\xi \in \mathbb{T}^3$. Next, observe that v_{q+1} must satisfy the divergence-free condition $\operatorname{div} v_{q+1} = 0$ and $v + w_o$ is not likely to fulfill this: we need to add a suitable correction w_c in order to satisfy it. Consider therefore a vector potential for v_q, namely write v_q as $\nabla \times z_q$ for some smooth z_q. Subsequently we would like to perturb z_q to a new

$$z_{q+1}(x,t) = z_q(x,t) + \frac{1}{\lambda_{q+1}} Z(v(x,t), R(x,t), \lambda_{q+1}x, \lambda_{q+1}t)\,.$$

Computing $v_{q+1} := \nabla \times z_{q+1}$ we get

$$v_{q+1}(x,t) = v_q(x,t) + \underbrace{(\nabla_\xi \times Z)(v(x,t), \tilde{R}(x,t), \lambda x, \lambda t)}_{(P)} + O\left(\frac{1}{\lambda}\right)\,.$$

The term (P) would correspond to w_o if we were able to find a vector potential Z for W which is *periodic in* ξ. This requires $\operatorname{div}_\xi W = 0$ and $\langle W \rangle = 0$, where we use the notation $\langle , \rangle$ to denote the average in the ξ variable.

Similar considerations (see for instance [50]) lead to the following set of conditions that we would like to impose on W:

- $\xi \mapsto W(v, R, \xi, \tau)$ is 2π-periodic with vanishing average, i.e.

$$\langle W \rangle := \frac{1}{(2\pi)^3} \int_{\mathbb{T}^3} W(v, R, \xi, \tau)\, d\xi = 0; \tag{H1}$$

- The average stress is given by R, i.e.

$$\langle W \otimes W \rangle = R\,; \tag{H2}$$

- The "cell problem" is satisfied:

$$\begin{cases} \partial_\tau W + v \cdot \nabla_\xi W + \operatorname{div}_\xi (W \otimes W) + \nabla_\xi P = 0 \\ \operatorname{div}_\xi W = 0\,, \end{cases} \tag{H3}$$

  where $P = P(v, R, \xi, \tau)$ is a suitable pressure;
- W is smooth in all its variables and satisfies the estimates

$$|W| \lesssim |R|^{1/2},\ |\partial_v W| \lesssim |R|^{1/2},\ |\partial_R W| \lesssim |R|^{-1/2}. \tag{H4}$$

As a consequence of (H1)-(H2) we obtain

$$\int_{\mathbb{T}^3} |v_{q+1}|^2\,dx \sim \int_{\mathbb{T}^3} |v_q|^2\,dx + \int_{\mathbb{T}^3} \langle |W|^2\rangle\,dx = \int_{\mathbb{T}^3} |v_q|^2\,dx + 3(2\pi)^3\rho_q(t)$$

and thus the total kinetic energy of the v_{q+1} is (up to small errors) $e_{q+1}(t)$.

Having defined the couple (v_{q+1}, p_{q+1}) we face the problem of finding a suitable stress tensor $\mathring{R}_{q+1}$. An important remark is that it is possible to select a good "elliptic operator" which solves the equations $\operatorname{div} \mathring{R} = f$. The relevant technical lemma is the following one.

Lemma 6.1 (The operator div^{-1}). *There exists a homogeneous Fourier-multiplier operator of order -1, denoted*

$$\operatorname{div}^{-1} : C^\infty(\mathbb{T}^3;\mathbb{R}^3) \to C^\infty(\mathbb{T}^3;\mathcal{S}_0^{3\times 3})$$

such that, for any $f \in C^\infty(\mathbb{T}^3;\mathbb{R}^3)$ with average $\fint_{\mathbb{T}^3} f = 0$ we have

(a) $\operatorname{div}^{-1} f(x)$ *is a symmetric trace-free matrix for each* $x \in \mathbb{T}^3$*;*
(b) $\operatorname{div}\operatorname{div}^{-1} f = f$.

Assuming the existence of an ideal profile W, the next stress tensor $\mathring{R}_{q+1}$ would then be defined through

$$\begin{aligned}
\mathring{R}_{q+1} &= -\operatorname{div}^{-1}\Big[\partial_t v_{q+1} + \operatorname{div}(v_{q+1}\otimes v_{q+1}) + \nabla p_{q+1}\Big]\\
&= \underbrace{-\operatorname{div}^{-1}\Big[\partial_t w_{q+1} + v_q\cdot\nabla w_{q+1}\Big]}_{=:\mathring{R}^{(1)}_{q+1}}\\
&\quad \underbrace{-\operatorname{div}^{-1}\Big[\operatorname{div}(w_{q+1}\otimes w_{q+1} - R_q) + \nabla(p_{q+1}-p_q)\Big]}_{=:\mathring{R}^{(2)}_{q+1}}\\
&\quad \underbrace{-\operatorname{div}^{-1}\Big[w_{q+1}\cdot\nabla v_q\Big]}_{=:\mathring{R}^{(3)}_{q+1}}
\end{aligned} \tag{14}$$

where div^{-1} is the operator of order -1 from Lemma 6.1. Since we are assuming that the size of the corrector w_c is negligible compared to w_o, we will discuss the corresponding terms where w_o replaces w_q.

The main issues are therefore

- to show that indeed it is possible to send δ_q to 0 as $q\uparrow\infty$ (so that the scheme converges)
- and to obtain a relation between δ_q and λ_q in the form of (10).

If we were able to find a "profile" W satisfying (H1)-(H2)-(H3)-(H4), then the iteration proposed so far would lead to a proof of the Onsager's conjecture. In order to see this first expand $W(v, R, \xi, \tau)$ as a Fourier series in ξ. We then could compute

$$\mathring{R}^{(3)} = \operatorname{div}^{-1}\left[w_o \cdot \nabla v_q\right] = \operatorname{div}^{-1} \sum_{k \in \mathbb{Z}^3, k \neq 0} c_k(x,t) e^{i\lambda_{q+1} k\cdot x}, \tag{15}$$

where the coefficients $c_k(x,t)$ vary much slower than the rapidly oscillating exponentials. When we apply the operator div^{-1} we can therefore treat the c_k as constants and gain a factor $\frac{1}{\lambda_{q+1}}$ in the outcome: a typically "stationary phase argument". Note that it is crucial that c_0 vanishes: this is in fact the content of condition (H1).

Using (H4) we can estimate the size of each term c_k as

$$\|c_k\|_0 \lesssim \|W\|_0 \|\nabla v_q\|_0 \lesssim \|R_q\|_0^{1/2} \|\nabla v_q\|_0.$$

Applying (9) and (11) we arrive at

$$\|\mathring{R}^{(3)}_{q+1}\|_0 \lesssim \frac{\delta_{q+1}^{1/2} \delta_q^{1/2} \lambda_q}{\lambda_{q+1}}. \tag{16}$$

In fact in our computations so far we are ignoring a lot of technical issues: the relevant estimates are much more complicated and affected by several other terms which we are neglecting.

Similar arguments for the two other error tensors $\mathring{R}^{(1)}_{q+1}$ and $\mathring{R}^{(2)}_{q+1}$ would lead to an estimate of type

$$\|\mathring{R}_{q+1}\|_0 \lesssim \frac{\delta_{q+1}^{1/2} \delta_q^{1/2} \lambda_q}{\lambda_{q+1}}. \tag{17}$$

Of course, this is just one of the estimates for $(v_{q+1}, p_{q+1}, R_{q+1})$ and similar ones should be obtained for all the other quantities (and for other norms). However, (17) already implies a relation between δ_q and λ_q. Indeed, comparing it with (11), the inductive step requires

$$\delta_{q+2} \sim \frac{\delta_{q+1}^{1/2} \delta_q^{1/2} \lambda_q}{\lambda_{q+1}}.$$

Assuming $\lambda_q \sim \lambda^q$ for some fixed $\lambda \gg 1$, this would lead to

$$\delta_q^{1/2} \sim \lambda^{-q/3} \sim \lambda_q^{-1/3}, \tag{18}$$

which gives $\theta_0 = 1/3$ as the critical Hölder regularity.

7 First Höder regularity and improvement

It tuns out that almost all conditions on the function $W = W(v, R, \xi, \tau)$ can be fulfilled, as shown in [27]. Let us first examine the simple case in which we set $v = 0$: it is then possible to construct a function $W_s(R, \xi) = W(0, R, \xi, \tau)$ satisfying the constraints (H1)-(H4). The basic building block is given by Beltrami flows. For the details we refer the reader to [27], but one important aspect of Beltrami flows is that one can construct several different W_s with the property that any linear combinations of them still satisfy (H1)-(H3)-(H4) (and a suitable version of (H2)). In fact W_s takes the form

$$W_s(R,\xi) = \sum_{k\in\Lambda} a_k(R) B_k e^{ik\cdot\xi}$$

where Λ is a subset of $\mathbb{Z}^3$ with the property that $|k|$ is a fixed constant for every $k \in \Lambda$ and B_k is related to k by a precise algebraic formula. Note in particular that two distinct profiles W_s^1 and W_s^2 whose linear combination is still a profile can be obtained by choosing two disjoint faimilies Λ's in the same sphere intersected with the lattice $\mathbb{Z}^3$.

Another aspect which is important about Beltrami flows is that the corresponding stationary profiles W_s are only defined for R in a suitably small cone $\mathcal{C}$ of tensors R, whose axis is the half-line $\{\lambda \mathrm{Id} : \lambda \in \mathbb{R}^+\}$.

Having obtained a profile $W(0,R,\xi,\tau) = W_s(R,\xi)$, it seems natural to extend W by imposing that $\partial_\tau W + v\cdot\nabla_\xi W = 0$, leading to the formula

$$W(v,R,\xi,\tau) = W_s(R,\xi - v\tau) = \sum_{k\in\Lambda} a_k(R) B_k e^{i(k-v\tau)\cdot\xi}\,. \tag{19}$$

However the latter fails to satisfy (H4), because $|\partial_v W(v,R,\xi,\tau)| \sim |R|^{1/2}|\tau|$. This is a serious problem: observing that τ is the "fast time" variable, in the construction (13) $\tau = \lambda_{q+1} t$, leading to an additional factor λ_{q+1} in the estimates for $\mathring{R}^{(1)}_{q+1}$ and $\mathring{R}^{(2)}_{q+1}$: this loss destroys any hope that the scheme might converge.

In [27] a "phase function" $\phi_k(v,\tau)$ was introduced to deal with the transport part of the cell problem. By considering W of the form

$$\sum_{|k|=\lambda_0} a_k(R)\phi_k(v,\tau) B_k e^{ik\cdot\xi} \tag{20}$$

the cell problem in (H3) leads to the equation

$$\partial_\tau \phi_k + i(v\cdot k)\phi_k = 0\,.$$

Since the exact solution $\phi_k(v,\tau) = e^{-i(v\cdot k)\tau}$ is incompatible with the requirement (H4), an approximation is used such that

$$\partial_\tau \phi_k + i(v\cdot k)\phi_k = O\left(\mu_q^{-1}\right), \qquad |\partial_v \phi_k| \lesssim \mu_q$$

for some new parameter μ_q. To be precise, the approximation involves a partition of unity over the space of velocities and the use of 8 distinct families $\Lambda^{(j)}$.

This leads to the following corrections to (H3) and (H4): (H3) is only satisfied approximately,

$$\partial_\tau W + v\cdot\nabla_\xi W + \mathrm{div}_\xi(W\otimes W) + \nabla_\xi P = O(\mu_q^{-1})$$

and in (H4) the second inequality is replaced by

$$|\partial_v W| \lesssim \mu_q |R|^{1/2}.$$

In [28] the approach above was subsequently used to show the first example of Hölder flows with prescribed energy profiles, more precisely:

Theorem 7.1. *Given any positive smooth function e on $[0,T]$ and any $\alpha < \frac{1}{10}$ there is a pair $(v,p) : \mathbb{T}^3 \times [0,T] \to \mathbb{R}^3 \times \mathbb{R}$ of C^α functions which solves (4) in the distributional sense and satisfies $\frac{1}{2}\int_{\mathbb{T}^3} |v|^2(x,t)\,dx = e(t)$.*

A further improvement was obtained in [12], following an idea first introduced by Isett in [34]. We change the Ansatz (20) on W and look for a perturbation w_o which has the form

$$w_o(x,t) = W_s(R_q(x,t), \lambda_{q+1}\Phi_q(x,t)) = \sum_{k\in\Lambda^{(1)}} a_k(R_q(x,t))B_k e^{i\lambda_{q+1}\Phi_q(x,t)}, \quad (21)$$

where Φ_q solves the transport equation

$$\partial_t \Phi_q + (v_q \cdot \nabla_x)\Phi_q = 0\,. \quad (22)$$

With (21), we would have

$$\mathring{R}^{(1)}_{q+1} = \sum_{k\in\Lambda^{(1)}} \nabla a_k(R_q)(\partial_t R_q + (v_q\cdot\nabla)R_q)e^{i\lambda_{q+1}\Phi_q}\,. \quad (23)$$

Assuming that $D\Phi_q(x,t)$ is not too far from the identity, the stationary phase argument leads to

$$\|\mathring{R}^{(1)}\|_0 \lesssim \delta_{q+1}^{3/2}\delta_q^{1/2}\lambda_q\lambda_{q+1}^{-1}\,. \quad (24)$$

In fact in the latter estimates we are also assuming that the advective derivative $\partial_t R_q + (v_q\cdot\nabla)R_q$ satisfies a better bound than the usual derivative DR_q. This is indeed correct, as first pointed out by Isett in [34], and intuitively it can be justified by observing that even the advective derivative $(\partial_t v_q + v_q\cdot\nabla)v_q$ satisfies a better bound than Dv_q.

However, since $\|Dv_q\|_0 \to \infty$, we expect the deformation matrix $D\Phi_q$ to be controllable only for short times. More precisely, by a well-known elementary estimate on ODEs, if $\Phi_q(x,t_0) = x$, then

$$\|D\Phi_q(\cdot,t) - \mathrm{Id}\|_0 \lesssim \|\nabla v_q\|_0 |t-t_0| \lesssim \delta_q^{1/2}\lambda_q|t-t_0| \quad (25)$$

for $|t-t_0| \lesssim (\delta_q^{1/2}\lambda_q)^{-1}$. The latter is a typical "CFL condition", cf. [19].

To handle this problem we proceed as in [12] and consider a partition of unity $(\chi_j)_j$ on the time interval $[0,T]$ such that the support of each χ_j is an interval I_j of size $\frac{1}{\mu_q}$ for some $\mu_q \gg 1$. In each time interval I_j we set $\Phi_{q,j}$ to be the solution of the transport equation (22) which satisfies

$$\Phi_{q,j}(x,t_j) = x,$$

where t_j is the center of the interval I_j. Recalling that $\|Dv_q\|_0 \lesssim \delta_q^{1/2}\lambda_q$, (25) leads to

$$\|D\Phi_{q,j}\|_0 = O(1) \quad \text{and} \quad \|D\Phi_{q,j} - \mathrm{Id}\|_0 \lesssim \frac{\delta_q^{1/2}\lambda_q}{\mu_q} \tag{26}$$

provided

$$\mu_q \geq \delta_q^{1/2}\lambda_q, \tag{27}$$

an estimate we will henceforth assume. Observe also that $|\partial_t \chi_j| \lesssim \mu_q$.

The new fluctuation will take the form

$$w_o = \sum_j \chi_j(t) \sum_{k\in\Lambda^{(i(j))}} a_k(R_q) B_k e^{i\lambda_{q+1}k\cdot\Phi_{q,j}} \tag{28}$$

where:

- $i(j)$ equals 1 if j is odd and 2 if j is even;
- $\Lambda^{(1)}$ and $\Lambda^{(2)}$ are two disjoint families.

The new Ansatz leads then to the following estimate

$$\|\mathring{R}_{q+1}\|_0 \lesssim \delta_{q+1}^{1/2}\mu_q\lambda_{q+1}^{-1} + \delta_{q+1}\delta_q^{1/2}\lambda_q\mu_q^{-1} \tag{29}$$

Optimizing in μ_q we then reach

$$\|\mathring{R}_{q+1}\|_0 \lesssim \delta_{q+1}^{3/4}\delta_q^{1/4}\lambda_q^{1/2}\lambda_{q+1}^{-1/2}, \tag{30}$$

namely

$$\delta_{q+2} \sim \delta_{q+1}^{3/4}\delta_q^{1/4}\lambda_q^{1/2}\lambda_{q+1}^{-1/2}.$$

The latter relation leads to a threshold $\theta_0 = \frac{1}{5}$ and hence to the following theorem

Theorem 7.2. *Given any positive smooth function e on $[0,T]$ and any $\alpha < \frac{1}{5}$ there is a pair $(v,p) : \mathbb{T}^3 \times [0,T] \to \mathbb{R}^3 \times \mathbb{R}$ of C^α functions which solves (4) in the distributional sense and satisfies $\frac{1}{2}\int_{\mathbb{T}^3} |v|^2(x,t)\,dx = e(t)$.*

8 First Onsager-critical construction

In [11] Buckmaster observed that, by choosing the cut-off functions χ_i appropriately in (28) it is possible to show that the solution produced in the proof of Theorem 7.2 enjoys $C^{1/3-\varepsilon}$ regularity at almost every time-slice. The idea is to make the cut-off flat on large portions of their supports while paying very steep time derivatives on small portions. The price to pay is thatthe "global" Hölder control gets much weaker: the solutions is just slightly better than continuous (i.e. it has a very small Hölder exponent, depending on ε). In [9], jointly with Buckmaster and Székelyhidi we exploited a quantitative version of the latter idea to reach the first nonconservative solutions up Onsager's threshold 1/3, albeit in a weaker form than as stated in his conjecture.

Theorem 8.1. *For every $\alpha < \frac{1}{3}$ there are a nontrivial continuous compactly supported solution $(v,p) : \mathbb{T}^3 \times \mathbb{R} \to \mathbb{R}^3 \times \mathbb{R}$ of (4) and an L^1 function $C : \mathbb{R} \to \mathbb{R}^+$ such that*

$$|v(x,t) - v(y,t)| \leq C(t)|x-y|^\alpha \qquad \forall t \in \mathbb{R}, \forall x,y \in \mathbb{T}^3.$$

9 h-principle

The Beltrami flows together with the transport Ansatz explained in the previous sections settle the issue of convergence (at least for Hölder exponents $\theta < 1/5$), but are not sufficient to conclude an h-principle statement which is a satisfactory counterpart of Theorem 1.2. The reason is that the stationary profiles W_s defined through Beltrami flows are only defined for R's belonging to a suitably small cone of tensors.

Nevertheless, there is a very simple set of stationary flows (which we will call "Mikado flows") based on pipe flow, which can generate all R. These flows were introduced by Daneri and Székelyhidi in [22].

Lemma 9.1. *For any compact subset $\mathcal{N}$ consisting of positive definite 3×3 matrices there exists a smooth vector field*

$$W_s : \mathcal{N} \times \mathbb{T}^3 \to \mathbb{R}^3, \quad i = 1, 2$$

such that, for every $R \in \mathcal{N}$

$$\begin{cases} \operatorname{div}_\xi (W_s(R,\xi) \otimes W(R,\xi)) = 0, \\ \operatorname{div}_\xi W_s(R,\xi) = 0, \end{cases} \tag{31}$$

and

$$\langle W_s \rangle = 0, \tag{32}$$
$$\langle W_s \otimes W_s \rangle = R. \tag{33}$$

In particular, in [22] the authors could prove the following h-principle result

Theorem 9.2. *Let $(\bar{v}, \bar{p}, \bar{R})$ be a smooth solution of*

$$\begin{cases} \partial_t \bar{v} + \operatorname{div}(\bar{v} \otimes \bar{v}) + \nabla \bar{p} = -\operatorname{div} \bar{R} \\ \operatorname{div} \bar{v} = 0 \end{cases} \tag{34}$$

on $\mathbb{T}^3 \times [0,T]$ such that $\bar{R}(x,t)$ is positive definite for all x, t. Then for any $\alpha < 1/5$ there exists a sequence $\{(v_k, p_k)\} \subset C^\alpha$ of weak solutions of (4) such that

$$v_k \overset{*}{\rightharpoonup} \bar{v} \quad \text{and} \quad v_k \otimes v_k \overset{*}{\rightharpoonup} \bar{v} \otimes \bar{v} + \bar{R} \quad \text{in } L^\infty$$

uniformly in time and furthermore for all $t \in [0,T]$

$$\int_{\mathbb{T}^3} v_k \otimes v_k \, dx = \int_{\mathbb{T}^3} (\bar{v} \otimes \bar{v} + \bar{R}) \, dx.$$

10 Isett's proof of Onsager's conjecture

The stationary profile W_s reached through the Mikado flows in [22] have not only the feature of being defined on any compact subset of positive definite R's, but they also yield improved estimates in several error terms in the iteration. On the other hand they are not compatible with the "patching in time" used in (28) and indeed in [22] they are only used for finitely many steps of the iteration, whereas the "tail" of the series $\sum_q w_q$ still consists of oscillatory perturbations whose building blocks are Beltrami flows.

In [35] Isett has been able to overcome this last obstruction by introducing a different "patching strategy". Isett's key idea can be easily explained as follows. Considered a given triple $(v_q, p_q, \mathring{R}_q)$ reached at a certain step of the iteration, satisfying all the estimates outlined in the previous subsection. The obstruction to using Mikado flows could be overcome if $\mathring{R}_q$ were supported in a union of disjoint time-stripes of the form $\mathbb{T}^3 \times [a_k, b_k]$, where $b_k - a_k \sim (\delta_q^{1/2}\lambda_q)^{-1}$, compatibly with the CFL condition (25). In this case there would be no need of "patching" the oscillatory perturbations, since they would be supported on disjoint time-stripes where the CFL condition holds and the flows $\Phi_{q,j}$ of the previous subsections are close to the identity.

In order to reach this ideal situation, Isett in [35] partitions the whole time interval in smaller intervals $[c_k, c_{k+1}]$ with size $\sim (\delta_q^{1/2}\lambda_q)^{-1}$. In intervals of comparable size it is possible to find exact solutions (z_k, r_k) of the incompressible Euler equations with $z_k(\cdot, c_k) = v_q(\cdot, c_k)$. Patching such solutions with a partition of unity one can obtain new velocity and pressure fields $(\tilde{v}_q, \tilde{p}_q)$ together with "separate" time-stripes where they are exact solutions of the Euler equations. The remaining regions consist of time-stripes where we have to find a new stress tensor $\tilde{R}_q$. Isett shows that such tensor can be found so that its size is not much larger than $\mathring{R}_q$: in fact it essentially satisfies the same estimates with worse constants.

Now there are no obstructions to apply the oscillatory perturbations of [22] to the new triple $(\tilde{v}_q, \tilde{p}_q, \tilde{R}_q)$ and therefore one can reach the following statement

Theorem 10.1. *For every $\alpha < \frac{1}{3}$ there is a nontrivial continuous compactly supported solution $(v,p) \in C^\alpha(\mathbb{T}^3 \times \mathbb{R})$ of* (4).

11 h-principle and Onsager's conjecture with dissipative solutions

In the previous "patching" of exact solutions of the Euler equations a canonical choice of the stress tensor $\tilde{R}_q$ would be

$$\tilde{R}_q := \operatorname{div}^{-1}[\partial_t \tilde{v}_q + (\tilde{v}_q \cdot \nabla)\tilde{v}_q + \nabla \tilde{p}_q]\,. \tag{35}$$

However [35] generates $\tilde{R}_q$ with a different, more complicated, procedure, since the author is not able to reach the desired estimate through the operator div^{-1}. A suboptimal outcome is that Theorem 10.1 does not produce "dissipative solutions".

This has been instead accomplished in [10], where in a joint work with Buckmaster, Székelyhidi and Vicol we derive appropriate estimates for the "canonical" $\tilde{R}_q$ as defined in (35). We can therefore derive the existence of dissipative solutions in the whole range of Hölder exponents of the second part of Onsager's conjecture. Indeed such a statement is obtained as a corollary of the exact counterpart of the h-principle result in [22]:

Theorem 11.1. *Let $(\bar{v}, \bar{p}, \bar{R})$ be a smooth solution of (34) on $\mathbb{T}^3 \times [0,T]$ such that $\bar{R}(x,t)$ is positive definite for all x, t. Then for any $\alpha < 1/3$ there exists a sequence $\{(v_k, p_k)\} \subset C^\alpha$ of weak solutions of (4) such that*

$$v_k \overset{*}{\rightharpoonup} \bar{v} \quad \text{and} \quad v_k \otimes v_k \overset{*}{\rightharpoonup} \bar{v} \otimes \bar{v} + \bar{R} \quad \text{in } L^\infty$$

uniformly in time and furthermore for all $t \in [0,T]$

$$\int_{\mathbb{T}^3} v_k \otimes v_k \, dx = \int_{\mathbb{T}^3} (\bar{v} \otimes \bar{v} + \bar{R}) \, dx.$$

Corollary 11.2. *For every $\alpha < \frac{1}{3}$ and every positive smooth $e : [0,T] \to \mathbb{R}$ there exists a solution $(v,p) \in C^{1/3}(\mathbb{T}^3 \times [0,T])$ of (4) such that*

$$\frac{1}{2} \int_{\mathbb{T}^3} |v(x,t)|^2 \, dx = e(t) .$$

References

[1] A. D. Alexandrov. *Intrinsic geometry of convex surfaces.* OGIZ, Moscow-Leningrad, 1948.

[2] Ju. F. Borisov. The parallel translation on a smooth surface. I. *Vestnik Leningrad. Univ.*, 13(7):160–171, 1958.

[3] Ju. F. Borisov. The parallel translation on a smooth surface. II. *Vestnik Leningrad. Univ.*, 13(19):45–54, 1958.

[4] Ju. F. Borisov. The parallel translation on a smooth surface. III. *Vestnik Leningrad. Univ.*, 14(1):34–50, 1959.

[5] Ju. F. Borisov. The parallel translation on a smooth surface. IV. *Vestnik Leningrad. Univ.*, 14(13):83–92, 1959.

[6] Ju. F. Borisov. $C^{1,\alpha}$-isometric immersions of Riemannian spaces. *Doklady*, 163:869–871, 1965.

[7] Yu.F. Borisov. Irregular $C^{1,\beta}$-surfaces with analytic metric. *Sib. Mat. Zh.*, 45(1):25–61, 2004.

[8] A. Bressan and F. Flores. On total differential inclusions. *Rend. Sem. Mat. Univ. Padova*, 92:9–16, 1994.

[9] T. Buckmaster, C. De Lellis, and L. Székelyhidi, Jr. Dissipative Euler flows with Onsager-critical spatial regularity. *Comm. Pure Appl. Math.*, 69(9):1613–1670, 2016.

[10] T. Buckmaster, C. De Lellis, L. Székelyhidi, Jr., and V. Vicol. Onsager's conjecture for admissible weak solutions. *ArXiv e-prints*, January 2017. To appear in *Comm. Pure Appl. Math.*

[11] T. Buckmaster. Onsager's conjecture almost everywhere in time. *Comm. Math. Phys.*, 333(3):1175–1198, 2015.
[12] T. Buckmaster, C. De Lellis, P. Isett, and L. Székelyhidi, Jr. Anomalous dissipation for 1/5-Hölder Euler flows. *Ann. of Math. (2)*, 182(1):127–172, 2015.
[13] C Burstin. Ein Beitrag zum Problem der Einbettung der Riemannschen Räume in euklidischen Räumen. *Rec. Math. Moscou*, 38(3-4):74–85, 1931.
[14] E. Cartan. Sur la possibilité de plonger un espace Riemannien donné dans un espace Euclidien. *Ann. Soc. Polon. Math.*, 6:1–7, 1928.
[15] A. Cellina. On the differential inclusion $x' \in [-1, +1]$. *Atti Accad. Naz. Lincei Rend. Cl. Sci. Fis. Mat. Natur. (8)*, 69(1-2):1–6 (1981), 1980.
[16] A. Cheskidov, P. Constantin, S. Friedlander, and R. Shvydkoy. Energy conservation and Onsager's conjecture for the Euler equations. *Nonlinearity*, 21(6):1233–1252, 2008.
[17] St. Cohn-Vossen. Zwei Sätze über die Starrheit der Eiflächen. *Nachrichten Göttingen*, 1927:125–137, 1927.
[18] P. Constantin, W. E, and E. S. Titi. Onsager's conjecture on the energy conservation for solutions of Euler's equation. *Comm. Math. Phys.*, 165(1):207–209, 1994.
[19] R. Courant, K. Friedrichs, and H. Lewy. On the partial difference equations of mathematical physics. *IBM J. Res. Develop. 11* (1967), 215–234.
[20] S. Conti, C. De Lellis, and L. Székelyhidi, Jr. h-principle and rigidity for $C^{1,\alpha}$ isometric embeddings. In *Nonlinear partial differential equations*, volume 7 of *Abel Symp.*, pages 83–116. Springer, Heidelberg, 2012.
[21] B. Dacorogna and P. Marcellini. General existence theorems for Hamilton-Jacobi equations in the scalar and vectorial cases. *Acta Math.*, 178:1–37, 1997.
[22] S. Daneri and L. Székelyhidi, Jr. Non-uniqueness and h-principle for H\"older-continuous weak solutions of the Euler equations. *ArXiv e-prints*, March 2016.
[23] C. De Lellis, D. Inauen, and L. Székelyhidi, Jr. A Nash-Kuiper theorem for $C^{1,\frac{1}{5}-\delta}$ immersions of surfaces in 3 dimensions. *ArXiv e-prints*, October 2015. To appear in *Rev. Mat. Iberoam.*
[24] C. De Lellis and L. Székelyhidi, Jr. The Euler equations as a differential inclusion. *Ann. of Math. (2)*, 170(3):1417–1436, 2009.
[25] C. De Lellis and L. Székelyhidi, Jr. On admissibility criteria for weak solutions of the Euler equations. *Arch. Ration. Mech. Anal.*, 195(1):225–260, 2010.
[26] C. De Lellis and L. Székelyhidi, Jr. The h-principle and the equations of fluid dynamics. *Bull. Amer. Math. Soc. (N.S.)*, 49(3):347–375, 2012.
[27] C. De Lellis and L. Székelyhidi, Jr. Dissipative continuous Euler flows. *Invent. Math.*, 193(2):377–407, 2013.
[28] C. De Lellis and L. Székelyhidi, Jr. Dissipative Euler flows and Onsager's conjecture. *J. Eur. Math. Soc. (JEMS)*, 16(7):1467–1505, 2014.
[29] C. De Lellis and László Székelyhidi, Jr. High dimensionality and h-principle in PDE. *Bull. Amer. Math. Soc. (N.S.)*, 54(2):247–282, 2017.
[30] G. L. Eyink. Energy dissipation without viscosity in ideal hydrodynamics. I. Fourier analysis and local energy transfer. *Phys. D*, 78(3-4):222–240, 1994.

[31] M. Gromov. *Partial differential relations*, volume 9 of *Ergebnisse der Mathematik und ihrer Grenzgebiete (3)*. Springer-Verlag, Berlin, 1986.

[32] M. Gromov. Geometric, algebraic, and analytic descendants of Nash isometric embedding theorems. *Bull. Amer. Math. Soc. (N.S.)*, 54(2):173–245, 2017.

[33] G. Herglotz. Über die Starrheit der Eiflächen. *Abh. Math. Semin. Hansische Univ.*, 15:127–129, 1943.

[34] P. Isett. Hölder Continuous Euler Flows in Three Dimensions with Compact Support in Time. *ArXiv e-prints*, November 2012.

[35] P. Isett. A Proof of Onsager's Conjecture. *ArXiv e-prints*, August 2016.

[36] M. Janet. Sur la possibilité de plonger un espace Riemannien donné à n dimensions dans un espace Euclidien à $\frac{n(n+1)}{2}$ dimensions. *C. R. Acad. Sci., Paris*, 183:942–943, 1926.

[37] B. Kirchheim, S. Müller, and V. Šverák. Studying nonlinear PDE by geometry in matrix space. In Stefan Hildebrandt and Hermann Karcher, editors, *Geometric analysis and Nonlinear partial differential equations*, pages 347–395. Springer-Verlag, 2003.

[38] N. H. Kuiper. On C^1-isometric imbeddings. I, II. *Nederl. Akad. Wetensch. Proc. Ser. A.* **58** = *Indag. Math.*, 17:545–556, 683–689, 1955.

[39] H. Lewy. On the existence of a closed convex surface realizing a given Riemannian metric. *Proc. Natl. Acad. Sci. USA*, 24:104–106, 1938.

[40] S. Müller and V. Šverák. Convex integration for Lipschitz mappings and counterexamples to regularity. *Ann. of Math. (2)*, 157(3):715–742, 2003.

[41] J. Nash. C^1 isometric imbeddings. *Ann. of Math. (2)*, 60:383–396, 1954.

[42] L. Onsager. Statistical hydrodynamics. *Nuovo Cimento (9)*, 6(Supplemento, 2(Convegno Internazionale di Meccanica Statistica)):279–287, 1949.

[43] A. V. Pogorelov. *Izgibanie vypuklyh poverhnosteĭ.* Gosudarstv. Izdat. Tehn.-Teor. Lit., Moscow-Leningrad, 1951.

[44] A. V. Pogorelov. *Extrinsic geometry of convex surfaces.* American Mathematical Society, Providence, R.I., 1973. Translations of Mathematical Monographs, Vol. 35.

[45] V. Scheffer. An inviscid flow with compact support in space-time. *J. Geom. Anal.*, 3(4):343–401, 1993.

[46] L. Schläfli. Nota alla memoria del sig. Beltrami, "Sugli spazii di curvatura costante". *Annali di Mat. (2)*, 5:178–193, 1871.

[47] A. Shnirelman. On the nonuniqueness of weak solution of the Euler equation. *Comm. Pure Appl. Math.*, 50(12):1261–1286, 1997.

[48] A. Shnirelman. Weak solutions with decreasing energy of incompressible Euler equations. *Comm. Math. Phys.*, 210(3):541–603, 2000.

[49] M. Spivak. *A comprehensive introduction to differential geometry. Vol. V. 2nd ed.* Berkeley: Publish or Perish, Inc., 1979.

[50] L. Székelyhidi Jr. Weak solutions of the Euler equations: non-uniqueness and dissipation. *Journées Équations aux derivées partielles*, 2016.

[51] H. Weyl. Über die Bestimmung einer geschlossenen konvexen Fläche durch ihr Linienelement. Zürich. Naturf. Ges. 61, 40-72, 1916.

[52] S. T. Yau. Open problems in geometry. In *Differential geometry: partial differential equations on manifolds (Los Angeles, CA, 1990)*, volume 54 of

Proc. Sympos. Pure Math., pages 1–28. Amer. Math. Soc., Providence, RI, 1993.

INSTITUT FÜR MATHEMATIK, UNIVERSITÄT ZÜRICH
CH-8057 ZÜRICH, SWITZERLAND
E-mail address: camillo.delellis@math.uzh.ch

Proceedings of the 7th International Congress of Chinese Mathematicians
ALM 43, pp. 173–213

Seismic Imaging and Optimal Transport

Björn Engquist and Yunan Yang

Abstract

Seismology has been an active science for a long time. It changed character about 50 years ago when the earth's vibrations could be measured on the surface more accurately and more frequently in space and time. The full wave field could be determined, and partial differential equations (PDE) started to be used in the inverse process of finding properties of the interior of the earth. We will briefly review earlier techniques but mainly focus on Full Waveform Inversion (FWI) for the acoustic formulation. FWI is a PDE constrained optimization in which the variable velocity in a forward wave equation is adjusted such that the solution matches measured data on the surface. The minimization of the mismatch is usually coupled with the adjoint state method, which also includes the solution to an adjoint wave equation. L^2 norm is the conventional objective function measuring the difference between simulated and measured data, but it often results in the minimization trapped in local minima. One way to mitigate this is by selecting another misfit function with better convexity properties. Here we propose using the quadratic Wasserstein metric (W_2) as a new misfit function in FWI. The optimal map defining W_2 can be computed by solving a Monge-Ampère equation. Theorems pointing to the advantages of using optimal transport over L^2 norm will be discussed, and a number of large-scale computational examples will be presented.

Mathematics Subject Classification: 65K10, 65Z05, 86A15, 86A22.

Keywords and Phrases: Seismic imaging, full-waveform inversion, optimal transport, Monge-Ampère equation.

1 Introduction

Earth Science is an early scientific subject. The efforts started as early as AD 132 in China when Heng Zhang invented the first seismoscope in the world (Figure 1a). The goal was to record that an earthquake had happened and to try to determine the direction of the earthquake. Substantial progress in seismology had to wait until about 150 years ago when seismological instruments started to record travel time.

With increasing sophistication in devices measuring the vibrations of seismic waves and in the availability of high-performance computing increasingly advances mathematical techniques could be used to explore the interior of the earth. The development started with calculations by hand based on geometrical optics and travel time measurement. It continued with a variety of wave equations when the equipment allowed for measuring wave fields and modern computers became available. As we will see below a wide range of mathematical tools are used today

(a)

(b)

Figure 1: (a) The first Seismoscope designed in AD 132 and (b) Modern seismic vibrator used in seismic survey

in seismic imaging, including partial differential equation (PDE) constrained optimization, advanced signal processing, optimal transport and the Monge-Ampère equation.

Since 19th-century modern seismographs were developed to record seismic signals, which are vibrations in the earth. In 1798 Henry Cavendish measured the density of the earth with less than 1% error compared with the number we can measure nowadays. Nearly one hundred years later, German physicist Emil Wiechert first discovered that the earth has a layered structure and the theory was further completed as the crust-mantle-core three-layer model in 1914 by one of his student Beno Gutenberg. In the meantime, people studied the waves including body waves and surface waves to better understand the earthquake. P-waves and S-waves were first clearly identified for their separate arrivals by English geologist Richard Dixon Oldham in 1897. The Murchison earthquake in 1929 inspired the Danish female seismologist and geophysicist Inge Lehmann to study the unexpected P-waves recorded by the seismographs. Later on, she proposed that the core of the earth has two parts: the solid inner core of iron and a liquid outer core of nickel-iron alloy, which was soon acknowledged by peer geophysicists worldwide.

We will see that measuring travel time plays a vital role in the development of modern techniques for the inverse problem of finding geophysical properties from measurements of seismic waves on the surface. The methods are often related to travel time tomography. They are quite robust and cost-efficient for achieving low-resolution information of the subsurface velocities. The forward problem is based on ray theory or geometric optics [12, 128].

The development of man-made seismic sources and advanced recording devices (Figure 1b) facilitate the research on the entire wavefields in time and space (Figure 2a) rather than merely travel time. This setup results in a more controlled setting and large amounts of data, which is needed for an accurate inverse process of estimating geophysical properties, for example, Figure 2b. The forward model-

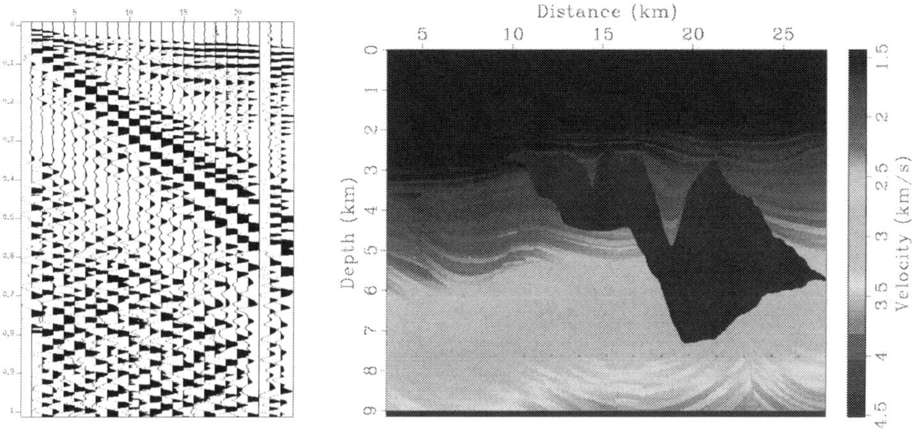

(a) (b)

Figure 2: (a) An example of the seismic data measured from the receivers and (b) Goal of inversion: geophysical properties as in the Sigsbee velocity model [5]

ing is a wave equation with many man-made sources and many receivers. The wave equation can vary from pure acoustic waves to anisotropic viscoelasticity. Even if there are various techniques in computational exploration seismology, there are two processes that currently stand out: reverse time migration (RTM) [6, 144] and full waveform inversion (FWI) [121, 124].

Migration techniques can be applied in both the time domain and the frequency domain following the early breakthroughs by Claerbout on imaging conditions [33, 34]. In RTM the computed forward wavefield starting from the source is correlated in time with the computed backward wavefield which is modeled with the measured data as the source term in the adjoint wave equation. The goal is to determine details of the reflecting surfaces as, for example, faults and sedimentary layers based on the measured data and a rough estimate of the geophysical properties. The least-squares reverse time migration (LSRTM) [43] is a new migration technique designed to improve the image quality generated by RTM. Reflectivity is regarded as a small perturbation in velocity, and the quantity is recovered through a linear inverse problem.

FWI is a high-resolution seismic imaging technique which recently gets great attention from both academia and industry [132]. The goal of FWI is to find both the small-scale and large-scale components which describe the geophysical properties using the entire content of seismic traces. A trace is the time history measured at a receiver. In this paper, we will consider the inverse problem of finding the wave velocity of an acoustic wave equation in the interior of a domain from knowing the Cauchy boundary data together with natural boundary conditions [36], which is implemented by minimizing the difference between computed and measured data on the boundary. It is thus a PDE-constrained optimization.

There are various kinds of numerical techniques that are used in seismic inversion, but FWI is increasing in popularity even if it is still facing three main computational challenges. First, the physics of seismic waves are complex, and we need more accurate forward modeling in inversion going from pure acoustic waves to anisotropic viscoelasticity [133]. Second, even as PDE-constrained optimization, the problem is highly non-convex. FWI requires more efficient and robust optimization methods to tackle the intrinsic nonlinearity. Third, the least-squares norm, classically used in FWI, suffers from local minima trapping, the so-called cycle skipping issues, and sensitivity to noise [114]. We will see that optimal transport based Wasserstein metric is capable of dealing with the last two limitations by including both amplitudes mismatches and travel time differences [47, 48].

We will introduce the mathematical formulation of these techniques in the following sections. The emphasis will be on FWI, but we will also summarize the state of the art of other standard imaging steps. Finally, we will relate FWI to RTM and LSRTM. These approaches all involve the interaction of the forward and the time-reversed wavefields, which is well known as the "imaging condition" in geophysics.

2 Seismic Imaging

Seismic data contains interpretable information about subsurface properties. Imaging predicts the spatial locations as well as specifies parameter values describing the earth properties that are useful in seismology. It is particularly important for exploration seismology which mainly focuses on prospecting for energy sources, such as oil, gas, coal. Seismic attributes contain both travel time records and waveform information to create an image of the subsurface to enable geological interpretation, and to obtain an estimate of the distribution of material properties in the underground. Usually, the problem is formulated as an inverse problem incorporating both physics and mathematics. Seismic inversion and migration are terms often used in this setting.

2.1 Seismic data

There are two types of seismic signals. Natural earthquakes propagate with substantial ultra-low frequency wave energy and penetrate deeply through the whole earth. Recorded by seismometers, the natural seismic waves are used to study earth structures. The other type of data is generated by man-made "earthquakes" to obtain an image of the sedimentary basins in the interior of the earth close to the surface. A wavefield has to be produced using suitable sources at appropriate locations, measured by receivers at other locations after getting reflected back from within the earth, and stored using recorders.

In this paper, we mainly discuss the second type of seismic events. The raw seismic data is not ideal to interpret and to create an accurate image of the subsurface. Recorded artifacts are related to the surface upon which the survey was performed, the instruments of receiving and recording and the noise generated by the procedure. We must remove or at least minimize these artifacts. Seismic data processing aims to eliminate or reduce these effects and to leave only the influences due to the structure of geology for interpretation. Typical data processing steps include but are not limited to deconvolution, demultiple, deghosting, frequency filtering, normal moveout (NMO) correction, dip moveout (DMO) correction, common midpoint (CMP) stack, vertical seismic profiling (VSP), etc [108, 142].

In the recent two decades, the availability of the increased computer power makes it possible to process each trace of the recorded common source gathers separately, aiming for a better image. We will discuss several primary imaging methods such as traveltime tomography, seismic migration, least squares migration and full waveform inversion (FWI).

2.2 Traveltime tomography

Most discoveries related to the structure of the earth were based on the assumption that seismic waves can be represented by rays, which is closely associated with geometric optics [107, 106, 145]. The primary advantages are its applicability to complex, isotropic and anisotropic, laterally varying layered media and its numerical efficiency in such computations. A critical observation is the travel time information of seismic arrivals. We can understand many arrival time observations with ray theory [26], which describes how short-wavelength seismic energy propagates.

As a background illustration, we will derive the ray tracing expressions in a 1D setting where the velocity only varies vertically [116]. Ray tracing in general 3D structure is more complicated but follows similar principles. Considering a laterally homogeneous earth model where velocity v only depends on depth, the ray parameter which is also called the horizontal slowness p, can be expressed in the following equation by the Snell's law:

$$p = s(z)\sin(\theta) = \frac{dT}{dX}, \tag{1}$$

where $s(z)$ $(= \frac{1}{v(z)})$ is the slowness, θ is the incidence angle, T is the travel time, X is the horizontal range. At the turning point depth z_p, $p = s(z_p)$, a constant for a given ray. The vertical slowness $\eta = \sqrt{s^2 - p^2}$.

When the velocity is a continuous function of depth, the surface to surface travel time $T(p)$ and the distance traveled $X(p)$ have the following expressions:

$$T(p) = 2\int_0^{z_p} \frac{s^2(z)}{\sqrt{s^2(z) - p^2}}dz = 2\int_0^{z_p} \frac{s^2(z)}{\eta}dz, \tag{2}$$

and

$$X(p) = 2p\int_0^{z_p} \frac{dz}{\sqrt{s^2(z) - p^2}} = 2p\int_0^{z_p} \frac{dz}{\eta}. \tag{3}$$

The expressions above are the forward problem in traveltime tomography. The seismologists are interested in inverting model parameter $s(z)$ from observed traveltime T and traveled distance X. Using integral transform pair, we can obtain

$$z(s) = -\frac{1}{\pi}\int_{s_0}^{s} \frac{X(p)}{\sqrt{p^2 - s^2(z)}}d(p) = \frac{1}{\pi}\int_0^{X(s)} \cosh^{-1}(p/s)dX, \tag{4}$$

which gives us the 1D velocity model.

Equation (4) is one example of the 1D velocity inversion problem at a given depth. There are limitations about traveltime tomography in general. First, the first arrivals are inherently nonunique. Second, the lateral velocity variations are not considered in this setting. If we divide the earth model into blocks, the 3D velocity inversion techniques can resolve some of the lateral velocity perturbations by using the travel time in each block. The problem can be formulated into a least-squares (L^2) inversion by minimizing the travel time residual between the predicted time and the observed time: $||t_{\text{obs}} - t_{\text{pred}}||_2^2$ [116, 146].

One limitation of ray theory is that it is applicable only to smooth media with smooth interfaces, in which the characteristic dimensions of inhomogeneities are considerably larger than the dominant wavelength of the considered waves. The ray method can yield distorted results and will fail at caustics or in general at so-called singular regions [28]. Moreover, much more information is available from the observed seismograms than travel times. To some extent, travel time tomography can be seen as phase-based inversion, and next, we will introduce waveform-based methods where the wave equation plays a significant role.

2.3 Reverse Time Migration

To overcome the difficulties of ray theory and further improve image resolutions, reverse time migration (RTM), least-squares reverse time migration (LSRTM) and full-waveform inversion (FWI) replace the semi-analytical solutions to the wave equation by fully numerical solutions including the full wavefield. Without loss of generality, we will explain all the methods in a simple acoustic setting:

$$\begin{cases} m(\mathbf{x})\frac{\partial^2 u(\mathbf{x},t)}{\partial t^2} - \triangle u(\mathbf{x},t) = s(\mathbf{x},t) \\ u(\mathbf{x},0) = 0 \\ \frac{\partial u}{\partial t}(\mathbf{x},0) = 0 \end{cases} \tag{5}$$

We assume the model $m(\mathbf{x}) = \frac{1}{c(\mathbf{x})^2}$ where $c(\mathbf{x})$ is the velocity, $u(\mathbf{x},t)$ is the wavefield, $s(\mathbf{x},t)$ is the source. It is a linear PDE but a nonlinear operator from model domain $m(\mathbf{x})$ to data domain $u(\mathbf{x},t)$.

Despite the fact that migration can be used to update velocity model [80, 110, 119], its chief purpose is to transform measured reflection data into an image of reflecting interfaces in the subsurface. There are two principal varieties of migration techniques: reverse time migration (RTM) which gives a modest resolution of the reflectivity [6, 143] and least-squares reverse-time migration (LSRTM) which typically yields a higher resolution of the reflectivity [43, 44].

Reverse-time migration is a prestack two-way wave-equation migration to illustrate complex structure, especially strong contrast geological interfaces such as environments involving salts. Conventional RTM uses an imaging condition which is the zero time-lag cross-correlation between the source and the receiver wavefields [33]:

$$R(\mathbf{x}) = \sum_{\text{shots}} \int_0^T u(\mathbf{x},t) \cdot v(\mathbf{x},t)dt, \tag{6}$$

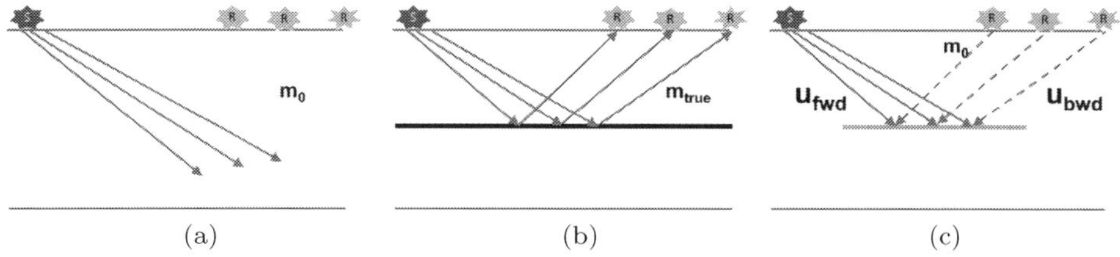

Figure 3: RTM: (a) Synthetic forward wavefield u_{fwd}, (b) True forward wavefield and (c) Reflectors generated as the backward wavefield u_{bwd} cross-correlated with u_{fwd}

where u is the source wavefield in (5) and v is the receiver wavefield which is the solution to the adjoint equation (7):

$$\begin{cases} m(\mathbf{x})\frac{\partial^2 v(\mathbf{x},t)}{\partial t^2} - \triangle v(x,t) = d(\mathbf{x},t)\delta(\mathbf{x}-\mathbf{x_r}) \\ v(\mathbf{x},T) = 0 \\ v_t(\mathbf{x},T) = 0 \end{cases} \tag{7}$$

Here T is the final recording time, d is the observed data from the receiver $\mathbf{x_r}$ and m is the assumed background velocity. The adjoint wave equation (7) is always solved backward in time from T to 0. Therefore it is also referred as backward propagation.

In classical RTM, the forward modeling typically does not contain reflection information. For example, it can be the paraxial approximation of the wave equation, which does not allow for reflections [36], or a smooth velocity model with unknown reflecting layers. As a summary, the conventional RTM consists three steps as Figure 3 shows:

1. Forward modeling of a wave field with a good velocity model to get u_{fwd};
2. Backpropagation of the measured data through the same model to get u_{bwd};
3. Cross-correlation the source wavefield u_{fwd} and receiver wavefield u_{bwd} based on an imaging condition (e.g., Equation (6)) to detect the reflecting interfaces.

RTM uses the entire solution of the wave equations instead of separating the downgoing or upgoing wavefields. Theoretically, RTM produces a more accurate image than ray-based methods since it does not rely on the asymptotic theory or migration using the one-way equation, which typically introduces modeling errors [113]. A good background velocity model that contains accurate information about the low-wavenumber components is also crucial for the quality of the image [55]. Recent advances in computation power make it possible to compute and store the solution of the wave equation efficiently, which significantly aids RTM to generate high-quality images [49].

2.4 Least-squares Reverse Time Migration

Least-squares reverse time migration (LSRTM) is a new migration method designed to improve the image quality generated by RTM. It is formulated as a linear inverse problem based on the Born approximation which we will describe briefly in this section. The wave equation (5) defines a nonlinear operator $\mathcal{F}$ from model domain to data domain that maps m to u. The Born approximation is a linearization of this map to the first order so that we can denote it as $L = \frac{\delta \mathcal{F}}{\delta m}$ [64, 129].

One can derive the Born approximation as follows [46]. If we denote the model $m(\mathbf{x})$ as the sum of a background model and a small perturbation:

$$m(\mathbf{x}) = m_0(\mathbf{x}) + \varepsilon m_1(\mathbf{x}), \tag{8}$$

the corresponding wavefield u also splits into two parts:

$$u(\mathbf{x}, t) = u_0(\mathbf{x}, t) + u_{sc}(\mathbf{x}, t), \tag{9}$$

where u satisfies (5), and u_0 solves the following equation:

$$\begin{cases} m_0(\mathbf{x})\frac{\partial^2 u_0(\mathbf{x},t)}{\partial t^2} - \triangle u_0(\mathbf{x}, t) = s(\mathbf{x}, t) \\ u_0(\mathbf{x}, 0) = 0 \\ \frac{\partial u_0}{\partial t}(\mathbf{x}, 0) = 0 \end{cases} \tag{10}$$

Subtracting (10) from (5) and using (8) , we derive an equation of u_{sc} with zero initial conditions:

$$m_0 \frac{\partial^2 u_{sc}(\mathbf{x}, t)}{\partial t^2} - \triangle u_{sc}(\mathbf{x}, t) = -\varepsilon m_1 \frac{\partial^2 u(\mathbf{x}, t)}{\partial t^2}. \tag{11}$$

We can write u_{sc} using Green's function G:

$$u_{sc}(\mathbf{x}, t) = -\varepsilon \int_0^t \int_{\mathbb{R}^n} G(\mathbf{x}, y; t-s) m_1(y) \frac{\partial^2 u}{\partial t^2}(y, s) dy ds. \tag{12}$$

As a result, the original wavefield u has an implicit relation:

$$u = u_0 - \varepsilon G m_1 \frac{\partial^2 u}{\partial t^2} = \left[I + \varepsilon G m_1 \frac{\partial^2}{\partial t^2} \right]^{-1} u_0 \tag{13}$$

The last term can be expanded in terms of Born series,

$$u = u_0 - \varepsilon \int_0^t \int_{\mathbb{R}^n} G(\mathbf{x}, y; t-s) m_1(y) \frac{\partial^2 u_0}{\partial t^2}(y, s) dy ds + \mathcal{O}(\varepsilon^2) \tag{14}$$

$$= u_0 + \varepsilon u_1 + \mathcal{O}(\varepsilon^2) \tag{15}$$

Therefore, we can approximate u_{sc} explicitly by εu_1 as $-\varepsilon G m_1 \frac{\partial^2 u_0}{\partial t^2}$, which is called the Born approximation. We also derive a linear map from m_1 to u_1:

$$\begin{cases} m_0 \frac{\partial^2 u_1(\mathbf{x},t)}{\partial t^2} - \triangle u_1(\mathbf{x}, t) = -m_1 \frac{\partial^2 u_0(\mathbf{x},t)}{\partial t^2} \\ u_1(\mathbf{x}, 0) = 0 \\ \frac{\partial u_1}{\partial t}(\mathbf{x}, 0) = 0 \end{cases} \tag{16}$$

Unlike (11), (16) is an explicit formulation with m_0 as the background velocity and u_0 as the background wavefiled which is the solution to (10).

It is convenient to denote the nonlinear forward map (5) as $\mathcal{F} : m \mapsto u$. A Taylor expansion of $u = \mathcal{F}(m)$ in the sense of calculus of variation, gives us:

$$u = u_0 + \varepsilon \frac{\delta \mathcal{F}}{\delta m}[m_0]m_1 + \frac{\varepsilon^2}{2} < \frac{\delta^2 \mathcal{F}}{\delta m^2}[m_0]m_1, m_1 > + \dots \tag{17}$$

The functional derivative $\frac{\delta \mathcal{F}}{\delta m} : m_1 \mapsto u_1$ is the linear operator (16), which we hereafter denote as L . The convergence of the Born series and the accuracy of the Born approximation can be proved mathematically [95, 96].

We assume there is an accurate background velocity model m_0. The Born modeling operator maps the reflectivity m_r to the scatted wavefield $d_r = \mathcal{F}(m) - \mathcal{F}(m_0)$:

$$Lm_r = d_r \tag{18}$$

Although L is linear, there is no guarantee that it is invertible [35]. Instead of computing L^{-1}, we seek the reflectivity model by minimizing the least-squares error between observed data d_r and predicted scattering wavefield:

$$J(m_r) = ||Lm_r - d_r||_2^2 \tag{19}$$

The normal least-squares solution to (19) is $m_r = (L^T L)^{-1} L^T d_r$ where L^T is the adjoint operator, but it is numerically expensive and unstable to invert the term $L^T L$ directly. Instead, the problem is solved in an iterative manner using optimization methods such as conjugate gradient descent (CG).

Another interesting way of approximating $(L^T L)^{-1}$ is to consider the problem as finding a non-stationary matching filter [56, 61]. Similar to RTM, we can get an image by doing one step of migration:

$$m_1 = L^T d_r. \tag{20}$$

One step of de-migration (Born modeling) based on m_1 generates data d_1

$$d_1 = Lm_1. \tag{21}$$

Finally, the re-migration step provides another image m_2

$$m_2 = L^T d_1. \tag{22}$$

Combining (20) to (22), the inverse Hessian operator $(L^T L)^{-1}$ behaves like a matching filter between m_1 and m_2 which we are able to produce from the observed data. It is also the filter between m_r and m_1 as (23) and (24) show below:

$$m_1 = (L^T L)^{-1} m_2 \tag{23}$$

$$m_r = (L^T L)^{-1} m_1 \tag{24}$$

Therefore, LSRTM can be seen as a process which first derives a filter to match the re-migration m_2 to the initial migration m_1 and then applies the filter back to the initial migrated image to give an estimate of the reflectivity. Seeking the reflectivity is equivalent to finding the best filter K by minimizing the misfit $J(K)$ in the image or model domain:

$$J(K) = ||m_1 - Km_2||_2^2. \tag{25}$$

The final reflectivity image $m_r \approx Km_1$. It is a single-iteration method which greatly reduces the computational cost of the iterative methods like CG.

A potentially better way of implementing the filter-based idea is to transform the image into curvelet domain [25] to improve the stability and structural consistency in the matching [134]. The formulation of obtaining the Hessian filter in curvelet domain is to minimize a misfit function $J(s)$ where

$$J(s) = ||C(m_1) - sC(m_2)||_2^2 + \varepsilon||s||_2^2, \tag{26}$$

where C is the curvelet domain transform operator, s is the matching filter and ε is the Tikhonov regularization parameter. The final reflectivity image $m_r \approx C^{-1}(|s|C(m_1))$, where C^{-1} is the inverse curvelet transform operator.

In general, least-squares reverse time migration (LSRTM) is still facing challenges. First of all, the image quality highly depends on the accuracy of the background velocity model m_0. Even a small error can make the two wavefields meet at a wrong location, which generates a blurred image or an incorrect reflectivity [83]. Another drawback is its high computational cost compared with other traditional migration techniques. In practice, LSRTM fits not only the data but also the noise in the data. Consequently, it boosts the high-frequency noise in the image during the iterative inversion [42, 147].

2.5 Inversion

The process of imaging through modeling the velocity structure is a form of inversion of seismic data [125], but in this paper, we regard inversion as a process of recovering the quantitative features of the geographical structure, that is, finding $m(\mathbf{x})$ in (5). Inversion is often used to build a velocity model iteratively until the synthetic data matches the actual recording [94].

Wave equation traveltime tomography [84] and the ray-based tomography in the earlier section are phase-like inversion methods [113]. Least-squares inversion is known as linearized waveform inversion [75, 122]. The migration method introduced earlier, LSRTM, can also be seen as a linear inverse problem. The background model m_0 is not updated after each iteration in least-squares inversion. Similar to the goal of migration, the model to be updated iteratively is the reflectivity distribution instead of the velocity model. One can interpret the process as a series of reverse time migrations, where the data residual is backpropagated into the model instead of the recorded data itself (Figure 3c).

If the background model m_0 is the parameter we invert for, the problem turns into a nonlinear waveform inversion, which is also called full-waveform inversion (FWI). Both the low-wavenumber and high-wavenumber components are updated simultaneously in FWI so that the final image has high resolution and high accuracy [133]. FWI is the primary focus of the paper. In the following sections, we

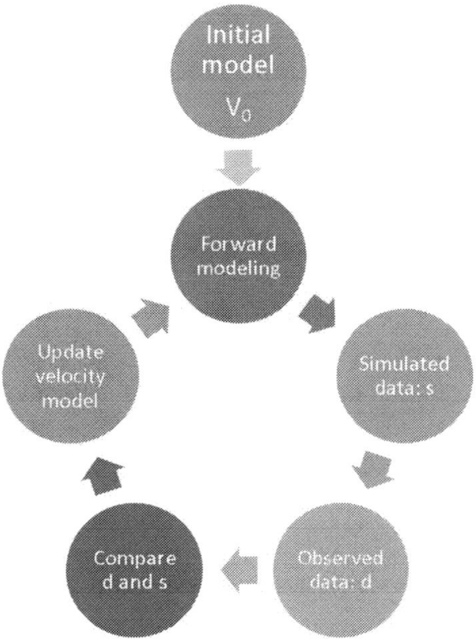

Figure 4: The framework of FWI as a PDE-constrained optimization

will further discuss the topic and especially the merit of using optimal transport based ideas to tackle the current limitations.

3 Full Waveform Inversion

FWI is a nonlinear inverse technique that utilizes the entire wavefield information to estimate the earth properties. The notion of FWI was first brought up three decades ago [74, 124] and has been actively studied as the computing power increases. As we will see, the mathematical formulation of FWI is PDE constrained optimization. Even inversion for subsurface elastic parameters using FWI has become increasingly popular in exploration applications [17, 93, 133]. Currently, FWI can achieve stunning clarity and resolution. Both academia and industry have been actively working on the innovative algorithms and software of FWI. However, this technique is still facing three main challenges.

First, the physics of seismic waves are complex, and we need more accurate forward modeling in inversion going from pure acoustic waves to anisotropic viscoelasticity. Recent developments focus on this multiparameter and multi-mode modeling. FWI strategies for simultaneous and hierarchical velocity and attenuation inversion have been investigated recently [105], but there is a dilemma. The more realistic with more parameters the models of the earth become, the more ill-posed and even non-unique will the inverse problem be.

Second, it is well known that the accuracy of FWI deteriorates from the lack of low frequencies, data noise, and poor starting model. The limitation is mainly due to the ill-posedness of the inverse problem which we treat as a PDE-constrained optimization. FWI is typically performed using local optimization methods in which the subsurface model is described by using a large number of unknowns, and the number of model parameters is determined a priori [123]. These methods typically only use the local gradient of the objective function. As a result, the inversion process is easily trapped in the local minima. Markov chain Monte Carlo (MCMC) based methods [109], particle swarm optimization [29], and many other global optimization methods [115] can avoid the pitfall theoretically, but they are not cost-efficient to handle practical large-scale inversion currently.

Third, it is relatively inexpensive to update the model through local optimization methods in FWI, but the convergence of the algorithm highly depends on the choice of a starting model. The research directions can be grouped into two main ideas to tackle this problem. One idea is to replace the conventional least-squares norm with other objective functions in optimization for a wider basin of attraction [48]. The other idea is to expand the dimensionality of the unknown model by adding non-physical coefficients. The additional coefficients may convexify the problem and fit the data better [13, 62].

The essential elements of FWI framework (Figure 4) includes forward modeling, the adjoint-state method (gradient calculation) and objective functions which will be introduced in the next three subsections.

3.1 Forward modeling

Wave-propagation modeling is the most significant step in seismic imaging. The earth is complex with various heterogeneity on many scales, and the real physics is far more complicated than the simple acoustic setting of this paper, but the industry standard is still the acoustic model in time or frequency domain. The current research of FWI covers multiple parameters inversion of seismic waveforms including anisotropic parameters, density, and attenuation factors [138] including viscoelastic modeling which is related to fractional Laplacian wave equations [111]. It should be noted that the more parameters in a model, the less well-posed is the inverse problem.

If we exclude the attenuation parameter, the general elastic wave equation is a realistic model. Based on the equation of conservation of momentum (Newton's law of dynamics) and Hooke's law for stress and strain tensors, we have the following elastic wave equation:

$$\rho \frac{\partial^2 u_i}{\partial t^2} = f_i + \frac{\partial \sigma_{ij}}{\partial x_j}, \tag{27}$$

$$\frac{\partial \sigma_{ij}}{\partial t} = c_{ijkl} \frac{\partial \varepsilon_{ij}}{\partial t} + \frac{\partial \widetilde{\sigma}_{ij}}{\partial t}, \tag{28}$$

where ρ is the density, $\mathbf{u}$ is the displacement vector, σ is the nine-component stress tensor (i,j = 1,2,3), $\widetilde{\sigma}$ is the internal stress, $\mathbf{f}$ is the outer body force, ε is the nine-component strain tensor which satisfies $\varepsilon_{ij} = \frac{1}{2}\left(\frac{\partial u_i}{\partial x_j} + \frac{\partial u_j}{\partial x_i}\right)$ and c_{ijkl} is the stiffness tensor containing twenty-one independent components.

One can classify the current numerical methods of complex wave propagation into three categories: direct methods, integral wave equation methods and asymptotic methods [65]. Direct methods include finite-difference method (FDM) [91], pseudospectral method [53], finite element method (FEM) [86], spectral element method (SEM) [71], discontinuous Galerkin method (DG) [67], etc. Integral wave equation methods include both boundary element method (BEM) [14] and the indirect boundary element methods (IBEM) [101] with a fast multipole method (FMM) [52] for efficiency. Asymptotic methods include geometrical optics, Gaussian beams [27] and frozen Gaussian beams [81].

3.2 Measure of mismatch

In seismic inversion, the misfit function, i.e. the objective function in the optimization process, is defined as a functional on the data domain. Common misfit functions include cross-correlation traveltime measurements [84, 87], amplitude variations [41] and waveform differences [124]. In both time [121] and frequency domain [102, 103], the least-squares norm has been the most widely used misfit function. For example, in time domain conventional FWI defines a least-squares waveform misfit as

$$d(f, g) = J(m) = \frac{1}{2} \sum_r \int |f(\mathbf{x_r}, t; m) - g(\mathbf{x_r}, t)|^2 \, dt, \tag{29}$$

where $\mathbf{x_r}$ are receiver locations, g is observed data, and f is simulated data which solves (5) with model parameter m. The time integral is carried out numerically as a sum. This formulation can also be extended to the case with multiple sources.

Real seismic data usually contains noise. As a result, denoising becomes a important step in seismic data processing. The L^2 norm is well known to be sensitive to noise [18]. Other norms have been proposed to mitigate this problem. For example, the L^1 norm [37, 121], the Huber criterion [57, 59] and the hybrid L^1/L^2 criterion [19] all demonstrated improved robustness to noise compared with conventional L^2 norm.

All the misfit functions above are point-by-point based objective functions which means they only accumulate the differences in amplitude at each fixed time grid point. There are global misfit functions that compare the simulated and measured signals not just pointwise. The Wasserstein metric is one such metric which we will discuss later. It is very robust with respect to noise, see Theorem 9 in section 5.

The oscillatory and periodic nature of waveforms lead to another main challenge in FWI: the cycle-skipping issue when implementing FWI as a local inversion scheme. If the true data and the initial synthetic data are more than half wavelength ($> \frac{\lambda}{2}$) away from each other, the first gradient can go in the wrong direction regarding the phase mismatch, but can nonetheless reduce the data misfit in the fastest manner [11]. Mathematically, it is related to the highly nonconvex and highly nonlinear nature of the inverse problem and results in finding only a local minima. Figure 5a displays two signals, each of which contains two Ricker wavelets and f is simply a shift of g. The L^2 norm between f and g is plotted in Figure 5b

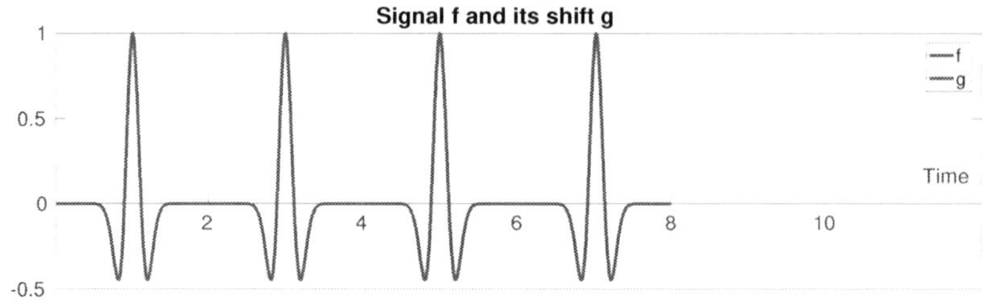

(a) Two signals

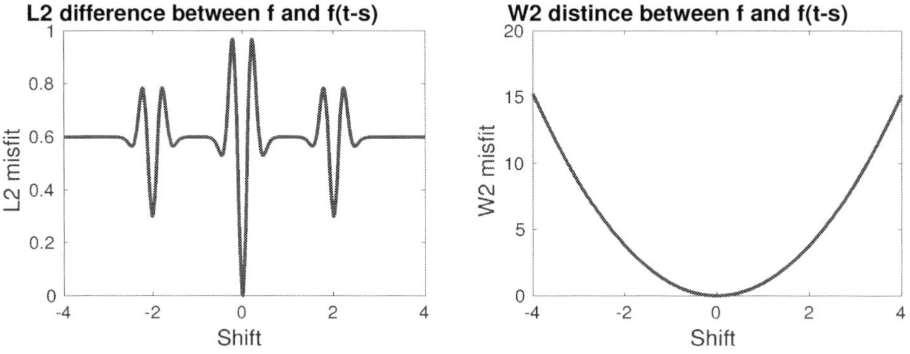

(b) L^2 sensitivity curve (c) W_2 sensitivity curve

Figure 5: (a) A signal consisting two Ricker wavelets (blue) and its shift (red) (b) L^2 norm of the difference between f and $f(t-s)$ in terms of shift s (c) W_2 norm between f and $f(t-s)$ in terms of shift s

as a function of the shift s. We observe many local minima and maxima in this simple two-event setting which again demonstrated the difficulty of the, so called, cycle-skipping issues [139].

The lower frequency components have a wider basin of attraction with the least-squares norm being the misfit function. Several hierarchical methods that invert from low frequencies to higher frequencies have been proposed in the literature to mitigate the cycle-skipping of the inverse problem [20, 69, 103, 117, 137]. Several other methods instead compare the integrated waveforms [63, 79] (Figure 6) and the waveform envelops [15, 82]. They share a similar idea with the hierarchical methods of taking advantage of the lower frequency components in the data.

A recently introduced class of misfit functions is based on optimal transport [30, 47, 48, 88, 89, 139, 140, 141]. As a useful tool from the theory of optimal transport, the Wasserstein metric computes the minimal cost of rearranging one distribution into another. The optimal transport based methods compare the observed and simulated data globally and thus include phase information. We will discuss these measures in section 4 and 5.

Other misfit functions with the idea of non-local comparison proposed in the

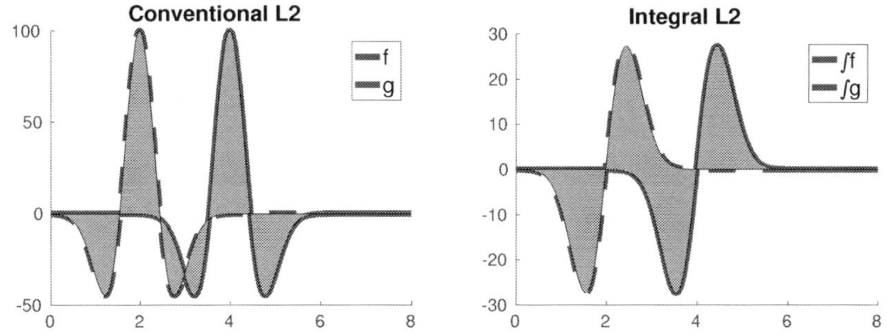

(a) Misfit for L^2 norm

(b) Misfit for Integral L^2 method

Figure 6: The shaded areas represent the mismatch each misfit function considers. (a) L^2: $\int(f-g)^2dt$. (b) Integral wavefields method: $\int(\int f - \int g)^2dt$. [140]

literature include filter based misfit functions [136, 148] as well as inversion using, so called, dynamic time warping [85] and the registration map [3]. The differential semblance optimization [120] exploits both phase and amplitude information of the reflections. Tomographic full waveform inversion [13] has some global convergence characteristics of wave-equation migration velocity analysis. In the filter based methods [136, 148], a filter is designed to minimize the L^2 difference between filtered simulated data and the observed data. The misfit is then a measure of how much the filter deviates from the identity. As we will see in the optimal transport based technique, this is done in one step where the optimal map directly determines the mapping of the simulated data. The optimal transport map is general and does not need to have the form of a convolution filter as in the filter based methods.

3.3 Adjoint-state method

Large-scale realistic 3D inversion is possible today. The advances in numerical methods and computational power allow for solving the 3D wave equations and compute the Fréchet derivative with respect to model parameters, which are needed in the optimization. In the adjoint-state method, one only needs to solve two wave equations numerically, the forward propagation and the backward adjoint wavefield propagation. Different misfit functions typically only affect the source term in the adjoint wave equation [100, 123].

Let us consider the misfit function $J(m)$ for computing the difference between predicted data f and observed data g where m is the model parameter, $F(m)$ is the forward modeling operator, $u(\mathbf{x},t)$ is the wavefield and $s(\mathbf{x},t)$ is the source. The predicted data f is the partial Cauchy boundary data of u which can be written as $f = Ru$ where R is a restriction operator only at the receiver locations. The wave equation (5) can be denoted as

$$F(m)u = s. \tag{30}$$

Taking first derivative regarding model m on both sides gives us:

$$\frac{\partial F}{\partial m} u + F \frac{\partial u}{\partial m} = 0. \tag{31}$$

Therefore,

$$\frac{\partial f}{\partial m} = -RF^{-1} \frac{\partial F}{\partial m} u. \tag{32}$$

By the chain rule, the gradient of misfit function J with respect to m is

$$\frac{\partial J}{\partial m} = \left(\frac{\partial f}{\partial m} \right)^T \frac{\partial J}{\partial f} \tag{33}$$

We can derive the following equation by plugging (32) into (33) :

$$\frac{\partial J}{\partial m} = -u^T \left(\frac{\partial F}{\partial m} \right)^T F^{-T} R^T \frac{\partial J}{\partial f} \tag{34}$$

Equation (34) is the adjoint-state method. The term $F^{-T} R^T \frac{\partial J}{\partial f}$ denotes the backward wavefield v generated by the adjoint wave equation whose source is the data residual $R^T \frac{\partial J}{\partial f}$. The gradient is similar to the usual imaging condition (6):

$$\frac{\partial J}{\partial m} = -\int_0^T \frac{\partial^2 u(\mathbf{x}, t)}{\partial t^2} v(\mathbf{x}, t) dt, \tag{35}$$

where v is the solution to the adjoint wave equation:

$$\begin{cases} m \frac{\partial^2 v(\mathbf{x},t)}{\partial t^2} - \triangle v(\mathbf{x}, t) = R^T \frac{\partial J}{\partial f} \\ v(\mathbf{x}, T) = 0 \\ v_t(\mathbf{x}, T) = 0 \end{cases} \tag{36}$$

Therefore F^T can be seen as the backward modeling operator which is similar to the adjoint wave equation (7) but with a different source term.

There are many other equivalent ways to formulate the adjoint-state method. One can refer to [46, 100] for more details.

In FWI, our aim is to find the model parameter m^* that minimizes the objective function, i.e. $m^* = \operatorname{argmin} J(m)$. For this PDE-constrained optimization, one can use the Fréchet derivative in a gradient-based iterative scheme to update the model m, such as steepest descent, conjugate gradient descent (CG), L-BFGS, Gauss-Newton method, etc. One can also derive the second-order adjoint equation for the Hessian matrix and use the full Newton's method in each iteration, but it is not practical regarding memory and current computing power. It is one of the current research interests to analyze and approximate the Hessian matrix in optimization [132].

4 Optimal Transport for FWI

Optimal transport has become a well-developed topic in mathematics since it was first brought up by Monge [92] in 1781. Due to its ability to incorporate both intensity and spatial information, optimal transport based metrics for modeling and signal processing have recently been adopted in a variety of applications including image retrieval, cancer detection, and machine learning [70]. In computer science, the metric is often called the "Earth Mover's Distance" (EMD).

The idea of using optimal transport for seismic inversion was first proposed in [47]. The Wasserstein metric is a concept based on optimal transportation [131]. Here, we transform our datasets of seismic signals into density functions of two probability distributions. Next, we find the optimal map between these two datasets and compute the corresponding transport cost as the misfit function in FWI. In this paper, we will focus on the quadratic cost function. The corresponding misfit is the quadratic Wasserstein metric (W_2). As Figure 5c shows, the convexity of W_2 is much better than the L^2 norm when comparing oscillatory seismic data with respect to shift.

Following the idea that changes in velocity cause a shift or "transport" in the arrival time, [48] demonstrated the advantageous mathematical properties of the quadratic Wasserstein metric (W_2) and provided rigorous proofs that laid a solid theoretical foundation for this new misfit function. We can apply W_2 as misfit function in two different ways: trace-by-trace comparison which is related to 1D optimal transport in the time dimension, and the entire dataset comparison in multiple dimensions. We will see that solving the Monge-Ampère equation in each iteration of FWI is a useful technique [141] for calculating the Wasserstein distance. An analysis of the 1D optimal transport approach and the conventional misfit functions such as L^2 norm and integral L^2 norm illustrated the intrinsic advantages of this transport idea [140].

4.1 Wasserstein metric

Let X and Y be two metric spaces with nonnegative Borel measures μ and ν respectively. Assume X and Y have equal total measure:

$$\int_X d\mu = \int_Y d\nu \tag{37}$$

Without loss of generality, we will hereafter assume the total measure to be one, i.e., μ and ν are probability measures.

Definition 1 (Mass-preserving map). *A transport map $T : X \to Y$ is mass-preserving if for any measurable set $B \in Y$,*

$$\mu(T^{-1}(B)) = \nu(B) \tag{38}$$

If this condition is satisfied, ν is said to be the push-forward of μ by T, and we write $\nu = T_{\#}\mu$

In another word, given two nonnegative densities $f = d\mu$ and $g = d\nu$, we are interested in the mass-preserving map T such that $f = g \circ T$. The transport cost function $c(x, y)$ maps pairs $(x, y) \in X \times Y$ to $\mathbb{R} \cup \{+\infty\}$, which denotes the cost of transporting one unit mass from location x to y. The most common choices of $c(x, y)$ include $|x - y|$ and $|x - y|^2$, which denote the Euclidean norms for vectors x and y hereafter. Once we find a mass-preserving map T, the cost corresponding to T is

$$I(T, f, g, c) = \int\limits_X c(x, T(x)) f(x)\, dx.$$

While there are many maps T that can perform the relocation, we are interested in finding the optimal map that minimizes the total cost

$$I(f, g, c) = \inf_{T \in \mathcal{M}} \int\limits_X c(x, T(x)) f(x)\, dx,$$

where $\mathcal{M}$ is the set of all maps that rearrange f into g.

Thus we have informally defined the optimal transport problem, the optimal map as well as the optimal cost, which is also called the Wasserstein distance:

Definition 2 (The Wasserstein distance)**.** *We denote by $\mathscr{P}_p(X)$ the set of probability measures with finite moments of order p. For all $p \in [1, \infty)$,*

$$W_p(\mu, \nu) = \left(\inf_{T_{\mu,\nu} \in \mathcal{M}} \int_{\mathbb{R}^n} |x - T_{\mu,\nu}(x)|^p \, d\mu(x) \right)^{\frac{1}{p}}, \quad \mu, \nu \in \mathscr{P}_p(X). \tag{39}$$

$\mathcal{M}$ is the set of all maps that rearrange the distribution μ into ν.

4.2 1D problem

In [141], we proposed two ways of using W_2 in FWI were proposed. One can either compute the misfit globally by solving a 2D or 3D optimal transport problem or compare data trace-by-trace with the 1D explicit formula, see Theorem 1 below. For the 1D approach, the corresponding misfit function in FWI becomes

$$J_1(m) = \sum_{r=1}^{R} W_2^2(f(\mathbf{x_r}, t; m), g(\mathbf{x_r}, t)), \tag{40}$$

where R is the total number of time history traces, g is the observed data, f is the simulated data, $\mathbf{x_r}$ are the receiver locations, and m is the model parameter. Mathematically it is W_2 metric in the time domain and L^2 norm in the spatial domain.

For f and g in one dimension, it is possible to exactly solve the optimal transportation problem [131] in terms of the cumulative distribution functions

$$F(x) = \int_{-\infty}^{x} f(t)\, dt, \quad G(y) = \int_{-\infty}^{y} g(t)\, dt. \tag{41}$$

In fact, the optimal map is just the unique monotone rearrangement of the density f into g. In order to compute the Wasserstein metric (W_p), we need the cumulative distribution functions F and G and their inverses F^{-1} and G^{-1} as the following theorem states:

Theorem 1 (Optimal transportation on $\mathbb{R}$). *Let $0 < f, g < \infty$ be two probability density functions, each supported on a connected subset of $\mathbb{R}$. Then the optimal map from f to g is $T = G^{-1} \circ F$.*

From the theorem above, we derive another formulation for the 1D quadratic Wasserstein metric:

$$\begin{aligned} W_2^2(f,g) &= \int_0^1 |F^{-1} - G^{-1}|^2 dy \\ &= \int_X |x - G^{-1}(F(x))|^2 f(x) dx. \end{aligned} \tag{42}$$

The corresponding Fréchet derive which is also the adjoint source term in the backward propagation is:

$$\begin{aligned} \frac{\partial W_2^2(f,g)}{\partial f} &= \left(\int_t^{T_0} -2(s - G^{-1}(F(s)) \frac{dG^{-1}(y)}{dy}\bigg|_{y=F(s)} f(s) ds \right) dt \\ &\quad + |t - G^{-1}(F(t))|^2 dt. \end{aligned} \tag{43}$$

This adjoint source term in the discrete 1D setting can be computed as

$$\left[U \operatorname{diag}\left(\frac{-2f(t)dt}{g(G^{-1} \circ F(t))} \right) \right] (t - G^{-1} \circ F(t)) dt + |t - G^{-1} \circ F(t)|^2 dt, \tag{44}$$

where U is the upper triangular matrix whose non-zero components are 1.

4.3 Monge-Ampère equation

This fully nonlinear partial differential equation plays an important role in computing the Wasserstein metric.

4.3.1 Introduction

In the previous section, we introduced the 1D optimal transport technique of comparing seismic data trace by trace and the explicit solution formula. Another option is a general optimal transport problem in all dimensions. In the global case we compare the full datasets and consider the whole synthetic data f and observed data g as objects with the general quadratic Wasserstein metric (W_2):

$$J_2(m) = W_2^2(f(\mathbf{x_r}, t; m), g(\mathbf{x_r}, t)). \tag{45}$$

The simple exact formula for 1D optimal transportation does not extend to optimal transportation in higher dimensions. Nevertheless, it can be computed by relying on two important properties of the optimal mapping $T(x)$: conservation of mass and cyclical monotonicity. From the definition of the problem, $T(x)$ maps f into g. If T is a sufficiently smooth map and $\det(\nabla T(x)) \neq 0$, the change of variables formula formally leads to the requirement

$$f(x) = g(T(x)) \det(\nabla T(x)). \tag{46}$$

The optimal map takes on additional structure in the special case of the cost function (i.e., $c(x, y) = |x - y|^2$): it is cyclically monotone [16, 68].

Definition 3 (Cyclical monotonicity). *We say that $T : X \to Y$ is cyclically monotone if for any $m \in \mathbb{N}^+$, $x_i \in X$, $1 \leq i \leq m$,*

$$\sum_{i=1}^{m} |x_i - T(x_i)|^2 \leq \sum_{i=1}^{m} |x_i - T(x_{i-1})|^2 \tag{47}$$

or equivalently

$$\sum_{i=1}^{m} \langle T(x_i), x_i - x_{i-1} \rangle \geq 0 \tag{48}$$

where $x_0 \equiv x_m$.

Additionally, a cyclically monotone mapping is formally equivalent to the gradient of a convex function [16, 68]. Making the substitution $T(x) = \nabla u(x)$ into the constraint (46) leads to the Monge-Ampère equation

$$\det(D^2 u(x)) = \frac{f(x)}{g(\nabla u(x))}, \quad u \text{ is convex.} \tag{49}$$

In order to compute the misfit between distributions f and g, we first compute the optimal map $T(x) = \nabla u(x)$ via the solution of this Monge-Ampère equation coupled to the non-homogeneous Neumann boundary condition

$$\nabla u(x) \cdot \nu = x \cdot \nu, \; x \in \partial X. \tag{50}$$

The squared Wasserstein metric is then given by

$$W_2^2(f, g) = \int_X f(x) \left| x - \nabla u(x) \right|^2 dx. \tag{51}$$

For the general Monge-Ampère equation, the uniqueness of the optimal map is not guaranteed. One need to discuss it in the context of a particular cost function and certain hypothesis. For example, the cyclical monotonicity is the key element in the proof of the following Brenier's theorem [16, 45] which gives an elegant result about the uniqueness of optimal transport map for the quadratic cost $|x - y|^2$:

Theorem 2 (Brenier's theorem). *Let μ and ν be two compactly supported probability measures on $\mathbb{R}^n$. If μ is absolutely continuous with respect to the Lebesgue measure, then*

1. *There is a unique optimal map T for the cost function $c(x, y) = |x - y|^2$.*
2. *There is a convex function $u : \mathbb{R}^n \to \mathbb{R}$ such that the optimal map T is given by $T(x) = \nabla u(x)$ for μ-a.e. x.*

Furthermore, if $\mu(dx) = f(x)dx$, $\nu(dy) = g(y)dy$, then T is differential μ-a.e. and

$$\det(\nabla T(x)) = \frac{f(x)}{g(T(x))}. \tag{52}$$

We are here considering the connection between the Monge-Ampère equation and optimal transport where the transport map is geometric in nature. The Monge-Ampère equation is of course also known for many other connections to geometry and mathematical physics. Let us mention a few examples. It arises naturally in many problems such as affine geometry [32], Riemannian geometry [1], isometric embedding [60], reflector shape design [135], etc. In the last century, treatments about this equation mostly came from the geometric problems above [21, 31, 72, 90, 126]. If we consider the following general Monge-Ampère equation:

$$\det(D^2 u(x)) = f(x, u, Du), \tag{53}$$

when $f = K(x)(1 + |Du|^2)^{(n+2)/2}$, the equation becomes the prescribed Gaussian curvature equation [45]. In affine geometry, an affine sphere in the graph satisfies the Monge-Ampère equation (53). The affine maximal surface satisfies a fourth-order equation which is related to the general Monge-Ampère equation:

$$\sum_{i,j=i}^{n} U^{ij} \partial_{x_i} \partial_{x_j} \left[\det(D^2 u)\right]^{-\frac{n+1}{n+2}} = 0, \tag{54}$$

where U^{ij} is the cofactor matrix of $D^2 u$ [127].

4.3.2 Weak solutions

Although the Monge-Ampère equation is a second-order PDE, there is no guarantee that the classical C^2 solution always exists. For the generalized Monge-Ampère equation (53) with homogeneous Dirichlet boundary condition $u = 0$ on $\partial\Omega$, it is well-known that there exists a classical convex solution $u \in C^2(\Omega) \cup C(\overline{\Omega})$, when f is strictly positive and sufficiently smooth [22, 23, 24]. When the assumptions no longer hold, we solve for two types of weak solutions instead: the Aleksandrov solution and the viscosity solution. One can refer to [58] for more details and proofs of the following definitions and theorems.

Let Ω be the open subset of $\mathbb{R}^d$ and $u : \Omega \to \mathbb{R}$. We denote $\mathcal{P}(\mathbb{R}^d)$ as the set of all subsets of $\mathbb{R}^d$.

Definition 4. *The normal mapping of u, or the subdifferential of u, is the set-valued mapping $\partial u : \Omega \to \mathcal{P}(\mathbb{R}^d)$ defined by*

$$\partial u(x_0) = \{p : u(x) \geq u(x_0) + p \cdot (x - x_0), \text{ for all } x \in \Omega\} \tag{55}$$

Given $V \in \Omega$, $\partial u(V) = \cup_{x \in V} \partial u(x)$.

Theorem 3 (Monge-Ampère measure). *If Ω is open and $u \in C(\Omega)$, then the class*

$$\mathcal{S} = \{V \subset \Omega : \partial u(V) \text{ is Lebesgue measurable}\}$$

is a Borel σ-algebra. The set function $Mu : \mathcal{S} \to \overline{\mathbb{R}}$ defined by

$$Mu(V) = |\partial u(V)| \tag{56}$$

is a measure, finite on compact sets, called the Monge-Ampère measure associated with the function u.

This is a measure generated by the the Monge-Ampère operator, which naturally defines the notion of the Aleksandrov solution.

Definition 5 (Aleksandrov solution). *Let ν be a Borel measure defined on Ω which is an open and convex subset of $\mathbb{R}^n$. The convex function u is a weak solution, in the sense of Aleksandrov, to the Monge-Ampère equation*

$$\det D^2 u = \nu \quad \text{in } \Omega \tag{57}$$

if the associated Monge-Ampère measure Mu definted in (56) is equal to ν.

Next we state one existence and uniqueness result for the Aleksandrov solution [2].

Theorem 4 (Existence and uniqueness of the Aleksandrov solution). *Consider the following Dirichlet problem of the Monge-Ampère equation*

$$\begin{aligned} \det D^2 u &= \nu \quad \text{in } \Omega \\ u &= g \quad \text{on } \partial\Omega, \end{aligned} \tag{58}$$

on a convex bounded domain $\Omega \in \mathbb{R}^d$ with boundary $\partial\Omega$. Assume that ν is a finite Borel measure and $g \in C(\partial\Omega)$ which can be extended to a convex function $\tilde{g} \in C(\bar{\Omega})$. Then the Monge-Ampère equation (58) has a unique convex Aleksandrov solution in $C(\overline{\Omega})$.

Aleksandrov's generalized solution corresponds to the curvature measure in the theory of convex bodies [127]. A finite difference scheme for computing Aleksandrov measure induced by $D^2 u$ in 2D was conducted in [98] with the solution u comes as a byproduct [50].

Another notion of weak solution is the viscosity solution which occurs naturally if f is continuous in (53).

Definition 6 (Viscosity solution). *Let $u \in C(\Omega)$ be a convex function and $f \in C(\Omega)$, $f \geq 0$. The function u is a viscosity subsolution (supersolution) of (53) in Ω if whenever convex function $\phi \in C^2(\Omega)$ and $x_0 \in \Omega$ are such that $(u - \phi)(x) \leq (\geq)(u - \phi)(x_0)$ for all x in the neighborhood of x_0, then we must have*

$$\det(D^2\phi(x_0)) \leq (\geq) f(x_0).$$

The function u is a viscosity solution if it is both a viscosity subsolution and supersolution.

We can relate these two notions of weak solution in the following proposition:

Proposition 5. *If u is a Aleksandrov (generalized) solution of (53) with f continuous, then u is also a viscosity solution.*

4.4 Numerical optimal transport in higher dimensions

In this section, we will summarize some of the current numerical methods for solving the optimal transport problems in higher dimensions. These methods are based on the equivalent or relaxed formulations of the original Monge's problem. In the end, we will introduce a monotone finite difference Monge-Ampère solver which is proved to converge to the viscosity solution to (49) [4, 51].

4.4.1 General methods

Optimal transport is a well-studied subject in mathematics while the computation techniques are comparatively underdeveloped. We will focus on analysis based methods. There are combinatorial techniques that typically are computationally costly in higher dimensions, for example, the Hungarian algorithm [73].

The definition (39) is the original static formulation of the optimal transport problem with a quadratic cost. It is an infinite dimensional optimization problem if we search for T directly. The non-symmetric nature of Monge's problem also generated difficulty because the map is unnecessarily bijective [76].

In the 40's, Kantorovich relaxed the constraints and formulated the dual problem [66]. Instead of searching for a map T, the transference plan γ is considered, which is also a measure supported by the product space $X \times Y$. The Kantorovich problem is the following:

$$\inf_{\gamma}\left\{\int_{X\times Y} c(x,y)d\gamma \mid \gamma \geq 0 \text{ and } \gamma \in \Pi(\mu,\nu)\right\}, \tag{59}$$

where $\Pi(\mu,\nu) = \{\gamma \in \mathcal{P}(X \times Y) \mid (P_X)\sharp\gamma = \mu, (P_Y)\sharp\gamma = \nu\}$. Here (P_X) and (P_Y) denote the two projections, and $(P_X)\sharp\gamma$ and $(P_Y)\sharp\gamma$ are two measures obtained by pushing forward γ with these two projections.

Consider $\varphi \in L^1(\mu)$ and $\psi \in L^1(\nu)$, the Kantorovich dual problem is formulated as the following [131]:

$$\sup_{\varphi,\psi}\left(\int_X \varphi\, d\mu + \int_Y \psi\, d\nu\right), \tag{60}$$

subject to $\varphi(x) + \psi(y) \leq c(x,y)$, for any $(x,y) \in X \times Y$.

The dual formulation is a linear optimization problem which is solvable by linear programming [40, 97, 112]. Kantorovich obtained the 1975 Nobel prize in economics for his contributions to resource allocation problems where he interpreted the dual problem as an economic equilibrium. Recently Cuturi introduced the entropy regularized optimal transport problem which enforces the desirable properties for optimal transference plan and convexifies the problem. There have been extremely efficient computational algorithms [38] which allow various applications in image processing, neuroscience, machine learning, etc [9, 39, 54, 118].

In the 90's, Benamou and Brenier derived an equivalent dynamic formulation [7] which has been one of the main tools for numerical computation. The

Benamou-Brenier formula identifies the squared quadratic Wasserstein metric between μ and ν by

$$W_2^2(\mu,\nu) = \inf \int_0^1 \int |v(t,x)|^2 \rho(t,x) dx dt, \tag{61}$$

where the infimum is taken among all the solutions of the continuity equation:

$$\begin{aligned} \frac{\partial \rho}{\partial t} + \nabla(v\rho) = 0, \\ \text{subject to} \quad \rho(0,x) = f,\ \rho(1,x) = g, \end{aligned} \tag{62}$$

In fact the infimum is taken among all Borel fields $v(t,x)$ that transports μ to ν continuously in time, satisfying the zero flux condition on the boundary. Many fast solvers based on this dynamic formulation has been proposed in literature [8, 77, 99]. They are used particularly in image registration, warping, texture mixing, etc.

4.4.2 The finite difference Monge-Ampère solver

As we have seen for the quadratic Wasserstein distance, the optimal map can be computed via the solution of a Monge-Ampère partial differential equation [10]. This approach has the advantage of drawing on the well-developed field of numerical partial differential equations (PDEs). We solve the Monge-Ampère equation numerically for the viscosity solution using an almost-monotone finite difference method relying on the following reformulation of the Monge-Ampère operator, which automatically enforces the convexity constraint [51]. The scientific reason for using monotone type schemes follows from the following theorem by Barles and Souganidis [4]:

Theorem 6 (Convergence of Approximation Schemes [4]). *Any consistent, stable, monotone approximation scheme to the solution of fully nonlinear second-order elliptic or parabolic PDE converges uniformly on compact subsets to the unique viscosity solution of the limiting equation, provided this equation satisfies a comparison principle.*

The numerical scheme of [10] uses the theory of [4] to construct a convergent discretization of the Monge-Ampère equation (49) as stated in Theorem 7. A variational characterization of the determinant on the left hand side which also involves the negative part of the eigenvalues was proposed as the following equation:

$$\begin{aligned} &\det(D^2 u) = \\ &\min_{\{v_1,v_2\}\in V} \{\max\{u_{v_1,v_1},0\}\max\{u_{v_2,v_2},0\} + \min\{u_{v_1,v_1},0\} + \min\{u_{v_2,v_2},0\}\} \end{aligned} \tag{63}$$

where V is the set of all orthonormal bases for $\mathbb{R}^2$.

Equation (63) can be discretized by computing the minimum over finitely many directions $\{\nu_1,\nu_2\}$, which may require the use of a wide stencil. In the low-order version of the scheme, the minimum in (63) is approximated using only two possible values. The first uses directions aligning with the grid axes.

$$\begin{aligned}MA_1[u] &= \max\{\mathcal{D}_{x_1x_1}u, \delta\}\max\{\mathcal{D}_{x_2x_2}u, \delta\}\\ &\quad + \min\{\mathcal{D}_{x_1x_1}u, \delta\} + \min\{\mathcal{D}_{x_2x_2}u, \delta\} - f/g\,(\mathcal{D}_{x_1}u, \mathcal{D}_{x_2}u) - u_0. \quad (64)\end{aligned}$$

Here dx is the resolution of the grid, $\delta > K\Delta x/2$ is a small parameter that bounds second derivatives away from zero, u_0 is the solution value at a fixed point in the domain, and K is the Lipschitz constant in the y-variable of $f(x)/g(y)$.

For the second value, we rotate the axes to align with the corner points in the stencil, which leads to

$$\begin{aligned}MA_2[u] &= \max\{\mathcal{D}_{vv}u, \delta\}\max\{\mathcal{D}_{v^\perp v^\perp}u, \delta\} + \min\{\mathcal{D}_{vv}u, \delta\} + \min\{\mathcal{D}_{v^\perp v^\perp}u, \delta\}\\ &\quad - f/g\left(\frac{1}{\sqrt{2}}(\mathcal{D}_v u + \mathcal{D}_{v^\perp}u), \frac{1}{\sqrt{2}}(\mathcal{D}_v u - \mathcal{D}_{v^\perp}u)\right) - u_0. \quad (65)\end{aligned}$$

Then the monotone approximation of the Monge-Ampère equation is

$$M_M[u] \equiv -\min\{MA_1[u], MA_2[u]\} = 0. \quad (66)$$

We also define a second-order approximation, obtained from a standard centred difference discretisation,

$$M_N[u] \equiv -\left((\mathcal{D}_{x_1x_1}u)(\mathcal{D}_{x_2x_2}u) - (\mathcal{D}_{x_1x_2}u^2)\right) + f/g\,(\mathcal{D}_{x_1}u, \mathcal{D}_{x_2}u) + u_0 = 0. \quad (67)$$

These are combined into an almost-monotone approximation of the form

$$M_F[u] \equiv M_M[u] + \epsilon S\left(\frac{M_N[u] - M_M[u]}{\epsilon}\right) \quad (68)$$

where ϵ is a small parameter and the filter S is given by

$$S(x) = \begin{cases} x & |x| \le 1\\ 0 & |x| \ge 2\\ -x+2 & 1 \le x \le 2\\ -x-2 & -2 \le x \le -1. \end{cases} \quad (69)$$

The Neumann boundary condition is implemented using standard one-sided differences. As described in [48, 51], the (formal) Jacobian $\nabla M_F[u]$ of the scheme can be obtained exactly. It is known to be sparse and diagonally dominant.

Theorem 7 (Convergence to Viscosity Solution [51, Theorem 4.4]). *Let the Monge-Ampère equation* (49) *have a unique viscosity solution and let $g > 0$ be Lipschitz continuous on $\mathbb{R}^d$. Then the solutions of the scheme* (68) *converge to the viscosity solution of* (49) *with a formal discretization error of $\mathcal{O}(Lh^2)$ where L is the Lipschitz constant of g and h is the resolution of the grid.*

Once the discrete solution u_h is computed, the squared Wasserstein metric is approximated via

$$W_2^2(f,g) \approx \sum_{j=1}^{n}(x_j - D_{x_j}u_h)^T \mathrm{diag}(f)(x_j - D_{x_j}u_h)dt, \tag{70}$$

where n is the dimension of the data f and g. Then the gradient of the discrete squared Wasserstein metric can be expressed as

$$\begin{aligned}\frac{\partial W_2^2(f,g)}{\partial f} &= \sum_{j=1}^{n}\left[-2\nabla M_F^{-1}[u_f]^T D_{x_j}^T \mathrm{diag}(f)\right](x_j - D_{x_j}u_f)dt \\ &+ \sum_{j=1}^{n}|x_j - D_{x_j}u_f|^2 dt,\end{aligned} \tag{71}$$

This term is the discretized version of the Fréchet derivative of the misfit function (45) with respect to the synthetic data f, i.e., the adjoint source $\frac{\partial J}{\partial f}$ in the adjoint wave equation (36).

5 Application of Optimal Transport to Seismic Inversion

In this section, we first review the good properties of the W_2 norm for the application of full-waveform inversion. We will also explain some details of the implementations and show numerical results of using optimal transport based metrics as the misfit function in FWI.

5.1 W_2 properties

As we demonstrated in [48], the squared Wasserstein metric has several properties that make it attractive as a choice for misfit function. One highly desirable feature is its convexity with respect to several parameterizations that occur naturally in seismic waveform inversion [141]. For example, variations in the wave velocity lead to simulated f that are derived from shifts,

$$f(x;s) = g(x + s\eta), \quad \eta \in \mathbb{R}^n, \tag{72}$$

or dilations,

$$f(x;A) = g(Ax), \quad A^T = A,\, A > 0, \tag{73}$$

applied to the observation g. Variations in the strength of a reflecting surface or the focusing of seismic waves can also lead to local rescalings of the form

$$f(x;\beta) = \begin{cases} \beta g(x), & x \in E \\ g(x), & x \in \mathbb{R}^n \backslash E. \end{cases} \tag{74}$$

Theorem 8 (Convexity of squared Wasserstein metric [48]). *The squared Wasserstein metric $W_2^2(f(m), g)$ is convex with respect to the model parameters m corresponding to a shift s in (72), the eigenvalues of a dilation matrix A in (73), or the local rescaling parameter β in (74).*

Another important property of optimal transport is the insensitivity to noise. All seismic data contains either natural or experimental equipment noise. For example, the ocean waves lead to extremely low-frequency data in the marine acquisition. Wind and cable motions also generate random noise.

Theorem 9 (Insensitivity to noise [48]). *Let f_{ns} be f with a piecewise constant additive noise of mean zero uniform distribution. The squared Wasserstein metric $W_2^2(f, f_{ns})$ is of $\mathcal{O}(\frac{1}{N})$ where N is the number of pieces of the additive noise in f_{ns}.*

The L^2 norm is known to be sensitive to noise since the misfit between clean and noisy data is calculated as the sum of squared noise amplitude at each sampling point.

5.2 Data normalization

In optimal transport theory, there are two main requirements for signals f and g: positivity and mass balance. Since these are not expected for seismic signals, some data pre-processing is needed before we can implement Wasserstein-based FWI. In [47, 48], the signals were separated into positive and negative parts $f^+ = \max\{f, 0\}$, $f^- = \max\{-f, 0\}$ and scaled by the total mass $\langle f \rangle = \int_X f(x)\, dx$. Inversion was accomplished using the modified misfit function

$$W_2^2\left(\frac{f^+}{\langle f^+ \rangle}, \frac{g^+}{\langle g^+ \rangle}\right) + W_2^2\left(\frac{f^-}{\langle f^- \rangle}, \frac{g^-}{\langle g^- \rangle}\right). \tag{75}$$

While this approach preserves the desirable theoretical properties of convexity to shifts and noise insensitivity, it is not easy to combine with the adjoint-state method and more realistic examples. We require the scaling function to be differentiable so that it is easy to apply the chain rule when calculating the Fréchet derivative for FWI backpropagation and also better suited for the Monge-Ampère and the wave equation solvers.

There are other ways to rescale the datasets so that they become positive. For example, we can square the data as $\tilde{f} = f^2$ or extract the envelope of the data. These methods preserve the convexity concerning simple shifts, but we have lost the uniqueness: $f^2 = g^2$ does not imply $f = g$. As a result, more local minima are present since the fact that the misfit $J(f^2, g^2)$ is decreasing does not necessarily indicate that f is approaching g, not to mention the non-unique issue of the inverse problem itself.

Typically, we first scale the data f to be positive as $\tilde{f}$ and then normalize to ensure mass balance as $\tilde{f}/ < \tilde{f} >$. We now introduce three normalization methods that are robust in realistic large-scale inversions: the linear scaling [141] (Figure 7a)

$$\tilde{f} = f + c_1, \quad c_1 \geq \max\{-f, -g\}, \tag{76}$$

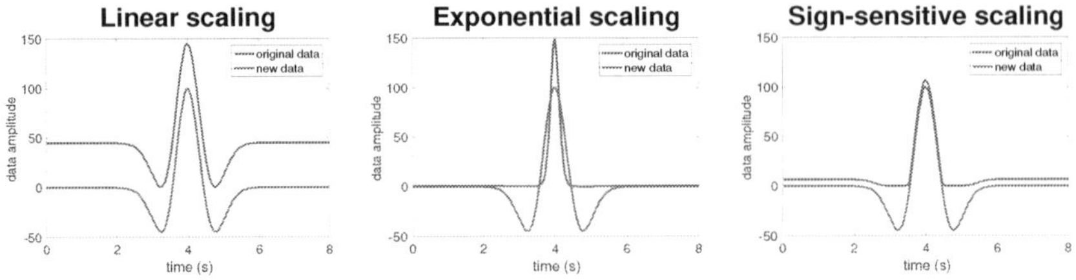

(a) Linear (b) Exponential (c) Sign-sensitive

Figure 7: (a) The linear, (b) the exponential and (c) the sign-sensitive scaling of a Ricker wavelet (Blue)

the exponential scaling [104] (Figure 7b)

$$\tilde{f} = \exp(c_2 f), \quad c_2 > 0, \tag{77}$$

and the sign-sensitive scaling (Figure 7c)

$$\tilde{f} = \begin{cases} f + \frac{1}{c_3}, & f \geq 0 \\ \frac{1}{c_3}\exp(c_3 f), & f < 0 \end{cases}, \quad c_3 > 0. \tag{78}$$

If c_2 in (77) and c_3 in (78) are large enough, these two scaling methods keep the convexity of W_2 norm regarding simple shifts as shown in Figure 5c. From Taylor expansion, we can see that the scalings are very close to the linear scaling when c_2 is small. One has to be careful with the exponential scaling (77) since it can easily become extremely large, but the sign-sensitive scaling (78) will not.

5.3 FWI with Kantorovich-Rubinstein norm

When the cost function $c(x, y)$ is the L^1 norm $|x - y|$, i.e. $p = 1$ in (39) with $f \geq 0$, $g \geq 0$, and $\int f = \int g$, the corresponding alternative W_1 distance has the following equivalent dual formulation:

$$W_1(f, g) = \max_{\varphi \in \text{Lip}_1} \int_X \varphi(x)(f(x) - g(x))dx, \tag{79}$$

where Lip_1 is the space of Lipschitz continuous functions with Lipschitz constant 1. However, seismic data f and g are oscillatory containing both positive and negative parts. If $\int f \neq \int g$, the value of (79) is always $+\infty$. Recently, [88, 89] introduced the following Kantorovich-Rubinstein (KR) norm in FWI which is a relaxation of the original W_1 distance by constraining the dual space:

$$\text{KR}(f, g) = \max_{\varphi \in \text{BLip}_1} \int_X \varphi(x)(f(x) - g(x))dx \tag{80}$$

Here BLip_1 is the space of bounded Lipschitz continuous functions with Lipschitz constant 1. One advantage of using KR norm in FWI is that there is no need to

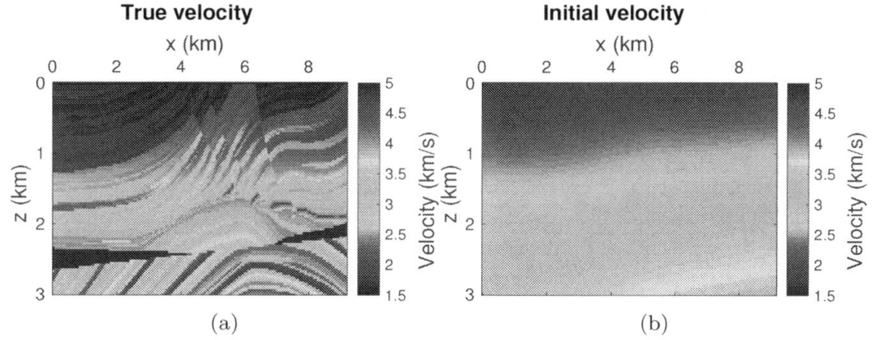

Figure 8: (a) True velocity and (b) inital velocity for full Marmousi model

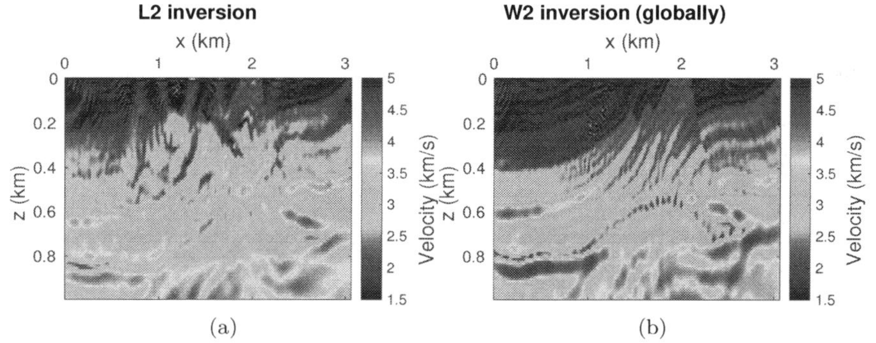

Figure 9: Inversion results of (a) L^2 and (b) global W_2 for the scaled Marmousi model

normalize the data to be positive and mass balanced. However, KR norm has no direct connection with optimal transport once we no longer require f and g to be probability measures [130]. When f and g are far apart which is very common when the initial velocity is rough, the maximum in (80) is achieved by "moving" f^+ to f^- and g^+ to g^-. The notion of transport is void in this case and convexity is lost.

5.4 Numerical results of global W_2

In the next two subsections, we provide numerical results for two approaches to using W_2 with linear normalization (76): trace-by-trace comparison and using the entire 2D datasets as objects. Here a trace is the time history measured at one receiver while the entire dataset consists of the time history of all the receivers. These are compared with results produced by using the standard least-squares norm L^2 to measure the misfit. More examples can be found in [141].

First, we use a scaled Marmousi model to compare the inversion between global W_2 and the conventional L^2 misfit function. Figure 8a is the P-wave velocity of the true Marmousi model, but in this experiment, we use a scaled model which is 1km in depth and 3km in width. The inversion starts from an initial model that is the true velocity smoothed by a Gaussian filter with a deviation of 40, which is highly smoothed and far from the true model (a scaled version of Figure 8b). We place 11 evenly spaced sources on top at 50m depth and 307 receivers on top at the same depth with a 10m fixed acquisition. The discretization of the forward wave equation is 10m in the x and z directions and 10ms in time. The source is a Ricker wavelet which is the second derivative of the Gaussian function with a peak frequency of 15Hz, and a bandpass filter is applied to remove the frequency components from 0 to 2Hz.

We use L-BFGS, a quasi-Newton method as the optimization algorithm [78]. Inversions are terminated after 200 iterations. Figure 9a shows the inversion result using the traditional L^2 least-squares method after 200 l-BFGS iterations. The inversion result of global W_2 (Figure 9b) avoids the problem of local minima suffered by the conventional L^2 metric, whose result demonstrates spurious high-frequency artifacts due to a point-by-point comparison of amplitude.

We solve the Monge-Ampère equation numerically in each iteration of the inversion. The drawback to the PDE approach is that data must be sufficiently regular for solutions to be well-defined and for the numerical approximation to be accurate. To remain robust on realistic examples, we use filters that effectively smooth the seismic data, which can lead to a loss of high-frequency information. For illustration in this paper, we perform computations using a Monge-Ampère solver for synthetic examples. Even in 2D, some limitations are apparent. This is expected to become even more of a problem in higher-dimensions and motivates our introduction of a trace-by-trace technique that relies on the exact 1D solution. The trace-by-trace technique is currently more promising for practical applications, as is evidenced in our computational examples in the next example.

5.5 Numerical results of trace-by-trace W_2

Recall that for the 1D trace-by-trace approach, the misfit function in FWI is

$$J_1(m) = \sum_{r=1}^{R} W_2^2(f(\mathbf{x_r}, t; m), g(\mathbf{x_r}, t)), \tag{81}$$

where R is the total number of traces, g is observed data, f is simulated data, $\mathbf{x_r}$ are receiver locations, and m is the model parameter. The adjoint source term for each single trace is

$$\frac{\partial W_2^2(f,g)}{\partial f} = \left(\int_t^{T_0} \frac{-2(s - G^{-1} \circ F(s))}{g(G^{-1} \circ F(s))} f(s) ds + |t - G^{-1}(F(t))|^2 \right) dt. \tag{82}$$

The next experiment is to invert the full Marmousi model by conventional L^2 and trace-by-trace W_2 misfit. Figure 8a is the P-wave velocity of the true Marmousi model, which is 3km in depth and 9km in width. The inversion starts

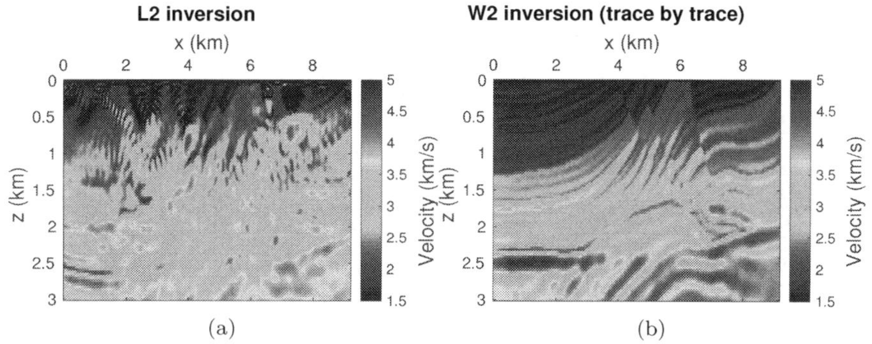

Figure 10: Inversion results of (a) L^2 and (b) trace-by-trace W_2 for the true Marmousi model

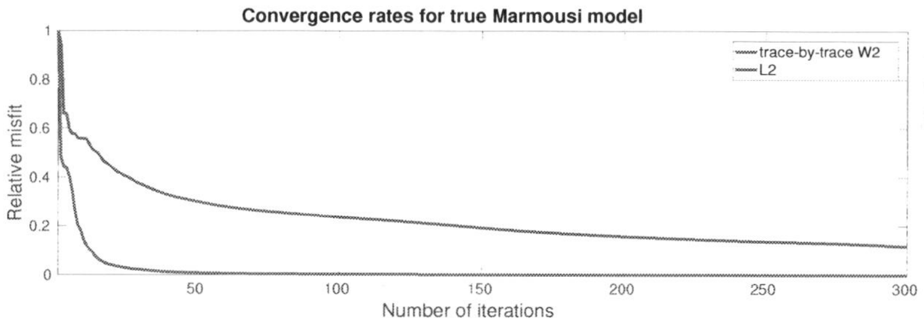

Figure 11: The convergence curves for trace-by-trace W_2 and L^2 based inversion of the full Marmousi model

from an initial model that is the true velocity smoothed by a Gaussian filter with a deviation of 40 (Figure 8b). The rest of the settings are the same as the previous section. Inversions are terminated after 300 L-BFGS iterations. Figure 10a shows the inversion result using the traditional L^2 least-squares method and figure 10b shows the final result using trace-by-trace W_2 misfit function. Again, the result of L^2 metric has spurious high-frequency artifacts while W_2 correctly inverts most details in the true model. The convergence curves in Figure 11 show that W_2 reduces the relative misfit to 0.1 in 20 iterations while L^2 converges slowly to a local minimum.

5.6 Insensitivity to noise

One of the good properties of the quadratic Wasserstein metric is the insensitivity to noise [48]. We repeat the previous experiment with a noisy reference by adding a uniform random iid noise to the data from the true velocity (Figure 12a). The signal-to-noise ratio (SNR) is -3.47 dB. In optimal transport, the effect of noise is essentially negligible due to the strong cancellation between the nearby positive and negative noise.

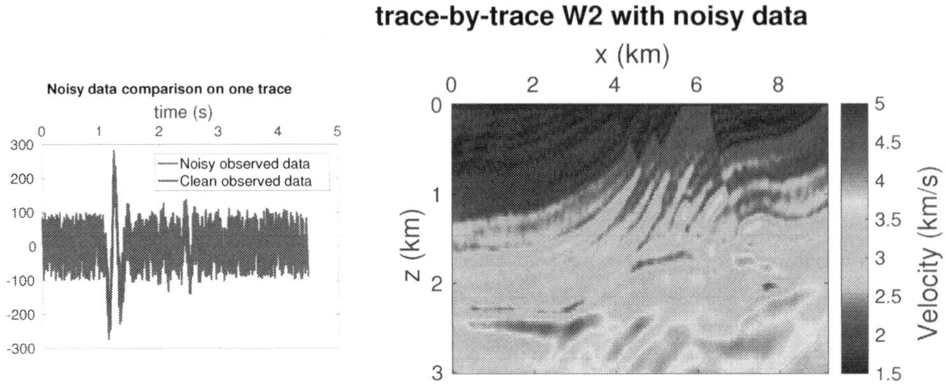

(a) (b)

Figure 12: (a) Noisy and clean data and (b) inversion result with the noisy data

All the settings remain the same as in the previous numerical experiment except the observed data. After 96 iterations, the optimization converges to a velocity presented in Figure 12b. Although the result has lower resolution than Figure 10b, it still recovers most features of Marmousi model correctly. Even when the noise is much larger than the signal, the quadratic Wasserstein metric still converges reasonably well.

References

[1] Thierry Aubin. *Some nonlinear problems in Riemannian geometry*. Springer Science & Business Media, 2013.

[2] Gerard Awanou. Discrete Aleksandrov solutions of the Monge-Ampère equation. *arXiv preprint arXiv:1408.1729*, 2014.

[3] Hyoungsu Baek, Henri Calandra, and Laurent Demanet. Velocity estimation via registration-guided least-squares inversion. *Geophysics*, 2014.

[4] Guy Barles and Panagiotis E Souganidis. Convergence of approximation schemes for fully nonlinear second order equations. *Asymptotic analysis*, 4(3):271–283, 1991.

[5] Vladimir Bashkardin, Sergey Fomel, Parvaneh Karimi, Alexander Klokov, and Xiaolei Song. Sigsbee model.

[6] Edip Baysal, Dan D. Kosloff, and John W. C. Sherwood. Reverse time migration. *Geophysics*, 48(11):1514–1524, Nov 1983.

[7] J.-D. Benamou and Y. Brenier. A computational fluid mechanics solution to the Monge-Kantorovich mass transfer problem. *Numer. Math.*, 84(3):375–393, 2000.

[8] Jean-David Benamou and Guillaume Carlier. Augmented lagrangian methods for transport optimization, mean field games and degenerate elliptic equations. *Journal of Optimization Theory and Applications*, 167(1):1–26,

2015.

[9] Jean-David Benamou, Guillaume Carlier, Marco Cuturi, Luca Nenna, and Gabriel Peyré. Iterative bregman projections for regularized transportation problems. *SIAM Journal on Scientific Computing*, 37(2):A1111–A1138, 2015.

[10] Jean-David Benamou, Brittany D Froese, and Adam M Oberman. Numerical solution of the optimal transportation problem using the Monge-Ampère equation. *Journal of Computational Physics*, 260:107–126, 2014.

[11] Wafik B. Beydoun and Albert Tarantola. First Born and Rytov approximations: Modeling and inversion conditions in a canonical example. *The Journal of the Acoustical Society of America*, 83(3):1045–1055, Mar 1988.

[12] Harmen Bijwaard, Wim Spakman, and E. Robert Engdahl. Closing the gap between regional and global travel time tomography. *Journal of Geophysical Research: Solid Earth*, 103(B12):30055–30078, Dec 1998.

[13] Biondo Biondi and Ali Almomin. Tomographic full waveform inversion (TFWI) by combining full waveform inversion with wave-equation migration velocity anaylisis. In *SEG Technical Program Expanded Abstracts 2012*, pages 1–5. Society of Exploration Geophysicists, 2012.

[14] Michel Bouchon and Francisco J Sánchez-Sesma. Boundary integral equations and boundary elements methods in elastodynamics. *Advances in geophysics*, 48:157–189, 2007.

[15] Ebru Bozdağ, Jeannot Trampert, and Jeroen Tromp. Misfit functions for full waveform inversion based on instantaneous phase and envelope measurements. *Geophysical Journal International*, 185(2):845–870, May 2011.

[16] Y. Brenier. Polar factorization and monotone rearrangement of vector-valued functions. *Comm. Pure Appl. Math.*, 44:375–417, 1991.

[17] Romain Brossier, Stéphane Operto, and Jean Virieux. Seismic imaging of complex onshore structures by 2D elastic frequency-domain full-waveform inversion. *Geophysics*, 74(6):WCC105–WCC118, 2009.

[18] Romain Brossier, Stéphane Operto, and Jean Virieux. Which data residual norm for robust elastic frequency-domain full waveform inversion? *Geophysics*, 75(3):R37–R46, 2010.

[19] Kenneth P. Bube and Robert T. Langan. Hybrid L^1/L^2 minimization with applications to tomography. *Geophysics*, 62(4):1183–1195, Jul 1997.

[20] Carey Bunks, Fatimetou M Saleck, S Zaleski, and G Chavent. Multiscale seismic waveform inversion. *Geophysics*, 60(5):1457–1473, 1995.

[21] L. Caffarelli, J. J. Kohn, L. Nirenberg, and J. Spruck. The Dirichlet problem for nonlinear second-order elliptic equations. II. Complex Monge-Ampère, and uniformaly elliptic, equations. *Communications on Pure and Applied Mathematics*, 38(2):209–252, Mar 1985.

[22] Luis A. Caffarelli. Interior a Priori Estimates for Solutions of Fully Non-Linear Equations. *The Annals of Mathematics*, 130(1):189, Jul 1989.

[23] Luis A. Caffarelli. Interior $W^{2,p}$ Estimates for Solutions of the Monge-Ampère Equation. *The Annals of Mathematics*, 131(1):135, Jan 1990.

[24] Luis A. Caffarelli. Some regularity properties of solutions of Monge Ampère equation. *Communications on Pure and Applied Mathematics*, 44(8-9):965–

969, Oct 1991.

[25] Emmanuel Candès, Laurent Demanet, David Donoho, and Lexing Ying. Fast discrete curvelet transforms. *Multiscale Modeling & Simulation*, 5(3):861–899, 2006.

[26] V Červený, IA Molotkov, and I Psencik. Ray method in seismology: Charles univ, 1977.

[27] Vlastislav Červený, Mikhail M Popov, and Ivan Pšenčík. Computation of wave fields in inhomogeneous media—Gaussian beam approach. *Geophysical Journal International*, 70(1):109–128, 1982.

[28] Vlastislav Červený and Ivan Pšenčík. Seismic, ray theory. pages 1244–1258. Springer, Dordrecht, 2011.

[29] Guiting Chen and Zhenli Wang. Robust full-waveform inversion based on particle swarm optimization. In *SEG Technical Program Expanded Abstracts 2017*, pages 1302–1306. Society of Exploration Geophysicists, Aug 2017.

[30] Jing Chen, Yifan Chen, Hao Wu, and Dinghui Yang. The quadratic Wasserstein metric for earthquake location. *arXiv preprint arXiv:1710.10447*, 2017.

[31] Shiu-Yuen Cheng and Shing-Tung Yau. On the regularity of the Monge-Ampère equation $det(\partial^2 u/\partial x_i \partial x_j) = f(x, u)$. *Communications on Pure and Applied Mathematics*, 30(1):41–68, Jan 1977.

[32] Shiu-Yuen Cheng and Shing-Tung Yau. Complete affine hypersurfaces. Part I. The completeness of affine metrics. *Communications on Pure and Applied Mathematics*, 39(6):839–866, Nov 1986.

[33] Jon F. Claerbout. Toward a unified theory of reflector mapping. *Geophysics*, 36(3):467–481, Jun 1971.

[34] Jon F Claerbout. *Imaging the earth's interior.* Blackwell scientific publications Oxford, 1985.

[35] Jon F Claerbout. *Earth soundings analysis: Processing versus inversion*, volume 6. Blackwell Scientific Publications London, 1992.

[36] Robert Clayton and Björn Engquist. Absorbing boundary conditions for acoustic and elastic wave equations. *Bulletin of the Seismological Society of America*, 67(6):1529–1540, 1977.

[37] E. Crase, A. Pica, M. Noble, J. McDonald, and A. Tarantola. Robust elastic nonlinear waveform inversion: Application to real data. *Geophysics*, 55(5):527–538, May 1990.

[38] Marco Cuturi. Sinkhorn distances: Lightspeed computation of optimal transport. In *Advances in neural information processing systems*, pages 2292–2300, 2013.

[39] Marco Cuturi and Arnaud Doucet. Fast computation of Wasserstein barycenters. In *International Conference on Machine Learning*, pages 685–693, 2014.

[40] Marco Cuturi and Gabriel Peyré. A smoothed dual approach for variational Wasserstein problems. *arXiv preprint arXiv:1503.02533*, 2015.

[41] F. A. Dahlen and Adam M. Baig. Fréchet kernels for body-wave amplitudes. *Geophysical Journal International*, 150(2):440–466, Aug 2002.

[42] Wei Dai, Hao Deng, Kun Jiao, and Denes Vigh. Least-squares reverse-time migration: An example of subsalt imaging. In *SEG Technical Program Ex-*

panded Abstracts 2017, pages 4427–4431. Society of Exploration Geophysicists, Aug 2017.

[43] Wei Dai, Paul Fowler, and Gerard T. Schuster. Multi-source least-squares reverse time migration. *Geophysical Prospecting*, 60(4):681–695, Jul 2012.

[44] Wei Dai and Gerard T. Schuster. Plane-wave least-squares reverse-time migration. *Geophysics*, 78(4):S165–S177, Jul 2013.

[45] Guido De Philippis and Alessio Figalli. The Monge-Ampère equation and its link to optimal transportation. Oct 2013.

[46] Laurent Demanet. Waves and imaging class notes -18.325. 2016.

[47] B. Engquist and B. D. Froese. Application of the Wasserstein metric to seismic signals. *Communications in Mathematical Sciences*, 12(5):979–988, 2014.

[48] Bjorn Engquist, Brittany D Froese, and Yunan Yang. Optimal transport for seismic full waveform inversion. *Communications in Mathematical Sciences*, 14(8):2309–2330, 2016.

[49] John Etgen, Samuel H. Gray, and Yu Zhang. An overview of depth imaging in exploration geophysics. *Geophysics*, 74(6):WCA5–WCA17, Nov 2009.

[50] Xiaobing Feng and Michael Neilan. Analysis of galerkin methods for the fully nonlinear monge-ampere equation. *Journal of Scientific Computing*, 47(3):303–327, 2011.

[51] B. D. Froese. A numerical method for the elliptic Monge-Ampère equation with transport boundary conditions. *SIAM J. Sci. Comput.*, 34(3):A1432–A1459, 2012.

[52] Hiroyuki Fujiwara. The fast multipole method for solving integral equations of three-dimensional topography and basin problems. *Geophysical Journal International*, 140(1):198–210, 2000.

[53] Takashi Furumura, BLN Kennett, and Hiroshi Takenaka. Parallel 3-D pseudospectral simulation of seismic wave propagation. *Geophysics*, 63(1):279–288, 1998.

[54] Alexandre Gramfort, Gabriel Peyré, and Marco Cuturi. Fast optimal transport averaging of neuroimaging data. In *International Conference on Information Processing in Medical Imaging*, pages 261–272. Springer, 2015.

[55] Samuel H. Gray. Seismic, migration. pages 1236–1244. Springer, Dordrecht, 2011.

[56] Antoine Guitton. Amplitude and kinematic corrections of migrated images for nonunitary imaging operators. *Geophysics*, 69(4):1017–1024, 2004.

[57] Antoine Guitton and William W. Symes. Robust inversion of seismic data using the Huber norm. *Geophysics*, 68(4):1310–1319, Jul 2003.

[58] Cristian E Gutiérrez. *The Monge-Ampère Equation*, volume 89. Birkhäuser, 2016.

[59] Taeyoung Ha, Wookeen Chung, and Changsoo Shin. Waveform inversion using a back-propagation algorithm and a huber function norm. *Geophysics*, 74(3):R15–R24, 2009.

[60] Qing. Han and Jiaxing Hong. *Isometric embedding of Riemannian manifolds in Euclidean spaces*. American Mathematical Society, 2006.

[61] Kaiming He, Jian Sun, and Xiaoou Tang. Guided image filtering. *IEEE*

Transactions on Pattern Analysis and Machine Intelligence, 35(6):1397–1409, Jun 2013.
[62] Guanghui Huang, Rami Nammour, and William Symes. Full-waveform inversion via source-receiver extension. *Geophysics*, 82(3):R153–R171, 2017.
[63] Guanghui Huang, Huazhong Wang, and Haoran Ren. Two new gradient precondition schemes for full waveform inversion. *arXiv preprint arXiv:1406.1864*, 2014.
[64] JA Hudson and JR Heritage. The use of the Born approximation in seismic scattering problems. *Geophysical Journal International*, 66(1):221–240, 1981.
[65] Ursula Iturrarán-Viveros and Francisco J. Sánchez-Sesma. Seismic Wave Propagation in Real Media: Numerical Modeling Approaches. pages 1200–1210. Springer, Dordrecht, 2011.
[66] Leonid Vital'evich Kantorovich. Mathematical methods of organizing and planning production. *Management Science*, 6(4):366–422, 1960.
[67] Martin Käser and Michael Dumbser. An arbitrary high-order discontinuous galerkin method for elastic waves on unstructured meshes—i. the two-dimensional isotropic case with external source terms. *Geophysical Journal International*, 166(2):855–877, 2006.
[68] M. Knott and C. S. Smith. On the optimal mapping of distributions. *Journal of Optimization Theory and Applications*, 43(1):39–49, 1984.
[69] Pierre Kolb, Francis Collino, and Patrick Lailly. Pre-stack inversion of a 1-D medium. *Proceedings of the IEEE*, 74(3):498–508, 1986.
[70] Soheil Kolouri, Serim Park, Matthew Thorpe, Dejan Slepčev, and Gustavo K Rohde. Transport-based analysis, modeling, and learning from signal and data distributions. *arXiv preprint arXiv:1609.04767*, 2016.
[71] Dimitri Komatitsch and Jeroen Tromp. Introduction to the spectral element method for three-dimensional seismic wave propagation. *Geophysical journal international*, 139(3):806–822, 1999.
[72] N. V. Krylov. On the general notion of fully nonlinear second-order elliptic equations. *Transactions of the American Mathematical Society*, 347(3):857–895, Mar 1995.
[73] Harold W Kuhn. The Hungarian method for the assignment problem. *Naval Research Logistics (NRL)*, 2(1-2):83–97, 1955.
[74] P. Lailly. The seismic inverse problem as a sequence of before stack migrations. In *Conference on inverse scattering: theory and application*, pages 206–220. Society for Industrial and Applied Mathematics, Philadelphia, PA, 1983.
[75] Patrick Lailly. Migration methods: partial but efficient solutions to the seismic inverse problem. *Inverse problems of acoustic and elastic waves*, 51:1387–1403, 1984.
[76] Bruno Lévy and Erica Schwindt. Notions of optimal transport theory and how to implement them on a computer. 2017.
[77] Wuchen Li, Ernest K Ryu, Stanley Osher, Wotao Yin, and Wilfrid Gangbo. A parallel method for earth mover's distance. *UCLA Comput. Appl. Math. Pub.(CAM) Rep*, pages 17–12, 2017.

[78] Dong C Liu and Jorge Nocedal. On the limited memory BFGS method for large scale optimization. *Mathematical programming*, 45(1):503–528, 1989.
[79] Jiangbo Liu, Hervé Chauris, and Henri Calandra. The normalized integration method-an alternative to full waveform inversion? In *25th Symposium on the Application of Geophpysics to Engineering & Environmental Problems*, 2012.
[80] Zhenyue Liu and Norman Bleistein. Migration velocity analysis: Theory and an iterative algorithm. *Geophysics*, 60(1):142–153, 1995.
[81] Jianfeng Lu and Xu Yang. Frozen gaussian approximation for high frequency wave propagation. *Communications in Mathematical Sciences*, 9(3):663–683, 2011.
[82] Jingrui Luo and Ru-Shan Wu. Seismic envelope inversion: Reduction of local minima and noise resistance. *Geophysical Prospecting*, 63(3):597–614, 2015.
[83] Simon Luo and Dave Hale. Least-squares migration in the presence of velocity errors. *Geophysics*, 79(4):S153–S161, 2014.
[84] Yi Luo and Gerard T Schuster. Wave-equation traveltime inversion. *Geophysics*, 56(5):645–653, 1991.
[85] Yong Ma and Dave Hale. Wave-equation reflection traveltime inversion with dynamic warping and full-waveform inversion. *Geophysics*, 78(6):R223–R233, 2013.
[86] Kurt J Marfurt. Accuracy of finite-difference and finite-element modeling of the scalar and elastic wave equations. *Geophysics*, 49(5):533–549, 1984.
[87] H. Marquering, F. A. Dahlen, and G. Nolet. Three-dimensional sensitivity kernels for finite-frequency traveltimes: the banana-doughnut paradox. *Geophysical Journal International*, 137(3):805–815, Jun 1999.
[88] L Métivier, R Brossier, Q Mérigot, E Oudet, and J Virieux. Measuring the misfit between seismograms using an optimal transport distance: application to full waveform inversion. *Geophysical Journal International*, 205(1):345–377, 2016.
[89] L Métivier, R Brossier, Q Mérigot, E Oudet, and J Virieux. An optimal transport approach for seismic tomography: application to 3D full waveform inversion. *Inverse Problems*, 32(11):115008, 2016.
[90] Hermann Minkowski. Volumen und Oberfläche. pages 146–192. Springer, Vienna, 1989.
[91] Peter Moczo, Johan OA Robertsson, and Leo Eisner. The finite-difference time-domain method for modeling of seismic wave propagation. *Advances in Geophysics*, 48:421–516, 2007.
[92] Gaspard Monge. Mémoire sur la théorie des déblais et de remblais. histoire de l'académie royale des sciences de paris. *avec les Mémoires de Mathématique et de Physique pour la même année*, pages 666–704, 1781.
[93] Peter Mora. Elastic wave-field inversion of reflection and transmission data. *Geophysics*, 53(6):750–759, Jun 1988.
[94] Peter Mora. Inversion = migration + tomography. *Geophysics*, 54(12):1575–1586, Dec 1989.
[95] F Natterer. An error bound for the Born approximation. *Inverse problems*,

20(2):447–452, 2004.

[96] Roger G Newton. *Scattering theory of waves and particles.* Springer Science & Business Media, 2013.

[97] Adam M Oberman and Yuanlong Ruan. An efficient linear programming method for optimal transportation. *arXiv preprint arXiv:1509.03668*, 2015.

[98] VI Oliker and LD Prussner. On the numerical solution of the equation \frac {{\partial^ 2 z}}{{\partial x^ 2}}\frac {{\partial^ 2 z}}{{\partial y^ 2}}- \left ({\frac {{\partial^ 2 z}}{{\partial x\partial y}}}\right)^ 2= f and its discretizations, i. *Numerische Mathematik*, 54(3):271–293, 1989.

[99] Nicolas Papadakis, Gabriel Peyré, and Edouard Oudet. Optimal transport with proximal splitting. *SIAM Journal on Imaging Sciences*, 7(1):212–238, 2014.

[100] R.-E. Plessix. A review of the adjoint-state method for computing the gradient of a functional with geophysical applications. *Geophysical Journal International*, 167(2):495–503, 2006.

[101] Tim Pointer, Enru Liu, and John A Hudson. Numerical modelling of seismic waves scattered by hydrofractures: application of the indirect boundary element method. *Geophysical Journal International*, 135(1):289–303, 1998.

[102] R. Gerhard Pratt. Inverse Theory Applied to Multi-Source Cross-Hole Tomography. Part 2: Elastic Wave-Equation. *Geophysical Prospecting*, 38(3):311–329, Apr 1990.

[103] R Gerhard Pratt and MH Worthington. Inverse theory applied to multi-source cross-hole tomography. Part 1: Acoustic wave-equation method. *Geophysical prospecting*, 38(3):287–310, 1990.

[104] Lingyun Qiu, Jaime Ramos-Martínez, Alejandro Valenciano, Yunan Yang, and Björn Engquist. Full-waveform inversion with an exponentially encoded optimal-transport norm. In *SEG Technical Program Expanded Abstracts 2017*, pages 1286–1290. Society of Exploration Geophysicists, 2017.

[105] Yingming Qu, Zhenchun Li, Jianping Huang, and Jinli Li. Viscoacoustic anisotropic full waveform inversion. *Journal of Applied Geophysics*, 136:484–497, Jan 2017.

[106] N Rawlinson and M Sambridge. Seismic traveltime tomography of the crust and lithosphere. *Advances in Geophysics*, 46:81–199, 2003.

[107] Nicholas Rawlinson, Gregory Austin Houseman, and Clive DN Collins. Inversion of seismic refraction and wide-angle reflection traveltimes for three-dimensional layered crustal structure. *Geophysical Journal International*, 145(2):381–400, 2001.

[108] Kabir Roy Chowdhury. Seismic data acquisition and processing. pages 1081–1097. Springer, Dordrecht, 2011.

[109] Malcolm Sambridge and Kerry Gallagher. Inverse Theory, Monte Carlo Method. pages 639–644. Springer, Dordrecht, 2011.

[110] Paul Sava and Biondo Biondi. Wave-equation migration velocity analysis. I. Theory. *Geophysical Prospecting*, 52(6):593–606, 2004.

[111] H Schiessel, R Metzler, A Blumen, and TF Nonnenmacher. Generalized viscoelastic models: their fractional equations with solutions. *Journal of physics A: Mathematical and General*, 28(23):6567, 1995.

[112] Bernhard Schmitzer. A sparse multiscale algorithm for dense optimal transport. *Journal of Mathematical Imaging and Vision*, 56(2):238–259, 2016.
[113] Gerard T. Schuster. Seismic Imaging, Overview. pages 1121–1134. Springer, Dordrecht, 2011.
[114] Gji Seismology, Ebru Bozdağ, Jeannot Trampert, and Jeroen Tromp. Misfit functions for full waveform inversion based on instantaneous phase and envelope measurements. *Geophys. J. Int*, 185:845–870, 2011.
[115] Mrinal K. Sen and Paul L. Stoffa. Inverse Theory, Global Optimization. pages 625–632. Springer, Dordrecht, 2011.
[116] Peter M Shearer. *Introduction to seismology.* Cambridge University Press, 2009.
[117] Laurent Sirgue and R Gerhard Pratt. Efficient waveform inversion and imaging: A strategy for selecting temporal frequencies. *Geophysics*, 69(1):231–248, 2004.
[118] Justin Solomon, Fernando De Goes, Gabriel Peyré, Marco Cuturi, Adrian Butscher, Andy Nguyen, Tao Du, and Leonidas Guibas. Convolutional Wasserstein distances: Efficient optimal transportation on geometric domains. *ACM Transactions on Graphics (TOG)*, 34(4):66, 2015.
[119] William W Symes. Migration velocity analysis and waveform inversion. *Geophysical prospecting*, 56(6):765–790, 2008.
[120] WW Symes and James J Carazzone. Velocity inversion by differential semblance optimization. *Geophysics*, 56(5):654–663, 1991.
[121] A. Tarantola. Inverse problems theory. *Methods for Data Fitting and Model Parameter Estimation. Elsevier, Southampton*, 1987.
[122] Albert Tarantola. Linearized inversion of seismic reflection data. *Geophysical prospecting*, 32(6):998–1015, 1984.
[123] Albert Tarantola. *Inverse problem theory and methods for model parameter estimation.* SIAM, 2005.
[124] Albert Tarantola and Bernard Valette. Generalized nonlinear inverse problems solved using the least squares criterion. *Reviews of Geophysics*, 20(2):219–232, 1982.
[125] Sven Treitel and Larry Lines. Past, present, and future of geophysical inversion—a new millennium analysis. *Geophysics*, 66(1):21–24, Jan 2001.
[126] Neil S Trudinger and Xu-Jia Wang. The Monge-Ampère equation and its geometric applications. *Handbook of geometric analysis*, 1:467–524, 2008.
[127] Neil S Trudinger and Xu-Jia Wang. The Monge-Ampère equation and its geometric applications. pages 467–524, 2008.
[128] Gunther Uhlmann. Travel time tomography. *J. Korean Math. Soc*, 38(4):711–722, 2001.
[129] Léon Van Hove. Correlations in space and time and Born approximation scattering in systems of interacting particles. *Physical Review*, 95(1):249, 1954.
[130] Anatoly Moiseevich Vershik. Long history of the Monge-Kantorovich transportation problem. *The Mathematical Intelligencer*, 35(4):1–9, 2013.
[131] C. Villani. *Topics in optimal transportation*, volume 58 of *Graduate Studies in Mathematics.* American Mathematical Society, Providence, RI, 2003.

[132] J. Virieux, A. Asnaashari, R. Brossier, L. Métivier, A. Ribodetti, and W. Zhou. 6. An introduction to full waveform inversion. In *Encyclopedia of Exploration Geophysics*, pages R1–1–R1–40. Society of Exploration Geophysicists, Jan 2014.

[133] Jean Virieux and Stéphane Operto. An overview of full-waveform inversion in exploration geophysics. *Geophysics*, 74(6):WCC1–WCC26, 2009.

[134] Ping Wang, Adriano Gomes, Zhigang Zhang, and Ming Wang. Least-squares rtm: Reality and possibilities for subsalt imaging. In *SEG Technical Program Expanded Abstracts 2016*, pages 4204–4209. Society of Exploration Geophysicists, Sep 2016.

[135] Xu-Jia Wang. On the design of a reflector antenna. *Inverse Problems*, 12(3):351–375, Jun 1996.

[136] Mike Warner and Lluís Guasch. Adaptive waveform inversion: Theory. In *SEG Technical Program Expanded Abstracts 2014*, pages 1089–1093. 2014.

[137] Arthur B Weglein, Fernanda V Araújo, Paulo M Carvalho, Robert H Stolt, Kenneth H Matson, Richard T Coates, Dennis Corrigan, Douglas J Foster, Simon A Shaw, and Haiyan Zhang. Inverse scattering series and seismic exploration. *Inverse problems*, 19(6):R27, 2003.

[138] Pengliang Yang, Romain Brossier, Ludovic Métivier, and Jean Virieux. A review on the systematic formulation of 3-D multiparameter full waveform inversion in viscoelastic medium. *Geophysical Journal International*, 207(1):129–149, 2016.

[139] Yunan Yang and Björn Engquist. Analysis of optimal transport related misfit functions in seismic imaging. In *International Conference on Geometric Science of Information*, pages 109–116. Springer, 2017.

[140] Yunan Yang and Björn Engquist. Analysis of optimal transport and related misfit functions in full-waveform inversion. *Geophysics*, 83(1):A7–A12, 2018.

[141] Yunan Yang, Björn Engquist, Junzhe Sun, and Brittany D Froese. Application of optimal transport and the quadratic Wasserstein metric to full-waveform inversion. *Geophysics*, 83(1):1–103, 2017.

[142] Öz Yilmaz. *Seismic Data Analysis.* Society of Exploration Geophysicists, Jan 2001.

[143] Kwangjin Yoon, Kurt J Marfurt, and William Starr. Challenges in reverse-time migration. In *SEG Technical Program Expanded Abstracts 2004*, pages 1057–1060. Society of Exploration Geophysicists, 2004.

[144] Kwangjin Yoon, Kurt J. Marfurt, and William Starr. Challenges in reverse-time migration. In *SEG Technical Program Expanded Abstracts 2004*, pages 1057–1060. Society of Exploration Geophysicists, Jan 2004.

[145] C. A. Zelt. Modelling strategies and model assessment for wide-angle seismic traveltime data. *Geophysical Journal International*, 139(1):183–204, Oct 1999.

[146] Colin A. Zelt. Traveltime tomography using controlled-source seismic data. pages 1453–1473. Springer, Dordrecht, 2011.

[147] Chong Zeng, Shuqian Dong, and Bin Wang. A guide to least-squares reverse time migration for subsalt imaging: Challenges and solutions. *Interpretation*, 5(3):SN1–SN11, Aug 2017.

[148] Hejun Zhu and Sergey Fomel. Building good starting models for full-waveform inversion using adaptive matching filtering misfit. *Geophysics*, 81(5):U61–U72, 2016.

Department of Mathematics and ICES
The University of Texas at Austin
1 University Station C1200, Austin, TX 78712, USA
E-mail address: engquist@math.utexas.edu

Department of Mathematics
The University of Texas at Austin
1 University Station C1200, Austin, TX 78712, USA
E-mail address: yunanyang@math.utexas.edu

Plenary Lectures

Proceedings of the 7th International
Congress of Chinese Mathematicians
ALM 43, pp. 217–235

Heterogeneously Localized Nonlocal Operators, Boundary Traces and Variational Problems

Qiang Du and Xiaochuan Tian

Abstract

It is a classical result of Sobolev spaces that any H^1 function has a well-defined $H^{1/2}$ trace on the boundary of a sufficient regular domain. We discuss its recent extensions given in [45] in some heterogeneously localized nonlocal function spaces. The new trace theorems are stronger and more general than the classical result. They can be established essentially for all functions having only square integrability away from the boundary or in any compact subset of interior domain. Yet, the heterogeneous localization offers the necessary regularity precisely at the boundary to have well-defined traces. A consequence is that we may study associated Dirichlet type boundary value problems, as well as the coupling of local and nonlocal equations through co-dimension-1 interfaces.

Mathematics Subject Classification: 46E35, 35A23, 47G10, 35Q74, 49N60.

Keywords and Phrases: Nonlocal operator, nonlocal function space, trace inequality, boundary value problem, heterogeneous localization, vanishing horizon.

1 Introduction

To study boundary value problems and interface problems, it is often necessary to study traces of associated function spaces. For standard second order elliptic problems defined on a spatial domain Ω, the associated function space is typically given by $H^1(\Omega)$, the standard Sobolev space of L^2 functions with square integrable derivatives. The following mathematical result is well-known: on a domain $\Omega \in \mathbb{R}^d$ with boundary $\partial\Omega$, a trace operator T on a subset Γ of $\partial\Omega$ can be extended continuously as a map from $H^1(\Omega)$ to $H^{1/2}(\Gamma)$. In other words, we have the classical trace inequality of Gagliardo [17]:

$$\|u\|_{H^{\frac{1}{2}}(\partial\Omega)} \le C\|u\|_{H^1(\Omega)}, \quad \forall\, u \in H^1(\Omega)$$

for a positive constant C depending only on Ω. Moreover, it is known that for suitably smooth domains, $H^{1/2}(\Gamma)$ functions can have $H^1(\Omega)$ interior extensions,

Supported in part by the U.S. NSF grants DMS-1558744, AFOSR MURI Center for Material Failure Prediction Through Peridynamics, and OSD/ARO/MURI W911NF-15-1-0562 on Fractional PDEs for Conservation Laws and Beyond: Theory, Numerics and Applications.

which is a key element in the formulation of boundary value problems of second order elliptic operators with Dirichlet type data. For generic $H^{1/2}(\Gamma)$ data, the $H^1(\Omega)$ regularity of the interior extension is the best possible.

A mathematical question underlying the study in [45] is to characterize some subspaces, denote by of $\mathcal{S}(\Omega)$, that are significantly larger than $H^1(\Omega)$, to have a continuous trace map into $H^{1/2}(\Gamma)$ and thus a stronger and more general trace inequality

$$\|u\|_{H^{\frac{1}{2}}(\partial\Omega)} \leq C\|u\|_{\mathcal{S}(\Omega)}, \quad \forall u \in \mathcal{S}(\Omega).$$

In this work, we present the main findings of [45] concerning some recent extensions of the classical trace theorem in the Sobolev space to some suitably defined nonlocal function spaces. To provide the background and motivation of such investigations, we first recall some elements of nonlocal operators with finite range of interactions and their connections with classical second order elliptic operators. Related works on the associated nonlocal and local (classical Sobolev) function spaces are also briefly discussed. Then the key technical results in [45] are reviewed, finally, we point out some implications on the well-posedness of boundary value problems associated with nonlocal elliptic operators and the coupling of local and nonlocal variational problems,

2 Background and motivation

In recent studies of nonlocal mechanical models and nonlocal diffusion, we have become interested in the study of nonlocal operators of the following form: for a given parameter $\delta > 0$ and a given density function (radially symmetric kernel) $\underline{\omega}_\delta = \underline{\omega}_\delta(s) = \underline{\omega}_\delta(|s|)$ compactly defined on $B_\delta(0) \subset \mathbb{R}^d$ the ball of radius δ at the origin),

$$\mathcal{L}_\delta u(\boldsymbol{x}) = d\int_{B_\delta(0)} \frac{u(\boldsymbol{x}+s) - 2u(\boldsymbol{x}) + u(\boldsymbol{x}-s)}{|s|^2}\underline{\omega}_\delta(\boldsymbol{s})d\boldsymbol{s} \quad \forall u = u(\boldsymbol{x}) : \mathbb{R}^d \to \mathbb{R}. \tag{1}$$

The operator $\mathcal{L}_\delta$ is said to be nonlocal since $\mathcal{L}_\delta u(\boldsymbol{x})$ depends not only on the properties of u at $\boldsymbol{x}$ but also its δ neighborhood. The positive parameter δ is commonly called the nonlocal horizon parameter that characterizes the range of nonlocal interactions.

The nonlocal operator $\mathcal{L}_\delta$ can be seen as a continuum average of finite difference operators with weights specified by the kernel $\underline{\omega}_\delta$. Meanwhile, formally by Taylor expansion, we have

$$\mathcal{L}_\delta u(\boldsymbol{x}) = \Delta u(\boldsymbol{x}) \int_{B_\delta(0)} \underline{\omega}_\delta(|\boldsymbol{s}|)d\boldsymbol{s} + O(\delta^2) \approx \Delta u(\boldsymbol{x}) = \mathcal{L}_0 u(\boldsymbol{x}), \quad \text{as } \delta \to 0,$$

Thus we may get the local differential operator $\mathcal{L}_0 = \Delta$ being the local limit. This can be made rigorous for kernels taking the form $\underline{\omega}_\delta(s) = \delta^{-d}\omega(s/\delta)$.

For a given spatial domain Ω, we may define a nonlocal space $\mathcal{S}(\Omega)$ that is given by the completion of $C^1(\Omega)$ with respect to the nonlocal norm

$$\| \cdot \|_{\mathcal{S}(\Omega)} = (\| \cdot \|^2_{L^2(\Omega)} + | \cdot |^2_{\mathcal{S}(\Omega)})^{1/2}, \tag{2}$$

where the associated nonlocal semi-norm $| \cdot |_{\mathcal{S}(\Omega)}$ is defined by

$$|u|^2_{\mathcal{S}(\Omega)} = \int_\Omega \int_{\Omega \cap B_\delta(\boldsymbol{x})} \gamma(\boldsymbol{x}, \boldsymbol{y}) \frac{(u(\boldsymbol{y}) - u(\boldsymbol{x}))^2}{|\boldsymbol{y} - \boldsymbol{x}|^2} d\boldsymbol{y} d\boldsymbol{x} \tag{3}$$

for some nonlocal interaction kernel $\gamma(\boldsymbol{x}, \boldsymbol{y})$.

For the nonlocal operator and nonlocal considered here with an associated nonlocal horizon parameter δ, a natural and popular choice of the kernel is in fact

$$\gamma(\boldsymbol{x}, \boldsymbol{y}) = \underline{\omega}_\delta(|\boldsymbol{x} - \boldsymbol{y}|) = \delta^{-d} \omega(|\boldsymbol{x} - \boldsymbol{y}|/\delta),$$

for a density $\omega = \omega(|\boldsymbol{x}|)$, though other forms may prove to also be useful, as shown later.

A nonlocal model $-\mathcal{L}_\delta u_\delta{=}f$ defines in $\mathbb{R}^d$ or Ω with a given data f and the unknown solution u can be seen as a more general model than differential/difference models or a bridge connecting them.

In recent years, there has been significant progress towards a systematic (or even axiomatic) development of a rigorous mathematical framework for nonlocal operators and nonlocal models. In [12], attempts have been made to develop a nonlocal vector calculus as the basic mathematical building blocks for nonlocal continuum models, mimicking the classical Newton's vector calculus for local models in the form of partial differential equations. A comparison of a few key ingredients and concepts is presented in the Table 1.

Newton's vector calculus	$\Leftrightarrow$	Nonlocal vector calculus
Local balance (PDE)	$\Leftrightarrow$	Nonlocal balance (PD)
Differential gradient/divergence	$\Leftrightarrow$	Nonlocal gradient/divergence
$-\nabla \cdot (K \nabla u) = f$	$\Leftrightarrow$	$-\mathscr{D} \cdot (\omega_\delta \mathscr{D}^* u) = f$
Boundary conditions	$\Leftrightarrow$	Volumetric constraints
Integral identities/integration by parts	$\Leftrightarrow$	Nonlocal integration by parts
$\int_\Omega u\Delta v - v\Delta u = \int_{\partial\Omega} u\partial_n v - v\partial_n u$	$\Leftrightarrow$	$\iint u\mathscr{D}(\mathscr{D}^* v) - v\mathscr{D}(\mathscr{D}^* u) = 0$

Table 1: Nonlocal vector calculus vesus Newton's vector calculus

For further development and applications of the framework to the analysis of various nonlocal models such as linear peridynamic bond-based and state-based models, peridynamic bending elasticity models, nonlinear hyperelastic peridynamic models, and nonlocal in time dynamic models, we refer to [11, 12, 29, 30, 31, 32, 13, 44]. Some of the main distinctions of these series of works include the theory developed for systems of equations involving vectors and tensor fields, the finite range of nonlocal interactions and the precise dependence of the mathematical estimates on the horizon δ, and the minimal regularity assumptions associated with the underlying function spaces.

Generically, nonlocal equations posed on a domain $\Omega \subset \mathbb{R}^d$ are complemented by nonlocal boundary conditions, or more precisely, constraints on a some domain with nonzero d-dimensional volume, hence leading to the so-called constrained value problems [11]. A pictorial depiction of nonlocal problems with a nonlocal horizon $\delta > 0$ and volumetric constraints is given in Figure.1, together with its possible differential equation limit in the form of a standard Dirichlet boundary value problem. We note that some key ingredients in the development of the well-posedness theory for linear nonlocal models are the nonlocal Kohn's inequalities and nonlocal Poincare inequalities, as well as compactness properties and nonlocal Gagliardo-Nirenberg inequalities, which serve as the nonlocal analog of their classical, local counterparts for linear PDEs.

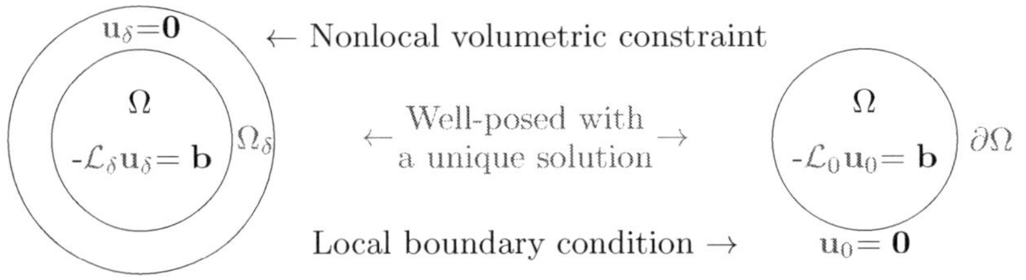

Figure 1: A nonlocal constrained value problem and a local boundary value problem.

In general, nonlocal models may allow more singular solutions in a function space $\mathcal{S}_\delta$ larger than H^1. The ability to produce more singular solutions is a major motivation for studyingf nonlocal models in subjects such as fracture mechanics and anomalous diffusion. For more studies on nonlocal spaces and associated nonlocal operators, we refer to [3, 5, 6, 11, 12, 16, 18, 24, 28, 31, 33, 37, 42] and references cited therein on related applications and mathematical analysis. Moreover, developing nonsmooth calculus [20] has also been an active subject of mathematical research with strong connections to geometry [15].

To avoid the use of such nonlocal constraints, the corresponding nonlocal operators need to be properly modified near the boundary, which is often the case for fractional differential equations [6]. For a more recent survey on the nonlocal elliptic equations, we refer to [38].

In order to have well-defined nonlocal problems on Ω with Dirichlet type data on part of its boundary $\partial\Omega$ of codimension-1, study of the trace map in $\mathcal{S}(\Omega)$ becomes natural and at the same time a necessity. However, it is known that in general $\mathcal{S}(\Omega)$ contains $H^1(\Omega)$ as a subspace and it is sufficient large so that it may contain more singular functions as desired by nonlocal models. Meanwhile, we know from localization results that as $\delta \to 0$, we may recover the classical H^1 space, that means that in the local limit, the trace property is attained together with the improvement in regularity homogeneously over the whole domain. This of course does not bode well with our goal, but it leads to the natural idea of heterogeneous localization. That is, we desire to define a nonlocal space which maintains the nonlocality inside the domain (thus allowing more singular functions to be included), but gets localized at the boundary (giving rise to the well-defined trace map). This is the path followed in [45].

3 New trace theorem in heterogeneously localized nonlocal spaces

The main findings of [45] are that the trace map exists and is continuous on a nonlocal function space $\mathcal{S}(\Omega)$ if the support of the kernel is heterogeneously localized as $\boldsymbol{x}$ approaches the boundary of Ω. By considering such a class of kernels, the study departs from many existing works corresponding to typical translate-invariant kernels (such as the celebrated work [5]), though in other settings, variable order and variable growth function spaces have been a popular subject with a rich history and much recent interest, see for instance [8, 34].

In [45], the class of kernels under consideration is given by

$$\gamma(\boldsymbol{x},\boldsymbol{y}) = \frac{1}{|\delta(\boldsymbol{x})|^{d+2}}\hat{\gamma}\left(\frac{|\boldsymbol{y}-\boldsymbol{x}|}{\delta(\boldsymbol{x})}\right) \tag{4}$$

where $\hat{\gamma} = \hat{\gamma}(s)$ is a non-increasing nonnegative function defined for $s \in (0,1)$ with a finite $d+1$ moment. The influence horizon $\delta = \delta(\boldsymbol{x})$ is a function defined on Ω that approaches zero when $\boldsymbol{x}$ approaches the boundary. A simple choice taken in [45] is

$$\delta(\boldsymbol{x}) = \sigma\,\mathrm{dist}(\boldsymbol{x},\Gamma), \quad \boldsymbol{x}\in\Omega\,, \tag{5}$$

for some $\sigma \in (0,1]$ and $\Gamma \subset \partial\Omega$. The associated nonlocal neighborhood $\mathcal{H}(\boldsymbol{x})$ is defined by

$$\mathcal{H}(\boldsymbol{x}) := \{\boldsymbol{y}\in\Omega : |\boldsymbol{y}-\boldsymbol{x}| \le \delta(\boldsymbol{x})\} = B_{\delta(\boldsymbol{x})}(\boldsymbol{x})\,.$$

A key observation of [45] is that, by allowing heterogeneous localization with vanishing effective neighborhood $\mathcal{H}(\boldsymbol{x})$ when $\boldsymbol{x}$ approaches the boundary $\partial\Omega$, we expect to have a well defined continuous trace map from the associated nonlocal space $\mathcal{S}(\Omega)$ to $H^{1/2}(\partial\Omega)$. This leads to the generalized trace theorem of [45] that is stated below.

Theorem 3.1 (General trace theorem). *Assume that Ω is a bounded simply connected Lipschitz domain in $\mathbb{R}^d$ $(d\ge 2)$ and $\Gamma = \partial\Omega$, and the nonlocal interaction kernel is taken as in* (4) *with the heterogeneously defined horizon given by* (5)*, then there exists a constant C depending only on Ω and Γ such that the trace map T for Γ satisfies*

$$\|Tu\|_{H^{\frac{1}{2}}(\Gamma)} \le C\|u\|_{\mathcal{S}(\Omega)}\,, \quad \forall u \in \mathcal{S}(\Omega)\,. \tag{6}$$

The following proposition has been established in [45], which is of independent interests on its own while showing the relation between the classical Sobolev space $H^1(\Omega)$ and the new heterogeneously localized nonlocal space $\mathcal{S}(\Omega)$. The result generalizes a similar conclusion for nonlocal spaces defined with a constant horizon and translation invariant kernel known in the literature, see [5].

Proposition 3.2. *For the nonlocal interaction kernel given in* (4) *and the heterogeneously defined horizon* $\delta(\boldsymbol{x}) = \sigma \mathrm{dist}(x, \Gamma)$ *with* $\sigma \in (0,1)$, *the space* $H^1(\Omega)$ *is continuously imbedded in* $\mathcal{S}(\Omega)$ *and there exists a constant* C *depending only on* σ *and* Ω *such that*

$$\|u\|_{\mathcal{S}(\Omega)} \le C\|u\|_{H^1(\Omega)}\,, \quad \forall u \in H^1(\Omega)\,. \tag{7}$$

Moreover, C *is independent of* σ *for* σ *small.*

The proof of (7) can be done for a smooth function $u \in C^1(\bar{\Omega}) \cap H^1(\Omega)$ first. One key observation is that

$$\begin{aligned} |u|^2_{\mathcal{S}(\Omega)} &\le \int_{\mathbb{R}^d}\int_{|\boldsymbol{h}|<\delta(\boldsymbol{x})} \frac{1}{|\delta(\boldsymbol{x})|^{d+2}}\hat{\gamma}(\frac{|\boldsymbol{h}|}{\delta(\boldsymbol{x})})|u(\boldsymbol{x}+\boldsymbol{h}) - u(\boldsymbol{x})|^2 d\boldsymbol{h}d\boldsymbol{x} \\ &= \int_{\mathbb{R}^d}\int_{|\boldsymbol{h}|<1} \frac{1}{|\delta(\boldsymbol{x})|^2}\hat{\gamma}(\boldsymbol{h})|u(\boldsymbol{x}+\delta(\boldsymbol{x})\boldsymbol{h}) - u(\boldsymbol{x})|^2 d\boldsymbol{h}d\boldsymbol{x} \\ &\le \int_{\mathbb{R}^d}\int_{|\boldsymbol{h}|<1} |\boldsymbol{h}|^2\hat{\gamma}(\boldsymbol{h})\int_0^1 |\nabla u(\boldsymbol{x}+t\delta(\boldsymbol{x})\boldsymbol{h})|^2 dt d\boldsymbol{h}d\boldsymbol{x}\,. \end{aligned}$$

For $\boldsymbol{y} = \boldsymbol{x} + t\delta(\boldsymbol{x})\boldsymbol{h}$, by noting that

$$\frac{\partial \boldsymbol{y}}{\partial \boldsymbol{x}} = I + t\nabla\delta(\boldsymbol{x}) \otimes \boldsymbol{h}\,,$$

and its inverse are uniformly bounded everywhere if $\|\nabla\delta\| = \sigma < 1$, with the bounds independent of σ if σ is small, we get the existence of a generic constant $C > 0$ such that

$$\begin{aligned} |u|^2_{\mathcal{S}(\Omega)} &\le C(\int_{|\boldsymbol{h}|<1} |\boldsymbol{h}|^2\hat{\gamma}(\boldsymbol{h})d\boldsymbol{h})|u|^2_{H^1(\mathbb{R}^d)} \\ &\le C\|u\|^2_{H^1(\Omega)}\,. \end{aligned}$$

The constant C may depend on Ω but is independent of σ for σ small. For a complete proof, we refer to [45].

By the Proposition 3.2, we have

$$\|Tu\|_{H^{\frac{1}{2}}(\Gamma)} \le C\|u\|_{\mathcal{S}(\Omega)} \le C\|u\|_{H^1(\Omega)}\,, \quad \forall u \in H^1(\Omega)\,,$$

thus, the trace inequality (6) is indeed a refinement of the classical trace inequality in the conventional Sobolev space $H^1(\Omega)$. The latter follows as a simple consequence.

Corollary 3.3 (Sobolev trace theorem)**.** *Assume that* Ω *is a bounded simply connected Lipschitz domain in* $\mathbb{R}^d$ $(d \ge 2)$ *and* $\Gamma = \partial\Omega$, *then there exists a constant* C *depending only on* Ω *and* Γ *such that the trace map* T *for* Γ *satisfies*

$$\|Tu\|_{H^{\frac{1}{2}}(\Gamma)} \le C\|u\|_{H^1(\Omega)}\,, \quad \forall u \in H^1(\Omega)\,. \tag{8}$$

4 A special trace theorem as an illustrative example

The full proof of the trace theorem 3.1 is presented in [45]. To help understanding what the result conveys and how it compares with other relevant works, it is suggestive to consider a special case. That is, in terms of Ω and Γ , we take a special stripe domain $\Omega = (0, r) \times \mathbb{R}^{d-1}$ and a portion of its boundary $\Gamma = \{0\} \times \mathbb{R}^{d-1}$ where r is any given positive constant, while for the kernel, we adopt a special kernel with $\hat{\gamma}$ to be the characteristic function $\chi_{(0,1)}$ for $|\boldsymbol{y} - \boldsymbol{x}| \in (0, 1)$ with the horizon $\delta(\boldsymbol{x})$ given by the first component of $\boldsymbol{x}$, that is,

$$\gamma(\boldsymbol{x}, \boldsymbol{y}) = \frac{\chi_{(0,1)}(|\boldsymbol{y} - \boldsymbol{x}|)}{|\delta(\boldsymbol{x})|^{d+2}}, \quad \delta(\boldsymbol{x}) = \mathrm{dist}(\boldsymbol{x}, \Gamma) = x_1, \quad \text{for } \boldsymbol{x} = (x_1, \tilde{\boldsymbol{x}}), \quad \tilde{\boldsymbol{x}} \in \mathbb{R}^{d-1}. \tag{9}$$

The geometric set up is depicted in Figure 2.

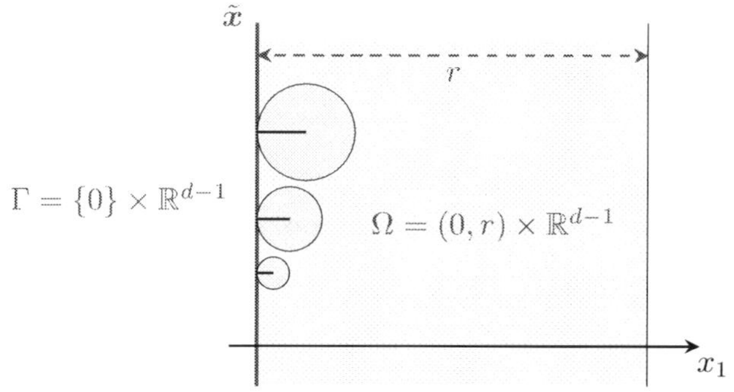

Figure 2: Depiction of the stripe geometry.

As shown in [45] , this special case serves as not only a helpful step towards proving the more general result but also an illustrative example.

Let us now take a look at the special form of the nonlocal (semi)-norm in this case. By definition, we get

$$|u|^2_{\mathcal{S}(\Omega)} = \int_\Omega \int_{\Omega \cap \{|\boldsymbol{y} - \boldsymbol{x}| < |x_1|\}} \frac{(u(\boldsymbol{y}) - u(\boldsymbol{x}))^2}{|x_1|^{2+d}} d\boldsymbol{y} d\boldsymbol{x}\,. \tag{10}$$

Clearly, the denominator x_1 penalizes the spatial variation only at $x_1 = 0$, thus $\mathcal{S}(\Omega)$ contains all functions that are in $L^2(\tilde{\Omega})$ for any domain $\tilde{\Omega}$ with its closure being a compact subset of Ω. Hence, functions in $\mathcal{S}(\Omega)$ are generally not expected to have regularity better than $L^2(\Omega')$ over the subdomain Ω', or may be significantly less regular away from the boundary than H^1 functions. Yet, as elucidated in [45], due to the shrinking horizon towards the boundary, the penalization of spatial variations provides enough regularity for the functions in $\mathcal{S}(\Omega)$ to have well-defined traces just on the boundary itself. Intuitively, this is a natural consequence of the localization of nonlocal interactions on the boundary.

In contrast, we recall standard norms associated with fractional Sobolev spaces on Ω.

$$|u|^2_{H^\alpha(\Omega)} = \int_\Omega \int_\Omega \frac{(u(\boldsymbol{y}) - u(\boldsymbol{x}))^2}{|\boldsymbol{y} - \boldsymbol{x}|^{2\alpha+d}} d\boldsymbol{y} d\boldsymbol{x} \,.$$

The regularity of the functions is effected by the denominator which vanishes at $\boldsymbol{x} = \boldsymbol{y}$.

We now state the trace theorem in the special form.

Theorem 4.1 (Special trace theorem). *For $\Omega = (0,r)\times\mathbb{R}^{d-1}$ and $\Gamma = \{0\}\times\mathbb{R}^{d-1}$, there exists a constant C depends only on d such that such that for any $u \in C^1(\bar{\Omega})\cap\mathcal{S}(\Omega)$,*

$$\|u\|_{L^2(\Gamma)} \le C\Big(r^{-1/2}\|u\|_{L^2(\Omega)} + r^{1/2}|u|_{\mathcal{S}(\Omega)}\Big)\,, \tag{11}$$

and for $d \ge 2$,

$$|u|_{H^{1/2}(\Gamma)} \le C\Big(r^{-1}\|u\|_{L^2(\Omega)} + |u|_{\mathcal{S}(\Omega)}\Big) \tag{12}$$

where the nonlocal semi-norm of $\mathcal{S}(\Omega)$ is as given in (10).

The special form allows us to give more quantitative estimates on the constants appearing in the trace inequality. Note that we are mainly interested in the small r dependence of the imbedding coefficients. For large r, the results remain true with uniformly bounded coefficients, that is, $\|u\|_{H^{1/2}(\Gamma)} \le C\|u\|_{\mathcal{S}(\Omega)}$ where C is a constant as $r \to \infty$.

In [45], it has also been noted that the need of the $L^2(\Omega)$ norm in (11) and (12) is similar to that for standard trace inequalities in Sobolev spaces. The dependence on the $L^2(\Omega)$ norm may be removed by considering L^p type norms for the trace with a suitable choice of p, just like the classical counterpart in Sobolev spaces.

5 More on the nonlocal spaces and norms

In the process of establishing the trace inequality, there are a number of results presented in [45] that are of independent interests. For example, a generalization of the classical Hardy's inequality has been shown there, with the one dimensional version given below.

Proposition 5.1 (Nonlocal Hardy-type inequality in one-dimension). *Let $\Omega = (0,r)$ for some $r > 0$, $\Gamma = \{0\}$, and a and b satisfy $0 \le a < b \le 1$. Then we have for any $u \in C^1(\bar{\Omega})$ with $u\,|_\Gamma = u(0) = 0$,*

$$\int_\Omega \frac{|u(x)|^2}{|x|^2}dx \le C_{a,b}\int_\Omega\int_{ax}^{bx}\frac{|u(y)-u(x)|^2}{|x|^3}dydx\,, \quad \textit{where } C_{a,b} = \frac{4(2+b+a)}{(b-a)(2-b-a)^2}. \tag{13}$$

We recall the classical Hardy's inequality is of the form:

$$\int_0^r \frac{|u(x)|^2}{\mathrm{dist}(x,\Gamma)^2}dx \leq C\int_0^r |u'(x)|^2 dx\,. \tag{14}$$

Obviously, in the above classiical Hardy's inequality, the requirement of $u' \in L^2$ is redundant away from $x = 0$ since a simple observation shows that the singular weight on the left hand side of (14) only becomes effective right on the boundary $x = 0$. Thus, though widely useful, the inequality itself is not sharp. On the other hand, the singular weight for the weighted L^2 norm is in perfect synchrony with the effects of singular weight in the nonlocal norm, and the nonlocal Hardy's inequality is in fact much more natural.

One can expect a higher dimensional nonlocal Hardy's inequality as well, which is also given in [45] and stated below.

Proposition 5.2 (Nonlocal Hardy's inequality). *Given a bounded Lipshitz domain Ω, there exists a constant $C > 0$ such that if $Tu = 0$ on $\partial\Omega$, then*

$$\int_\Omega \frac{|u(\boldsymbol{x})|^2}{(dist(\boldsymbol{x},\partial\Omega))^2}d\boldsymbol{x} \leq C|u|^2_{\mathcal{S}(\Omega)}\,. \tag{15}$$

We note that the nonlocal space $\mathcal{S}(\Omega)$ in the above proposition is defined the same way as in the new trace theorem 3.1. The one-dimensional nonlocal Hardy's inequality (13) can be proved via elementary means, but the proof for the more general multidimensional case is somewhat more involved and it is shown with the help of another estimate on the variations of u along the normal direction. discussed later. Such an estimate requires careful studies. Hence, in the same spirit of norms of directional derivatives in classical, local function spaces, the following definition is introduced in [45] as a nonlocal analog that refines our understanding of how the nonlocal norm $\|\cdot\|_{\mathcal{S}(\Omega)}$ provides control on the function variation in different directions. This not only helps proving the nonlocal Hardy's inequality, but also plays important roles in proving the new trace theorems.

Definition 5.3. *On the domain $\Omega = (0,r)\times\mathbb{R}^{d-1}$, we define in the following two directional nonlocal semi-norms $|\cdot|_n$ and $|\cdot|_t$, standing for normal and tangential directions respectively with reference to the boundary segment $\Gamma = \{0\}\times\mathbb{R}^{d-1}$,*

$$|u|_n^2 = \int_{\mathbb{R}^{d-1}}\int_0^r \fint_{ax_1}^{bx_1} \frac{|u(y_1,\bar{\boldsymbol{x}}) - u(x_1,\bar{\boldsymbol{x}})|^2}{|x_1|^2} dy_1 dx_1 d\bar{\boldsymbol{x}} \tag{16}$$

$$|u|_t^2 = \int_{\mathbb{R}^{d-1}}\int_0^r \fint_{B_{cx_1}(\bar{\boldsymbol{x}})} \frac{|u(x_1,\bar{\boldsymbol{y}}) - u(x_1,\bar{\boldsymbol{x}})|^2}{|x_1|^2} d\bar{\boldsymbol{y}} dx_1 d\bar{\boldsymbol{x}} \tag{17}$$

where $0 \leq a < b \leq 1$ and $0 < c < 1$ are constants, and $\fint$ represents the integral average over a domain, that is, the integral over the domain divided by the domain volume.

Some heuristic comments are provided in [45] to offer insight into the above definitions. For a smooth function $u = u(\boldsymbol{x})$, we may approximately have, in an informal manner, that $|\cdot|_n$ and $|\cdot|_t$ mimic norms of directional directives, and they are thus named nonlocal normal and tangential directional semi-norms. In comparison, we may also informally see that $|\cdot|^2_{\mathcal{S}(\Omega)}$ involves nonlocal variations in all directions like a total gradient semi-norm.

While the norms of classical (local) directional derivatives are obviously bounded by that of the total gradient, a challenging step in [45] is to establish a nonlocal analog, stated in the the following lemma, that shows that $|\cdot|_n$ and $|\cdot|_t$ are indeed controlled by the original semi-norm $|\cdot|_{\mathcal{S}(\Omega)}$ as given by (10).

Lemma 5.4. *Let $\Omega = (0, r) \times \mathbb{R}^{d-1}$ for some $r > 0$, a, b and c satisfy $0 \leq a < b \leq 1$, $0 < c < 1$ and $(a-1)^2 + c^2 \leq 1$. Then there exists a constant C depending only on a, b and c such that for any $u \in \mathcal{S}(\Omega)$,*

$$|u|_n \leq C|u|_{\mathcal{S}(\Omega)}\,, \tag{18}$$

$$|u|_t \leq C|u|_{\mathcal{S}(\Omega)}\,. \tag{19}$$

The above inequalities are established as consequences of the following inequalities

$$|u|_n^2 \leq c_1|u|_t^2 + C|u|^2_{\mathcal{S}(\Omega)} \tag{20}$$

$$|u|_t^2 \leq c_2|u|_n^2 + C|u|^2_{\mathcal{S}(\Omega)} \tag{21}$$

where $c_1c_2 < 1$.

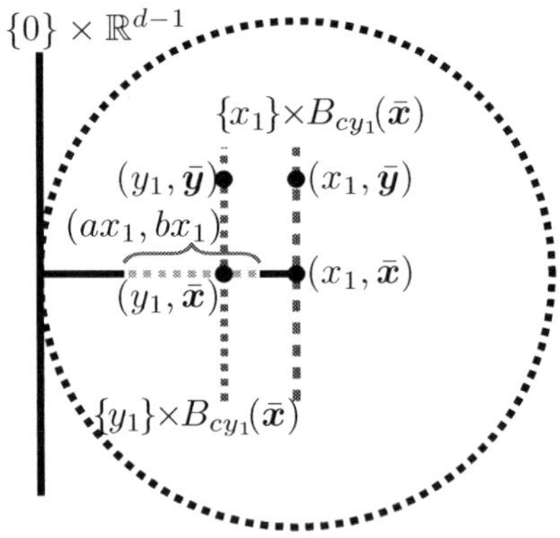

Figure 3: Depiction of geometry used in the proof of Lemma 5.4 in [45].

The Figure 3 can help explaining how the above estimates are achieved with some suitable choices of a, b and c. Geometrically, the three constants determines the geometric picture in Figure 3. The cyan horizontal dotted line shows the range of $(y_1, \bar{\boldsymbol{x}})$, the blue vertical dotted line for $(y_1, \bar{\boldsymbol{y}})$, and the red vertical dashed line for $(x_1, \bar{\boldsymbol{y}})$. The key to choose these positions is to make sure that $(y_1, \bar{\boldsymbol{y}})$ stays in the effective neighborhood of $(x_1, \bar{\boldsymbol{x}})$, which is bounded by the black dashed circle. This essentially becomes the issue of how to pick the constants a, b and c.

6 More about the proof of theorem 4.1 on the stripe domain

To show the bound on the $H^{1/2}$ semi-norm in the inequality (12), we essentially need to prove the following inequality

$$\int_{\mathbb{R}^{d-1}} \int_{B_{r/2}(\bar{\boldsymbol{x}})} \frac{|u(0,\bar{\boldsymbol{y}}) - u(0,\bar{\boldsymbol{x}})|^2}{|\bar{\boldsymbol{y}} - \bar{\boldsymbol{x}}|^d} d\bar{\boldsymbol{y}} d\bar{\boldsymbol{x}} \leq C|u|^2_{\mathcal{S}(\Omega)} \,. \tag{22}$$

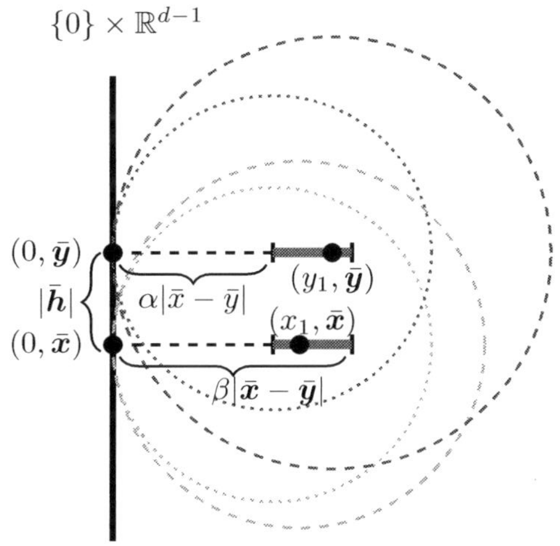

Figure 4: Depiction of geometry used in the proof of Theorem 4.1.

The idea is to make the interior extension using

$$|u(0,\bar{\boldsymbol{y}}) - u(0,\bar{\boldsymbol{x}})| \leq |u(0,\bar{\boldsymbol{y}}) - u(y_1,\bar{\boldsymbol{y}})| + |u(y_1,\bar{\boldsymbol{y}}) - u(x_1,\bar{\boldsymbol{x}})| + |u(x_1,\bar{\boldsymbol{x}}) - u(0,\bar{\boldsymbol{x}})|$$

and to control all the terms associated with the right hand side by the nonlocal norm. The blue and red solid horizontal line segments in Figure 4 show the possible positions of $(x_1, \bar{\boldsymbol{x}})$ and $(y_1, \bar{\boldsymbol{y}})$ respectively. The key is to determine the end points of these line segments so that any $(y_1, \bar{\boldsymbol{y}})$ on the red solid line should remain in the effective neighborhood (shown as balls bounded by dotted and dashed spheres, with the dotted ones giving the smallest neighborhoods) of any $(x_1, \bar{\boldsymbol{x}})$ on the other blue solid line and vice versa. The term associated with $|u(y_1, \bar{\boldsymbol{y}}) - u(x_1, \bar{\boldsymbol{x}})|$ is most similar to the nonlocal norm, while the other two terms involves essentially variations in the normal direction that can be estimated with the help of Lemma 5.4.

Concerning the process of interior extension, it is worthy noting that the problem of relating boundary estimates and interior estimates appears often in the study of PDE boundary value problems, such as in Kellogg's theorem for deriving C^α regularity estimates up to the boundary with prescribed C^α data

[23], and in deriving interior regularity estimates from the coincidence set for free boundary problems [26].

Indeed, the idea of relating boundary points to interior points in order to get an estimate of boundary from those in the interior leads to a popular approach to establish the classical trace theorem, see for example, [25, chapter 15]. However, a new challenge in our work here in the nonlocal case, unlike the straightforward constructions in the classical case, is that the interior points need to be carefully chosen to make the nonlocal norm $\|u\|_{\mathcal{S}(\Omega)}$ coming into play. The lemma 5.4 provides us analogies of estimates on tangential and normal derivatives that are important for the derivation. The details are very involved and can be found in [45].

7 Generalizations

7.1 More general kernels

Although much of our discussion so far is focused on the choice that $\hat{\gamma}$ takes on a constant value over its support, the new nonlocal trace theorem is valid for more general nonlocal interactions. The special choice of $\hat{\gamma}$ not only avoids technical complication but also corresponds to some of the weakest nonlocal norms among those associated with popular kernels used in the literature. For example, for a typical fractional power law kernel $\hat{\gamma}(s) = 1/s^\lambda$ [2, 9], we have the rescaled fractional type kernel

$$\gamma^\lambda(\boldsymbol{x},\boldsymbol{y}) = \frac{c_\lambda}{|\delta(\boldsymbol{x})|^{d+2-\lambda}} \cdot \frac{1}{|\boldsymbol{y}-\boldsymbol{x}|^\lambda} \quad \text{for } \boldsymbol{y} \in \mathcal{H}(\boldsymbol{x}), \quad \lambda \in [0, d+2)\,. \tag{23}$$

For such kernels, it is easy to make the following comparison of norms.

Lemma 7.1. *For γ^λ defined in (23) with $\lambda \in (0, d+2)$, there is a constant $C > 0$ such that the following inequality holds for any function u,*

$$\int_\Omega \int_{\Omega\cap\mathcal{H}(\boldsymbol{x})} \gamma^0(\boldsymbol{x},\boldsymbol{y})(u(\boldsymbol{y})-u(\boldsymbol{x}))^2 d\boldsymbol{y}d\boldsymbol{x} \le C \int_\Omega \int_{\Omega\cap\mathcal{H}(\boldsymbol{x})} \gamma^\lambda(\boldsymbol{x},\boldsymbol{y})(u(\boldsymbol{y})-u(\boldsymbol{x}))^2 d\boldsymbol{y}d\boldsymbol{x}\,.$$

The lemma shows that the nonlocal norm defined with the constant kernel associated with $\lambda = 0$ indeed gives the weakest norm among ones corresponding to a large class of kernels either associated with (23) or are bounded below and above by such kernels when $\boldsymbol{x}$ and $\boldsymbol{y}$ are close by.

7.2 More general choices of horizon

To generalize the choices of the variable horizon, we may consider, for example, the case $\delta(\boldsymbol{x}) = \sigma \text{dist}(\boldsymbol{x}, \Gamma)$ where $\sigma \in (0, \sigma_0]$ for $\sigma_0 > 0$. On this matter, we define some notations first.

$$|u|^2_{\delta(\boldsymbol{x}),r} = \int_{\Omega_r} \int_{\Omega_r\cap\mathcal{H}(\boldsymbol{x})} \frac{1}{|\delta(\boldsymbol{x})|^{d+2}} \hat{\gamma}\left(\frac{|\boldsymbol{y}-\boldsymbol{x}|}{\delta(\boldsymbol{x})}\right)(u(\boldsymbol{y}) - u(\boldsymbol{x}))^2 d\boldsymbol{y}d\boldsymbol{x}\,.$$

where $\Omega_r = (0, r) \times \mathbb{R}^{d-1}$. The next lemma shows that the smaller σ is, the larger the nonlocal norm we can get.

Lemma 7.2. *Let $\delta(\boldsymbol{x}) = \sigma\, dist(\boldsymbol{x}, \Gamma)$, where $\sigma \in [\frac{1}{2}, 1)$ and $\Gamma = \{0\} \times \mathbb{R}^{d-1}$, then there exists a constant C depending only on d such that the following inequality holds for any $r > 0$ and $\alpha \in (0, 1]$,*

$$|u|^2_{\delta(\boldsymbol{x}),r/2} \le C\Big(\frac{1+\sigma}{1-\sigma}\Big)^{d+2} |u|^2_{\alpha\delta(\boldsymbol{x}),r}\,. \tag{24}$$

We may then get an extension to more general horizons given by $\delta(\boldsymbol{x}) = \sigma \text{dist}(\boldsymbol{x}, \Gamma)$.

Proposition 7.3. *The results of Theorem 3.1 and Theorem 4.1 remain valid for influence horizon of the form $\delta(\boldsymbol{x}) = \sigma\, dist(\boldsymbol{x}, \Gamma)$ where $\sigma \in (0, \sigma_0]$ for some $\sigma_0 > 0$. Moreover, the embedding constant C depends only on Ω, Γ and σ_0.*

A horizon proportional to the distance to the boundary is a specific choice that can be further generalized. One instance is that $\delta(\boldsymbol{x})$ is proportional to $\text{dist}(\boldsymbol{x}, \Gamma)$ for $\boldsymbol{x}$ only on a boundary layer of finite positive width but remains constant elsewhere. A possible form of such a $\delta(\boldsymbol{x})$ could be $\delta(\boldsymbol{x}) = \min\{\sigma \text{dist}(x, \Gamma), \eta\}$, for some $\eta > 0$ to be specified. Another possibility is to have $\delta(\boldsymbol{x})$ vanishes in other nonlinear ways as $\boldsymbol{x}$ approaches the boundary. In particular, smooth variations in $\delta(\boldsymbol{x})$ may be useful to regularity control. Similar results can be shown in these cases and they follow naturally from the fact that it is the nonlocal interaction in the boundary layer, rather than the interior of the domain, that provides the essential control on the $H^{\frac{1}{2}}$ trace.

The discussion on the general form of $\delta(\boldsymbol{x})$ is meaningful since it is important in many applications to note that the imbedding constant in (7) does not depend on σ and η, just like the constants appearing in the new nonlocal trace inequalities. For example, for the coupled PDE and nonlocal model depicted in Fig. 6, we may recover a coupled PDE models in the local limit. This again implies that the nonlocal trace theorems are refinement and improvement of the classical trace theorems in $H^1(\Omega)$.

7.3 More general domains and trace theorems on portions of the domain boundary

Concerning extending the trace theorem from the special stripe domain to more general Liphshitz domain, one can establish theorem 3.1 using partition of unity techniques.

As for generalization of the trace theorem to portions of the domain boundary. we note that the classical, local trace inequality on $\partial\Omega$ automatically implies the same result for the trace on a subset Γ of $\partial\Omega$. This is not, however, straightforward for the nonlocal space involving the Γ dependent horizon, and thus the Γ dependent nonlocal kernel.

To demonstrate that similar results remain valid, a possible route is pointed out in [45]. That is, one can consider first a special domain in the form of a (rectangular) section of the stripe domain, for instance, $\Omega = (0, r) \times (a, b) \times \mathbb{R}^{d-2}$ and $\Gamma = \{0\} \times (a, b) \times \mathbb{R}^{d-2}$. By a suitable extension in the second variable from the interval (a, b) to the whole real line, the result in Theorem 4.1 can be utilized for the whole stripe domain to get the desired result on its subsection. One may then employ similar partition of unity techniques and domain transformations to more general domains and more general subset of their boundary.

7.4 Other generalizations

In terms of further generalizations of the trace theorems, we note that although the results of this paper are only shown for the L^2 or the Hilbert space setting, it is not surprising that they can be generalized to the L^p and other more general Banach spaces. With the choices of more general kernels, one may also consider nonlocal extensions of trace results in fractional $W^{s,p}$ type spaces. Extensions of the notion of trace may also go beyond co-dimensional one manifolds to other more general subdomains or sets. Furthermore, the position-dependent and heterogeneous feature in the nonlocal norms may be related for the study of more general Morrey, Campanato, Besov and Lizorkin-Triebel spaces, possibly of variable order and growth conditions, to obtain new type of spaces and the associated trace maps [22, 35]. In addition, connections with the study of Sobolev and other function spaces on metric measure spaces may also be explored [20, 21]. Mathematically, one may also ask questions concerning optimal constants in the trace inequality, as in the classical case [14]. Moreover, while it is known that the H^1 space gives the smallest Sobolev space with continuous $H^{1/2}$ boundary trace map, we now see much larger spaces can also preserve the same property, even in spaces like what we define here whose functions may only be in L^2 over any compact subset away from the boundary. Thus, the issue of how large such a space can be, as communicated to us by Luis Caffarelli, becomes very interesting to study.

Another direction, motivated naturally by interests in nonlocal mechanics, is to consider analogous results for spaces of vector fields such as those studied in [30, 31]. Likewise, one may investigate high order extensions as well, following the discussions of high order nonlocal spaces likes ones in [44], which were relevant to the studies of beams and shells.

Furthermore, the results presented here are indicative of the conceptually simple observation on the improved regularity of functions in nonlocal spaces associated with a vanishing nonlocal horizon, either uniformly across the domain of interest, or when approaching a codimension-1 surface. In the former case we recover the limit of nonlocal spaces being the classical Sobolev space, as in [5, 37], while in the latter case we obtain the analogue and extension of the classical trace theorem. One may further investigate regularity estimates, multiscale analysis and homogenization issues associated with nonlocal problems having a heterogeneous choice of variable horizon and nonlocal interaction kernels.

8 Variational problems associated with heterogeneously localized nonlocal operators

8.1 Boundary value problems for heterogeneously localized nonlocal models

Generically, as alluded earlier, nonlocal equations posed on a domain $\Omega \subset \mathbb{R}^d$ are complemented by nonlocal boundary conditions. More precisely, constraints on a some domain with nonzero d-dimensional volume, hence leading to so-called constrained value problems [11], as illustrated in Figure 1. To avoid the use of such nonlocal constraints, the nonlocal operators need to be properly modified near the boundary, which is often the case for fractional differential equations [6]. For a more recent survey on the nonlocal elliptic equations, we refer to [38].

With heterogeneous localization of the nonlocal interactions, we can have well-posed boundary value problem with Dirichlet data on the boundary only, thanks to the trace theorem and nonlocal Poincare's inequality. This is depicted in Figure 5 that shows the striking similarity with the Figure 1 concerning the local boundary condition for both the heterogeneously localized nonlocal model and the local PDE model. Furthermore, we may also consider local Neumann type traction boundary conditions and Robin type mixed boundary conditions, imposed all just on the boundary.

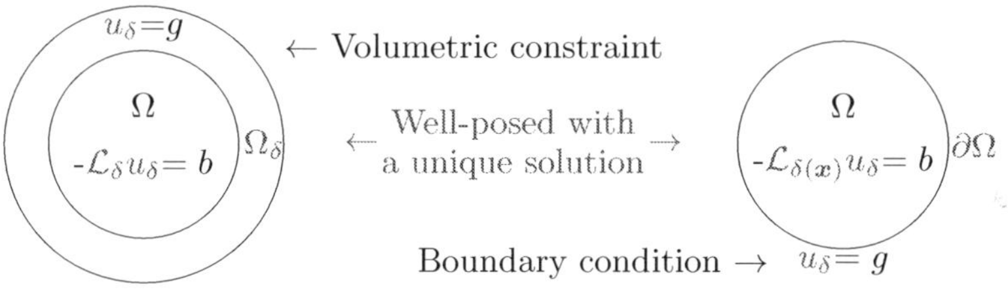

Figure 5: Nonlocal problems with either volumetric constraints or local boundary conditions.

One may further establish the convergence of numerical methods, for example, conforming finite element methods as long as the finite element spaces are dense subspaces of the nonlocal energy space. The latter is assured if we have the density in the classical Sobolev space H^1, a result that is reminiscent and consistent to the findings on asymptotically compatible schemes for the numerical solution of nonlocal problems involving a constant horizon [43].

8.2 Coupled local and nonlocal models

Another consequence of a well-defined trace map with the trace belonging to a space similar to that for standard Sobolev spaces would allow a seamless coupling between a classical, local PDE (for instance the Poisson equation $-\Delta u = f$) on one side Ω_- of a codimension-1 interface Γ with a nonlocal equation (say the variational equation $-\mathcal{L}u = f$ associated with the nonlocal energy) on the other side Ω_+ of Γ, see Fig. 6 for an illustration (the circular domains depict domains of nonlocal interactions associated with a heterogeneously defined horizon parameter). The study on transmission conditions and the well-posedness of the coupled local and nonlocal models can be readily obtained from a variational formulation

$$\min_{u_-=u_+ \text{ on } \Gamma} E_{\text{local}}(u_-)+E_{\text{nonlocal}}(u_+) = |\nabla u_-|^2_{L^2(\Omega_-)}+|u|^2_{\mathcal{S}(\Omega_+)}-2(u,f)_{L^2(\Omega_-\cup\Omega_+)}\,.$$

Furthermore, having heterogeneous horizon allows one to harvest the flexibility in working with nonlocal interactions on a wide range of scales so that more effective numerical simulations can be carried out, along the lines of asymptotically compatible schemes [43].

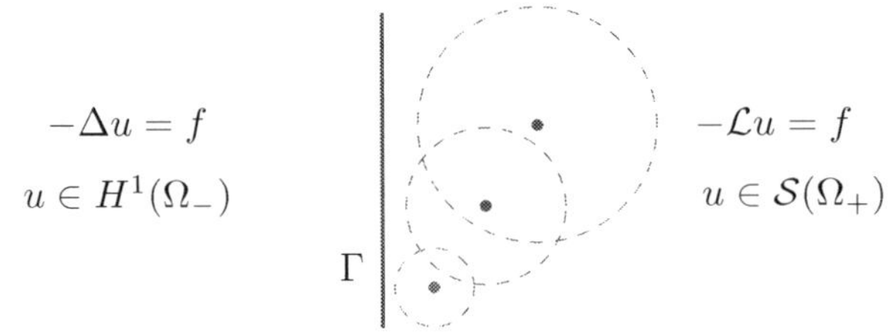

Figure 6: A PDE model (in Ω_-) is coupled with a nonlocal model (in Ω_+) using suitably defined boundary trace and transmission condition on Γ.

9 Conclusion

Nonlocality is intrinsic and ubiquitous in nature. In theory and modeling, it is also a generic feature of multiscale analysis and model reduction as it has often been (knowingly or implicitly) encoded in many earlier works (such as the Mori-Zwanzig formalism, Duhamel principle, ...). Nonlocal continuum models are arguably more general than local continuum models and discrete models and can also serve as a bridge connecting different models. Meanwhile, developing a systematic/axiomatic mathematical framework for nonlocal models is challenging and important in many important applications (that often involve defects/anomalies). Moreover, the study of nonlocal continuum models can be linked to many other concepts in mathematics, data and computational sciences as well as applications, like the combinatorial hodge theory, graph Laplacian, diffusion map, Levy flights, SPH, RKPM, etc [4, 19, 27, 36, 41, 48]. Hence, there are many good reasons to get better mathematical understanding of nonlocal models and nonlocality.

The main focus of the work presented here (based on [45]) is the heterogeneous localization of nonlocal models. This study leads to interesting mathematical extensions of classical results in functional analysis, and it could also be of practical significance for developing coupled local and nonlocal models that can retain nonlocal features wherever necessary and utilize effective local models wherever feasible, hence achieving effectiveness while maintaining generality. As a guidance for practical mathematical modeling, our findings can be summarized into a simple slogan, that is, think nonlocal, act local.

Acknowledgement

The first author would like to thank the organizing committee of ICCM 2016 for the invitation to speak at the conference.

References

[1] D. Adams and L. Hedberg. Function spaces and potential theory. *Grundlehren der Mathematischen Wissenschaften*, 314, 1996.

[2] M. Allen, L. Caffarelli, and A. Vasseur. A parabolic problem with a fractional-time derivative. *Archive for Rational Mechanics and Analysis*, 221(2): 603-630, 2015.

[3] F. Andreu, J. M. Mazón, J. D. Rossi, and J. Toledo. *Nonlocal diffusion problems*, V.165, *Mathematical Surveys and Monographs.* American Mathematical Society, 2010.

[4] L. Bartholdi, T. Schick, N. Smale and S. Smale. Hodge theory on metric spaces. *Foundations of Computational Mathematics*, 12(1):1-48, 2012.

[5] J. Bourgain, H. Brezis, and P. Mironescu. *Another look at Sobolev spaces*, 439–455, IOS Press, Amsterdam, 2001.

[6] L. Caffarelli and L. Silvestre. An extension problem related to the fractional Laplacian. *Communications in Partial Differential Equations*, 32:1245–1260, 2007.

[7] E.B. Davies. A review of Hardy inequalities. In *The Maz'ya anniversary collection*, pp.55–67. Springer, 1999.

[8] L. Diening, Maximal functions on generalized $L^{p(\cdot)}$ spaces, *Math. Inequalities and Applications*, 7:245-253, 2004.

[9] K. Diethelm. *The analysis of fractional differential equations: an application-oriented exposition using differential operators of Caputo type*, Springer, 2010.

[10] Z. Ding. A proof of the trace theorem of Sobolev spaces on Lipschitz domains. *Proceedings of the American Mathematical Society*, 124(2): 591-600. 1996.

[11] Q. Du, M. Gunzburger, R. Lehoucq, and K. Zhou. Analysis and approximation of nonlocal diffusion problems with volume constraints. *SIAM Review*, 56:676–696, 2012.

[12] Q. Du, M. Gunzburger, R. Lehoucq, and K. Zhou. A nonlocal vector calculus, nonlocal volume-constrained problems, and nonlocal balance laws. *Math. Mod. Meth. Appl. Sci.*, 23:493–540, 2013.

[13] Q. Du, J. Yang, Z. Zhou, Analysis for a nonlocal-in-time parabolic equation, Discrete and Continuous Dynamical Systems–Series B 22 (2) (2017) 339–368

[14] J. Escobar. Sharp constant in a Sobolev trace inequality. Indiana University Mathematics Journal, 37.3: 687-698, 1988.

[15] C. Fefferman, S. Ivanov, V. Kurylev, M. Lassas and H. Narayanan. *Recon-*

struction and interpolation of manifolds I: The geometric Whitney problem. arXiv preprint arXiv:1508.00674, 2015.

[16] M. Felsinger, M. Kassmann, and P. Voigt. The Dirichlet problem for nonlocal operators. *Mathematische Zeitschrift*, 279:1–31, 2013.

[17] E. Gagliardo. Caratterizzazioni delle tracce sulla frontiera relative ad alcune classi di funzioni in n variabili. *Rendiconti del seminario matematico della università di Padova*, 27:284–305, 1957.

[18] G. Gilboa and S. Osher. Nonlocal operators with applications to image processing. *Multiscale Modeling Simulation*, 7:1005–1028, 2008.

[19] R. Gingold and J. Monaghan. Smoothed particle hydrodynamics: theory and application to non-spherical stars, *Monthly Notices Royal Astronomical Society*, 181:375-389, 1977.

[20] J. Heinonen. Nonsmooth calculus, *Bulletin of the American Mathematical Society* 44(2):163–232, 2007.

[21] J. Heinonen, P. Koskela, N. Shanmugalingam and J. Tyson. *Sobolev spaces on metric measure spaces*, Cambridge University Press, 2015.

[22] J. Johnsen, S. Hansen and W. Sickel. Anisotropic Lizorkin-Triebel spaces with mixed norms — traces on smooth boundaries. *Mathematische Nachrichten* 288, 1327-1359, 2015.

[23] O. Kellogg. On the derivatives of harmonic functions on the boundary. *Transactions of the American Mathematical Society*, 33(2):486–510, 1931.

[24] J. Klafler and I.M. Sokolov. Anomalous diffusion spreads its wings. *Physics world*, 18(8):29, 2005.

[25] G. Leoni. A first course in Sobolev spaces. Vol. 105. *Providence, RI: American Mathematical Society*, 2009.

[26] F.-H. Lin. Lectures on Elliptic Free Boundary Problems, in *Lectures on the analysis of nonlinear partial differential equations*, 4, 113-191, Higher Education Press and International Press, Beijing-Boston, 2016

[27] W. Liu, Y. Chen, S. Jun, J. Chen, T. Belytschko, C. Pan, R. Uras and C. Chang. Overview and applications of the reproducing kernel particle methods, *Archives of Computational Methods in Engineering*, 3:3-80, 1996.

[28] Y. Lou, X. Zhang, S. Osher, and A. Bertozzi. Image recovery via nonlocal operators. *J. Sci. Comput.*, 42:185-197, 2010.

[29] T. Mengesha and Q. Du, The bond-based peridynamic system with Dirichlet-type volume constraint, *Proceeding of Royal Soc. Edinburgh A*, 144:161-186, 2014,

[30] T. Mengesha and Q. Du. Nonlocal constrained value problems for a linear peridynamic Navier equation. *Journal of Elasticity*, 116(1):27–51, 2014.

[31] T. Mengesha and Q. Du. On the variational limit of a class of nonlocal functionals related to peridynamics. *Nonlinearity*, 28(11):3999-4035, 2015.

[32] T. Mengesha and Q. Du, Characterization of function spaces of vector fields and an application in nonlinear peridynamics, *Nonlinear Analysis: Theory, Methods & Applications*, 140: 82-111, 2016.

[33] R. Metzler and J. Klafter. The random walk's guide to anomalous diffusion: a fractional dynamics approach. *Physics Reports*, 339(1):1-77, 2000.

[34] E. Nakai, Singular and fractional integral operators on Campanato spaces

with variable growth conditions. *Revista Matemática Complutense* 23:355-381, 2010.

[35] S. Nakamura, T. Noi and Y. Sawano, Generalized Morrey spaces and trace operator, *Science China Mathematics*, 59(2):281-336, 2016.

[36] B. Nadler, G. Lafon, R.B. Coifman and I.G, Kevrekidis, Diffusion maps, spectral clustering and reaction coordinates of dynamical systems. *Applied and Computational Harmonic Analysis*, 21:113-127, 2006.

[37] A. C. Ponce. An estimate in the spirit of Poincaré's inequality. *J. Eur. Math. Soc*, 6(1):1–15, 2004.

[38] X. Ros-Oton. Nonlocal elliptic equations in bounded domains: a survey. *preprint arXiv:1504.04099*, 2015.

[39] S.A. Silling. Reformulation of elasticity theory for discontinuities and long-range forces. *Journal of the Mechanics and Physics of Solids*, 48(1):175–209, 2000.

[40] S. A. Silling, D. J. Littlewood, and P. Seleson. Variable horizon in a peridynamic medium. Technical report, No. SAND2014-19088, Sandia National Laboratories (SNL-NM), Albuquerque, NM (United States), 2014. 124(2):591–600, 1996.

[41] A. Singer. From graph to manifold Laplacian: The convergence rate. *Applied and Computational Harmonic Analysis* 21.1:128-134, 2006.

[42] X. Tian and Q. Du. Nonconforming discontinuous Galerkin methods for nonlocal variational problems. *SIAM J. Numerical Analysis*, 53:762-781, 2015.

[43] X. Tian and Q. Du. Asymptotically compatible schemes and applications to robust discretization of nonlocal models. *SIAM J. Numerical Analysis*, 52:1641-1665, 2014.

[44] X. Tian and Q. Du. A class of high order nonlocal operators. *Archive for Rational Mechanics and Analysis*, 222(3):1521-1553, 2016.

[45] X. Tian and Q. Du. Trace theorems for some nonlocal energy spaces with heterogeneous localization, *SIAM J. Math. Analysis*, 49(2): 1621-1644, 2017.

[46] H. Triebel. Theory of function spaces. II, volume 84 of Monographs in Mathematics. *Birkhäuser Verlag*, 1992.

[47] H. Triebel. Function spaces and wavelets on domains. European Mathematical Society, 2007.

[48] V. Zaburdaev, S. Denisov and J. Klafter. Lévy walks. *Reviews of Modern Physics* 87.2:483-530, 2015.

Department of Applied Physics and Applied Mathematics
Columbia University, New York, NY 10027, USA
E-mail address: qd2125@columbia.edu; xt2156@columbia.edu

Proceedings of the 7th International Congress of Chinese Mathematicians
ALM 43, pp. 237–253

Deformations and Rigidity of Local Systems

Lei Fu

Abstract

Let X be a smooth connected projective algebraic curve over an algebraically closed field, and let S be a finite nonempty closed subset in X. We study deformations of lisse $\overline{\mathbb{F}}_\ell$-sheaves on $X-S$. The universal deformation space is a formal scheme. Its generic fiber has a rigid analytic space structure. By a dimension counting of this rigid analytic space, we prove a conjecture of Katz which says that a lisse irreducible $\overline{\mathbb{Q}}_\ell$-sheaf $\mathcal{F}$ on $X-S$ is rigid if and only if $\dim H^1(X, j_*\mathcal{E}nd(\mathcal{F})) \leq 2g$, where $j: X-S \to X$ is the open immersion, and g is the genus of X.

Mathematics Subject Classification: 14D15, 14G22.

Keywords and Phrases: Deformation of Galois representations, formal scheme, rigid analytic space.

Introduction

Let X be a complex smooth projective connected curve, and let S be a nonempty finite subset of $X(\mathbb{C})$. A local system $\mathcal{F}$ on the complex manifold $(X-S)^{\mathrm{an}}$ is a locally constant sheaf of finite dimensional complex vector spaces. Fix a base point x of $(X-S)^{\mathrm{an}}$. The fiber functor $\mathcal{F} \to \mathcal{F}_x$ defines an equivalence from the category of local systems on $(X-S)^{\mathrm{an}}$ to the category of finite dimensional complex representations of the fundamental group $\pi_1((X-S)^{\mathrm{an}}, x)$. We say $\mathcal{F}$ is irreducible if the corresponding representation of $\pi_1((X-S)^{\mathrm{an}}, x)$ is irreducible.

For any point $s \in S$, let D_s be a small disc around s in X^{an}, let $D_s^* = D_s - \{s\}$, let γ_s be the loop starting from x, going to D_s^* along a path, turning once around s in the counterclockwise direction, and returning back to x along the same path. We define the *local monodromy transformation* of $\mathcal{F}$ around s to be the image of $[\gamma_s]$ under the representation $\pi_1((X-S)^{\mathrm{an}}, x) \to \mathrm{GL}(\mathcal{F}_x)$ defined by the local system $\mathcal{F}$. Let $\mathcal{G}$ be another local system on $(X-S)^{\mathrm{an}}$. We say $\mathcal{F}$ and $\mathcal{G}$ have the same local monodromy at s if $\mathcal{F}|_{D_s^*} \cong \mathcal{G}|_{D_s^*}$. This is equivalent to saying the local monodromy transformations for $\mathcal{F}$ and for $\mathcal{G}$ are conjugate. We say $\mathcal{F}$ is *physically rigid* if for any local system $\mathcal{G}$ on $(X-S)^{\mathrm{an}}$ with the same local monodromy as $\mathcal{F}$ at each $s \in S$, we have $\mathcal{F} \cong \mathcal{G}$.

I would like to thank P. Deligne, Y. Hu and J. Xie for interesting discussions. I am also thankful for H. Esnault who invited me to visit Essen in 2005 and suggested the problem studied in this paper. This research is supported by the NSFC.

The notion of physical rigidity is reasonable only if the genus g of X is 0. Indeed, if $g \geq 1$, then from the fact that the maximal abelian quotient $\pi_1(X^{\mathrm{an}}, x)^{\mathrm{ab}}$ of $\pi_1(X^{\mathrm{an}}, x)$ is isomorphic to $\mathbb{Z}^{2g}$, we can find a character $\chi : \pi_1(X^{\mathrm{an}}, x) \to \mathbb{C}^*$ such that $\chi^r \neq 1$ for all r. Let $\mathcal{L}_\chi$ be the local system on X corresponding to the representation χ. We have $\mathcal{F}|_{D_s^*} \cong (\mathcal{F}|_{D_s^*}) \otimes (\mathcal{L}_\chi|_{D_s^*})$ for all s since $\mathcal{L}_\chi$ has trivial local monodromy around s. However, $\mathcal{F}$ is not isomorphic to $\mathcal{F} \otimes (L_\chi|_{(X-S)^{\mathrm{an}}})$ since

$$\det(\mathcal{F} \otimes \mathcal{L}_\chi|_{(X-S)^{\mathrm{an}}}) \cong \det(\mathcal{F}) \otimes (\mathcal{L}_\chi|_{(X-S)^{\mathrm{an}}})^{\otimes r}$$

is not isomorphic to $\det(\mathcal{F})$ by our choice of χ, where r is the rank of $\mathcal{F}$. So $\mathcal{F}$ cannot be physically rigid. We will extend the notion of rigidity to higher genus curves at the end of the introduction.

Assume $X = \mathbb{P}^1$. One can show that any rank 1 local system is physically rigid. In [5], Katz shows that any physically rigid local system can be constructed from rank 1 local systems by taking middle convolutions and tensor products. The local system formed by the solutions of the Gauss hypergeometric differential equation and studied by Riemann is rigid. The following theorem of Katz ([5, 1.1.2]]) gives a cohomological criterion for a local system to be physically rigid.

Theorem 0.1 (Katz). *Let S be a nonempty finite subset of $\mathbb{P}^1$, let $j : (\mathbb{P}^1 - S)^{\mathrm{an}} \hookrightarrow (\mathbb{P}^1)^{\mathrm{an}}$ be the inclusion, and let $\mathcal{F}$ be a local system on $(\mathbb{P}^1 - S)^{\mathrm{an}}$. If $\mathcal{F}$ is irreducible and $H^1((\mathbb{P}^1)^{\mathrm{an}}, j_*\mathcal{E}nd(\mathcal{F})) = 0$, then $\mathcal{F}$ is physically rigid.*

Proof. We have

$$H^0((\mathbb{P}^1)^{\mathrm{an}}, j_*\mathcal{E}nd(\mathcal{F})) \cong \mathrm{End}(\mathcal{F}).$$

Since $\mathcal{F}$ is irreducible, we have $\mathrm{End}(\mathcal{F}) \cong \mathbb{C}$ by Schur's lemma. Hence

$$\dim H^0((\mathbb{P}^1)^{\mathrm{an}}, j_*\mathcal{E}nd(\mathcal{F})) = 1.$$

The pairing

$$\mathcal{E}nd(\mathcal{F}) \times \mathcal{E}nd(\mathcal{F}) \to \mathbb{C}, \quad (\phi, \psi) \mapsto \mathrm{Tr}(\psi \circ \phi)$$

defines a self duality on $\mathcal{E}nd(\mathcal{F})$, and it induces a perfect pairing

$$H^0((\mathbb{P}^1)^{\mathrm{an}}, j_*\mathcal{E}nd(\mathcal{F})) \times H^2((\mathbb{P}^1)^{\mathrm{an}}, j_*\mathcal{E}nd(\mathcal{F})) \to \mathbb{C}$$

by Poincaré duality. So we have

$$\dim H^2((\mathbb{P}^1)^{\mathrm{an}}, j_*\mathcal{E}nd(\mathcal{F})) = 1.$$

Combining with the assumption that $H^1((\mathbb{P}^1)^{\mathrm{an}}, j_*\mathcal{E}nd(\mathcal{F})) = 0$, we get

$$\chi((\mathbb{P}^1)^{\mathrm{an}}, j_*\mathcal{E}nd(\mathcal{F})) = 2.$$

Recall that if two local systems $\mathcal{H}_1$ and $\mathcal{H}_2$ on $(\mathbb{P}^1 - S)^{\mathrm{an}}$ have the same local monodromy at each $s \in S$, then $j_*\mathcal{H}_1$ and $j_*\mathcal{H}_2$ have the same Euler characteristic. Let $\mathcal{G}$ be a local systems on $(\mathbb{P}^1 - S)^{\mathrm{an}}$ having the same local monodromy as $\mathcal{F}$ at each $s \in S$. Then $\mathcal{E}nd(\mathcal{F})$ and $\mathcal{H}om(\mathcal{F}, \mathcal{G})$ have the same local monodromy. So we have

$$\chi((\mathbb{P}^1)^{\mathrm{an}}, j_*\mathcal{H}om(\mathcal{F}, \mathcal{G})) = 2.$$

In particular, we have

$$\dim H^0((\mathbb{P}^1)^{\mathrm{an}}, j_*\mathcal{H}om(\mathcal{F},\mathcal{G})) + \dim H^2((\mathbb{P}^1)^{\mathrm{an}}, j_*\mathcal{H}om(\mathcal{F},\mathcal{G})) \geq 2 \qquad (0.1)$$

We have

$$H^0((\mathbb{P}^1)^{\mathrm{an}}, j_*\mathcal{H}om(\mathcal{F},\mathcal{G})) = \mathrm{Hom}(\mathcal{F},\mathcal{G}).$$

Again by Poincaré duality, $H^2((\mathbb{P}^1)^{\mathrm{an}}, j_*\mathcal{H}om(\mathcal{F},\mathcal{G}))$ is dual to $H^0((\mathbb{P}^1)^{\mathrm{an}}, j_*\mathcal{H}om(\mathcal{G},\mathcal{F})) \cong \mathrm{Hom}(\mathcal{G},\mathcal{F})$. Combined with the inequality (0.1), we get

$$\dim \mathrm{Hom}(\mathcal{F},\mathcal{G}) + \dim \mathrm{Hom}(\mathcal{G},\mathcal{F}) \geq 2.$$

So either $\mathrm{Hom}(\mathcal{F},\mathcal{G})$ or $\mathrm{Hom}(\mathcal{G},\mathcal{F})$ is nonzero. Since $\mathcal{F}$ is irreducible and $\mathcal{G}$ necessarily has the same rank as $\mathcal{F}$, any nonzero element in $\mathrm{Hom}(\mathcal{F},\mathcal{G})$ or $\mathrm{Hom}(\mathcal{G},\mathcal{F})$ must be an isomorphism. Hence $\mathcal{F} \cong \mathcal{G}$. So $\mathcal{F}$ is physically rigid. □

Katz also shows that the converse of the above theorem is true ([5, 1.1.2]). Slightly more generally, we have the following.

Theorem 0.2 (Katz). *Let S be a nonempty finite subset of $\mathbb{P}^1$, let $j : (\mathbb{P}^1 - S)^{\mathrm{an}} \hookrightarrow (\mathbb{P}^1)^{\mathrm{an}}$ be the inclusion, and let $\mathcal{F}$ be a local system on $(\mathbb{P}^1 - S)^{\mathrm{an}}$. Suppose* $\mathrm{End}(\mathcal{F})$ *consists of only scalar multiplications, (this condition holds if $\mathcal{F}$ is irreducible by Schur's lemma), and suppose $\mathcal{F}$ is physically rigid. Then $H^1((\mathbb{P}^1)^{\mathrm{an}}, j_*\mathcal{E}nd(\mathcal{F})) = 0$.*

Proof. Let $\rho_0 : \pi_1((\mathbb{P}^1 - S)^{\mathrm{an}}, x) \to \mathrm{GL}(r, \mathbb{C})$ be the representation corresponding to $\mathcal{F}$. Note that $\pi_1((\mathbb{P}^1 - S)^{\mathrm{an}}, x)$ is the group generated by $[\gamma_s]$ $(s \in S)$ with the relation $\prod_{s\in S}[\gamma_s] = 1$. Let $A_s^{(0)} = \rho_0([\gamma_s])$, and let U be the closed subvariety of $\prod_{s\in S}\mathrm{GL}(r,\mathbb{C}) \times \prod_{s\in S}\mathrm{GL}(r,\mathbb{C})$ so that $((A_s)_{s\in S}, (P_s)_{s\in S}) \in \prod_{s\in S}\mathrm{GL}(r,\mathbb{C}) \times \prod_{s\in S}\mathrm{GL}(r,\mathbb{C})$ lies in U if and only if

$$\prod_{s\in S} A_s = I, \quad P_s^{-1}A_sP_s = A_s^{(0)}.$$

We have an action

$$\mathrm{SL}(n, C) \times U \to U,$$
$$\Big(P, ((A_s)_{s\in S}, (P_s)_{s\in S})\Big) \mapsto ((PA_sP^{-1})_{s\in S}, (PP_s)_{s\in S}).$$

Let $\mathcal{M} = \mathrm{SL}(n,\mathbb{C})\backslash U$ be the quotient space. Let $\mathcal{G}$ be a local system on $(\mathbb{P}^1 - S)^{\mathrm{an}}$ with the same local monodromy as $\mathcal{F}$ at each $s \in S$, let $\rho : \pi_1((\mathbb{P}^1 - S)^{\mathrm{an}}, x) \to \mathrm{GL}(r,\mathbb{C})$ be the representation corresponding to $\mathcal{G}$, and let $A_s = \rho([\gamma_s])$. Then there exists $P_s \in \mathrm{GL}(r, C)$ such that $P_s^{-1}A_sP_s = A_s^{(0)}$. The triple $((A_s)_{s\in S}, (P_s)_{s\in S})$ gives rise to a point in U, and hence a point in $\mathcal{M}$. We may think column vectors of P_s as a basis or a *frame* of $\mathbb{C}^n$, and with respect to this frame, the matrix of $\rho^{(0)}([\gamma_s])$ is exactly the matrix $A_s^{(0)}$. We can call $\mathcal{M}$ the *moduli space of framed local systems with the same local monodromy as $\mathcal{F}$.*

Note that for a point $((A_s)_{s\in S}, (P_s)_{s\in S})$ in U, the component $(A_s)_{s\in S}$ is completely determined by the component $(P_s)_{s\in S}$ since we have

$$A_s = P_s A_s^{(0)} P_s^{-1}.$$

The component $(P_s)_{s\in S}$ is subject to the relation

$$\prod_{s\in S} (P_s A_s^{(0)} P_s^{-1}) = I.$$

So U is isomorphic to the closed subvariety U' of $\prod_{s\in S} \mathrm{GL}(r, \mathbb{C})$ defined by this equation. The group $\mathrm{SL}(n, C)$ acts on U' by left multiplication, and $\mathcal{M} \cong \mathrm{SL}(n, C)\backslash U'$. Note that U' is the fiber over I of the morphism

$$\prod_{s\in S} \mathrm{GL}(r, \mathbb{C}) \to \mathrm{SL}(r, \mathbb{C}), \quad (P_s)_{s\in S} \mapsto \prod_{s\in S} (P_s A_s^{(0)} P_s^{-1}).$$

(This map has image $\mathrm{SL}(r, \mathbb{C})$ since $\prod_{s\in S} A_s^{(0)} = I$.) One can show that

$$\dim U' \geq \dim\Big(\prod_{s\in S} \mathrm{GL}(r, \mathbb{C})\Big) - \dim(\mathrm{SL}(r, \mathbb{C})) = (|S| - 1)r^2 + 1.$$

Hence

$$\dim \mathcal{M} \geq \dim U' - \dim(\mathrm{SL}(r, \mathbb{C})) = (|S| - 2)r^2 + 2. \tag{0.2}$$

Let V be the closed subvariety of $\prod_{s\in S} \mathrm{GL}(r, \mathbb{C})$ so that $(P_s)_{s\in S} \in \prod_{s\in S} \mathrm{GL}(r, \mathbb{C})$ lies in V if and only if

$$P_s^{-1} A_s^{(0)} P_s = A_s^{(0)}.$$

We have a morphism

$$V \to U, \quad (P_s)_{s\in S} \mapsto ((A_s^{(0)})_{s\in S}, (P_s)_{s\in S}).$$

Composed with the projection $U \mapsto \mathrm{SL}(r, \mathbb{C})\backslash U$, we get a morphism

$$f : V \to \mathcal{M}.$$

Given a point $((A_s)_{s\in S}, (P_s)_{s\in S})$ in U, let $\rho : \pi_1((\mathbb{P}^1 - S)^{\mathrm{an}}, x) \to \mathrm{GL}(r, \mathbb{C})$ be the representation defined by $\rho([\gamma_s]) = A_s$, and let $\mathcal{G}$ be the local system corresponding to ρ. Since $P_s^{-1} A_s P_s = A_s^{(0)}$, $\mathcal{G}$ has the same local monodromy as $\mathcal{F}$. Since $\mathcal{F}$ is physically rigid, we have $\mathcal{G} \cong \mathcal{F}$. So there exists $P \in \mathrm{GL}(r, \mathbb{C})$ such that $P\rho P^{-1} = \rho^{(0)}$. In particular, we have $PA_sP^{-1} = A_s^{(0)}$ for all s. Moreover, we can find such a P lying in $\mathrm{SL}(r, \mathbb{C})$. It follows that $((A_s)_{s\in S}, (P_s)_{s\in S})$ and $((A_s^{(0)})_{s\in S}, (PP_s))$ define the same point in $\mathcal{M}$. This shows that $f : V \to \mathcal{M}$ is surjective. So we have

$$\dim V \geq \dim \mathcal{M}. \tag{0.3}$$

Let $Z_i(s) = \{P \in \mathfrak{gl}(n, \mathbb{C}) | A_s^{(0)} P = P A_s^{(0)}\}$. Note that $Z_i(s) \cong (j_* \mathcal{E}nd(\mathcal{F}))_s$. We have

$$\dim V = \sum_{s\in S} \dim Z_i(s) = \sum_{s\in S} \dim\, (j_*\mathcal{E}nd(\mathcal{F}))_s. \tag{0.4}$$

From (0.2)-(0.4), we get

$$\sum_{s\in S} \dim\,(j_*\mathcal{E}nd(\mathcal{F}))_s \geq (|S|-2)r^2+2.$$

By the Euler-Poincaré formula, we have

$$\chi((\mathbb{P}^1)^{\mathrm{an}}, j_*\mathcal{E}nd(\mathcal{F})) = (2-|S|)r^2 + \sum_{s\in S} \dim\,(j_*\mathcal{E}nd(\mathcal{F}))_s.$$

So we have

$$\chi((\mathbb{P}^1)^{\mathrm{an}}, j_*\mathcal{E}nd(\mathcal{F})) \geq 2.$$

As in Theorem 0.1, we have

$$\dim H^0((\mathbb{P}^1)^{\mathrm{an}}, j_*\mathcal{E}nd(\mathcal{F})) = \dim H^2((\mathbb{P}^1)^{\mathrm{an}}, j_*\mathcal{E}nd(\mathcal{F})) = 1.$$

We thus get

$$\dim H^1((\mathbb{P}^1)^{\mathrm{an}}, j_*\mathcal{E}nd(\mathcal{F})) \leq 0.$$

So $H^1((\mathbb{P}^1)^{\mathrm{an}}, j_*\mathcal{E}nd(\mathcal{F}))=0$. □

The above discussion can be extended to ℓ-adic sheaves on curves over an arbitrary field. Let k be an algebraically closed field of characteristic p, let X be a smooth connected projective curve over k, let S be a finite nonempty closed subset of X, let ℓ be a prime number distinct from p, and let E be a finite extension of $\mathbb{Q}_\ell$. For any $s\in S$, let η_s be the generic point of the strict henselization of X at s. A lisse E-sheaf $\mathcal{F}_E$ on $X-S$ is called *physically rigid* if for any lisse $\overline{\mathbb{Q}}_\ell$-sheaf $\mathcal{G}$ on $X-S$ with the property $(\mathcal{F}_E\otimes_E\overline{\mathbb{Q}}_\ell)|_{\eta_s}\cong\mathcal{G}|_{\eta_s}$ for all $s\in S$, we have $\mathcal{F}_E\otimes_E\overline{\mathbb{Q}}_\ell\cong\mathcal{G}$. Let η be the generic point of X, let $\bar{\eta}_s$ and $\bar{\eta}$ be geometric points over η_s and η, respectively, and let $\bar{\eta}_s\to\bar{\eta}$ be morphisms so that the diagrams

$$\begin{array}{ccc}\bar{\eta}_s & \to & \bar{\eta}\\ \downarrow & & \downarrow\\ \eta_s & \to & \eta\end{array}$$

commute. Then we have canonical homomorphisms $\mathrm{Gal}(\bar{\eta}_s/\eta_s)\to\pi_1(X-S,\bar{\eta})$, where $\pi_1(X-S,\bar{\eta})$ is the étale fundamental group of $X-S$. The lisse E-sheaf $\mathcal{F}_E$ on $X-S$ corresponds to a representation

$$\rho_E:\pi_1(X-S,\bar{\eta})\to\mathrm{GL}(E^r).$$

$\mathcal{F}$ is physically rigid if and only if for any representation $\rho':\pi_1(X-S,\bar{\eta})\to\mathrm{GL}(\overline{\mathbb{Q}}_\ell^r)$ such that $\rho|_{\mathrm{Gal}(\bar{\eta}_s/\eta_s)}\cong\rho'|_{\mathrm{Gal}(\bar{\eta}_s/\eta_s)}$ for all $s\in S$, we have $\rho\cong\rho'$ as $\overline{\mathbb{Q}}_\ell$-representations.

To get a good notion of physical rigidity, we have to assume $X=\mathbb{P}^1$. Indeed, the abelian-pro-ℓ quotient of $\pi_1(X,\bar{\eta})$ of X is isomorphic to $\mathbb{Z}_\ell^{2g}$, where g is the genus of X. If $g\geq 1$, then there exists a character $\chi:\pi_1(X,\bar{\eta})\to\overline{\mathbb{Q}}_\ell^*$ such that χ^n are nontrivial for all n. So there exists a lisse $\overline{\mathbb{Q}}_\ell$-sheaf $\mathcal{L}_\chi$ of rank 1 on X such that $\mathcal{L}_\chi^{\otimes n}$ are nontrivial for all n. For any lisse E-sheaf $\mathcal{F}_E$ on $X-S$, the lisse

$\overline{\mathbb{Q}}_\ell$-sheaf $\mathcal{G} = (\mathcal{F} \otimes_E \overline{\mathbb{Q}}_\ell) \otimes_{\overline{\mathbb{Q}}_\ell} (\mathcal{L}_\chi|_{X-S})$ is not isomorphic to $\mathcal{F}_E \otimes_E \overline{\mathbb{Q}}_\ell$ since they have non-isomorphic determinant, but $(\mathcal{F}_E \otimes_E \overline{\mathbb{Q}}_\ell)|_{\eta_s} \cong \mathcal{G}|_{\eta_s}$ for each $s \in X$ since $\chi|_{\mathrm{Gal}(\bar{\eta}_s/\eta_s)}$ is unramfied and hence trivial. Thus $\mathcal{F}_E$ is not physically rigid.

Examples.

1. Rank 1 lisse sheaves on $\mathbb{P}^1 - S$ are physically rigid.

2. Kloosterman sheaves and more generally hypergeometric geometric sheaves on $\mathbb{G}_m$ defined by Katz [6] are physically rigid.

3. Let $\mathcal{F}$ be an irreducible lisse sheaf on $\mathbb{A}^1 - S$ so that the Deligne-Fourier transform $FT(j_*\mathcal{F}[1])$ is isomorphic to $j'_*\mathcal{F}'[1]$ for some lisse sheaf $\mathcal{F}'$ on $\mathbb{A}^1 - S'$, where $j : \mathbb{A}^1 - S \hookrightarrow \mathbb{A}^1$ and $j' : \mathbb{A}^1 - S' \hookrightarrow \mathbb{A}^1$ are open immersions. If $\mathcal{F}$ is physically rigid, then $\mathcal{F}'$ is physically rigid.

4. Starting with a rank 1 sheaf $\mathcal{L}$ on $\mathbb{P}^1 - S$, iterate the following types of operations on sheaves:

(a) $\mathcal{F} \mapsto f^*\mathcal{F}$ for any automorphism $f : \mathbb{P}^1 \to \mathbb{P}^1$.

(b) $\mathcal{F} \mapsto \mathcal{F}'$, where $j'_*\mathcal{F}' = FT(j_*\mathcal{F})$.

Then we end up with a rigid sheaf. Katz ([5]) proves that any tamely ramified physically rigid sheaf can be obtained from tamely ramified rank 1 sheaf by iterating the above operations.

Under the assumption that $\mathrm{End}(\mathcal{F}_E) = E$, one can develop a theory of deformations ([2]) of $\mathcal{F}_E$ with undeformed local monodromy so that $H^1(X, j_*\mathcal{E}nd(\mathcal{F}_E))$ is the Zariski tangent space, where $j : X - S \hookrightarrow X$ is the open immersion. The same proof as in Theorem 0.1 shows the following.

Theorem 0.3. *Suppose $\mathcal{F}_E$ is an irreducible lisse sheaf on $\mathbb{P}^1 - S$. If $H^1(X, j_*\mathcal{E}nd(\mathcal{F}_E)) = 0$, then $\mathcal{F}_E$ is physically rigid.*

It was conjectured that the converse of this theorem should be true. In this paper we sketch a proof the conjecture. Details can be found in [3]. Our theorem is as follows.

Theorem 0.4. *Suppose that $X = \mathbb{P}^1$, that $\mathrm{End}(\mathcal{F}_E)$ consists of scalar multiplications, and that $\mathcal{F}_E$ is physically rigid. Then $H^1(X, j_*\mathcal{E}nd(\mathcal{F}_E)) = 0$, where $j : X - S \hookrightarrow X$ is the canonical open immersion.*

We first give a heuristic argument. Let

$$\mathcal{M} = \{(\rho, (P_s)_{s\in S}) \mid \rho : \pi_1(X - S, \bar{\eta}) \to \mathrm{GL}(\overline{\mathbb{Q}}_\ell^r) \text{ is a representation}, \quad P_s \in \mathrm{GL}(\overline{\mathbb{Q}}_\ell^r), \\ P_s^{-1}\rho|_{\mathrm{Gal}(\bar{\eta}_s/\eta_s)}P_s = \rho_E|_{\mathrm{Gal}(\bar{\eta}_s/\eta_s)} \text{ for all } s \in S\}/\sim,$$

The column vectors of $P_s \in \mathrm{GL}(\overline{\mathbb{Q}}_\ell^r)$ can be regarded as a basis, that is, a frame of $\overline{\mathbb{Q}}_\ell^r$. In the definition of $\mathcal{M}$, two tuples $(\rho^{(i)}, (P_s^{(i)})_{s\in S})$ $(i = 1, 2)$ are equivalent if there exists an isomorphism of representations $P : \overline{\mathbb{Q}}_\ell^r \to \overline{\mathbb{Q}}_\ell^r$ from $\rho^{(1)}$ to $\rho^{(2)}$ transforming the frame $P_s^{(1)}$ to the frame $P_s^{(2)}$ for each $s \in S$, or equivalently, there exists $P \in \mathrm{GL}(\overline{\mathbb{Q}}_\ell^r)$ such that

$$(\rho^{(1)}, (P_s^{(1)})_{s\in S}) = (P^{-1}\rho^{(2)}P, (P^{-1}P_s^{(2)})_{s\in S}).$$

Let's pretend that $\mathcal{M}$ is the moduli space of framed lisse $\overline{\mathbb{Q}}_\ell$-sheaves on $X-S$ whose local monodromy at s is isomorphic to $\rho_E|_{\mathrm{Gal}(\bar{\eta}_s/\eta_s)}$. Suppose $\mathcal{F}_E$ is physically rigid. Then from the condition $P_s^{-1}\rho|_{\mathrm{Gal}(\bar{\eta}_s/\eta_s)}P_s = \rho_E|_{\mathrm{Gal}(\bar{\eta}_s/\eta_s)}$, we get $\rho \cong \rho_E$, that is, there exists $P \in \mathrm{GL}(\overline{\mathbb{Q}}_\ell^r)$ such that $P^{-1}\rho P = \rho_E$. We have

$$(\rho, (P_s)_{s\in S}) \sim (P^{-1}\rho P, (P^{-1}P_s)_{s\in S}) = (\rho_E, (P^{-1}P_s)_{s\in S}).$$

Hence any element in $\mathcal{M}$ is equivalent to a tuple of the form $(\rho_E, (P_s)_{s\in S})$ with

$$P_s^{-1}\rho_E|_{\mathrm{Gal}(\bar{\eta}_s/\eta_s)}P_s = \rho_E|_{\mathrm{Gal}(\bar{\eta}_s/\eta_s)},$$

that is, $P_s \in \mathrm{Aut}_{\mathrm{Gal}(\bar{\eta}_s/\eta_s)}(\overline{\mathbb{Q}}_\ell^r)$, where $\overline{\mathbb{Q}}_\ell^r$ is endowed with the $\mathrm{Gal}(\bar{\eta}_s/\eta_s)$-module structure via ρ_E. Moreover, for any scalar matrix $P = \lambda^{-1}I$, we have

$$(\rho_E, (P_s)_{s\in S}) \sim (P^{-1}\rho_E P, (P^{-1}P_s)_{s\in S}) = (\rho_E, (\lambda P_s)_{s\in S}).$$

It follows that

$$\dim \mathcal{M} \leq \sum_{s\in S} \dim \mathrm{End}_{\mathrm{Gal}(\bar{\eta}_s/\eta_s)}(\overline{\mathbb{Q}}_\ell^r) - 1,$$

where $\overline{\mathbb{Q}}_\ell^r$ is provided with the $\mathrm{Gal}(\bar{\eta}_s/\eta_s)$-module structure via the representation ρ_E. One might hope that by calculating the dimension of $\mathcal{M}$, this leads to a proof of the theorem.

However, since the étale fundamental group is a pro-finite group, the moduli space of ℓ-adic representations of the étale fundamental group does not exist in the usual sense. We will put a rigid analytic space structure on a piece of $\mathcal{M}$ so that the above estimate on its dimension holds. Our idea is to study the framed deformation of the torsion sheaf obtained from $\mathcal{F}_E$. Our treatment is similar to Mazur's theory of deformations of Galois representations ([8]) and Kisin's theory of framed deformations of Galois representations ([7]). The universal deformation space is a formal scheme. Its generic fiber is a rigid analytic space. We use this rigid analytic space as a substitute for the moduli space of framed ℓ-adic local systems. A counting dimension argument on this rigid analytic space proves Theorem 0.4.

In the case where X has higher $g \geq 1$, we may modify the concept of physical rigidity as follows: The lisse E-sheaf $\mathcal{F}_E$ on $X - S$ is called *rigid* if for any lisse $\overline{\mathbb{Q}}_\ell$-sheaf $\mathcal{G}$ on $X - S$ with the property $\det(\mathcal{F} \otimes_E \overline{\mathbb{Q}}_\ell) \cong \det(\mathcal{G})$ and $(\mathcal{F}_E \otimes_E \overline{\mathbb{Q}}_\ell)|_{\eta_s} \cong \mathcal{G}|_{\eta_s}$ for all $s \in S$, we have $\mathcal{F}_E \otimes_E \overline{\mathbb{Q}}_\ell \cong \mathcal{G}$. Equivalently, one says $\mathcal{F}_E$ is rigid if for any representation $\rho' : \pi_1(X - S, \bar{\eta}) \to \mathrm{GL}(\overline{\mathbb{Q}}_\ell^r)$ with the property $\det(\rho) = \det(\rho')$ and $\rho|_{\mathrm{Gal}(\bar{\eta}_s/\eta_s)} \cong \rho'|_{\mathrm{Gal}(\bar{\eta}_s/\eta_s)}$ for all $s \in S$, we have $\rho \cong \rho'$. Note that in the case where $X = \mathbb{P}^1$, this notion coincides with physically rigidity. Indeed, if $\rho|_{\mathrm{Gal}(\bar{\eta}_s/\eta_s)} \cong \rho'|_{\mathrm{Gal}(\bar{\eta}_s/\eta_s)}$ holds, we have $\det(\rho)|_{\mathrm{Gal}(\bar{\eta}_s/\eta_s)} = \det(\rho')|_{\mathrm{Gal}(\bar{\eta}_s/\eta_s)}$. It follows that $\det(\rho)\det(\rho)^{-1}$ is unramified everywhere. Since $\mathbb{P}^1$ is simply connected, we must have $\det(\rho)\det(\rho)^{-1} = 1$. Hence the condition $\rho|_{\mathrm{Gal}(\bar{\eta}_s/\eta_s)} \cong \rho'|_{\mathrm{Gal}(\bar{\eta}_s/\eta_s)}$ for all $s \in S$ implies the condition $\det(\rho) = \det(\rho')$ in the case $X = \mathbb{P}^1$.

We have the following generalization of Theorem 0.4.

Theorem 0.5. *Suppose that $\mathcal{F}_E$ is rigid, and that* $\mathrm{End}(\mathcal{F}_E)$ *consists of scalar multiplications. Then we have* $\dim H^1(X, j_*\mathcal{E}nd(\mathcal{F}_E)) \leq 2g$, *where g is the genus of X.*

Corollary 0.6. *Suppose that $\mathcal{F}_E$ is rigid, and that* $\mathrm{End}(\mathcal{F}_E)$ *consists of scalar multiplications. If $g \geq 2$, then the rank of $\mathcal{F}_E$ is ≤ 1.*

Proof. Let r be the rank of $\mathcal{F}_E$. On the one hand, by Theorem 0.5, we have

$$\chi(X, j_*\mathcal{E}nd(\mathcal{F}_E)) = 2 - \dim H^1(X, j_*\mathcal{E}nd(\mathcal{F}_E)) \geq 2 - 2g. \tag{0.5}$$

On the other hand, by the Grothendieck-Ogg-Shafarevich formula, we have

$$\begin{aligned}\chi(X, j_*\mathcal{E}nd(\mathcal{F}_E)) &= (2 - 2g - |S|)r^2 + \sum_{s\in S}\Big(\dim\,(j_*\mathcal{E}nd(\mathcal{F}_E))_{\bar{s}} - \mathrm{sw}_s(\mathcal{E}nd(\mathcal{F}_E))\Big) \\ &= (2-2g)r^2 + \sum_{s\in S}\Big(\Big(\dim\,(j_*\mathcal{E}nd(\mathcal{F}_E)\Big)_{\bar{s}} - r^2) - \mathrm{sw}_s(\mathcal{E}nd(\mathcal{F}_E))\Big),\end{aligned}$$

where sw_s is the Swan conductor at s. We have $\dim\,(j_*\mathcal{E}nd(\mathcal{F}_E))_{\bar{s}} - r^2 \leq 0$. It follows that

$$\chi(X, j_*\mathcal{E}nd(\mathcal{F}_E)) \leq (2-2g)r^2. \tag{0.6}$$

From (0.5)-(0.6), we get

$$2 - 2g \leq (2-2g)r^2.$$

Since $g \geq 2$, we must have $r \leq 1$. □

The above corollary shows that we have interesting higher rank rigid ℓ-adic lisse sheaves only when $g \leq 1$. For the case $g = 0$, Katz studies tamely ramified rigid ℓ-adic sheaf ([5]) and rigid hypergeometric sheaves ([6]) on $\mathbb{P}^1$ in detail. For the case $g = 1$, Katz ([5, 1.3-4]) classifies rigid complex local systems on punctured elliptic curves.

Go back to the case where the ground field is $\mathbb{C}$. By the Riemann-Hilbert correspondence, the category of complex local systems on $(X - S)^{\mathrm{an}}$ is isomorphic to the category of integrable meromorphic connections on X with regular singularity at each $s \in S$ and holomorphic elsewhere. It is natural to extend the above discussion to integrable meromorphic connections. This is done in [1]. By the Simpson correspondence, the category of complex local systems on X^{an} is isomorphic to the category of Higgs bundles on X with zero Chern class. It is also natural to extend the above discussion to Higgs bundles on punctured curves. We hope to return to this problem in the future.

1 Deformation theory

In the following, we take Λ to be either a finite extension E of $\mathbb{Q}_\ell$, or the integer ring $\mathcal{O}$ of such E. Let $\mathfrak{m}$ be the maximal ideal of Λ, and let $\kappa = \Lambda/\mathfrak{m}$ be the residue field of Λ. Denote by $\mathcal{C}_\Lambda$ the category of Artinian local Λ-algebras with same residue field κ as Λ. Morphisms in $\mathcal{C}_\Lambda$ are homomorphisms of Λ-algebras.

Using the fact that the maximal ideal of an Artinian local ring coincides with its nilpotent radical, one can check that morphisms in $\mathcal{C}_\Lambda$ are necessarily local homomorphisms, and they induce the identity homomorphism on the residue field. If A is an object in $\mathcal{C}_\Lambda$, we denote by $\mathfrak{m}_A$ the maximal ideal of A. For any $A \in \operatorname{ob} \mathcal{C}$, a homomorphism $\rho : \pi_1(X - S, \bar{\eta}) \to \mathrm{GL}(A^r)$ is called a representation if it is continuous. Here, if $\Lambda = \mathcal{O}$, then A is finite and we put the discrete topology on $\mathrm{GL}(A^r)$. If $\Lambda = E$, then A is a finite dimensional vector space over E, and we put the topology induced from the ℓ-adic topology on $\mathrm{GL}(A^r)$.

Suppose that S is nonempty and that we are given a representation $\rho_\Lambda : \pi_1(X - S, \bar{\eta}) \to \mathrm{GL}(\Lambda^r)$. Let $\lambda : \pi_1(X - S, \bar{\eta}) \to \Lambda^*$ be the determinant of ρ_Λ and let $\rho_0 : \pi_1(X - S, \bar{\eta}) \to \mathrm{GL}(\kappa^r)$ be $\rho_\Lambda \mod \mathfrak{m}_\Lambda$. Fix a frame $P_{0,s} \in \mathrm{GL}(\kappa^r)$ for each $s \in S$ with the property

$$P_{0,s}^{-1} \rho_0|_{\mathrm{Gal}(\bar{\eta}_s/\eta_s)} P_{0,s} = \rho_0|_{\mathrm{Gal}(\bar{\eta}_s/\eta_s)}.$$

For any $A \in \operatorname{ob} \mathcal{C}_\Lambda$, denote the composite

$$\pi_1(X - S, \bar{\eta}) \xrightarrow{\lambda} \Lambda^* \to A^*$$

also by λ. Define

$$\begin{aligned} D_S^{\square}(A) = \{(\rho, (P_s)_{s\in S}) \mid \rho : \pi_1(X - S, \bar{\eta}) \to \mathrm{GL}(A^r) \text{ is a representation,} \\ P_s \in \mathrm{GL}(A^r),\ \ \rho \mod \mathfrak{m}_A = \rho_0,\ P_s \mod \mathfrak{m}_A = P_{s,0}\}/\sim, \end{aligned}$$

where two tuples $(\rho^{(i)}, (P_s^{(i)})_{s\in S})$ $(i = 1, 2)$ are equivalent if there exists $P \in \mathrm{GL}(A^r)$ such that

$$(\rho^{(1)}, (P_s^{(1)})_{s\in S}) = (P^{-1}\rho^{(2)}P, (P^{-1}P_s^{(2)})_{s\in S}).$$

Define

$$D_S^{\square,\lambda}(A) = \{(\rho, (P_s)_{s\in S}) \in D_S^{\square}(A) \mid \det(\rho) = \lambda\},$$

and for any $s \in S$, define

$$\begin{aligned} D_s^{\square}(A) &= \{\rho \mid \rho : \mathrm{Gal}(\bar{\eta}_s/\eta_s) \to \mathrm{GL}(A^r) \text{ is a representation,} \\ &\qquad \rho \mod \mathfrak{m}_A = \rho_0|_{\mathrm{Gal}(\bar{\eta}_s/\eta_s)}\}, \\ D_s^{\square,\lambda}(A) &= \{\rho \in D_s^{\square}(A) \mid \det(\rho) = \lambda|_{\mathrm{Gal}(\bar{\eta}_s/\eta_s)}\}. \end{aligned}$$

For any morphism $A' \to A$ in $\mathcal{C}_\Lambda$, define $D(A') \to D(A)$ to be the map induced by $\mathrm{GL}(A'^r) \to \mathrm{GL}(A^r)$ for $D = D_S^{\square}, D_S^{\square,\lambda}, D_s^{\square}, D_s^{\square,\lambda}$. Then $D_S^{\square}, D_S^{\square,\lambda}, D_s^{\square}, D_s^{\square,\lambda}$ are functors from the category $\mathcal{C}_\Lambda$ to the category of sets. We have canonical morphisms of functors

$$D_S^{\square} \to \prod_{s\in S} D_s^{\square}, \quad D_S^{\square,\lambda} \to \prod_{s\in S} D_s^{\square,\lambda}$$

given by

$$(\rho, (P_s)_{s\in S}) \to (P_s^{-1}\rho|_{\mathrm{Gal}(\bar{\eta}_s/\eta_s)} P_s)_{s\in S}.$$

Finally, let F (resp. F^λ) be the subfunctor of $D_S^\square$ (resp. $D_S^{\square,\lambda}$) defined by

$$F(A) = \{(\rho, (P_s)_{s\in S}) \in D_S^\square(A) | P_s^{-1}\rho|_{\mathrm{Gal}(\bar\eta_s/\eta_s)}P_s = \rho_\Lambda|_{\mathrm{Gal}(\bar\eta_s/\eta_s)}\},$$
$$(\text{resp. } F^\lambda(A) = \{(\rho, (P_s)_{s\in S}) \in D_S^{\square,\lambda}(A) | P_s^{-1}\rho|_{\mathrm{Gal}(\bar\eta_s/\eta_s)}P_s = \rho_\Lambda|_{\mathrm{Gal}(\bar\eta_s/\eta_s)}\}).$$

Note that F (resp. F^λ) can be thought as the fiber of the morphism of functors

$$D_S^\square \to \prod_{s\in S} D_s^\square \quad (\text{resp. } D_S^{\square,\lambda} \to \prod_{s\in S} D_s^{\square,\lambda})$$

over $(\rho_\Lambda|_{\mathrm{Gal}(\bar\eta_s/\eta_s)})_{s\in S}$. For any morphism $A' \to A$ and $A'' \to A$ in $\mathcal{C}_\Lambda$, one can verify that the canonical map

$$D(A' \times_A A'') \to D(A') \times_{D(A)} D(A'')$$

is bijective if $A'' \to A$ is surjective for each of the functors $D = D_S^\square, D_S^{\square,\lambda}, D_s^\square, D_s^{\square,\lambda}$, F, F^λ. Proposition 1.1 below shows that each $D(\kappa[\epsilon])$ is finite dimensional. So the functors $D_S^\square, D_S^{\square,\lambda}, D_s^\square, D_s^{\square,\lambda}, F, F^\lambda$ are pro-representable by Schlessinger's criteria [9, Theorem 2.11]. In [3, Proposition 1.1], we prove the following.

Proposition 1.1. *Let $\kappa[\epsilon]$ be the ring of dual numbers. We have*

$$\begin{aligned}
\dim D_S^\square(\kappa[\epsilon]) &= -\chi(\pi_1(X-S,\bar\eta), \mathrm{Ad}(\rho_0)) + |S|r^2 \\
&= -\chi(X-S, \mathcal{E}nd(\mathcal{F}_0)) + |S|r^2, \\
\dim D_S^{\square,\lambda}(\kappa[\epsilon]) &= -\chi(\pi_1(X-S,\bar\eta), \mathrm{Ad}^{(0)}(\rho_0)) + |S|r^2 - 1 \\
&= -\chi(X-S, \mathcal{E}nd^{(0)}(\mathcal{F}_0)) + |S|r^2 - 1, \\
\dim D_s^\square(\kappa[\epsilon]) &= -\chi(\mathrm{Gal}(\bar\eta_s/\eta_s), \mathrm{Ad}(\rho_0)) + r^2 \\
&= -\chi(\eta_s, \mathcal{E}nd(\mathcal{F}_0|_{\eta_s})) + r^2, \\
\dim D_s^{\square,\lambda}(\kappa[\epsilon]) &= -\chi(\mathrm{Gal}(\bar\eta_s/\eta_s), \mathrm{Ad}^{(0)}(\rho_0)) + r^2 - 1 \\
&= -\chi(\eta_s, \mathcal{E}nd^{(0)}(\mathcal{F}_0|_{\eta_s})) + r^2 - 1, \\
\dim F(\kappa[\epsilon]) &= \dim H_c^1(X, \mathcal{E}nd(\mathcal{F}_0)).
\end{aligned}$$

where $\mathrm{Ad}(\rho_0)$ *is the κ-vector space of $r\times r$ matrices with entries in κ on which $\pi_1(X-S,\bar\eta)$ and $\mathrm{Gal}(\bar\eta_s/\eta_s)$ act by the composition of ρ_0 with the adjoint representation of $\mathrm{GL}(\kappa^r)$, $\mathrm{Ad}^{(0)}(\rho_0)$ is the subspace of $\mathrm{Ad}(\rho_0)$ consisting of matrices of trace 0, $\mathcal{F}_0$ is the lisse κ-sheaf on $X-S$ corresponding to the representation $\rho_0 : \pi_1(X-S,\bar\eta) \to \mathrm{GL}(\kappa^r)$, and $\mathcal{E}nd^{(0)}(\mathcal{F}_0)$ is the subsheaf of $\mathcal{E}nd(\mathcal{F}_0)$ formed by sections of trace 0.*

In [3, Proposition 2.2], we develop an obstruction theory, and we prove the following.

Lemma 1.2. *Let $A' \to A$ be an epimorphism in $\mathcal{C}_\Lambda$ such that its kernel $\mathfrak{a}$ has the property $\mathfrak{m}_{A'}\mathfrak{a} = 0$. Let $(\rho, (P_s)_{s\in S})$ be a tuple whose equivalent class is an element in $D_S^\square(A)$, let $\rho_s = P_s^{-1}\rho|_{\mathrm{Gal}(\bar\eta_s/\eta_s)}P_s$, and let $\rho'_s : \mathrm{Gal}(\bar\eta_s/\eta_s) \to \mathrm{GL}(A'^r)$ be representations such that*

$$\rho'_s \mod \mathfrak{a} = \rho_s$$

for all $s \in S$. Suppose all elements in $\mathrm{End}_{\pi_1(X-S,\bar{\eta})}(\kappa^r)$ are scalar multiplications and suppose $\det(\rho)$ can be lifted to a representation $\lambda' : \pi_1(X - S, \bar{\eta}) \to \mathrm{GL}(A')$ with the property $\lambda'|_{\mathrm{Gal}(\bar{\eta}_s/\eta_s)} = \det \rho'_s$. Then there exists a tuple $(\rho', (P'_s)_{s\in S})$ whose equivalent class is an element in $D_S^{\Box}(A')$ so that $(\rho', (P'_s)_{s\in S})$ is a lifting $(\rho, (P_s)_{s\in S})$, and $P_s'^{-1}\rho'|_{\mathrm{Gal}(\bar{\eta}_s/\eta_s)}P'_s = \rho'_s$ for all $s \in S$.

Proposition 1.3. *Suppose all elements in $\mathrm{End}_{\pi_1(X-S,\bar{\eta})}(\kappa^r)$ are scalar multiplications.*

(i) If $X = \mathbb{P}^1$, then the morphism of functors

$$D_S^{\Box,\lambda} \to \prod_{s\in S} D_s^{\Box,\lambda}, \quad (\rho, (P_s)_{s\in S}) \mapsto (P_s^{-1}\rho|_{\mathrm{Gal}(\bar{\eta}_s/\eta_s)}P_s)_{s\in S}$$

is smooth. In particular, its fiber F^λ is smooth. We have $F = F^\lambda$ in this case, and hence F is also smooth.

(ii) For general X, suppose r is prime to $\mathrm{char}\,\kappa$ and suppose κ contains all r-th roots of unity. Then the morphism of functors

$$D_S^{\Box,\lambda} \to \prod_{s\in S} D_s^{\Box,\lambda}, \quad (\rho, (P_s)_{s\in S}) \mapsto (P_s^{-1}\rho|_{\mathrm{Gal}(\bar{\eta}_s/\eta_s)}P_s)_{s\in S}$$

is smooth. In particular, its fiber F^λ is smooth.

Proof. We prove (ii) and leave the proof of (i) to the reader. Keep the notation in Lemma 1.2, and suppose furthermore that $\det(\rho) = \lambda$ and $\det(\rho'_s) = \lambda|_{\mathrm{Gal}(\bar{\eta}_s/\eta_s)}$ so that the equivalent class of $(\rho, (P_s)_{s\in S})$ is an element of $D_S^{\Box,\lambda}(A)$, and $(\rho'_s)_{s\in S}$ is an element in $\prod_{s\in S} D_s^{\Box,\lambda}(A')$. There is no obstruction to lifting $\lambda : \pi_1(X-S, \bar{\eta}) \to A^*$ since λ takes its value in the ground coefficient ring Λ. By Lemma 1.2, we can lift $(\rho, (P_s)_{s\in S})$ to $(\rho', (P'_s)_{s\in S})$ such that $P_s'^{-1}\rho'|_{\mathrm{Gal}(\bar{\eta}_s/\eta_s)}P'_s = \rho'_s$. We have

$$\det(\rho')|_{\mathrm{Gal}(\bar{\eta}_s/\eta_s)} = \det(\rho'_s) = \lambda|_{\mathrm{Gal}(\bar{\eta}_s/\eta_s)}.$$

So $\lambda^{-1}\det(\rho')$ is unramified at each $s \in S$. It is also unramified outside S. So $\lambda^{-1}\det(\rho')$ defines a character

$$\chi : \pi_1(X, \bar{\eta}) \to A'^*.$$

Note that we have

$$\lambda^{-1}\det(\rho') \mod \mathfrak{a} = \lambda^{-1}\det(\rho) = 1.$$

So the image of χ lies in the subgroup $1+\mathfrak{m}_{A'}$ of A'^*. This subgroup has a filtration

$$1 + \mathfrak{m}_{A'} \supset 1 + \mathfrak{m}_{A'}^2 \supset \cdots.$$

For each i, we have an isomorphism of groups

$$\mathfrak{m}_{A'}^i/\mathfrak{m}_{A'}^{i+1} \cong (1 + \mathfrak{m}_{A'}^i)/(1 + \mathfrak{m}_{A'}^{i+1}),$$

and $\mathfrak{m}_{A'}^i/\mathfrak{m}_{A'}^{i+1}$ is the underlying abelian group of a finite dimensional vector space over κ. Any profinite subgroup of a finite dimensional κ-vector space must be a pro-ℓ-group. It follows that the representation $\chi : \pi_1(X, \bar{\eta}) \to A'^*$ must factor through the maximal abelian pro-ℓ-quotient Γ of $\pi_1(X, \bar{\eta})$. By [4, X 3.10], the pro-ℓ-completion of $\pi_1(X, \bar{\eta})$ is isomorphic to the pro-ℓ-completion of the group with generators s_i, t_i $(1 \leq i \leq g = \mathrm{genus}(X))$ and with one relation

$$(s_1 t_1 s_1^{-1} t_1^{-1}) \cdots (s_g t_g s_g^{-1} t_g^{-1}) = 1.$$

It follows that $\Gamma \cong \mathbb{Z}_\ell^{2g}$. Since $(r, \mathrm{char}\,\kappa) = 1$, and κ contains all r-th roots of unities, any character of

$$\mathbb{Z}_\ell^{2g} \to 1 + \mathfrak{m}_{A'}$$

is the r-th power of another such character. So we can find a character

$$\chi' : \pi_1(X, \bar{\eta}) \to 1 + \mathfrak{m}_{A'}$$

such that $\chi'^r = \chi$. Then

$$\det(\rho' \chi'^{-1}) = \det(\rho') \chi'^{-r} = \det(\rho') \chi^{-1} = \lambda.$$

So $(\rho' \chi'^{-1}, (P_s)_{s \in S})$ defines an element of $D_S^{\square, \lambda}(A')$ lifting $(\rho, (P_s)_{s \in S})$ in $D_S^{\square, \lambda}(A)$. This proves our assertion. □

Proposition 1.4. *Suppose that r is prime to* $\mathrm{char}\,\kappa$, *that κ contains all r-th roots of unity, and that* $\mathrm{End}_{\pi_1(X-S, \bar{\eta})}(\kappa^r)$ *consists of scalar multiplications. Then*

$$\dim F^\lambda(\kappa[\epsilon]) = \dim H_c^1(X - S, \mathcal{E}nd(\mathcal{F}_0)) - 2g,$$

where $\mathcal{F}_0$ is the lisse κ-sheaf on $X - S$ corresponding to the representation ρ_0.

Proof. By Proposition 1.3 (ii), the canonical morphism $D_S^{\square, \lambda} \to \prod_{s \in S} D_s^{\square, \lambda}$ is smooth. Combined with Proposition 1.1, we have

$$\begin{aligned}
&\dim F^\lambda(\kappa[\epsilon]) \\
&= \dim D_S^{\square, \lambda}(\kappa[\epsilon]) - \sum_{s \in S} \dim D_s^{\square, \lambda}(\kappa[\epsilon]) \\
&= -\chi(X - S, \mathcal{E}nd^{(0)}(\mathcal{F}_0)) + |S| r^2 - 1 - \sum_{s \in S} \Big(- \chi(\eta_s, \mathcal{E}nd^{(0)}(\mathcal{F}_0|_{\eta_s})) + r^2 - 1 \Big) \\
&= -\Big(\chi(X, Rj_* \mathcal{E}nd^{(0)}(\mathcal{F}_0)) - \sum_{s \in S} \dim(Rj_* \mathcal{E}nd^{(0)}(\mathcal{F}_0))_{\bar{s}} \Big) + |S| - 1 \\
&= -\chi(X, j_! \mathcal{E}nd^{(0)}(\mathcal{F}_0)) + |S| - 1 \\
&= -\chi(X, j_! \mathcal{E}nd(\mathcal{F}_0)) + \chi(X, j_! \mathbb{Q}_\ell) + |S| - 1 \\
&= \Big(\dim H_c^1(X - S, \mathcal{E}nd(\mathcal{F}_0)) - 1 \Big) + (2 - 2g - S|) + |S| - 1 \\
&= \dim H_c^1(X - S, \mathcal{E}nd(\mathcal{F}_0)) - 2g.
\end{aligned}$$

□

2 Proof of Theorems 0.4 and 0.5

As in §1, we take Λ to be either a finite extension E of $\mathbb{Q}_\ell$, or the integer ring $\mathcal{O}$ of such E. Let $\mathfrak{m}$ be the maximal ideal of Λ, and let $\kappa = \Lambda/\mathfrak{m}$ be the residue field of Λ. Suppose we are given a representation $\rho_\Lambda : \pi_1(X - S, \bar{\eta}) \to \mathrm{GL}(\Lambda^r)$. Let $\rho_0 : \pi_1(X - S, \bar{\eta}) \to \mathrm{GL}(\kappa^r)$ be the representation obtained from ρ_Λ by passing to residue field. Suppose we are given $P_{0,s} \in \mathrm{GL}(\kappa^r)$ for each $s \in S$ with the property

$$P_{0,s}^{-1}\rho_0|_{\mathrm{Gal}(\bar{\eta}_s/\eta_s)}P_{0,s} = \rho_0|_{\mathrm{Gal}(\bar{\eta}_s/\eta_s)}.$$

In application, we often take $P_{0,s}$ to be the identity matrix I for all $s \in S$. In this case, we denote the data $(\rho_0, (P_{0,s})_{s\in S})$ by $(\rho_0, (I)_{s\in S})$. For any $A \in \mathrm{ob}\,\mathcal{C}_\Lambda$, recall that $F(A)$ is the set of deformations $(\rho, (P_s)_{s\in S})$ of the data $(\rho_0, (P_{0,s})_{s\in S})$ with the prescribed local monodromy $\rho_\Lambda|_{\mathrm{Gal}(\bar{\eta}_s/\eta_s)}$. More precisely, we have

$$\begin{aligned} F(A) = \{(\rho, (P_s)_{s\in S}) \mid\ & \rho : \pi_1(X - S, \bar{\eta}) \to \mathrm{GL}(A^r) \text{ is a representation, } P_s \in \mathrm{GL}(A^r), \\ & \rho \mod \mathfrak{m}_A = \rho_0, \quad P_s \mod \mathfrak{m}_A = P_{0,s}, \\ & P_s^{-1}\rho|_{\mathrm{Gal}(\bar{\eta}_s/\eta_s)}P_s = \rho_\Lambda|_{\mathrm{Gal}(\bar{\eta}_s/\eta_s)} \text{ for all } s \in S\}/\sim, \end{aligned}$$

where two tuples $(\rho^{(i)}, (P_s^{(i)})_{s\in S})$ $(i = 1, 2)$ are equivalent if there exists $P \in \mathrm{GL}(A^r)$ such that

$$(\rho^{(1)}, (P_s^{(1)})_{s\in S}) = (P^{-1}\rho^{(2)}P, (P^{-1}P_s^{(2)})_{s\in S}).$$

Let $\lambda = \det(\rho_\Lambda)$. Recall that $F^\lambda(A)$ is the set of deformations $(\rho, (P_s)_{s\in S})$ of the data $(\rho_0, (P_{0,s})_{s\in S})$ with the prescribed determinant λ and the prescribed local monodromy $\rho_\Lambda|_{\mathrm{Gal}(\bar{\eta}_s/\eta_s)}$, that is, $F^\lambda(A)$ consists of those equivalent classes of $(\rho, (P_s)_{s\in S})$ in $F(A)$ with $\det(\rho) = \lambda$. We summarize properties of F and F^λ as follows, which follow from Propositions 1.1, 1.3 and 1.4.

Proposition 2.1. *Let $\mathcal{F}_0$ be the lisse κ-sheaf on $X - S$ corresponding to the representation ρ_0.*

(i) The functor F is pro-representable and

$$\dim F(\kappa[\epsilon]) = \dim H_c^1(X - S, \mathcal{E}nd(\mathcal{F}_0)).$$

If $X = \mathbb{P}^1$ and $\mathrm{End}(\mathcal{F}_0)$ consists of scalar multiplications, then F is smooth.

(ii) The functor F^λ is pro-representable. Suppose the rank r of $\mathcal{F}_0$ is prime to char κ *and suppose κ contains all r-th root of unity. Then F^λ is smooth and*

$$\dim F(\kappa[\epsilon]) = \dim H_c^1(X - S, \mathcal{E}nd(\mathcal{F}_0)) - 2g.$$

As an example, let's make the pro-representability of F more explicit as follows: There is a complete noetherian local Λ-algebra $R(\rho_\Lambda)$ with residue field κ and a universal tuple $(\rho_{\mathrm{univ}}, (P_{\mathrm{univ},s})_{s\in S})$ such that $\rho_{\mathrm{univ}} : \pi_1(X - S, \bar{\eta}) \to \mathrm{GL}(R(\rho_\Lambda)^r)$ is a representation, $P_{\mathrm{univ},s} \in \mathrm{GL}(R(\rho_\Lambda)^r)$,

$$\begin{aligned} \rho_{\mathrm{univ}} \mod \mathfrak{m}_{R(\rho_\Lambda)} &= \rho_0, \\ P_{\mathrm{univ},s} \mod \mathfrak{m}_{R(\rho_\Lambda)} &= P_{0,s}, \\ P_{\mathrm{univ},s}^{-1}\rho_{\mathrm{univ}}|_{\mathrm{Gal}(\bar{\eta}_s/\eta_s)}P_{\mathrm{univ},s} &= \rho_\Lambda|_{\mathrm{Gal}(\bar{\eta}_s/\eta_s)}, \end{aligned}$$

and such that for any $(\rho, (P_s)_{s\in S}) \in F(A)$, there exists a unique local Λ-algebra homomorphism $R(\rho_\Lambda) \to A$ which brings $(\rho_{\text{univ}}, (P_{\text{univ},s})_{s\in S})$ to a tuple equivalent to $(\rho, (P_s)_{s\in S})$.

Let E be a finite extension of $\mathbb{Q}_\ell$, and let $\mathcal{O}$ be its integer ring. From now on, we denote the residue field of $\mathcal{O}$ by κ. Suppose $\mathcal{F}_E$ is a lisse E-sheaf on $X - S$ of rank r. Choose a torsion free lisse $\mathcal{O}$-sheaf $\mathcal{F}_\mathcal{O}$ such that $\mathcal{F}_E \cong \mathcal{F}_\mathcal{O} \otimes_\mathcal{O} E$. Let $\mathcal{F}_0 = \mathcal{F}_\mathcal{O} \otimes_\mathcal{O} \kappa$, let $\rho_\mathcal{O} : \pi_1(X - S, \bar\eta) \to \mathrm{GL}(\mathcal{O}^r)$ be the representation corresponding to the sheaf $\mathcal{F}_\mathcal{O}$, and let $\rho_E : \pi_1(X - S, \bar\eta) \to \mathrm{GL}(E^r)$ and $\rho_0 : \pi_1(X - S, \bar\eta) \to \mathrm{GL}(\kappa^r)$ be the representations obtained from $\rho_\mathcal{O}$ by passing to the fraction field and the residue field of $\mathcal{O}$, respectively. Note that ρ_E and ρ_0 are also the representations corresponding to the E-sheaf $\mathcal{F}_E$ and the κ-sheaf $\mathcal{F}_0$, respectively. Take $\Lambda = \mathcal{O}$. Consider the universal deformation ring $R(\rho_\mathcal{O})$ of the functor $F : \mathcal{C}_\mathcal{O} \to (\text{Sets})$ for the data $(\rho_0, (I)_{s\in S})$, where $P_{0,s} = I$ for all $s \in S$. Applying the universal property to $(\rho_\mathcal{O}, (I)_{s\in S}) \mod \mathfrak{m}_\mathcal{O}^n$, we get an $\mathcal{O}$-algebra homomorphism $\varphi_0 : R(\rho_\mathcal{O}) \to \mathcal{O}$ which brings $(\rho_{\text{univ}}, (P_{\text{univ},s})_{s\in S})$ to the tuple $(\rho_\mathcal{O}, (I)_{s\in S})$. Let $\mathfrak{p}_0$ be the kernel of φ. It lies in the generic fiber of $\operatorname{Spec} R(\rho_\mathcal{O}) \to \operatorname{Spec} \mathcal{O}$. The next proposition shows that we can recover the universal deformation ring $R(\rho_E)$ of the functor $F : \mathcal{C}_E \to (\text{Sets})$ for the data $(\rho_E, (I)_{s\in S})$ from the universal deformation ring $R(\rho_\mathcal{O})$.

Proposition 2.2. *Notation as above. We have* $R(\rho_\mathcal{O})^\wedge_{\mathfrak{p}_0} \cong R(\rho_E)$.

Corollary 2.3. *Suppose* $X = \mathbb{P}^1$ *and* $\mathrm{End}(\mathcal{F}_E)$ *consists of scalar multiplications. Then*

$$\dim R(\rho_\mathcal{O})^\wedge_{\mathfrak{p}_0} = \dim H^1_c(X - S, \mathrm{End}(\mathcal{F}_{\mathbb{Q}_\ell})).$$

Proof. Indeed, by Proposition 2.1 (i) applied to the case $\Lambda = E$, $R(\rho_E)$ is a formally smooth E-algebra. Its dimension coincides with the dimension of its tangent space, which is $\dim H^1_c(X - S, \mathrm{End}(\mathcal{F}_E))$ by Proposition 2.1 (i). On the other hand, we have $R(\rho_\mathcal{O})^\wedge_{\mathfrak{p}_0} \cong R(\rho_E)$ by Proposition 2.2. Our assertion follows. □

Proof of Theorem 0.4. Let $a_1, \ldots, a_n$ be a family of generators for $\mathfrak{p}_0$. We have an epimorphism of $\mathcal{O}$-algebras

$$\mathcal{O}[[T_1, \ldots, T_n]] \to R(\rho_\mathcal{O}), \quad T_i \mapsto a_i.$$

Let $\{f_1, \ldots, f_m\}$ be a family of generators for the kernel of this epimorphism so that

$$R(\rho_\mathcal{O}) \cong \mathcal{O}[[T_1, \ldots, T_n]]/(f_1, \ldots, f_m).$$

Let $\mathcal{M}_0$ be the rigid analytic subvariety $f_1 = \cdots = f_m = 0$ of the n-dimensional open unit polydisc. It can be considered as the generic fiber of the formal scheme $\operatorname{Spf} R(\rho_\mathcal{O})$ over $\operatorname{Spf} \mathcal{O}$. Points in $\mathcal{M}_0$ are of the form $t = (t_1, \ldots, t_n)$ with $t_i \in \overline{\mathbb{Q}}_\ell$, $|t_i| < 1$ $(i = 1, \ldots, n)$ and $f_j(t_1, \ldots, t_n) = 0$ $(j = 1, \ldots, m)$. Such a point defines an $\mathcal{O}$-algebra homomorphism

$$\varphi_t : R(\rho_\mathcal{O}) \cong \mathcal{O}[[T_1, \ldots, T_n]]/(f_1, \ldots, f_m) \to \mathcal{O}_{\overline{\mathbb{Q}}_\ell}, \quad f(T_1, \ldots, T_n) \mapsto f(t_1, \ldots, t_n),$$

where $\mathcal{O}_{\overline{\mathbb{Q}}_\ell}$ is the ring of integers of $\overline{\mathbb{Q}}_\ell$. It brings $(\rho_{\mathrm{univ}}, (P_{\mathrm{univ},s})_{s\in S})$ to a tuple $(\rho_t, (P_{t,s})_{s\in S})$ so that $\rho_t : \pi_1(X - S, \bar{\eta}) \to \mathrm{GL}(\mathcal{O}^r_{\overline{\mathbb{Q}}_\ell})$ is a representation, $P_{t,s} \in \mathrm{GL}(\mathcal{O}^r_{\overline{\mathbb{Q}}_\ell})$, and

$$\rho_t \mod \mathfrak{m}_{\mathcal{O}_{\overline{\mathbb{Q}}_\ell}} = \rho_0, \quad P_{t,s} \mod \mathfrak{m}_{\mathcal{O}_{\overline{\mathbb{Q}}_\ell}} = I, \quad P_{t,s}^{-1}\rho_t|_{\mathrm{Gal}(\bar{\eta}_s/\eta_s)}P_{t,s} = \rho_{\mathcal{O}}|_{\mathrm{Gal}(\bar{\eta}_s/\eta_s)},$$

$\mathcal{M}_0$ can be thought as a piece of the module space $\mathcal{M}$. Suppose ρ_E is physically rigid. Then the third equation above implies that ρ_t is isomorphic to ρ_E as $\overline{\mathbb{Q}}_\ell$-representations, that is, there exists $P \in \mathrm{GL}(\overline{\mathbb{Q}}_\ell^r)$ such that $P^{-1}\rho_t P = \rho_E$. Deligne shows that there exists an integer N such that for those t with

$$|t_i| \leq |\lambda^N| \ (i = 1, \ldots, n)$$

where λ is a uniformizer for $\mathcal{O}$, we can choose P so that $P \in \mathrm{GL}(\mathcal{O}^r_{\overline{\mathbb{Q}}_\ell})$ and $P \equiv I \mod \mathfrak{m}_{\mathcal{O}_{\overline{\mathbb{Q}}_\ell}}$. Let V be the subdomain of $\mathcal{M}_0$ defined by $|t_i| \leq |\lambda^N|$ $(i = 1, \ldots, n)$. Then for any $t \in V$, we have an equivalence of tuples

$$(\rho_t, (P_{t,s})_{s\in I}) \sim (P^{-1}\rho_t P, (P^{-1}P_{t,s})_{s\in I}) = (\rho_{\mathcal{O}}, (P^{-1}P_{t,s})_{s\in I}).$$

Let $\mathcal{N}$ be the rigid analytic space whose points are given by

$$\mathcal{N} = \{(P_s)_{s\in I} | P_s \in \mathrm{Aut}_{\mathrm{Gal}(\bar{\eta}_s/\eta_s)}(\mathcal{O}^r_{\overline{\mathbb{Q}}_\ell}), \ P_s \equiv I \mod \mathfrak{m}_{\overline{\mathbb{Q}}_\ell}\}/\sim,$$

where two tuples $(P_s^{(i)})$ $(i = 1, 2)$ are equivalent if and only if there exists a scalar u such that $P_s^{(1)} = uP_s^{(2)}$ for all $s \in S$, and $\mathcal{O}^r_{\overline{\mathbb{Q}}_\ell}$ is provided with the $\mathrm{Gal}(\bar{\eta}_s/\eta_s)$-module structure via ρ_E. We have

$$\dim \mathcal{N} = \sum_{s\in S} \dim \mathrm{End}_{\mathrm{Gal}(\bar{\eta}_s/\eta_s)}(\overline{\mathbb{Q}}_\ell^r) - 1.$$

We have a morphism

$$f : \mathcal{N} \to \mathcal{M}_0$$

which maps the point $(P_s)_{s\in I}$ in $\mathcal{N}$ to the point $(\rho_{\mathcal{O}}, (P_s)_{s\in I})$ in $\mathcal{M}_0$. The above argument shows that f induces a surjective map $f^{-1}(V) \to V$. Xie proves that this implies $\dim V \leq \dim f^{-1}(V)$. We thus have

$$\dim V \leq \sum_{s\in S} \dim \mathrm{End}_{\mathrm{Gal}(\bar{\eta}_s/\eta_s)}(\overline{\mathbb{Q}}_\ell^r) - 1.$$

One can show that $\widehat{\mathcal{O}}_{\mathcal{M}_0,t_0} \cong R(\rho_{\mathcal{O}})^\wedge_{\mathfrak{p}_0}$, where t_0 is the origin of the open polydisc which lies in V. Hence

$$\dim R(\rho_{\mathcal{O}})^\wedge_{\mathfrak{p}_0} \leq \sum_{s\in S} \dim \mathrm{End}_{\mathrm{Gal}(\bar{\eta}_s/\eta_s)}(\overline{\mathbb{Q}}_\ell^r) - 1.$$

Combined with Corollary 2.3, we have

$$\dim H^1_c(X - S, \mathcal{E}nd(\mathcal{F}_E)) \leq \sum_{s\in S} \dim H^0(\eta_s, \mathcal{E}nd(\mathcal{F}_E|_{\eta_s})) - 1. \tag{1.1}$$

Note that $j_!\mathcal{E}nd(\mathcal{F}_E)$ is a subsheaf of $j_*\mathcal{E}nd(\mathcal{F}_E)$, the quotient sheaf is a sky-scrapper sheaf supported on S, and

$$\Big(j_*\mathcal{E}nd(\mathcal{F}_E)/j_!\mathcal{E}nd(\mathcal{F}_E)\Big)_{\bar{s}} \cong H^0(\eta_s, \mathcal{E}nd(\mathcal{F}_E|_{\eta_s}))$$

for any $s \in S$. We have $H^0_c(X - S, \mathcal{E}nd(\mathcal{F}_E)) = 0$ since $X - S$ is an affine curve. Moreover, we have $H^0(X, j_*\mathcal{E}nd(\mathcal{F}_E)) \cong \mathrm{End}(\mathcal{F}_E)$. So we have a long exact sequence

$$\begin{aligned} 0 \to \mathrm{End}(\mathcal{F}_E) \to & \bigoplus_{s\in S} H^0(\eta_s, \mathcal{E}nd(\mathcal{F}_E|_{\eta_s})) \\ & \to H^1_c(X - S, \mathcal{E}nd(\mathcal{F}_E)) \to H^1(X, j_*\mathcal{E}nd(\mathcal{F}_E)) \to 0. \end{aligned}$$

It follows that

$$\begin{aligned} \dim H^1(X, j_*\mathcal{E}nd(\mathcal{F}_E)) = \dim H^1_c(X - S, \mathcal{E}nd(\mathcal{F}_E)) - \sum_{s\in S} \dim H^0(\eta_s, \mathcal{E}nd(\mathcal{F}_E|_{\eta_s})) \\ + \dim \mathrm{End}(\mathcal{F}_E). \end{aligned}$$

So by the inequality (1.1), we have

$$\dim H^1(X, j_*\mathcal{E}nd(\mathcal{F}_E)) \leq \dim \mathrm{End}(\mathcal{F}_E) - 1.$$

Since $\mathrm{End}(\mathcal{F}_E)$ consists of scalar multiplications, we have $\dim \mathrm{End}(\mathcal{F}_E) = 1$. So $H^1(X, j_*\mathcal{E}nd(\mathcal{F}_E)) = 0$. This proves Theorem 0.4. □

The proof of Theorem 0.5 is similar. We first enlarge E so that it contains all r-th roots of unity. Then we use the same argument as the proof of Theorem 0.4, using the functor functor F^λ instead of F. We leave the details to the reader.

References

[1] S. Bloch and H. Esnault, Local Fourier transform and rigidity for $\mathcal{D}$-modules, *Asian J. Math.* 8 (2004), 587-606.

[2] L. Fu, Deformation of ℓ-adic sheaves with undeformed local monodromy, *J. Number Theory* 133 (2013), 675-691.

[3] L. Fu, Deformations and rigidity of ℓ-adic sheaves, arXiv:1611.03964.

[4] A. Grothendieck, *Revêtements Étales et Groupe Fondamental* (SGA 1), Lecture Notes in Math. 224, Springer-Verlag (1971).

[5] N. Katz, *Rigid Local Systems*, Annals of Math. Studies 139, Princeton University Press (1996).

[6] N. Katz, *Exponential sums and differential equations*, Annals of Math. Studies 124, Princeton University Press (1990).

[7] M. Kisin, Moduli of finite flat group schemes and modularity, *Ann. of Math.* 170 (2009), 1085-1180.

[8] B. Mazur, Deforming Galois representations, *Galois groups over* $\mathbb{Q}$, 385-437, Math. Sci. Res. Inst. Publ., 16, Springer, New York, 1989.

[9] M. Schlessinger, Functors on Artin rings, *Trans. A.M.S.* 130 (1968), 208-222.

Yau Mathematical Sciences Center, Tsinghua University
Beijing, China
E-mail address: leifu@mail.tsinghua.edu.cn

Proceedings of the 7th International Congress of Chinese Mathematicians
ALM 43, pp. 255–266

On the Cocenters of p-adic Groups

Xuhua He

Abstract

In this report, we give an introduction to a new approach towards the understanding the complex and modular representations of p-adic groups. This is based on my talk at ICCM 2016. The main references are the series of papers [8], [9], [5].

Mathematics Subject Classification: 22E50, 20C08.

Keywords and Phrases: Hecke algebras, p-adic groups, cocenter.

1 Motivation: finite groups

1.1

To explain some basic ideas of the cocenter program, we start with the representation theory of finite groups.

Let G be a finite group. A fundamental result in representation theory is that the number of irreducible complex representations of G equals the number of conjugacy classes of G. This important identity can be reformulated in the following way.

Let $\mathfrak{R}_{\mathbb{C}}(G)$ be the Grothendieck group of finite dimensional complex representations of G. Then it has a natural basis given by the set of irreducible complex representations. In particular, the number of irreducible complex representations of G is just the rank of $\mathfrak{R}_{\mathbb{C}}(G)$.

On the other hand, let $H_{\mathbb{C}} = \mathbb{C}[G]$ be the group algebra of G. Let $[H_{\mathbb{C}}, H_{\mathbb{C}}]$ be the commutator of $H_{\mathbb{C}}$, the subspace of H spanned by $[f, f'] = ff' - f'f$ for $f, f' \in H_{\mathbb{C}}$. Let $\bar{H}_{\mathbb{C}} = H_{\mathbb{C}}/[H_{\mathbb{C}}, H_{\mathbb{C}}]$ be the cocenter of $H_{\mathbb{C}}$. It is easy to see that given a conjugacy class $\mathcal{O}$ of G, the image of g in $H_{\mathbb{C}}$ is independent of the choice of the element $g \in \mathcal{O}$. Thus we have a well-defined element $[\mathcal{O}]$ in $\bar{H}_{\mathbb{C}}$ for each conjugacy class $\mathcal{O}$ of G. One may check that $\{[\mathcal{O}]\}$ forms a basis of $\bar{H}_{\mathbb{C}}$. In particular, $\dim \bar{H}_{\mathbb{C}}$ equals the number of conjugacy classes of G. Now we have the equality

$$\dim \bar{H}_{\mathbb{C}} = \operatorname{rank} \mathfrak{R}_{\mathbb{C}}(G). \tag{a}$$

We can make this equality more explicit via the trace map. For any $g \in G$ and representation π of G, we have $Tr_{\mathbb{C}}(g, \pi) \in \mathbb{C}$. The trace map extends linearly to

Partially supported by NSF DMS-1463852 and DMS-1128155 (from IAS).

the group algebra $H_{\mathbb{C}}$ and the Grothendieck group $\mathfrak{R}_{\mathbb{C}}(G)$ and induces a bijection

$$Tr_{\mathbb{C}} : \bar{H}_{\mathbb{C}} \to \mathfrak{R}_{\mathbb{C}}(G)^*, \qquad g \mapsto (\pi \mapsto Tr_{\mathbb{C}}(g, \pi)). \tag{b}$$

1.2

In practice, such bijection can be written down explicitly via the character table, an invertible square matrix with rows indexed by the irreducible representations and the columns indexed by the conjugacy classes of G and the entries given by the characters.

Now we give an example.

Let $G = S_3$, the permutation group of $\{1, 2, 3\}$. It has three conjugacy classes: $\mathcal{O}_1 = \{id\}$, $\mathcal{O}_2 = \{(12), (13), (23)\}$, $\mathcal{O}_3 = \{(123), (132)\}$. It has three irreducible complex representations, the trivial representation *triv*, the sign representation sgn and the irreducible 2-dimensional representation σ. The character table of S_3 is

S_3	$\mathcal{O}_1$	$\mathcal{O}_2$	$\mathcal{O}_3$
triv	1	1	1
sgn	1	-1	1
σ	2	0	-1

The determinant of this matrix is 6 and the matrix is invertible.

1.3

One may also consider the modular representations.

Let R be an algebraically closed field of characteristic l. Then the number of irreducible representations of G over R, in general, does not equal to the number of conjugacy classes of G. One needs to replace the conjugacy classes by the l-regular conjugacy classes, the conjugacy classes of G whose elements have order prime to l.

We may still relate the cocenters and representations in the following way.

Let $\mathfrak{R}_R(G)$ be the Grothendieck group of finite dimension representations of G over R. Let H_R be the group algebra $R[G]$ and $\bar{H}_R$ be its cocenter. We have the trace map

$$Tr_R : \bar{H}_R \to \mathfrak{R}_R(G).$$

The trace map is always surjective, but not injective in general. However, the kernel is easy to describe: it is spanned by the differences $[\mathcal{O}] - [\mathcal{O}']$, where $\mathcal{O}$ is a conjugacy class of G and $\mathcal{O}'$ is the associated l-regular conjugacy class.

We may look again at the example $G = S_3$. Instead of complex representations, we consider the representations over R instead, where R is an algebraic closed field of characteristic 2. In this case, trivial representation and the sign representation are the same. On the other hand, $\mathcal{O}_2$ is not 2-regular and the associated 2-regular conjugacy class is $\mathcal{O}_1$. Over R, $\mathcal{O}_1$ and $\mathcal{O}_2$ has the same character values. If we delete the row *sgn* and the column $\mathcal{O}_2$ from the character table in §1.2, we get the character table for mod-2 representations.

S_3	$\mathcal{O}_1$	$\mathcal{O}_3$
triv	1	1
σ	2	-1

The determinant is $-3 \in R$ and the matrix is invertible.

1.4

It is also worth mentioning that the structure of the cocenter is independent of the choice of the field. In fact, one may consider $H = \mathbb{Z}[G]$. Then for any field R, we have $H_R = H \otimes_{\mathbb{Z}} R$. We define the commutator $[H, H]$ as the $\mathbb{Z}$-submodule of H generated by $[f, f']$ for $f, f' \in H$ and define the cocenter $\bar{H} = H/[H, H]$. Then

(1) $\bar{H}$ is a free $\mathbb{Z}$-module with basis $\{[\mathcal{O}]\}$.

(2) For any R, $\bar{H}_R = \bar{H} \otimes_{\mathbb{Z}} R$.

(3) For any R, $\operatorname{rank} \mathfrak{R}_R(G) \leqslant \operatorname{rank} \bar{H}$.

2 p-adic groups and their Hecke algebras

2.1

Now we move to the p-adic groups. More precisely, let F be a nonarchimedean local field with residue characteristic p (i.e. a finite extension of $\mathbb{Q}_p$ or $\mathbb{F}_q((\epsilon))$), where q is a power of p. Let $\mathbb{G}$ be a connected reductive group over F and $G = \mathbb{G}(F)$ be its F-points. Note that we do not think of G just as an abstract group, but we also take into account of its topology. A key fact is that every neighborhood of the identity of G contains an open compact subgroup.

On the representation side, we do not consider arbitrary representation of G, but we are mainly interested in the smooth admissible representations:

- "Smooth" means that the stabilizer of every point in the representation is an open subgroup of G;
- "Admissible" means that given an open compact subgroup $\mathcal{K}$ of G, the subspace of $\mathcal{K}$-fixed points is finite dimensional.

2.2

Since we are mainly interested in the smooth admissible representations, we will not use the group algebra of G as in the finite group case, but use the Hecke algebra instead.

By definition, the Hecke algebra is the convolution algebra of functions on G that are

- locally constant (which corresponds to the smoothness condition of the representations).
- compactly supported (which guarantees the convolution product of any two such functions make sense).

Usually one considers the Hecke algebra of $\mathbb{C}$-valued functions, which is related to the complex representations. One may also consider the Hecke algebra of R-valued functions for algebraically closed field of characteristic $l \neq p$, which is related to the mod-l representations. In both cases, the category of smooth admissible representations of G is equivalent to the category of (non-degenerate admissible) representations of H (see [16, 4.4]).

Note that there is no Haar measure of G over characteristic p. Thus the Hecke algebra does not exist in characteristic p. To study the mod-p representations of G, one uses the pro-p Iwahori Hecke algebra instead. This is also a very interesting topics, but it is beyond the scope of this note. We refer to [1] in this directions.

2.3

Let R be an algebraically closed field of characteristic not equal to p. Let $\mathfrak{R}_R(G)$ be the Grothendieck group of smooth admissible representations of G over R and H_R be the Hecke algebra of locally constant, compactly supported R-valued functions on G. Let $\bar{H}_R$ be its cocenter. Then we have the trace map

$$Tr_R : \bar{H}_R \to \mathfrak{R}_R(G)^*.$$

As motivated by the case of finite groups, we would like to understand the structure $\bar{H}_R$ and the trace map and we expect that the study will eventually lead us to a better understand of the representation theory of p-adic groups.

2.4

As in the finite group case, we would like to study the integral form of the Hecke algebras and their cocenters. By the discussion above, we have to allow p to be invertible. So by the integral form, we mean the Hecke algebra H of locally constant, compactly supported, $\mathbb{Z}[\frac{1}{p}]$-valued functions on G. We have

$$H_R = H \otimes_{\mathbb{Z}[\frac{1}{p}]} R, \quad \bar{H}_R = \bar{H} \otimes_{\mathbb{Z}[\frac{1}{p}]} R.$$

We also expect that $\bar{H}$ is a free $\mathbb{Z}[\frac{1}{p}]$-module. This is proved by Vignéras if $\text{char}(F) = 0$.

Recall that the cocenter of the group algebra has a standard basis given by the conjugacy classes of the group. However, such a simple and nice description does not apply to the Hecke algebra H. The reason basically comes from the "locally constant" condition as any open subset of G intersects infinitely many conjugacy classes.

Another major difficulty to study the cocenter is that it is of infinite rank, which makes the connection between the cocenter and the representations complicated. The simple yet powerful counting argument for finite groups does not work for p-adic groups.

To overcome these difficulties, we introduce a decomposition of G into subsets, each of which is a nice (geometric) union of conjugacy classes. We discuss in details in the next section.

3 Newton decomposition

3.1

We first take a short digression and discuss the finite reductive group. Let $\mathbb{G}$ be a connected reductive group over $\mathbb{F}_q$. Then $\mathbb{G}(\mathbb{F}_q)$ is a finite reductive group. However, to study the group $\mathbb{G}(\mathbb{F}_q)$ and its representations, it is very useful to think it in a more geometric way as follows.

Let σ be the Frobenius of $\bar{\mathbb{F}}_q$ over $\mathbb{F}_q$. Then we have the algebraic group $\mathbb{G}(\bar{\mathbb{F}}_q)$, which has an algebraic variety structure. The finite reductive group $\mathbb{G}(\mathbb{F}_q)$ can be viewed as the subgroup of Frobenius-fixed points. This geometric point of view is essential in the study of representations of finite reductive groups, e.g. in the Deligne-Lusztig theory and Lusztig's theory of character sheaves.

3.2

We will view the p-adic group G in a similar way. We first introduce a few more notations.

Let $\breve{F}$ be the completion of maximal unramified extension of F and σ be the Frobenius of $\breve{F}$ over F. Let $\breve{G} = \mathbb{G}(\breve{F})$. Then we have $G = \breve{G}^\sigma$.

We consider the σ-twisted conjugation action on $\breve{G}$ defined by $g \cdot_\sigma g' = gg'\sigma(g)^{-1}$. Since G is σ-fixed, each conjugacy class of G is contained in a single σ-conjugacy class of $\breve{G}$ under the natural embedding $G \hookrightarrow \breve{G}$. The map from the set of conjugacy classes of G to the set of σ-conjugacy classes of $\breve{G}$ is far from being one-to-one. However, as we explained in §2.4, we are mainly interested in a nice union of conjugacy classes of G, not each individual ones. As we will see later, the fibers of this map actually provide the nice unions we are looking for.

3.3

The σ-conjugacy classes on $\breve{G}$ are classified by Kottwitz in [14] and [15]. Roughly speaking, a σ-conjugacy class is determined by the Newton point. Newton points are certain dominant rational coweights. We will not give the precise definition here, but provide an example of GL_n instead.

The Newton points for GL_n are $(a_1, \cdots, a_n) \in \mathbb{Q}^n$ satisfying

- The dominance condition: $a_1 \geqslant a_2 \geqslant \cdots \geqslant a_n$.
- The integrality condition: for any $r \in \mathbb{Q}$, $r\sharp\{i; a_i = r\} \in \mathbb{Z}$.

In the case where $n = 2$, the Newton points are (a, b) for integers $a \geqslant b$ and $(a + \frac{1}{2}, a + \frac{1}{2})$ for $a \in \mathbb{Z}$. The σ-conjugacy class of $GL_2(F)$ corresponding to the Newton point (a, b) for integers $a \geqslant b$ has a standard representative $\begin{pmatrix} \epsilon^a & 0 \\ 0 & \epsilon^b \end{pmatrix}$ and the σ-conjugacy class of $GL_2(F)$ corresponding to the Newton point $(a+\frac{1}{2}, a+\frac{1}{2})$ for $a \in \mathbb{Z}$ has a standard representative $\begin{pmatrix} 0 & \epsilon^a \\ \epsilon^{a+1} & 0 \end{pmatrix}$. Here ϵ is the uniformizer.

3.4

Now we have the decomposition

$$\breve{G} = \sqcup_\nu [\nu]$$

into σ-conjugacy classes. Intersecting this decomposition with G, we obtain

$$G = \sqcup_\nu G(\nu), \qquad \text{where } G(\nu) = G \cap [\nu].$$

The subset $G(\nu)$ of G is called the *Newton stratum* of G associated to the Newton point ν.

This definition works for all the split groups. For quasi-split groups, the construction also works under some modification. However, there are some problems with this definition for non-quasi-split groups, as the Newton points lies in the vector space associated to the apartment of the Bruhat-Tits building. Although the Bruhat-Tits buildings and the apartments of G over F and over $\breve{F}$ are compatible, the special vertices used to determine the vector spaces do not match.

In [8], we give another definition which works in the general case and coincide with the above definition in the split case. As the definition is a bit technical, we skip the details.

3.5

By definition, each Newton stratum is a union of conjugacy classes of G. For any ν, let $H(\nu) \subset H$ consisting of functions supported in $G(\nu)$ and $\bar{H}(\nu)$ be the image of $H(\nu)$ in the cocenter $\bar{H}$.

A key feature of the Newton strata is that they are all admissible, i.e., for any open compact subset X of G, there exists an open compact subgroup $\mathcal{K}$ of G such that $G(\nu) \cap X$ is stable under the left and right multiplication. Note that the admissibility implies openness. Therefore $H(\nu)$ is nonempty for each ν.

In fact, the admissibility condition is stronger than the open condition (for example, $\mathcal{K} - \{1\}$ is open but not admissible for an open compact subgroup $\mathcal{K}$ of G). The admissibility of Newton strata guarantees that the Newton strata works well with the "locally constant, compactly supported" condition of the Hecke algebras.

The first main result we establish in [8] is

Theorem 3.1. *We have the Newton decompositions*

$$H = \oplus_\nu H(\nu), \qquad \bar{H} = \oplus_\nu \bar{H}(\nu).$$

3.6

In order to study representation theory, we also need to consider the Hecke algebra at a given level.

Let $\mathcal{K}$ be an open compact subgroup of G and $H(G, \mathcal{K})$ be the Hecke algebra of compactly supported, $\mathcal{K}$-biinvariant functions on G. Then we have $H = \varinjlim_{\mathcal{K}} H(G, \mathcal{K})$.

However, given an open compact subgroup $\mathcal{K}$, the Newton stratum $G(\nu)$ is not stable under the left/right multiplication by $\mathcal{K}$. In other words, there is no Newton decomposition for the Hecke algebra at a given level:

$$H(G,\mathcal{K}) \neq \oplus_{\nu} H(G,\mathcal{K};\nu).$$

Here $H(G,\mathcal{K};\nu) \subset H$ consisting of compactly supported, $\mathcal{K}$-biinvariant functions supported in $G(\nu)$.

But quite amazingly, the cocenter of $H(G,\mathcal{K})$ (for "good" K) does have Newton decomposition.

Theorem 3.2. *Let $\mathcal{I}_n$ be the n-th congruent subgroup of the Iwahori subgroup $\mathcal{I}$. Then*

$$\bar{H}(G,\mathcal{I}_n) = \oplus_{\nu} \bar{H}(G,\mathcal{I}_n;\nu).$$

As $\{\mathcal{I}_n\}_{n\in\mathbb{N}}$ form a fundamental system of neighborhoods of 1, we have $H = \varinjlim_n H(G,\mathcal{I}_n)$. Therefore for each ν, $\bar{H}(\nu) = \varinjlim_n \bar{H}(G,\mathcal{I}_n;\nu)$. The structure of $\bar{H}(G,\mathcal{I}_n;\nu)$ for each ν and n will help us to understand the structure of the Newton component $\bar{H}(\nu)$ of the cocenter $\bar{H}$ and the representation theory of G.

4 Iwahori-Matsumoto presentation

Theorem 3.2 is a major result in the cocenter program. In this section, we will explain some ideas behind the proof. This will also give some insight on the Newton components of the cocenter of Hecke algebras, as well as lead to some important applications.

4.1

The key ingredient in the proof is the Iwahori-Matsumoto presentation of cocenter.

We first give a quick review of history:

- Bruhat decomposition $G = \sqcup_{w\in\tilde{W}} \mathcal{I}w\mathcal{I}$, where $\tilde{W}$ is the Iwahori-Weyl group;
- The original Iwahori-Matsumoto presentation [12] is for the affine Hecke algebra $H(G,\mathcal{I})$: $H(G,\mathcal{I})$ has a basis $\{T_w\}$ for $w \in \tilde{W}$;
- For the cocenter of affine Hecke algebra, the I-M presentation is established in [10]: The cocenter $\bar{H}(G,I)$ has a basis $\{T_{\mathcal{O}}\}$, where $\mathcal{O}$ runs over conjugacy classes of $\tilde{W}$ and $T_{\mathcal{O}}$ is the image of T_w in the cocenter for any Minimal length representative $w \in \mathcal{O}$.

4.2

The previous study is on the Iwahori level. In order to study H (i.e. the Hecke algebra with arbitrarily deep level), we first establish the Iwahori-Matsumoto presentation for $H(G,\mathcal{I}_n)$ for $n \in \mathbb{N}$. Then we use the reduction method development in [10] and [7] to study the cocenter of $\bar{H}(G,\mathcal{I}_n)$. Finally we show that the reduction method on $\bar{H}(G,\mathcal{I}_n)$ is compatible with the reduction method we used to establish the Newton decomposition $G = \sqcup_{\nu} G(\nu)$.

Theorem 4.1. *Let $\tilde{W}_{\min}$ be the set of elements in $\tilde{W}$ that are of minimal length in their conjugacy class.*

(1) For any n,

$$\bar{H}(G,\mathcal{I}_n) = \sum_{w \in \tilde{W}_{\min}} \bar{H}(G,\mathcal{I}_n)_w,$$

where $\bar{H}(G,\mathcal{I}_n)_w$ is the image in the cocenter of $\mathcal{I}_n$-biinvariant functions supported in $\mathcal{I}w\mathcal{I}$.

(2) For any n and Newton point ν, we have

$$\bar{H}(G,\mathcal{I}_n;\nu) = \sum \bar{H}(G,\mathcal{I}_n)_w,$$

where w runs over the elements in $\tilde{W}_{\min}$ associated to the Newton point ν in the sense of [8].

As a consequence, we deduce Newton decomposition for the cocenter of $H(G,\mathcal{I}_n)$.

4.3

As we mentioned before, one major difficulty is that $\dim \bar{H} = \infty$. We need some finiteness results.

Conjecture 4.2 (Howe). *Let X be a compact subset of G and $J(X)$ be the set of invariant distributions supported in $G \cdot X$. Then for any open compact subgroup K of G,*

$$\dim J(X) \mid_{H(G,K)} < \infty.$$

It is conjectured by Howe in [11] in 1973, proved by Clozel in [4] for $\mathrm{char}(F) = 0$ and by Barbasch-Moy in [2].

Now we give a new proof of it, based on the Iwahori-Matsumoto presentation. The outline is as follows.

- Any open compact subgroup contains $\mathcal{I}_n$ for some n.
- Any compact subset of G is contained in a finite union of Newton strata $G(\nu)$.
- By definition, $J(G(\nu)) \mid_{H(G,\mathcal{I}_n)} = \bar{H}(G,\mathcal{I}_n;\nu)^*$.
- For any ν, there are only finitely many $w \in \tilde{W}_{\min}$ associated to it.
- For any w, $\dim \bar{H}(G,\mathcal{I}_n)_w \leqslant \dim H(G,\mathcal{I}_n)_w = \sharp(\mathcal{I}_n \backslash IwI/\mathcal{I}_n)$ is finite.
- The space $\bar{H}(G,\mathcal{I}_n;\nu) = \sum_{w \in \tilde{W}_{\min}, \nu_w = \nu} \bar{H}(G,\mathcal{I}_n)_w$ is a finite sum of finite dimensional spaces and is thus finite dimensional.

5 Induction and restriction functors

5.1

Now we discuss some relations between the cocenter and representations.

A large family of representations comes from inductions, more precisely, the parabolic inductions.

Let M be a (standard) proper Levi subgroup of G. We have $i_M : \mathfrak{R}_{\mathbf{k}}(M) \to \mathfrak{R}_{\mathbf{k}}(G)$, the induction functor. In the other direction, we have $r_M : \mathfrak{R}_{\mathbf{k}}(G) \to \mathfrak{R}_{\mathbf{k}}(M)$, the Jacquet functor.

We should have the induction and restriction functors on the cocenter side.

- The restriction map $\bar{r}_M : \bar{H}(G) \to \bar{H}(M)$ which is adjoint to i_M.
- The induction map $\bar{i}_M : \bar{H}(M) \to \bar{H}(G)$ which is adjoint to r_M.

The restriction map $\bar{r}_M$ can be written down explicitly via Van Dijk's formula. The induction map $\bar{i}_M$ is more complicated. For affine Hecke algebras, such map can be written down explicitly via Bernstein-Lusztig presentation (which leads to an algebra homomorphism $H(M, \mathcal{I} \cap M) \to H(G, \mathcal{I})$). However, there is no such presentation for the Hecke algebra at a deeper level or for the whole Hecke algebra H.

5.2

For any algebraically closed field R of characteristic not equal to p, one may use the adjunction formula

$$Tr_R^M(f, r_M(\pi)) = Tr_R^G(\bar{i}_M(f), \pi)$$

to characterize the induction map $\bar{i}_M$. This works well if $R = \mathbb{C}$ based on the spectral density theorem of Kazhdan [13] which asserts that the trace map $Tr_{\mathbb{C}} : \bar{H}_{\mathbb{C}} \to \mathfrak{R}_{\mathbb{C}}(G)^*$ is injective. However, if R is of positive characteristic, the trace map Tr_R may not be injective and thus the map $\bar{i}_{M,R}$ is not uniquely determined by the adjunction formula.

In [9], we give an explicit formula for $\bar{i}_M$ for the "dominant part" of $\bar{H}(M)$. The basic idea is as follows. Note that $\bar{H}(M)$ is spanned by the image of $\delta_{m\mathcal{K}_M}$, where $m \in M$, $\mathcal{K}_M \subset M$ is an open compact subgroup and δ is the characteristic function. We would like to enlarge the open subset $m\mathcal{K}_M$ by multiplying a suitable open compact subgroup $\mathcal{K}$ of G so that it becomes an open subset of G. The most difficult part is that this map is independent of the choice of the open compact subgroups $\mathcal{K}$ and that this map (on the level of Hecke algebra) induces a map on the cocenter. The main result in [9] is

Theorem 5.1. *Let ν be a Newton point of G and $M = M_\nu$ be the centralizer of ν, a standard Levi subgroup of G. Then the map*

$$\delta_{m\mathcal{K}_M} \mapsto \delta_{P_\nu}(m)^{-\frac{1}{2}} \frac{\mu_M(\mathcal{K}_M)}{\mu_G(\mathcal{K}_M\mathcal{K})} \delta_{m\mathcal{K}_M\mathcal{K}} + [H, H]$$

for sufficiently small open compact subgroup $\mathcal{K}$ of G gives a well-defined map $\bar{i}_\nu : \bar{H}(M; \nu) \to \bar{H}$ which is adjoint to the Jacquet functor r_M.

Moreover, the image of $\bar{i}_\nu$ equals $\bar{H}(G; \nu)$.

5.3

We would like to point out that our formula does not have the whole $\bar{H}(M)$ as its domain. The problem is that we use the Newton points of G, which are certain G-dominant rational coweight. The Newton decomposition for M uses M-dominant rational coweight. A G-dominant coweight is a priori M-dominant, but not vice versa.

Nevertheless, the explicit formula for $\bar{i}_\nu$ covers exactly the whole cocenter $\bar{H}$ as the image. This provides a very useful way to deduce the study of the cocenter and the representations of G into the rigid part, which we will discuss in the next subsection.

Another important feature of the induction map $\bar{i}_M$ is that it is compatible with the Newton decomposition. For any Newton point ν, the induction map $\bar{i}_{M_\nu}$ sends $\bar{H}(M_\nu;\nu)$ onto $\bar{H}(G;\nu)$. In fact, we expect that such map is an isomorphism. This is proved in [9] in the case $\mathrm{char}(F)=0$.

5.4

In [8], we introduced the rigid cocenter $\bar{H}^{rig}=\oplus\bar{H}(\nu)$, where ν runs over rational central coweights. For any standard Levi subgroup M, we define the $+$-rigid part $\bar{H}(M)^{+,rig}=\oplus\bar{H}(M;\nu)$, where ν runs over rational dominant coweights with $M=M_\nu$. We then have the well-defined map $\bar{i}^+_M=\oplus_v\bar{i}_v:\bar{H}(M)^{+,rig}\to\bar{H}$ and the decomposition of the cocenter $\bar{H}$ into $+$-rigid parts:

$$\bar{H}=\oplus_{M \text{ is a standard Levi subgroup}}\,\bar{i}^+_M(\bar{H}(M)^{+,rig}).$$

This asserts that the rigid cocenters of Levi subgroups form the "building blocks" of the whole cocenter $\bar{H}$. In [8], we give an explicit description of the rigid cocenter in terms of generators (the cocenters of the Hecke algebras of parahoric subgroups) and the relations (the so-called (P,Q,x)-graphs). In [9], we also show that that the decomposition into $+$-rigid part is compatible with the trace map.

6 Trace Paley-Wiener theorem

6.1

We have the trace map $Tr_{\mathbf{k}}:\bar{H}_{\mathbf{k}}\to\mathfrak{R}_{\mathbf{k}}(G)^*$. Unlike the finite group case, this map is not surjective. One may see it from the following example.

Let T be a one-dimensional torus. Then any irreducible representation is a character. A character of T is called *unramified* if the maximal open compact subgroup T_0 acts trivially. Note that $T/T_0\cong\mathbb{Z}$ but the unramified characters are parameterized by $\mathbf{k}^*$.

In fact, the set of irreducible representations of G has a natural algebraic variety structure and $Tr(f,-)$ for any $f\in H$ must be an algebraic function on this variety.

Bernstein, Deligne and Kazhdan in [3] introduced the notion of good forms. By definition, $f\in\mathfrak{R}_{\mathbf{k}}(G)^*$ is good if

1. For any standard Levi M, $\sigma \in \mathfrak{R}_{\mathbf{k}}(M)$, $\psi \mapsto f(i_M(\psi\sigma))$ is regular on unramified char ψ of M.
2. There exists an open compact subgroup $\mathcal{K}$ such that for any $\pi \in \mathfrak{R}_{\mathbf{k}}(G)$ with $\pi^{\mathcal{K}} = \{0\}$, we have $f(\pi) = 0$.

It is easy to see that the image of the trace map must be the good forms. It is an important problem if every good form comes from Hecke algebras. We have the following result

Theorem 6.1 (Trace Paley-Wiener Theorem). *Under a mild assumption on the characteristic of* $\mathbf{k}$, *the image of* $Tr_{\mathbf{k}} : \bar{H}_{\mathbf{k}} \to \mathfrak{R}_{\mathbf{k}}(G)^*$ *is the space of good forms.*

6.2

This result is first proved by Bernstein, Deligne and Kazhdan for complex representations. It is proved very recently by Henniart and Lemaire in [6] for complex ω-representations coming from the theory of twisted endoscopy. The basic strategy consists of two key ingredients:

- Use the A-operator of [3] to reduce the study to discrete series and elliptic cocenter of $\bar{H}$.
- Obtain some "finiteness result" for the discrete series.

In [3] and [6], the second part is obtained by studying the tempered modules. This is the reason that only complex representations are considered there.

In [5], we follow the first part of the strategy as in [3]. This is the reason that we need to assume that the order of the relative Weyl group is invertible in $\mathbf{k}$. But our approach to the second part is completely different.

We first show that the elliptic cocenter is contained in the rigid cocenter, using the compatibility of the Newton decomposition with the induction and restriction map. Then we show that modulo the action of the connected center of G, there are only finitely many discrete series of a given level by applying the Howe's conjecture to the rigid cocenter. Our approach is algebraic, and thus works over positive characteristic as well as over $\mathbb{C}$. We did not study the mod-l ω-representations but we believe under suitable modification, our method can be applied to that situation as well.

References

[1] N. Abe, G. Henniart, F. Herzig and M.-F. Vignéras, *A classification of irreducible admissible mod p representations of p-adic reductive groups*, J. Amer. Math. Soc. **30** (2017), no. 2, 495–559.

[2] D. Barbasch and A. Moy, *A new proof of the Howe conjecture*, J. Amer. Math. Soc. **13** (2000), no. 3, 639–650.

[3] J. Bernstein, P. Deligne, D. Kazhdan, *Trace Paley-Wiener theorem for reductive p-adic groups*, J. d'Analyse Math. **47** (1986), 180–192.

[4] L. Clozel, *Orbital integrals on p-adic groups: a proof of the Howe conjecture*, Ann. of Math. (2) **129** (1989), no. 2, 237–251.

[5] D. Ciubotaru and X. He, *Cocenters of p-adic groups, III: Elliptic cocenter and rigid cocenter*, arXiv:1703.00378.

[6] G. Henniart and B. Lemaire, *La transformée de Fourier pour les espaces tordus sur un groupe réductif p-adique*, Astérisque **386** (2017), ix+366 pp.

[7] X. He, *Geometric and homological properties of affine Deligne-Lusztig varieties*, Ann. of Math. **179** (2014), 367–404.

[8] X. He, *Cocenters of p-adic groups, I: Newton decomposition*, Forum Math. Pi **6** (2018), e2, 27 pp.

[9] X. He, *Cocenters of p-adic groups, II: induction map*, arXiv:1611.06825.

[10] X. He and S. Nie, *Minimal length elements of extended affine Weyl group*, Compos. Math. **150** (2014), 1903–1927.

[11] R. Howe, *Two conjectures about reductive p-adic groups*, Proc. of A.M.S. Symposia in Pure Math. XXVI (1973), 377–380.

[12] N. Iwahori and H. Matsumoto, *On some Bruhat decomposition and the structure of the Hecke rings of p-adic Chevalley groups*, Inst. Hautes Études Sci. Publ. Math. **25** (1965), 5–48.

[13] D. Kazhdan, *Cuspidal geometry of p-adic groups*, J. Analyse Math. **47** (1986), 1–36.

[14] R. Kottwitz, *Isocrystals with additional structure*, Compositio Math. **56** (1985), 201–220.

[15] R. Kottwitz, *Isocrystals with additional structure. II*, Compositio Math. **109** (1997), 255–339.

[16] M.-F. Vignéras, *Représentations l-modulaires d'un groupe réductifp-adique avec $l \neq p$*, Progress in Math. 137, Birkhäuser, 1996.

[17] M.-F. Vignéras, *Intégrales orbitales modulo l pour un groupe réductif p-adique*, Algebraic geometry: Hirzebruch 70 (Warsaw, 1998), Amer. Math. Soc., Providence, RI, 1999, 241, 327–338.

Department of Mathematics, University of Maryland
College Park, MD 20742, USA
and Institute for Advanced Study
Princeton, NJ 08540, USA
E-mail address: xuhuahe@math.umd.edu

Proceedings of the 7th International Congress of Chinese Mathematicians
ALM 43, pp. 267–302

Metric Schottky Problem and Satake Compactifications of Moduli Spaces

Lizhen Ji

Abstract

In this paper, we discuss several results on the interaction and similarity between moduli spaces and locally symmetric spaces, in particular the Schottky problem from the metric geometric perspective: to understand the metric properties of the Jacobian (period) map $J : \mathcal{M}_g \to \mathcal{A}_g$ and the Jacobian locus $J(\mathcal{M}_g)$ inside the moduli space $\mathcal{A}_g$ of principally polarized abelian varieties of dimension g, when $\mathcal{A}_g$ is viewed as a noncompact locally symmetric space of finite volume with respect to the canonical locally symmetric Riemannian metric, hence as a complete noncompact metric space. We discuss several results about the location, density, size, shape and distortion of $J(\mathcal{M}_g)$ in $\mathcal{A}_g$. The period map J can be used to construct the universal moduli space of Riemann surfaces. We will also discuss Satake compactifications of locally symmetric spaces and how to relate them to compactifications as moduli spaces of flat tori, and give moduli interpretations of some compactifications of symmetric spaces.

Contents

1 Introduction: moduli spaces versus locally symmetric spaces

The moduli spaces of Riemann surfaces have been studied extensively since Riemann's introduction of Riemann surfaces and the moduli space $\mathcal{M}_g$ of compact Riemann surfaces; and locally symmetric spaces have been studied even earlier since Gauss' work on binary quadratic forms and their reduction in terms of the action of the modular group $\mathrm{SL}(2,\mathbb{Z})$ on the upper half-plane, which was explained in his classical book *Disquisitiones Arithmeticae.*

In this first section, we will explain the history of moduli spaces of Riemann surfaces, and its similarity with arithmetic locally symmetric spaces.

1.1 The origin and history

For each nonnegative integer g, let Σ_g denote a compact Riemann surface of genus g, or equivalently, a smooth projective curve of genus g over the field $\mathbb{C}$ of complex numbers.

The notion of Riemann surfaces was introduced by Riemann in his thesis in 1851.[1] Six years later in a very influential paper titled *Theory of abelian functions*, [Rie, p. 111, Section I, §12], Riemann defined the *moduli space of Riemann surfaces* $\mathcal{M}_g$ as the set of biholomorphic equivalence classes of compact Riemann surfaces Σ_g,

$$\mathcal{M}_g = \{\Sigma_g\}/\sim .$$

He also counted the number of complex parameters needed to describe points in $\mathcal{M}_g$.

When $g = 0$, $\mathcal{M}_0$ consists of one point and corresponds to the complex projective line $\mathbb{C}P^1 = S^2$. This fact is now usually stated as a part of the uniformization theorem for Riemann surfaces and is not obvious. Actually, it was proved earlier by Clebsch and Schwartz (see [dSG, p. xxvi] for references) and can be interpreted as one of the first examples of rigidity results.

When $g = 1$, Riemann proved [Rie, p. 113] that "the number of moduli for a class is 1." Compact Riemann surfaces of genus 1 are usually called elliptic curves due to the fact that they arise from the inversion of elliptic integrals.

It is now well-known that by the uniformization theorem, every compact Riemann surface of genus 1 is biholomorphic to a quotient $\Lambda\backslash\mathbb{C}$, where Λ is a lattice of $\mathbb{C}$. This fact is not obvious either and was first proved by Clebsch in 1865 (see [dSG, p. xxvi]), though it is very reasonable given the work on the inversion of elliptic integrals by Abel and Jacobi, and also by Gauss. Once such a uniformization of compact Riemann surfaces of genus 1 is known, then it is easy

[1] On [Rie, p. 4], Riemann wrote, "The position of the point O is no longer considered as being in the plane A, but in a surface T spread out over the plane. We choose this wording since it is inoffensive to speak of one surface lying on another, to leave open the possibility that the position of O can extend more than once over a given part of the plane. However, in such a case we suppose that the portions of the surface lying upon one another do not connect along a line. Thus a folding of the surface, or the splitting of the surface into superimposed parts, does not occur."

to identify $\mathcal{M}_1$ with $\mathrm{SL}(2,\mathbb{Z})\backslash\mathbb{H}^2$, which had been used by Gauss, Lagrange to parametrize lattices in $\mathbb{R}^2 \cong \mathbb{C}$ or equivalently positive definite binary quadratic forms.

For $g \geq 2$, Riemann found that the number of *effective complex parameters* needed to describe an equivalence class of compact Riemann surfaces Σ_g of genus g as branching coverings of $\mathbb{C}P^1$ is equal to $3g-3$.[2] These parameters are called *moduli* by Riemann. The name moduli was motivated by the common use of *modulus* k in describing families of elliptical integrals by people such as Gauss, Legendre, Jacobi, and Abel. See [MM, Chap. 2], for example. Consequently, $\mathcal{M}_g$ is called the *moduli space*, or *Riemann's moduli space,* since an equivalence class of Riemann surfaces is the best parameter to describe the equivalence class and is called the *moduli* of the equivalence class.

The upper half plane $\mathbb{H}^2$ is a basic symmetric space of noncompact type since it is simply connected and has constant negative curvature. It can be written as the following homogeneous space

$$\mathbb{H}^2 = \mathrm{SL}(2,\mathbb{R})/\mathrm{SO}(2),$$

where $\mathrm{SL}(2,\mathbb{R})$ acts on $\mathbb{H}^2$ by Mobius transformations, and $\mathrm{SO}(2)$ is the stabilizer of the point i.

The group $\mathrm{SL}(2,\mathbb{Z})$ is an arithmetic subgroup of the Lie group $\mathrm{SL}(2,\mathbb{R})$, and hence $\mathrm{SL}(2,\mathbb{Z})\backslash\mathbb{H}^2$ is an *arithmetic locally symmetric space* and admits an important generalization: locally symmetric spaces $\Gamma\backslash X$ associated with arithmetic subgroups, where G is a noncompact semisimple Lie group with finitely many connected components, $K \subset G$ a maximal compact subgroup, $X = G/K$ a symmetric space of noncompact type, and $\Gamma \subset G$ an arithmetic subgroup. One important example is $G = \mathrm{SL}(n,\mathbb{R})$, $K = \mathrm{SO}(n)$, and $\Gamma = \mathrm{SL}(n,\mathbb{Z})$. Another important example is $G = \mathrm{Sp}(2g,\mathbb{R})$, $K = \mathrm{U}(g)$, $\Gamma = \mathrm{Sp}(2g,\mathbb{Z})$, which will give rise to the Siegel upper halfspace $\mathfrak{h}_g$ and the Siegel modular variety $\mathcal{A}_g = \mathrm{SL}(n,\mathbb{Z})\backslash\mathfrak{h}_g$ used later in this paper.

In summary, the moduli space $\mathcal{M}_1 = \mathrm{SL}(2,\mathbb{Z})\backslash\mathbb{H}^2$ of compact Riemann surfaces Σ_1 of genus 1 admits two natural and important generalizations:

1. The moduli space $\mathcal{M}_g$ of compact Riemann surfaces of genus $g \geq 1$,
2. Arithmetic locally symmetric spaces $\Gamma\backslash X$ of noncompact type.

1.2 Similarity between moduli spaces and locally symmetric spaces

The common root of the spaces $\mathcal{M}_g$ and $\Gamma\backslash X$ in $\mathrm{SL}(2,\mathbb{Z})\backslash\mathbb{H}^2$ suggests the possibility of similar properties enjoyed by these two classes of generalizations. To discuss more concretely such properties, we need analogues of $\mathbb{H}^2$ and $\mathrm{SL}(2,\mathbb{Z})$

[2]On [Rie, p. 111, Section I, §12], Riemann wrote, "The remaining $3p-3$ branch points in these systems of similarly branching μ-valued functions can therefore be assigned any given values; and thus a class of systems of similarly branching functions with connectivity $2p+1$, and the corresponding class of algebraic equations, depends on $3p-3$ continuous variables, which we shall call the moduli of the class."

for the moduli space $\mathcal{M}_g$, which are given by the Teichmüller space $\mathcal{T}_g$ and the mapping class group Mod_g of compact surfaces.

In fact, after the above work of Riemann in 1857, several people including Klein, Poincaré, Hurwitz, Fricke, Severi, Brouwer and Koebe tried to understand better the moduli space $\mathcal{M}_g$. For some of them, the motivation was to prove the uniformization theorem for Riemann surfaces coming from algebraic curves by the method of continuity.

As defined by Riemann, $\mathcal{M}_g$ is only a set of equivalence classes without any other structure,[3] though his count of complex parameters, or moduli, suggested that $\mathcal{M}_g$ is a complex space of dimension $3g-3$, which became known as *Riemann's problem of moduli.* It was only Teichmüller who, around 1940, raised explicitly in [Tei] the question of defining a topology, an *intrinsic* complex structure and algebraic variety structure on $\mathcal{M}_g$. In doing this, he introduced the notion of fine moduli spaces, which was formulated in the modern language of representability of moduli functors. See [J2] [J4] for more details.

Recall that a *marked compact Riemann surface* is a compact Riemann surface Σ_g together with a homotopy class of an orientation preserving diffeomorphism from Σ_g to a fixed compact oriented surface S_g of genus g. Then the set of equivalence classes of marked Riemann surfaces of genus g gives the Teichmüller space $\mathcal{T}_g$. Teichmüller used the dilatation of extremal quasi-conformal maps between marked Riemann surfaces to define the Teichmüller metric, and hence a topology, on $\mathcal{T}_g$ and showed that $\mathcal{T}_g$ is homeomorphic to $\mathbb{R}^{6g-6}$ for $g \geq 2$.

To undo the markings on Riemann surfaces, we need the action on $\mathcal{T}_g$ of the *mapping class group* defined by

$$\mathrm{Mod}_g = \mathrm{Diff}^+(S_g)/\mathrm{Diff}^0(S_g),$$

where $\mathrm{Diff}^+(S_g)$ is the group of orientation preserving diffeomorphisms of S_g and $\mathrm{Diff}^0(S_g)$ its identity component. Then Mod_g acts on $\mathcal{T}_g$ by changing markings of Riemann surfaces, and hence by definition, we have the identification

$$\mathcal{M}_g = \mathrm{Mod}_g \backslash \mathcal{T}_g.$$

It is known that Mod_g acts on $\mathcal{T}_g$ properly by homomorphisms and hence $\mathcal{M}_g$ has a natural Hausdorff topology. It is also known that $\mathcal{T}_g$ is a fine moduli space of marked compact Riemann surfaces of genus g and hence admits a *natural* complex manifold structure, and Mod_g acts on $\mathcal{T}_g$ by biholomorphic maps. Consequently, $\mathcal{M}_g$ is a complex orbifold. When $g = 1$, $\mathrm{Mod}_1 = \mathrm{SL}(2,\mathbb{Z})$, $\mathcal{T}_1 = \mathbb{H}^2$, and the action of Mod_1 on $\mathcal{T}_1$ is the standard action of $\mathrm{SL}(2,\mathbb{Z})$ on $\mathbb{H}^2$ by fractional linear transformations.

Since $\mathcal{T}_g$ is an analogue of symmetric spaces of noncompact type, and Mod_g is an analogue of arithmetic groups, this realization of $\mathcal{M}_g$ as a quotient of $\mathcal{T}_g$

[3]Given the work of Riemann on Riemannian geometry in his paper titled *The hypotheses on which geometry is based* [Rie, pp. 257-272], one may argue that he might have in mind a topology on $\mathcal{M}_g$ and even an orbifold structure on $\mathcal{M}_g$, since his choices of effective and independent parameters are locally determined up to finite many possibilities. The parameters give a manifold structure and the finite ambiguity gives the orbifold structure.

suggests a stronger analogy between $\mathcal{M}_g$ and arithmetic locally symmetric spaces $\Gamma\backslash X$.

The correspondence between these two types of spaces and groups can be summarized in the following table.

$\mathcal{T}_1 = \mathbb{H}^2$, Teichmüller space	$\mathbb{H}^2 = \mathcal{T}_1$, Poincaré upper half-plane
$\mathrm{Mod}_1 = \mathrm{SL}(2,\mathbb{Z})$, mapping class group	$\mathrm{SL}(2,\mathbb{Z}) = \mathrm{Mod}_1$, modular group
$\mathcal{M}_1 = \mathrm{SL}(2,\mathbb{Z})\backslash\mathbb{H}^2$, moduli space	$\mathrm{SL}(2,\mathbb{Z})\backslash\mathbb{H}^2 = \mathcal{M}_1$, modular curve
$\mathcal{T}_g$, Teichmüller space	$X = G/K$, symmetric spaces of noncompact
Mod_g, mapping class group	Γ, arithmetic subgroups
$\mathcal{M}_g$, moduli space of Riemann surfaces	$\Gamma\backslash X$, arithmetic locally symmetric spaces
$\Omega\mathcal{M}_g$, moduli space of abelian differentials on Riemann surfaces	$\Gamma\backslash G$, homogeneous spaces of semisimple Lie groups
Compactifications of Teichmüller space $\mathcal{T}_g$	Compactifications of symmetric space $X = G/K$
Compactifications of moduli space $\mathcal{M}_g$	Compactifications of locally symmetric space $\Gamma\backslash X$
Curve complexes of surfaces	Tits buildings of linear algebraic groups

In the seventh row, the space $\Omega\mathcal{M}_g$ is the moduli space of holomorphic 1-forms on compact Riemann surfaces,

$$\Omega\mathcal{M}_g = \{(\Sigma_g, \omega)\},$$

where ω is a nonzero holomorphic differential (i.e., an abelian differential) on a compact Riemann surface Σ_g. Hence $\Omega\mathcal{M}_g$ is also called the space of abelian differentials. Unlike $\mathcal{M}_g$, the Lie group $\mathrm{SL}(2,\mathbb{R})$ acts on $\Omega\mathcal{M}_g$, and hence it resembles the homogeneous space $\Gamma\backslash G$. (Note that from the point of view of representation theory, $\Gamma\backslash G$ is a more natural space than $\Gamma\backslash X$, which only admits correspondences but not actions of Lie groups.)

Objects in the same row share several common properties. In particular, there is a lot of interaction between moduli spaces, in particular $\mathcal{M}_g$, and locally symmetric spaces $\Gamma\backslash X$.

The traffic usually goes from locally symmetric spaces to moduli spaces since there is more group structures on locally symmetric spaces and hence more tools to study them. In particular, problems and solutions for moduli spaces are sometimes suggested by those of locally symmetric spaces and homogeneous spaces.

On the other hand, there are also direct interactions between them. The main part of this paper deals with metric properties of the Jacobian map $J : \mathcal{M}_g \to \mathcal{A}_g$, which is a direct bridge between these two classes of spaces.

We will also consider implication in the opposite direction: viewing locally symmetric spaces as moduli spaces and understanding their properties and compactifications using such moduli interpretations. Similarly, we will also discuss moduli interpretations of some compactifications of symmetric spaces. The table of contents gives a clear indication of topics discussed.

1.3 Tits buildings versus curve complexes

The analogy between two types of spaces and groups in the previous subsection is sometimes more than formal and can be unexpectedly fruitful. We describe one important example here. See [J3] for more examples and references.

As mentioned before, both the Teichmüller space $\mathcal{T}_g$ and the symmetric spaces X are contractible. It turns out that the action of Mod_g on $\mathcal{T}_g$ gives a model of universal spaces for proper actions of Mod_g. Similarly, X is a model of universal spaces for proper actions of an arithmetic subgroup Γ. Consequently, cohomological properties of Mod_g and Γ can be understood by using these actions. For example, Borel and Serre [BoS] proved that any arithmetic subgroup Γ of a semisimple Lie group G is a virtual generalized Poincaré duality group. When $\Gamma\backslash X$ is compact, Γ is a virtual Poincaré duality group, and this follows from the Poincaré duality of compact manifolds. When $\Gamma\backslash X$ is noncompact, to prove this duality property of Γ, they constructed a compactification $\overline{\Gamma\backslash X}^{BS}$ of $\Gamma\backslash X$ which is a compact real analytic manifold with corners with an explicit description of the homotopy type of the boundary. Then the Poincaré-Lefschetz duality implies the duality property of the group.

The Borel-Serre compactification $\overline{\Gamma\backslash X}^{BS}$ [BoS] was constructed in two steps:

1. Construct a partial compactification $\overline{X}^{BS}$ of the symmetric space X which is a real analytic manifold with corners, whose boundary $\partial\overline{X}^{BS}$ is homotopy equivalent to the rational Tits building of the linear algebraic group G and hence to a bouquet of infinitely many spheres.
2. Show that the action of Γ on X extends to a proper analytic action on $\overline{X}^{BS}$, and the quotient $\Gamma\backslash\overline{X}^{BS}$ is a real analytic manifold with corners and gives the desired compactification $\overline{\Gamma\backslash X}^{BS}$.

Motivated by this, Harvey constructed in [Harv] a Borel-Serre partial compactification $\overline{\mathcal{T}_g}^{BS}$ of $\mathcal{T}_g$ and hence a Borel-Serre compactification of the moduli space $\mathcal{M}_g$, which is different from the known Deligne-Mumford compactification of $\mathcal{M}_g$ obtained by adding stable Riemann surfaces at infinity [DM]. To describe the topology of the boundary $\partial\overline{\mathcal{T}_g}^{BS}$, Harvey introduced the notion of *curve complex* $\mathcal{C}(S_g)$ of a surface S_g, whose vertices are homotopy equivalence classes of essential simple closed curves in S_g, and whose simplexes are homotopy equivalence classes of *disjoint* essential simple closed curves in S_g. The curve complex $\mathcal{C}(S_g)$ plays the similar role in describing degeneration of compact Riemann surfaces or the structure at infinity of Teichmüller space $\mathcal{T}_g$ as the Tits buildings do in describing the structure at infinity of symmetric spaces X, and it is also homotopy equivalent to a bouquet of infinitely many spheres. Consequently, this property was similarly used to prove that Mod_g is a virtual generalized Poincaré duality group [Har], but not a virtual Poincaré duality group [IJ].

It turns out that the curve complex $\mathcal{C}(S_g)$ is of fundamental importance in many problems in low dimensional topology and geometric group theory. See [Min] and its references. The influence of $\mathcal{C}(S_g)$ is beyond the original expectation, or the mere formal analogy with the Tits buildings when it was first introduced in [Har]. See [J6] [J7] for more discussions and references.

2 Jacobian varieties and Jacobian map

In the previous sections, we discussed the analogy between moduli spaces and locally symmetric spaces. One particular locally symmetric space which is most directly related to the moduli space $\mathcal{M}_g$ is probably the Siegel modular variety $\mathcal{A}_g$, and their interaction is provided by the *period* (or *Jacobian*) *map* of Riemann surfaces. This was started in the work of Gauss, Abel and Jacobi on periods of elliptic integrals and elliptic functions.

A high dimensional generalization of the upper half plane $\mathbb{H}^2$ is the *Siegel upper half space* of degree $g \geq 1$,

$$\mathfrak{h}_g = \{X + iY \mid X, Y \text{ are } g \times g \text{ real symmetric matrices}, Y > 0\}.$$

When $g = 1$, $\mathfrak{h}_1 = \mathbb{H}^2$. It is known that $\mathfrak{h}_g$ is a Hermitian symmetric space of noncompact type. The symplectic group $\mathrm{Sp}(2g, \mathbb{R})$ acts holomorphically and transitively on $\mathfrak{h}_g$, and the stabilizer of the point $iI_g \in \mathfrak{h}_g$ can be identified with $\mathrm{U}(g)$. Consequently, $\mathfrak{h}_g$ can be written as the homogeneous space

$$\mathfrak{h}_g = \mathrm{Sp}(2g, \mathbb{R})/\mathrm{U}(g).$$

The *Siegel modular group* $\mathrm{Sp}(2g, \mathbb{Z})$ acts properly and holomorphically on $\mathfrak{h}_g$, and the quotient space is a complex orbifold and called the *Siegel modular variety*, and is usually denoted by

$$\mathcal{A}_g = \mathrm{Sp}(2g, \mathbb{Z})\backslash\mathfrak{h}_g.$$

Since $\mathrm{Sp}(2g, \mathbb{Z})$ is an arithmetic subgroup and the $\mathbb{Q}$-rank of $\mathrm{Sp}(2g)$ is equal to g and hence positive, $\mathcal{A}_g$ is a noncompact locally symmetric space of finite volume. (Or we can see the noncompactness of $\mathcal{A}_g$ from degeneration of abelian varieties once A_g is interpreted as the moduli space of principally polarized abelian varieties.)

2.1 Abelian varieties

For any lattice $\Lambda \subset \mathbb{C}^g$, the quotient $\Lambda\backslash\mathbb{C}^g$ is a compact complex torus. When $g \geq 2$, a compact complex torus $\Lambda\backslash\mathbb{C}^g$ may not be realized as a projective variety, i.e., may not be embedded into a complex projective space $\mathbb{C}P^n$. It is a general fact that a compact complex manifold M can be realized as a projective variety if and only if there is a Hodge class on M, i.e., a closed, positive (1,1)-form which represents an integral cohomology class of M. Such a Hodge class is called a *polarization* of M. For a compact complex torus $V = \Lambda\backslash\mathbb{C}^g$, polarizations can be described more explicitly. Specifically, a polarization on V amounts to a positive definite Hermitian form H on $\mathbb{C}^g$ such that its imaginary part $E = \mathrm{Im}\, H$ takes integral values on $\Lambda \times \Lambda$. (Note that the relation is similar to that between a Hermitian metric and the Kähler form, and the Hermitian form is needed to construct the theta function of the abelian variety [Mu2, p 267].) These two conditions on H and E correspond to the two conditions of a Hodge class ω: positivity and integrality. Since the form E is skew-symmetric and integral, its determinant $\det E = d^2$ for a positive integer d. When $d = 1$, the polarization H is called a *principal polarization.* In this case, there exists a suitable basis $\lambda_1, \cdots, \lambda_{2g}$ of the lattice Λ, called a symplectic basis, and a basis $e_1, \cdots, e_g$ of the complex vector space $\mathbb{C}^g$ such that the *period matrix* of $\lambda_1, \cdots, \lambda_{2g}$ with respect to $e_1, \cdots, e_g$ is of the form

$$(I_g, Z),$$

and the $g \times g$-matrix Z satisfies the condition

$$Z \in \mathfrak{h}_g,$$

(see [GH, pp. 303-305]). Let $x_1, \cdots, x_{2g}$ be the real coordinates dual to the basis $\lambda_1, \cdots, \lambda_{2g}$, and $z_1, \cdots, z_g$ the complex coordinates dual to the basis $e_1, \cdots, e_g$. Then the Hodge class ω can be expressed in these two coordinates as

$$\omega = \sum_{\alpha=1}^{g} dx_\alpha \wedge dx_{g+\alpha} = \frac{\sqrt{-1}}{2} \sum_{\alpha,\beta=1}^{g} (\mathrm{Im}\ Z)^{-1}_{\alpha\beta} dz_\alpha \wedge \bar{z}_\beta,$$

where the first expression shows the integrality condition, and the second expression gives the positivity of ω. We also note that the Hermitian form is given by

$$H(\mathbf{z}, \mathbf{w}) = \mathbf{z}^t (\mathrm{Im}\ Z)^{-1} \bar{\mathbf{w}}.$$

For every $Z \in \mathfrak{h}_g$, $\mathbb{Z}^n + Z\mathbb{Z}^n$ is a lattice of $\mathbb{C}^n$, and the compact torus $\mathbb{Z}^n + Z\mathbb{Z}^n\backslash\mathbb{C}^n$ has a natural principal polarization, since with respect to obvious choices of bases of the lattice and the complex space, the period (I_g, Z) satisfies the desired property. Conversely, the above discussion shows that every abelian variety with a principal polarization is of this form. Since the choice of such a symplectic basis $\lambda_1, \cdots, \lambda_{2g}$ of a lattice Λ is well-defined up to action of $\mathrm{Sp}(2g, \mathbb{Z})$, the quotient space $\mathcal{A}_g = \mathrm{Sp}(2g, \mathbb{Z})\backslash\mathfrak{h}_g$ is canonically identified with the *moduli space of principally polarized abelian varieties* of dimension g.

2.2 Jacobian varieties

The bridge between the two spaces $\mathcal{M}_g$ and $\mathcal{A}_g$ is the period (or Jacobian) map $J : \mathcal{M}_g \to \mathcal{A}_g$, which can be defined in several equivalent ways. We will use complex analytic ones.

The most direct and classical one is in terms of periods of Riemann surfaces. Pick any symplectic basis $a_1, b_1, \cdots, a_g, b_g$ of $H_1(\Sigma_g, \mathbb{Z}) \cong \mathbb{Z}^{2g}$:

$$a_i \cdot a_j = 0, \quad b_i \cdot b_j = 0, \quad a_i \cdot b_j = \delta_{ij}, \quad i, j = 1, \cdots, g.$$

Let $\omega_1, \cdots, \omega_g$ be a basis of the space $H^0(\Sigma_g, \Omega^1)$ of holomorphic 1-forms on Σ_g normalized by the condition

$$\int_{a_i} \omega_j = \delta_{ij}, \quad i, j = 1, \cdots, g.$$

Then

$$\pi_{ij} = \int_{b_j} \omega_i, \quad i, j = 1, \cdots, g,$$

are called the *periods*, and they form the *period matrix* of the Riemann surface Σ_g with respect to the chosen basis,

$$\Pi(\Sigma_g) = (\pi_{ij}).$$

The Riemann bilinear relations imply that

$$\Pi(\Sigma_g) \in \mathfrak{h}_g.$$

Though the period $\Pi(\Sigma_g)$ depends on the choice of symplectic basis $a_1, b_1, \cdots, a_g, b_g$ of $H_1(\Sigma_g, \mathbb{Z})$, its image in $\mathrm{Sp}(2g, \mathbb{Z})\backslash\mathfrak{h}_g = \mathcal{A}_g$ does not and hence only depends on the Riemann surface Σ_g. This defines the *period map*

$$\Pi : \mathcal{M}_g \to \mathcal{A}_g.$$

As mentioned above, $\mathcal{A}_g$ can be identified with the moduli space of principally polarized abelian varieties of dimension g. Another basis-free way to define the period map without explicitly using the period is as follows. The Hodge decomposition shows that integration of holomorphic 1-forms on 1-cycles gives an embedding $H_1(\Sigma_g, \mathbb{Z})$ into the dual space $H^0(\Sigma_g, \Omega^1)^*$, and the image is a lattice. Then the quotient $H_1(\Sigma_g, \mathbb{Z})\backslash H^0(\Sigma_g, \Omega^1)^*$ is a compact complex torus and is called the *Jacobian variety* of Σ_g and denoted by $J(\Sigma_g)$:

$$J(\Sigma_g) = H_1(\Sigma_g, \mathbb{Z})\backslash H^0(\Sigma_g, \Omega^1)^*.$$

The theta divisor Θ of $H_1(\Sigma_g, \mathbb{Z})\backslash H^0(\Sigma_g, \Omega^1)^*$ (or its associated line bundle) gives a principal polarization of $J(\Sigma_g)$. In constructing the theta divisor, the period matrix Z and the fact that $Z \in \mathfrak{h}_g$ are needed. We note that for the Jacobian variety $J(\Sigma_g)$, the cohomology class of the principal polarization (or the skew-symmetric form) corresponds to the intersection

$$H_1(\Sigma_g, \mathbb{Z}) \times H_1(\Sigma_g, \mathbb{Z}) \to H_0(\Sigma_g, \mathbb{Z}) \cong \mathbb{Z}.$$

(We note that this intersection form is unimodular and hence gives another argument that the polarization is principal.)

In view of the above discussion, the period map $\Pi : \mathcal{M}_g \to \mathcal{A}_g$ is also called the *Jacobian map* and denoted by

$$J : \mathcal{M}_g \to \mathcal{A}_g.$$

Fix a basepoint z_0 in Σ_g, for any $\omega \in \mathbb{H}^0(\Sigma_g, \Omega^1)$, the integratal $\int_{z_0}^{z} \omega$ defines the *Abel-Jacobi map*,

$$\Sigma_g \to H_1(\Sigma_g, \mathbb{Z})\backslash H^0(\Sigma_g, \Omega^1)^* = J(\Sigma_g),$$

which extends to the space of divisors on Σ_g, in particular, to the symmetric product $\Sigma_g^{(g-1)}$ of Σ_g. The image of $\Sigma_g^{(g-1)}$ in $J(\Sigma_g)$ is a divisor and is equal to a translate of the theta divisor Θ, according to a theorem of Riemann [GH, p. 338].

The Abel-Jacobi map $\Sigma_g \to J(\Sigma_g)$ is classical and has played an important role in the development of complex analysis, complex geometry and algebraic geometry. The Jacobian map $J : \mathcal{M}_g \to A_g$ has also played an important role in several subjects, in particular, in the construction and study of the moduli space $\mathcal{M}_g$. For example, it can be used to construct $\mathcal{M}_g$ as a quasi-projective algebraic variety. (See [Mu1, Chap 6]). It has also been used to construct cohomology classes of $\mathcal{M}_g$ by pulling back cohomology classes of $\mathcal{A}_g$.

3 The classical Schottky problem

Right after the work of Riemann in 1857 on abelian functions and periods of Riemann surfaces, there have been many results and questions about the Jacobian map $J : \mathcal{M}_g \to \mathcal{A}_g$, in particular about its image $J(\mathcal{M}_g)$, called the *Jacobian locus*.

The Torelli theorem says that J is injective. The map J can be lifted to a holomorphic map between two complex manifolds which cover $\mathcal{M}_g$ and $\mathcal{A}_g$ respectively:

$$J : \mathcal{T}_g \to \mathfrak{h}_g.$$

In fact, the complex structure on $\mathcal{T}_g$ and hence on $\mathcal{M}_g$ was defined by Rauch, Ahlfors and Bers so that J is a holomorphic map. It turns out that J fails to be an immersion on the locus of hyperelliptic Riemann surfaces in $\mathcal{T}_g$. Though J is not an embedding, there exists a unique complex structure on $\mathcal{T}_g$ such that J is a holomorphic map. On the other hand, this does not automatically imply that this complex structure defined by Rauch, Ahlfors and Bers is the intrinsic complex structure on $\mathcal{T}_g$ as the fine moduli space of marked Riemann surfaces. See [J3] and [J4] for references and some discussion on this important yet subtle point.

When $g = 1$, $\mathcal{M}_1 = \mathcal{A}_1 = \mathrm{SL}(2, \mathbb{Z})\backslash\mathbb{H}^2$, and for every compact Riemann surface Σ_1, its Jacobian $J(\Sigma_1)$ is biholomorphic to Σ_1. Therefore, J is the identity map.

When $g \geq 2$, $\dim_{\mathbb{C}} \mathcal{M}_g = 3g - 3$ and $\dim_{\mathbb{C}} \mathcal{A}_g = \frac{g(g+1)}{2}$. Consequently, for $g \geq 4$, $\dim \mathcal{M}_g < \dim \mathcal{A}_g$. Since abelian varieties can degenerate, or the lattices Λ in complex tori $\Lambda\backslash\mathbb{C}^g$ can degenerate, $\mathcal{A}_g$ is noncompact. In [Sa], Satake constructed a topological compactification $\overline{\mathcal{A}_g}^S$ of $\mathcal{A}_g$ by adding at infinity the smaller dimensional spaces: $\mathcal{A}_{g-1}, \cdots, \mathcal{A}_1$ and a point $\{\infty\}$, and put a complex analytic structure on the compactification $\overline{\mathcal{A}_g}^S$ as a ringed space. He also stated that $\overline{\mathcal{A}_g}^S$ is a compact complex analytic space, i.e., locally isomorphic to the locus in some $\mathbb{C}^n$ defined by the vanishing of finitely many holomorphic functions. Baily [Ba1] proved this and a stronger result that $\overline{\mathcal{A}_g}^S$ is a normal projective variety, and consequently $\mathcal{A}_g$ is a quasi-projective variety. In [Ba2], Baily proved that the image $J(\mathcal{M}_g)$ is an algebraic subvariety of $\mathcal{A}_g$. Once $\mathcal{M}_g$ was constructed as a quasi-projective variety in [Mu1], the Jacobian map J becomes a morphism between algebraic varieties.

In summary, for $g = 2, 3$, $J(\mathcal{M}_g)$ is a Zariski dense algebraic subvariety of $\mathcal{A}_g$ (i.e., the complement is a proper subvariety of positive codimension), and for $g \geq 4$, $J(\mathcal{M}_g)$ is proper algebraic subvariety of $\mathcal{A}_g$ of positive codimension.

The above discussion suggests the problem to determine the subvariety $J(\mathcal{M}_g)$ in $\mathcal{A}_g$, for example, to find defining equations for it. This is the classical *Schottky problem* raised by Schottky [Sc1] [Sc2] in 1888 (see also [FR]). Briefly, since the number of effective complex parameters of $\mathcal{M}_g$ is equal to $3g - 3$ as Riemann's count shows, the $\frac{g(g+1)}{2}$ entries of the period $\Pi(\Sigma_g)$ of a compact Riemann surface Σ_g should satisfy $\frac{(g-2)(g-3)}{2} = \frac{g(g+1)}{2} - (3g-3)$ relations between them, and Schottky exhibited one such relation when $g = 4$. (This is all needed since $\frac{(g-2)(g-3)}{2} = 1$ for $g = 4$.) This problem was raised implicitly earlier by Riemann. In [Rie, p. 93], Riemann talked about special theta functions coming from Jacobian varieties. See also [Lew, Theorem 9]. The above problem is a concrete way to characterize the Jacobian locus $J(\mathcal{M}_g)$ as an algebraic subvariety of $\mathcal{A}_g$. Another version of the Schottky problem is to *characterize Jacobian varieties $J(\Sigma_g)$ among principally polarized abelian varieties in $\mathcal{A}_g$*.

The latter version of Schottky problem was solved by Shiota [Sh] by proving a conjecture of Novikov, which states that a principally polarized abelian variety $V \in \mathcal{A}_g$ is isomorphic to a Jacobian variety $J(\Sigma_g)$ if and only if the following two conditions are satisfied:

1. The theta divisor Θ of V is irreducible.
2. The theta function of V satisfies the KP-equation, a nonlinear partial differential equation.

There have been a lot of work on the Schottky problems using different approaches, for example, in terms of the singularities of the theta divisor Θ, and many surveys have been written. An introductory yet comprehensive one is by Mumford [Mu2]. Other surveys include [Do], [Gr] and [KS]. These approaches in algebraic geometry are not effective in the sense that if a principally polarized abelian variety V is given explicitly, one cannot really use any of these characterizations to decide if it is a Jacobian variety. There arises a need to view the problem from a different perspective.

4 The metric geometric Schottky problem

In the previous section, we explained that the moduli spaces $\mathcal{M}_g$ and $\mathcal{A}_g$, and the map $J : \mathcal{M}_g \to \mathcal{A}_g$ are basic in algebraic geometry, and hence the Schottky problem on characterizing the image $J(\mathcal{M}_g)$ as a subvariety is a natural and important problem.

In this section, we describe some problems and results on understanding the image $J(\mathcal{M}_g)$ as a subset of $\mathcal{A}_g$ from a different perspective – the metric geometric perspective. There are several reasons for doing this:

1. Besides being an important algebraic variety, $\mathcal{A}_g$ is also a noncompact locally symmetric space of finite volume and hence carries a complete Riemannian metric (or distance). Locally symmetric spaces are special Riemannian manifolds, and $\mathcal{A}_g$ is even special among locally symmetric spaces. Hence it is natural and important to understand the subspace $J(\mathcal{M}_g)$ as a metric subspace.
2. To use such a metric geometric perspective to find explicit points in $\mathcal{A}_g$ which do not lie in $J(\mathcal{M}_g)$, i.e., explicit principally polarized abelian varieties which are not Jacobian varieties.
3. As we shall explain, such a geometric metric perspective is closely related to some recent conjectures and results on the Jacobian map $J : \mathcal{M}_g \to \mathcal{A}_g$ and the image $J(\mathcal{M}_g)$ from the theory of arithmetic geometry.

To explain the metric geometric Schottky problem more precisely, we need to start with metrics on $\mathcal{A}_g$. We note that the Siegel upper half space $\mathfrak{h}_g$ is a Hermitian symmetric space and hence admits canonical symmetric Riemannian metrics. Since $\mathfrak{h}_g$ is an irreducible symmetric space, such symmetric metrics are unique up to scaling. When $g = 1$, $\mathfrak{h}_g = \mathbb{H}^2$, and the symmetric metric on $\mathfrak{h}_g$ can be scaled so that it is equal to the hyperbolic metric on $\mathbb{H}^2$. It is known that $\mathfrak{h}_1 = \mathbb{H}^2$ can be naturally embedded into $\mathfrak{h}_g$ as a totally geodesic submanifold through the base-point iI_g. We normalize the symmetric metric on $\mathfrak{h}_g$ such that its restriction to this copy of $\mathfrak{h}_1$ is equal to the hyperbolic metric on $\mathbb{H}^2$.

Let d be the distance on the Siegel upper half-space $\mathfrak{h}_g$ induced from the symmetric Riemannian metric. The induced distance on the quotient space $\mathcal{A}_g = \mathrm{Sp}(2g,\mathbb{Z})\backslash\mathfrak{h}_g$ is also denoted by d. Then the metric space $(\mathcal{A}_g, d)$ is a noncompact complete metric space, and it also has finite volume as a Riemannian orbifold.

Definition 4.1 *The metric geometric Schottky problem* aims to understand the Jacobian locus $J(\mathcal{M}_g)$ as a metric subspace of $(\mathcal{A}_g, d)$. More specifically, we can ask the following questions:

1. The location of $J(\mathcal{M}_g)$ inside $(\mathcal{A}_g, d)$ and how it sits in $\mathcal{A}_g$. As a noncompact locally symmetric space, $\mathcal{A}_g$ has a thick-thin decomposition, and the closely related decomposition into a compact core and boundary neighborhood of the infinite. We can understand the subset $J(\mathcal{M}_g)$ with respect to these decompositions.
2. The density (or denseness) of $J(\mathcal{M}_g)$ inside $(\mathcal{A}_g, d)$ in the metric geometric sense, but not in the sense of measure, since for $g \geq 4$, $J(\mathcal{M}_g)$ is a subvariety

of positive codimension and hence of measure 0. For example, how dense is it in the sense of large scale geometry (or coarse geometry) introduced and emphasized by Gromov? How large a locally sub-symmetric space inside $\mathcal{A}_g$ does $J(\mathcal{M}_g)$ span?
3. The metric distortion of $J(\mathcal{M}_g)$ inside $(\mathcal{A}_g, d)$ as a metric subspace. Though there is a lot of similarity between $\mathcal{M}_g$ and $\mathcal{A}_g$, it is known that when $g \geq 2$, $\mathcal{M}_g$ is not a locally symmetric space (for example, $\mathcal{M}_g$ is not biholomorphic to any Hermitian locally symmetric space, or isometric to any locally symmetric space when $\mathcal{M}_g$ is endowed with any metric.) Hence the image $J(\mathcal{M}_g)$ is not a totally geodesic subspace. The question is how distorted it is with respect to the metric d of the ambient space $\mathcal{A}_g$.

Problem 1 is related to a result of Buser and Sarnak [BoS] to be discussed below. The large scale geometric version of Schottky problem in Problem 2 was raised by Farb [Fa, Prob. 4.11] and will be discussed later as well. Problem 3 was also raised by Farb [Fa, Prob. 4.12] and will also be discussed.

Problems 2 and 3 are also related to the Coleman-Oort conjecture on the location of $J(\mathcal{M}_g)$ with respect to special subvarieties (defined below), and the André-Oort conjecture on relations between special subvarieties of $\mathcal{A}_g$ and CM (complex multiplication) points. In turn, the Coleman-Oort conjecture was motivated by the Manin-Mumford conjecture for subvarieties of abelian varieties and the Coleman conjecture for Jacobian varieties. We will explain briefly these conjectures.

The *Manin-Mumford conjecture* says that if an algebraic subvariety of an abelian variety contains a Zariski dense set of torsion points, then it is a translate of an abelian subvariety by a torsion point, since the latter obviously has a Zariski dense set of torsion points.

By definition, a *CM point* in $\mathcal{A}_g$ corresponds to a principally polarized abelian variety V whose endormorphism ring $\mathrm{End}(V)$ is large in the sense that $\mathrm{End}(V) \otimes \mathbb{Q}$ contains a commutative semisimple algebra of rank $2 \dim V = 2g$. This definition was motivated by the definition of elliptic curves with complex multiplications. In some sense, CM points in $\mathcal{A}_g$ correspond to the most symmetric abelian varieties.

The *Coleman conjecture* says that for g sufficiently large, there are only finitely many compact Riemann surfaces Σ_g in $\mathcal{M}_g$ such that their Jacobians $J(\Sigma_g)$ are CM points in $\mathcal{A}_g$, i.e., the image $J(\mathcal{M}_g)$ misses most CM points in $\mathcal{A}_g$. (We note that CM points are Zariski dense in $\mathcal{A}_g$).

The *André-Oort conjecture* states that any algebraic subvariety of $\mathcal{A}_g$ which contains a Zariski dense set of CM points is a special subvariety. By definition, a special subvariety is a Shimura variety of $\mathcal{A}_g$ which contains at least one CM point, in particular, it is a quotient of a Hermitian symmetric space X by a congruence subgroup. It is known that every special subvariety contains a dense set of CM points.

Remark 4.2 When a Shimura curve $\Gamma\backslash\mathbb{H}^2$ is a special variety contained in $J(\mathcal{M}_g)$ and the inverse image in $\mathcal{T}_g$ under the projection map $\mathcal{T}_g \to \mathcal{M}_g = J(\mathcal{M}_g)$ is a holomorphic geodesic with respect to the Teichmüller metric on $\mathcal{T}_g$, then $\Gamma\backslash\mathbb{H}^2$ is called a Shimura-Teichmüller curve in [Mo]. Special varieties of dimension 0 are the CM points in $\mathcal{A}_g$.

The previous three conjectures motivated the *Coleman-Oort conjecture* which says that when $g \gg 1$, there is no special subvariety of positive dimension of $\mathcal{A}_g$ which is contained in $J(\mathcal{M}_g)$.

There has been a lot of recent work on these conjectures in algebraic geometry and arithmetic geometry. See the recent survey [MoO] for more details on these conjectures and references on some results.

We note that the Coleman-Oort conjecture suggests that $J(\mathcal{M}_g)$ is different from locally symmetric subspaces of $\mathcal{A}_g$ and hence also supports Problem 3 in the metric geometric Schottky problem in (4.1) above.

In the rest of this section, we will define more precisely the above problems in the Schottky problem from the metric geometric perspective, and describe some results, for example, Theorem 4.9 and Theorem 4.21.

4.1 Problem 1: Location of the Jacobian locus $J(\mathcal{M}_g)$

Since the image $J(\mathcal{M}_g)$ is a proper subvariety of $\mathcal{M}_g$, a generic or randomly chosen point in $\mathcal{M}_g$ should not lie in $J(\mathcal{M}_g)$, i.e., should not be a Jacobian variety. On the other hand, whenever we choose a point, we do not know if it is generic. As pointed out before, the solutions to the Schottky problem by Shiota and others in algebraic geometry do not give effective methods either. One of the motivating problems in the paper [BoS] by Buser and Sarnak was to find an explicit principally polarized abelian variety which is not a Jacobian variety. For this purpose, they took a differential geometric point of view on the location of $J(\mathcal{M}_g)$.

Their starting point is that $\mathcal{A}_g$ is a metric space besides being an algebraic variety. Buser and Sarnak then defined a "distance" to the "boundary" $\partial \mathcal{A}_g$, and showed that $J(\mathcal{A}_g)$ is contained in an exponentially small "tubular" neighborhood of $\partial \mathcal{A}_g$ as $g \to +\infty$. By picking a point outside such a neighborhood, they found an explicit principally polarized abelian variety which is not a Jacobian variety.

As mentioned above, the complete symmetric Riemannian metric on $\mathfrak{h}_g$ induces a distance function d on the quotient space $\mathcal{A}_g = \mathrm{Sp}(2g, \mathbb{Z}) \backslash \mathfrak{h}_g$. Since the metric d on $\mathcal{A}_g$ is complete, the distance from any interior point of $\mathcal{A}_g$ to any "boundary point" in the "boundary" $\partial \mathcal{A}_g$ is equal to infinite.[4]

To explain the "distance" to the "boundary" $\partial \mathcal{A}_g$ in [BoS], we need to interpret $\mathcal{A}_g$ as a different moduli space and apply some basic results from the geometry of numbers.

Note that with respect to the standard Euclidean metric on $\mathbb{R}^n$, the lattice $\mathbb{Z}^n$ in $\mathbb{R}^n$ is unimodular, i.e., the quotient $\mathbb{Z}^n \backslash \mathbb{R}^n$ has volume 1. Any element $M \in \mathrm{SL}(n, \mathbb{R})$ maps $\mathbb{Z}^n$ to another unimodular lattice $M\mathbb{Z}^n$ in $\mathbb{R}^n$, and any unimodular lattice in $\mathbb{R}^n$ is of this form. Since $\mathrm{SL}(n, \mathbb{Z})$ is the stabilizer of the lattice $\mathbb{Z}^n$ in $\mathrm{SL}(n, \mathbb{R})$ under the above action, the space of unimodular lattices in $\mathbb{R}^n$ can be identified with the homogeneous space $\mathrm{SL}(n, \mathbb{Z}) \backslash \mathrm{SL}(n, \mathbb{R})$. If we identify

[4] The boundary $\partial \mathcal{A}_g$ is not precisely defined unless we have a compactification. Here it means an abstract boundary at infinity of $\mathcal{A}_g$.

lattices up to isometry, then the space of unimodular lattices in $\mathbb{R}^n$ can be identified with the locally symmetric space $\mathrm{SL}(n,\mathbb{Z})\backslash\mathrm{SL}(n,\mathbb{R})/\mathrm{SO}(n)$. Equivalently, $\mathrm{SL}(n,\mathbb{Z})\backslash\mathrm{SL}(n,\mathbb{R})/\mathrm{SO}(n)$ is also the *moduli space* $\mathcal{F}_n$ of oriented compact flat tori of volume 1 and dimension n.

For any $Z \in \mathfrak{h}_g$, the lattice $\mathbb{Z}^g + Z\mathbb{Z}^g$ in $\mathbb{C}^g = \mathbb{R}^{2g}$ is not unimodular in general. By multiplying by a suitable constant c, we obtain a unimodular lattice $c(\mathbb{Z}^g + Z\mathbb{Z}^g)$ and a compact flat torus $c(\mathbb{Z}^g + Z\mathbb{Z}^g)\backslash\mathbb{R}^{2g}$ of volume 1. Therefore, we have a map

$$\mathcal{A}_g \to \mathcal{F}_{2g} = \mathrm{SL}(2g,\mathbb{Z})\backslash\mathrm{SL}(2g,\mathbb{R})/\mathrm{SO}(2g). \tag{4.1}$$

For any compact flat torus $T = \Lambda\backslash\mathbb{R}^n$ of volume 1, i.e., Λ is a unimodular lattice, let $\ell(T)$ be the shortest length of closed geodesics of T, which is also equal to the minimum norm of nonzero vectors in Λ and is often called the *systole* of $\mathcal{T}$. We note that $\frac{1}{2}\ell(T)$ is equal to the injectivity radius of the flat torus T and it is essentially the density of lattice sphere packing based on Λ.

A basic result in the geometry of number is Mahler's compactness criterion which says that a sequence of flat tori T_i in $\mathcal{F}_n$, $i \geq 1$, diverges to the infinity if and only if $\lim_{i\to\infty} \ell(T_i) = 0$. See [Ca, Chap 5] for example. Therefore, $\ell(T)$ can be thought of as a "defining function" of, or a "distance" to the "boundary" of $\mathcal{F}_n$.[5] (The boundary here is not exactly the boundary of any specific compactification of $\mathcal{F}_n$, though $\mathcal{F}_n$ can be compactified in several different ways.) The function $\ell(T)$ induces a function on $\mathcal{A}_g$ through the map in Equation 4.1, and it can also be viewed as a "distance" function to its "boundary" $\partial\mathcal{A}_g$.

Mahler's compactness criterion implies that the following maximum values exist:

$$\gamma_{2g} = \max_{T\in\mathcal{F}_{2g}} \ell(T)^2,$$

$$\delta_{2g} = \max_{T\in\mathcal{A}_g} \ell(T)^2.$$

It follows from a generalization of Mahler's compactness criterion for compact hyperbolic surfaces by Mumford [Mu3] that the following maximum also exists:

$$\eta_{2g} = \max_{T\in J(\mathcal{M}_g)} \ell(T)^2.$$

By the theory of geometry of numbers [BoS, p. 29], we have the following bounds: for $g \gg 1$,

$$\frac{g}{\pi e} \leq \gamma_{2g} \leq \frac{4g}{\pi e}. \tag{4.2}$$

It was proved in [BoS, p. 29] that for $g \gg 1$,

$$\frac{g}{\pi e} \leq \delta_{2g}. \tag{4.3}$$

On the other hand, it was proved in [BoS, p. 29] that much smaller bounds hold for the Jacobians: for all $g \geq 1$,

$$c\log g \leq \eta_{2g} \leq \frac{3}{\pi}\log(4g+3), \tag{4.4}$$

[5]If we lift this function to the symmetric space $\mathfrak{h}_g$, it is related to Buseman functions on $\mathfrak{h}_g$ as a simply connected and nonpositively complete Riemannian manifold.

where c is a small positive constant.

Remark 4.3 To understand the bounds in Equation 4.2, it is helpful to compare it with the radius r of a ball $B^n(r)$ in $\mathbb{R}^n$ of volume 1 when $n \gg 1$. It is known that for any fixed r, when n is large, the volume of the ball $B^n(r)$ has the following asymptotics:

$$\mathrm{vol}(B^n(r)) \sim \frac{1}{\sqrt{n\pi}}(\frac{2\pi e}{n})^{\frac{n}{2}} r^n.$$

Consequently, if the volume of $B^n(r)$ is fixed to be V, then as $n \to +\infty$,

$$r \sim (\pi n)^{\frac{1}{2n}} \sqrt{\frac{n}{2\pi e}} V^{\frac{1}{n}}.$$

In particular, when $V = 1$, as $n \to +\infty$,

$$r^2 \sim (\pi n)^{\frac{1}{n}} \frac{n}{2\pi e} \sim \frac{n}{2\pi e}.$$

Now given a flat torus $V = \Lambda\backslash\mathbb{R}^n$, its injectivity radius is equal to $r = \frac{1}{2}\ell(V)$, and hence the ball $B^n(r)$ can be isometrically embedded into V. For a flat torus V_0 achieving the maximum value in Equation 4.2, the above asymptotic formula and Equation 4.2 show that $\mathrm{vol}(B^n(r))$ is comparable to $\mathrm{vol}(V_0)$, and hence $B^n(r)$ is comparable to, or occupies a definite proportion of the whole manifold V_0. Consequently, the torus V_0 is very round in a certain sense, for example, far away from the cube lattice $\mathbb{Z}^n\backslash\mathbb{R}^n$, whose shortest length of geodesics is equal to 1, much shorter than n for large values of n.

Using this upper bound, Buser and Sarnak defined a subset in $\mathcal{A}_g$,

$$N_g = \{V \in \mathcal{A}_g \mid \ell(V)^2 \leq \frac{3}{\pi}\log(4g+3)\},$$

which can be viewed as a "tubular neighborhood" of the boundary $\partial\mathcal{A}_g$, since $\ell(V)$ can be viewed as a defining equation of the "boundary" of $\mathcal{A}_g$. The upper bound in Equation 4.4 implies that

$$J(\mathcal{M}_g) \subset N_g.$$

They [BoS, p. 29, Equation (1.15)] showed that

Proposition 4.4 *With respect to the Riemannian metric induced from the symmetric space $\mathfrak{h}_g$, as $g \to +\infty$,*

$$\frac{\mathrm{vol}(N_g)}{\mathrm{vol}(\mathcal{A}_g)} = O(g^{-vg})$$

for every $v < 1$. Consequently, the Jacobian locus $J(\mathcal{M}_g)$ is contained in an exponentially small part N_g of $\mathcal{A}_g$ near its "boundary" $\partial\mathcal{A}_g$.

Using the upper bound in Equation 4.4, Buser and Sarnak [BoS, p. 29, p. 52] showed that

Corollary 4.5 *The Leech lattice Λ_{24} in $\mathbb{C}^{12} = \mathbb{R}^{24}$, or rather its associated compact complex torus $\Lambda_{24}\backslash\mathbb{C}^{12}$, does not belong to the subset N_g for $g = 12$, and hence it is not a Jacobian variety.*

Remark 4.6 Recall that the Leech lattice Λ_{24} is the unique unimodular lattice such that the square of the length of each vector in Λ_{24} is an even integer, and the length of its nonzero vectors is at least 2. (Note that the lengths of nonzero vectors may not be integers). In general, it is difficult to construct explicitly unimodular lattices with large minimal length, i.e., with high sphere packing density. Since the Leech lattice Λ_{24} has a large sphere packing density, this is one natural candidate to check.

4.2 Problem 2(a): Coarse density of the Jacobian locus $J(\mathcal{M}_g)$

Motivated by the above results in [BoS], Farb posed a version of the Schottky problem in the sense of large scale geometry, as developed by Gromov, in [Fa, Prob. 4.11]. Basically, he asked what $J(\mathcal{M}_g)$ looks like from far away inside the unbounded metric space $(\mathcal{A}_g, d)$. We will explain the precise meaning of this.

To describe the large scale shape of an unbounded metric space (M, d), Gromov introduced the notion of *tangent cone at infinity* $T_\infty(M, d)$ of (M, d), which is also called the *asymptotic cone* and denoted by $\mathrm{Cone}_\infty(M, d)$.

For any two bounded subsets A and B of a metric space (X, d), the Hausdorff distance between A and B is

$$d_X^H(A, B) = \inf\{\varepsilon \mid A \subset N_\varepsilon(B), B \subset N_\varepsilon(A)\},$$

where $N_\varepsilon(B)$ and $N_\varepsilon(A)$ are ε-neighborhoods of B and A respectively. For any two bounded metric spaces (X_1, d_1) and (X_2, d_2), the Gromov-Hausdorff distance between them is defined by

$$d^{GH}((X_1, d_1), (X_2, d_2)) = \inf_{(Z,d)} d_Z^H(X_1, X_2),$$

where (Z, d) ranges over all metric spaces such that (X_1, d_1) and (X_2, d_2) can be isometrically embedded into them. The Gromov-Hausdorff distance gives the notion of Gromov-Hausdorff convergence and of Gromov-Hausdorff limit of a sequence of bounded metric spaces.

Clearly, the Gromov-Hausdorff convergence cannot be applied directly to convergence of a sequence of unbounded metric spaces (M_n, d_n) to a metric space (M_∞, d_∞) as $n \to \infty$. For this purpose, we need to reduce them to bounded metric spaces. One way to do this is to introduce a base point x_n in each space M_n and consider metric balls around x_n. We call a metric space with a chosen base point a *pointed metric space.* Then we say that a sequence of *pointed metric spaces* (M_n, x_n, d_n) *converges in the sense of Gromov-Hausdorff* to a limit pointed metric space $(M_\infty, x_n, d_\infty)$ if for every (large) radius $r > 0$, the Gromov-Hausdorff distance between the ball $B_r(x_n)$ in M_n of radius r with center x_n and the ball $B_r(x_\infty)$ in M_∞ goes to 0 as $n \to +\infty$.

Given any metric space (M, d), pick a base point x_0. For any $\varepsilon > 0$, scale down the metric d to εd to get a family of pointed metric spaces $(M, x_0, \varepsilon d)$. Then the *tangent cone at infinity* $T_\infty(M, d)$ is defined to be the Gromov-Hausdorff limit of the family of pointed metric spaces $(M, x_0, \varepsilon d)$,

$$T_\infty(M, d) = \lim_{\varepsilon \to 0}(M, x_0, \varepsilon d).$$

In general, we need to use ultrafilters to ensure the existence of the Gromov-Hausdorff limit. On the other hand, if the limit exists, it is clearly independent of the choice of the base point x_0. It is also clear that $T_\infty(M, d)$ is a metric cone, and hence is also called the *asymptotic cone* of (M, d).

Remark 4.7 The name of the tangent cone at infinity comes from the fact that for a Riemannian manifold (M, ds^2), for any base point $x_0 \in M$, the tangent space $T_{x_0}M$ with the Riemannian norm can be obtained as the Gromov-Hausdorff limit of the family of pointed Riemannian manifolds $(M, x_0, \frac{1}{\varepsilon}ds^2)$ as $\varepsilon \to 0$.

The asymptotic cone of arithmetic locally symmetric spaces $\Gamma\backslash X$ was determined in [Hat] [JM] [Le]. It is a metric cone over a finite simplicial complex which is the quotient of the rational Tits building of the algebraic group $\mathbf{G}$ under the action of Γ. For $\mathcal{A}_g$ endowed with the locally symmetric metric d, the quotient of the rational Tits building of $\mathrm{Sp}(2g, \mathbb{C})$ by $\mathrm{Sp}(2g, \mathbb{Z})$ is one simplex, which is a spherical section of a positive Weyl chamber. Consequently, $\mathrm{Cone}_\infty(\mathcal{A}_g, d)$ is isometric to a Weyl chamber for the Lie group $\mathrm{Sp}(2g, \mathbb{R})$, which is a g-dimensional cone in $\mathbb{R}^g$.

In the above notation, Farb's problem [Fa, Prob. 4.11] can be stated precisely as follows:

Problem 4.8 (Coarse Schottky Problem) *Describe, as a subset of the asymptotic cone* $\mathrm{Cone}_\infty(\mathcal{A}_g)$, *the subset determined by the Jacobian locus* $J(\mathcal{M}_g)$ *in* $\mathcal{A}_g$.

In other words, the problem asks for a description of all points in $\mathrm{Cone}_\infty(\mathcal{A}_g)$ which are limits of points in $J(\mathcal{M}_g)$.

In a joint work with Leuzinger [JL, Theorem 1.2], we proved the following result.

Theorem 4.9 *The Jacobian locus* $J(\mathcal{M}_g)$ *is coarsely dense in* $\mathcal{A}_g$ *in the sense that there exists a constant* c_g *such that for every point* $p \in \mathcal{A}_g$, *there exists a point* $q \in J(\mathcal{M}_g)$ *such that the distance* $d(p, q) \leq c_g$. *Consequently, every point in* $\mathrm{Cone}_\infty(\mathcal{A}_g)$ *is the limit of a sequence of points in* $J(\mathcal{M}_g)$.

Clearly, there exists a *minimum constant* c_g satisfying the density condition in Theorem 4.9, which is denoted by $c_{g,\min}$. The result of [BoS] mentioned above immediately suggests the following conclusion.

Proposition 4.10 *The minimal constant* $c_{g,\min}$ *in the above theorem converges to* $+\infty$ *as* $g \to +\infty$.

Proposition 4.10 was stated in [JL] without any explanation. It is reasonable but not entirely obvious. One problem is that even though the volume of $\mathcal{A}_g$ is much bigger than the neighborhood N_g of $J(\mathcal{M}_g)$, $\mathcal{A}_g$ might be very "flat" and might be contained in a somewhat bigger tubular neighborhood of $J(\mathcal{M}_g)$. One possible and reasonable explanation is to show that when we increase the radius of the tubular neighborhood of $J(\mathcal{M}_g)$, the volume of the tubular neighborhood increases in a controlled way. Since this seems to be complicated, we give another explanation, or rather a proof.

By the reduction theory for $\mathrm{Sp}(2g,\mathbb{Z})$ acting on the Siegel upper half space $\mathfrak{h}_g$, a fundamental set (or a coarse fundamental domain) for $\mathrm{Sp}(2g,\mathbb{Z})$ is given by a Siegel set Ω. By Equation 4.3, there exist points x in the Siegel set Ω whose associated flat tori T of volume 1 have the minimal length $\ell(T) \geq \frac{g}{e\pi}$. By the relation between the points in $\mathfrak{h}_g$ and positive definite quadratic forms of determinant 1 in [BoS, p. 27], a point $Z = X + iY \in \mathfrak{h}_g$, where X, Y are real matrices, corresponds to the positive definite symmetric matrix

$$\begin{pmatrix} I_g & 0 \\ X & I_g \end{pmatrix} \begin{pmatrix} Y^{-1} & 0 \\ 0 & Y \end{pmatrix} \begin{pmatrix} I_g & X \\ 0 & I_g \end{pmatrix}.$$

It is also known that there is an 1-1 correspondence between the space of positive definite symmetric matrices of determinant 1 and the space of a flat tori of volume 1, where a flat torus given by $M\mathbb{Z}^n$, M being a $n \times n$-matrix, corresponds to the positive definite matrix $M^t\, M$. From this, we can see that flat tori corresponding to points in a ball $B_\beta(x)$ of radius β with center x in the Siegel upper half-space $\mathfrak{h}_g$ have the minimal length $\geq \ell(T) - \beta'$, where β' is a constant depending only on β. The upper bound in Equation 4.4 on the minimal lengths of Jacobian varieties implies that for any fixed β, the ball $B_\beta(x)$ does not contain any point of $J(\mathcal{M}_g)$ when $g \gg 1$. This implies that the constant $c_{g,\min} \to +\infty$ as $g \to +\infty$.

The above discussion naturally raises the following problem:

Problem 4.11 *Find the order of the minimal constant $c_{g,\min}$ going to ∞ as $g \to +\infty$.*

The gap between the bounds in Equations 4.3 and 4.4 and the relation between $\mathfrak{h}_g$ and the space of flat tori of volume 1 in [BoS, p. 27] might suggest that $c_{g,min}$ is of the order $\log g$, if we consider the relation between the Poincaré upper half plane $\mathbb{H}^2$ and the moduli space of flat tori of volume 1. On the other hand, an explanation through Siegel sets as fundamental sets suggests the following conjecture.

Conjecture 4.12 *As $g \to +\infty$, the minimal constant $c_{g,min}$ goes to ∞ like g^2.*

Since all minimal parabolic subgroups of $\mathrm{Sp}(2g,\mathbb{C})$ defined over $\mathbb{Q}$ are conjugate, the reduction theory for the arithmetic group $\mathrm{Sp}(2g,\mathbb{Z})$ implies that one Siegel set

$$\Omega = U \times A_{P,t} \subset \mathfrak{h}_g$$

associated with a minimal parabolic subgroup P of $\mathrm{Sp}(2g,\mathbb{C})$ gives a fundamental set for the action of $\mathrm{Sp}(2g,\mathbb{Z})$ on $\mathfrak{h}_g$. (This was used above in the proof of Proposition 4.10). Note that in describing the Siegel set, we have used the horospherical decomposition of $\mathfrak{h}_g$ with respect to the minimal parabolic subgroup

$$\mathfrak{h}_g = N_P \times A_P,$$

where N_P is the unipotent radical of P, and A_P is the split component of P. In the Siegel set Ω, $U \subset N_P$ is a bounded subset, and $A_{p,t}$ is a translate of the positive Weyl chamber.

Let P_∞ be the subgroup of $\mathrm{GL}(2g, \mathbb{C})$ consisting of upper triangular matrices, and N_∞ the subgroup of P_∞ consisting of those with 1s along the diagonal. Then P_∞ is a minimal subgroup of $\mathrm{GL}(2g, \mathbb{C})$ defined over $\mathbb{Q}$, and N_∞ is its unipotent radical. Then one minimal parabolic subgroup of $\mathrm{Sp}(2g, \mathbb{C})$ is the subgroup

$$P = \mathrm{Sp}(2g, \mathbb{C}) \cap P_\infty, \tag{4.5}$$

and its unipotent radical is

$$N_P = \mathrm{Sp}(2g, \mathbb{C}) \cap N_\infty. \tag{4.6}$$

In the above notation, the Siegel set Ω for $\mathrm{Sp}(2g, \mathbb{Z})$ can be described more explicitly. The subset U in the unipotent radical N_P in Equation 4.6 is given by

$$U = \{X \in N_P \mid \text{for every upper triangular entry } X_{ij}, |X_{ij}| \leq \frac{1}{2}\},$$

and $A_{P,t}$ is a shift of the positive Weyl chamber by a constant t close to 1 which is independent of g.

For points in $J(\mathcal{M}_g)$, its $A_{P,t}$-parts almost fill up the space $A_{P,t}$ up to a distance of the order $\log g$, but the diameter of U in the invariant metric d is of the order g^2, since the number of nonzero entries of $X \in U$ is of the order g^2. Therefore, for a tubular neighborhood of $J(\mathcal{M}_g)$ to cover the whole space $\mathcal{A}_g$ (or the Siegel set Ω), the radius $c_{g,\min}$ should be of the order g^2.

We also note that in the case of $g = 1$, the lattice with the largest sphere packing density occurs at a point in the Siegel set with the maximal N_P-component.

4.3 Problem 2(b): Locally symmetric spaces spanned by the Jacobian locus $J(\mathcal{M}_g)$

In the previous subsection, we studied the coarse density of the Jacobian locus $J(\mathcal{M}_g)$ in $\mathcal{A}_g$. There are other ways to measure the size and density of a metric subspace. Since we view $\mathcal{A}_g$ as a locally symmetric space, it is natural to measure the size of $J(\mathcal{M}_g)$ in terms of locally symmetric spaces which are contained in $\mathcal{A}_g$. Partially motivated by the Coleman-Oort conjecture, we can also ask the following questions:

Question 4.13 *Which locally symmetric spaces are contained in $J(\mathcal{M}_g)$? What is the highest dimension of a locally symmetric subspace contained in $J(\mathcal{M}_g)$?*

Question 4.14 *Is there a proper locally symmetric subspace of $\mathcal{A}_g$ which contains $J(\mathcal{M}_g)$?*

The Coleman-Oort conjecture deals with special subvarieties, and hence Hermitian locally symmetric spaces. On the other hand, in Question 4.13, we allow all locally symmetric spaces, including non-Hermitian locally symmetric spaces of $\mathcal{A}_g$. Question 4.14 can be viewed as the opposite of the first question. In both questions, we want to bound $J(\mathcal{M}_g)$ in terms of locally symmetric subspaces in $\mathcal{A}_g$.

In a symmetric space of noncompact type, any submanifold is totally geodesic if and only if it is a symmetric space. Therefore, locally symmetric spaces of $\mathcal{A}_g$ can be viewed as totally geodesic subspaces. Since totally geodesic subspaces have no distortion as metric subspaces, the two questions above also fit into Problem 3 on metric distortion of the Jacobian locus $J(\mathcal{M}_g)$ in $\mathcal{A}_g$.

Before explaining Questions 4.13 and 4.14 more precisely, we want to make some comments on the internal symmetric space structure of the spaces $\mathfrak{h}_g$ and $\mathcal{A}_g$.

In any symmetric space of noncompact type X, any totally geodesic (i.e., convex) complete subspace $Y \subset X$ is also a symmetric space which is of nonpositive curvature, which could be a product of a symmetric space of noncompact type with a flat subspace. For example, a complete geodesic X is an isometric copy of $\mathbb{R}^1$ embedded in it. If the rank of X is equal to r, $\mathbb{R}^n$ can be isometrically embedded in X. For $X = \mathfrak{h}_g$, $g \geq 2$, $\mathfrak{h}_1 \times \mathfrak{h}_1$ can be isometrically embedded into $\mathfrak{h}_g$, and consequently $\mathfrak{h}_1 \times \mathbb{R}^1$ can also be isometrically embedded into $\mathfrak{h}_g$, where $\mathbb{R}^1$ is a geodesic in $\mathfrak{h}_1 = \mathbb{H}^2$.

If X is a symmetric subspace of $\mathfrak{h}_g$, its image in $\mathcal{A}_g$ is a quotient of X but may not be an embedded locally symmetric space (or a finite quotient of locally symmetric spaces). For example, non-closed geodesics in $\mathcal{A}_g$ can accumulate at some points. If there exists a discrete subgroup Γ such that the map $X \to \mathcal{A}_g$ factor through a map $\Gamma\backslash X \to \mathcal{A}_g$ which is finite-to-one, then the image of X in $\mathcal{A}_g$ can be viewed as an embedded locally symmetric space. In this case, the convexity of X in $\mathfrak{h}_g$ suggests some convexity of its image in $\mathcal{A}_g$ too.

Let $X \subset \mathfrak{h}_g$ be a symmetric subspace contained in $\mathfrak{h}_g$. Then the image of X under the projection $\pi : \mathfrak{h}_g \to \mathrm{Sp}(2g, \mathbb{Z})\backslash\mathfrak{h}_g$ is a locally symmetric space in $\mathcal{A}_g$ when X satisfies certain rationality condition.

We are interested in the following situation. Let $\mathbf{G} \subset \mathrm{Sp}(2g, \mathbb{C})$ be a linear semisimple algebraic subgroup defined over $\mathbb{Q}$. Let $G = \mathbf{G}(\mathbb{R})$ the associated semisimple Lie group. Let $B \in \mathrm{Sp}(2g, \mathbb{Q})$, and $x_0 = B(iI_g) \in \mathfrak{h}_g$. Then the orbit of G in $\mathfrak{h}_g$ through x_0 is a realization of the symmetric space $X = G/K$ associated with G,

$$X \hookrightarrow \mathfrak{h}_g,$$

where K is a suitable maximal compact subgroup of G. This induces a map $X \to \mathcal{A}_g$. Then there exists an arithmetic subgroup $\Gamma \subset \mathbf{G}(\mathbb{Q})$ such that the map $X \to \mathcal{A}_g$ factors through the projection $X \to \Gamma\backslash X$ and the map $\Gamma\backslash X \to \mathcal{A}_g$ is a finite-to-one map. For convenience, we denote the image of $\Gamma\backslash X$ in $\mathcal{A}_g$ also by $\Gamma\backslash X$. We note that this setup is also natural in view of the fact that $J(\mathcal{M}_g)$ is an algebraic subvariety defined over $\mathbb{Q}$ of $\mathcal{A}_g$.

Let $\Gamma\backslash X$ be a locally symmetric space. We call it *irreducible* if there is no finite cover of $\Gamma\backslash X$ which splits isometrically as a nontrivial product, which is equivalent to the condition that the linear algebraic group **G** which defines the arithmetic subgroup Γ is not a product of two semisimple algebraic groups defined over $\mathbb{Q}$ up to finite index.

Using the results in [JL], we can prove the following.

Proposition 4.15 *There is no proper irreducible locally symmetric space $\Gamma\backslash X$ of $\mathcal{A}_g$ which contains $J(\mathcal{M}_g)$.*

Proof. Since the locally symmetric space $\Gamma\backslash X$ is irreducible and $\mathrm{Sp}(2g, \mathbb{C})$ is split over $\mathbb{Q}$, the $\mathbb{Q}$-rank of $\Gamma\backslash X$, or the linear semisimple algebraic group **G**, is strictly less than the $\mathbb{Q}$-rank of $\mathrm{Sp}(2g, \mathbb{C})$ (or of $\mathcal{A}_g$), which is equal to g. Since the map $\Gamma\backslash X \to \mathcal{A}_g$ is a uniform finite-to-one map, the image of $\Gamma\backslash X$ is within a bounded distance of a finite union of Weyl chambers of the rational root systems of **G**. Since these chambers are of strictly smaller dimension than g, while g is the dimension of the asymptotic cone $\mathrm{Cone}_\infty(\mathcal{A}_g)$, the image of $\Gamma\backslash X$ in $\mathrm{Cone}_\infty(\mathcal{A}_g)$ is a proper subset of positive codimension which is a finite union of polyhedral cones. This contradicts Theorem 4.9 (i.e., [JL, Theorem 1.2]), which says that the image of $J(\mathcal{M}_g)$ in $\mathrm{Cone}_\infty(\mathcal{A}_g)$ is equal to the whole space.

Remark 4.16 The above result shows that $J(\mathcal{M}_g)$ fills up $\mathcal{A}_g$ or is full in $\mathcal{A}_g$ in some sense. In [MoO, Remark 4.5], it was proved that when $g > 3$, $\mathcal{M}_g$ is not contained in any proper special subvariety of $\mathcal{A}_g$. It also mentioned that it is a known fact that $\mathcal{M}_g$ is not isomorphic to any special variety.

We give an explanation for the fact that $\mathcal{M}_g$ is not biholomorphic to any Hermitian locally symmetric space $\Gamma\backslash X$ when $g > 1$. If not, there exists a biholomorphic map $f : \mathcal{M}_g \to \Gamma\backslash X$. Note that $\mathcal{M}_g$ is a complex orbifold and $\mathcal{T}_g$ is the universal covering space of $\mathcal{M}_g$ as an orbifold. Similarly, $\Gamma\backslash X$ is also a complex orbifold and X is the universal covering space of the orbifold $\Gamma\backslash X$. Then the biholomorphic map f lifts to a biholomorphic map $\tilde{f} : \mathcal{T}_g \to X$. By Royden's theorem, the automorphism group of $\mathcal{T}_g$ is essentially given by the mapping class group Mod_g and is hence discrete. This implies that $\mathcal{T}_g$ is not biholomorphic to the Hermitian symmetric space X.

Remark 4.17 It is reasonable to believe that the assumption that $\Gamma\backslash X$ is irreducible in Proposition 4.15 is not needed. One reason is the result of [MoO, Remark 4.5] mentioned in the previous remark. Note that Jacobian varieties are irreducible, i.e., they are not products of principally polarized abelian varieties of lower dimension. It is not clear if this can be helpful.

The same argument proves a more general statement.

Proposition 4.18 *There is no proper irreducible locally symmetric space $\Gamma\backslash X$ of $\mathcal{A}_g$ such that $J(\mathcal{M}_g)$ is contained in a bounded neighborhood of $\Gamma\backslash X$, i.e., there does not exist any constant c such that for every $p \in J(\mathcal{M}_g)$, there exists $q \in \Gamma\backslash X$, $d(p, q) \leq c$.*

Note that for this stronger result, the condition of irreducibility of $\Gamma\backslash X$ is necessary. By the proof of [JL, Theorem 1.1], the closure of $J(\mathcal{M}_g)$ in $\mathcal{A}_g$ contains the reducible $\Gamma\backslash X_\infty = \mathrm{SL}(2,\mathbb{Z})\backslash\mathbb{H}^2 \times \cdots \times \mathrm{SL}(2,\mathbb{Z})$ (g of them), and a suitable tubular neighborhood of $\Gamma\backslash X_\infty$ contains $J(\mathcal{M}_g)$.

4.4 Problem 3: Metric distortion of the Jacobian locus $J(\mathcal{M}_g)$

As mentioned before, one version of the Schottky problem, motivated by the work of Schottky for the case of genus $g = 4$, is to identify algebraic equations which define $J(\mathcal{M}_g)$, i.e., to understand its structure as an algebraic subvariety. In the above discussion, we have viewed $\mathcal{A}_g$ as a locally symmetric space (or Riemannian orbifold). Consequently, one natural problem is to understand metric properties of $J(\mathcal{M}_g)$ with respect to locally symmetric subspaces in $\mathcal{A}_g$, for example, whether $J(\mathcal{M}_g)$ has certain convexity property. One aspect of such problems was discussed in the previous section. Now we discuss it from another perspective.

In [Fa, Prob. 4.12], Farb raised the following problem.

Problem 4.19 (Distortion of the Schottky locus) *Compute the distortion of the Schottky locus in $\mathcal{A}_g$.*

Then he commented that "A naive guess might be that it is exponential." The reason is that it is expected the distortion of $J(\mathcal{M}_g)$ in $\mathcal{A}_g$ is severe and hence more than polynomials. (We will explain the meaning of this sentence below). In [JL, §8], some remarks were made to suggest that $J(\mathcal{M}_g)$ is folded inside $\mathcal{A}_g$, a phenomenon of very different nature from the usual distortion such as a circle in the plane, or a parabola in the plane $\mathbb{R}^2$. We will give a definition of distortion function, the property of folding, and a proof of this result later in Theorem 4.21.

We also note that in the discussion of the Coleman-Oort conjecture, it is expected $J(\mathcal{M}_g)$ is distorted with respect to the algebraic subgroup structures of $\mathcal{A}_g$. Hence the results and discussions here are consistent.

To define metric distortion and to solve Problem 4.19, we need to introduce some notations. Given a metric space (M, d) and a path connected subspace N, there are two induced metrics on N. The first one is the *restriction* of the distance function $d : M \times M \to [0, +\infty)$ to $N \times N$. Denote this metric on N also by d. The second one is the *induced length metric* ℓ, which is defined as follows. For any continuous path $\sigma : [0,1] \to N$, define its length $\ell(\sigma)$ by

$$\ell(\sigma) = \sup_{0=t_0<t_1<\cdots t_n=1} \sum_{i=1}^{n} d(\sigma(t_{i-1}), \sigma(t_i)),$$

where the supremum is over all partitions of $[0,1]$. For any two points $p, q \in N$, consider all continuous paths σ connecting them, and define the *induced length distance* between them by

$$\ell(p,q) = \inf_{\sigma} \ell(\sigma).$$

We note that the length distance $\ell(p,q)$ may not be finite.

Clearly, for any two points $p, q \in N$,

$$\ell(p,q) \geq d(p,q).$$

Hence the ratio $\ell(p,q)/d(p,q) \geq 1$, and it measures the distortion of the embedding

$$(N,d) \to (M,d)$$

of metric spaces. If M is a complete Riemannian manifold and N is a totally geodesic subspace, then $\ell = d$, and the ratio is equal to 1.

One simple example of metric distortion to keep in mind is the circle embedded in $\mathbb{R}^2$. In this case,

$$1 \leq \ell(p,q)/d(p,q) \leq \frac{\pi}{2},$$

where the upper bound is achieved when p and q are the end points of a diameter.

If there exists a polynomial function P such that

$$\ell(p,q) \leq P(d(p,q)),$$

then we say the distortion of N inside M is *at most of the polynomial P*. If

$$\ell(p,q) \sim P(d(p,q))$$

for all $p, q \in N$ with $d(p,q)$ sufficiently large, then we say that the distortion *is of the order of the polynomial P*. Similarly, we can define exponential distortion of N inside M. (Note that in [Fa, p. §3.2], only the upper bound is imposed in defining distortion.)

Remark 4.20 In geometric group theory, a basic notion is the word metric for a finitely generated group Γ with respect to a finite symmetric generating set S. Denote the word metric on Γ, or its associated Cayley graph if one prefers a connected metric space, by d_S. Given a finitely generated group Γ and a finitely generated subgroup Γ', let d_S and $d_{S'}$ be their word metrics. Comparison between the two metrics d_S and $d_{S'}$ on Γ', i.e., bounding $d_{S'}$ in terms of d_S, gives the notion of distortion of Γ' in Γ. In this case, the distortion is usually either polynomial or exponential.

An example of a polynomial distortion of metric spaces is a parabola in $\mathbb{R}^2$, and an example of exponential distortion is to take M to be the upper half plane and N to be a horocircle, for example, $\{x + i1 \mid x \in \mathbb{R}\}$.

Note that even for Riemannian manifolds M and its submanifolds N, the distortion may not be described by any function as defined above, i.e., there may not exist a function $f : [0, +\infty) \to [0, +\infty)$ which is usually increasing, such that

$$\ell(p,q) \sim f(d(p,q)), \quad \text{or} \quad \ell(p,q) \leq f(d(p,q)), \tag{4.7}$$

for all $p, q \in N$.

One such example is obtained by taking $M = \mathbb{R}^2$, and N the image of *folding* $\mathbb{R}$ around the origin so that two half rays are like

$$\{(|x|, \frac{x}{x^2+1}) \mid x \in \mathbb{R}\}.$$

For $x \gg 0$, take $p = (x, \frac{x}{x^2+1}), q = (x, \frac{-x}{x^2+1})$, then

$$\ell(p, q) \sim 2x, \quad \text{but} \quad d(p, q) \sim \frac{2}{x}.$$

Therefore, when p, q move to the infinite along two branches,

$$d(p, q) \to 0, \quad \text{but } \ell(p, q) \to \infty.$$

Consequently, there is no way to bound $\ell(p, q)$ by $f(d(p, q))$, where f is an increasing function, no matter how fast it grows.

On the other hand, if we take $p' = (x, \frac{x}{x^2+1})$, and $q' = (2x, \frac{2x}{(2x)^2+1})$, then

$$\ell(p', q') \sim x, \quad \text{but} \quad d(p', q') \sim x.$$

We can also find sequences of pairs of points p_j, q_j such that $d(p_j, q_j)$ can approach any finite nonnegative value, while $\ell(p_j, q_j)$ goes to infinite when $j \to +\infty$. All these confirm the statement above that there is no distortion function f as in Equation 4.7. Such a folding is the opposite of totally geodesic embedding.

With the above notation, we can give a precise answer to Problem 4.19 raised by Farb [Fa, Prob. 4.12]. This result is motivated on some speculations in [JL, §8].

Theorem 4.21 *When $g \gg 1$, there exist a sequence of pairs of points $p_j, q_j \in J(\mathcal{M}_g)$ and a positive constant $\varepsilon_0 > 0$ such that $\ell(p_j, q_j) \geq \varepsilon_0$, while $d(p_j, q_j) \to 0$ as $j \to +\infty$.*

The above result shows that the metric subspace $J(\mathcal{M}_g)$ is folded in $\mathcal{A}_g$. The idea of the proof is this. We construct two families of Riemann surfaces $\Sigma_{g,t}, \Sigma'_{g,t}$, $t \in (0, 1]$, such that their Jacobian varieties $J(\Sigma_{g,t})$ and $J(\Sigma'_{g,t})$ converge to the same limiting point in $\mathcal{A}_g$ as $t \to 0$, while the length of any continuous path $\sigma : [0, 1] \to J(\mathcal{M}_g)$ between $J(\Sigma_{g,t})$ and $J(\Sigma'_{g,t})$ is uniformly bounded away from zero. This is achieved by proving that that any path σ must go through the interior of $J(\mathcal{M}_g)$ (or $\mathcal{M}_g$) deeply enough. This implies that $\lim_{t \to 0} d(J(\Sigma_{g,t}), J(\Sigma'_{g,t})) = 0$, while there exists a positive constant δ such that $\ell(J(\Sigma_{g,t}), J(\Sigma'_{g,t})) \geq \delta$ for all $t \in (0, 1]$. See [J8] for detail.

Remark 4.22 The above discussion also raises the problem of understanding the induced metric on $\mathcal{M}_g$ and $\mathcal{T}_g$ by pulling back the locally symmetric metric on $\mathcal{A}_g$ by the Jacobian map $J : \mathcal{M}_g \to \mathcal{A}_g$. It is not complete, though in some directions, geodesic rays can be extended infinitely.

Remark 4.23 Right before [Fa, Prob. 4.12], Farb also raised the problem on computing the second fundamental form of the subspace $J(\mathcal{M}_g) \subset \mathcal{A}_g$ in order to understand the distortion of $J(\mathcal{M}_g)$. This has been carried out in [CPT], and it was used in [CFG] to bound the dimension of germs of totally geodesic submanifolds contained in $J(\mathcal{M}_g)$. In some sense, computation of the second fundamental form describes the local distortion, while the distortion function deals with more global distortion.

Remark 4.24 Note that in raising [Fa, Prob. 4.12], Farb was motivated by distortion of word metrics for groups and subgroups in [Fa, §3.2], and folding phenomenon probably does not occur in such a situation.

5 Universal moduli space of Riemann surfaces and its partial compactification

The modular curve $\mathrm{SL}(2,\mathbb{Z})\backslash\mathbb{H}^2$ is noncompact. It is can be compactified by adding a point at infinity. The compactification admits the structure of a compact Riemann surface, biholomorphic to $\mathbb{C}P^1$. In fact, the j function (or j-invariant of elliptic curves) gives a bijection $j : \mathrm{SL}(2,\mathbb{Z})\backslash\mathbb{H}^2 \to \mathbb{C}$, and we can compactify it to $\mathbb{C}P^1$ by adding a point at infinity.

It is known that $\mathcal{M}_g$ is noncompact, and many arithmetic locally symmetric spaces such as $\mathcal{A}_g$ are also noncompact.

As mentioned before, Satake [Sa] and Baily [Ba1] compactified $\mathcal{A}_g$ to a normal projective variety $\overline{\mathcal{A}_g}^S$. Baily and Borel generalized this to all arithmetic Hermitian locally symmetric spaces and constructed the Baily-Borel compactification in [BB]. Because of this, $\overline{\mathcal{A}_g}^S$ is also called the Sataki-Baily-Borel compactification and denoted by $\overline{\mathcal{A}_g}^{SBB}$.

The compactification $\overline{\mathcal{A}_g}^S$ was used by Baily [Ba2] to construct the first compactification of $\mathcal{M}_g$, which is the closure of $J(\mathcal{M}_g)$ in $\overline{\mathcal{A}_g}^S$. This is highly singular and different from the later better known Deligne-Mumford compactification $\overline{\mathcal{M}_g}^{DM}$ constructed by Deligne and Mumford in [DM].

The Deligne-Mumford compactification $\overline{\mathcal{M}_g}^{DM}$ is a complex orbifold and is a modular compactification in the sense whose ideal points correspond to stable projective algebraic curves with nodal points which are of the same Euler characteristic as compact Riemann surfaces of genus g and can fit together with compact Riemann surfaces into flat families of Riemann surfaces.

On the other hand, the Satake compactification $\overline{\mathcal{A}_g}^S$ is not a modular compactification. But it has one important feature that its ideal points correspond to compact Riemann surfaces and their unions of total genus less than g. No singular Riemann surfaces are introduced in the boundary of the compactification. This fact is important in applications of Riemann surfaces to the Plateau problem for minimal surfaces in $\mathbb{R}^3$. The introduction of [JJ] gives more detail.

In [JJ], we used $\mathcal{A}_g$ and its compactification $\overline{\mathcal{A}_g}^S$ to construct a universal moduli space $\mathcal{M}_\infty$ and a partially compactified universal moduli space $\overline{\mathcal{M}_\infty}^{SBB}$ satisfying the following properties:

1. The universal moduli space $\mathcal{M}_\infty$ is a union of all moduli spaces $\mathcal{M}_g$, $g \geq 0$.
2. The partial compactification $\overline{\mathcal{M}_\infty}^{SBB}$ is connected stratified complex analytic space such that the closure of each stratum is a projective variety.
3. For every $g \geq 0$, the compactification $\overline{\mathcal{M}_g}^{SBB}$ is embedded into $\overline{\mathcal{M}_\infty}^{SBB}$, and $\overline{\mathcal{M}_\infty}^{SBB}$ is the union of these subsets $\overline{\mathcal{M}_g}^{SBB}$, $g \geq 0$.
4. There is a natural measure on $\overline{\mathcal{M}_\infty}^{SBB}$ which induces compatible measures on different strata, and with respect to the induced measure, every stratum has finite volume.

In constructing these spaces, the Jacobian map $J : \mathcal{M}_g \to \mathcal{A}_g$, the compactification $\overline{\mathcal{A}_g}^{SBB}$, a universal moduli space of principally polarized abelian varieties $\mathcal{A}_\infty$, and its partial compactification $\overline{\mathcal{A}_\infty}^{SBB}$ play a crucial role. This is another instance of application of the interaction between $\mathcal{M}_g$ and $\mathcal{A}_g$.

Besides applications to the Plateau problem for minimal surfaces as mentioned above, such a universal moduli space $\mathcal{M}_\infty$ and its partial compactification $\overline{\mathcal{M}_\infty}^{SBB}$ can be applied to possibly construct a non-perturbative string theory. This problem was raised in [Jo, p. 82]. See also the paper [JJ] for additional explanation on potential applications to string theory.

6 Satake compactifications of moduli spaces of flat tori

The discussions in the previous section show applications of compactifications of locally symmetric spaces to moduli spaces. In this section, we give an example of identifying locally symmetric spaces as moduli spaces and then constructing compactifications of locally symmetric spaces using such moduli interpretations.

One of the most basic example of locally symmetric spaces is

$$\mathrm{SL}(n,\mathbb{Z})\backslash\mathrm{SL}(n,\mathbb{R})/\mathrm{SO}(n), \ n \geq 2.$$

It admits finitely many Satake compactifications. Briefly, given a symmetric space $X = G/K$ of noncompact type, for every projectively faithful finite dimensional representation $\rho : G \to SL(n,\mathbb{C})$, there exists a Satake compactification $\overline{X}_\rho^S$ of the symmetric space X. This Satake compactification $\overline{X}_\rho^S$ only depends on the relative position of the highest weight μ_ρ of ρ, i.e., the Weyl chamber face which contains μ_ρ. Consequently, there are only finitely many non-isomorphic Satake compactifications of X, which are partially ordered. For example, when μ_ρ is contained in the interior of the Weyl chamber, $\overline{X}_\rho^S$ is the maximal Satake compactification, denoted by $\overline{X}_{\max}^S$, and it dominates all other Satake compactifications of X. When μ_ρ is contained in a 1-dimensional Weyl chamber face, $\overline{X}_\rho^S$ is a minimal Satake compactification. Note that there are in general more than one minimal Satake compactifications when the rank of X is greater than or equal to 2.

Under suitable conditions on the representation ρ, the Satake compactification of symmetric space $\overline{X}^S_\rho$ induces a compactification $\overline{\Gamma\backslash X}^S_\rho$ of an arithmetic locally symmetric space $\Gamma\backslash X$. See [BJ] for more detail.

The standard representation of $\mathrm{SL}(n,\mathbb{R})$ given by the inclusion $\mathrm{SL}(n,\mathbb{R}) \to \mathrm{SL}(n,\mathbb{C})$ gives a minimal Satake compactification of the symmetric space $\mathrm{SL}(n,\mathbb{R})/\mathrm{SO}(n)$, denoted by $\overline{\mathrm{SL}(n,\mathbb{R})/\mathrm{SO}(n)}^S_{\mathrm{min,\ st}}$, which induces a minimal Satake compactification of the locally symmetric space $\mathrm{SL}(n,\mathbb{Z})\backslash\mathrm{SL}(n,\mathbb{R})/\mathrm{SO}(n)$, denoted by $\overline{\mathrm{SL}(n,\mathbb{Z})\backslash\mathrm{SL}(n,\mathbb{R})/\mathrm{SO}(n)}^S_{\mathrm{min,\ st}}$.

It is known that the space $\mathrm{SL}(n,\mathbb{Z})\backslash\mathrm{SL}(n,\mathbb{R})/\mathrm{SO}(n)$ can be interpreted as the moduli space of flat tori of dimension n, though not in a unique way. Unlike the hyperbolic metric on compact Riemann surfaces of genus $g \geq 2$, there is no canonical normalization of flat metric on tori. Due to different scaling of flat metrics on tori, we have several different moduli interpretations to $\mathrm{SL}(n,\mathbb{Z})\backslash\mathrm{SL}(n,\mathbb{R})/\mathrm{SO}(n)$. The basic point of this section is that these different moduli interpretations lead to different compactifications.

As mentioned before, $\mathrm{SL}(n,\mathbb{Z})\backslash\mathrm{SL}(n,\mathbb{R})/\mathrm{SO}(n)$ can be identified with the moduli space of n-dimensional oriented flat tori of volume 1. This is the best known and most common normalization and hence moduli interpretation. On the other hand, there are other ways to normalize it:

1. For each flat torus of volume 1 $\Lambda\backslash\mathbb{R}^n$, we can scale the metric so that its diameter is equal to 1.
2. For each flat torus of volume 1 $\Lambda\backslash\mathbb{R}^n$, we can scale the metric so that its injectivity radius is equal to 1.

We have the following result.

Theorem 6.1 *For every sequence of n-dimensional flat tori T_j of diameter 1, there exists a subsequence T_j' such that it converges in the Gromov-Hausdorff distance to a flat torus of dimension* **less than or equal to** *n. When $\mathrm{SL}(n,\mathbb{Z})\backslash\mathrm{SL}(n,\mathbb{R})/\mathrm{SO}(n)$ is identified with the moduli space of flat tori of diameter 1, its completion in the Gromov-Hausdorff distance is compact, and it is isomorphic to the minimal Satake compactification* $\overline{\mathrm{SL}(n,\mathbb{Z})\backslash\mathrm{SL}(n,\mathbb{R})/\mathrm{SO}(n)}^S_{\mathrm{min,\ st}}$.

Theorem 6.2 *For every sequence of n-dimensional flat tori T_j of injectivity radius 1, there exists a subsequence T_j' such that with respect any choice of base point on T_j', it converges in the sense of Gromov-Hausdorff for pointed metric spaces to a flat manifold* **of dimension** *n which may not be compact but has injectivity radius equal to 1. When $\mathrm{SL}(n,\mathbb{Z})\backslash\mathrm{SL}(n,\mathbb{R})/\mathrm{SO}(n)$ is identified with the moduli space of flat tori of injectivity radius 1, we can construct a compactification by adding limits of convergent sequences in the above sense, and the resulting compactification is isomorphic to another minimal Satake compactification of $\mathrm{SL}(n,\mathbb{Z})\backslash\mathrm{SL}(n,\mathbb{R})/\mathrm{SO}(n)$.*

Theorem 6.3 *For every sequence of n-dimensional flat tori T_j of volume 1, there exists a subsequence T_j' such that with respect any choice of base point on T_j',* *it converges in the sense of Gromov-Hausdorff for pointed metric spaces to a flat manifold of dimension* **less than or equal to** *n which may not be compact. When $\mathrm{SL}(n,\mathbb{Z})\backslash\mathrm{SL}(n,\mathbb{R})/\mathrm{SO}(n)$ is identified with the moduli space of flat tori of volume 1, we can construct a compactification by adding limits of convergent sequences in the above sense, and the resulting compactification of $\mathrm{SL}(n,\mathbb{Z})\backslash\mathrm{SL}(n,\mathbb{R})/\mathrm{SO}(n)$ is not isomorphic to any Satake compactification.*

The above results are essentially contained in [BJ, Chap 12, III. 19], though formulated in different languages.

For the locally symmetric space $\mathcal{A}_g$, probably the most important compactification is the minimal Satake compactification $\overline{\mathcal{A}_g}^S$ in [Sa] [Ba2] which is a normal projective variety. If we identify $\mathcal{A}_g$ with the moduli space of principally polarized abelian varieties, we can also consider the flat metrics associated with the underlying compact complex tori and rescale them to normalize either the volume, the diameter, or the injectivity radius. Each such normalization leads to a compactification of $\mathcal{A}_g$, but none leads to this special Satake compactification $\overline{\mathcal{A}_g}^S$.

Remark 6.4 When a moduli space is given by a locally symmetric space, the objects under classification are usually linear. This is supported by examples. This also shows that $\mathcal{M}_g$ is probably richer and more difficult than $\Gamma\backslash X$.

Remark 6.5 As mentioned before, the geometry of numbers is important for understanding several locally symmetric spaces, and attempts have been made to establish similar results for $\mathcal{M}_g$ or the action of Mod_g on $\mathcal{T}_g$, but with only partial success so far. See [SS] for some results and references.

7 Moduli interpretation of the hyperbolic plane and its compactification

When Teichmuller first defined a topology on Teichmuller space $\mathcal{T}_g$, he used the Teichmuller metric, which is defined in terms of the dilatation of extremal quasi-conformal maps between marked Riemann surfaces.

It is well-known that when $g = 1$, the Teichmüller space $\mathcal{T}_1$ can be canonically identified with the upper halfplane $\mathbb{H}^2$, where each point $\tau \in \mathbb{H}^2$ determines a lattice $\Lambda_\tau = \mathbb{Z} + \tau\mathbb{Z}$ and hence a compact Riemann surface of genus 1, $\Lambda_\tau\backslash\mathbb{C}$, together with a choice of a basis $1, \tau$ of

$$\Lambda_\tau = H_1(\Lambda_\tau\backslash\mathbb{C}, \mathbb{Z}) \cong \pi_1(\Lambda_\tau\backslash\mathbb{C}).$$

It is probably less well-known that Teichmüller also showed that when $g = 1$, the Teichmuller metric is equal to the hyperbolic metric on $\mathbb{H}^2 = \mathcal{T}_1$. This gives a *moduli interpretation* of $\mathbb{H}^2$ in terms of complex structures of compact Riemann surfaces of genus 1.

When $g \geq 2$, each compact Riemann surface Σ_g of genus g admits a canonical hyperbolic metric which is conformal to the complex structure. Thurston defined the Thurston metric d^{Th} on $\mathcal{T}_g$ in [Th] using the minimal Lipschitz constants of Lipschitz maps between marked hyperbolic surfaces. The important point is that this metric only depends on the hyperbolic geometry of surfaces. It turns out that the Thurston metric is not symmetric, i.e., for two points $p, q \in \mathcal{T}_g$, $d^{Th}(p,q) \neq d^{Th}(q,p)$ in general.

One problem is to generalize the Thurston metric to the Teichmüller space $\mathcal{T}_1$. In this case, each point in $\mathcal{T}_1 = \mathbb{H}^2$ can be represented by a flat torus whose metric is well-defined up to scalar multiples together with a choice of a basis of the first homology group, which is a *marking of the torus*. As we discussed flat tori of general dimension in the previous section, there are different normalizations of the flat metrics of tori. For the purpose of defining the Thurston metric on $\mathcal{T}_1$, it turns out that the normalization of area equal to 1 seems to be the most suitable.

In [GJ, Theorem 1.4], we prove the following property of the Thurston metric on the Teichmüller space $\mathcal{T}_1$.

Theorem 7.1 *Identify the Teichmüller space $\mathcal{T}_1$ with the space of marked flat tori of dimension 2 of area 1. Then the analogue of Thurston metric on the Teichmüller space $\mathcal{T}_1$ is symmetric. When $\mathcal{T}_1$ is canonically identified with the upper half plane $\mathbb{H}^2$, the Thurston metric is equal to the hyperbolic metric on $\mathbb{H}^2$. Consequently, the hyperbolic plane $\mathbb{H}^2$ can be canonically realized through the moduli space of marked oriented flat tori of dimension 2 of volume 1.*

In the above theorem, the normalization of tori by volume 1 is crucial. There are other normalizations, such as that in [BPT], but they lead to non-symmetric metrics, which are not equal to the hyperbolic metric on $\mathbb{H}^2$.

Remark 7.2 It is perhaps worthwhile to point out that for every two marked compact Riemann surfaces of genus $g \geq 2$, there exists a unique extremal quasi-conformal map between them. This is a crucial result in the Teichmüller theory. On the other hand, there is no such uniqueness result for extremal Lipschitz maps between marked flat tori of area 1. See [GJ, Proposition 3.8].

After describing the above moduli interpretation of the upper half-plane $\mathbb{H}^2$, we want to conclude with a moduli interpretation of the well-known compactification $\overline{\mathbb{H}^2}$ of $\mathbb{H}$.

The most common way to compactify $\mathbb{H}^2$ is to take the closure of $\mathbb{H}^2$ in $\mathbb{C}P^1 = \mathbb{C} \cup \{\infty\}$, and the compactification is equal to

$$\overline{\mathbb{H}^2} = \mathbb{H}^2 \cup \mathbb{R} \cup \{\infty\},$$

and its boundary $\partial\mathbb{H}^2$ is a circle $S^1 = \mathbb{R} \cup \{\infty\}$

One problem is to interpret this compactification and its boundary. If we put the Poincaré metric on $\mathbb{H}^2$, then the boundary $\partial\mathbb{H}^2$ can be identified with the set of equivalence classes of unit speed oriented geodesics in $\mathbb{H}^2$. The topology of the compactification $\overline{\mathbb{H}^2}$ can also be described in terms of the hyperbolic geometry. This gives a purely geometric interpretation of the compactification $\overline{\mathbb{H}^2}$.

If we consider harmonic functions on $\mathbb{H}^2$, we can also identify the boundary $\partial\mathbb{H}^2$ with the Martin boundary and also with the Poisson boundary of $\mathbb{H}^2$, and the compactification $\overline{\mathbb{H}^2}$ can also be identified with the Martin compactification of $\mathbb{H}^2$ in potential theory. This is an analysis way to interpret the compactification $\overline{\mathbb{H}^2}$.

If we identify $\mathbb{H}^2$ with the homogeneous space $\mathrm{SL}(2,\mathbb{R})/\mathrm{SO}(2)$, there are also group theoretical ways to interpret the compactification $\overline{\mathbb{H}^2}$. See the book [GJT] for some detail.

In the spirit of this paper, we want to give a moduli interpretation of the compactification $\overline{\mathbb{H}^2}$. To motivate the result, we start with special boundary points, the so-called rational points. As explained above, $\mathbb{H}^2$ can be identified with the moduli space of marked flat tori of area 1 or some other normalizations. We want to interpret the boundary points in $\partial\overline{\mathbb{H}^2}$ in terms of some geometric structures.

We start with the rational boundary points $\mathbb{Q} \cup \{\infty\} \subset \partial\overline{\mathbb{H}^2}$. We can define a partial compactification

$$\mathbb{H}^2 \cup \mathbb{Q} \cup \{\infty\},$$

by constructing a topology on it such that for each boundary point $q \in \mathbb{Q} \cup \{\infty\}$, unions of the point $\{q\}$ with horodisks based at q form a basis of neighborhoods of q. Then the modular group $\mathrm{SL}(2,\mathbb{Z})$ acts continuously (though not properly) on it with a compact quotient

$$\mathrm{SL}(2,\mathbb{Z})\backslash\mathbb{H}^2 \cup \{\infty\}.$$

This can also be obtained by adding one point at infinity to the well-known fundamental domain

$$\Omega = \{z \in \mathbb{H}^2 \mid |z| > 1, -\frac{1}{2} < \mathrm{Re}(z) < \frac{1}{2}\}$$

of $\mathrm{SL}(2,\mathbb{Z})$ acting on $\mathbb{H}^2$.

The question is how to identify this point ∞ with some geometric objects. There are at least three geometric objects to describe this point ∞. When a sequence of flat tori T_n corresponds to points in the fundamental domain Ω and goes to infinity, we have three different normalizations of metrics on T_n and hence three limits:

1. Normalize the area of T_n to be equal to 1. Then T_n converges in the sense of Gromov-Hausdorff for pointed metric spaces to a real line $\mathbb{R}$.
2. Normalize the diameter of T_n to be equal to 1. Then T_n converges in the Gromov-Hausdorff distance to a circle S^1 of diameter 1.
3. Normalize the injectivity radius of T_n to be equal to 1. Then T_n converges in the sense of Gromov-Hausdorff for pointed metric spaces to an infinite flat cylinder of injectivity radius 1.

Since the rational boundary points $\mathbb{Q} \cup \{\infty\} \subset \partial\mathbb{H}^2$ are mapped to the unique boundary point ∞ of $\mathrm{SL}(2,\mathbb{Z})\backslash\mathbb{H}^2$, these geometric objects also give moduli interpretations to all rational boundary points. One question is which one of the above three normalizations is the most natural one? The answer depends one's perspective.

The normalization of flat metrics on tori by area 1 is the most natural choice in view of the Thurston metric above. Since the interior points of $\mathbb{H}^2$ correspond to 2-dimensional flat tori, adding 1-dimensional objects, a real line $\mathbb{R}$ or a circle S^1 may not seem satisfying. This excludes Options (1) and (2). In Option (3), we add an infinite flat cylinder. In taking the Gromov-Hausdorff limit for pointed metric spaces, we need to specify a base point. Because of this, we lose one symmetry of $S^1 \times S^1$ for tori in Option (3).

Besides the above problems, there is a more serious issue:

Question 7.3 *How to assign geometric objects to irrational boundary points in $\partial\mathbb{H}^2$?*

Our answer is to use measured foliations on the standard torus

$$\mathbb{Z}^2\backslash\mathbb{R}^2 = S^1 \times S^1.$$

This idea was motivated by Thurson's work on compactifications of the Teichmüller space $\mathcal{T}_g$, for $g \geq 2$, using measured foliations on surfaces (see [FLP]).

Definition 7.4 *A measured foliation on the standard torus $\mathbb{Z}^2\backslash\mathbb{R}^2 = S^1 \times S^1$ consists of a direction V, or a nonzero vector in $\mathbb{R}^2$, and a measure on lines perpendicular to V. Two measured foliations are projectively equivalent if they have the same direction and their transverse measures are multiples of each other.*

Proposition 7.5 *The boundary points of the compactification $\overline{\mathbb{H}^2}$ correspond bijectively to equivalence classes of measured foliations on the standard torus $\mathbb{Z}^2\backslash\mathbb{R}^2 = S^1 \times S^1$.*

Proof. Besides the three identifications of the compactification $\overline{\mathbb{H}^2}$, there is another one which comes from identifying points of $\mathbb{H}^2$ with positive definite quadratic forms on $\mathbb{R}^2$ and the boundary points with projective equivalence classes of positive semidefinite quadratic forms. This is the Satake compactification of $\mathbb{H}^2$ as a symmetric space. For each semidefinite, but not definite, quadratic form on $\mathbb{R}^2$, its kernel, or the zero eigenspace, defines a direction in $\mathbb{R}^2$, and the form restricts to a measure on lines orthogonal to it. They project and give a measured foliation on the standard torus $\mathbb{Z}^2\backslash\mathbb{R}^2$. Conversely, every measure foliation on $\mathbb{Z}^2\backslash\mathbb{R}^2$ can arise this way.

Remark 7.6 To relate these measured foliations to the limiting geometric structures we obtained for the rational boundary points $\mathbb{Q} \cup \{\infty\}$ of $\partial\mathbb{H}^2$, we note that when the direction of a measured foliation is rational, we can collapse the rational direction and keep the orthogonal lines which all coincide and become one line, and a measure on it. This becomes a circle with a measure or a metric on it. On the other hand, if we keep the direction and make a positive length on it, we need to blow up the metric on the orthogonal lines, which gives an infinite flat cylinder.

Remark 7.7 A similar moduli interpretation works for the boundary of the minimal Satake compactification $\overline{\mathrm{SL}(n,\mathbb{R})/\mathrm{SO}(n)}^S_{\mathrm{min,\ st}}$ of the symmetric space $\mathrm{SL}(n,\mathbb{R})/\mathrm{SO}(n)$ associated with the standard representation by using measured foliations on the standard torus $\mathbb{Z}^n\backslash\mathbb{R}^n$, where a direction is replaced by a linear subspace.

References

[Ba1] W. Baily, Satake's compactification of V_n. Amer. J. Math. 80 (1958) 348–364.

[Ba2] W. Baily, On the moduli of Jacobian varieties. Ann. of Math. (2) 71 (1960) 303–314.

[BB] W. Baily, A. Borel, Compactification of arithmetic quotients of bounded symmetric domains. Ann. of Math. (2) 84 (1966) 442–528.

[BPT] A. Belkhirat, A. Papadopoulos, M. Troyanov, Thurston's weak metric on the Teichmüller space of the torus. Trans. Am. Math. Soc. 357 (2005) 3311–3324.

[BJ] A. Borel, L. Ji, Compactifications of symmetric and locally symmetric spaces. Mathematics: Theory & Applications. Birkhäuser Boston, Inc., Boston, MA, 2006. xvi+479 pp.

[BoS] A. Borel, J.-P. Serre, Corners and arithmetic groups. Comment. Math. Helv. 48 (1973), 436–491.

[BuS] P. Buser, P. Sarnak, On the period matrix of a Riemann surface of large genus. Invent. Math. 117 (1994), no. 1, 27–56.

[Ca] J. Cassels, *An introduction to the geometry of numbers.* Corrected reprint of the 1971 edition. Classics in Mathematics. Springer-Verlag, Berlin, 1997. viii+344 pp.

[CFG] E. Colombo, P. Frediani, A. Ghigi, On totally geodesic submanifolds in the Jacobian locus. Internat. J. Math. 26 (2015), no. 1, 1550005, 21 pp.

[CPT] E. Colombo, G. Pirola, A. Tortora, Hodge-Gaussian maps. Ann. Scuola Norm. Sup. Pisa Cl. Sci. (4) 30 (2001), no. 1, 125–146.

[DM] P. Deligne, D. Mumford, The irreducibility of the space of curves of given genus. Inst. Hautes Études Sci. Publ. Math. No. 36 (1969) 75–109.

[dSG] H. de Saint-Gervais, *Uniformization of Riemann surfaces. Revisiting a hundred-year-old theorem.* Translated from the 2010 French original by Robert G. Burns. Heritage of European Mathematics. European Mathematical Society (EMS), Zürich, 2016. xxx+482 pp.

[Do] R. Donagi, The Schottky problem. *Theory of moduli* (Montecatini Terme, 1985), 84–137, Lecture Notes in Math., 1337, Springer, Berlin, 1988.

[Fa] B. Farb, Some problems on mapping class groups and moduli space, in *Problems on Mapping Class Groups and Related Topics*, Proc. Sympos. Pure Math. 74, Amer. Math. Soc., Providence (2006), 11–55.

[FR] H. Farkas, H. Rauch, Two kinds of theta constants and period relations on a Riemann surface. Proc. Nat. Acad. Sci. U.S.A. 62 (1969) 679–686.

[FLP] A. Fathi, F. Laudenbach, V. Poénaru, *Thurston's work on surfaces.* Translated from the 1979 French original by Djun M. Kim and Dan Margalit. Mathematical Notes, 48. Princeton University Press, Princeton, NJ, 2012. xvi+254 pp.

[GJ] M. Greenfield, L. Ji, A new modular characterization of the hyperbolic plane, preprint, arXiv:1707.00818, 2017.

[GH] P. Griffiths, J. Harris, *Principles of algebraic geometry.* Pure and Applied Mathematics. Wiley-Interscience [John Wiley & Sons], New York, 1978.

xii+813 pp.

[Gr] S. Grushevsky, The Schottky problem. *Current developments in algebraic geometry*, 129–164, Math. Sci. Res. Inst. Publ., 59, Cambridge Univ. Press, Cambridge, 2012.

[GJT] Y. Guivarc'h, L. Ji, J.C. Taylor, *Compactifications of symmetric spaces.* Progress in Mathematics, 156. Birkhäuser Boston, Inc., Boston, MA, 1998. xiv+284 pp.

[Har] J. Harer, The virtual cohomological dimension of the mapping class group of an orientable surface. Invent. Math. 84 (1986), no. 1, 157–176.

[Hat] T. Hattori, Asymptotic geometry of arithmetic quotients of symmetric spaces, Math. Z. 222 (1996), 247–277.

[Harv] W. Harvey, Boundary structure of the modular group. *Riemann surfaces and related topics*: Proceedings of the 1978 Stony Brook Conference (State Univ. New York, Stony Brook, N.Y., 1978), pp. 245–251, Ann. of Math. Stud., 97, Princeton Univ. Press, Princeton, N.J., 1981.

[IJ] N. Ivanov, L. Ji, Infinite topology of curve complexes and non-Poincaré duality of Teichmüller modular groups. Enseign. Math. (2) 54 (2008), no. 3–4, 381–395.

[J1] L. Ji, A tale of two groups: arithmetic groups and mapping class groups. *Handbook of Teichmüller theory.* Volume III, 157–295, IRMA Lect. Math. Theor. Phys., 17, Eur. Math. Soc., Zürich, 2012.

[J2] L. Ji, The story of Riemann's moduli space. ICCM Not. 3 (2015), no. 2, 46–73.

[J3] L. Ji, A tale of two groups: arithmetic groups and mapping class groups. *Handbook of Teichmüller theory.* Volume III, 157–295, IRMA Lect. Math. Theor. Phys., 17, Eur. Math. Soc., Zürich, 2012.

[J4] L. Ji, A brief history on finding the right complex structure on Teichmüller space and moduli space, preprint 2016.

[J5] L. Ji, The greatest common quotient of Borel-Serre and the toroidal compactifications of locally symmetric spaces. Geom. Funct. Anal. 8 (1998), no. 6, 978–1015.

[J6] L. Ji, Curve complexes versus Tits buildings: structures and applications. *Buildings, finite geometries and groups*, 93–152, Springer Proc. Math., 10, Springer, New York, 2012.

[J7] L. Ji, From the Borel-Serre compactification to curve complex of surfaces, preprint, 2017.

[J8] L. Ji, Metric distortion in the geometric Schottky problem, in preparation, 2017.

[JJ] L. Ji, J. Jost, Universal moduli spaces of Riemann surfaces. J. Geom. Phys. 114 (2017), 124–137.

[JL] L. Ji, E. Leuzinger, The asymptotic Schottky problem. Geom. Funct. Anal. 19 (2010), no. 6, 1693–1712.

[JM] L. Ji, R. MacPherson, Geometry of compactifications of locally symmetric spaces, Ann. Inst. Fourier, Grenoble 52 (2002), 457–559.

[Jo] J. Jost, *Geometry and physics.* Springer-Verlag, Berlin, 2009. xiv+217 pp.

[KS] I. Krichever, T. Shiota, Soliton equations and the Riemann-Schottky problem. *Handbook of moduli.* Vol. II, 205–258, Adv. Lect. Math. (ALM), 25, Int. Press, Somerville, MA, 2013.

[Le] E. Leuzinger, Tits geometry, arithmetic groups, and the proof of a conjecture of Siegel, J. Lie Theory 14 (2004), 317–338.

[Lew] J. Lewittes, Riemann surfaces and the theta function. Acta Math. 111 (1964) 37–61.

[MM] H. McKean, V. Moll, *Elliptic curves. Function theory, geometry, arithmetic.* Cambridge University Press, Cambridge, 1997. xiv+280 pp.

[Min] Y. Minsky, Curve complexes, surfaces and 3-manifolds. International Congress of Mathematicians. Vol. II, 1001–1033, Eur. Math. Soc., Zürich, 2006.

[Mo] M. Möller, Shimura and Teichmüller curves. J. Mod. Dyn. 5 (2011), no. 1, 1–32.

[MoO] B. Moonen, F. Oort, The Torelli locus and special subvarieties. *Handbook of moduli.* Vol. II, 549–594, Adv. Lect. Math. (ALM), 25, Int. Press, Somerville, MA, 2013.

[Mu1] D. Mumford, *Geometric invariant theory.* Ergebnisse der Mathematik und ihrer Grenzgebiete, Neue Folge, Band 34 Springer-Verlag, Berlin-New York 1965 vi+145 pp.

[Mu2] D. Mumford, *Curves and their Jacobians.* The University of Michigan Press, Ann Arbor, Mich., 1975. vi+104 pp. It also appears as an appendix of *The red book of varieties and schemes.* Second, expanded edition. Lecture Notes in Mathematics, 1358. Springer-Verlag, Berlin, 1999. x+306 pp.

[Mu3] D. Mumford, A remark on Mahler's compactness theorem. Proc. Amer. Math. Soc. 28 (1971) 289–294.

[Rie] B. Riemann, *Collected papers.* Translated from the 1892 German edition by Roger Baker, Charles Christenson and Henry Orde. Kendrick Press, Heber City, UT, 2004. x+555 pp.

[Sa] I. Satake, On the compactification of the Siegel space. J. Indian Math. Soc. (N.S.) 20 (1956), 259–281.

[SS] P. Schmutz Schaller, Geometry of Riemann surfaces based on closed geodesics. Bull. Amer. Math. Soc. (N.S.) 35 (1998), no. 3, 193–214.

[Sc1] F. Schottky, Ueber specielle Abelsche Functionen vierten Ranges. J. Reine Angew. Math. 103 (1888), 185–203.

[Sc2] F. Schottky, Über die Moduln der Thetafunctionen. Acta Math. 27 (1903), no. 1, 235–288.

[Sh] T. Shiota, Characterization of Jacobian varieties in terms of soliton equations. Invent. Math. 83 (1986), no. 2, 333–382.

[Tei] O. Teichmüller, Veränderliche Riemannsche Flächen. Deutsche Math. 7, (1944). 344–359. An English translation in "Variable Riemann surfaces." Translated from the German by Annette A'Campo-Neuen. IRMA Lect. Math. Theor. Phys., 19, Handbook of Teichmüller theory. Vol. IV, 787–803, Eur. Math. Soc., Zürich, 2014.

[Th] W.P. Thurston, Minimal stretch maps between hyperbolic surfaces. Preprint (1985) available at http://arxiv.org/abs/math.GT/9801039.

Department of Mathematics, University of Michigan
Ann Arbor, MI 48109, USA
E-mail address: lji@umich.edu

Proceedings of the 7th International
Congress of Chinese Mathematicians
ALM 43, pp. 303–325

Cohomology of Automorphic Bundles

Kai-Wen Lan

Abstract

In this survey article, we review some recent works (by the author and his collaborators Junecue Suh, Michael Harris, Richard Taylor, Jack Thorne, and Benoît Stroh) on the cohomology of automorphic bundles over locally symmetric varieties and some related geometric objects.

Mathematics Subject Classification: Primary 11G18; Secondary 11F75, 14F17, 11F33, 11F80.

Keywords and Phrases: Locally symmetric varieties, Shimura varieties, compactifications, automorphic bundles, cohomology, vanishing theorems, construction of Galois representations.

1 Classical story: modular curves and modular forms

Let us begin by briefly reviewing the classical story of modular curves and modular forms. For more details and references, we shall refer the readers to the first two sections of the survey article [Lan12b], which had a similar starting point. Nevertheless, our goals in this article are more general, with less emphasis on the good reduction integral models of PEL-type Shimura varieties.

1.1 Classical modular forms

The group $\mathrm{SL}_2(\mathbb{R})$ acts on the *Poincaré upper-half plane*

$$\mathcal{H} = \{z \in \mathbb{C} : \mathrm{im}(z) > 0\}$$

by the usual *Möbius transformation*

$$\gamma = \begin{pmatrix} a & b \\ c & d \end{pmatrix} \in \mathrm{SL}_2(\mathbb{R}) : z \mapsto \gamma(z) := \frac{az+b}{cz+d}.$$

This is induced by the (transitive) action of $\mathrm{SL}_2(\mathbb{C})$ on the *projective coordinates* of $\mathbb{P}^1(\mathbb{C})$:

$$\gamma = \begin{pmatrix} a & b \\ c & d \end{pmatrix} \in \mathrm{SL}_2(\mathbb{C}) : \begin{pmatrix} z \\ 1 \end{pmatrix} \mapsto \begin{pmatrix} az+b \\ cz+d \end{pmatrix} \sim \begin{pmatrix} \gamma(z) \\ 1 \end{pmatrix}.$$

Note that $\sim$ is given by division by the factor $(cz+d)$.

Suppose $\Gamma \subset \mathrm{SL}_2(\mathbb{Z})$ is a finite index subgroup (which is then an *arithmetic subgroup* of $\mathrm{SL}_2(\mathbb{Q})$). Suppose $k \in \mathbb{Z}$ is an integer.

Definition 1.1. *A* **classical modular form** *of level Γ and weight k is a holomorphic function $f : \mathcal{H} \to \mathbb{C}$ satisfying the following two conditions:*

1. (**automorphy condition**) *$f(\gamma(z)) = (cz+d)^k f(z)$ for all $\gamma \in \Gamma$.*
2. (**growth condition**) *$(cz+d)^{-k} f(\gamma(z))$ stays bounded as $\mathrm{im}(z) \to +\infty$, for all $\gamma \in \mathrm{SL}_2(\mathbb{Z})$.*

We say that f is a **cusp form** *if, instead of (2), it satisfies the following:*

3. (**cuspidal condition**) *$(cz+d)^{-k} f(\gamma(z)) \to 0$ as $\mathrm{im}(z) \to +\infty$, for all $\gamma \in \mathrm{SL}_2(\mathbb{Z})$.*

We shall denote the space of modular forms (resp. cusp forms) of level Γ and weight k by $M_k(\Gamma; \mathbb{C})$ (resp. $S_k(\Gamma; \mathbb{C})$). (These might be zero spaces.)

1.2 Modular curves and a geometric definition of modular forms

We can redefine $M_k(\Gamma; \mathbb{C})$ and $S_k(\Gamma; \mathbb{C})$ geometrically. The group Γ acts naturally on $\mathcal{H}$ and on

$$\mathcal{H}^* := \mathcal{H} \cup \mathbb{P}^1(\mathbb{Q}),$$

and we have the open and compactified modular curves

$$Y_\Gamma := \Gamma \backslash \mathcal{H} \hookrightarrow X_\Gamma := \Gamma \backslash \mathcal{H}^* \tag{1.2}$$

(with a suitable topology on $\Gamma \backslash \mathcal{H}^*$). The pullback of $\mathcal{O}_{\mathbb{P}^1(\mathbb{C})}(1)$ to $\mathcal{H}$ descends to a line bundle ω on Y_Γ, which extends to a line bundle ω on X_Γ such that

$$M_k(\Gamma; \mathbb{C}) \cong H^0(X_\Gamma, \omega^{\otimes k}) \tag{1.3}$$

and

$$S_k(\Gamma; \mathbb{C}) \cong H^0(X_\Gamma, \omega^{\otimes k}(-\infty)), \tag{1.4}$$

where $(-\infty)$ means vanishing at the *cusps* $X_\Gamma - Y_\Gamma = \Gamma \backslash \mathbb{P}^1(\mathbb{Q})$.

By the *Kodaira–Spencer isomorphism* and *Serre duality*, we have

$$\omega^2(-\infty) \cong \Omega^1_{X_\Gamma/\mathbb{C}}$$

and

$$\overline{S_k(\Gamma; \mathbb{C})} \cong \overline{H^0(X_\Gamma, \omega^{\otimes k}(-\infty))} \cong H^1(X_\Gamma, \omega^{\otimes(2-k)}),$$

where the complex conjugations (denoted by overlines) are induced by integrations on modular curves. When $k \geq 2$, we have the *Eichler–Shimura isomorphism*

$$H^1(Y_\Gamma, \mathrm{Sym}^{k-2}(\mathbb{C}^{\oplus 2})) \cong M_k(\Gamma; \mathbb{C}) \oplus \overline{S_k(\Gamma; \mathbb{C})}, \tag{1.5}$$

where $\mathrm{Sym}^{k-2}(\mathbb{C}^{\oplus 2})$ abusively denotes the local system (over Y_Γ) associated with the algebraic representation (denoted by the same symbols) of SL_2 over $\mathbb{C}$ (induced by the standard action of $\mathrm{SL}_2(\mathbb{C})$ on $\mathbb{C}^2$), which can be rewritten as

$$H^1(Y_\Gamma, \mathrm{Sym}^{k-2}(\mathbb{C}^{\oplus 2})) \cong H^0(X_\Gamma, \omega^{\otimes k}) \oplus H^1(X_\Gamma, \omega^{\otimes(2-k)}). \tag{1.6}$$

When $k = 2$, this is a consequence of the comparison isomorphism

$$H^1(Y_\Gamma, \mathbb{C}) \cong H^1(X_\Gamma, \Omega^\bullet_{X_\Gamma/\mathbb{C}}(\log\infty))$$

(where the right-hand side denotes the first hypercohomology of the log de Rham complex $\Omega^\bullet_{X_\Gamma/\mathbb{C}}(\log\infty)$), and the degeneration of Hodge spectral sequence:

$$H^1(X_\Gamma, \Omega^\bullet_{X_\Gamma/\mathbb{C}}(\log\infty)) \cong H^0(X_\Gamma, \Omega^1_{X_\Gamma/\mathbb{C}}(\log\infty)) \oplus H^1(X_\Gamma, \mathscr{O}_{X_\Gamma}).$$

The previous isomorphism (1.6) for general $k > 2$ is a similar consequence for nontrivial coefficient systems (using mixed Hodge theory).

1.3 Integral models and some applications

In this subsection, let us assume that Γ is a *congruence subgroup* of $\mathrm{SL}_2(\mathbb{Z})$, which means Γ contains the principal congruence subgroup

$$\Gamma(N) := \ker(\mathrm{SL}_2(\mathbb{Z}) \to \mathrm{SL}_2(\mathbb{Z}/N\mathbb{Z}))$$

for some integer $N \geq 1$. For any rational prime p and any field isomorphism $\mathbb{C} \cong \overline{\mathbb{Q}}_p$, we can compare $H^1(Y_\Gamma, \mathrm{Sym}^{k-2}(\mathbb{C}^{\oplus 2}))$ with $H^1_{\text{ét}}(Y_\Gamma, \mathrm{Sym}^{k-2}(\overline{\mathbb{Q}}_p^{\oplus 2}))$, where $\mathrm{Sym}^{k-2}(\overline{\mathbb{Q}}_p^{\oplus 2})$ abusively denotes the (lisse) $\overline{\mathbb{Q}}_p$-étale sheaf associated with the algebraic representation (denoted by the same symbols) of SL_2 over $\overline{\mathbb{Q}}_p$, and such a comparison is compatible with varying Γ (and with the *Hecke actions* on these spaces). An important fact is that Y_Γ admits a model over some number field K, and hence $\mathrm{Gal}(\overline{K}/K)$ acts on $H^1_{\text{ét}}(Y_\Gamma, \mathrm{Sym}^{k-2}(\overline{\mathbb{Q}}_p^{\oplus 2}))$. Very roughly speaking (we are being intentionally vague here), this provides the main source of *Galois representations* attached to cusp forms of weight $k \geq 2$. By using also some *integral model* of X_Γ (over the integers $\mathcal{O}_K$), we obtain information about the restriction of such Galois representations to $\mathrm{Gal}(\overline{K}_v/K_v)$ at nonarchimedean places v. By using integral models differently, we can define the integral versions $M_k(\Gamma; \mathcal{O}_K)$ and $S_k(\Gamma; \mathcal{O}_K)$, and study *congruences* between modular forms. Then we can extend the attachment of Galois representations to the case of the *low weight* $k = 1$ (see [DS74]). (For simplicity, we mentioned only the semisimple algebraic group SL_2 in the above, but for the consideration of Hecke and Galois actions, it is important to also introduce the reductive algebraic group GL_2.)

In what follows, our goals will be to explain the generalizations of modular curves and modular forms to higher dimensions, and to provide a survey of some recent results on such generalizations due to this author and his collaborators Junecue Suh, Michael Harris, Richard Taylor, Jack Thorne, and Benoît Stroh.

2 Locally symmetric varieties and automorphic bundles

Here is an overview of the generalizations we shall explain:

classical story	generalizations
SL_2 or GL_2	more general algebraic groups
Γ	arithmetic subgroups
Poincaré upper-half plane $\mathcal{H}$	Hermitian symmetric domains
open modular curve $\Gamma\backslash\mathcal{H}$	locally symmetric varieties or Shimura varieties
compactified modular curve $\Gamma\backslash\mathcal{H}^*$	various compactifications
ω^k and $\mathrm{Sym}^k(\mathbb{C}^{\oplus 2})$	automorphic (vector) bundles (two kinds)
M_k = sections of ω^k	sections and also cohomology of automorphic bundles
Eichler–Shimura isomorphism	Faltings's dual BGG spectral sequence (degeneration by mixed Hodge theory)
integral models	integral models
⋮	⋮

2.1 Locally symmetric varieties and their compactifications

Let us start with some generalization of the Poincaré upper-half plane. For simplicity of exposition, consider the following setup:

- $\mathcal{H}$ is a *Hermitian symmetric domain* of dimension d, and
- G is a simply-connected connected semisimple algebraic group over $\mathbb{Q}$

such that $\mathcal{H} \cong \mathrm{G}(\mathbb{R})/K$ for some maximal compact subgroup K of G. (The setup here can be generalized to the case where $\mathcal{H}$ is a finite disjoint union of Hermitian symmetric domains with a transitive action of $\mathrm{G}(\mathbb{R})$, where G is a possibly disconnected reductive algebraic group satisfying some conditions. This more general setup will be tacitly allowed when we discuss about Shimura varieties and their integral models.) Then there is the *Borel embedding*

$$\mathcal{H} = \mathrm{G}(\mathbb{R})/K \hookrightarrow \mathcal{H}^\vee = \mathrm{G}(\mathbb{C})/\mathrm{P}(\mathbb{C})$$

for some maximal parabolic subgroup P of $\mathrm{G}_\mathbb{C}$, with a Levi subgroup M such that $\mathrm{M}(\mathbb{C}) = K_\mathbb{C}$ in $\mathrm{G}(\mathbb{C})$. (See, e.g., [Hel01, Ch. III, Sec. 7], [AMRT75, Ch. III, Sec. 2.1], and [Mil90, Sec. III.1].) This generalizes the canonical embedding of the Poincaré upper-half plane into the projective line $\mathbb{P}^1(\mathbb{C})$.

Let Γ be any arithmetic subgroup of $\mathrm{G}(\mathbb{Q})$ which we assume to be *neat* (so that it acts freely on $\mathcal{H}$), and let

$$X := \Gamma\backslash\mathcal{H}.$$

Then we have some useful compactifications of X, generalizing (1.2) in two ways:

$$\begin{array}{ccc} X & \hookrightarrow & X^{\mathrm{tor}} \\ & \searrow & \downarrow \\ & & X^{\mathrm{min}} \end{array}$$

In this diagram, the *minimal compactification* X^{min} due to Satake and Baily–Borel (see [BB66]) is canonical, normal, and projective. This shows that X is quasi-projective and canonically algebraic. But the generally rather singular X^{min} is not a good starting point for generalizing modular forms. On the other hand, the *toroidal compactifications* X^{tor} due to Mumford and others (see [AMRT75] and [AMRT10]) are noncanonical, but still canonically depend on certain *cone decompositions* (or *fans*), and can be chosen to be projective and smooth with boundaries given by simple normal crossings divisors. For simplicity, we shall always assume that the toroidal compactifications we consider have these nice properties. (This is harmless for our purpose.) They are useful for applications of mixed Hodge theory, and for defining and studying generalizations of modular forms.

2.2 Automorphic bundles and canonical extensions

To define the desired generalizations of modular forms, we need the automorphic (vector) bundles over X, together with their canonical and subcanonical extensions over X^{tor}.

Let $\mathrm{Rep}_{\mathbb{C}}(\mathrm{G}_{\mathbb{C}})$ (resp. $\mathrm{Rep}_{\mathbb{C}}(\mathrm{P})$, resp. $\mathrm{Rep}_{\mathbb{C}}(\mathrm{M})$) denote the category of finite dimensional representations of $\mathrm{G}_{\mathbb{C}}$ (resp. P, resp. M) over $\mathbb{C}$. Given any object W in $\mathrm{Rep}_{\mathbb{C}}(\mathrm{P})$, we can naturally define a vector bundle

$$\underline{W} := (\mathrm{G}(\mathbb{C}) \times W)/\mathrm{P}(\mathbb{C})$$

over $\mathcal{H}^\vee = \mathrm{G}(\mathbb{C})/\mathrm{P}(\mathbb{C})$, which pulls back to $\mathcal{H}$ and descends to $X = \Gamma\backslash\mathcal{H}$, still denoted by $\underline{W}$. By abuse of notation, we shall denote the corresponding sheaf of sections over X by the same symbol $\underline{W}$. Given any object W in $\mathrm{Rep}_{\mathbb{C}}(\mathrm{M})$, we can pull it back to an object W in $\mathrm{Rep}_{\mathbb{C}}(\mathrm{P})$ via the canonical homomorphism $\mathrm{P} \twoheadrightarrow \mathrm{M}$ and define a vector bundle $\underline{W}$ (and its sheaf of sections) over X. On the other hand, given any object V in $\mathrm{Rep}_{\mathbb{C}}(\mathrm{G}_{\mathbb{C}})$, its restriction $V|_{\mathrm{P}}$ defines a vector bundle $\underline{V} := \underline{(V|_{\mathrm{P}})}$ over X, and the $\mathrm{Lie}\,\mathrm{G}_{\mathbb{C}}$ action on V defines an integrable connection

$$\nabla : \underline{V} \to \underline{V} \otimes_{\mathscr{O}_X} \Omega^1_{X/\mathbb{C}}. \tag{2.1}$$

Such $\underline{W}$ and $(\underline{V}, \nabla)$ will be called *automorphic bundles* over X.

According to Mumford and Harris (see [Mum77] and [Har89]), we have a *canonical extension*

$$\underline{W}^{\mathrm{can}}$$

of $\underline{W}$ to a vector bundle over X^{tor}, for each $\underline{W}$ as above. As explained in [Har89, Sec. 4], Deligne's canonical extension $(\underline{V}^{\mathrm{can}}, \nabla^{\mathrm{can}})$ of $(\underline{V}, \nabla)$ over X^{tor} (see [Del70]) is compatible with the above in the sense that $\underline{V}^{\mathrm{can}} = (\underline{V|_{\mathrm{P}}})^{\mathrm{can}}$. Following [Har90, Sec. 2], we introduce the *subcanonical extensions*

$$\underline{W}^{\mathrm{sub}} := \underline{W}^{\mathrm{can}}(-D)$$

and

$$\underline{V}^{\mathrm{sub}} := \underline{V}^{\mathrm{can}}(-D)$$

of $\underline{W}$ and $\underline{V}$ over X^{tor}, respectively, where

$$D := (X^{\mathrm{tor}} - X)_{\mathrm{red}}$$

is the (reduced) boundary divisor. Then we obtain integrable connections

$$\nabla : \underline{V}^{\mathrm{can}} \to \underline{V}^{\mathrm{can}} \otimes_{\mathscr{O}_{X^{\mathrm{tor}}}} \Omega^1_{X^{\mathrm{tor}}/\mathbb{C}}(\log D)$$

and

$$\nabla : \underline{V}^{\mathrm{sub}} \to \underline{V}^{\mathrm{sub}} \otimes_{\mathscr{O}_{X^{\mathrm{tor}}}} \Omega^1_{X^{\mathrm{tor}}/\mathbb{C}}(\log D)$$

with *log poles* along D, extending (2.1), for each $\underline{V}$ as above.

The coherent sheaf cohomology groups $H^i(X^{\mathrm{tor}}, \underline{W}^{\mathrm{can}})$ and $H^i(X^{\mathrm{tor}}, \underline{W}^{\mathrm{sub}})$ are called the *coherent cohomology* of X^{tor} of weight W, which are natural generalizations of modular forms (cf. (1.3) and (1.4)). (Such cohomology depends only on X and W, but not on the choice of X^{tor}.) The hypercohomology of the de Rham complex of $\underline{V}^{\mathrm{can}}$ (resp. $\underline{V}^{\mathrm{sub}}$), defined by the connection above with log poles, computes the de Rham cohomology $H^\bullet_{\mathrm{dR}}(X, \underline{V})$ (resp. the compactly supported $H^\bullet_{\mathrm{dR,c}}(X, \underline{V})$).

2.3 Cohomological weights and dual BGG decompositions

Let us compatibly choose weights $\mathbf{X}_{\mathrm{G}_\mathbb{C}} = \mathbf{X}_{\mathrm{M}}$ and positive roots $\Phi^+_{\mathrm{G}_\mathbb{C}}$ and Φ^+_{M}. Then $\Phi^+_{\mathrm{G}_\mathbb{C}} \supset \Phi^+_{\mathrm{M}}$ and $\mathbf{X}^+_{\mathrm{G}_\mathbb{C}} \subset \mathbf{X}^+_{\mathrm{M}}$. For each $\mu \in \mathbf{X}^+_{\mathrm{G}_\mathbb{C}}$ and $\nu \in \mathbf{X}^+_{\mathrm{M}}$, we shall denote by $V_\mu \in \mathrm{Rep}_\mathbb{C}(\mathrm{G}_\mathbb{C})$ and $W_\nu \in \mathrm{Rep}_\mathbb{C}(\mathrm{M})$ the irreducible representations of highest weights μ and ν, respectively. For simplicity, we shall make various minor adjustments in the notation and terminologies without explicitly introducing them. For example, we shall say that $H^i(X^{\mathrm{tor}}, \underline{W}^{\mathrm{can}}_\nu)$ and $H^i(X^{\mathrm{tor}}, \underline{W}^{\mathrm{sub}}_\nu)$ are coherent cohomology *of weight* ν, rather than of weight W_ν.

Consider $\rho = \rho_{\mathrm{G}_\mathbb{C}} := \frac{1}{2}\sum_{\alpha \in \Phi^+_{\mathrm{G}_\mathbb{C}}} \alpha$, $\rho_{\mathrm{M}} := \frac{1}{2}\sum_{\alpha \in \Phi^+_{\mathrm{M}}} \alpha$, and $\rho^{\mathrm{M}} := \rho - \rho_{\mathrm{M}}$. Consider the Weyl groups $W = W_{\mathrm{G}_\mathbb{C}} \supset W_{\mathrm{M}}$, and consider the minimal length representatives $W^{\mathrm{M}} := \{w \in W : w(X^+_{\mathrm{G}_\mathbb{C}}) \subset X^+_{\mathrm{M}}\}$ of $W_{\mathrm{M}} \backslash W$. Consider the usual *dot action* of W on $\mathbf{X}$ given by $w \cdot \mu = w(\mu + \rho) - \rho$.

Definition 2.2. *We say* $\nu \in X^+_{\mathrm{M}}$ *is (de Rham)* **cohomological** *if there exist* $\mu = \mu(\nu) \in \mathbf{X}^+_{\mathrm{G}_\mathbb{C}}$ *and* $w = w(\nu) \in W^{\mathrm{M}}$ *such that* $W_\nu \cong W^\vee_{w \cdot \mu}$.

This notion is justified by Faltings's *dual BGG spectral sequence* (see [Fal83, Sec. 3 and 7] and [FC90, Ch. VI, Sec. 5]) and its degeneracy due to the theory of *mixed Hodge modules* (see [Sai90] and [HZ01, Cor. 4.2.3]):

$$H^i_{\mathrm{dR}}(X, \underline{V}^\vee_\mu) \cong \bigoplus_{w \in W^{\mathrm{M}}} H^{i-l(w)}(X^{\mathrm{tor}}, (\underline{W}^{\mathrm{can}}_{w \cdot \mu})^\vee) \tag{2.3}$$

and

$$H^i_{\mathrm{dR,c}}(X, \underline{V}^\vee_\mu) \cong \bigoplus_{w \in W^{\mathrm{M}}} H^{i-l(w)}(X^{\mathrm{tor}}, (\underline{W}^{\mathrm{sub}}_{w \cdot \mu})^\vee). \tag{2.4}$$

We shall call such isomorphisms *dual BGG decompositions*. The de Rham cohomology at the left-hand sides of (2.3) and (2.4) can be compared with the Betti (singular) and étale cohomology with coefficients in the local systems given by the corresponding sheaves of horizontal sections, while the coherent cohomology on the right-hand sides of (2.3) and (2.4) can be viewed as generalizations of modular forms. But only coherent cohomology of cohomological weights can contribute to the de Rham cohomology via the dual BGG decompositions.

Example 2.5 (G = SL_2; modular curve case).

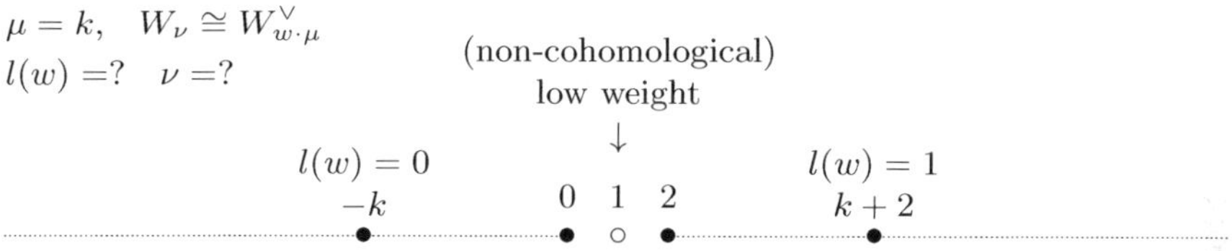

In this case, since $\underline{V}_k = \underline{V}^\vee_k = \mathrm{Sym}^k(\mathbb{C}^{\oplus 2})$ for $k \geq 0$ and $\underline{W}^{\mathrm{can}}_k = (\underline{W}^{\mathrm{can}}_{-k})^\vee = \omega^k$ for all k, the dual BGG decomposition (2.3) can be identified with the Eichler–Shimura isomorphism (1.6) (for weights ≥ 2):

$$H^1_{\mathrm{dR}}(X, \underline{V}_k) \cong H^0(X^{\mathrm{tor}}, \underline{W}^{\mathrm{can}}_{k+2}) \oplus H^1(X^{\mathrm{tor}}, \underline{W}^{\mathrm{can}}_{-k}).$$

Note that *weight* 1 *modular forms* do not contribute to the de Rham cohomology (but can be studied by congruences with forms of cohomological weights).

Example 2.6 (G = $\mathrm{SL}_{2,F}$ with $F/\mathbb{Q}$ real quadratic; Hilbert modular surfaces).

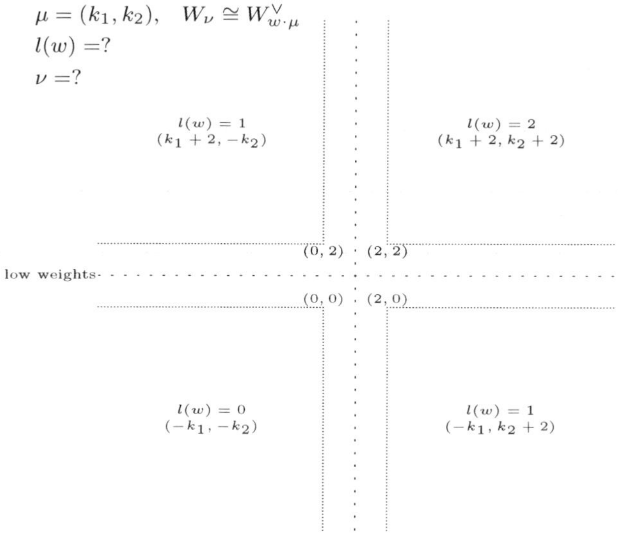

In this case, the dual BGG decomposition gives

$$H^2_{\mathrm{dR}}(X, \underline{V}_{(k_1,k_2)}) \cong H^0(X^{\mathrm{tor}}, \underline{W}^{\mathrm{can}}_{(k_1+2,k_2+2)}) \oplus H^1(X^{\mathrm{tor}}, \underline{W}^{\mathrm{can}}_{(-k_1,k_2+2)}) \\ \oplus H^1(X^{\mathrm{tor}}, \underline{W}^{\mathrm{can}}_{(k_1+2,-k_2)}) \oplus H^2(X^{\mathrm{tor}}, \underline{W}^{\mathrm{can}}_{(-k_1,-k_2)}).$$

Example 2.7 ($\mathrm{G} = \mathrm{Sp}_4$; Siegel threefolds).

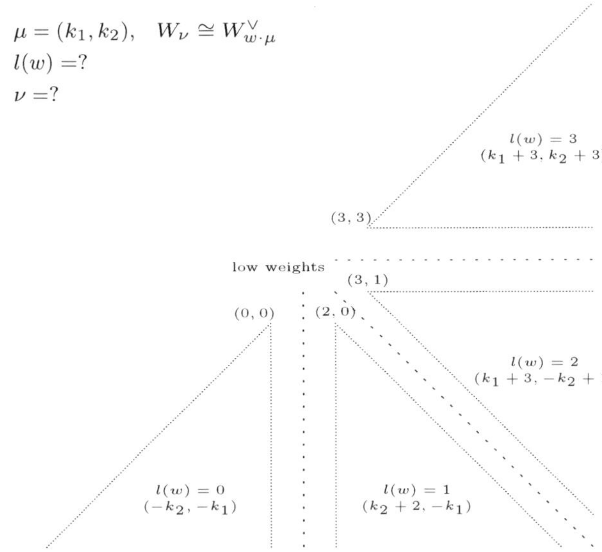

In this case, the dual BGG decomposition gives

$$H^3_{\mathrm{dR}}(X, \underline{V}_{(k_1,k_2)}) \cong H^0(X^{\mathrm{tor}}, \underline{W}^{\mathrm{can}}_{(k_1+3,k_2+3)}) \oplus H^1(X^{\mathrm{tor}}, \underline{W}^{\mathrm{can}}_{(k_1+3,-k_2+1)}) \\ \oplus H^2(X^{\mathrm{tor}}, \underline{W}^{\mathrm{can}}_{(k_2+2,-k_1)}) \oplus H^3(X^{\mathrm{tor}}, \underline{W}^{\mathrm{can}}_{(-k_2,-k_1)})$$

3 Vanishing theorems

In this section, we present some vanishing theorems for the cohomology of automorphic bundles over general locally symmetric varieties, in the author's collaborations with Junecue Suh and Benoît Stroh, together with some examples.

3.1 Positivity of automorphic line bundles

By [BB66], the canonical bundle $\underline{W}_{2\rho^{\mathrm{M}}} \cong \Omega^d_{X/\mathbb{C}}$ of X, where $d = \dim_{\mathbb{C}}(X)$, is *ample*. By [Mum77, Prop. 3.4 b)], $\underline{W}^{\mathrm{can}}_{2\rho^{\mathrm{M}}} \cong \Omega^d_{X^{\mathrm{tor}}/\mathbb{C}}(\log D)$, where $D := (X^{\mathrm{tor}} - X)_{\mathrm{red}}$, descends to an ample line bundle over X^{min}. Hence, although the automorphic line bundle $\underline{W}_{2\rho^{\mathrm{M}}}$ over X is ample, its canonical extension $\underline{W}^{\mathrm{can}}_{2\rho^{\mathrm{M}}}$ over X^{tor} is not ample unless the canonical morphism $X^{\mathrm{tor}} \to X^{\mathrm{min}}$ is an isomorphism.

One might guess that the subcanonical extension $\underline{W}^{\mathrm{sub}}_{2\rho^{\mathrm{M}}}$ over X^{tor} (defined by projective and smooth cone decompositions) is ample, but subcanonical extensions are not preserved by tensor powers, and what can indeed be shown to be ample are more complicated. By Tai's work (see [AMRT75, Ch. IV, Sec. 2]) and the observation first made in [LS11, property (5) preceding (2.1)] (see also [LS13, Prop. 4.2(5)]), there exist an integer N_0 and a (possibly nonreduced) normal crossings divisor D' with $D'_{\mathrm{red}} = D$ such that $(\underline{W}^{\mathrm{can}}_{2\rho^{\mathrm{M}}})^{\otimes N}(-D')$ is ample for all $N \geq N_0$, and this turned out to be the most useful positivity property for our purpose. (As a consequence, although $\underline{W}^{\mathrm{can}}_{2\rho^{\mathrm{M}}}$ is generally not ample, it is still *nef and big*.)

Definition 3.1. *We say* $\nu \in \mathbf{X}^+_{\mathrm{M}}$ *is* **positive parallel** *if* $\dim_{\mathbb{C}} W_\nu = 1$ *and if the pullback of* ν *to each* $\mathbb{Q}$*-simple factor* G' *of* $\mathrm{G}_{\mathbb{C}}$ *is a positive rational multiple of* $\rho^{\mathrm{M}'}$*, for the corresponding factor* M' *of* M.

Example 3.2. Here are some examples of low ranks:

G	all positive parallel weights
SL_2	$\nu = k$ for $k \in \mathbb{Z}_{\geq 1}$
$\mathrm{Res}_{F/\mathbb{Q}} \mathrm{SL}_2$, $F/\mathbb{Q}$ real quadratic	$\nu = k(1,1)$ for $k \in \mathbb{Z}_{\geq 1}$
$\mathrm{SL}_2 \times \mathrm{SL}_2$	$\nu = (k_1, k_2)$ for $k_1, k_2 \in \mathbb{Z}_{\geq 1}$
Sp_4	$\nu = k(1,1)$ for $k \in \mathbb{Z}_{\geq 1}$

Remark 3.3. See [Lan16c, Sec. 3] for a fairly complete description of the smallest positive parallel weights in all cases.

3.2 Vanishing for coherent cohomology

Theorem 3.4 (see [LS12, Thm. 8.7 and 8.20], [LS13, Thm. 8.13 and 8.23], and [Lan16c, Thm. 4.1]). *Let* $\nu \in X^+_{\mathrm{M}}$.

1. *If there exists a positive parallel weight* ν_- *in* X^+_{M} *such that* $\nu + \nu_-$ *is cohomological, then* $H^i(X^{tor}, \underline{W}^{can}_\nu) = 0$ *for* $i < d - l(w(\nu + \nu_-))$.
2. *If there exists a positive parallel weight* ν_+ *in* X^+_{M} *such that* $\nu - \nu_+$ *is cohomological, then* $H^i(X^{tor}, \underline{W}^{sub}_\nu) = 0$ *for* $i > d - l(w(\nu - \nu_+))$.
3. *If there exist positive parallel weights* ν_+ *and* ν_- *in* X^+_{M} *such that* $\nu - \nu_+$ *and* $\nu + \nu_-$ *are both cohomological, then the* **interior cohomology**
$$H^i_{int}(X^{tor}, \underline{W}^{can}_\nu) := \mathrm{im}(H^i(X^{tor}, \underline{W}^{sub}_\nu) \to H^i(X^{tor}, \underline{W}^{can}_\nu))$$
satisfies $H^i_{int}(X^{tor}, \underline{W}^{can}_\nu) = 0$ *for* $i \notin [d - l(w(\nu + \nu_-)), d - l(w(\nu - \nu_+))]$.

Remark 3.5. Using the dual BGG decompositions (2.3) and (2.3) (which replace the usual Hodge decompositions), Theorem 3.4 can be viewed as a Kodaira-type vanishing theorem for the coherent cohomology of automorphic bundles, although (in noncompact cases) it did not follow from vanishing results readily available in the literature of algebraic geometry. (See [LS13, Sec. 1–3] and [Suh].)

3.3 Vanishing for de Rham cohomology

Theorem 3.6 (see [LS12, Thm. 8.16], [LS13, Thm. 8.18], and [Lan16c, Thm. 4.10]). *Suppose* $\mu \in \mathbf{X}^+_{\mathrm{G}_{\mathbb{C}}}$ *is* **sufficiently regular** *in the sense that, for each*

$\alpha \in \Phi^+_{G_{\mathbb{C}}}$, *which comes from some* $\mathbb{C}$*-simple factor of* $G_{\mathbb{C}}$, *we have:*

$$\langle \mu, \alpha^\vee \rangle \geq \begin{cases} 0, & \text{if the factor is compact (i.e., roots are all in } \Phi_M\text{);} \\ 1, & \text{if the factor is not compact and not of types B or C;} \\ 2, & \text{if the factor is not compact but is of types B or C.} \end{cases}$$

Then:

1. $H^i_{dR}(X, \underline{V}^\vee_\mu) = 0$ *for* $i < d$.
2. $H^i_{dR,c}(X, \underline{V}^\vee_\mu) = 0$ *for* $i > d$.
3. *The* **interior cohomology**

$$H^i_{dR,int}(X, \underline{V}^\vee_\mu) := \mathrm{im}(H^i_{dR,c}(X, \underline{V}^\vee_\mu) \to H^i_{dR}(X, \underline{V}^\vee_\mu))$$

satisfies $H^i_{dR,int}(X, \underline{V}^\vee_\mu) = 0$ *for* $i \neq d$.

Remark 3.7. This follows from the vanishing for coherent cohomology (using the dual BGG decompositions (2.3) and (2.3)), and (in noncompact cases) reproves many Hermitian cases of Li and Schwermer's result (see [LS04, Cor. 5.6]). (See [Lan16c, Rem. 4.16] for a more complete documentation of what were known earlier.) It is new (and beyond methods involving either automorphic or Galois representations) when Γ is not a congruence subgroup, although our understanding of the cohomology of automorphic bundles for such general Γ is still very limited.

3.4 Some examples

Example 3.8. This is an example for Theorem 3.4, which is about vanishing for the coherent cohomology, when $G = \mathrm{Res}_{F/\mathbb{Q}} \mathrm{SL}_2$ with $F/\mathbb{Q}$ real quadratic (which is the case of Hilbert modular surfaces). In the following diagram, starting with each weight represented by a bullet $\bullet$, we shift it by both positive and negative multiples of $(1,1)$ (which is the smallest positive parallel weight, as we have seen in Example 3.2), and consider the first cohomological weights we encounter in both directions, represented by two circles $\circ$. Then we record the Weyl lengths $l(w)$ associated with such cohomological weights, and determine the bounds $d - l(w)$ for vanishing degrees (with $d = 2$). For example, if we start with the weight $(2,-1)$, then the first cohomological weight we encounter after shifting by a positive integral multiple of $(1,1)$ is $(3,0)$, which is of Weyl length $l(w) = 1$, and hence we obtain the vanishing of the cohomology of $\underline{W}^{\mathrm{can}}_{(2,-1)}$ over X^{tor} below degree $d - l(w) = 1$. On the other hand, the first cohomological weight we encounter after shifting by a negative integral multiple of $(1,1)$ is $(0,-3)$, which is of Weyl length $l(w) = 0$, and hence we obtain the vanishing of the cohomology of $\underline{W}^{\mathrm{sub}}_{(2,-1)}$ over X^{tor} above degree $d - l(w) = 2$ (which is useless because the cohomology of a surface always vanishes above degree $d = 2$). The cases for the weights $(3,3)$ and $(-1,0)$ are similar.

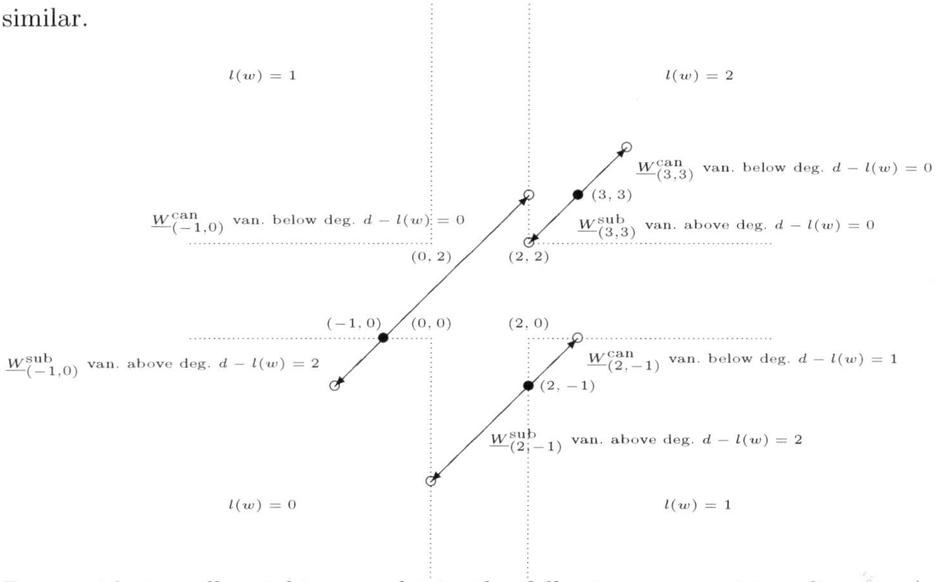

By considering all weights, we obtain the following summarizing diagram (see [Lan16c, Ex. 4.18] for more details), in which an interval $[a,b]$ (or simply $[a] = [a,a]$) means $H^i(X^{\mathrm{tor}}, \underline{W}^{\mathrm{can}}_\nu) = 0$ for $i < a$, $H^i(X^{\mathrm{tor}}, \underline{W}^{\mathrm{sub}}_\nu) = 0$ for $i > b$, and $H^i_{\mathrm{int}}(X^{\mathrm{tor}}, \underline{W}^{\mathrm{can}}_\nu) = 0$ for $i \notin [a,b]$:

•: regular and concentrate in one degree
∘: $[0,2]$ (useless)

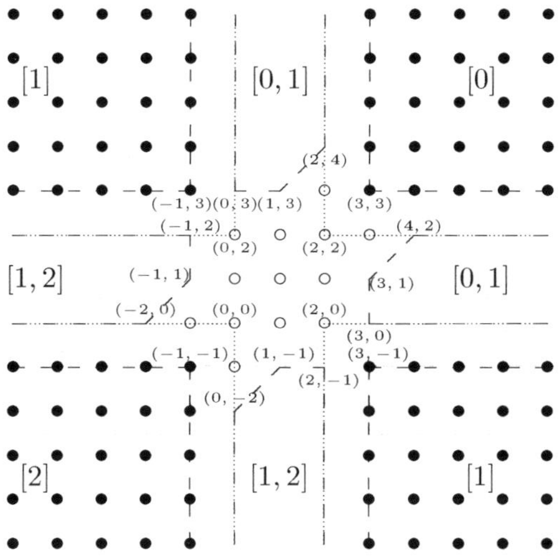

Example 3.9. This is an example for Theorem 3.4, which is about vanishing for the de Rham cohomology, again when $\mathrm{G} = \mathrm{Res}_{F/\mathbb{Q}} \mathrm{SL}_2$ with $F/\mathbb{Q}$ real quadratic. In the following diagram, the bullets • denote the weights of the dual BGG pieces of the sufficiently regular weight $\mu = (1,1)$, and the upshot is that these weights remain cohomological (with the same associated Weyl length) when shifted by the smallest positive parallel weight $(1,1)$ and the opposite $(-1,-1)$.

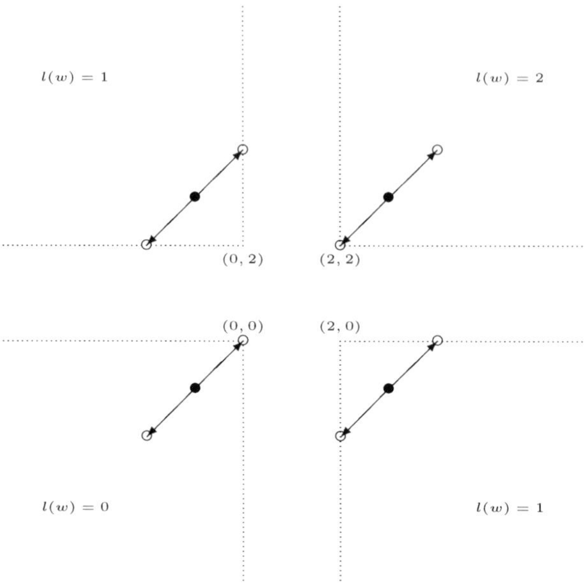

In the following diagram, we apply the Weyl group action and move all weights to the cohomological region of Weyl length zero. Thus, the point of being sufficiently regular is that the weight then remains (up to duality) in the dominant chamber after shifting by the whole Weyl orbit of the smallest positive parallel weight.

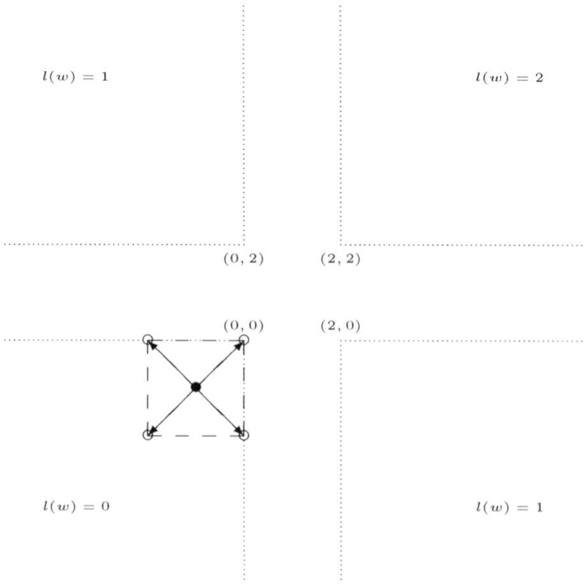

Example 3.10. This is an example for Theorem 3.4 when $\mathrm{Lie}\, G_{\mathbb{R}} \cong \mathfrak{su}_{2,1}$ (which is the case of Picard modular surfaces; see [Lan16c, Ex. 4.19] for more details). (In the following diagram, we visualize $(k_1, k_2; k_3)$ mod $(1, 1; 1)$ by $(k_1 - k_3, k_2 - k_3)$.)

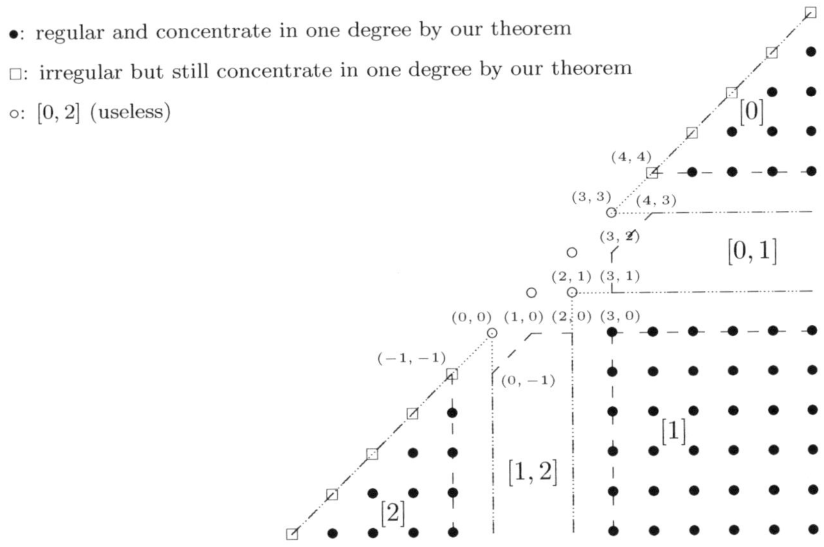

Example 3.11. This is an example for Theorem 3.4 when $G = \mathrm{Sp}_4$ (which is the case of Siegel threefolds; see [Lan16c, Ex. 4.17] for more details).

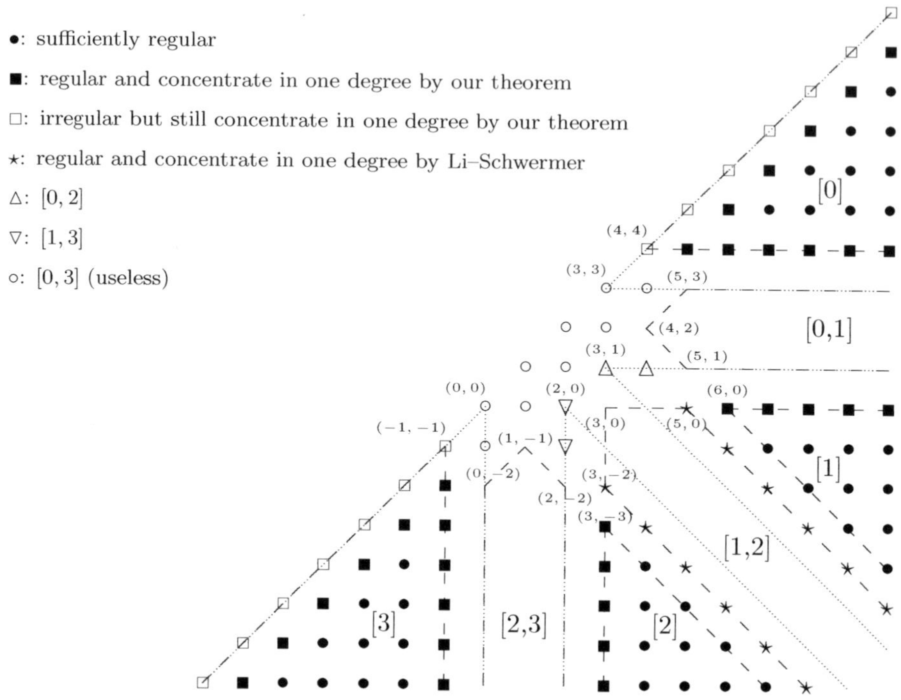

Example 3.12. This is an example for Theorem 3.4 when $\mathrm{Lie}\,\mathrm{G}_{\mathbb{R}} \cong \mathfrak{e}_{7(-25)}$. In this case, we embed $\Phi_{\mathrm{G}_{\mathbb{C}}}$ in $\mathbb{R}^7$ (with Killing form induced, up to scaling, by the Euclidean inner product) with positive simple roots in $\Phi^+_{\mathrm{G}_{\mathbb{C}}}$ given by

$$\alpha_1 = (1,-1,0,0,0,0,0),\ \ \alpha_2 = (0,1,-1,0,0,0,0),$$
$$\alpha_3 = (0,0,1,-1,0,0,0),\ \ \alpha_4 = (0,0,0,1,-1,0,0),$$
$$\alpha_5 = (0,0,0,0,1,-1,0),\ \ \alpha_6 = (0,0,0,0,1,1,0),$$
$$\alpha_7 = (-\tfrac{1}{2},-\tfrac{1}{2},-\tfrac{1}{2},-\tfrac{1}{2},-\tfrac{1}{2},-\tfrac{1}{2},\tfrac{\sqrt{2}}{2}),$$

with $\alpha_1 \notin \Phi_{\mathrm{M}}$. Then we have (see [Lan16c, Ex. 5.46, 5.47, and 5.48]):

ν	cohomological?	$\mu(\nu)$ regular?	$H^i_{\mathrm{int}} = 0$ for?
$(10,10,9,7,4,0,26\sqrt{2})$	yes	yes	$i \neq 6$
$(-14,8,3,2,1,0,\sqrt{2})$	yes	no	$i \notin [25,27]$
$(-7,5,5,2,1,0,3\sqrt{2})$	no	undefined	$i \notin [23,24]$

3.5 Relative vanishing

Let $\pi : X^{\mathrm{tor}} \to X^{\mathrm{min}}$ denote the canonical morphism. A rather unpredicted recent discovery about the cohomology of automorphic bundles is the following:

Theorem 3.13 (see [LS14] and [Lan16c, Thm. 4.5]). *For every* $\nu \in \mathbf{X}^+_{\mathrm{M}}$, *we have* $R^i\pi_*\underline{W}^{sub}_\nu = 0$ *for all* $i > 0$.

Remark 3.14. It is not true in general that $R^i\pi_*\underline{W}^{\mathrm{can}}_\nu = 0$ for all $i > 0$.

Remark 3.15. The first proofs of results like Theorem 3.13 (see, for example, [HLTT16, Thm. 5.4], [Lan17a, Sec. 8.2], and [Lan16b, Thm. 3.9 and Rem. 10.1]) were based on detailed analyses of the formal fibers of π. The later (shorter) proofs in [LS14] and [Lan16c, Thm. 4.5] were based on (various versions of) Theorem 3.4.

3.6 Higher Koecher's principle

The relative vanishing of Theorem 3.13 implies the following generalization of the classical Koecher's principle, whose discovery was surprisingly late, given how fundamental the assertion seems to be:

Theorem 3.16 (higher Koecher's principle; see [Lan16b, Thm. 2.5] and [Lan16c, Thm. 4.7]). *Let* $\nu \in \mathbf{X}^+_{\mathrm{M}}$. *Let*

$$c_X := \mathrm{codim}_{\mathbb{C}}(X^{min} - X, X^{min}).$$

Let us denote by $j^{tor} : X \hookrightarrow X^{tor}$ *and* $j^{min} : X \hookrightarrow X^{min}$ *the canonical morphisms. Then the canonical morphism (induced by* j^{tor}*)*

$$R^i\pi_*\underline{W}^{can}_\nu \to R^i j^{min}_* \underline{W}_\nu$$

is an isomorphism for $i < c_X - 1$, *and is injective for* $i = c_X - 1$. *Hence, by the Leray spectral sequence, for any open* U *in* X^{min}, *the canonical restriction*

$$H^i(\pi^{-1}(U), \underline{W}^{can}_\nu) \to H^i((j^{min})^{-1}(U), \underline{W}_\nu)$$

is bijective (resp. injective) for all $i < c_X - 1$ *(resp.* $i = c_X - 1$*).*

Remark 3.17. This theorem implies its complex analytic analogue, by the same argument as in [Lan16b, Sec. 3], using GAGA (see [Ser56] and [Gro71, XII]) and the results on local cohomology in [Gro68, VIII] and [Siu70].

Remark 3.18. When $i = 0$, $U = X^{\mathrm{min}}$, and $c_X > 1$, this specializes to the classical Koecher's principle, which asserts that a section of $\underline{W}_\nu$ over $(j^{\mathrm{min}})^{-1}(U) = X$ automatically extends to a section of $\underline{W}^{\mathrm{can}}_\nu$ over all of $\pi^{-1}(U) = X^{\mathrm{tor}}$, or in other words that the *growth condition* is unnecessary in the definition of such (generalizations of) modular forms. (See [Lan16b, Sec. 2].)

Example 3.19. Applying Theorem 3.16 with $U = X^{\mathrm{min}}$, we see that the canonical restriction map

$$H^i(X^{\mathrm{tor}}, \underline{W}^{\mathrm{can}}_\nu) \to H^i(X, \underline{W}_\nu) \tag{3.20}$$

is bijective if $i < c_X - 1$, and injective if $i = c_X - 1$. For the values of c_X, we have the following summarizing table:

G (most split form)	$\mathrm{rk}_{\mathbb{Q}}$	$\dim_{\mathbb{C}}(X)$	$\dim_{\mathbb{C}}(X^{\min} - X)$	c_X
SL_2	1	1	0	1
$\mathrm{SL}_{2,F}$, $F/\mathbb{Q}$ totally real	1	$d = [F : \mathbb{Q}]$	0	d
Sp_4	2	3	1	2
Sp_{2n}, $n \geq 1$	n	$\frac{1}{2}n(n+1)$	$\frac{1}{2}n(n-1)$	n
$\mathrm{Lie}\, \mathrm{G}_{\mathbb{R}} \cong \mathfrak{su}_{2,1}$	1	2	0	2
$\mathrm{Lie}\, \mathrm{G}_{\mathbb{R}} \cong \mathfrak{su}_{n,1}$, $n \geq 1$	1	n	0	n
$\mathrm{Lie}\, \mathrm{G}_{\mathbb{R}} \cong \mathfrak{su}_{a,b}$, $ab \geq 1$	$\min(a, b)$	ab	$(a-1)(b-1)$	$a+b-1$
$\mathrm{Lie}\, \mathrm{G}_{\mathbb{R}} \cong \mathfrak{so}_{1,2}$	1	1	0	1
$\mathrm{Lie}\, \mathrm{G}_{\mathbb{R}} \cong \mathfrak{so}_{n,2}$, $n \geq 2$	2	n	1	$n-1$
$\mathrm{Lie}\, \mathrm{G}_{\mathbb{R}} \cong \mathfrak{so}^*_{2n}$, $n \geq 2$	$\lfloor \frac{n}{2} \rfloor$	$\frac{1}{2}n(n-1)$	$\frac{1}{2}(n-2)(n-3)$	$2n-3$
$\mathrm{Lie}\, \mathrm{G}_{\mathbb{R}} \cong \mathfrak{e}_{6(-14)}$	2	16	5	11
$\mathrm{Lie}\, \mathrm{G}_{\mathbb{R}} \cong \mathfrak{e}_{7(-25)}$	3	27	10	17

Remark 3.21. If $i = c_X - 1$, then (3.20) can actually fail to be surjective, up to shifting ν by a sufficiently large multiple of $2\rho^{\mathrm{M}}$ (by the same argument as in [Lan16b, Sec. 9]). This suggests that there are some mysterious *fake modular forms* in degree $c_X - 1$. But we have no idea what they are—the proof of their existence is based on [Gro68, VIII, Prop. 3.2], a general fact in the theory of local cohomology.

3.7 Vanishing over integral models

To prove analogues of the above theorems with coefficients over integers or their reductions modulo integers, we can no longer just work with the complex analytic locally symmetric varieties—we need certain good integral models. (We are being vague here about what "good" means.)

So far all known constructions of good integral models rely on an important coincidence: When $\mathrm{G} = \mathrm{Sp}_{2n}$ or better GSp_{2n} (to avoid introducing roots of unity into the base rings), X is a moduli space of principally polarized abelian varieties with some level structures over $\mathbb{C}$. This led to the construction of good integral models of X for the *PEL-* and *Hodge-type* Shimura varieties, by taking normalizations over moduli of *abelian schemes* with PEL structures (i.e., polarizations, endomorphism structures, and level structures) and with certain Hodge tensors. (See, for example, [Kot92] and [Lan13] for the construction of *smooth PEL moduli*; see [Kis10] and [KMP16] for the construction of good reduction integral models of Hodge-type Shimura varieties; and see [Lan16a] for the construction by normalization in all PEL-type cases, allowing bad reduction.) We now know how to also construct good integral models of X^{tor} and $X^{\min}$ in such cases, allowing arbitrary ramifications, levels, polarization degrees, and isogeny collections in all PEL-type cases. (See [Lan13], [Lan12c], [Lan12a], [Lan16a], [Lan15], [Lan17b], and [Lan18] for various constructions and comparisons in PEL-type cases; and see [MP15] for a different approach in Hodge-type cases, with some restriction on the levels at p allowing all hyperspecial and many parahoric ones.) The construction of good reduction integral models of X extends to the *abelian-type* cases—see [Kis10] and

[KMP16] again. In [KP15], the construction is further extended to allow bad reductions with parahoric levels at p (under the assumptions that $p > 2$ and that G splits over a tamely ramified extension of $\mathbb{Q}_p$). (The construction for X^{tor} and $X^{\min}$ should also extend to abelian-type cases with parahoric levels at p, but is not carried out yet.)

At least in PEL-type cases, the automorphic bundles $\underline{W}_\nu$ and their extensions $\underline{W}_\nu^{\mathrm{can}}$ and $\underline{W}_\nu^{\mathrm{sub}}$ can also be defined over even bad reduction integral models of X. In good reduction cases, we can also consider $(\underline{V}_\mu, \nabla)$ over smooth integral models of X, their extensions $(\underline{V}_\mu^{\mathrm{can}}, \nabla)$ and $(\underline{V}_\mu^{\mathrm{sub}}, \nabla)$ over smooth integral models X^{tor} with log poles over $D = (X^{\mathrm{tor}} - X)_{\mathrm{red}}$, and their (log) de Rham cohomology; and we have p-torsion or p-integral analogues of Theorems 3.4 and 3.6 when p is unramified and larger than a bound (dependent on ν or μ but) independent of the level away from p. The short exact sequence

$$0 \to \underline{W}_{\mathbb{Z}_{(p)}}^{\mathrm{can}} \xrightarrow{p} \underline{W}_{\mathbb{Z}_{(p)}}^{\mathrm{can}} \to \underline{W}_{\mathbb{Z}/p\mathbb{Z}}^{\mathrm{can}} \to 0$$

induces a long exact sequence

$$\cdots \to H^{i-1}(\underline{W}_{\mathbb{Z}/p\mathbb{Z}}^{\mathrm{can}}) \to H^i(\underline{W}_{\mathbb{Z}_{(p)}}^{\mathrm{can}}) \xrightarrow{p} H^i(\underline{W}_{\mathbb{Z}_{(p)}}^{\mathrm{can}}) \to H^i(\underline{W}_{\mathbb{Z}/p\mathbb{Z}}^{\mathrm{can}}) \to \cdots .$$

Hence, vanishing mod p induces not just vanishing over $\mathbb{Z}_{(p)}$, but also *freeness* and *liftability*. By *crystalline comparison* (see [LS12, Sec. 5] and [LS13, Sec. 9], and the references there), under certain assumptions on p (and on the local properties of the integral models), the vanishing of (log) de Rham cohomology mod p implies the vanishing of the corresponding étale and Betti cohomology with coefficients mod p (which then also implies the corresponding freeness and liftability assertions). (For more details, see [LS12], [LS13], and [LP]; see also the survey article [Lan12b].)

On the other hand, the relative vanishing in Theorem 3.13 has analogues over integral models of PEL-type Shimura varieties even in bad reduction cases (see, for example, [Lan17b, Thm. 8.6] and [Lan18, Thm. 4.4.9]), which are important because, roughly speaking, many new techniques for congruences involve the consideration of the sheaves $\pi_* \underline{W}_\nu^{\mathrm{sub}}$ over $X^{\min}$. (There is no restriction at all on the residue characteristics involved. We do not need them to be larger than any bounds dependent on, for example, the levels or weights.) However, in such generality, we have to resort to some detailed analyses of formal fibers of π, because we no longer expect any Kodaira-type vanishing as in Theorem 3.4 to be true (cf. Remark 3.15). Thus, the nature of the relative vanishing as in Theorem 3.13 remains mysterious.

4 Application to the construction of Galois representations

4.1 Conjectural framework and historical developments

Let us begin with the following conjectural framework (incorporating conjectures due to Langlands, Clozel [Clo90], Fontaine–Mazur [FM97], and some others).

Given any prime number p, any field isomorphism $\iota : \overline{\mathbb{Q}}_p \xrightarrow{\sim} \mathbb{C}$, any number field F, and any $n \in \mathbb{Z}_{\geq 1}$, it is conjectured that there is a natural bijection:

$$\left\{\begin{matrix}\text{irreducible algebraic cuspidal}\\ \text{automorphic representations}\\ \pi \cong \otimes'_v \pi_v \text{ of } \mathrm{GL}_n(\mathbb{A}_F)\end{matrix}\right\}_{/\cong} \leftrightarrow \left\{\begin{matrix}\text{irreducible algebraic}\\ \text{continuous representations}\\ r : \mathrm{Gal}(\overline{F}/F) \to \mathrm{GL}_n(\overline{\mathbb{Q}}_p)\end{matrix}\right\}_{/\cong}$$

The term *algebraic* at the left-hand side means the Harish-Chandra parameter of the archimedean component $\pi_\infty = \otimes_{v|\infty} \pi_v$ is integral after shifting by the half-sum of positive roots. The term *algebraic* at the right-hand side means r is unramified at all but finitely many places and de Rham at all $v|p$. (These are often called *geometric* because they are satisfied by the p-adic étale cohomology of varieties over F.) The bijection is *natural* in the sense that it matches the local L-factors at both sides—or it suffices to know the *local-global compatibility* at all finite places $v \nmid p$ over which F, π_v, and r_v are all unramified, by matching (under ι) the Satake parameters of π_v with the Frobenius eigenvalues of r_v (up to some normalization).

Here is a summary of some historical developments:

- When $n = 1$, the conjecture is known, thanks to *class field theory.*
- When $n > 1$, the conjecture is often studied in two directions: *attaching Galois representations* r to automorphic representations π, and the converse (i.e., *modularity* or *automorphy* problems).
- When F is CM or totally real, for cohomological (i.e., regular algebraic) and *polarized* π, one can construct r by traditional methods: by realizing r (up to base change, patching, and twisting) in the étale cohomology of *Shimura varieties* when π_∞ has sufficient regular weights (which is often proved by trace formula techniques), and by extending such r to other cohomological weights by congruences. (One can further extend the construction to cases where π_∞ is some holomorphic limits of discrete series, realized in H^0, by congruences; see, for example, [Tay91].)
- The modularity or automorphy problems are often studied by the so-called *Taylor–Wiles method.* For the method to work in higher dimensions, one often needs the cohomology to behave like free modules over some auxiliary Hecke algebras, and this requires rather strong vanishing results.

4.2 Removal of polarizability condition

Theorem 4.1 (Harris–Lan–Taylor–Thorne; see [HLTT16, Introduction, Thm. A]). *Let p, $\iota : \overline{\mathbb{Q}}_p \xrightarrow{\sim} \mathbb{C}$, and n be as above, and F totally real or CM. Suppose π is a* **cohomological** *cuspidal automorphic representation of* $\mathrm{GL}_n(\mathbb{A}_F)$ *(which is then necessarily algebraic). Then there is a unique semisimple representation*

$$r = r_{p,\iota}(\pi) : \mathrm{Gal}(\overline{F}/F) \to \mathrm{GL}_n(\overline{\mathbb{Q}}_p)$$

such that, if $\ell \neq p$ is a prime above which F and π are both unramified and if $v|\ell$ is a place of F, then r is unramified at v and satisfies the local-global compatibility

$$r|_{W_{F_v}}^{ss} = \iota^{-1} \operatorname{rec}_{F_v}(\pi_v |\det|_v^{\frac{1-n}{2}}).$$

Remark 4.2. The most important point of Theorem 4.1 is that we impose *no polarizability condition* on π (i.e., not requiring π to be conjugate self-dual up to character twists). Then r generally does not occur in the p-adic étale cohomology of any Shimura variety!

Instead, we constructed r using *p-adic limits* of Galois representations which do occur in the p-adic étale cohomology of some Shimura varieties. The main ideas can be briefly (and somewhat imprecisely) summarized as follows:

1. We may assume that $n > 1$. Also, by patching, we may assume that F is CM and satisfies certain simplifying assumptions.
2. The starting point is Skinner's idea that, to construct $r = r_{p,\iota}(\pi)$, it suffices to construct $R_t = r_{p,\iota}(\pi|\det|^t) \oplus r_{p,\iota}(\pi|\det|^t)^{c,\vee}$ for sufficiently many t.
3. Each R_t should correspond to Eisenstein series for "GU(n,n)" (with Levi $\mathrm{GL}_{n,F} \times \mathrm{G_m}$). But nothing along that line really works.
4. Rather, we construct them as *overconvergent cusp forms*, which are global sections of $\pi_* \underline{W}_\nu^{\mathrm{sub},\dagger}$ (i.e., the overconvergent version of $\pi_* \underline{W}_\nu^{\mathrm{sub}}$) over the affinoid (multiplicative-type) ordinary locus $X^{\mathrm{ord},\mathrm{min},\dagger}$ of $X^{\mathrm{min},\dagger}$ for some $\nu \in \mathbf{X}_{\mathrm{M}}^+$, for some "GU$(n,n)$" Shimura varieties X. (Here the superscripts ^{ord} and † mean the *dagger spaces* attached to the tubes defined by certain p-integral models with only *ordinary loci* in characteristic p.) They are naturally p-adic limits of cusp forms whose associated Galois representations were already known (thanks to historical developments in the *polarizable case*, because we are working on Shimura varieties associated with unitary similitude groups over CM fields).
5. Using the mysterious relative vanishing $R^i \pi_* \underline{W}_\nu^{\mathrm{sub},\dagger} = 0$ for $i > 0$, it suffices to construct sections of $\underline{W}_\nu^{\mathrm{sub},\dagger}$ over the (non-affinoid) ordinary locus $X^{\mathrm{ord},\mathrm{tor},\dagger}$ of $X^{\mathrm{tor},\dagger}$, and we achieved this using the Hodge spectral sequence of certain *rigid cohomology of* $\overline{X}^{ord}$ *with compact support along the partial boundary* $\overline{X}^{ord,tor} - \overline{X}^{ord}$, or of the analogue with X and X^{tor} replaced with certain Kuga families (which are self-fiber-products of the universal abelian schemes over the original X) and their toroidal compactifications. (Here the overlines and the superscripts ^{ord} mean the characteristic p fibers of certain integral models with only ordinary loci, the same ones used in the definition of the dagger spaces above.)
6. The key observation is the following: The partial boundary $\overline{X}^{\mathrm{ord},\mathrm{tor}} - \overline{X}^{\mathrm{ord}}$ and its analogue for Kuga families can be arranged to be simple normal crossings divisors such that the incidence relations of their smooth irreducible components are essentially the same as in characteristic zero, which encode the interior cohomology of the (real analytic) locally symmetric manifold associated with $\mathrm{GL}_{n,F}$ (with coefficients in local systems associated with finite-dimensional representations of $\mathrm{GL}_{n,F}$ of polynomial weights). Then a Frobenius-weight argument (based on [Ber97, Thm. 3.1] and [Chi98, Thm. 2.2]) shows that such interior cohomology contributes to the rigid cohomology mentioned above. (Note that we are really realizing the interior cohomology of a real analytic manifold in the rigid cohomology of some characteristic p algebraic variety.) But all π considered in Theorem 4.1 (or more precisely

their nonarchimedean components) contribute up to determinant twists (and ι) to such interior cohomology!

Remark 4.3. In [Sch15], Scholze reproved Theorem 4.1 using a much more advanced method, and also treated the analogue for torsion cohomological classes—in fact, the consideration of torsion classes is crucial in his argument. Nevertheless, to show that the Galois representation r constructed in Theorem 4.1 is indeed *algebraic*, at least with current techniques, it still seems easier to use the results in [HLTT16].

There have been many other exciting developments since [HLTT16] and [Sch15], but it is not easy to give a fair overview of all of them, and we decided to stop here due to limitation of time and energy.

Acknowledgements

The author is partially supported by the National Science Foundation under agreement No. DMS-1352216 and by an Alfred P. Sloan Research Fellowship. He would also like to thank the anonymous referee for a careful reading and many helpful comments.

References

[AMRT75] A. Ash, D. Mumford, M. Rapoport, and Y. Tai, *Smooth compactification of locally symmetric varieties*, Lie Groups: History Frontiers and Applications, vol. 4, Math Sci Press, Brookline, Massachusetts, 1975, TEXified and republished as [AMRT10].

[AMRT10] ———, *Smooth compactification of locally symmetric varieties*, 2nd ed., Cambridge Mathematical Library, Cambridge University Press, Cambridge, New York, 2010.

[BB66] W. L. Baily, Jr. and A. Borel, *Compactification of arithmetic quotients of bounded symmetric domains*, Ann. Math. (2) **84** (1966), no. 3, 442–528.

[Ber97] P. Berthelot, *Finitude et pureté cohomologique en cohomologie rigide*, Invent. Math. **128** (1997), 329–377.

[Chi98] B. Chiarellotto, *Weights in rigid cohomology applications to unipotent F-isocrystals*, Ann. Sci. Ecole Norm. Sup. (4) **31** (1998), no. 5, 683–715.

[Clo90] L. Clozel, *Motifs et formes automorphes: applications du principe de fonctorialité*, in Clozel and Milne [CM90a], pp. 77–159.

[CM90a] L. Clozel and J. S. Milne (eds.), *Automorphic forms, Shimura varieties, and L-functions. Volume I*, Perspectives in Mathematics, vol. 10, Proceedings of a Conference held at the University of Michigan, Ann Arbor, July 6–16, 1988, Academic Press Inc., Boston, 1990.

[CM90b] L. Clozel and J. S. Milne (eds.), *Automorphic forms, Shimura varieties, and L-functions. Volume II*, Perspectives in Mathematics, vol. 11, Proceedings of a Conference held at the University of Michigan, Ann Arbor, July 6–16, 1988, Academic Press Inc., Boston, 1990.

[CY97] J. Coates and S.T. Yau (eds.), *Elliptic curves, modular forms & Fermat's last theorem*, Proceedings of a conference held in the Institute of Mathematics of the Chinese Univerisity of Hong Kong, International Press, Cambridge, Massachusetts, 1997.

[Del70] P. Deligne, *Equations différentielles à points singuliers réguliers*, Lecture Notes in Mathematics, vol. 163, Springer-Verlag, Berlin, Heidelberg, New York, 1970.

[DS74] P. Deligne and J.-P. Serre, *Formes modulaires de poids 1*, Ann. Sci. Ecole Norm. Sup. (4) **7** (1974), no. 4, 507–530.

[Fal83] G. Faltings, *On the cohomology of locally symmetric hermitian spaces*, in Malliavin [Mal83], pp. 55–98.

[FC90] G. Faltings and C.-L. Chai, *Degeneration of abelian varieties*, Ergebnisse der Mathematik und ihrer Grenzgebiete, 3. Folge, vol. 22, Springer-Verlag, Berlin, Heidelberg, New York, 1990.

[FM97] J.-M. Fontaine and B. Mazur, *Geometric Galois representations*, in Coates and Yau [CY97], pp. 190–227.

[Gro68] A. Grothendieck (ed.), *Cohomologie locale des faiseaux cohérents et théorèmes de Lefschetz locaux et globaux* (*SGA 2*), North-Holland Publishing Co., Amsterdam, 1968.

[Gro71] A. Grothendieck (ed.), *Revêtements étales et groupe fondamental* (*SGA 1*), Lecture Notes in Mathematics, vol. 224, Springer-Verlag, Berlin, Heidelberg, New York, 1971.

[Har89] M. Harris, *Functorial properties of toroidal compactifications of locally symmetric varieties*, Proc. London Math. Soc. (3) **59** (1989), 1–22.

[Har90] ———, *Automorphic forms and the cohomology of vector bundles on Shimura varieties*, in Clozel and Milne [CM90b], pp. 41–91.

[Hel01] S. Helgason, *Differential geometry, Lie groups, and symmetric spaces*, Graduate Studies in Mathematics, vol. 34, American Mathematical Society, Providence, Rhode Island, 2001.

[HLTT16] M. Harris, K.-W. Lan, R. Taylor, and J. Thorne, *On the rigid cohomology of certain Shimura varieties*, Res. Math. Sci. **3** (2016), article no. 37, 308 pp.

[HZ01] M. Harris and S. Zucker, *Boundary cohomology of Shimura varieties III. Coherent cohomology on higher-rank boundary strata and applications to Hodge theory*, Mémoires de la Société Mathématique de France. Nouvelle Série, vol. 85, Société Mathématique de France, Paris, 2001.

[JPYY12] L. Ji, Y. S. Poon, L. Yang, and S.-T. Yau (eds.), *Fifth International Congress of Chinese Mathematicians*, AMS/IP Studies in Advanced Mathematics, vol. 51, part 1, Proceedings of the Fifth International Congress of Chinese Mathematicians held at Tsinghua University, Beijing, in December 2010, American Mathematical Society, Providence, Rhode Island, and International Press, Cambridge, MA, 2012.

[Kis10] M. Kisin, *Integral models for Shimura varieties of abelian type*, J.

Amer. Math. Soc. **23** (2010), no. 4, 967–1012.

[KMP16] W. Kim and K. Madapusi Pera, *2-adic integral canonical models*, Forum Math. Sigma **4** (2016), e28, 34.

[Kot92] R. E. Kottwitz, *Points on some Shimura varieties over finite fields*, J. Amer. Math. Soc. **5** (1992), no. 2, 373–444.

[KP15] M. Kisin and G. Pappas, *Integral models of Shimura varieties with parahoric level structure*, preprint, 2015.

[Lan12a] K.-W. Lan, *Comparison between analytic and algebraic constructions of toroidal compactifications of PEL-type Shimura varieties*, J. Reine Angew. Math. **664** (2012), 163–228.

[Lan12b] ———, *Geometric modular forms and the cohomology of torsion automorphic sheaves*, in Ji et al. [JPYY12], pp. 183–208.

[Lan12c] ———, *Toroidal compactifications of PEL-type Kuga families*, Algebra Number Theory **6** (2012), no. 5, 885–966.

[Lan13] ———, *Arithmetic compactification of PEL-type Shimura varieties*, London Mathematical Society Monographs, vol. 36, Princeton University Press, Princeton, 2013, errata available on the author's website.

[Lan15] ———, *Boundary strata of connected components in positive characteristics*, Algebra Number Theory **9** (2015), no. 9, 1955–2054, an appendix to the article "Families of nearly ordinary Eisenstein series on unitary groups" by Xin Wan.

[Lan16a] ———, *Compactifications of PEL-type Shimura varieties in ramified characteristics*, Forum Math. Sigma **4** (2016), e1, 98 pp.

[Lan16b] ———, *Higher Koecher's principle*, Math. Res. Lett. **23** (2016), no. 1, 163–199.

[Lan16c] ———, *Vanishing theorems for coherent automorphic cohomology*, Res. Math. Sci. **3** (2016), article no. 39, 43 pp.

[Lan17a] ———, *Compactifications of PEL-type Shimura varieties and Kuga families with ordinary loci*, World Scientific, Singapore, 2017.

[Lan17b] ———, *Integral models of toroidal compactifications with projective cone decompositions*, Int. Math. Res. Not. IMRN **2017** (2017), no. 11, 3237–3280.

[Lan18] ———, *Compactifications of splitting models of PEL-type Shimura varieties*, Trans. Amer. Math. Soc. **370** (2018), no. 4, 2463–2515.

[LP] K.-W. Lan and P. Polo, *Dual BGG complexes for automorphic bundles*, to appear in Math. Res. Lett.

[LS04] J.-S. Li and J. Schwermer, *On the Eisenstein cohomology of arithmetic groups*, Duke Math. J. **123** (2004), no. 1, 141–169.

[LS11] K.-W. Lan and J. Suh, *Liftability of mod p cusp forms of parallel weights*, Int. Math. Res. Not. IMRN **2011** (2011), 1870–1879.

[LS12] ———, *Vanishing theorems for torsion automorphic sheaves on compact PEL-type Shimura varieties*, Duke Math. J. **161** (2012), no. 6, 1113–1170.

[LS13] ———, *Vanishing theorems for torsion automorphic sheaves on general PEL-type Shimura varieties*, Adv. Math. **242** (2013), 228–286.

[LS14] K.-W. Lan and B. Stroh, *Relative cohomology of cuspidal forms on*

PEL-type Shimura varieties, Algebra Number Theory **8** (2014), no. 8, 1787–1799.

[Mal83] M.-P. Malliavin (ed.), *Séminaire d'algèbre Paul Dubreil et Marie–Paule Malliavin*, Lecture Notes in Mathematics, vol. 1029, Proceedings, Paris 1982 (35ème Année), Springer-Verlag, Berlin, Heidelberg, New York, 1983.

[Mil90] J. S. Milne, *Canonical models of (mixed) Shimura varieties and automorphic vector bundles*, in Clozel and Milne [CM90a], pp. 283–414.

[MP15] K. Madapusi Pera, *Toroidal compactifications of integral models of Shimura varieties of Hodge type*, preprint, 2015.

[Mum77] D. Mumford, *Hirzebruch's proportionality theorem in the non-compact case*, Invent. Math. **42** (1977), 239–272.

[Sai90] M. Saito, *Mixed Hodge modules*, Publ. Res. Inst. Math. Sci. **26** (1990), no. 2, 221–333.

[Sch15] P. Scholze, *On torsion in the cohomology of locally symmetric varieties*, Ann. Math. (2) **182** (2015), no. 3, 946–1066.

[Ser56] J.-P. Serre, *Geométrie algébrique et géométrie analytique*, Ann. Inst. Fourier. Grenoble **6** (1955–1956), 1–42.

[Siu70] Y.-T. Siu, *Analytic sheaves of local cohomology*, Trans. Amer. Math. Soc. **148** (1970), 347–366.

[Suh] J. Suh, *Vanishing theorems for mixed hodge modules and applications*, to appear in J. Eur. Math. Soc.

[Tay91] R. Taylor, *Galois representations associated to Siegel modular forms of low weight*, Duke Math. J. **63** (1991), no. 2, 281–332.

School of Mathematics, University of Minnesota
Minneapolis, MN 5545, USA
E-mail address: kwlan@math.umn.edu

Proceedings of the 7th International Congress of Chinese Mathematicians
ALM 43, pp. 327–348

New Developments in Mean Curvature Flow of Arbitrary Codimension Inspired by Yau Rigidity Theory

Li Lei and Hong-Wei Xu

Abstract

In this survey, we will focus on the mean curvature flow theory with sphere theorems, and discuss the recent developments on the convergence theorems for the mean curvature flow of arbitrary codimension inspired by the Yau rigidity theory of submanifolds. Several new differentiable sphere theorems for submanifolds are obtained as consequences of the convergence theorems for the mean curvature flow. It should be emphasized that Theorem 4.1 is an optimal convergence theorem for the mean curvature flow of arbitrary codimension, which implies the first optimal differentiable sphere theorem for submanifolds with positive Ricci curvature. Finally, we present a list of unsolved problems in this area.

Mathematics Subject Classification: 53C24, 53C20, 53C40, 53C42.

Keywords and Phrases: Mean curvature flow of arbitrary codimension, curvature and topology of submanifolds, convergence theorems, sphere theorems, rigidity theory.

1 Huisken's classical theorems for MCF of hypersurfaces

Let (M, g) be a closed Riemannian n-manifold, and let $F_t : M^n \to N^{n+q}$ be a one-parameter family of smooth submanifolds immersed in an $(n+q)$-dimensional Riemannian manifold (N, h). We say that $M_t = F_t(M)$ is a solution of the mean curvature flow if F_t satisfies

$$\begin{cases} \frac{\partial}{\partial t}F(x,t) = H(x,t), \\ F(x,0) = F_0(x), \end{cases} \tag{1.1}$$

where $F(x,t) = F_t(x)$, $H(x,t)$ is the mean curvature vector, and $F_0(M)$ is the initial submanifold immersed in N.

In 1984, Huisken first studied the mean curvature flow for compact hypersurfaces in the Euclidean and proved the following convergence theorem [19].

Research supported by the National Natural Science Foundation of China, Grant Nos. 11531012, 11371315.

Theorem 1.1. *Let M_0 be an n-dimensional ($n \geq 2$) closed hypersurface in $\mathbb{R}^{n+1}$. Assume that M_0 is uniformly convex. Then the mean curvature flow with initial value M_0 has a smooth solution on a finite time interval $0 \leq t < T$, and converge to a single round point as $t \to T$.*

Afterwards, Huisken [20] generalized Theorem 1.1 to the case of the mean curvature flow of compact hypersurfaces in a Riemannian manifold N^{n+1} with bounded geometry, which is stated as follows.

Theorem 1.2. *Let $n \geq 2$ and N^{n+1} be a complete Riemannian manifold which satisfies uniform bounds*

$$-K_1 \leq \bar{K}_N \leq K_2, \quad |\bar{\nabla}\bar{R}| \leq L^2, \quad \mathrm{inj}(N) \geq i_N,$$

for nonnegative constants K_1, K_2, L and positive constant i_N. Let M_0 be a closed hypersurface immersed in N, and suppose that M_0 satisfies the pinching condition

$$|H|h_{ij} > nK_1 g_{ij} + \frac{n^2}{|H|} L g_{ij}.$$

Then the mean curvature flow with initial value M_0 has a smooth solution on a finite time interval $0 \leq t < T$, and converge to a single round point as $t \to T$.

Motivated by the rigidity theorem for hypersurfaces with constant mean curvature in a sphere due to Okumura [40], Huisken [21] proved the following convergence theorem for the mean curvature flow in a sphere.

Theorem 1.3. *Let $n \geq 2$ and $\mathbb{F}^{n+1}(c)$ be a spherical space form with positive constant curvature c. Let M_0 be a closed hypersurface immersed in $\mathbb{F}^{n+1}(c)$, and suppose that M_0 satisfies the pinching condition*

$$|A|^2 < \begin{cases} \frac{3}{4}|H|^2 + \frac{4}{3}c, & n = 2, \\ \frac{1}{n-1}|H|^2 + 2c, & n \geq 3. \end{cases} \tag{1.2}$$

Then one of the following holds:
(i) The mean curvature flow with initial value M_0 has a smooth solution M_t on a finite time interval $0 \leq t < T$ and the M_t's converge uniformly to a single round point as $t \to T$.
(ii) The mean curvature flow with initial value M_0 has a smooth solution M_t for all $0 \leq t < \infty$ and the M_t's converge in the C^∞-topology to a smooth totally geodesic hypersurface M_∞.

During the past three decades, there are many progresses on the theory of mean curvature flows. Most of the results focus on the mean curvature flow of hypersurfaces. On the other hand, fruitful results on the mean curvature flow of submanifolds of higher codimension were obtained by several geometers. Mean curvature flow of surfaces in 4-manifolds, Lagrangian mean curvature flow of higher codimension, graphic mean curvature flow, mean curvature flow with convex Gauss image were investigated by Chen, Li, Smoczyk, Wang, Xin and others [8, 50, 51, 52, 53, 54, 55, 56].

In the present article, we will introduce the recent developments on the convergence theorems for the mean curvature flow of arbitrary codimension inspired by the Yau rigidity theory of submanifolds with parallel mean curvature. Several new differentiable sphere theorems for submanifolds are obtained as consequences of the convergence theorems for the mean curvature flow. Finally, we present a list of conjectures in this area.

2 Yau rigidity theory of submanifolds and its developments

More than forty years ago, Okumura [39, 40] investigated compact submanifolds with parallel mean curvature and flat normal bundle in $\mathbb{S}^{n+q}$ whose squared norm of the second fundamental form satisfies $|A|^2 < 2 + \frac{|H|^2}{n-1}$. Meanwhile, Chen-Okumura [7] proved a rigidity theorem for compact submanifolds with parallel mean curvature in Euclidean spaces under pinching condition $|A|^2 < \frac{|H|^2}{n-1}$. In 1975, Yau [65] established the rigidity theory of submanifolds with parallel mean curvature, which includes the following important results.

Theorem 2.1. *Let M be an n-dimensional compact submanifold with parallel mean curvature in $\mathbb{S}^{n+q}$ with $q > 1$. If $|A|^2 < n[3 + n^{\frac{1}{2}} - (q-1)^{-1}]^{-1}$, then M lies in a totally geodesic sphere $\mathbb{S}^{n+1}$.*

Theorem 2.2. *Let M be an n-dimensional oriented compact minimal submanifold in $\mathbb{S}^{n+q}$. If $K_M \geq \frac{q-1}{2q-1}$, then either M is the totally geodesic sphere, the standard immersion of the product of two spheres, or the Veronese surface in $\mathbb{S}^4$.*

Notice that Yau's pinching constant $\frac{q-1}{2q-1}$ is better than the pinching constant $\frac{1}{2}$ given by Simon [48]. Moreover, Yau's pinching constant above is the best possible in the case where $q = 1$, or $n = 2$ and $q = 2$. Inspired by the Yau rigidity theorem, Ejiri [11] obtained a rigidity theorem for compact and simply connected minimal submanifolds in a sphere with $Ric_M \geq n-2$. The Yau parameter method was first introduced in the proof of the Yau rigidity theorem, which has many applications in the study of the Chern conjecture and other rigidity problems.

Many other important results were obtained by Yau [65]. For example, he obtained the geometric structure theorem for non-negatively curved submanifolds with parallel mean curvature in spaces forms, the classification theorem for surfaces with parallel mean curvature in spaces forms, and the codimension reduction theorem for submanifolds in a conformally flat manifold.

The Yau rigidity theory of submanifolds plays a very important role in the study of geometry, topology and curvature flows of submanifolds. During the past four decades, the Yau rigidity theory has been developed by several geometers [17]. In 1984, Cheng, Li, and Yau [9] proved the following volume gap theorem for minimal submanifolds.

Theorem 2.3. *Let M be an n-dimensional compact minimal submanifold in $\mathbb{S}^{n+q}$. Then there exists an explicit positive constant ε_n depending only on n such that if $Vol(M) < Vol(\mathbb{S}^n) + \varepsilon_n$, then M is congruent to the great sphere $\mathbb{S}^n$.*

The possible generalization of the Cheng-Li-Yau theorem will give a proof of the higher dimensional version of the Willmore conjecture on the total mean curvature for closed submanifolds in Euclidean spaces.

In 1990, Xu [57] proved the generalized Simons-Lawson-Chern-do Carmo-Kobayashi theorem for compact submanifolds with parallel mean curvature in a sphere.

Theorem 2.4. *Let M be an n-dimensional oriented compact submanifold with parallel mean curvature in $\mathbb{S}^{n+q}$. If $|A|^2 \le C_0(n,q,|H|)$, then M is either a totally umbilic sphere $\mathbb{S}^n(n/\sqrt{|H|^2+n^2})$, a Clifford hypersurface in an $(n+1)$-sphere, or the Veronese surface in $\mathbb{S}^4(2/\sqrt{|H|^2+4})$. Here the constant $C_0(n,q,|H|)$ is defined by*

$$C_0(n,q,|H|) := \begin{cases} \alpha(n,|H|), & q=1, \ \text{or } q=2 \text{ and } |H|\neq 0, \\ \frac{n}{2-\frac{1}{q}}, & q\ge 2 \text{ and } |H|=0, \\ \min\left\{\alpha(n,|H|), \frac{n^2+|H|^2}{(2-\frac{1}{q-1})n}+\frac{|H|^2}{n}\right\}, & q\ge 3 \text{ and } |H|\neq 0, \end{cases}$$

$$\alpha(n,|H|) = n + \frac{n}{2(n-1)}|H|^2 - \frac{n-2}{2(n-1)}\sqrt{|H|^4+4(n-1)|H|^2}.$$

In [29, 30], Li and Li improved Simons' pinching constant for n-dimensional compact minimal submanifolds in $\mathbb{S}^{n+q}$ to $\max\{\frac{n}{2-1/q}, \frac{2}{3}n\}$. Using Li-Li's matrix inequality, Xu [58] improved the pinching constant $C_0(n,q,|H|)$ in Theorem 2.4 to

$$C(n,q,|H|) = \begin{cases} \alpha(n,|H|), & q=1, \ \text{or } q=2 \text{ and } |H|\neq 0, \\ \min\left\{\alpha(n,|H|), \frac{2n}{3}+\frac{5}{3n}|H|^2\right\}, & \text{otherwise.} \end{cases}$$

This is the best pinching constant for n-dimensional compact submanifolds with parallel mean curvature in $\mathbb{S}^{n+q}$ up to date. In fact, Theorem 2.4 is stronger than the rigidity results in [1, 3, 42]. In [59], Xu obtained an optimal rigidity theorem for complete submanifolds with parallel mean curvature in hyperbolic spaces.

Let $\mathbb{F}^{n+q}(c)$ be an $(n+q)$-dimensional simply connected space form with constant curvature c. Put

$$\alpha(n,|H|,c) = nc + \frac{n}{2(n-1)}|H|^2 - \frac{n-2}{2(n-1)}\sqrt{|H|^4+4(n-1)c|H|^2}.$$

If $c>0$, we have $\min_{|H|}\alpha(n,|H|,c) = 2\sqrt{n-1}c$.

Using the techniques in stable currents and algebraic topology, Shiohama and Xu [46] proved the following optimal topological sphere theorem.

Theorem 2.5. *Let $M^n (n\ge 4)$ be an oriented complete submanifold in $\mathbb{F}^{n+q}(c)$ with $c\ge 0$. Suppose that $\sup_M(|A|^2-\alpha(n,|H|,c))<0$. Then M is homeomorphic to a sphere.*

After the work by Shen [43] and Lin-Xia [32], Xu [57] proved the following theorem.

Theorem 2.6. *Let M be an n-dimensional compact submanifold with parallel mean curvature in the space form $\mathbb{F}^{n+q}(c)$ with $c \geq 0$. Denote by $\mathring{A}$ the traceless second fundamental form of M. If*

$$\int_M |\mathring{A}|^n dM < C_1(n),$$

where $C_1(n)$ is an explicit positive constant depending only on n, then M is a totally umbilic sphere.

Applying the Morse theory of submanifolds, Shiohama and Xu [47] proved the following topological sphere theorem for compact submanifolds.

Theorem 2.7. *Let M be an n-dimensional compact submanifold in the space form $\mathbb{F}^{n+q}(c)$ with $c \geq 0$. There exists a positive constant $C_2(n)$ depending only on n such that if*

$$\int_M |\mathring{A}|^n dM < C_2(n),$$

then M is homeomorphic to $\mathbb{S}^n$.

After the work due to Ejiri [11], Shen [45], Li [31] and Xu-Tian [63], Xu-Gu [61] proved the following pinching theorem for submanifolds with pinched Ricci curvatures.

Theorem 2.8. *Let M be an $n(\geq 3)$-dimensional oriented compact submanifold with parallel mean curvature in the space form $\mathbb{F}^{n+q}(c)$. If*

$$Ric_M \geq (n-2)(c + \frac{|H|^2}{n^2}),$$

where $n^2c + |H|^2 > 0$, then M is either a totally umbilic sphere, a Clifford hypersurface $\mathbb{S}^m(r) \times \mathbb{S}^m(r)$ in an $(n+1)$-sphere with $n = 2m$, or $\mathbb{C}P^2(\frac{4}{3}c + \frac{1}{12}|H|^2)$ in $\mathbb{S}^7(\frac{4}{\sqrt{16c+|H|^2}})$.

Remark 2.1. When n is even, the pinching condition in Theorem 2.8 is optimal. Furthermore, Gu-Tian-Xu [61, 63] obtained the sharp topological and differentiable sphere theorems for submanifolds with pinched Ricci curvatures.

Using the DDVV inequality verified by Ge-Tang [12] and Lu [37] and the Yau parameter method, Gu-Xu [14] proved the following rigidity theorem for minimal submanifolds in spheres.

Theorem 2.9. *Let M be an n-dimensional oriented compact minimal submanifold in the unit sphere $\mathbb{S}^{n+q}$. If*

$$K_M \geq \frac{q \cdot sgn(q-1)}{2(q+1)},$$

then M is either a totally geodesic sphere, the standard immersion of the product of two spheres, or the Veronese surface in $\mathbb{S}^4$. Here $sgn(\cdot)$ is the standard sign function.

Remark 2.2. When $2 < q < n$, the pinching constant in Theorem 2.9 is better than ones given by Yau [65] and Itoh [22].

Combing Theorems 2.2, 2.9, and rigidity results in [14, 22, 44], we present a general version of the Yau rigidity theorem for submanifolds with parallel mean curvature in spaces forms.

Generalized Yau rigidity theorem. *Let M be an n-dimensional oriented compact submanifold with parallel mean curvature in $\mathbb{F}^{n+q}(c)$, where $n^2c + |H|^2 > 0$. Set $k(m,n) = \min\{m \cdot sgn(m-1), n\}$. Then we have*

(i) if $|H| = 0$ and

$$K_M \geq \frac{k(q,n)c}{2[k(q,n)+1]},$$

then M is either a totally geodesic sphere, the standard immersion of the product of two spheres, or the Veronese submanifold in $\mathbb{F}^{n+d}(c)$, where $d = \frac{1}{2}n(n+1) - 1$;

(ii) if $|H| \neq 0$ and

$$K_M \geq \frac{k(q-1,n)(c + \frac{1}{n^2}|H|^2)}{2[k(q-1,n)+1]},$$

then M is either a totally umbilical sphere, a Clifford hypersurface in an $(n+1)$-sphere, a product of three spheres in an $(n+2)$-sphere, or the Veronese submanifold in $\mathbb{F}^{n+d}(c + \frac{1}{n^2}|H|^2)$, where $d = \frac{1}{2}n(n+1) - 1$.

Remark 2.3. Motivated by the generalized Yau rigidity theorem, Gu and Xu [16] verified a sharp differentiable sphere theorem for submanifolds with positive sectional curvature.

3 MCF meets Ricci flow: positive sectional curvature

Using the Ricci flow and stable currents [5, 6, 18, 24], Xu and Zhao [64] initiated the study of differentiable pinching problems for submanifolds, and proved the following differentiable sphere theorem.

Theorem 3.1. *Let M be an $n(\geq 4)$-dimensional oriented complete submanifold in the unit sphere $\mathbb{S}^{n+q}$. Then*
(i) if $n = 4, 5, 6$ and $\sup_M(|A|^2 - \alpha(n, |H|)) < 0$, then M is diffeomorphic to $\mathbb{S}^n$;
(ii) if $n \geq 7$ and $|A|^2 < 2\sqrt{2}$, then M is diffeomorphic to $\mathbb{S}^n$.

In [60], Xu and Gu proved an optimal differentiable sphere theorem for submanifolds in the space form $\mathbb{F}^{n+q}(c)$ with constant curvature $c \geq 0$.

Theorem 3.2. *Let M be an n-dimensional oriented complete submanifold in $\mathbb{F}^{n+q}(c)$ with $c \geq 0$. If $\sup_M \left(|A|^2 - \frac{|H|^2}{n-1} - 2c\right) < 0$, then M is diffeomorphic to $\mathbb{S}^n$.*

Meanwhile, Andrews-Baker [2] proved the following convergence theorem for the mean curvature flow of arbitrary codimension in Euclidean spaces, which implies a differentiable sphere theorem for submanifolds in Euclidean spaces.

Theorem 3.3. *Let M_0 be an n-dimensional compact submanifold immersed in $\mathbb{R}^{n+q}$. If M_0 has $|H| \neq 0$ everywhere and satisfies*

$$|A|^2 \leq \begin{cases} \frac{4}{3n}|H|^2, & n = 2, 3, \\ \frac{1}{n-1}|H|^2, & n \geq 4, \end{cases} \tag{3.1}$$

then the mean curvature flow has a smooth solution $F : M \times [0, T) \to \mathbb{R}^{n+q}$ on a finite maximal time interval, and M_t converges uniformly to a point $q \in \mathbb{R}^{n+q}$ as $t \to T$. The rescaled maps $\widetilde{F}_t = \frac{F_t - x}{\sqrt{2n(T-t)}}$ converge in C^∞ to a limiting embedding $\widetilde{F}_T$ such that $\widetilde{F}_T(M)$ is the unit n-sphere in some $(n+1)$-dimensional subspace of $\mathbb{R}^{n+q}$.

In [15], Gu and Xu obtained a refined version of Theorem 3.2. Later, Baker [4] proved a sharp convergence theorem for the mean curvature flow of submanifolds in spheres.

Theorem 3.4. *Let M_0 be an n-dimensional compact submanifold immersed in $\mathbb{S}^{n+q}$. If M_0 satisfies*

$$|A|^2 \leq \begin{cases} \frac{4}{3n}|H|^2 + \frac{2(n-1)}{3}, & n = 2, 3, \\ \frac{1}{n-1}|H|^2 + 2, & n \geq 4, \end{cases} \tag{3.2}$$

then either M_t shrinks uniformly to a round point $p \in \mathbb{S}^{n+q}$ as $t \to T < \infty$, or M_t converges to a totally geodesic sphere in $\mathbb{S}^{n+q}$ as t tends to infinity.

In [34], Liu-Xu-Ye-Zhao proved the following sharp convergence theorem for the mean curvature flow of submanifolds in hyperbolic spaces.

Theorem 3.5. *Let $F_0 : M^n \to \mathbb{F}^{n+q}(c)$ be an n-dimensional closed submanifold in a hyperbolic space with constant curvature $c < 0$. Assume F_0 satisfies*

$$|A|^2 \leq \begin{cases} \frac{4}{3n}|H|^2 + \frac{n}{2}c, & n = 2, 3, \\ \frac{1}{n-1}|H|^2 + 2c, & n \geq 4. \end{cases} \tag{3.3}$$

Then the mean curvature flow with F_0 as initial value converges to a round point in finite time. In particular, M is diffeomorphic to $\mathbb{S}^n$.

More generally, Xu and Gu [60] proved the following differentiable sphere theorem, which generalized the Brendle-Schoen [6] differentiable sphere theorem for manifolds with strictly 1/4-pinched curvatures in the pointwise sense to the cases of submanifolds in a Riemannian manifold with codimension $q(\geq 0)$.

Theorem 3.6. *Let M^n be an n-dimensional complete submanifold in an $(n+q)$-dimensional Riemannian manifold N^{n+q}. If*

$$|A|^2 < \frac{8}{3}\left(\overline{K}_{\min} - \frac{1}{4}\overline{K}_{\max}\right) + \frac{|H|^2}{n-1},$$

then M is diffeomorphic to a space form. In particular, if M is simply connected, then M is diffeomorphic to $\mathbb{S}^n$ or $\mathbb{R}^n$.

Let N^{n+q} be an $(n+q)$-dimensional complete Riemannian manifold. Suppose that N satisfies

(1) the sectional curvature $-K_1 \leq K_N \leq K_2$ for $K_1, K_2 \geq 0$;

(2) $|\bar{\nabla}\bar{R}| \leq L$ for $L \geq 0$;

(3) the injectivity radius $inj(N) \geq i(N) > 0$.

N is called a Riemannian manifold with bounded geometry. We call K_1, K_2, L, and i_N the bound constants.

In 2012, Liu-Xu-Zhao [35] proved the following convergence theorem.

Theorem 3.7. *Let $F_0 : M^n \to N^{n+q}$ be an n-dimensional closed submanifold in an $(n+q)$-dimensional complete Riemannian manifold with bounded geometry with bound constants K_1, K_2, L, and i_N. There is an explicit constant b_0 depending on n, q, K_1, K_2 and L such that if F_0 satisfies*

$$|A|^2 < \begin{cases} \frac{4}{3n}|H|^2 - b_0, & n = 2, 3, \\ \frac{1}{n-1}|H|^2 - b_0, & n \geq 4, \end{cases} \tag{3.4}$$

then the mean curvature flow with F_0 as initial value contracts to a round point in finite time. In particular, M is diffeomorphic to $\mathbb{S}^n$.

In [33], Liu-Xu-Ye-Zhao obtained some convergence theorems for the mean curvature flow of closed submanifolds in Euclidean spaces under suitable integral curvature pinching conditions.

4 MCF with positive Ricci curvature

There are very few optimal differentiable spheres in curvature and topology of manifolds. In 1995, Grove and Wilhelm [13] first verified an optimal differentiable sphere theorem for a class of Riemannian manifolds with positive sectional curvature. Using the Ricci flow and stable currents, Xu and Gu [60] proved an optimal differentiable sphere theorem for certain submanifolds with positive sectional curvature. A challenging problem is: is it possible to prove an optimal differentiable sphere theorem for certain manifolds with positive Ricci curvature?

In 2015, the authors [25] first proved the following optimal convergence theorem for the mean curvature flow of arbitrary codimension in hyperbolic spaces whose initial submanifold has positive Ricci curvature.

Theorem 4.1. *Let $F_0 : M \to \mathbb{F}^{n+q}(c)$ be an n-dimensional ($n \geq 6$) complete submanifold in a hyperbolic space with negative constant curvature c. If M satisfies*

$$\sup_M(|A|^2 - \alpha(n, |H|, c)) < 0, \quad \textit{where} \ \ n^2c + |H|^2 > 0,$$

then the mean curvature flow with initial value F_0 has smooth solution $F_t(\cdot)$, and $F_t(\cdot)$ converges to a round point in finite time. In particular, M is diffeomorphic to $\mathbb{S}^n$.

Remark 4.1. Since $\alpha(n, |H|, c) > \frac{1}{n-1}|H|^2 + 2c$, our theorem above improves Theorem 3.5 for $n \geq 6$. Note that almost all initial submanifolds in the convergence results possess positive curvature. The pinching condition in Theorem 4.1 implies that the Ricci curvature of the initial submanifold is positive, but does not imply positivity of the sectional curvature. The following example shows that the pinching condition in Theorem 4.1 is optimal for arbitrary $n(\geq 6)$. Therefore, Theorem 4.1 is an optimal convergence theorem for the mean curvature flow of arbitrary codimension, which implies the first optimal differentiable sphere theorem for submanifolds with positive Ricci curvature.

EXAMPLE. Let λ, μ be positive constants satisfying $\lambda\mu = -c$ and $\lambda > \sqrt{-c}$, where $c < 0$. For $n \geq 3$, we consider the submanifold $M = \mathbb{F}^{n-1}(c + \lambda^2) \times \mathbb{F}^1(c + \mu^2) \subset \mathbb{H}^{n+q}(c)$. Then M is a complete submanifold with parallel mean curvature, which satisfies $|H| \equiv (n-1)\lambda + \mu > n\sqrt{-c}$ and $|A|^2 \equiv (n-1)\lambda^2 + \mu^2 = \alpha(n, |H|, c)$.

The key ingredient of the proof of Theorem 4.1 is to establish the elaborate estimates for the pinching quantity $\mathring{\alpha} = \alpha(n, |H|, c) - \frac{1}{n}|H|^2$, because our pinching condition is sharper than that in Theorem 3.5. Using the properties of $\mathring{\alpha}$ and the evolution equations, we first derive that $|\mathring{A}|^2 < \mathring{\alpha}$ is preserved along the mean curvature flow. Applying a new auxiliary function $f_\sigma = |\mathring{A}|^2/\mathring{\alpha}^{1-\sigma}$, we deduce that $|\mathring{A}|^2 \leq C_0|H|^{2(1-\sigma)}$ via the De Giorgi iteration. We then obtain an estimate for $|\nabla H|$. Finally, using estimates for $|\nabla H|$ and the Ricci curvature, we show that $\mathrm{diam}(M_t) \to 0$ and $|H|_{\min}/|H|_{\max} \to 1$ as $t \to T$. This implies the flow shrinks to a round point.

Proof of Theorem 4.1. We split the proof into several steps.

Step 1. We first show that the pinching condition in Theorem 4.1 is preserved under the mean curvature flow.

Suppose that M_0 is an $n(\geq 6)$-dimensional complete submanifold satisfying

$$\sup_M(|A|^2 - \alpha(n, |H|, c)) < 0, \qquad |H|^2 + n^2c > 0.$$

From a theorem due to Shiohama-Xu [46], we see that M_0 is actually a compact submanifold.

For a positive integer n and a negative constant c, we define a function $\mathring{\alpha} : (-n^2c, +\infty) \to \mathbb{R}$ by

$$\mathring{\alpha}(x) := nc + \frac{n^2 - 2n + 2}{2(n-1)n}x - \frac{n-2}{2(n-1)}\sqrt{x^2 + 4(n-1)cx}.$$

Then $|A|^2 < \alpha(n, |H|, c)$ is equivalent to $|\mathring{A}|^2 < \mathring{\alpha}(|H|^2)$.

Note that $\mathring{\alpha}(|H|^2) \to 0$ as $|H|^2 \to -n^2c$. This implies that if $|\mathring{A}|^2 < \mathring{\alpha}(|H|^2)$ remains true along the flow, then $|H|^2 + n^2c > 0$ also remains true.

We consider $U = |\mathring{A}|^2 - \mathring{\alpha}(|H|^2)$. Along the mean curvature flow, U satisfies

$$\begin{aligned}
&(\partial_t - \Delta)\, U \\
=&(\partial_t - \Delta)\, |\mathring{A}|^2 - \mathring{\alpha}'(|H|^2) \cdot (\partial_t - \Delta)\, |H|^2 \\
&+ \mathring{\alpha}''(|H|^2)\big|\nabla |H|^2\big|^2 \\
\leq& - 2|\nabla A|^2 + 2\left(\tfrac{1}{n} + \mathring{\alpha}'(|H|^2)\right) |\nabla H|^2 + \mathring{\alpha}''(|H|^2) \cdot |\nabla |H|^2|^2 \\
&+ 2|\mathring{A}|^2 \left(|A|^2 - nc\right) - 2\mathring{\alpha}'(|H|^2)|H|^2 \left(|A|^2 + nc\right).
\end{aligned}$$

For convenience, we denote $\mathring{\alpha}(|H|^2)$, $\mathring{\alpha}'(|H|^2)$ and $\mathring{\alpha}''(|H|^2)$ by $\mathring{\alpha}$, $\mathring{\alpha}'$ and $\mathring{\alpha}''$, respectively.

Using the inequality $|\nabla A|^2 \geq \frac{3}{n+2}|\nabla H|^2$ on submanifolds of space forms, we obtain

$$\begin{aligned}
&(\partial_t - \Delta)\, U \\
\leq& \; 2U\left(2\mathring{\alpha} + \tfrac{1}{n}|H|^2 - nc - \mathring{\alpha}' \cdot |H|^2 + U\right) \\
&+ 2\left[-\tfrac{2(n-1)}{n(n+2)} + \mathring{\alpha}' + 2|H|^2\mathring{\alpha}''\right] |\nabla H|^2 \\
&+ 2\left[\mathring{\alpha} \cdot \left(\mathring{\alpha} + \tfrac{1}{n}|H|^2 - nc\right) - \mathring{\alpha}' \cdot |H|^2 \cdot \left(\mathring{\alpha} + \tfrac{1}{n}|H|^2 + nc\right)\right].
\end{aligned}$$

To apply the maximum principle for the parabolic partial differential equation, we need to prove the two expressions in the square brackets of the above formula are non-positive.

By some computations, we prove the following.

Lemma 4.2. *If $x > -n^2c$, then $\mathring{\alpha}$ has the following properties.*

(i) $2x\mathring{\alpha}''(x) + \mathring{\alpha}'(x) < \frac{2(n-1)}{n(n+2)}$,

(ii) $x\mathring{\alpha}'(x) \cdot \left(\mathring{\alpha}(x) + \frac{1}{n}x + nc\right) \equiv \mathring{\alpha}(x) \cdot \left(\mathring{\alpha}(x) + \frac{1}{n}x - nc\right)$.

It follows from Lemma 4.2 that $(\partial_t - \Delta)\, U \leq 0$ at the point where $U = 0$. Hence, if U is initially negative, then $U < 0$ is preserved along the mean curvature flow. Therefore, we obtain the following.

Lemma 4.3. *If M_0 satisfies $|\mathring{A}|^2 < \mathring{\alpha}$ and $|H|^2 + n^2c > 0$, then there exist small positive constants ε and δ, such that for all time $t \in [0, T_{max})$, we have*

$$|\mathring{A}|^2 < \mathring{\alpha} - \varepsilon|H|^2 \quad \text{and} \quad |H|^2 + n^2c > \delta.$$

Step 2. We next prove a pinching estimate for the traceless second fundamental form, which guarantees that M_t becomes spherical along the mean curvature flow.

Lemma 4.4. *There are positive constants C_0 and σ independent of t such that*

$$|\mathring{A}|^2 \leq C_0|H|^{2(1-\sigma)}$$

holds along the mean curvature flow.

To prove the above lemma, we introduce an auxiliary function:

$$f_\sigma = \frac{|\mathring{A}|^2}{\mathring{\alpha}^{1-\sigma}},$$

where $\sigma \in (0,1)$. We need to prove that f_σ is bounded.

Computing the evolution equation of f_σ, we get

$$\frac{\partial}{\partial t} f_\sigma \le \Delta f_\sigma + \frac{2}{|\mathring{A}|}|\nabla f_\sigma||\nabla H| - 12\varepsilon \frac{f_\sigma}{|\mathring{A}|^2}|\nabla H|^2 - 4c f_\sigma + \sigma |H|^2 f_\sigma.$$

Because of the existence of term $\sigma|H|^2 f_\sigma$ in the inequality above, the maximum principle doesn't work here. To deal with this term, we need the following estimate.

Lemma 4.5. *If a submanifold in* $\mathbb{H}^{n+q}(c)$ *satisfies* $|\mathring{A}|^2 < \mathring{\alpha} - \varepsilon|H|^2$ and $|H|^2 + n^2 c > 0$, then we have

$$\Delta|\mathring{A}|^2 \ge 2\left\langle \mathring{A}, \nabla^2 H\right\rangle + \frac{\varepsilon}{2}|H|^2|\mathring{A}|^2.$$

From Lemma 4.5, we obtain the following inequality.

$$\Delta f_\sigma \ge \frac{2 f_\sigma}{|\mathring{A}|^2}\left\langle \mathring{A}, \nabla^2 H\right\rangle + \frac{\varepsilon}{2}|H|^2 f_\sigma - (1-\sigma)\frac{f_\sigma}{\mathring{\alpha}}\Delta\mathring{\alpha} - \frac{2}{|\mathring{A}|}|\nabla f_\sigma||\nabla H|.$$

Integrating both sides of the inequality above and making use of the divergence theorem, we get

$$\frac{\varepsilon}{2}\int_{M_t}|H|^2 f_\sigma^p \mathrm{d}\mu_t \le \int_{M_t}\left(\frac{3p f_\sigma^{p-1}}{|\mathring{A}|}|\nabla f_\sigma||\nabla H| + \frac{5 f_\sigma^p}{|\mathring{A}|^2}|\nabla H|^2\right)\mathrm{d}\mu_t.$$

Then we get the following estimate for the time derivative of the integral of f_σ.

$$\begin{aligned}
&\frac{\mathrm{d}}{\mathrm{d}t}\int_{M_t} f_\sigma^p \mathrm{d}\mu_t \\
&= p\int_{M_t} f_\sigma^{p-1}\frac{\partial f_\sigma}{\partial t}\mathrm{d}\mu_t - \int_{M_t} f_\sigma^p |H|^2 \mathrm{d}\mu_t \\
&\le p\int_{M_t} f_\sigma^{p-2}\left[-(p-1)|\nabla f_\sigma|^2 + \left(2 + \frac{6\sigma p}{\varepsilon}\right)\frac{f_\sigma}{|\mathring{A}|}|\nabla f_\sigma||\nabla H|\right. \\
&\quad \left. - \left(12\varepsilon - \frac{10\sigma}{\varepsilon}\right)\frac{f_\sigma^2}{|\mathring{A}|^2}|\nabla H|^2\right]\mathrm{d}\mu_t - 4cp\int_{M_t} f_\sigma^p \mathrm{d}\mu_t.
\end{aligned}$$

Note that the expression in the square bracket of the above formula is a quadratic polynomial. By choosing suitable constants σ and p, we make this quadratic polynomial non-positive.

Hence we get

$$\frac{\mathrm{d}}{\mathrm{d}t}\int_{M_t} f_\sigma^p \mathrm{d}\mu_t \leq -4cp\int_{M_t} f_\sigma^p \mathrm{d}\mu_t.$$

It can be proved that the maximal existence time of the mean curvature flow is finite. Therefore, we obtain

Lemma 4.6. *There exist a constant C independent of t, such that*

$$\left(\int_{M_t} f_\sigma^p d\mu_t\right)^{\frac{1}{p}} \leq C.$$

Then we show that f_σ is bounded along the mean curvature flow via a Stampacchia iteration procedure.

Step 3. We establish a gradient estimate for the mean curvature flow, which will be used to compare the mean curvature at different points of the submanifold.

Lemma 4.7. *For every $\eta > 0$, there exists a constant C_η independent of t such that for all $t \in [0, T_{\max})$, the following inequality holds*

$$|\nabla H|^2 \leq \eta |H|^4 + C_\eta.$$

Proof. Consider the function

$$f = |\nabla H|^2 + (N_1 + N_2|A|^2)|\mathring{A}|^2 - \eta|H|^4,$$

where N_1 and N_2 are sufficiently large positive constants independent of t.

By the evolution equations of $|H|^4$ and $|\nabla H|^2$, we obtain

$$\frac{\partial}{\partial t} f \leq \triangle f + C_5.$$

The assertion follows from the maximum principle and the definition of f. □

Step 4. Now we finish the proof of Theorem 4.1.

By an estimate for Ricci curvature due to Shiohama-Xu [46], we have

$$\mathrm{Ric}_{M_t} \geq \frac{n-1}{n}\left(nc + \frac{2}{n}|H|^2 - |A|^2 - \frac{n-2}{\sqrt{n(n-1)}}|H||\mathring{A}|\right).$$

By the preserved pinching condition $|\mathring{A}|^2 < \mathring{\alpha} - \varepsilon|H|^2$ and the following identity

$$\frac{n-2}{\sqrt{n(n-1)}}\sqrt{x\mathring{\alpha}(x)} \equiv \frac{1}{n}x - \mathring{\alpha}(x) + nc,$$

we see that there exists a positive constant ε_0 independent of t such that $\mathrm{Ric}_{M_t} \geq \varepsilon_0|H|^2$.

Since $T_{\max}$ is finite, $\max_{M_t}|A|^2 \to \infty$ as $t \to T_{\max}$. At a time t close to $T_{\max}$, let x be a point on M_t where $|H|$ achieves its maximum. From the gradient estimate of H, along all geodesics of length $l = (2\eta|H|_{\max})^{-1}$ starting from x, we have $|H| > (1-\eta)|H|_{\max}$. With η small enough, the Ricci curvature of M_t

satisfies $\mathrm{Ric}_{M_t} > (n-1)\pi^2/l^2$ on these geodesics. Then from Myers' theorem, these geodesics can reach any point of M_t.

Thus we have $|H|_{\min} > (1-\eta)|H|_{\max}$ and $\mathrm{diam}(M_t) \leq (2\eta|H|_{\max})^{-1}$. So, $\frac{\max_{M_t}|H|}{\min_{M_t}|H|} \to 1$ and $\mathrm{diam}(M_t) \to 0$ as $t \to T_{\max}$. Hence M_t's converge to a single point o as $t \to T_{\max}$.

Take a rescaling around o such that the total area of the expanded submanifolds are fixed. Then the rescaled immersions converge to a totally umbilical immersion as $t \to T_{\max}$.

Recently, Lei-Xu [26] proved the following sharp convergence theorem for the mean curvature flow of hypersurfaces in spheres.

Theorem 4.8. *Let $F_0 : M^n \to \mathbb{F}^{n+1}(c)$ be an n-dimensional ($n \geq 3$) closed hypersurface immersed in a spherical space form. If F_0 satisfies*

$$|A|^2 < \gamma_1(n, |H|, c),$$

then the mean curvature flow with initial value F_0 has a unique smooth solution $F : M \times [0, T) \to \mathbb{F}^{n+1}(c)$, and F_t converges to a round point in finite time, or converges to a totally geodesic hypersurface as $t \to \infty$. Here $\gamma_1(n, H, c)$ is an explicit positive scalar defined by

$$\gamma_1(n, H, c) = \min\{\alpha(H^2), \beta(H^2)\},$$

where

$$\alpha(x) = nc + \frac{n}{2(n-1)}x - \frac{n-2}{2(n-1)}\sqrt{x^2 + 4(n-1)cx},$$

$$\beta(x) = \alpha(x_0) + \alpha'(x_0)(x - x_0) + \frac{1}{2}\alpha''(x_0)(x - x_0)^2,$$

$$x_0 = y_n c, \quad y_n = 4(1-n) + \frac{2(n^2-4)}{\sqrt{2n-5}}\cos\left(\frac{1}{3}\arctan\frac{n^2-4n+6}{2(n-1)\sqrt{2n-5}}\right).$$

Remark 4.2. The scalar $\gamma_1(n, |H|, c)$ satisfies (i) $\gamma_1(n, H, c) > \frac{1}{n-1}H^2 + 2c$; (ii) $\gamma_1(n, |H|, c) > \frac{9}{5}\sqrt{n-1}c$; (iii) $\gamma_1(n, |H|, c) = \alpha(n, |H|, c)$, for $|H|^2 \geq y_n c$. Thus Theorem 4.8 substantially improves the famous convergence theorem due to Huisken [21].

Now we state the idea of the proof of Theorem 4.8.

Step 1. We first prove that the pinching condition in Theorem 4.8 is preserved under the mean curvature flow.

Suppose that M_0 is an n-dimensional ($n \geq 3$) closed hypersurface satisfying $|A|^2 < \gamma_1(|H|^2)$. Here γ_1 is an explicit function defined by

$$\gamma_1(x) = \begin{cases} \alpha(x), \ x \geq x_0, \\ \beta(x), \ 0 \leq x < x_0. \end{cases}$$

Consider $U = |A|^2 - \gamma_1(H^2)$. Computing the evolution equation of U, we get

$$\begin{aligned}
&\left(\frac{\partial}{\partial t} - \Delta\right) U \\
&\leq 2U\left(2\gamma_1(H^2) - nc - \gamma_1'(H^2)H^2 + U\right) \\
&\quad +2\left[-\frac{3}{n+2} + \gamma_1'(H^2) + 2H^2\gamma_1''(H^2)\right]|\nabla H|^2 \\
&\quad +2\Big[2cH^2 + \gamma_1(H^2)^2 - nc\gamma_1(H^2) - (\gamma_1(H^2) + nc)H^2\gamma_1'(H^2)\Big].
\end{aligned}$$

To apply the maximum principle, we need to prove the two expressions in the square brackets of the above formula are non-positive.

By some complicated calculations, we get the following lemma.

Lemma 4.9. *For $x \geq 0$, the function γ_1 has the following properties.*
(i) $2x\gamma_1''(x) + \gamma_1'(x) \leq \frac{3}{n+2}$,
(ii) $(\gamma_1(x) + nc)x\gamma_1'(x) \geq 2cx + \gamma_1(x)^2 - nc\gamma_1(x)$.

It follows from Lemma 4.9 that $(\partial_t - \Delta)U \leq 0$ at the point where $U = 0$. Applying the maximum principle, we prove that if U is initially negative, then $U < 0$ is preserved along the mean curvature flow.

Step 2. We prove a pinching estimate for the traceless second fundamental form, which guarantees that M_t becomes spherical along the mean curvature flow.

Lemma 4.10. *There are positive constants C_0 and σ independent of t such that*

$$|\mathring{A}|^2 \leq C_0(H^2 + c)^{1-\sigma}\,\mathrm{e}^{-2\sigma ct}$$

holds along the mean curvature flow.

To prove the lemma above, we introduce an auxiliary function:

$$f_\sigma = |\mathring{A}|^2/\gamma_1(H^2)^{1-\sigma},$$

where $\sigma \in (0,1)$. We need to show that f_σ decays exponentially.

Let p be a sufficiently large number. Computing the evolution equation of f_σ^p, we get

$$\frac{\mathrm{d}}{\mathrm{d}t}\int_{M_t} f_\sigma^p \mathrm{d}\mu_t \leq -3\sigma cp\int_{M_t} f_\sigma^p \mathrm{d}\mu_t.$$

Thus we obtain

Lemma 4.11. *There exist a constant C independent of t, such that*

$$\left(\int_{M_t} f_\sigma^p d\mu_t\right)^{\frac{1}{p}} \leq C\,\mathrm{e}^{-3\sigma ct}.$$

Let $g_\sigma = f_\sigma\,\mathrm{e}^{2\sigma ct}$. Then we show that g_σ is bounded along the mean curvature flow via a Stampacchia iteration procedure. This proves the lemma.

Step 3. We establish a gradient estimate for the mean curvature flow, which will be used to compare the mean curvature at different points of the hypersurface.

Lemma 4.12. *For every $\eta > 0$, there exists a constant C_η independent of t such that for all $t \in [0, T_{\max})$, there holds*

$$|\nabla H|^2 < [(\eta H)^4 + C(\eta)^2]\, \mathrm{e}^{-\sigma ct}.$$

Proof. Consider the function

$$f = \Big[|\nabla H|^2 + (N_1 + N_2|A|^2)|\mathring{A}|^2\Big]\, \mathrm{e}^{\sigma ct} - \eta |H|^4,$$

where N_1 and N_2 are sufficiently large positive constants independent of t. By the evolution equations of $|H|^4$ and $|\nabla H|^2$, we obtain

$$\frac{\partial}{\partial t} f \leq \triangle f + C_5\, \mathrm{e}^{-\sigma ct}.$$

The assertion follows from the maximum principle and the definition of f. □

Step 4. Now we finish the proof of Theorem 4.8. We have two cases as follows.

Case (i). Suppose that $T_{\max}$ is finite. By an estimate for Ricci curvature due to Shiohama-Xu [46], we obtain $\mathrm{Ric}_{M_t} \geq \varepsilon_0(H^2 + c)$. Since $T_{\max}$ is finite, $\max_{M_t} |A|^2 \to \infty$ as $t \to T_{\max}$. Using the gradient estimate for H and Myers' theorem, we obtain $\frac{\max_{M_t} |H|}{\min_{M_t} |H|} \to 1$ and $\mathrm{diam}(M_t) \to 0$ as $t \to T_{\max}$. Hence M_t's converge to a single point o as $t \to T_{\max}$.

We take a rescaling around o. Then the rescaled immersions converge to a totally umbilical immersion as $t \to T_{\max}$.

Case (ii). Suppose that $T_{\max} = \infty$. It follows from the gradient estimate of H that $H^2 < C\,\mathrm{e}^{-\sigma ct}$. Noting that $|\mathring{A}|^2 \leq C_0 H^{2(1-\sigma)} \cdot \mathrm{e}^{-2\sigma ct}$, we get $|h|^2 \leq C\,\mathrm{e}^{-\sigma ct}$. Since $|h| \to 0$ as $t \to \infty$, M_t converges to a totally geodesic hypersurface.

This completes the proof of Theorem 4.8.

More recently, Lei-Xu [27] proved a sharp convergence theorem for mean curvature flow ($n \geq 6$) of arbitrary codimension in spheres, which substantially improves the convergence theorem for mean curvature flow of arbitrary codimension in spheres due to Baker [4].

Theorem 4.13. *Let $F_0 : M^n \to \mathbb{F}^{n+q}(c)$ be an n-dimensional ($n \geq 6$) closed submanifold immersed in a spherical space form. If F_0 satisfies*

$$|A|^2 < \gamma(n, |H|, c),$$

then the mean curvature flow with initial value F_0 has a unique smooth solution $F : M \times [0, T) \to \mathbb{F}^{n+q}(c)$, and F_t converges to a round point in finite time, or converges to a totally geodesic sphere as $t \to \infty$. Here $\gamma(n, |H|, c)$ is an explicit positive scalar defined by

$$\gamma(n, |H|, c) = \min\{\alpha(|H|^2), \beta(|H|^2)\},$$

where

$$\alpha(x) = nc + \frac{n}{2(n-1)}x - \frac{n-2}{2(n-1)}\sqrt{x^2 + 4(n-1)cx},$$

$$\beta(x) = \alpha(x_0) + \alpha'(x_0)(x - x_0) + \frac{1}{2}\alpha''(x_0)(x - x_0)^2,$$

$$x_0 = y_n c, \quad y_n = \frac{2n+2}{n-4}\sqrt{n-1}\left(\sqrt{n-1} - \tfrac{n-4}{2n+2}\right)^2.$$

Remark 4.3. The scalar $\gamma(n, |H|, c)$ satisfies the following: (i) $\gamma(n, H, c) > \frac{1}{n-1}H^2 + 2c$; (ii) $\gamma(n, |H|, c) > \frac{7}{6}\sqrt{n-1}c$; (iii) $\gamma(n, |H|, c) = \alpha(n, |H|, c)$, for $|H|^2 \geq z_n c$. Hence Theorem 4.13 substantially improves Theorem 3.4.

In [41], Pipoli and Sinestrari obtained the following convergence theorem for mean curvature flow of small codimension in the complex projective space.

Theorem 4.14. *Let $F_0 : M^n \to \mathbb{CP}^{\frac{n+q}{2}}$ be a closed submanifold of dimension n and codimension q in the complex projective space with Fubini-Study metric. Let $F : M^n \times [0, T) \to \mathbb{CP}^{\frac{n+q}{2}}$ be the mean curvature flow with initial value F_0. If F_0 satisfies*

$$|h|^2 < B_0(n, q, |H|) := \begin{cases} \frac{1}{n-1}|H|^2 + 2, & q = 1 \text{ and } n \geq 5, \\ \frac{1}{n-1}|H|^2 + \frac{n-3-4q}{n}, & 2 \leq q < \frac{n-3}{4}, \end{cases}$$

then F_t converges to a round point in finite time, or converges to a totally geodesic submanifold as $t \to \infty$. In particular, M is diffeomorphic to either $\mathbb{S}^n$ or $\mathbb{CP}^{n/2}$.

Most recently, Lei-Xu [28] proved a new convergence theorem for mean curvature flow of arbitrary codimension in complex projective spaces, which substantially improves the convergence theorem for mean curvature flow of small codimension in complex projective spaces due to Pipoli-Sinestrari [41]. Precisely, we prove the following theorem.

Theorem 4.15. *Let $F_0 : M^n \to \mathbb{CP}^{\frac{n+q}{2}}$ be an n-dimensional closed submanifold in $\mathbb{CP}^{\frac{n+q}{2}}$. Let $F : M^n \times [0, T) \to \mathbb{CP}^{\frac{n+q}{2}}$ be the mean curvature flow with initial value F_0. If F_0 satisfies*

$$|h|^2 < B(n, q, |H|) := \begin{cases} \varphi(|H|^2), & q = 1 \text{ and } n \geq 3, \\ \frac{1}{n-1}|H|^2 + 2 - \frac{3}{n}, & 2 \leq q < n-4, \\ \psi(|H|^2), & q \geq n-4 \geq 2, \end{cases}$$

then F_t converges to a round point in finite time, or converges to a totally geodesic submanifold as $t \to \infty$. In particular, M is diffeomorphic to $\mathbb{S}^n$ or $\mathbb{CP}^{n/2}$.

Here $\varphi(|H|^2)$ and $\psi(|H|^2)$ are given by

$$\varphi(|H|^2) = 2 + a_n + \left(b_n + \tfrac{1}{n-1}\right)|H|^2 - \sqrt{b_n^2|H|^4 + 2a_n b_n |H|^2},$$

$$\psi(|H|^2) = \frac{9}{n^2 - 3n - 3} + \frac{n^2 - 3n}{n^3 - 4n^2 + 3}|H|^2 - \frac{3\sqrt{|H|^4 + \frac{2}{n}(n-1)(n^2-3)|H|^2 + 9(n-1)^2}}{n^3 - 4n^2 + 3},$$

where $a_n = 2\sqrt{(n^2 - 4n + 3)b_n}$, $b_n = \min\left\{\frac{n-3}{4n-4}, \frac{2n-5}{n^2+n-2}\right\}$.

Remark 4.4. In particular, (i) if $q = 1$ and $n \geq 5$, we have $B(n, q, |H|) > B_0(n, q, |H|)$ and $B(n, q, |H|) > \sqrt{2(n-3)}$; (ii) if $2 \leq q < \frac{n-3}{4}$, we have $B(n, q, |H|) - B_0(n, q, |H|) = 1 + \frac{4q}{n}$. Therefore, Theorem 4.15 substantially improves Theorem 4.14.

5 Problem section

In this section, we give several unsolved problems on the convergence theorems for the mean curvature flow and sphere theorems for submanifolds.

Denote by $\mathbb{F}^{n+q}(c)$ the $(n+q)$-dimensional complete simply connected space form of constant sectional curvature c. Let M be an n-dimensional oriented compact submanifold in $\mathbb{F}^{n+q}(c)$. Based on the discussions in [14, 17, 26, 27, 33, 36, 38, 46, 47, 61, 62], we present the following conjectures.

Conjecture 5.1 (Liu-Xu-Ye-Zhao [33]). *Let $F_0 : M \to \mathbb{F}^{n+q}(c)$ be an n-dimensional closed submanifold in $\mathbb{F}^{n+q}(c)$ with $c > 0$. Suppose F_0 satisfies*

$$|A|^2 < \alpha(n, |H|, c),$$

then the mean curvature flow with F_0 as initial value has a unique solution $F : M \times [0, T) \to \mathbb{F}^{n+q}(c)$ on a maximal time interval, and either
(i) $T < \infty$ and M_t converges to a round point as $t \to T$; or
(ii) $T = \infty$ and M_t converges to a totally geodesic submanifold in $\mathbb{F}^{n+q}(c)$ as $t \to \infty$.

In particular, M is diffeomorphic to the standard n-sphere.

Conjecture 5.2 (see also [33]). *Let $F_0 : M \to \mathbb{F}^{n+q}(c)$ be an n-dimensional closed submanifold in $\mathbb{F}^{n+q}(c)$. Then there exists a positive constant C_n depending only on n, such that if M satisfies*

$$\int_M |\mathring{A}|^n dM < C_n,$$

then the mean curvature flow with F_0 as initial value has a unique solution $F : M \times [0, T) \to \mathbb{F}^{n+q}(c)$ on a maximal time interval, and either
(i) $T < \infty$ and M_t converges to a round point as $t \to T$; or
(ii) $c > 0$, $T = \infty$ and M_t converges to a totally geodesic submanifold in $\mathbb{F}^{n+q}(c)$ as $t \to \infty$.

In particular, M is diffeomorphic to the standard n-sphere.

Conjecture 5.3. *Let $F_0 : M \to \mathbb{R}^{n+q}$ be an n-dimensional closed submanifold in $\mathbb{R}^{n+q}$. There exists a positive constant C_n depending only on n, such that if M satisfies*

$$\int_M |H|^n dM < n^n Vol(\mathbb{S}^n) + C_n,$$

then the mean curvature flow with F_0 as initial value has a unique solution $F : M \times [0, T) \to \mathbb{R}^{n+q}$ on a finite maximal time interval, and M_t converges to a round point as $t \to T$. In particular, M is diffeomorphic to the standard n-sphere.

A challenging problem is: what is the best pinching constant in Conjecture 5.3? In particular, we have the following stronger version of the Willmore theorem verified by Marques and Neves [38].

Conjecture 5.4. *Let $F_0 : M^2 \to \mathbb{R}^3$ be a closed surface in $\mathbb{R}^3$. If M satisfies*

$$\int_M |H|^2 dM < 8\pi^2,$$

then the mean curvature flow with F_0 as initial value has a unique solution $F : M \times [0, T) \to \mathbb{R}^3$ on a finite maximal time interval, and M_t converges to a round point as $t \to T$. In particular, M is diffeomorphic to $\mathbb{S}^2$.

Conjecture 5.5 (Xu-Gu [61]). *Let $F_0 : M \to \mathbb{F}^{n+q}(c)$ be an $n(\geq 3)$-dimensional closed submanifold in an $(n+q)$-dimensional space form $\mathbb{F}^{n+q}(c)$ with $n^2c+|H|^2 > 0$. If the Ricci curvature of M satisfies*

$$Ric_M > (n-2)(c + \frac{1}{n^2}|H|^2),$$

then the mean curvature flow with F_0 as initial value has a unique solution $F_t(\cdot)$ on a maximal time interval, and M_t converges to a round point in finite time, or $c > 0$ and M_t converges to a totally geodesic sphere as $t \to \infty$. In particular, M is diffeomorphic to $\mathbb{S}^n$.

Conjecture 5.6 (Gu-Xu [14]). *Let $F_0 : M \to \mathbb{F}^{n+q}(c)$ be an n-dimensional closed submanifold in an $(n+q)$-dimensional space form $\mathbb{F}^{n+q}(c)$ with $n^2c+|H|^2 > 0$. Set $k(q,n) = \min\{q \cdot sgn(q-1), n\}$. If the sectional curvature of M satisfies*

$$K_M > \frac{k(q,n)(c + \frac{1}{n^2}|H|^2)}{2[k(q,n)+1]},$$

then the mean curvature flow with F_0 as initial value has a unique solution $F_t(\cdot)$ on a maximal time interval, and M_t converges to a round point in finite time, or $c > 0$ and M_t converges to a totally geodesic sphere as $t \to \infty$. In particular, M is diffeomorphic to $\mathbb{S}^n$.

Even for $q = 1$ and $c = 1$, the above problem is still open.

Conjecture 5.7. *Let $F_0 : M \to \mathbb{S}^{n+q}$ be an n-dimensional closed submanifold in an $(n+q)$-dimensional unit sphere. If $|h(u,u)|^2 < \frac{1}{3}$ for any $u \in UM$, then the mean curvature flow with F_0 as initial value has a unique solution $F_t(\cdot)$ on a maximal time interval, and M_t converges to a round point in finite time, or converges to a totally geodesic sphere as $t \to \infty$. In particular, M is diffeomorphic to $\mathbb{S}^n$.*

Conjecture 5.8 (Xu-Huang-Zhao [62]). *Let $F_0 : M \to \mathbb{S}^{n+q}$ be an n-dimensional closed submanifold in an $(n+q)$-dimensional unit sphere. Set*

$$\tau(x) = \max_{u,v \in U_xM, u \perp v} |h(u,u) - h(v,v)|^2.$$

If $\tau(x) < \frac{4}{3}$ for all $x \in M$, then the mean curvature flow with F_0 as initial value has a unique solution $F_t(\cdot)$ on a maximal time interval, and M_t converges to a round point in finite time, or converges to a totally geodesic sphere as $t \to \infty$. In particular, M is diffeomorphic to $\mathbb{S}^n$.

Notice that if $|h(u,u)|^2 < \frac{1}{3}$ for any $u \in U_xM$, then $\tau(x) < \frac{4}{3}$. This implies that Conjecture 5.8 is stronger than Conjecture 5.7.

References

[1] H. Alencar and M. do Carmo, *Hypersurfaces with constant mean curvature in spheres*, Proc. Amer. Math. Soc., **120**(1994), 1223-1229.

[2] B. Andrews and C. Baker, *Mean curvature flow of pinched submanifolds to spheres,* J. Differential Geom., **85**(2010), 357-395

[3] K. O. Araújo and K. Tenenblat, *On submanifolds with parallel mean curvature vector*, Kodai Math. J., **32**(2009), 59-76.

[4] C. Baker, *The mean curvature flow of submanifolds of high codimension*, arXiv:1104.4409v1.

[5] S. Brendle, *A general convergence result for the Ricci flow in higher dimensions*, Duke Math. J., **145**(2008), 585-601.

[6] S. Brendle and R. Schoen, *Manifolds with 1/4-pinched curvature are space forms,* J. Amer. Math. Soc., **22**(2009), 287-307.

[7] B. Y. Chen and M. Okumura, *Scalar curvature, inequality and submanifold.* Proc. Amer. Math. Soc., **38** (1973), 605-608.

[8] J. Y. Chen and J. Y. Li, *Singularity of mean curvature flow of Lagrangian submanifolds*, Invent. Math., **156**(2004), 25-51.

[9] S. Y. Cheng, P. Li and S. T. Yau, *Heat equations on minimal submanifold and their applications*, Amer. J. Math., **106**(1984), 1033-1065.

[10] S. S. Chern, M. do Carmo and S. Kobayashi, *Minimal submanifolds of a sphere with second fundamental form of constant length,* in Functional Analysis and Related Fields, Springer-Verlag, New York(1970).

[11] N. Ejiri, *Compact minimal submanifolds of a sphere with positive Ricci curvature,* J. Math. Soc. Japan, **31** (1979), 251-256.

[12] J. Q. Ge and Z. Z. Tang, *A proof of the DDVV conjecture and its equality case*, Pacific J. Math., **237**(2008), 87-95.

[13] K. Grove and F. Wilhelm, *Hard and soft packing radius theorems*, Ann. of Math., **142**(1995), 213-237.

[14] J. R. Gu and H. W. Xu, *On Yau rigidity theorem for minimal submanifolds in spheres*, Math. Res. Lett., **19**(2012), 511-523.

[15] J. R. Gu and H. W. Xu, *The sphere theorems for manifolds with positive scalar curvature*, J. Differential Geom., **92**(2012), 507-545.

[16] J. R. Gu and H. W. Xu, *A sharp differentiable pinching theorem for submanifolds in space forms*, Proc. Amer. Math. Soc., **144**(2016), 337-346.

[17] J. R. Gu, H. W. Xu, Z. Y. Xu and E. T. Zhao, *A survey on rigidity problems in geometry and topology of submanifolds*, Proceedings of the 6th International Congress of Chinese Mathematicians, ALM 37, pp. 79-99, 2016.

[18] R. Hamilton, *Three manifolds with positive Ricci curvature*, J. Differential Geom., **17**(1982), 255-306.

[19] G. Huisken, *Flow by mean curvature of convex surfaces into spheres*, J. Differential Geom., **20**(1984), 237-266.

[20] G. Huisken, *Contracting convex hypersurfaces in Riemannian manifolds by their mean curvature*, Invent. Math., **84**(1986), 463-480.

[21] G. Huisken, *Deforming hypersurfaces of the sphere by their mean curvature,* Math. Z., **195** (1987), 205-219.

[22] T. Itoh, *On Veronese manifolds*, J. Math. Soc. Japan., **27**(1975), 497-506.

[23] B. Lawson, *Local rigidity theorems for minimal hypersurfaces,* Ann. of Math., **89**(1969), 187-197.

[24] B. Lawson and J. Simons, *On stable currents and their application to global problems in real and complex geometry,* Ann. of Math., **98**(1973), 427-450.

[25] L. Lei and H. W. Xu, *An optimal convergence theorem for mean curvature flow of arbitrary codimension in hyperbolic spaces*, arXiv:1503.06747, 2015.

[26] L. Lei and H. W. Xu, *A new version of Huisken's convergence theorem for mean curvature flow in spheres*, arXiv:1505.07217, 2015.

[27] L. Lei and H. W. Xu, *Mean curvature flow of arbitrary codimension in spheres and sharp differentiable sphere theorem*, arXiv:1506.06371v2, 2015.

[28] L. Lei and H. W. Xu, *Mean curvature flow of arbitrary codimension in complex projective spaces*, arXiv:1605.07963, 2016.

[29] A. M. Li and J. M. Li, *An inequality for matrices and its applications in differential geometry,* Adv. Math.(China), **20**(1991), 375-376.

[30] A. M. Li and J. M. Li, *An intrinsic rigidity theorem for minimal submanifolds in a sphere,* Arch. Math., **58**(1992), 582-594.

[31] H. Z. Li, *Curvature pinching for odd-dimensional minimal submanifolds in a sphere*, Publ. Inst. Math. (Beograd), **53**(1993), 122-132.

[32] J. M. Lin and C. Y. Xia, *Global pinching theorem for even dimensional minimal submanifolds in a unit sphere*, Math. Z., **201**(1989), 381-389.

[33] K. F. Liu, H. W. Xu, F. Ye and E. T. Zhao, *The extension and convergence of mean curvature flow in higher codimension,* Trans. Amer. Math. Soc., **370**(2018), 2231-2262.

[34] K. F. Liu, H. W. Xu, F. Ye and E. T. Zhao, *Mean curvature flow of higher codimension in hyperbolic spaces*, Comm. Anal. Geom., **21**(2013), 651-669.

[35] K. F. Liu, H. W. Xu and E. T. Zhao, *Mean curvature flow of higher codimension in Riemannian manifolds*, arXiv:1204.0107, 2012.

[36] K. F. Liu, H. W. Xu and E. T. Zhao, *Some recent results on mean curvature flow of arbitrary codimension*, Proceedings of the 6th International Congress of Chinese Mathematicians, ALM 37, pp. 115-131, 2016.

[37] Z. Lu, *Normal scalar curvature conjecture and its applications*, J. Funct. Anal., **261**(2011), 1284-1308.

[38] F. C. Marques and A. Neves, *Min-max theory and the Willmore conjecture*, Ann. of Math., **179**(2014), 683-782.

[39] M. Okumura, *Submanifolds and a pinching problem on the second fundamental tensor*, Trans. Amer. Math. Soc., **178**(1973), 285-291.

[40] M. Okumura, *Hypersurfaces and a pinching problem on the second fundamen-*

tal tensor, Amer. J. Math., **96**(1974), 207-213.

[41] G. Pipoli and C. Sinestrari, *Mean curvature flow of pinched submanifolds of* $\mathbb{CP}^n$, arXiv:1502.00519v3, 2015.

[42] W. Santos, *Submanifolds with parallel mean curvature vector in spheres*, Tohoku Math. J., **46**(1994), 403-415.

[43] C. L. Shen, *A global pinching theorem for minimal hypersurfaces in sphere*, Proc. Amer. Math. Soc., **105**(1989), 192-198.

[44] Y. B. Shen, *Submanifolds with nonnegative sectional curvature*, Chinese Ann. Math. SerB, **5**(1984), 625-632.

[45] Y. B. Shen, *Curvature pinching for three-dimensional minimal submanifolds in a sphere*, Proc. Amer. Math. Soc., **115**(1992), 791-795.

[46] K. Shiohama and H. W. Xu, *The topological sphere theorem for complete submanifolds,* Compositio Math., **107**(1997), 221-232.

[47] K. Shiohama and H. W. Xu, *Rigidity and sphere theorems for submanifolds*, Kyushu J. Math. I, **48**(1994), 291-306; II, **54**(2000), 103-109.

[48] U. Simon, *Submanifolds with parallel mean curvature vector and the curvature of minimal submanifolds of spheres*, Arch. Math., (Basel) **29**(1977), 106-112.

[49] J. Simons, *Minimal varieties in Riemannian manifolds,* Ann. of Math., **88**(1968), 62-105.

[50] K. Smoczyk, *Mean curvature flow in higher codimension: introduction and survey,* Global differential geometry, 231-274, Springer Proc. Math., 17, Springer, Heidelberg, 2012.

[51] M. T. Wang, *Mean curvature flow of surfaces in einstein four-manifolds*, J. Differential Geom., **57**(2001), 301-338.

[52] M. T. Wang, *Long time existence and convergence of graphic mean curvature flow in arbitrary codimension*, Invent. Math., **148**(2002), 525-543.

[53] M. T. Wang, *Gauss maps of the mean curvature flow*, Math. Res. Lett., **10**(2003), 287-299.

[54] M. T. Wang, *A convergence result of the lagrangian mean curvature flow*, Third International Congress of Chinese Mathematicians. Part 1, 2, 291-295, AMS/IP Stud. Adv. Math., 42, pt. 1, 2, Amer. Math. Soc., Providence, RI, 2008.

[55] M. T. Wang, *Lectures on mean curvature flows in higher codimensions*, Handbook of geometric analysis. No. 1, 525-543, Adv. Lect. Math. (ALM), 7, Int. Press, Somerville, MA, 2008.

[56] Y. L. Xin, *Mean curvature flow with bounded Gauss image*, Results Math., **59**(2011), 415-436.

[57] H. W. Xu, *Pinching theorems, global pinching theorems and eigenvalues for Riemannian submanifolds,* Ph.D. dissertation, Fudan University, 1990.

[58] H. W. Xu, *A rigidity theorem for submanifolds with parallel mean curvature in a sphere,* Arch. Math., **61**(1993), 489-496.

[59] H. W. Xu, *Rigidity of submanifolds with parallel mean curvature in space forms*, preprint, 1993.

[60] H. W. Xu and J. R. Gu, *An optimal differentiable sphere theorem for complete manifolds,* Math. Res. Lett., **17**(2010), 1111-1124.

[61] H. W. Xu and J. R. Gu, *Geometric, topological and differentiable rigidity of*

submanifolds in space forms, Geom. Funct. Anal., **23**(2013), 1684-1703.

[62] H. W. Xu, F. Huang and E. T. Zhao, *Geometric and differentiable rigidity of submanifolds in spheres*, J. Math. Pures Appl., **99**(2013), 330-342.

[63] H. W. Xu and L. Tian, *A differentiable sphere theorem inspired by rigidity of minimal submanifolds*, Pacific J. Math., **254**(2011), 499-510.

[64] H. W. Xu and E. T. Zhao, *Topological and differentiable sphere theorems for complete submanifolds*, Comm. Anal. Geom., **17**(2009), 565-585.

[65] S. T. Yau, *Submanifolds with constant mean curvature I, II,* Amer. J. Math., **96, 97**(1974, 1975), 346-366, 76-100.

CENTER OF MATHEMATICAL SCIENCES, ZHEJIANG UNIVERSITY
HANGZHOU 310027, CHINA
E-mail address: lei-li@zju.edu.cn; xuhw@zju.edu.cn

Proceedings of the 7th International
Congress of Chinese Mathematicians
ALM 43, pp. 349–372

Effective Batalin-Vilkovisky Quantization and Geometric Applications

Si Li

Abstract

We explain the effective renormalization method of quantum field theory in the Batalin-Vilkovisky formalism and illustrate its mathematical applications by three geometric examples: (1) Topological quantum mechanics and algebraic index, (2) Elliptic curve and higher genus mirror symmetry, (3) Calabi-Yau geometry and integrable hierarchy. This note is an expansion of author's talk at ICCM 2016.

1 Introduction

Quantum field theory has been a rich source of mathematical structures. Typically, quantum field theory deals with infinite dimensional geometry, a fact that lies behind many of its nontrivial consequences. A famous example is the mysterious mirror symmetry conjecture between symplectic and complex geometries. It has a clean physics interpretation as a version of infinite dimensional Fourier transform. Still, we do not have a good mathematical framework to understand why mirror symmetry holds although many nontrivial statements have been checked. Nevertheless, asymptotic analysis can always be performed with the help of the celebrated idea of renormalization and effective field theory. The purpose of this article is to explain some aspects of effective renormalization method of quantum field theory and show how it leads to powerful applications to geometric problems.

A physics system is usually described by a map

$$S : \mathcal{E} \to \mathbb{R}.$$

Here $\mathcal{E}$ is called the *space of fields* and S is the *action functional.* Here are some typical examples.

- $\mathcal{E} = C^\infty(X)$. *Scalar field theory.*
- $\mathcal{E} = \{\text{connections on a vector bundle } V \to X\}$. *Gauge theory.*
- $\mathcal{E} = \mathrm{Map}(\Sigma, X)$. *$\sigma$-model.*
- $\mathcal{E} = \{\text{metrics on } X\}$. *Gravity.*

In all examples above, $\mathcal{E}$ is an infinite dimensional space carrying a delicate topology. Classical physics is decribed by the critical locus

$$\mathrm{Crit}(S) = \{\delta S = 0\}$$

which usually leads to interesting differential equations (*equations of motion*). One convenient way to approach quantum physics is by Feynman's "path integral"

$$\langle \mathcal{O} \rangle := \int_{\mathcal{E}} \mathcal{O} e^{S/\hbar}.$$

$\mathcal{O}$ is called an *observable*. $\langle \mathcal{O} \rangle$ is the *correlation function*. Again, we encounter the mathematical challenge for such ∞-dim integral. On the other hand, $\hbar$-asymptotic analysis has a solid foundation usually called *perturbative field theory*. Although modern physics is deeply rooted in understanding nonperturbative aspects of quantum field theory, the $\hbar$-asymptotic analysis gives very useful physics information as well as new mathematical phenomenons. We illustrate this by three geometric applications.

1. **Topological quantum mechanics and index theorem**. Quantum field theory in one dimension is quantum mechanics. We first explore the well-studied example of topological quantum mechanics and explain the result of [27] on its connection with Fedosov's deformation quantization [18]:

   $$\boxed{\text{low energy effective theories of topological quantum mechanics}}\ .$$
   $$\Downarrow$$
   $$\boxed{\text{Fedosov's abelian connections on the Weyl bundle}}$$

   This fact together with a version of Atiyah-Bott localization computation leads to a geometric proof of the *algebraic index theorem*, which was first formulated by Fedosov [19] and Nest-Tsygan [38] as the algebraic analogue of Atiyah-Singer index theorem. This result is analogous to the physics interpretation of Atiyah-Singer index theorem [2,22,44,45] via semi-classical analysis of supesymmetric quantum mechanics, as well as the symbol calculus of pseudodifferential operators [20,23,39].
2. **Elliptic curve and higher genus mirror symmetry**. We consider chiral deformations of free CFT's in the Batalin-Vilkovisky (BV) formalism in two dimension [33]. We explain that

   $$\boxed{\text{BV quantization of 2d chiral theory}}$$
   $$\Downarrow$$
   $$\boxed{\text{Integrable evolution of vertex algebra}}$$

   where the evolution is parametrized by the relevant background symmetry. This can be viewed as the 2d analogue of the above 1d quantum mechanical result. As an application, we show how this construction leads to an exact solution [33] of quantum B-model (BCOV theory [5,12]) in complex one dimension that solves the higher genus mirror symmetry conjecture on elliptic curves [32,33].

3. **Calabi-Yau geometry and integrable hierarchy**. Topological string on Calabi-Yau geometry usually leads to integrable hierarchies. We propose a geometric interpretation of this mysterious phenomenon in terms of B-twisted topological string field theory (BCOV theory [5, 12]). Let X be a Calabi-Yau geometry (either a compact Calabi-Yau or a Landau-Ginzburg model). We consider $X \times \mathbb{C}$ which is again Calabi-Yau but with dimension increased by one. Compactifying along X,

$$\boxed{\text{BCOV theory on } X \times \mathbb{C}} \Downarrow \boxed{\text{2d chiral theory on } \mathbb{C}}$$

we get an effective 2d chiral theory on $\mathbb{C}$. We explain the proposal in [28] that the BV master equation for the effective 2d chiral theory induced from that on $X \times \mathbb{C}$ leads to integrable hierarchies.

$$\boxed{\text{BV master equation}} \Longrightarrow \boxed{\text{Integrable hierarchy}}$$

In all examples above, we work with Batalin-Vilkovisky (BV or BRST-BV) quantization scheme, which arises in physics as a general framework to quantize theories with gauge symmetries [4]. In this article we work with Costello's formalism[10] of homotopic effective renormalization which is a convenient framework for our geometric problems. We introduce the basic philosophy and notions of BV quantization in the asymptotic analysis of quantum gauge theories, and then explain with more details on the above examples.

Conventions

In this article we work with the category of $\mathbb{Z}$-graded vector spaces over a field of characteristic 0.

- Let $V = \bigoplus_{m\in\mathbb{Z}} V_m$ be a $\mathbb{Z}$-graded k-vector space. Given $a \in V_m$, we let $\bar{a} = m$ be its degree.
- $V[n]$ denotes the degree shifting such that $V[n]_m = V_{n+m}$.
- All tensors are in the category of graded vector spaces. Koszul sign rule is assumed. $\mathrm{Sym}^m(V)$ and $\wedge^m(V)$ denote the graded symmetric product and graded skew-symmetric product respectively.
- $V[\hbar]$, $V[[\hbar]]$ and $V((\hbar))$ denote polynomial series, formal power series and Laurent series respectively in a variable $\hbar$ valued in V.
- Given $A, B \in \mathrm{Hom}(V,V)$, the commutator $[A,B] \in \mathrm{Hom}(V,V)$ always means the graded commutator: $[A,B] = AB - (-1)^{\bar{A}\bar{B}}BA$.

2 Batalin-Vilkovisky quantization

2.1 A prototype of BV formalism

Calculus is nicely packaged into the framework of de Rham theory: let X be an oriented manifold, $C \subset X$ be a submanifold, α be a smooth differential form, then there is a natural integration pairing between α and C

$$\int_C \alpha.$$

If α is closed, i.e. $d\alpha = 0$, and C has no boundary, then Stokes' theorem tells that the above integration is invariant under continuous deformations of C.

The BV approach to the above problem is to consider the super manifold $T^*X[-1]$, where we have graded the cotangent fiber direction to be odd of degree 1. It carries a natural (-1)-shifted symplectic structure (which we denote by ω), and we can identify its structure sheaf via polyvector fields

$$\mathcal{O}_{T^*X[-1]} = \mathrm{PV}^{-\bullet}(X), \quad \text{where } \mathrm{PV}^{-k}(X) := \Gamma(X, \wedge^k T_X).$$

The induced Poisson bracket on $\mathcal{O}_{T^*X[-1]}$ coincides with the Schouten-Nijenhuis bracket.

Let us fix a chosen volume form Ω on X. It allows us to identify via contracting with Ω

$$\mathrm{PV}^{-\bullet}(X) \overset{\lrcorner\Omega}{\longleftrightarrow} \Omega^{n-\bullet}(X), \quad n = \dim(X).$$

Under this identification, the de Rham differential d gives rise to a differential

$$\Delta : \mathrm{PV}^{-\bullet}(X) \to \mathrm{PV}^{1-\bullet}(X).$$

Δ will be called the BV operator, which is the divergence operator with respect to the volume form Ω. The triple $(T^*X[-1], \omega, \Delta)$ is an example of BV manifold [43]. In general, a BV manifold M carries a natural integration pairing between its functions and lagrangian super submanifolds [43]:

$$\int_{\mathcal{L}} : \mathcal{O}_M \to \mathbb{R}.$$

If $\mathcal{L}$ has no boundary, and μ is Δ-closed, then $\int_{\mathcal{L}} \mu$ is invariant under continuous deformations of $\mathcal{L}$ and a version of Stokes' theorem holds. This is the BV analogue of de Rham theory.

In our case

$$\int_{\mathcal{L}} : \mathcal{O}_{T^*X[-1]} \to \mathbb{R},$$

if $\mathcal{L} = N^*_{C/X}[-1] \subset T^*X[-1]$ is the (shifted) conormal bundle of a submanifold C inside X, then

$$\int_{\mathcal{L}} \mu = \int_C \mu \lrcorner \Omega.$$

In particular, $\int_{\mathcal{L}} \Delta\mu = \int_C d(\mu \lrcorner \Omega) = 0$ if $\partial C = \emptyset$.

In our geometric situation above, the de Rham theory and BV formalism are completely equivalent. There is one subtle difference when we move on to discuss infinite dimensional geometry of quantum field theories. In de Rham theory, the integration map

$$\int_X : \Omega^\bullet(X) \to \mathbb{R}$$

can be identified with the cohomology

$$\int_X \cong H^n : \Omega^\bullet(X) \to H^n(X) \cong \mathbb{R}.$$

Here n is the dimension of X. In the BV approach, the BV integration

$$\int_X : \mathrm{PV}^{-\bullet} \to \mathbb{R}$$

can be identified with H^0. When $n = \infty$ as in quantum field theory, H^0 behaves better than H^n. The problem of integration is transferred to construct Δ, which has a convenient formulation at least in perturbative quantum field theory. This is one of the folklore advantage of working with BV.

2.2 BV algebra and master equation

Let us first introduce some algebraic structures that are relevant to the above discussion.

Definition 2.1. A differetial Batalin-Vilkovisky (BV) algebra is a triple $(\mathcal{A}, Q, \Delta)$ where

- $\mathcal{A}$ is a $\mathbb{Z}$-graded commutative associative unital algebra,
- $\Delta : \mathcal{A} \to \mathcal{A}$ is a second-order operator of degree 1 such that $\Delta^2 = 0$,
- $Q : \mathcal{A} \to \mathcal{A}$ is a derivation of degree 1 such that $Q^2 = 0$ and $[Q, \Delta] = 0$.

Here Δ is called the BV operator. Δ being "second-order" means the following: let us define the *BV bracket* $\{-, -\}$ as the failure of Δ to be a derivation

$$\{a, b\} := \Delta(ab) - (\Delta a)b - (-1)^{\bar{a}} a\Delta b.$$

Then $\{-, -\}$ defines a Lie bracket of degree 1 (Gerstenhaber algebra) such that Δ is compatible with $\{-, -\}$ via a graded version of Leibniz rule. In our discussion above, the triple is $(\mathrm{PV}^{-\bullet}(X), 0, \Delta)$.

Definition 2.2. Let $(\mathcal{A}, Q, \Delta)$ be a differential BV algebra. An element $I = I_0 + I_1\hbar + \cdots \in \mathcal{A}[[\hbar]]$ of degree 0 is said to satisfy *quantum master equation (QME)* if

$$QME: \quad \boxed{(Q + \hbar\Delta)e^{I/\hbar} = 0}.$$

This is equivalent to

$$\boxed{QI + \hbar\Delta I + \frac{1}{2}\{I, I\} = 0}.$$

The leading $\hbar$-order I_0 satisfies

$$CME: \quad \boxed{QI_0 + \frac{1}{2}\{I_0, I_0\} = 0}$$

which is called the *classical master equation (CME)*.

Definition 2.3. A solution I of quantum master equation gives rise to a differential

$$Q + \hbar\Delta + \{I, -\} : \mathcal{A}[[\hbar]] \to \mathcal{A}[[\hbar]].$$

The cohomology

$$\mathrm{Obs}^q = \boxed{H^\bullet(\mathcal{A}[[\hbar]], Q + \hbar\Delta + \{I, -\})}$$

is called the *quantum observables*. Similarly, its classical limit

$$\mathrm{Obs}^c = \boxed{H^\bullet(\mathcal{A}, Q + \{I_0, -\})}$$

is called the *classical observables*.

Here we give some mathematical and physical explanations of the above definitions.

1. Classical master equation: $Q + \{I_0, -\}$ squares zero and describes the infinitesimal gauge transformations (BRST). Mathematically, this defines a homotopic Lie-algebra (L_∞-algebra) $\mathfrak{g}$ if $\mathcal{A} = \prod_{n \geq 0} \mathrm{Sym}^n(\mathfrak{g}[1]^*)$.
2. Quantum master equation: the quantum gauge consistency condition for the measure $e^{I/\hbar}$.
3. A quantum observable is represented by an element $\mathcal{O} \in \mathcal{A}[[\hbar]]$ such that

$$(Q + \hbar\Delta)(\mathcal{O}e^{I/\hbar}) = 0.$$

   This is the analogue of a closed differential form in the BV setting.
4. Quantum correlation function is given by a BV integral

$$\langle \mathcal{O} \rangle = \int_{\mathcal{L}} \mathcal{O}e^{I/\hbar}.$$

   $\mathcal{L}$ is a lagrangian super subspace of $\mathcal{E}$ related to a *gauge fixing condition*. In particular, for a quantum observable $\mathcal{O}$, $\langle \mathcal{O} \rangle$ is invariant under continuous deformations of the gauge fixing condition $\mathcal{L}$ and independent of the choice of representatives. This is the original motivation for the BV description of quantum gauge consistency.

Example 2.4 (Singularity theory)**.** We compare BV formalism in the finite dimensional case with classical singularity theory. Let $f : \mathbb{C}^n \to \mathbb{C}$ be a polynomial in n variables with an isolated critical point at the origin. Let $\{z^i\}_{i=1}^n$ be holomorphic coordinates on $\mathbb{C}^n$. Let $\{\theta_i\}_{i=1}^n$ be Grassmann variables of degree -1

$$\theta_i\theta_j = -\theta_j\theta_i, \quad \forall 1 \leq i, j \leq n.$$

Let $\mathbb{C}\{z\}$ be the germ of holomorphic functions at 0. We consider the triple $(\mathcal{A} = \mathbb{C}\{z^i\}[\theta_i], Q = 0, \Delta)$ where

$$\Delta = \sum_{i=1}^n \frac{\partial}{\partial\theta_i}\frac{\partial}{\partial z^i}.$$

It is easy to check that $\Delta^2 = 0$, and its expression explains the name "second order".

- $f = f(z^i)$ gives a solution of QME: $\Delta e^{f/\hbar} = 0$ since f does not depend on θ_i's.
- The space of classical observables is given by the Jacobian algebra or Milnor ring
$$\mathrm{Obs}^c = H^\bullet(\mathcal{A}, \{f, -\}) = \mathrm{Jac}(f) := \mathbb{C}\{z^i\}/(\partial_i f).$$
- The space of quantum observables is isomorphic to a formal completion of the Brieskorn lattice [41]
$$\mathrm{Obs}^q = H^\bullet(\mathcal{A}[[\hbar]], \hbar\Delta + \{f, -\}) \overset{d^n z}{\cong} \frac{\Omega^n(\mathbb{C}^n)_0[[\hbar]]}{(\hbar d + df\wedge)\Omega^{n-1}(\mathbb{C}^n)_0[[\hbar]]}$$
Here $d^n z = dz^1 \wedge \cdots \wedge dz^n$, $\Omega^k(\mathbb{C}^n)_0$ is the germ of holomorphic k-forms at the origin 0.
- The BV integration models the oscillatory integral
$$\langle \mathcal{O} \rangle = \int_{\mathcal{L}} d^n z \mathcal{O} e^{f/\hbar}$$
where $\mathcal{L}$ is a Lefschetz thimble.

The Brieskorn lattice (or the quantum observables) plays the important role of Hodge filtration in singularity theory (see [3, 31] for an exposition). The above finite dimensional model is the effective theory of topological Landau-Ginzburg B-model. In general, a quantum field theory can be viewed as a version of infinite dimensional singularity theory, and the $\hbar$-filtration is related to the Hodge filtration.

2.3 (-1)-shifted symplectic geometry

A general class of differential BV algebras arises from odd symplectic geometry. Let us start with a finite dimensional toy model. Let (V, Q) be a finite dimensional dg vector space. The differential $Q : V \to V$ induces a differential on various tensors of V, V^*, still denoted by Q. Let
$$\omega \in \wedge^2 V^*, \quad Q(\omega) = 0,$$
be a Q-compatible symplectic structure such that $\deg(\omega) = -1$. Then ω induces an isomorphism
$$V^* \simeq V[1].$$
Let $K = \omega^{-1} \in \mathrm{Sym}^2(V)$ be the Poisson kernel of degree 1 under the identification via ω
$$\wedge^2 V^* \simeq \mathrm{Sym}^2(V)[2],$$
where we have used the canonical isomorphism $\wedge^2(V[1]) \simeq \mathrm{Sym}^2(V)[2]$. Let us define the formal power series ring
$$\mathcal{O}(V) := \prod_{n \geq 0} \mathrm{Sym}^n(V^*).$$

Then $(\mathcal{O}(V), Q)$ is a graded-commutative dga. Let Δ_K denote the second order operator

$$\Delta_K : \mathrm{Sym}^n(V^*) \to \mathrm{Sym}^{n-2}(V^*)$$

by contracting with the kernel $K \in \mathrm{Sym}^2(V)$ using the pairing between V and V^*. It is straight-forward to see that $(\mathcal{O}(V), Q, \Delta_K)$ defines a differential BV algebra.

Remark 2.5. The relevant data defining the BV operator is K instead of ω. In fact, for any Q-closed element $K \in \mathrm{Sym}^2(V)$ of degree 1, the triple $(\mathcal{O}(V), Q, \Delta_K)$ defines a differential BV algebra. In particular, the above construction is generalized to the case when K is degenerate, i.e. the Poisson case.

Let P be a degree 0 element of $\mathrm{Sym}^2(V)$, and consider the new BV kernel

$$K_P := K + Q(P).$$

We say that K_P is homologous to K. We can form a new differential BV algebra $(\mathcal{O}(V), Q, \Delta_{K_P})$. Let ∂_P denote the second order operator on $\mathcal{O}(V)$ by contracting with P, which is defined similarly to Δ_K associated to K. The following simple algebraic lemma is key to Costello's homotopic renormalization theory [10].

Lemma 2.6. *The following equation holds formally as operators on* $\mathcal{O}(V)[[\hbar]]$

$$(Q + \hbar\Delta_{K_P})\, e^{\hbar\partial_P} = e^{\hbar\partial_P}\,(Q + \hbar\Delta_K)\,,$$

i.e., the following diagram commutes

$$\begin{array}{ccc} \mathcal{O}(V)[[\hbar]] & \xrightarrow{Q+\hbar\Delta_K} & \mathcal{O}(V)[[\hbar]] \\ {\scriptstyle \exp(\hbar\partial_P)}\downarrow & & \downarrow{\scriptstyle \exp(\hbar\partial_P)} \\ \mathcal{O}(V)[[\hbar]] & \xrightarrow[Q+\hbar\Delta_{K_P}]{} & \mathcal{O}(V)[[\hbar]] \end{array}$$

This lemma is a simple consequence of the observation that $[Q, \partial_P] = \Delta_K - \Delta_{K_P}$.

Definition 2.7. Let

$$\mathcal{O}^+(V) := \mathrm{Sym}^{\geq 3}(V^*)) \oplus \hbar\mathcal{O}(V)[[\hbar]]$$

denote the subspace of $\mathcal{O}(V)[[\hbar]]$ consisting of functions that are at least cubic modulo $\hbar$.

Corollary 2.8. *Let* $I \in \mathcal{O}^+(V)$ *be a solution of quantum master equation in* $(\mathcal{O}(V)[[\hbar]], Q, \Delta_K)$. *Let* I_P *be formally defined by the equation*

$$e^{I_P/\hbar} = e^{\hbar\partial_P} e^{I/\hbar}.$$

Then I_P *is a well-defined element of* $\mathcal{O}^+(V)$ *that solves the quantum master equation in* $(\mathcal{O}(V)[[\hbar]], Q, \Delta_{K_P})$.

The real content of the above formal definition of I_P is that

$$I_P = \sum_{\Gamma:\text{connected}} \frac{\hbar^{g(\Gamma)}}{|\text{Aut}(\Gamma)|} W_\Gamma(P, I)$$

where the summation is over all connected Feynman graphs with P being the propagator and I being the vertex (see [6] for an exposition on Feynman graph techniques). Here $g(\Gamma)$ denotes the genus of Γ, and $|\text{Aut}(\Gamma)|$ denotes the cardinality of the automorphism group of Γ. I being at least cubic implies that the above infinite graph sum leads to a well-defined element of $\mathcal{O}^+(V)$. The statement of I_P being a solution of the quantum master equation is a direct consequence of Lemma 2.6.

Definition 2.9. Given $P \in \text{Sym}^2(V)$ of degree 0, we define the *homotopic renormalization group flow* (HRG) operator $W(P, -)$ by

$$W(P, -) : \mathcal{O}^+(V) \to \mathcal{O}^+(V), \quad \boxed{e^{W(P,I)/\hbar} := e^{\hbar \partial_P} e^{I/\hbar}}.$$

Lemma 2.6 says that HRG links QME for differential BV algebras whose BV kernels are homologous.

2.4 Effective BV quantization

Quantum field theories always have infinite dimensional space of fields playing the role of V in the previous subsection. Typically, a classical field theory in the BV formalism consists of $(\mathcal{E}, Q, \omega, S_0)$ where

1. fields $\mathcal{E} = \Gamma(X, E)$ for a graded bundle E on a manifold X.
2. $Q : \mathcal{E} \to \mathcal{E}$ is a differential operator on E that makes $(\mathcal{E}, Q)$ into an elliptic complex.
3. ω: local symplectic pairing of degree (-1)

$$\omega(\alpha, \beta) = \int_X \langle \alpha, \beta \rangle, \quad \forall \alpha, \beta \in \mathcal{E}.$$

   where $(-, -)$ is a degree -1 skew-symmetric pairing on $\mathcal{E}$ valued in the density line bundle on X.
4. A classical action $S_0 = \omega(Q(-), -) + I_0$ satisfying the classical master equation

$$\{S_0, S_0\} = 0, \quad \text{equivalently } QI_0 + \frac{1}{2}\{I_0, I_0\} = 0.$$

   Here the quadratic term $\omega(Q(-), -)$ is called the *free part*, while I_0 is called the *interaction*.

Example 2.10 (Chern-Simons theory). Let X be a 3-manifold. $\mathfrak{g}$ be a Lie algebra with a Killing pairing Tr.

- $\mathcal{E} = \Omega^\bullet(X, \mathfrak{g})[1]$, and $Q = d$ is the de Rham differential.
- $\omega(\alpha, \beta) = \int_X \text{Tr}(\alpha \wedge \beta), \quad \alpha, \beta \in \mathcal{E}.$

- Let $[-,-]$ be the bracket on $\mathcal{E}$ induced from $\mathfrak{g}$. The Chern-Simons functional in the BV formalism is

$$CS(A) = \frac{1}{2}\int_X \operatorname{Tr}(A \wedge dA) + \frac{1}{6}\int_X \operatorname{Tr}(A \wedge [A, A]), \quad A \in \mathcal{E}$$

- Classical master equation $\Longleftrightarrow$ $\{\Omega^\bullet(X, \mathfrak{g}), d, [-,-]\}$ being a differential graded Lie algebra.

The space of fields $\mathcal{E}$ carries a natural topology, and V^* is replaced by the continuous dual of $\mathcal{E}$

$$\mathcal{E}^* = \operatorname{Hom}(\mathcal{E}, \mathbb{R}),$$

i.e., distributions. Then formal functions $\mathcal{O}(\mathcal{E})$ are defined in terms of the completed tensor product.

$$(\mathcal{E}^*)^{\otimes k} = \mathcal{E}^* \hat{\otimes} \cdots \hat{\otimes} \mathcal{E}^*$$

are distributions on the bundle $E \boxtimes \cdots \boxtimes E$ over $X \times \cdots \times X$. $\operatorname{Sym}^k(\mathcal{E}^*)$ is defined similarly (taking care of the sign of permutations). Then

$$\mathcal{O}(\mathcal{E}) := \prod_{k \geq 0} \operatorname{Sym}^k(\mathcal{E}^*)$$

is the analogue of $\mathcal{O}(V)$.

Since ω is given by integration, the Poisson kernel $K_0 = \omega^{-1}$ is given by δ-function. In particular, K_0 is *singular*. Therefore the naive analogue of the finite dimensional model

$$\Delta_{K_0} : \mathcal{O}(\mathcal{E}) \to \mathcal{O}(\mathcal{E})$$

is *ill-defined* since we can not pair a distribution with another distribution. This difficulty is related to the ultra-violet problem in quantum field theory. Note that the BV bracket $\{-,-\}$ is in fact well-defined for *local functionals*, i.e., functionals of the form $\int_X \mathcal{L}$ by integrating a langrangian density. In particular, the classical master equaition $QI_0 + \frac{1}{2}\{I_0, I_0\} = 0$ is well-defined for a local interaction I_0.

In [10], Costello uses a smooth kernel K_r that is homologous to K_0 to define the renormalized BV operator (the existence of such a homologous smooth kernel is a consequence of elliptic regularity). Since smooth functions can be paired with distributions, the BV operator (defined similarly as the finite dimensional case)

$$\Delta_{K_r} : \mathcal{O}(\mathcal{E}) \to \mathcal{O}(\mathcal{E})$$

is well-defined. In this way we obtain a family of differential BV algebras

$$\mathrm{BV}[r] := (\mathcal{O}(\mathcal{E}), Q, \Delta_{K_r}) .$$

Any two such smooth kernels K_{r_1} and K_{r_2} are homologous again by a smooth kernel, which well-defines a homotopic renormalization group flow between $\mathrm{BV}[r_1]$ and $\mathrm{BV}[r_2]$. This motivates the following definition.

Definition 2.11 ([10]). An effective solution of quantum master equation is an assignement $I[r] \in \mathcal{O}(\mathcal{E})[[\hbar]]$ for each smooth kernel K_r that is homologous to K_0, satisfying

- Renormalized quantum master equation
$$(Q + \hbar\Delta_{K_r})e^{I[r]/\hbar} = 0.$$
- Two different $I[r]$'s are related by the homotopic renormalization group flow equation.

If we start with a classical local interaction I_0 satisfying classical master equation, then we require that $I[r]$ becomes asymptotically local as $r \to 0$ and the classical unrenormalized limit $\lim_{r\to 0}\lim_{\hbar\to 0} I[r] = I_0$.

There is a parallel observable theory in such a homotopic renormalization framework developed by Costello-Gwilliam [11]. Here is a picture to illustrate the situation:

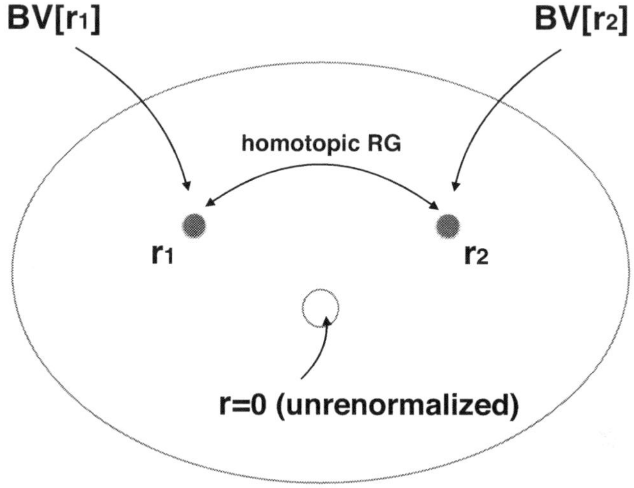

Remark 2.12. Typically for $L > 0$, let K_L be the kernel for the heat operator $e^{-L[Q,Q^\dagger]}$ associated to a generlized Laplacian $[Q, Q^\dagger]$ constructed from the elliptic operator Q and its adjoint $Q^\dagger$. Then $K_0 = \lim_{L\to 0} K_L$ is the δ-function which is homologous with K_L by
$$\int_0^L du(Q^\dagger \otimes 1)K_u.$$
For $0 < \epsilon < L$, K_ϵ and K_L are homologous by the smooth kernel
$$P_\epsilon^L = \int_\epsilon^L du(Q^\dagger \otimes 1)K_u$$

which is called the *regularized propagator*. It leads to the formalism of homotopic renormalization group flow used in [10].

3 Applications

3.1 Topological quantum mechanics and index theorem

The simplest nontrivial example is when the underlying manifold X is one-dimensional. This corresponds to quantum mechanical models. We give a complete description of the geometry of BV quantization for the low energy effective theory of topological quantum mechanical models.

Let us start with a linear model. Let $X = S^1$. Let V be a graded vector space with a degree 0 symplectic pairing

$$(-,-) : \wedge^2 V \to \mathbb{R}.$$

The space of fields of our local model will be

$$\mathcal{E} = \Omega^\bullet(S^1) \otimes_{\mathbb{R}} V.$$

Let us denote by X_{dR} the super-manifold whose underlying topological space is X and whose structure sheaf is the de Rham complex on X. We can identify $\varphi \in \mathcal{E}$ with a map $\hat{\varphi}$ between super-manifolds

$$\hat{\varphi} : X_{dR} \to V.$$

The differential $Q = d_{S^1}$ on $\mathcal{E}$ is the de Rham differential on $\Omega^\bullet(S^1)$. The (-1)-shifted symplectic pairing is given by the composition of the symplectic pairing and the integration map

$$\omega(\varphi_1, \varphi_2) := \int_{S^1} (\varphi_1, \varphi_2), \quad \varphi_1, \varphi_2 \in \mathcal{E}.$$

This model is just the AKSZ-formalism [1] applied to the one-dimensional σ-model. Given $I \in \mathcal{O}(V)$ of degree k, it induces an element $\hat{I} \in \mathcal{O}(\mathcal{E})$ of degree $k-1$ via

$$\hat{I}(\varphi) := \int_{S^1} \hat{\varphi}^*(I), \quad \forall \varphi \in \mathcal{E}.$$

We choose the standard flat metric on S^1 and work with the heat kernel regularization as in Remark 2.12. We have a regularized BV kernel K_L for each $L > 0$. K_ϵ and K_L are homologous by the regularized propagator P_ϵ^L for $0 < \epsilon < L \leq \infty$. The homotopic renormalization group flow operator $W(P_\epsilon^L, -)$ connects the corresponding solutions of quantum master equations.

Theorem 3.1 ([27]). *Given $I \in \mathcal{O}^+(V)$ of degree 1, the limit*

$$\hat{I}[L] = \lim_{\epsilon \to 0} W(P_\epsilon^L, \hat{I})$$

exists as a degree 0 *element of* $\mathcal{O}(\mathcal{E})[[\hbar]]$. *The family* $\{\hat{I}[L]\}_{L>0}$ *gives an effective solution of quantum master equation if and only if*

$$[I, I]_\star = 0.$$

Here $\mathcal{O}(V)[[\hbar]]$ *inherits a natural Moyal-Weyl product* $\star$ *from the linear symplectic form* $(-,-)$. $[-,-]_\star$ *is the commutator with respect to the Moyal-Weyl product.*

This theorem illustrates an explicit relationship between BV quantization in one dimension and the Weyl quantization. It can be globalized on a symplectic manifold (X, ω) as follows. For each point $p \in X$, the tangent space T_pX is a linear symplectic space and $\mathcal{O}(T_pX)$ has a canonical Weyl quantization. It glues to a bundle of algebra $Weyl(T_X)$ with fiberwise Moyal-Weyl product $\star$

$$\begin{array}{ccc} \mathcal{O}(T_X) & \overset{\text{quantization}}{\rightsquigarrow} & Weyl(T_X) \\ \downarrow & & \downarrow \\ X & & X \end{array}$$

Let ∇ be a torsion-free connection on T_X that is compatible with ω (called a symplectic connection). ∇ naturally induces a connection on $Weyl(T_X)$. In [18], Fedosov shows that ∇ can be modified to be flat by adding a "quantum correction" $\gamma \in \Omega^1(X, Weyl(T_X))$. Here γ satisfies

$$\nabla\gamma + \frac{1}{2\hbar}[\gamma, \gamma]_\star = \omega_\hbar$$

where $\omega_\hbar = -\omega + \sum_{k\geq 1} \hbar^k \omega_k$ and $\omega_k' s$ are closed 2-forms that classify the equivalence class of deformation quantizations on X. Since $\omega_\hbar$ is valued in the central element of $Weyl(T_X)$, the curvature of the quantum corrected connection $\nabla + \frac{1}{\hbar}[\gamma, -]_\star$ vanishes. Fedosov calls this an abelian connection. Let

$$\sigma : \Gamma(X, Weyl(T_X)) \to C^\infty(X)[[\hbar]]$$

be the symbol map by projecting to the constant component. Let $\Gamma^{flat}(X, Weyl(T_X))$ be the flat sections of the Weyl bundle under an abelian connection $\nabla + \frac{1}{\hbar}[\gamma, -]_\star$. Fedosov further shows that

$$\sigma : \Gamma^{flat}(X, Weyl(T_X)) \to C^\infty(X)[[\hbar]]$$

is a $\hbar$-linear isomorphism, which induces a star product (deformation quantization) on $C^\infty(X)[[\hbar]]$ via the fiberwise Moyal-Weyl product on $\Gamma^{flat}(X, Weyl(T_X))$.

We consider the topological σ-model of AKSZ type

$$S^1_{dR} \to X.$$

We consider the low energy effective theory $\mathcal{E}$ which describes the formal neighborhood of constant maps. It can be viewed as a sheafification of our linear model V by gluing the tangent spaces $\{T_pX\}$.

Theorem 3.2 (GLL). *Solutions of effective quantum master equation for $\mathcal{E}$ corresponds to Fedosov's abelian connections. The associative algebra of local quantum observables is identified with Fedosov's deformation quantization.*

We refer to [27] for more precise notions in the above theorem. Physically, this theorem says that the low energy effective theory of topological quantum mechanics is described by Fedosov's abelian connections and deformation quantization. There is a natural lagrangian super submanifold $\mathcal{L}$ inside $\mathcal{E}$, which induces a correlation function

$$\int_{\mathcal{L}} : \mathrm{Obs}^q \to \mathbb{R}((\hbar)).$$

Restricting to local quantum observables, it can be identified with the unique normalized trace map on the deformation quantized algebra [18, 38]

$$\mathrm{Tr} : C^\infty(X)[[\hbar]] \to \mathbb{R}((\hbar)).$$

Combining with Atiyah-Bott localization method, we can compute

$$\int_{BV} 1 = \int_X e^{-\omega_\hbar/\hbar} \hat{A}(X).$$

This is the simplest version of *algebraic index theorem* which was first formulated by Fedosov [19] and Nest-Tsygan [38] as the algebraic analogue of Atiyah-Singer type index theorem. It leads to the Atiyah-Singer index theorem with the help of the symbol calculus of pseudodifferential operators [20, 23, 39]. Our formulation is closely related to the physics interpretation of Atiyah-Singer index theorem [2, 22, 44, 45] via semi-classical analysis of supesymmetric quantum mechanics. When the symplectic manifold X is the total space of a cotangent bundle, there is a closely related result [25] where the quantization is one-loop exact.

3.2 2d chiral theory in the BV formalism

We move on to consider quantum field theories in two dimension. For simplicity, we consider a flat surface

$$\Sigma = \mathbb{C}, \mathbb{C}^*, \text{or } E_\tau = \mathbb{C}/\mathbb{Z} \oplus \mathbb{Z}\tau$$

where τ is an element of the upper half plane.

Here are some examples of free conformal field theory (CFT) in two dimension.

- free boson: $\int_\Sigma \partial\phi \wedge \bar{\partial}\phi$. ϕ a scalar field.
- bc-system: $\int_\Sigma b \wedge \bar{\partial} c$. b, c is a pair of fermions.
- $\beta\gamma$-system: $\int_\Sigma \beta \wedge \bar{\partial}\gamma$. β, γ is a pair of bosons.

We consider effective BV quantization in two dimension for chiral deformation of free CFT of the form:

$$S = \text{free CFT} + I.$$

Here I is a local functional of the form

$$I = \int d^2 z \mathcal{L}^{hol}(\partial_z \phi, b, c, \beta, \gamma)$$

where $\mathcal{L}^{hol}$ is a lagragian density involving only holomorphic derivatives of $\partial_z\phi, b, c, \beta, \gamma$.

There is an analogue of Theorem 3.1 describing the effective BV quantization, where the Weyl algebra is replaced by the *vertex algebra*. We give a brief discussion on vertex algebras and refer to [21, 29] for details.

A *vertex algebra* is a vector space $\mathcal{V}$ with the structure of *state-field correspondence* (and other axioms like vacuum, locality, etc.)

$$\begin{aligned} &\mathcal{V} \to End(\mathcal{V})[[z, z^{-1}]] \\ &A \to Y(A, z) = \sum_n A_{(n)} z^{-n-1} \end{aligned}$$

For simplicity, we will just write $A(z)$ for the operator $Y(A, z)$ associated to $A \in \mathcal{V}$. It defines the *operator product expansion* (OPE)

$$A(z)B(w) = \sum_{n\in\mathbb{Z}} \frac{(A_{(n)} \cdot B)(w)}{(z-w)^{n+1}}$$

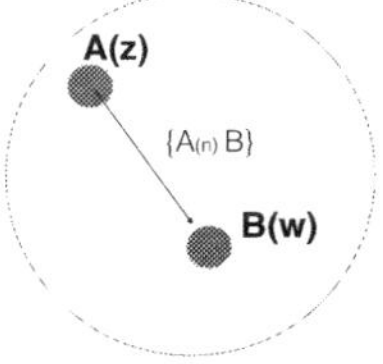

which is the 2d analogue of "associative product". We often write its singular part as

$$A(z)B(w) \sim \sum_{n\geq 0} \frac{(A_{(n)} \cdot B)(w)}{(z-w)^{n+1}}.$$

Given a vertex algebra $\mathcal{V}$, we can associate a Lie algebra $\oint \mathcal{V}$ from its Fourier modes. As a vector space

$$\oint \mathcal{V} := \mathrm{Span}_{\mathbb{C}} \left\{ \oint dz z^k A(z) := A_{(k)} \right\}_{A\in\mathcal{V}, k\in\mathbb{Z}}.$$

The Lie bracket is determined by the OPE (Borcherds commutator formula)

$$[A_{(m)}, B_{(n)}] = \sum_{j\geq 0} \binom{m}{j} \left(A_{(j)} B\right)_{m+n-j}.$$

Remark 3.3. For any $A \in \mathcal{V}$, the residues $A_{(0)} = \oint dz A(z)$ generates an infinitesimal automorphism of the vertex algebra. See for example [21, Corollary 3.3.8].

In our BV setup for chiral deformation of free CFT's, the space of fields has the form

$$\Omega^{0,\bullet}(E_\tau) \otimes h$$

of Dolbeault complex valued in some vector space h. h generates a vertex algebra $\mathcal{V}$, and a chiral lagrangian $\mathcal{L}^{hol}$ gives rise to an operator $\oint \mathcal{L}^{hol} \in \oint \mathcal{V}$. The following theorem is a 2d analogue of Theorem 3.1. We refer to [33] for details.

Theorem 3.4 ([33]). *We consider the following two dimensional quantum field theory with fields $\Omega^{0,\bullet}(E_\tau) \otimes h$ and*

$$S = \textit{free CFT} + I$$

where $I = \int d^2z \mathcal{L}^{hol}$ *is a chiral deformation described above. The differential* $Q = \bar\partial + \delta : \Omega^{0,\bullet} \otimes h \to \Omega^{0,\bullet} \otimes h$ *is the sum of* $\bar\partial$ *and a holomorphic differential operator* δ*.*

1. *The theory is UV-finite.*
2. *Solutions of effective quantum master equations*

$$\Longleftrightarrow \delta \oint dz \mathcal{L}^{hol} + \frac{1}{2\hbar}\left[\oint dz\mathcal{L}^{hol}, \oint dz \mathcal{L}^{hol}\right] = 0,$$

where $\oint dz\mathcal{L}^{hol}$ *is viewed as an operator on the corresponding vertex algebra.*

Remark 3.5. Remark 3.3 implies that we can globalize the above construction to obtain a 2d analogue of Theorem 3.2. For example, chiral de Rham complex [37] could be obtained in this way. See also [24]. It would be interesting to obtain a general algebraic index theorem for chiral vertex operators in this way.

We will explore applications of (2) in later subsections. Here we first explain (1) on the UV finiteness. The UV property for chiral deformations is known to physicists via the method of point-splitting regularization. It has a nice cohomological interpretation in the our current framework. We give an example to illustrate the basic idea. A full development will appear in a forthcoming paper [34].

Consider the following chiral deformation of free boson on the elliptic curve E_τ

$$S = \frac{1}{2}\int_{E_\tau} \partial\phi \wedge \bar\partial\phi + \frac{1}{3!}\int_{E_\tau} \frac{d^2z}{\operatorname{im}\tau}(\partial_z\phi)^3.$$

Let us look at a two-loop diagram

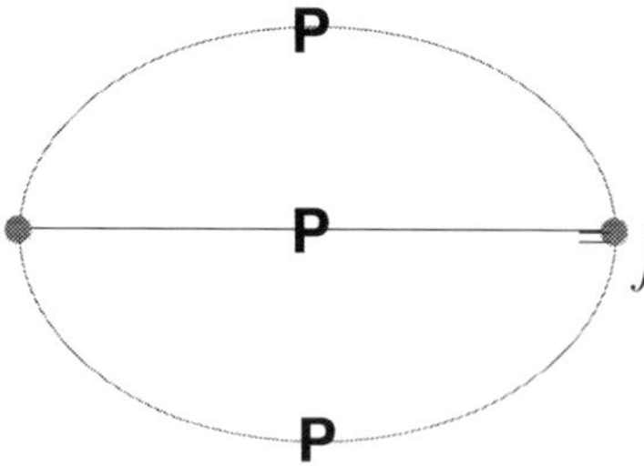

$$\int_{E_\tau} \frac{d^2z}{\operatorname{im}\tau}\mathbf{P}(z;\tau)^3.$$

Here the propagator is given by the $\mathbf{P}(z;\tau) = \int_0^\infty \partial_z^2 h_t$ where z is the linear holomorphic coordinate on $\mathbb{C}$, and $h_t(z) = \frac{1}{4\pi t}\sum\limits_{\lambda\in\mathbb{Z}+\mathbb{Z}\tau} e^{-|z+\lambda|^2/4t}$ is the heat kernel function on E_τ with respect to the flat metric $\frac{i}{2}dz\wedge d\bar z$. $\mathbf{P}(z;\tau)$ can be expressed as

$$\mathbf{P}(z;\tau) = \mathcal{P}(z,\tau) + \frac{\pi^2}{3}E_2^*$$

where $\mathcal{P}$ is the Weierstrass $\mathcal{P}$-function. E_2 is the second Eisenstein series, and $E_2^* = E_2 - \frac{3}{\pi}\frac{1}{\operatorname{im}\tau}$.

Naively, $\mathcal{P}$ has a pole of order 2, and the above integral looks like divergent. However, our homotopic renormalization scheme leads to a well-defined renormalized value, which can be described as follows. Consider the exact sequence of sheaves on E_τ

$$0 \to \mathbb{C} \to \mathcal{M} \xrightarrow{d} \Omega^1_{II} \to 0$$

where $\mathcal{M}$ is the sheaf of meromorphic functions, and Ω^1_{II} is the sheaf of abelian differentials of second kind, i.e. meromorphic 1-forms without order 1 pole. The integrand

$$\frac{d^2z}{\operatorname{im}\tau}\mathbf{P}(z;\tau)^3 = -d\bar{z} \wedge \frac{\mathbf{P}(z;\tau)^3 dz}{\operatorname{im}\tau}$$

can be viewed as an element of $H^1(E_\tau, \Omega^1_{II})$. Then the renormalized value of the above Feynman integral is given by the boundary map

$$H^1(E_\tau, \Omega^1_{II}) \to H^2(E_\tau, \mathbb{C}) \xrightarrow{\int} \mathbb{C}.$$

Using this we find

$$\frac{1}{\pi^6}\int_{E_\tau}\frac{d^2z}{\operatorname{im}\tau}\mathbf{P}^3 = \frac{2^2}{3^3 5}E_6 + \frac{2}{3^2 5}E_4E_2^* - \frac{2}{3^3}(E_2^*)^3$$

Under the $\bar{\tau} \to \infty$ limit, which amounts to replacing $E_2^* \to E_2$, the above integral reduces to an A-cycle integral computed by Douglas [17]. This explanation of renormalization works in great generality [34].

3.3 BCOV theory and higher genus B-model

The closed string field theory is described in terms of BV master equation [46]. This is one of the places where BV quantization plays an essential role. We are interested in mirror symmetry between topological strings. The leading cubic approximation of string field action in the B-twisted topological sector on Calabi-Yau 3-fold is proposed in [5], called the *Kodaira-Spencer gauge theory*. This is generalized in [12] where we have found the full B-twisted topological string field action on Calabi-Yaus of arbitary dimension that satisfies the classical master equation. We call this *BCOV theory* and propose (as a generalization of [5]) that the BV quantization of BCOV theory leads to quantum B-model that is mirror to the A-model Gromov-Witten theory of counting higher genus curves. As an application of the effective BV quantization method, we carry out this program explicitly for elliptic curves following [33]. See [7–9, 26] for another categorical approach to this example.

3.3.1 classical BCOV theory

Let X be a compact Calabi-Yau manifold of dimension d. Ω_X will be a fixed holomorphic volume form. By abuse of notation, we let

$$\mathrm{PV}(X) = \bigoplus_{0\leq i,j\leq d} \mathrm{PV}^{i,j}(X), \quad \mathrm{PV}^{i,j}(X) = \Omega^{0,j}(X, \wedge^i T_X)$$

be the space of polyvector fields on X. Here T_X is the holomorphic tangent bundle, and $\Omega^{0,j}(X, \wedge^i T_X)$ is the space of smooth $(0,j)$-forms valued in $\wedge^i T_X$. Ω_X induces an identification between the space of polyvector fields and differential forms

$$\mathrm{PV}^{i,j}(X) \overset{\lrcorner\Omega_X}{\cong} \Omega^{d-i,j}(X)$$
$$\alpha \to \alpha \lrcorner \Omega_X$$

where $\lrcorner$ is the contraction, and $\Omega^{i,j}(X)$ denotes smooth differential forms of type (i,j). The holomorphic de Rham differential ∂ on forms defines an operator on $\mathrm{PV}(X)$ via the above isomorphism, denoted again by

$$\partial : \mathrm{PV}^{i,j}(X) \to \mathrm{PV}^{i-1,j}(X)$$

i.e.

$$(\partial\alpha)\lrcorner\Omega_X := \partial(\alpha\lrcorner\Omega_X), \quad \alpha \in \mathrm{PV}(X).$$

The definition of ∂ doesn't depend on the choice of Ω_X on compact Calabi-Yau manifolds.

Definition 3.6 ([12]). The space of fields for BCOV theory on X is the complex

$$\mathcal{E} := (\mathrm{PV}(X)[[t]][2], Q = \bar{\partial} + t\partial)$$

where t is a formal variable of degree 2 representing the gravitational descendants. [2] is the degree shifting such that $\mathrm{PV}^{1,1}(X)$ lies in degree 0.

The BV kernel of BCOV theory is in fact degenerate, hence does not come from a (-1)-symplectic pairing. Nevertheless, the framework of effective BV quantization still works (see Remark 2.5). Consider the trace map

$$\mathrm{Tr} : \mathrm{PV}(X) \to \mathbb{C}, \quad \mu \to \int_X (\mu\lrcorner\Omega_X) \wedge \Omega_X.$$

The BV kernel is a $(\mathrm{PV}(X \times X) =) \, \mathrm{PV}(X) \otimes \mathrm{PV}(X)$-valued distribution representing the integral kernel (with respect to Tr) of the operator $\partial : \mathrm{PV}(X) \to \mathrm{PV}(X)$. In other words, it is given by

$$(\partial \otimes 1)\delta$$

where δ is the δ-function representing the integral kernel of the identity operator with respect to Tr.

In particular, $\mathrm{PV}(X)t^k$ for $k > 0$ does not appear in the BV kernel: they are not dynamical hence are called *background fields*. We see that BCOV theory is a highly degenerate BV theory with a huge number of background fields.

The classical interaction of BCOV theory is given by

$$I_0(\mu) = \mathrm{Tr}\left\langle e^{\mu}\right\rangle_0, \quad \mu \in \mathcal{E},$$

where

$$\left\langle \mu_1 t^{k_1}, \cdots, \mu_n t^{k_n}\right\rangle_0 := \binom{n-3}{k_1, \cdots, k_n} \mu_1 \wedge \cdots \wedge \mu_n, \quad \mu_i \in \mathrm{PV}(X)$$

and $\langle e^{\mu}\rangle_0$ is understood as $\sum\limits_{n\geq 3} \frac{1}{n!}\langle \mu, \cdots, \mu\rangle_0$. It is shown in [12] that I_0 satisfies the classical master equation

$$QI_0 + \frac{1}{2}\{I_0, I_0\} = 0.$$

The effective BV quantization of classical BCOV theory leads to higher genus B-model invariants which is expected to be identified with the A-model Gromov-Witten invariants of the mirror Calabi-Yau [5,12]. So far we have only established quantizations of BCOV theory on a few special Calabi-Yau geometries [13, 33]. Elliptic curve is an example where we have fully established the quantum BCOV theory.

Remark 3.7. The above setup is naturally genearlized to the Landau-Ginzburg case when X is a non-compact Calabi-Yau equipped with a holomorphic function $f : X \to \mathbb{C}$ (superpotential). There $\mathcal{E} = \mathrm{PV}_c(X)[[t]]$ is given by compactly supported polyvector fields, and $Q = \bar{\partial} + t\partial + df\lrcorner$. When $X = \mathbb{C}^n$ and f is a holomorphic function with an isolated critical point, the genus zero Landau-Ginzburg twisted BCOV theory is equivalent [35, 36] to K. Saito's theory of primitive forms [42].

3.3.2 The elliptic curve example

We carry out the effective BV quantization of BCOV theory explicitly in the case of elliptic curves.

Let τ be an element of the upper half plane. Let $\mathrm{E}_\tau = \mathbb{C}/(\mathbb{Z} \oplus Z\tau)$ be the corresponding elliptic curve. z denotes the linear holomorphic coordinate. The space of fields of BCOV theory on E_τ is

$$\mathcal{E} = \Omega^{0,\bullet}(\mathrm{E}_\tau, \mathcal{O}_{\mathrm{E}_\tau})[[t]] \oplus \Omega^{0,\bullet}(\mathrm{E}_\tau, T_{\mathrm{E}_\tau}[1])[[t]].$$

The differential is given by

$$Q = \bar{\partial} + t\partial,$$

where $\partial : T_{\mathrm{E}_\tau} \to \mathcal{O}_{\mathrm{E}_\tau}$ is the divergence operator with respect to the holormophic volume form dz.

Let us first describe the BV bracket concretely. Let us parametrize a field $\varphi \in \mathcal{E}$ by

$$\varphi = \sum_{k\geq 0} b_k t^k + \eta_k t^k, \quad b_k \in \Omega^{0,\bullet}(\mathrm{E}_\tau, \mathcal{O}_{\mathrm{E}_\tau}), \quad \eta_k \in \Omega^{0,\bullet}(\mathrm{E}_\tau, T_{\mathrm{E}_\tau}[1]).$$

Then the only nontrivial BV bracket is given by

$$\{b_0(z), b_0(w)\} = \partial_z \delta^2(z-w)$$

where $\delta^2(z-w)$ is the δ-function with respect to the volume form $\frac{i}{2}dz \wedge d\bar{z}$. All other fields $b_{>0}, \eta_\bullet$ are background fields. This theory can be identified with a chiral deformation of the free chiral boson ϕ, where

$$b_0 = \partial_z \phi, \quad \text{with OPE} \quad b_0(z)b_0(w) \sim \frac{\hbar}{(z-w)^2}.$$

In particular, Theorem 3.4 applies and the effective BV quantization amounts to find an operator I on the vertex algebra satisfying

$$(t\partial)I + \frac{1}{2\hbar}[I, I] = 0$$

and reduces to our classical BCOV interaction I_0 modulo $\hbar$. In [33], a canonical solution of quantum master equation is found using boson-fermion correspondence. The full solution is determined by its restriction to

$$b_{>0} = 0, \quad \eta_\bullet \in \mathbb{C}[d\bar{z}]\partial_z \quad (\text{i.e., } b_{>0} \text{ vanishes and } \eta_\bullet\text{=constants}).$$

Such subsector represents $t\partial$-cohomology of the fields, and is called the stationary sector [32] which is mirror to a similar notion in the A-model. It determines the full generating function on elliptic curves with the help of Virasoro constraint [40].

Restricting to the stationary subsector, the quantum corrected chiral interaction is given by

$$I = \sum_{k\geq 0} \eta_k \int d^2z \frac{W^{(k+2)}(b_0)}{k+2}$$

where

$$W^{(k)}(b_0) = \sum_{\sum_{i\geq 1} ik_i = k} \frac{k!}{\prod_i k_i!} : \prod_i \left(\frac{1}{i!}(\sqrt{\hbar}\partial)^{i-1} b_0\right)^{k_i} :_B$$

comes from the bosonic realization of the $W_{1+\infty}$-algebra. In the stationary subsector, quantum master equation is equivalent to

$$\left[\oint dz \frac{W^{(k+2)}}{k+2}, \oint dz \frac{W^{(m+2)}}{m+2}\right] = 0, \quad k, m \geq 0,$$

representing an integrable hierarchy of commuting Hamiltonians.

The character on the Heisenberg vertex algebra (generated by b_0)

$$\operatorname{Tr} q^{L_0 - \frac{1}{24}} e^{\frac{1}{\hbar}\sum\limits_{k\geq 0} \oint dz \eta_k \frac{W^{(k+2)}}{k+2}}, \quad q = e^{2\pi i \tau},$$

coincides with the A-model stationary Gromov-Witten invariants [40] under the boson-fermion correspondence. This can be viewed as a full generalization of [14] who considers the cubic interaction $W^{(3)}$.

3.4 Calabi-Yau geometry and integrable hierarchy

Topological string on Calabi-Yau manifolds is often found to be related to certain integrable hierarchies. In this section we propose a uniform interpretation in terms of BCOV theory. The classical aspects of this construction is explored in [28].

Let X be a Calabi-Yau geometry. It can be either a compact Calabi-Yau, or a noncompact Calabi-Yau with a Landau-Ginzburg superpotential $f : X \to \mathbb{C}$ having isolated critical points. We consider $X \times \mathbb{C}$, which is again a Calabi-Yau manifold. Here we equip $\mathbb{C}$ with the trivial volume form dz. We consider BCOV theory on $X \times \mathbb{C}$, with a parameter $\epsilon \neq 0$

$$\mathcal{E}^\epsilon = \big(\mathrm{PV}(X \times \mathbb{C})[[t]], Q_\epsilon = \bar{\partial} + t(\partial_X + \epsilon \partial_{\mathbb{C}})\big)$$

Here ∂_X ($\partial_{\mathbb{C}}$ respectively) is the divergence operator with respect to the Calabi-Yau volume form on X ($\mathbb{C}$ respectively). The BV kernel is the integral kernel of the operator $(\partial_X + \epsilon\partial_{\mathbb{C}})$ on $\mathrm{PV}(X \times \mathbb{C})$. It is easy to see that the same classical interaction satisfies the classical master equation. Now we consider the process of integrating out massive modes on X

$$\begin{array}{ll} \text{B- model (BCOV theory } \mathcal{E}^\epsilon\text{) on} & X \times \mathbb{C} \\ & \Big\downarrow \text{integration out over } X \\ \text{Effective 2d chiral theory on} & \mathbb{C} \end{array}$$

Mathematically, this process is a partial homological perturbation along component of X (this requires a choice of splitting of the Hodge filtration on X). We end up with an effective 2d chiral theory on $\mathbb{C}$, whose space of fields is

$$\mathrm{PV}(\mathbb{C}) \otimes_{\mathbb{C}} \mathrm{H}_X[[t]]$$

where

$$H_X = \begin{cases} H^\bullet(\mathrm{PV}(X), \bar{\partial}) & \text{compact Calabi-Yau model} \\ \mathrm{Jac}(f) & \text{Landau-Ginzburg model.} \end{cases}$$

$H_X[[t]]$ represents the zero modes of BCOV theory on X. The effective theory has again a huge number of background symmetry, which contains the abelian subalgebra

$$\partial_z \otimes \mathrm{H}_X[[t]].$$

Their associated Noether currents give rise to infinite many commutating operators $\{I_\alpha^{\hbar,\epsilon}\}_{\alpha \in \mathrm{H}_X[[t]]}$. This is our interpretation of integrable hierarchy.

Observe that the effective quantum parameter on $\mathbb{C}$ is $\hbar\epsilon$. In particular, the limit $\epsilon \to 0$ describes a family of classical master equations parametrized by $\hbar$.

Let us first consider the limit

$$I_{k,\mu} = \lim_{\substack{\hbar \to 0 \\ \epsilon \to 0}} I^{\hbar,\epsilon}_{t^k\mu}, \quad \mu \in H_X.$$

This can be described as follows. Recall that H_X carries a nondegenerate pairing $\langle -,-\rangle$. Our effective vertex algebra is generated by H_X with OPE

$$a(z)b(w) \sim \frac{\hbar\epsilon \langle a,b\rangle}{(z-w)^2}, \quad a,b \in H_X.$$

The classical limit leads to the Poisson bracket

$$\{a(z), b(w)\} = \langle a,b\rangle \, \partial_z \delta(z-w).$$

Let $F_0^X \in \mathcal{O}(H_X[[t]])$ be the genus zero generating function on $H_X[[t]]$ of BCOV theory (B-model). Let

$$J_{k,\mu} := \partial_{t^k\mu} F_0^X |_{\mathbb{H}_X} \in \mathcal{O}(\mathbb{H}_X)$$

be the generating function with one descendant insertion $t^k\mu$ and all others primary insertions $t^0 H_X$. Then

$$I_{k,\mu} = \oint dz J_{k,\mu}(b(z)), \quad b \in H_X$$

and classical master equation implies that they all commute under the Poisson bracket above [28]. This is precisely Dubrovin's classical integrable hierarchy associated to a Frobenius manifold [15]. The limit

$$\lim_{\epsilon\to 0} I_\alpha^{\hbar,\epsilon}$$

gives a dispersion deformation of such a classical integrable hierarchy which is contructed and studied extensively by Dubrovin-Zhang [16] for semi-simple Frobenius manifold. Topological string on $X \times \mathbb{C}$ suggests that such a dispersion deformation and a further quantization should exist in general. One possible approach is to establish quantum BCOV theory on $X \times \mathbb{C}$. We hope to explore this in the future.

References

[1] M. Alexandrov, A. Schwarz, O. Zaboronsky, and M. Kontsevich, *The geometry of the master equation and topological quantum field theory*, Internat. J. Modern Phys. A **12** (1997), no. 7, 1405–1429.

[2] L. Alvarez-Gaumé, *Supersymmetry and the Atiyah-Singer index theorem*, Comm. Math. Phys. **90** (1983), no. 2, 161–173.

[3] V. I. Arnold, S. M. Gusein-Zade, and A. N. Varchenko, *Singularities of differentiable maps. Volume 2*, Modern Birkhäuser Classics, Birkhäuser/Springer, New York, 2012. Monodromy and asymptotics of integrals; Translated from the Russian by Hugh Porteous and revised by the authors and James Montaldi; Reprint of the 1988 translation.

[4] I. A. Batalin and G. A. Vilkovisky, *Gauge algebra and quantization*, Phys. Lett. B **102** (1981), no. 1, 27–31.

[5] M. Bershadsky, S. Cecotti, H. Ooguri, and C. Vafa, *Kodaira-Spencer theory of gravity and exact results for quantum string amplitudes*, Comm. Math. Phys. **165** (1994), no. 2, 311–427.

[6] D. Bessis, C. Itzykson, and J. B. Zuber, *Quantum field theory techniques in graphical enumeration*, Adv. in Appl. Math. **1** (1980), no. 2, 109–157.

[7] A. Căldăraru and J. Tu, *Computing a categorical Gromov-Witten invariant*, arXiv:1706.09912 [math.AG].

[8] K. Costello, *The partition function of a topological field theory*, J. Topol. **2** (2009), no. 4, 779–822.
[9] ———, *Topological conformal field theories and Calabi-Yau categories*, Adv. Math. **210** (2007), no. 1, 165–214.
[10] ———, *Renormalization and effective field theory*, Mathematical Surveys and Monographs, vol. 170, American Mathematical Society, Providence, RI, 2011.
[11] K. Costello and O. Gwilliam, *Factorization algebras in quantum field theory. Vol. 1*, New Mathematical Monographs, vol. 31, Cambridge University Press, Cambridge, 2017.
[12] K. Costello and S. Li, *Quantum BCOV theory on Calabi-Yau manifolds and the higher genus B-model*, arXiv:1201.4501 [math.QA].
[13] ———, *Quantization of open-closed BCOV theory, I*, arXiv:1505.06703 [hep-th].
[14] R. Dijkgraaf, *Mirror symmetry and elliptic curves*, The moduli space of curves (Texel Island, 1994), Progr. Math., vol. 129, Birkhäuser Boston, Boston, MA, 1995, pp. 149–163.
[15] B. Dubrovin, *Geometry of 2D topological field theories*, Integrable systems and quantum groups (Montecatini Terme, 1993), Lecture Notes in Math., vol. 1620, Springer, Berlin, 1996, pp. 120–348.
[16] B. Dubrovin and Y. Zhang, *Normal forms of hierarchies of integrable PDEs, Frobenius manifolds and Gromov - Witten invariantsv.* arXiv:math/0108160 [math.DG].
[17] M. R. Douglas, *Conformal field theory techniques in large N Yang-Mills theory*, Quantum field theory and string theory (Cargèse, 1993), NATO Adv. Sci. Inst. Ser. B Phys., vol. 328, Plenum, New York, 1995, pp. 119–135.
[18] B. V. Fedosov, *A simple geometrical construction of deformation quantization*, J. Differential Geom. **40** (1994), no. 2, 213–238.
[19] ———, *Deformation quantization and index theory*, Mathematical Topics, vol. 9, Akademie Verlag, Berlin, 1996.
[20] ———, *Pseudo-differential operators and deformation quantization*, Quantization of singular symplectic quotients, Progr. Math., vol. 198, Birkhäuser, Basel, 2001, pp. 95–118.
[21] E. Frenkel and D. Ben-Zvi, *Vertex algebras and algebraic curves*, 2nd ed., Mathematical Surveys and Monographs, vol. 88, American Mathematical Society, Providence, RI, 2004.
[22] D. Friedan and P. Windey, *Supersymmetric derivation of the Atiyah-Singer index and the chiral anomaly*, Nuclear Phys. B **235** (1984), no. 3, 395–416.
[23] E. Getzler, *Pseudodifferential operators on supermanifolds and the Atiyah-Singer index theorem*, Comm. Math. Phys. **92** (1983), no. 2, 163–178.
[24] V. Gorbounov, O. Gwilliam, and B. Williams, *Chiral differential operators via Batalin-Vilkovisky quantization.* arXiv:1610.09657 [math.QA].
[25] O. Gwilliam and R. Grady, *One-dimensional Chern-Simons theory and the $\hat{A}$ genus*, Algebr. Geom. Topol. **14** (2014), no. 4, 2299–2377.
[26] M. Kontsevich and Y. Soibelman, *Notes on A_∞-algebras, A_∞-categories and non-commutative geometry*, Homological mirror symmetry, Lecture Notes in Phys., vol. 757, Springer, Berlin, 2009, pp. 153–219.
[27] R. E. Grady, Q. Li, and S. Li, *Batalin–Vilkovisky quantization and the algebraic index*, Adv. Math. **317** (2017), 575–639.
[28] W. He, S. Li, and P. Yoo, *Calabi-Yau geometry and integrable hierarchy*, in preparation.
[29] V. Kac, *Vertex algebras for beginners*, 2nd ed., University Lecture Series, vol. 10, American Mathematical Society, Providence, RI, 1998.
[30] M. Kaneko and D. Zagier, *A generalized Jacobi theta function and quasimodular forms*, The moduli space of curves (Texel Island, 1994), Progr. Math., vol. 129, Birkhäuser Boston, Boston, MA, 1995, pp. 165–172.
[31] V. S. Kulikov, *Mixed Hodge structures and singularities*, Cambridge Tracts in Mathematics, vol. 132, Cambridge University Press, Cambridge, 1998.
[32] S. Li, *BCOV theory on the elliptic curve and higher genus mirror symmetry*, arXiv:1112.4063 [math.QA].
[33] ———, *Vertex algebras and quantum master equaton.* arXiv:1612.01292[Math.QA].
[34] S. Li and J. Zhou, *Chiral CFTs on the Tate curve*, in preparation.
[35] C. Li, S. Li, and K. Saito, *Primitive forms via polyvector fields*, arxiv: 1311.1659[math.AG].
[36] Changzheng Li, Si Li, Kyoji Saito, and Yefeng Shen, *Mirror symmetry for exceptional unimodular singularities*, J. Eur. Math. Soc. (JEMS) **19** (2017), no. 4, 1189–1229.

[37] F. Malikov, V. Schechtman, and A. Vaintrob, *Chiral de Rham complex*, Comm. Math. Phys. **204** (1999), no. 2, 439–473.
[38] R. Nest and B. Tsygan, *Algebraic index theorem*, Comm. Math. Phys. **172** (1995), no. 2, 223–262.
[39] ———, *Formal versus analytic index theorems*, Internat. Math. Res. Notices **11** (1996), 557–564.
[40] A. Okounkov and R. Pandharipande, *Virasoro constraints for target curves*, Invent. Math. **163** (2006), no. 1, 47–108.
[41] K. Saito, *The higher residue pairings $K_F^{(k)}$ for a family of hypersurface singular points*, Singularities, Part 2 (Arcata, Calif., 1981), Proc. Sympos. Pure Math., vol. 40, Amer. Math. Soc., Providence, RI, 1983, pp. 441–463.
[42] ———, *Period mapping associated to a primitive form*, Publ. Res. Inst. Math. Sci. **19** (1983), no. 3, 1231–1264.
[43] A. Schwarz, *Geometry of Batalin-Vilkovisky quantization*, Comm. Math. Phys. **155** (1993), no. 2, 249–260.
[44] E. Witten, *Supersymmetry and Morse theory*, J. Differential Geom. **17** (1982), no. 4, 661–692.
[45] ———, *Index of Dirac operators*, Quantum fields and strings: a course for mathematicians, Vol. 1, 2 (Princeton, NJ, 1996/1997), Amer. Math. Soc., Providence, RI, 1999, pp. 475–511.
[46] B. Zwiebach, *Closed string field theory: quantum action and the Batalin-Vilkovisky master equation*, Nuclear Phys. B **390** (1993), no. 1, 33–152.

YAU MATHEMATICAL SCIENCES CENTER, TSINGHUA UNIVERSITY, BEIJING, CHINA
E-mail address: sili@mail.tsinghua.edu.cn

Proceedings of the 7th International Congress of Chinese Mathematicians
ALM 43, pp. 373–394

Global Regularities of 2-D Density Patches for Inhomogeneous Incompressible Viscous Flow

Xian Liao and Ping Zhang

Abstract

In this paper, we explain our recent results in [30, 31] concerning the propagation of regularities of the density patch problem for two dimensional viscous inhomogeneous incompressible flow. We first establish the global well-posedness of 2-D inhomogeneous incompressible Navier-Stokes system with initial density of the form $\eta_1 \mathbf{1}_{\Omega_0} + \eta_2 \mathbf{1}_{\Omega_0^c}$, for any pair of positive constants (η_1, η_2), and with Ω_0 a simply connected $W^{k+2,p}(\mathbb{R}^2)$-domain. We then show that the time evolved domain $\Omega(t)$ also belongs to the same regularity class $W^{k+2,p}(\mathbb{R}^2)$ for any order $k \geq 0$ and for any positive time $t > 0$.

Mathematics Subject Classification: 35Q30, 76D03.

Keywords and Phrases: Inhomogeneous incompressible Navier-Stokes equations, density patch, striated distributions, Littlewood-Paley theory.

1 Introduction

We consider the following two-dimensional inhomogeneous incompressible Navier-Stokes equations:

$$\begin{cases} \partial_t \rho + \operatorname{div}(\rho v) = 0, \qquad (t,x) \in \mathbb{R}^+ \times \mathbb{R}^2, \\ \partial_t(\rho v) + \operatorname{div}(\rho v \otimes v) - \Delta v + \nabla \pi = 0, \\ \operatorname{div} v = 0, \\ (\rho, v)|_{t=0} = (\rho_0, v_0). \end{cases} \tag{NS}$$

Here the unknown ρ is a positive function which represents the density of the fluid at time $t \geq 0$ and point $x \in \mathbb{R}^2$, $v = (v_1, v_2) \in \mathbb{R}^2$ stands for the velocity field of the fluid, and $\pi \in \mathbb{R}$ represents the pressure at time t and point x, which guarantees the incompressibility of the fluid. Notice that when the density function $\rho = 1$ is positive constant function, the system (NS) becomes the classic homogeneous Navier-Stokes system, which has been studied extensively since the pioneer work by J. Leray [27] in 1934. Here we are concerned with variable density function case and this system (NS) is a mathematical model describing an incompressible fluid with *nonconstant* density. Some basic examples are mixtures of incompressible and non-reactant flows, flows with complex structure such as blood flow or model of rivers, and fluids containing a melted substance.

The system (NS) has three major basic features. Let us explain them in a general Euclidean space with dimension $d \geq 2$:

- The incompressibility of the convection velocity in the density transport equation ensures that

$$\begin{aligned} &\|\rho(t)\|_{L^\infty} = \|\rho_0\|_{L^\infty} \quad \text{and} \\ &\text{meas}\big\{x \in \mathbb{R}^d \mid a \leq \rho(t,x) \leq b\big\} \text{ is independent of } t \geq 0, \end{aligned} \tag{1.1}$$

for any pair of real numbers (a, b);
- The kinetic energy is formally conserved

$$\frac{1}{2}\int_{\mathbb{R}^d} \rho(t,x)|v(t,x)|^2 \mathrm{d}x + \int_0^t \|\nabla v(t',\cdot)\|_{L^2(\mathbb{R}^d)}^2 \mathrm{d}t' = \frac{1}{2}\int_{\mathbb{R}^d} \rho_0(x)|v_0(x)|^2 \mathrm{d}x; \tag{1.2}$$

- If (ρ, v, π) is a solution of (NS) on $[0, T] \times \mathbb{R}^d$, then $(\rho, v, \pi)_\lambda$ defined by

$$(\rho, v, \pi)_\lambda(t, x) \stackrel{\text{def}}{=} \big(\rho(\lambda^2 t, \lambda x), \lambda v(\lambda^2 t, \lambda x), \lambda^2 \pi(\lambda^2 t, \lambda x)\big) \tag{1.3}$$

is also a solution of (NS) on $[0, T/\lambda^2] \times \mathbb{R}^d$. This scaling invariance property naturally leads to the notion of *critical regularity*.

Based on the energy estimate (1.2), A.V. Kazhikhov [25] proved the global-in-time existence of weak solutions of the inhomogeneous incompressible Navier-Stokes system (NS). See also J. Simon [36] for the case including vacuum and P.-L. Lions [32] for the variable viscosity coefficient case.

In the case of smooth initial data with no vacuum, the existence of strong unique solutions goes back to the work of O.A. Ladyzhenskaya and V.A. Solonnikov [26]. Motivated by (1.3), R. Danchin [13] established the well-posedness of (NS) in the whole space $\mathbb{R}^d$ in the so-called *critical functional framework* for small perturbations of some positive constant density. This result was extended to more general Besov spaces in [1, 2, 3, 4, 33].

In all those aforementioned works, the density has to be at least in the Besov space $\dot{B}^{\frac{d}{p}}_{p,\infty}(\mathbb{R}^d)$, and within this framework it is not possible to consider the density function with discontinuities across a hypersurface. In fact, the Besov regularity of the characteristic function of a smooth domain is only $\dot{B}^{\frac{1}{p}}_{p,\infty}(\mathbb{R}^d)$. Therefore, those results do not apply to a mixture composed of two separate fluids with different densities.

The first breakthrough toward this was made by R. Danchin and P.B. Mucha [14], where they essentially proved the global well-posedness of (NS) with initial density which can be discontinuous across a C^1 interface with a sufficiently small jump. Later on, essentially under the mere condition that ρ_0 is bounded and bounded away from zero, J. Huang, M. Paicu and the second author [24] obtained the global existence of solutions to (NS) in a *critical functional framework*.

Moreover, the uniqueness was also obtained under slightly stronger regularity assumptions on the velocity field. Later, R. Danchin and the second author extended this result to the half-space setting in [17].

In the general case where $0 < c_0 \le \rho_0 \in L^\infty(\mathbb{R}^d)$ and $v_0 \in H^2(\mathbb{R}^d)$, R. Danchin and P.B. Mucha [15] proved the local well-posedness of the strong solutions to the system (NS) and later [11, 34] improved this well-posedness results to less regularity assumptions on the initial velocity. Recently R. Danchin and P.B. Mucha [16] succeeded in including also the vacuum where $0 \le \rho_0 \in L^\infty(\Omega)$ and $v_0 \in H_0^1(\Omega)$, $\Omega \subset \mathbb{R}^d$, $d = 2, 3$ some bounded smooth domain.

It is natural to study the boundary regularity propagation of the interface of the fluids. On this issue, P.-L. Lions proposed the following question in [32]: suppose that the initial density takes the form $\rho_0 = \mathbf{1}_D$ for some smooth domain D, then by Theorem 2.1 of [32], there exists at least one global weak solution (ρ, v) of (NS) such that for all $t \ge 0$, $\rho(t) = \mathbf{1}_{D(t)}$ for some set $D(t)$ with $\mathrm{vol}(D(t)) = \mathrm{vol}(D)$. Then is the boundary regularity of D preserved by the time evolution or not?

As a first step toward this question, we [29] considered the case away from vacuum with the initial density $\rho_0(x) = \eta_1 \mathbf{1}_{\Omega_0} + \eta_2 \mathbf{1}_{\Omega_0^c}$ and confirmed the propagation of the $W^{k+2,p}(\mathbb{R}^2)$, $p \in (2,4)$-boundary regularities of the domain Ω_0 of *any order* $k \ge 0$ [1] under the small jump condition: $|\eta_1 - \eta_2|$ is sufficiently small. For any pair of positive numbers (η_1, η_2), we [30, 31] considered the low regularity case $k = 0, 1$ and the high regularity case $k \ge 2$ respectively. The three dimensional case was considered by Y. Liu and the first author in [28]. Recently [16, 18, 20] studied the propagation of the low boundary regularity of Hölder type $C^{1,\alpha}$ (which corresponds to $k = 0$ in the our L^p-framework) and as [16] investigated the vacuum case, the aforementioned P.-L. Lions' question has a complete answer for the constant viscosity coefficient case in the low regularity setting.

Let us explain our setting more precisely.

Let Ω_0 be a simply connected bounded domain with boundary regularity $W^{k+2,p}(\mathbb{R}^2)$, $k \ge 0$, $p \in (2,4)$, that is, we can parametrize $\partial\Omega_0$ as

$$\gamma : \mathbb{S}^1 \mapsto \partial\Omega_0 \text{ via } s \mapsto \gamma(s) \text{ with } \partial_s \gamma(s) = \nabla^\perp h_0(\gamma(s)), \quad \gamma \in (W^{k+2,p}(\mathbb{R}^2))^2, \tag{1.4}$$

where the real-valued function $h_0 \in W^{2,p}(\mathbb{R}^2)$ is taken such that $\partial\Omega_0 = h_0^{-1}(0)$ and that $\nabla^\perp h_0$, $\nabla^\perp = (-\partial_2, \partial_1)^T$ does not vanish on $\partial\Omega_0$.

Let $X_0 = \nabla^\perp h_0$ such that its restriction on $\partial\Omega_0$ is tangential: $X_0(\gamma(s)) = \partial_s\gamma(s)$. Let $\partial_{X_0} v = X_0 \cdot \nabla v$ denote the directional derivative of v along X_0. Then we easily calculate

$$\partial_s^2 \gamma(s) = \partial_s(X_0(\gamma(s))) = X_0(\gamma(s)) \cdot \nabla X_0(\gamma(s)) = (\partial_{X_0} X_0)(\gamma(s)),$$

and repeating this calculation gives $\partial_s^\ell \gamma(s) = (\partial_{X_0}^{\ell-1} X_0)(\gamma(s))$. Hence by the domain assumption we know that

$$X_0 = \nabla^\perp h_0 \in (W^{1,p}(\mathbb{R}^2))^2, \quad \partial_{X_0}^{\ell-1} X_0 \in (W^{2,p}(\mathbb{R}^2))^2, \quad \ell = 1, \cdots, k. \tag{1.5}$$

[1]The case $k = 0$ is implied by the well-posedness results with $v \in L^1_{\mathrm{loc}}(\mathbb{R}^+; W^{2,p}(\mathbb{R}^2))$ directly.

For any $\eta_1, \eta_2 > 0$, we take the initial density ρ_0 and the initial velocity v_0 of (NS) as

$$\rho_0 = \eta_1 \mathbf{1}_{\Omega_0} + \eta_2 \mathbf{1}_{\Omega_0^c}, \quad v_0 \in (L^2 \cap \dot{B}^{s_0}_{2,1}(\mathbb{R}^2))^2 \quad \text{and} \quad \partial^\ell_{X_0} v_0 \in (L^2 \cap \dot{B}^{s_\ell}_{2,1}(\mathbb{R}^2))^2, \tag{1.6}$$

for some $s_0 \in (0,1)$, $s_\ell = s_0 - \epsilon\ell/k$ with some fixed $\epsilon \in (0, s_0)$, $\ell = 1, \cdots, k$, $p \in (2, 2/(1-s_k))$. In particular, if the initial scalar vorticity $\omega_0 = \mathbf{1}_{\Omega_0}$, then the regularity condition (1.6) for the initial velocity field v_0 holds. This idea of considering conormal distributions or striated distributions was used by J.-Y. Chemin [8, 9] to prove the global regularities of the two-dimensional vortex patch for ideal flow (see also [6, 7, 10, 12, 19, 21, 22, 35]), and it also plays an essential role in this paper.

The main result of this paper is the following:

Theorem 1.1. *Let the initial data (ρ_0, v_0) be given by (1.4)-(1.5)-(1.6) for any pair of positive constants (η_1, η_2) and for any $k \geq 0$. Then the two dimensional inhomogeneous incompressible Navier-Stokes system* (NS) *has a unique global solution (ρ, v) so that*

$$\rho(t,x) = \eta_1 \mathbf{1}_{\Omega(t)} + \eta_2 \mathbf{1}_{\Omega(t)^c}$$

with a simply connected $W^{k+2,p}(\mathbb{R}^2)$-domain $\Omega(t)$ for any $t > 0$. Furthermore, the Besov regularities for the velocity field are also conserved.

In the above, we have taken the following definition of Besov norms (see e.g. [5]):

Definition 1.1. *Consider a smooth radial function φ on $\mathbb{R}$, supported in $[3/4, 8/3]$ such that $\sum_{j\in\mathbb{Z}} \varphi(2^{-j}\tau) = 1$ for any $\tau > 0$. We denote*

$$\Delta_j a = \mathcal{F}^{-1}(\varphi(2^{-j}|\xi|)\widehat{a}(\xi)), \quad j \in \mathbb{Z}.$$

Let $(p, r, \kappa) \in [1, +\infty]^3$ and $s \in \mathbb{R}$. The Besov norms are defined as

$$\|a\|_{\dot{B}^s_{p,r}} \stackrel{\text{def}}{=} \left\| \left(2^{js}\|\Delta_j a\|_{L^p}\right)_j \right\|_{\ell^r(\mathbb{Z})}, \|a\|_{\widetilde{L}^\kappa_T(\dot{B}^s_{p,r})} \stackrel{\text{def}}{=} \left\| \left(2^{js}\|\Delta_j a\|_{L^\kappa_t([0,T];L^p)}\right)_j \right\|_{\ell^r(\mathbb{Z})}.$$

When $p = r = 2$, the Besov spaces $\dot{B}^s_{p,r}$ coincide with the classical homogeneous Sobolev spaces $\dot{H}^s$. Moreover, the embedding $H^s \hookrightarrow L^2 \cap \dot{B}^{s_0}_{2,1}$ holds whenever $0 < s_0 < s$. We have to take Besov norms instead of (simpler) Sobolev norms for the velocity vector field v because we have not obtained the required L^2-estimate for $\partial_t v$ in the Sobolev setting (see Subsection 2.1.3 for details). Nevertheless, in view of the above embedding, we still have the regularity propagation result in the Sobolev framework.

2 Proof of the low regularity case $k = 0, 1$

If the velocity vector field $v(t, \cdot)$ is Lipschitzian, we can define $\psi(t, \cdot)$ as the associated inhomogeneous incompressible Navier-Stokes flow by the following

$$\begin{cases} \frac{d}{dt}\psi(t,x) = v(t,\psi(t,x)), \\ \psi(0,x) = x. \end{cases} \tag{2.1}$$

Then by view of the initial setting (1.4)-(1.5), the evolved domain $\Omega(t) = \psi(t, \Omega_0)$ is the level set of the real-valued function $h(t,x) = h_0(\psi^{-1}(t,x))$ and the tangential vector field $X(t) = (X^1(t), X^2(t)) = \nabla^\perp h(t,x)$ evolves along this flow as $X(t, \psi(t,x)) = X_0(x) \cdot \nabla\psi(t,x)$ and solves the following Cauchy problem

$$\begin{cases} \partial_t X + v \cdot \nabla X = X \cdot \nabla v, \text{ i.e. } [D_t, \partial_X] = 0 \text{ with } D_t = \partial_t + v\cdot\nabla,\ \partial_X v = X \cdot \nabla v, \\ X(0,x) = X_0(x). \end{cases} \tag{X}$$

Notice that the divergence free condition for the initial vector field X_0: $\operatorname{div} X_0 = \operatorname{div} \nabla^\perp h_0 = 0$ is conserved by this flow: $\operatorname{div} X(t,\cdot) = 0$.

In order to show that $\Omega(t) = \psi(t, \Omega_0)$ is still of class $W^{k+2,p}(\mathbb{R}^2)$, $k = 0, 1$, it suffices to show

$$\psi(t,\cdot) - \mathrm{Id} \in (W^{2,p}(\mathbb{R}^2))^2, \quad X(t,\cdot) \in (W^{k+1,p}(\mathbb{R}^2))^2,$$

and to this end we are going to prove $v \in (L^1_{\mathrm{loc}}(W^{2,p}(\mathbb{R}^2)))^2$ and when $k = 1$, $X \in (L^\infty_{\mathrm{loc}}(W^{2,p}(\mathbb{R}^2)))^2$ locally in time. The analysis depends heavily on the energy estimates for the velocity field v and its material derivative $D_t v$ and tangential derivative $\partial_X v$: We first derive a priori estimates for the velocity field v in Subsection 2.1 which implies Theorem 1.1 for the case $k = 0$, and then derive the a priori estimates for $D_t v$ in Subsection 2.2 which will be used to derive the a priori estimates for $\partial_X v$ and X in Subsection 2.3 when $k = 1$.

2.1 The velocity vector field v

Recalling the $L^2(\mathbb{R}^2)$-estimate (1.2) for v, we now consider the a priori estimates for its $H^2(\mathbb{R}^2)$-norm and $\dot{B}^s_{2,1}(\mathbb{R}^2)$-norm. We introduce the time weight $\sigma(t) = \min\{1, t\}$ in order to compensate the initial low regularity assumption and in the following context, the norm $\|\cdot\|_{L^r}$ will always denote the norm $\|\cdot\|_{L^r(\mathbb{R}^2)}$ in the space variable, while $\|\cdot\|_{L^q_t(L^r)}$ is simply the space-time norm $\|\cdot\|_{L^q([0,t];L^r(\mathbb{R}^2))}$.

2.1.1 H^1-estimate for v

We show the H^1-estimate for the general system of Navier-Stokes type as follows [2]

Lemma 2.1. *Let $(u, \nabla\Pi)$ be a smooth enough solution of*

$$\begin{cases} \rho\partial_t u + \rho v \cdot \nabla u - \Delta u + \nabla\Pi = f, \quad (t,x) \in \mathbb{R}^+ \times \mathbb{R}^2; \\ \operatorname{div} u = \operatorname{div} g. \end{cases} \tag{u}$$

[2] In the following context the unknowns such as $\partial_X v$ are not divergence-free.

Then we have

$$\begin{aligned}\frac{d}{dt}\|\nabla u(t)\|_{L^2}^2 &+ \|(\partial_t u, \nabla^2 u, \nabla\Pi)\|_{L^2}^2 \\ &\leq C\big(\|v\|_{L^4}^4\|\nabla u\|_{L^2}^2 + \|(\partial_t g, \nabla\operatorname{div} g, f)\|_{L^2}^2\big).\end{aligned} \tag{2.2}$$

Proof. By taking $L^2(\mathbb{R}^2)$ inner product of the equation (u) with $u_t := \partial_t u$, we obtain

$$\int_{\mathbb{R}^2}\rho|u_t|^2\,dx + \frac{1}{2}\frac{d}{dt}\|\nabla u(t)\|_{L^2}^2 = \int_{\mathbb{R}^2}\big(f - \nabla\Pi - \rho v\cdot\nabla u\big)|u_t\,dx.$$

It follows from the 2-D interpolation inequality

$$\|a\|_{L^r} \leq C\|a\|_{L^2}^{\frac{2}{r}}\|\nabla a\|_{L^2}^{1-\frac{2}{r}} \quad \forall\, r\in(2,\infty), \tag{2.3}$$

that

$$\begin{aligned}\Big|\int_{\mathbb{R}^2}\rho v\cdot\nabla u|u_t\,dx\Big| \leq& C\|v\|_{L^4}\|\nabla u\|_{L^4}\|\sqrt{\rho}u_t\|_{L^2} \\ \leq& C\|v\|_{L^4}\|\nabla u\|_{L^2}^{\frac{1}{2}}\|\nabla^2 u\|_{L^2}^{\frac{1}{2}}\|\sqrt{\rho}u_t\|_{L^2}.\end{aligned}$$

On the other hand, we get, by taking the space divergence operator to (u), that

$$\Delta\Pi = \operatorname{div}\big(f - \rho u_t - \rho v\cdot\nabla u\big) + \Delta\operatorname{div} u,$$

which together with the condition $\operatorname{div} u = \operatorname{div} g$ yields

$$\|\nabla\Pi\|_{L^2} \leq C\big(\|\nabla\operatorname{div} g\|_{L^2} + \|\sqrt{\rho}u_t\|_{L^2} + \|v\|_{L^4}\|\nabla u\|_{L^4} + \|f\|_{L^2}\big),$$

from which and (u), we deduce that $\|\nabla^2 u\|_{L^2}$ satisfies the same estimate. This together with (2.3) ensures

$$\|\nabla^2 u\|_{L^2} + \|\nabla\Pi\|_{L^2} \leq C\big(\|\nabla\operatorname{div} g\|_{L^2} + \|\sqrt{\rho}u_t\|_{L^2} + \|v\|_{L^4}^2\|\nabla u\|_{L^2} + \|f\|_{L^2}\big). \tag{2.4}$$

Hence we get

$$\Big|\int_{\mathbb{R}^2}\rho v\cdot\nabla u|u_t\,dx\Big| \leq C\big(\|v\|_{L^4}^4\|\nabla u\|_{L^2}^2 + \|\nabla\operatorname{div} g\|_{L^2}^2 + \|f\|_{L^2}^2\big) + \frac{1}{6}\|\sqrt{\rho}u_t\|_{L^2}^2.$$

While we get, by using integration by parts and (2.4), that

$$\begin{aligned}\Big|\int_{\mathbb{R}^2}\nabla\Pi|u_t\,dx\Big| =& \Big|\int_{\mathbb{R}^2}\nabla\Pi|g_t\,dx\Big| \leq \|\nabla\Pi\|_{L^2}\|g_t\|_{L^2} \\ \leq& C\big(\|v\|_{L^4}^4\|\nabla u\|_{L^2}^2 + \|(g_t, \nabla\operatorname{div} g, f)\|_{L^2}^2\big) + \frac{1}{6}\|\sqrt{\rho}u_t\|_{L^2}^2.\end{aligned}$$

As a result, it comes out

$$\frac{d}{dt}\|\nabla u(t)\|_{L^2}^2 + \|\sqrt{\rho}u_t\|_{L^2}^2 \leq C\big(\|v\|_{L^4}^4\|\nabla u\|_{L^2}^2 + \|(g_t, \nabla\operatorname{div} g, f)\|_{L^2}^2\big).$$

This together with (2.4) leads to (2.2). □

In particular, taking $u = v$ and $f = g = 0$, multiplying (2.2) by 1 or $\sigma(t) = \min\{1, t\}$ and then integrating the resulting inequality over $[0, t]$, gives rise to the H^1-estimates for v:

$$\|\nabla v\|_{L_t^\infty(L^2)}^2 + \big\|(\partial_t v, \nabla^2 v, \nabla\pi)\big\|_{L_t^2(L^2)}^2 \leq C\|\nabla v_0\|_{L^2}^2 \exp\big(C\|v_0\|_{L^2}^4\big), \quad \forall t > 0, \tag{v:H1}$$

and

$$\|\sigma^{\frac{1}{2}}\nabla v\|_{L_t^\infty(L^2)}^2 + \big\|\sigma^{\frac{1}{2}}(\partial_t v, \nabla^2 v, \nabla\pi)\big\|_{L_t^2(L^2)}^2 \leq C\|v_0\|_{L^2}^2 \exp\big(C\|v_0\|_{L^2}^4\big), \quad \forall t > 0. \tag{v:H1t}$$

2.1.2 L^2-estimate for $\partial_t v$

We now consider a priori estimates for the $L^2(\mathbb{R}^2)$-norm of $\partial_t u$ when $\operatorname{div} u = 0$:

Lemma 2.2. *Let $(u, \nabla\Pi)$ be a smooth enough solution of (u) with $g = 0$ and $\tau(t)$ be a non-negative Lipschitz function. Then we have*

$$\begin{aligned} \|\tau^{\frac{1}{2}} u_t\|_{L_t^\infty(L^2)}^2 + \|\tau^{\frac{1}{2}}\nabla u_t\|_{L_t^2(L^2)}^2 &\leq \exp\left(C\exp(C\|v_0\|_{L^2}^4)\right)\Big(\|\tau^{\frac{1}{2}} f_t\|_{L_t^1(L^2)}^2 \\ &+ \|\tau^{\frac{1}{2}}\sigma^{-\frac{1}{2}} f\|_{L_t^2(L^2)}^2 + \|\tau^{\frac{1}{2}}\sigma^{-\frac{1}{2}}\nabla u\|_{L_t^\infty(L^2)}^2 + \int_0^t \big(\tau' + \tau\sigma^{-1}\big)\|u_t\|_{L^2}^2\,dt'\Big). \end{aligned} \tag{2.5}$$

Sketchy proof. We first get, by taking ∂_t to the equation (u), that

$$\rho\partial_t^2 u + \rho v\cdot\nabla u_t - \Delta u_t + \nabla\Pi_t = -\rho_t(u_t + v\cdot\nabla u) - \rho v_t\cdot\nabla u + f_t.$$

Taking $L^2(\mathbb{R}^2)$ inner product of the above equation with u_t gives

$$\begin{aligned} &\frac{1}{2}\frac{d}{dt}\|\sqrt{\rho}u_t(t)\|_{L^2}^2 + \|\nabla u_t\|_{L^2}^2 \\ &= -\int_{\mathbb{R}^2}\big(\rho_t(u_t + v\cdot\nabla u) + \rho v_t\cdot\nabla u\big)|u_t\,dx + \int_{\mathbb{R}^2} f_t|u_t\,dx. \end{aligned} \tag{2.6}$$

By using the transport equation of (NS) and integrating by parts, one has

$$\begin{aligned} \Big|\int_{\mathbb{R}^2}\rho_t u_t|u_t\,dx\Big| = 2\Big|\int_{\mathbb{R}^2}\rho v\cdot\nabla u_t|u_t\,dx\Big| &\leq C\|v\|_{L^4}\|\nabla u_t\|_{L^2}\|u_t\|_{L^4} \\ &\leq \frac{1}{6}\|\nabla u_t\|_{L^2}^2 + C\|v\|_{L^4}^4\|u_t\|_{L^2}^2. \end{aligned}$$

Similarly, we get, by using integration by parts, (2.3) and (2.4), that

$$\begin{aligned} \Big|\int_{\mathbb{R}^2}\rho_t v\cdot\nabla u|u_t\,dx\Big| &\leq \frac{1}{6}\|\nabla u_t\|_{L^2}^2 + C\big(1 + \|v\|_{L^2}^2\big)\|\nabla v\|_{L^2}^2\|u_t\|_{L^2}^2 \\ &+ C\left(\|v_t\|_{L^2}^2 + \|v\|_{L^2}^2\|\nabla v\|_{L^2}^4\right)\|\nabla u\|_{L^2}^2 + C\|\nabla v\|_{L^2}^2\|f\|_{L^2}^2, \end{aligned}$$

and

$$\Big|\int_{\mathbb{R}^2} \rho v_t \cdot \nabla u | u_t \, dx\Big| \le \frac{1}{6}\|\nabla u_t\|_{L^2}^2 + \|v_t\|_{L^2}^2 \|\nabla u\|_{L^2}^2 \\ + \sigma^{-1}(t)\big(\|u_t\|_{L^2}^2 + \|f\|_{L^2}^2\big) + C\left(\|v\|_{L^4}^4 + \sigma(t)\|v_t\|_{L^2}^2\right)\|u_t\|_{L^2}^2.$$

Inserting the above estimates into (2.6) gives rise to

$$\frac{d}{dt}\|\sqrt{\rho} u_t(t)\|_{L^2}^2 + \|\nabla u_t\|_{L^2}^2 \le C\left((1+\|v\|_{L^2}^2)\|\nabla v\|_{L^2}^2 + \sigma(t)\|v_t\|_{L^2}^2\right)\|u_t\|_{L^2}^2 \\ + C\left(\|v_t\|_{L^2}^2 + \|v\|_{L^2}^2\|\nabla v\|_{L^2}^4\right)\|\nabla u\|_{L^2}^2 \\ + \sigma^{-1}(t)\|u_t\|_{L^2}^2 + C\big(\|\sigma^{\frac{1}{2}}\nabla v\|_{L^2}^2 + 1\big)\sigma^{-1}(t)\|f\|_{L^2}^2 + \|f_t\|_{L^2}\|u_t\|_{L^2}.$$

Multiplying the above inequality by $\tau(t)$ and then applying Gronwall's inequality and Young's inequality leads to (2.5). □

In particular, taking $u = v$, $f = 0$ and $\tau = \sigma^2$ in (2.5) and using (v:H1t) gives rise to the time weighted $\dot{H}^2$-estimate of v:

$$\|\sigma(\partial_t v, \nabla^2 v, \nabla\pi)\|_{L_t^\infty(L^2)}^2 + \|\sigma\nabla\partial_t v\|_{L_t^2(L^2)}^2 \le \exp\left(C\exp\big(C\|v_0\|_{L^2}^4\big)\right), \quad \forall t>0. \tag{v:H2t}$$

2.1.3 $\dot{B}^s_{2,1}$ estimate of v

We use the idea of interpolation to get the $\dot{B}^s_{2,1}(\mathbb{R}^2)$-estimate for v: we consider the evolution of the dyadic piece $\Delta_q v_0$ by use of the energy estimates (1.2), (v:H1) and then show the dyadic piece $\Delta_j v$ decaying as 2^{-js} which gives the $\dot{B}^s_{2,1}$ estimate.

Lemma 2.3. *Let $(u, \nabla\Pi)$ be the solution of (u) with $f = g = 0$ and with initial data $u_0 \in \dot{B}^s_{2,1}$ for some $s \in (0,1)$. Then there exists C_0 depending only on the initial data such that for any $t > 0$,*

$$\|u\|_{\widetilde{L}_t^\infty(\dot{B}^s_{2,1})} + \|\nabla u\|_{\widetilde{L}_t^2(\dot{B}^s_{2,1})} \le C\|u_0\|_{\dot{B}^s_{2,1}}\exp\big(C\|v_0\|_{L^2}^4\big), \tag{2.7}$$

$$\|\sigma^{\frac{1}{2}}\nabla u\|_{\widetilde{L}_t^\infty(\dot{B}^s_{2,1})} + \|\sigma^{\frac{1}{2}} u_t\|_{\widetilde{L}_T^2(\dot{B}^s_{2,1})} \le C_0, \tag{2.8}$$

$$\|\sigma^{\frac{1-s}{2}} u\|_{\widetilde{L}_t^\infty(\dot{B}^1_{2,1})} + \big\|\sigma^{\frac{1-s}{2}}(u_t, \nabla^2 u, \nabla\Pi)\big\|_{L_t^2(L^2)} \le C_0, \tag{2.9}$$

$$\big\|\sigma^{1-\frac{s}{2}}(u_t, \nabla^2 u, \nabla\Pi)\big\|_{L_t^\infty(L^2)} + \|\sigma^{1-\frac{s}{2}}\nabla u_t\|_{L_t^2(L^2)} \le C_0. \tag{2.10}$$

Proof. Let $(u_q, \nabla\Pi_q)$ be determined by

$$\begin{cases} \rho\partial_t u_q + \rho v\cdot\nabla u_q - \Delta u_q + \nabla\Pi_q = 0, \quad \operatorname{div} u_q = 0, \\ u_q|_{t=0} = \Delta_q u_0. \end{cases} \tag{2.11}$$

Then by the uniqueness of solution to (u) with $f = g = 0$, we have

$$u = \sum_{q\in\mathbb{Z}} u_q \quad \text{and} \quad \nabla\Pi = \sum_{q\in\mathbb{Z}} \nabla\Pi_q.$$

It follows from Definition 1.1, Bernstein type inequality and (1.2), (v:H1) that

$$\|u_q\|_{L_t^\infty(L^2)}^2 + \|\nabla u_q\|_{L_t^2(L^2)}^2 \leq C\|\Delta_q u_0\|_{L^2}^2 \leq C d_q^2 2^{-2qs}\|u_0\|_{\dot{B}_{2,1}^s}^2,$$

and

$$\begin{aligned}\|\nabla u_q\|_{L_t^\infty(L^2)}^2 + \big\|(\partial_t u_q, \nabla^2 u_q, \nabla\Pi_q)\big\|_{L_t^2(L^2)}^2 &\leq C\|\Delta_q u_0\|_{\dot{H}^1}^2 \exp\big(C\|v_0\|_{L^2}^4\big)\\ &\leq C d_q^2 2^{2q(1-s)}\|u_0\|_{\dot{B}_{2,1}^s}^2 \exp\big(C\|v_0\|_{L^2}^4\big).\end{aligned} \tag{2.12}$$

Here and in what follows, (d_q) denotes a generic element in $\ell^1(\mathbb{Z})$ such that $\|(d_q)\|_{\ell^1(\mathbb{Z})} = 1$. Exploiting the idea used in the proof of the characterization of Besov space with positive index (see Lemma 2.88 of [5] for instance) gives for any $j \in \mathbb{Z}$ that (noticing $s \in (0,1)$)

$$\begin{aligned}\|\Delta_j u\|_{L_t^\infty(L^2)} + \|\Delta_j \nabla u\|_{L_t^2(L^2)} \leq& \sum_{q>j}\Big(\|\Delta_j u_q\|_{L_t^\infty(L^2)} + \|\Delta_j \nabla u_q\|_{L_t^2(L^2)}\Big)\\ &+ 2^{-j}\sum_{q\leq j}\Big(\|\Delta_j \nabla u_q\|_{L_t^\infty(L^2)} + \|\Delta_j \nabla^2 u_q\|_{L_t^2(L^2)}\Big)\\ \leq& C\sum_{q>j}\Big(\|u_q\|_{L_t^\infty(L^2)} + \|\nabla u_q\|_{L_t^2(L^2)}\Big)\\ &+ C2^{-j}\sum_{q\leq j}\Big(\|\nabla u_q\|_{L_t^\infty(L^2)} + \|\nabla^2 u_q\|_{L_t^2(L^2)}\Big)\\ \leq& C d_j 2^{-js}\|u_0\|_{\dot{B}_{2,1}^s}\exp\big(C\|v_0\|_{L^2}^4\big),\end{aligned}$$

which together with Definition 1.1 ensures the Inequality (2.7).

Similarly, we deduce (2.8) from (v:H1t) and (v:H2t). Then we derive from the interpolation between (2.7) and (2.8) that

$$\|\sigma^{\frac{1-s}{2}} u\|_{\widetilde{L}_t^\infty(\dot{B}_{2,1}^1)} \leq C\|u\|_{\widetilde{L}_t^\infty(\dot{B}_{2,1}^s)}^s \|\sigma^{\frac{1}{2}}\nabla u\|_{\widetilde{L}_t^\infty(\dot{B}_{2,1}^s)}^{1-s} \leq C_0.$$

Since by (v:H1t) we have

$$\big\|\sigma^{\frac{1}{2}}\partial_t u_q\big\|_{L_t^2(L^2)}^2 \leq C_0 d_q^2 2^{-2qs},$$

we take interpolation between the above and (2.12) to obtain

$$\int_{\mathbb{R}^+} \sigma(t')^{1-s}\|\partial_t u_q\|_{L^2}^2\, dt \leq C_0 d_q^2,$$

which implies

$$\Big(\int_{\mathbb{R}^+} \sigma(t)^{1-s}\|\partial_t u\|_{L^2}^2\, dt\Big)^{\frac{1}{2}} \leq \sum_{q\in\mathbb{Z}}\Big(\int_{\mathbb{R}^+} \sigma(t')^{1-s}\|\partial_t u_q\|_{L^2}^2\, dt\Big)^{\frac{1}{2}} \leq C_0\sum_{q\in\mathbb{Z}} d_q \leq C_0. \tag{vt}$$

The same estimate holds for $\big\|\sigma^{\frac{1-s}{2}}(\nabla^2 u,\nabla\Pi)\big\|_{L^2_t(L^2)}$ by use of (2.4).

Finally, taking $f=0$ and $\tau(t)=\sigma(t)^{2-s}$ in Lemma 2.2 gives rise to

$$\|\sigma^{1-\frac{s}{2}}u_t\|^2_{L^\infty_t(L^2)}+\|\sigma^{1-\frac{s}{2}}\nabla u_t\|^2_{L^2_t(L^2)}$$
$$\le C_0\exp\big(C\exp(C\|v_0\|^4_{L^2})\big)\Big(\|\sigma^{\frac{1-s}{2}}\nabla u\|^2_{L^\infty_t(L^2)}+\|\sigma^{\frac{1-s}{2}}u_t\|^2_{L^2_t(L^2)}\Big).$$

This together with (2.9) gives (2.10). □

Let us remark that in order to show the convergence of $\|\sigma^{\frac{1-s}{2}}\partial_t u\|_{L^2_t(L^2)}$ in (vt) we have to assume $d_q\in\ell^r(\mathbb{Z})$, $r=1$: this is why we assume Besov regularity ($r=1$) instead of Sobolev regularity ($r=2$). We emphasis that as stated in (2.5), the estimate of $\|\sigma^{\frac{1-s}{2}}\partial_t u\|_{L^2_t(L^2)}$ is crucial for us to deal with L^2-energy estimate of $\partial_t u$ (and hence to deal with higher regularities, see e.g. the definition of $\mathfrak{b}_0$ in (b0) below).

2.1.4 L^r-type estimates for v

To conclude from all the above estimates and the interpolation inequality (2.3), we know that v is Lipschitzian and for any $r\in[2,\infty)$,

$$\begin{aligned}&\big\|\sigma^{\frac{1-s_0}{2}}v\big\|_{L^\infty_t(L^\infty)}+\big\|\sigma^{\frac{1-s_0}{2}(1-\frac{2}{r})}v\big\|_{L^\infty_t(L^r)}+\big\|\sigma^{\frac{1-s_0}{2}}\nabla v\big\|_{L_t^{\frac{2r}{r-2}}(L^r)}\\&+\big\|\sigma^{\left(1-\frac{1}{r}-\frac{s_0}{2}\right)}\nabla v\big\|_{L^\infty_t(L^r)}+\big\|\sigma^{\left(1-\frac{1}{r}-\frac{s_0}{2}\right)}(\partial_t v,\nabla^2 v,\nabla\pi)\big\|_{L^2_t(L^r)}+\big\|\sigma^{\frac{1-s_0}{2}}\nabla v\big\|_{L^2_t(L^\infty)}\\&+\big\|\sigma^{\left(1-\frac{s_0}{2}\right)}(\partial_t v,\nabla^2 v,\nabla\pi)\big\|_{L_t^{\frac{2r}{r-2}}(L^r)}+\|\sigma^{1-\frac{s_0}{2}}\nabla\partial_t v\|_{L^2_t(L^2)}\le C_0,\end{aligned}\tag{v:Lr}$$

where C_0 only depends on the initial data and r. In particular, $v\in(L^1_{\rm loc}(\mathbb{R}^+;W^{2,p}(\mathbb{R}^2)))^2$ (since $p\in(2,\frac{2}{1-s_0})$) and hence the $W^{2,p}$-boundary regularity of Ω_0 is propagated.

Proof of Theorem 1.1 when $k=0$. By mollifying the initial data satisfying (1.6), we first construct the global approximate solutions to (NS). Then along the same line to the proof of the above estimates, we obtain similar uniform estimates for such approximate solutions. Finally a standard compactness argument completes the existence part of Theorem 1.1. The uniqueness of such solutions has been proved in [34]. We skip the details here. □

2.2 The material derivative $D_t v$

In order to propagate higher order regularities, the first idea coming up is to derive the energy estimate of $\nabla\partial_t v$. In order to do this, we first apply ∂_t to the momentum equation of (NS) to get

$$\rho D_t v_t-\Delta v_t+\nabla\pi_t=-\rho_t D_t v-\rho v_t\cdot\nabla v,\quad D_t=\partial_t+v\cdot\nabla,\quad v_t=\partial_t v,$$

and then test the above equation by v_{tt}: We need to deal with such terms as

$$\int_{\mathbb{R}^2} \rho_t D_t v | v_{tt}\, dx = -\int_{\mathbb{R}^2} v \cdot \nabla \rho D_t v | v_{tt}\, dx,$$

which is impossible to go through with non-Lipschitz density function ρ.

The idea to overcome this difficulty is to apply the operator $D_t = \partial_t + v \cdot \nabla$ instead of ∂_t to the momentum equation of (NS), which gives

$$\begin{aligned} \rho D_t^2 v - \Delta D_t v + \nabla D_t \pi =& F_D(v, \pi) \quad \text{with} \\ F_D(u, \Pi) \stackrel{\text{def}}{=}& -2\nabla v_\alpha \cdot \partial_\alpha \nabla u - \Delta v \cdot \nabla u + \nabla v_\alpha \partial_\alpha \Pi. \end{aligned} \tag{Dv}$$

Here and in what follows, repeated indices of α means summation of α from 1 to 2. We remark that the advantage of the operator D_t is that $D_t \rho = 0$ so that difficult terms mentioned before will not appear anymore. This idea was used by D. Hoff [23].

2.2.1 H^1 estimate for $D_t v$

Note that due to $\operatorname{div} v = 0$, one has

$$\begin{aligned} &\operatorname{div} D_t v = \operatorname{div} \mathfrak{a}_0 = \partial_\alpha v \cdot \nabla v_\alpha, \quad \mathfrak{a}_0 = v \cdot \nabla v, \\ &\text{and} \quad \operatorname{div} D_t^2 v = \operatorname{div} \mathfrak{b}_0, \quad \mathfrak{b}_0 \stackrel{\text{def}}{=} v \cdot (\nabla v_t + D_t \nabla v) + D_t v \cdot \nabla v. \end{aligned} \tag{b0}$$

Lemma 2.4. *Let* $(\mathrm{w}, \nabla \mathrm{q})$ *be a smooth enough solution of the following system:*

$$\begin{cases} \rho D_t^2 \mathrm{w} - \Delta D_t \mathrm{w} + \nabla \mathrm{q} = \mathrm{F}, \\ \operatorname{div} D_t \mathrm{w} = \operatorname{div} \mathfrak{a} \quad \textit{and} \quad \operatorname{div} D_t^2 \mathrm{w} = \operatorname{div} \mathfrak{b}. \end{cases} \tag{w}$$

Then for any $s \in (0,1)$, *we have*

$$\begin{aligned} &\|\sigma^{1-\frac{s}{2}} D_t \mathrm{w}\|^2_{L^\infty_t(L^2) \cap L^2_t(\dot H^1)} + \|\sigma^{\frac{3-s}{2}} \nabla D_t \mathrm{w}\|^2_{L^\infty_t(L^2)} + \|\sigma^{\frac{3-s}{2}} (D_t^2 \mathrm{w}, \nabla^2 D_t \mathrm{w}, \nabla \mathrm{q})\|^2_{L^2_t(L^2)} \\ &\leq C \exp\left(C\|v_0\|^2_{L^2}\right) \Big(\|\sigma^{\frac{1-s}{2}} (D_t \mathrm{w}, \mathfrak{a})\|^2_{L^2_t(L^2)} + \|\sigma^{1-\frac{s}{2}} \nabla \mathfrak{a}\|^2_{L^2_t(L^2)} \\ &\qquad + \|\sigma^{\frac{3-s}{2}} (\nabla \operatorname{div} \mathfrak{a}, \mathfrak{b}, \mathrm{F})\|^2_{L^2_t(L^2)} \Big). \end{aligned} \tag{2.13}$$

Sketchy proof. Taking $L^2(\mathbb{R}^2)$ inner product of (w) with $D_t \mathrm{w}$, multiplying by $\sigma(t)^{2-s}$ and then integrating over $[0,t]$ leads to

$$\begin{aligned} &\frac{1}{2} \|\sigma^{1-\frac{s}{2}} \sqrt{\rho} D_t \mathrm{w}(t)\|^2_{L^2} + \|\sigma^{1-\frac{s}{2}} \nabla D_t \mathrm{w}\|^2_{L^2_t(L^2)} \\ &= \frac{2-s}{2} \|\sigma^{\frac{1-s}{2}} \sqrt{\rho} D_t \mathrm{w}\|^2_{L^2_t(L^2)} + \int_0^t \int_{\mathbb{R}^2} \sigma^{2-s} (\mathrm{F} - \nabla \mathrm{q}) | \mathrm{D_t w}\, \mathrm{dx}\, \mathrm{dt'}. \end{aligned} \tag{2.14}$$

It is easy to observe that

$$\Big|\int_0^t \int_{\mathbb{R}^2} \sigma^{2-s} \mathrm{F} | D_t \mathrm{w}\, dx\, dt'\Big| \leq \|\sigma^{\frac{3-s}{2}} \mathrm{F}\|^2_{L^2_t(L^2)} + \|\sigma^{\frac{1-s}{2}} D_t \mathrm{w}\|^2_{L^2_t(L^2)},$$

and due to $\operatorname{div} D_t\mathrm{w} = \operatorname{div}\mathfrak{a}$,

$$\Big|\int_0^t\int_{\mathbb{R}^2}\sigma^{2-s}\nabla\mathrm{q}|D_t\mathrm{w}\,dx\,dt'\Big| = \Big|\int_0^t\int_{\mathbb{R}^2}\sigma^{2-s}\nabla\mathrm{q}|\mathfrak{a}\,dx\,dt'\Big| \\ \leq \varepsilon\big\|\sigma^{\frac{3-s}{2}}\nabla\mathrm{q}\big\|^2_{L^2_t(L^2)} + C_\varepsilon\|\sigma^{\frac{1-s}{2}}\mathfrak{a}\|^2_{L^2_t(L^2)}.$$

Similarly, by taking the $L^2(\mathbb{R}^2)$ inner product of (w) with $D_t^2\mathrm{w}$, we get

$$\|\sqrt{\rho}D_t^2\mathrm{w}\|^2_{L^2} + \frac{1}{2}\frac{d}{dt}\|\nabla D_t\mathrm{w}\|^2_{L^2} \\ \leq \int_{\mathbb{R}^2}(\mathrm{F}-\nabla\mathrm{q})|D_t^2\mathrm{w}\,dx + C\|\nabla v\|_{L^2}\|\nabla D_t\mathrm{w}\|_{L^2}\|\nabla^2 D_t\mathrm{w}\|_{L^2}.$$

Multiplying the above equality by $\sigma(t)^{3-s}$ and then integrating over $[0,t]$ leads to

$$\frac{1}{2}\big\|\sigma^{\frac{3-s}{2}}\nabla D_t\mathrm{w}(t)\big\|^2_{L^2} + \big\|\sigma^{\frac{3-s}{2}}\sqrt{\rho}D_t^2\mathrm{w}\big\|^2_{L^2_t(L^2)} \leq \frac{3-s}{2}\big\|\sigma^{1-\frac{s}{2}}\nabla D_t\mathrm{w}\big\|^2_{L^2_t(L^2)} \\ + \frac{1}{4}\|\sigma^{\frac{3-s}{2}}\sqrt{\rho}D_t^2\mathrm{w}\|_{L^2_t(L^2)} + C\|\sigma^{\frac{3-s}{2}}\mathrm{F}\|^2_{L^2_t(L^2)} + \varepsilon\|\sigma^{\frac{3-s}{2}}\nabla\mathrm{q}\|^2_{L^2_t(L^2)} \\ + C_\varepsilon\|\sigma^{\frac{3-s}{2}}\mathfrak{b}\|^2_{L^2_t(L^2)} + \varepsilon\|\sigma^{\frac{3-s}{2}}\nabla^2 D_t\mathrm{w}\|^2_{L^2} + C_\varepsilon\int_0^t\|\nabla v\|^2_{L^2}\big\|\sigma^{\frac{3-s}{2}}\nabla D_t\mathrm{w}\big\|^2_{L^2}\,dt'. \tag{2.15}$$

To deal with the estimate of $\nabla\mathrm{q}$, in view of (w), we write

$$-\Delta\big(D_t\mathrm{w} - \nabla\Delta^{-1}\operatorname{div} D_t\mathrm{w}\big) + \nabla\mathrm{q} = -\rho D_t^2\mathrm{w} + \nabla\operatorname{div} D_t\mathrm{w} + \mathrm{F},$$

such that (noticing $\operatorname{div}\big(D_t\mathrm{w} - \nabla\Delta^{-1}\operatorname{div} D_t\mathrm{w}\big) = 0$)

$$\big\|\sigma^{\frac{3-s}{2}}(\nabla^2 D_t\mathrm{w}, \nabla\mathrm{q})\big\|_{L^2_t(L^2)} \leq C\big\|\sigma^{\frac{3-s}{2}}(\sqrt{\rho}D_t^2\mathrm{w}, \nabla\operatorname{div}\mathfrak{a}, \mathrm{F})\big\|_{L^2_t(L^2)}. \tag{2.16}$$

Inserting (2.16) into (2.14), (2.15) and choosing sufficiently small ε yields (2.13). □

In particular, we can estimate $\mathfrak{a}_0, \mathfrak{b}_0, F_D$ in (b0) and (Dv) by use of (v:Lr) such that the following $\dot{H}^1$-estimate for $D_t v$ holds:

$$\big\|\sigma^{\frac{3-s_0}{2}}(\nabla D_t v, D_t\nabla v, \nabla v_t)\big\|_{L^\infty_t(L^2)} + \big\|\sigma^{\frac{3-s_0}{2}}(D_t^2 v, \nabla^2 D_t v, \nabla D_t\pi)\big\|_{L^2_t(L^2)} \leq C_0. \tag{2.17}$$

2.2.2 L^r-type estimate for $v, D_t v$

By use of (v:Lr), (2.17), we deduce more estimates which will play an important role in the following context:

$$\big\|\sigma^{\left(\frac{3}{2}-\frac{1}{r}-\frac{s_0}{2}\right)}\big(D_t v, v_t, v\otimes\nabla v, \nabla^2 v, \nabla\pi\big)\big\|^2_{L^\infty_t(L^r)} \\ + \|\sigma^{1-\frac{s_0}{2}}\nabla v\|^2_{L^\infty_t(L^\infty)} + \int_0^t\mathfrak{B}_0(t')\,dt' \leq C_0, \quad \forall t>0, \tag{v':Lr}$$

where $r \in [2, +\infty)$ and

$$\begin{aligned}
\mathfrak{B}_0(t) &= \|\nabla v(t)\|_{L^2}^2 + \big\|\sigma^{(1-\frac{1}{r}-\frac{s_0}{2})}(v_t, \nabla^2 v, \nabla\pi, D_t v, v\otimes\nabla v)(t)\big\|_{L^r}^2 \\
&+ \|\sigma^{1-\frac{s_0}{2}}\nabla v_t(t)\|_{L^2}^2 + \|\sigma^{\frac{1-s_0}{2}}\nabla v(t)\|_{L^\infty\cap L^{\frac{2p}{p-2}}}^2 + \big\|\sigma^{1-\frac{s_0}{2}}(D_t v, v\otimes\nabla v, v_t)(t)\big\|_{L^\infty}^2 \\
&+ \big\|\sigma^{\frac{3-s_0}{2}}\big(D_t^2 v, \nabla^2 D_t v, D_t\nabla^2 v, \nabla D_t\nabla v, \nabla D_t\pi, D_t\nabla\pi\big)(t)\big\|_{L^2}^2 \\
&+ \big\|\sigma^{(\frac{3}{2}-\frac{1}{r}-\frac{s_0}{2})}(\nabla D_t v, D_t\nabla v, \nabla(v\otimes\nabla v), \nabla v_t)(t)\big\|_{L^r}^2 + \|\sigma^{\frac{3-s_0}{2}}\mathfrak{b}_0(t)\|_{L^2}^2.
\end{aligned}$$

Here, we denote $(f\otimes g) = (f^j g^k)_{jk}$ for $f = (f^j)_j$ and $g = (g^k)_k$.

2.3 The vector field X and the tangential derivative $\mathrm{v}^1 \overset{\text{def}}{=} \partial_X v$

In this subsection we consider the case $k = 1$ and show $X \in (L^\infty_{\text{loc}}(\mathbb{R}^+); W^{2,p}(\mathbb{R}^2))^2$ by applying the above estimates for $v, D_t v$ and the following estimates for $\mathrm{v}^1, D_t\mathrm{v}^1$ to the transport equation (X).

2.3.1 H^1 estimate for $\mathrm{v}^1 = \partial_X v$

Thanks to $\operatorname{div} v = \operatorname{div} X = 0$, we have

$$\operatorname{div}\mathrm{v}^1 = \partial_\alpha X\cdot\nabla v_\alpha = \operatorname{div} g_1, \quad g_1 = v\cdot\nabla X.$$

It is also easy to observe the commutators

$$[\partial_X; \partial_i]f = -\partial_i X\cdot\nabla f \quad\text{and}\quad [\partial_X; \partial_i^2]f = -\partial_i^2 X\cdot\nabla f - 2\partial_i X\cdot\nabla\partial_i f.$$

By taking ∂_X to the momentum equation of (NS), we find that (with $\pi^1 \overset{\text{def}}{=} \partial_X\pi$)

$$\begin{aligned}
&(\mathrm{v}^1, \nabla\pi^1) \text{ solves (u) with the data } (f, g, u_0) \text{ replaced by } (F_1, g_1, \partial_{X_0}v_0): \\
&F_1(v,\pi) \overset{\text{def}}{=} -(\Delta X\cdot\nabla v + 2\partial_\alpha X\cdot\nabla\partial_\alpha v) + \nabla X^\alpha\partial_\alpha\pi.
\end{aligned} \tag{v1}$$

In order to estimate v^1, we split $(\mathrm{v}^1, \nabla\pi^1)$ into two pieces

$$\mathrm{v}^1 = \mathrm{v}_{11} + \mathrm{v}_{12} \quad\text{and}\quad \nabla\pi^1 = \nabla\mathrm{p}_{11} + \nabla\mathrm{p}_{12},$$

with $(\mathrm{v}_{11}, \nabla\mathrm{p}_{11})$ and $(\mathrm{v}_{12}, \nabla\mathrm{p}_{12})$ solving the following systems respectively

$$\rho\partial_t\mathrm{v}_{11} + \rho v\cdot\nabla\mathrm{v}_{11} - \Delta\mathrm{v}_{11} + \nabla\mathrm{p}_{11} = 0, \quad \operatorname{div}\mathrm{v}_{11} = 0, \quad \mathrm{v}_{11}|_{t=0} = \partial_{X_0}v_0, \tag{v11}$$

and

$$\rho\partial_t\mathrm{v}_{12} + \rho v\cdot\nabla\mathrm{v}_{12} - \Delta\mathrm{v}_{12} + \nabla\mathrm{p}_{12} = F_1(v,\pi), \quad \operatorname{div}\mathrm{v}_{12} = \operatorname{div} g_1, \quad \mathrm{v}_{12}|_{t=0} = 0. \tag{v12}$$

Then we deduce from Lemma 2.3 that

$$\|\mathrm{v}_{11}\|_{\widetilde{L}^\infty_t(\dot{B}^{s_1}_{2,1})} + \|\nabla\mathrm{v}_{11}\|_{\widetilde{L}^2_t(\dot{B}^{s_1}_{2,1})} \le C\|\partial_{X_0}v_0\|_{\dot{B}^{s_1}_{2,1}}\exp\big(C\|v_0\|_{L^2}^4\big), \tag{v11:Bs}$$

$$\|\sigma^{\frac{1-s_1}{2}}\mathrm{v}_{11}\|_{\widetilde{L}^\infty_t(\dot{B}^1_{2,1})} + \big\|\sigma^{\frac{1-s_1}{2}}(\partial_t\mathrm{v}_{11}, \nabla^2\mathrm{v}_{11}, \nabla\mathrm{p}_{11})\big\|_{L^2_t(L^2)} \le C_0. \tag{v11:B1t}$$

We next get $\dot{H}^1$-estimate for v_{12} which depends on g_1: under the non-divergence free condition, we prefer to consider the case of *zero*-initial data (that is why we subtract v_{11} from v^1) and see below how the zero-initial data assumption helps to get the estimate for $\|\sigma^{-s/2}\nabla u\|_{L^2_t(L^2)}$.

Lemma 2.5. *Let* $(u, \nabla\Pi)$ *be a smooth solution of* (u) *with initial data*

$$u_0 = 0.$$

Then for any $s \in (0,1)$ *and* $\delta \in (s,1)$, *one has*

$$\begin{aligned} &\|\sigma^{\frac{1-s}{2}}\nabla u\|^2_{L^\infty_t(L^2)} + \big\|\sigma^{\frac{1-s}{2}}(u_t, \nabla^2 u, \nabla\Pi)\big\|^2_{L^2_t(L^2)} \\ &\le C_0\Big(\frac{\langle t\rangle}{\delta - s}\|g(0)\|^2_{\mathcal{B}^\delta} + \|\sigma^{-\frac{1-(s_0-s)}{2}}\nabla g\|^2_{L^1_t(L^2)} + \big\|\sigma^{-\frac{s}{2}}(f, g_t)\big\|^2_{L^1_t(L^2)} \\ &\quad + \|\sigma^{-\frac{s}{2}}\nabla g\|^2_{L^2_t(L^2)} + \big\|\sigma^{\frac{1-s}{2}}(g_t, \nabla\operatorname{div} g, f)\big\|^2_{L^2_t(L^2)}\Big), \quad \langle t\rangle = (1+t^2)^{1/2}. \end{aligned} \tag{2.18}$$

Sketchy proof. We multiply (2.2) by σ^{1-s} and then apply Gronwall's inequality to get

$$\begin{aligned} \|\sigma^{\frac{1-s}{2}}\nabla u\|^2_{L^\infty_t(L^2)} + \big\|\sigma^{\frac{1-s}{2}}(u_t, \nabla^2 u, \nabla\Pi)\big\|^2_{L^2_t(L^2)} &\le C\exp\big(C\|v\|^4_{L^4_t(L^4)}\big) \\ &\times \int_0^t \big(\sigma^{-s}\|\nabla u\|^2_{L^2} + \sigma^{1-s}\|(g_t, \nabla\operatorname{div} g, f)\|^2_{L^2}\big)\,dt'. \end{aligned} \tag{2.19}$$

To handle the term $\int_0^t \sigma^{-s}\|\nabla u\|^2_{L^2}\,dt'$, let us take

$$\mathfrak{y} = u - g, \quad \mathfrak{y} = \mathfrak{y}_1 + \mathfrak{y}_2 \quad \text{and} \quad \nabla\Pi = \nabla\Pi_1 + \nabla\Pi_2,$$

with $(\mathfrak{y}_1, \nabla\Pi_1)$ and $(\mathfrak{y}_2, \nabla\Pi_2)$ solving respectively

$$\rho\partial_t\mathfrak{y}_1 + \rho v\cdot\nabla\mathfrak{y}_1 - \Delta\mathfrak{y}_1 + \nabla\Pi_1 = 0, \quad \operatorname{div}\mathfrak{y}_1 = 0, \quad \mathfrak{y}_1|_{t=0} = -g(0), \tag{y1}$$

and

$$\begin{cases} \rho\partial_t\mathfrak{y}_2 + \rho v\cdot\nabla\mathfrak{y}_2 - \Delta\mathfrak{y}_2 + \nabla\Pi_2 = f - (\rho\partial_t g + \rho v\cdot\nabla g - \Delta g), \quad \operatorname{div}\mathfrak{y}_2 = 0, \\ \mathfrak{y}_2|_{t=0} = 0. \end{cases} \tag{y2}$$

For any $\delta \in (0,1)$, it follows $\|\sigma^{\frac{1-\delta}{2}}\nabla\mathfrak{y}_1\|_{L^\infty_t(L^2)} \le C_0\|g(0)\|_{\mathcal{B}^\delta}$, which in particular implies

$$\|\sigma^{-\frac{s}{2}}\nabla\mathfrak{y}_1\|^2_{L^2_t(L^2)} \le \|\sigma^{\frac{1-\delta}{2}}\nabla\mathfrak{y}_1\|^2_{L^\infty_t(L^2)}\|\sigma^{-1+\delta-s}\|_{L^1_t} \le \frac{C_0\langle t\rangle}{\delta - s}\|g(0)\|^2_{\mathcal{B}^\delta}, \quad s \in (0,\delta). \tag{y1:H1}$$

Whereas for any fixed $t_0 \in (0,1)$, we denote $\sigma_{t_0}(t) = \begin{cases} \sigma(t) & \text{if } t \ge t_0, \\ t_0 & \text{if } t \le t_0. \end{cases}$ Then we get, by first taking the L^2 inner product between (y2) and $\mathfrak{y}_2$, then multiplying by

$\sigma_{t_0}^{-s}(t)$ and finally integrating the resulting inequality over $[0,t]$, that

$$\begin{aligned}
&\|\sigma_{t_0}^{-\frac{s}{2}}\sqrt{\rho}\mathfrak{y}_2(t)\|_{L^2}^2+\int_0^t\sigma_{t_0}^{-s}\|\nabla\mathfrak{y}_2\|_{L^2}^2\,dt'\\
&=-s\int_0^t\sigma_{t_0}^{-1-s}\sigma_{t_0}'\|\sqrt{\rho}\mathfrak{y}_2\|_{L^2}^2\,dt'+\int_0^t\int_{\mathbb{R}^2}\sigma_{t_0}^{-s}\left(f-\rho\partial_t g-\rho v\cdot\nabla g+\Delta g\right)|\mathfrak{y}_2\,dx\,dt'\\
&\leq C\big\|\sigma_{t_0}^{-\frac{s}{2}}(f,g_t)\big\|_{L_t^1(L^2)}^2+C\|\sigma^{\frac{1-s_0}{2}}v\|_{L_t^\infty(L^\infty)}^2\big\|\sigma^{-\frac{1-(s_0-s)}{2}}\nabla g\big\|_{L_t^1(L^2)}^2\\
&\quad+\frac{1}{2}\|\sigma_{t_0}^{-\frac{s}{2}}\sqrt{\rho}\mathfrak{y}_2\|_{L_t^\infty(L^2)}^2+\frac{1}{2}\big\|\sigma_{t_0}^{-\frac{s}{2}}\nabla g\big\|_{L_t^2(L^2)}^2+\frac{1}{2}\|\sigma_{t_0}^{-\frac{s}{2}}\nabla\mathfrak{y}_2\|_{L_t^2(L^2)}^2,
\end{aligned}$$

from which, (v:Lr) and (y1:H1), we infer

$$\begin{aligned}
\int_0^t\sigma_{t_0}^{-s}\|\nabla u\|_{L^2}^2dt'\leq C_0\Big(&\frac{\langle t\rangle}{\delta-s}\|g(0)\|_{\mathcal{B}^\delta}^2+\big\|\sigma^{-\frac{s}{2}}(f,g_t)\big\|_{L_t^1(L^2)}^2\\
&+\big\|\sigma^{-\frac{1-(s_0-s)}{2}}\nabla g\big\|_{L_t^1(L^2)}^2+\|\sigma^{-\frac{s}{2}}\nabla g\|_{L_t^2(L^2)}^2\Big).
\end{aligned}$$

Applying Fatou's Lemma, we deduce that as $t_0\to 0_+$ in the above inequality, there holds (2.18) from (2.19). □

In particular, by estimating $\nabla g_1,\partial_t g_1,\nabla\operatorname{div} g_1,F_1$ precisely in virtue of (v:Lr), (v11:B1t) we obtain

$$\big\|\sigma^{\frac{1-s_1}{2}}\nabla \mathrm{v}_{12}\big\|_{L_t^\infty(L^2)}^2+\big\|\sigma^{\frac{1-s_1}{2}}(\partial_t\mathrm{v}_{12},\nabla^2\mathrm{v}_{12},\nabla p_{12})\big\|_{L_t^2(L^2)}^2\leq\mathfrak{G}_{1,X}(t),$$

with $\mathfrak{G}_{1,X}(t)=\exp\big(C_0\langle t\rangle^2\big)\Big(1+\int_0^t\big(\mathfrak{B}_0(t')+\sigma(t')^{-1+s_0-s_1}\big)\|\nabla X(t')\|_{W^{1,p}}^2\,dt'\Big),$

and hence one gets the H^1-estimate for v^1 in terms of X:

$$\begin{aligned}
&\|\mathrm{v}^1\|_{L_t^\infty(L^2)\cap L_t^2(\dot H^1)}+\|\sigma^{\frac{1-s_1}{2}}\nabla\mathrm{v}^1\|_{L_t^\infty(L^2)}+\big\|\sigma^{\frac{1-s_1}{2}}\big(\partial_t\mathrm{v}^1,\nabla^2\mathrm{v}^1,\nabla\pi^1\big)\big\|_{L_t^2(L^2)}\\
&\leq\mathfrak{G}_{1,X}(t).
\end{aligned}\tag{v1:H1}$$

2.3.2 H^1 estimate for $D_t\mathrm{v}^1$

We apply the operator D_t to the v^1-equation to get

$$\rho D_t^2\mathrm{v}^1-\Delta D_t\mathrm{v}^1+\nabla D_t\pi^1=F_D(\mathrm{v}^1,\pi^1)+D_tF_1(v,\pi)\stackrel{\text{def}}{=}F_{1D},\tag{Dv1}$$

where F_D and F_1 are given in (Dv) and (v1) respectively, and we thus obtain

$$\begin{aligned}
F_{1D}=&-2\partial_\alpha v\cdot\nabla\partial_\alpha(\mathrm{v}^1)-\Delta v\cdot\nabla(\mathrm{v}^1)+\nabla v_\alpha\partial_\alpha(\pi^1)\\
&-D_t(\Delta X\cdot\nabla v+2\partial_\alpha X\cdot\nabla\partial_\alpha v)+D_t(\nabla X^\alpha\partial_\alpha\pi).
\end{aligned}$$

A direct calculation shows

$$\begin{aligned}
&\operatorname{div}(D_t\mathrm{v}^1)=\operatorname{div}\mathfrak{a}_1,\quad\mathfrak{a}_1=2\mathrm{v}^1\cdot\nabla v+v_t\cdot\nabla X+v\partial_\alpha X\cdot\nabla v_\alpha-(v\cdot\nabla X)\cdot\nabla v,\\
&\operatorname{div}(D_t^2\mathrm{v}^1)=\operatorname{div}\mathfrak{b}_1,\quad\mathfrak{b}_1=D_t^2v\cdot\nabla X+\partial_X\mathfrak{b}_0-\mathfrak{b}_0\cdot\nabla X.
\end{aligned}$$

A careful use of all the above estimates yields

$$\|\sigma^{\frac{1-s_1}{2}}\mathfrak{a}_1\|_{L^2_t(L^2)}^2+\|\sigma^{1-\frac{s_1}{2}}\nabla\mathfrak{a}_1\|_{L^2_t(L^2)}^2+\|\sigma^{\frac{3-s_1}{2}}\nabla\operatorname{div}\mathfrak{a}_1\|_{L^2_t(L^2)}^2\le\mathfrak{G}_{1,X}(t),$$
$$\|\sigma^{\frac{3-s_1}{2}}\mathfrak{b}_1\|_{L^2_t(L^2)}^2\le\varepsilon\|\sigma^{\frac{3-s_1}{2}}D_t^2\mathrm{v}^1\|_{L^2_t(L^2)}^2+C_\varepsilon\mathfrak{G}_{1,X}(t),\tag{a1b1}$$

and then Lemma 2.4 implies

$$\|\sigma^{1-\frac{s_1}{2}}D_t\mathrm{v}^1\|_{L^\infty_t(L^2)\cap L^2_t(\dot H^1)}^2+\|\sigma^{\frac{3-s_1}{2}}\nabla D_t\mathrm{v}^1\|_{L^\infty_t(L^2)}^2$$
$$+\|\sigma^{\frac{3-s_1}{2}}(D_t^2\mathrm{v}^1,\nabla^2D_t\mathrm{v}^1,\nabla D_t\pi^1)\|_{L^2_t(L^2)}^2\le\mathfrak{G}_{1,X}(t).\tag{Dv1:H1}$$

2.3.3 $W^{2,p}$ estimate for X

We use L^p energy estimate to (X) to obtain

$$\frac{d}{dt}\|X\|_{W^{2,p}}\le C\big(\|\nabla v\|_{L^\infty}\|X\|_{W^{2,p}}+\|\Delta v\|_{L^p}\|\nabla X\|_{L^\infty}\big)+\|\Delta\mathrm{v}^1\|_{L^p}.\tag{2.20}$$

To deal with the estimate of $\|\Delta\mathrm{v}^1\|_{L^p}$, we rewrite the equation (v1) as

$$-\Delta\big(\mathrm{v}^1-\nabla\Delta^{-1}(\partial_\alpha v\cdot\nabla X^\alpha)\big)+\nabla\pi^1=-\rho D_t\mathrm{v}^1+\nabla(\partial_\alpha v\cdot\nabla X^\alpha)+F_1(v,\pi),$$

which ensures that for any $r\in[2,p]$

$$\|(\nabla^2\mathrm{v}^1,\nabla\pi^1)\|_{L^r}\le C\Big(\|(D_t\mathrm{v}^1,D_tv)\|_{L^r}+\big(\|\nabla v\|_{L^{\frac{rp}{p-r}}}+\|(\nabla^2v,\nabla\pi)\|_{L^r}\big)\|\nabla X\|_{W^{1,p}}\Big).\tag{2.21}$$

Inserting (2.21) for $r=p$ into (2.20), and then integrating over $[0,t]$, we obtain

$$\|X(t)\|_{W^{2,p}}\le\|X_0\|_{W^{2,p}}+C\|(D_t\mathrm{v}^1,D_tv)\|_{L^1_t(L^p)}$$
$$+C\int_0^t\big(\|\nabla v\|_{L^\infty}+\|(\nabla^2v,\nabla\pi)\|_{L^p}\big)\|X\|_{W^{2,p}}\,dt'.$$

Note from the hypothesis that $p\in\big(2,2/(1-s_1)\big)$, which implies

$$\frac{2p}{p+2}(1-\frac{s_1}{2})<1.$$

This together with the interpolation inequality (2.3) and (v':Lr), (Dv1:H1) ensures

$$\|(D_t\mathrm{v}^1,D_tv)\|_{L^1_t(L^p)}^2\le\|\sigma^{1-\frac{s_1}{2}}(D_t\mathrm{v}^1,D_tv)\|^2_{L_t^{\frac{2p}{p-2}}(L^p)}\|\sigma^{-(1-\frac{s_1}{2})}\|^2_{L^{\frac{2p}{p+2}}([0,t])}\le\mathfrak{G}_{1,X}(t),$$

that is,

$$\|X(t)\|_{W^{2,p}}^2$$
$$\le\|X_0\|_{W^{2,p}}^2+\exp(C_0\langle t\rangle^2)\Big(C_0+\int_0^t(\mathfrak{B}_0(t')+\sigma(t')^{-1+\frac{s_0-s_1}{2}})\|X(t')\|_{W^{2,p}}^2\,dt'\Big).$$

Then by virtue of (v':Lr), we get, by applying Gronwall's inequality, that

$$\|X\|_{L_t^\infty(W^{2,p})} \leq C_0 \exp\big(\exp(C_0\langle t\rangle^2)\big)\,. \tag{X:W2p}$$

Hence the righthand side of (v1:H1) and (Dv1:H1) can be replaced by $C_0 \exp\exp(C_0\langle t\rangle^2)$. Furthermore, we can follow the same strategy in the proof of Lemma 2.3 to propogate the Besov regularity of v^1.

Similarly as the derivation of (v:Lr), (v':Lr), we have the following L^r-type estimates of v^1 which will be useful to estimate $\partial_X^2 v$ and $\partial_X X$ in the next step:

$$\begin{aligned}
&\|\mathrm{v}^1\|_{L_t^\infty(L^2)} + \|\sigma^{\frac{1-s_1}{2}}\mathrm{v}^1\|_{L_t^\infty(L^{\frac{2p}{p-2}})} + \|\sigma^{\frac{1+\varepsilon_0-s_1}{2}}(\mathrm{v}^1, X_t)\|_{L_t^\infty(L^\infty)} \\
&+ \big\|\sigma^{\frac{1-s_1}{2}}(\partial_X\nabla v, \nabla\mathrm{v}^1, \nabla X_t)\big\|_{L_t^\infty(L^2)} + \big\|\sigma^{(1-\frac{s_1}{2})}(\partial_X v_t, \partial_t\mathrm{v}^1, D_t\nabla^2 X)\big\|_{L_t^\infty(L^2)} \\
&+ \big\|\sigma^{(\frac{1}{2}+\frac{1}{p}-\frac{s_1}{2})}(\partial_X\nabla v, \nabla\mathrm{v}^1, D_t\nabla X)\|_{L_t^\infty(L^{\frac{2p}{p-2}})} + \|\sigma^{1-\frac{s_1}{2}}(\partial_X\nabla v, \nabla\mathrm{v}^1, D_t\nabla X)\|_{L_t^\infty(L^\infty)} \\
&+ \sum_{i+j+l=1} \big\|\sigma^{(\frac{3}{2}-\frac{1}{r}-\frac{s_1}{2})}(\partial_X^i\nabla\partial_X^j\nabla\mathrm{v}^l, D_t\mathrm{v}^i, \partial_X^i\nabla\pi^j)\big\|_{L_t^\infty(L^2\cap L^p)} \\
&+ \sum_{i+j=1}\Big(\big\|\sigma^{(\frac{1}{2}+\frac{1}{p}-\frac{s_1}{2})}D_t\mathrm{v}^1(t)\|^2_{L^{\frac{2p}{p-2}}} + \|\sigma^{1-\frac{s_1}{2}}(\partial_X^i D_t\nabla\mathrm{v}^j, \partial_X^i\nabla D_t\mathrm{v}^j)(t)\big\|^2_{L^2} \\
&+ \big\|\sigma^{(1+\frac{1}{p}-\frac{s_1}{2})}(\partial_X^i D_t\nabla\mathrm{v}^j, \partial_X^i\nabla D_t\mathrm{v}^j)(t)\big\|^2_{L^{\frac{2p}{p-2}}} \\
&+ \big\|\sigma^{\frac{3-s_1}{2}}\big(D_t^2\mathrm{v}^i, \partial_X^i D_t\nabla^2\mathrm{v}^j, \nabla\partial_X^i\nabla D_t\mathrm{v}^j, \partial_X^i D_t\nabla\pi^j\big)(t)\big\|^2_{L^2}\Big) \leq C_0\exp\exp(C_0\langle t\rangle^2).
\end{aligned} \tag{S1}$$

We remark here that the estimate (S1) controls more terms than (v1:H1), (Dv1:H1) and (X:W2p): For example, terms such as $\partial_X\nabla^2 v, \nabla\partial_X\nabla v$, $\nabla^2\partial_X v$ rather than the mere term $\nabla^2\mathrm{v}^1$ are involved in (S1).

3 Outline of the proof for general case $k \geq 2$

In the following context, we shall use the following notations:

$$\begin{aligned}
&\mathrm{v}^\ell \overset{\text{def}}{=} \partial_X^\ell v,\ \pi^\ell \overset{\text{def}}{=} \partial_X^\ell \pi,\ \mathrm{X}^\ell \overset{\text{def}}{=} \partial_X^\ell X, \\
&\mathcal{H}_\ell(t) \overset{\text{def}}{=} \exp\cdots\exp\big(C_0(1+t^2)\big) \text{ with } (\ell+1)\text{-times } \exp\,.
\end{aligned}$$

Recall the parametrization of the initial boundary $\partial\Omega_0$: $\gamma : \mathbb{S}^1 \mapsto \partial\Omega_0$, $\gamma \in (W^{k+2,p}(\mathbb{R}^2))^2$, $k \geq 2$. In order to show that the evolved boundary $\partial\Omega(t)$ is also of regularity $W^{k+2,p}$, we have to prove $\psi(t,\gamma(\cdot)) \in (W^{k+2,p}(\mathbb{R}^2))^2$ where ψ denotes the inhomogeneous incompressible Navier-Stokes flow (2.1). It is easy to check that when $k=1$

$$\partial_s(\psi(t,\gamma(s))) = \partial_s\gamma(s)\cdot\nabla\psi(t,\gamma(s)) = (X_0\cdot\nabla\psi)(t,\gamma(s)) = X(t,\psi(t,\gamma(s))).$$

We repeat this calculation to obtain $\partial_s^\ell(\psi(t,\gamma(s))) = (\partial_X^{\ell-1}X)(t,\psi(t,\gamma(s)))$. Since we have shown $\psi - \mathrm{Id}, X \in (L^\infty(\mathbb{R}^+; W^{2,p}(\mathbb{R}^2)))^2$ in Section 2, in order to show that $\Omega(t)$ is of class $W^{k+2,p}$, $k \geq 2$, it suffices to prove

$$\mathrm{X}^{\ell-1} = \partial_X^{\ell-1} X \in (L^\infty(\mathbb{R}^+; W^{2,p}(\mathbb{R}^2)))^2,\ \ell = 2, \cdots, k.$$

We now derive the equations for $(\mathrm{v}^\ell, \nabla\pi^\ell, \mathrm{X}^{\ell-1})$. We apply $\partial_X^{\ell-1}$ to (X) to get

$$\partial_t \mathrm{X}^{\ell-1} + v \cdot \nabla \mathrm{X}^{\ell-1} = \mathrm{v}^\ell, \quad \mathrm{X}^{\ell-1}(0,\cdot) = \partial_{X_0}^{\ell-1} X_0 \in (W^{2,p}(\mathbb{R}^2))^2. \tag{Xell}$$

By taking $\partial_X^{\ell-1}$ with $\ell \geq 2$ to (v1) and using $\partial_X \rho = 0$, we write

$$\rho \partial_t \mathrm{v}^\ell + \rho v \cdot \nabla \mathrm{v}^\ell - \Delta \mathrm{v}^\ell + \nabla \pi^\ell = F_\ell(v, \pi), \tag{vell}$$

where the source term $F_\ell(v,\pi)$ is determined inductively by

$$\begin{aligned} F_\ell(v,\pi) &= F_1(\mathrm{v}^{\ell-1}, \pi^{\ell-1}) + \partial_X F_{\ell-1}(v,\pi) \\ &= \sum_{i=0}^{\ell-1} \partial_X^i \left(-2\partial_\alpha X \cdot \nabla \partial_\alpha \mathrm{v}^{\ell-1-i} - \Delta X \cdot \nabla \mathrm{v}^{\ell-1-i} + \nabla X \cdot \nabla \pi^{\ell-1-i}\right). \\ &= \sum_{i=0}^{\ell-1} \sum_{j=0}^{i} C_i^j \Big(-2\partial_X^j \partial_\alpha X \cdot \partial_X^{i-j} \nabla \partial_\alpha \mathrm{v}^{\ell-1-i} \\ &\quad - \partial_X^j \Delta X \cdot \partial_X^{i-j} \nabla \mathrm{v}^{\ell-1-i} + \partial_X^j \nabla X \cdot \partial_X^{i-j} \nabla \pi^{\ell-1-i}\Big). \end{aligned}$$

By applying the operator D_t to (vell) and using $D_t \rho = 0$, we obtain

$$\rho D_t^2 \mathrm{v}^\ell - \Delta D_t \mathrm{v}^\ell + \nabla D_t \pi^\ell = F_D(\mathrm{v}^\ell, \pi^\ell) + D_t F_\ell(v,\pi) \stackrel{\text{def}}{=} F_{\ell D}, \tag{Dvell}$$

where $F_D(\cdot,\cdot)$ is given by (Dv) and

$$\begin{aligned} F_{\ell D} = &- 2\partial_\alpha v \cdot \nabla \partial_\alpha \mathrm{v}^\ell - \Delta v \cdot \nabla \mathrm{v}^\ell + \nabla v \cdot \nabla \pi^\ell \\ &+ \sum_{i=0}^{\ell-1} D_t \partial_X^i \Big(-2\partial_\alpha X \cdot \nabla \partial_\alpha \mathrm{v}^{\ell-1-i} - \Delta X \cdot \nabla \mathrm{v}^{\ell-1-i} + \nabla X \cdot \nabla \pi^{\ell-1-i}\Big). \end{aligned}$$

We proceed with induction argument to show $\mathrm{X}^{l-1} \in (L^\infty(\mathbb{R}^+; W^{2,p}(\mathbb{R}^2)))^2$ for $l = 2, \cdots, k$: provided with the estimate $(S_{\ell-1})$ for $(\mathrm{v}^{\ell-1}, \nabla\pi^{\ell-1}, \mathrm{X}^{\ell-2})$ in the $(\ell-1)$-step, we will show in the ℓ-step

(I) H^1-estimates for $\mathrm{v}^\ell, D_t \mathrm{v}^\ell$ and $W^{2,p}$-estimate for $\mathrm{X}^{\ell-1}$;
(II) the estimate (S_ℓ) for $(\mathrm{v}^\ell, \nabla\pi^\ell, \mathrm{X}^{\ell-1})$.

Here the estimate (S_ℓ) is similar as (S1) but with 1 replaced by ℓ and with the righthand side replaced by $\mathcal{H}_\ell(t)$. See [30] for precise definition of (S_ℓ).

We first say some words from (I) to (II). The (S1)-type estimate (S_ℓ) involves more information than the mere bounds of H^1-norms of $(\mathrm{v}^\ell, D_t \mathrm{v}^\ell)$ and $W^{2,p}$-norm of $\mathrm{X}^{\ell-1}$: terms of general forms such as $\partial_X^i \nabla \partial_X^j \nabla \mathrm{v}^{\ell-i-j}, \partial_X^i D_t \nabla^2 \mathrm{v}^{\ell-i}$ rather than terms of the simple forms $\nabla^2 \mathrm{v}^\ell, \nabla^2 D_t \mathrm{v}^\ell$ are also involved in (S_ℓ). On one side, these estimates for such general terms are really needed in the proof of (I). On the other side, these general terms appear due to the non-commutativity of the operators ∂_X, D_t and ∂_t, ∇. Therefore, from (I) to (II) it requires the estimates for the commutators such as

$$\|\partial_X^i \nabla \mathrm{X}^{\ell-1-i} - \nabla \mathrm{X}^{\ell-1}\|_{L_t^\infty(W^{1,p})} + \|\partial_X^i \nabla \partial_X^j \nabla \mathrm{X}^{\ell-1-i-j} - \nabla^2 \mathrm{X}^{\ell-1}\|_{L_t^\infty(L^p)},$$
$$\big\|\sigma^{\left(1-\frac{1}{r}-\frac{s_{\ell-1}}{2}\right)}(\partial_X^i \nabla \mathrm{v}^{\ell-i} - \nabla \mathrm{v}^\ell)\big\|_{L_t^\infty(L^r)}$$
$$+ \big\|\sigma^{\left(\frac{3}{2}-\frac{1}{r}-\frac{s_{\ell-1}}{2}\right)}(\partial_X^i \nabla \partial_X^j \nabla \mathrm{v}^{\ell-i-j} - \nabla^2 \mathrm{v}^\ell)\big\|_{L_t^\infty(L^r)},$$
$$\big\|\sigma(t)^{\frac{3-s_{\ell-1}}{2}}(\partial_X^i D_t \nabla^2 \mathrm{v}^{\ell-i} - \nabla^2 D_t \mathrm{v}^\ell,\ \nabla \partial_X^i \nabla D_t \mathrm{v}^{\ell-i} - \nabla^2 D_t \mathrm{v}^\ell)(t)\big\|_{L^2}^2, \cdots$$

We need to show that these commutators can be controlled by $\mathcal{H}_{\ell-1}(t)$ in the previous step and hence (II) holds true whenever (I) holds.

We will apply Lemmas 2.3, 2.4 and 2.5 to show (I), following the procedure in Section 2.3. Indeed, similar as in Subsection 2.3.1, we decompose

$$\mathrm{v}^\ell = \mathrm{v}_{\ell 1} + \mathrm{v}_{\ell 2},$$

such that $\mathrm{v}_{\ell 1}$ satisfies (v11:Bs)-(v11:B1t) with obvious modifications. Calculate from inductive assumption $\operatorname{div} \mathrm{v}^{\ell-1} = \operatorname{div} g_{\ell-1}$ that

$$\operatorname{div} \mathrm{v}^\ell = \operatorname{div}(\partial_X \mathrm{v}^{\ell-1}) = \operatorname{div}(\mathrm{v}^{\ell-1} \cdot \nabla X + \mathcal{E}_X g_{\ell-1}), \quad \mathcal{E}_X(u) \stackrel{\text{def}}{=} \partial_X u - (\nabla X)^T \cdot u,$$

and hence $\operatorname{div} \mathrm{v}^\ell = \operatorname{div} g_\ell$ with

$$g_\ell = \sum_{i=0}^{\ell-1} \mathcal{E}_X^i(\mathrm{v}^{\ell-1-i} \cdot \nabla X) = \sum_{i=0}^{\ell-1} (\partial_X - (\nabla X)^T \cdot)^i (\mathrm{v}^{\ell-1-i} \cdot \nabla X)$$
$$= \sum_{i=0}^{\ell-1} \sum_{i_1+\cdots+i_r=i} \partial_X^{i_1}(\mathrm{v}^{\ell-1-i} \cdot \nabla X) \cdot \partial_X^{i_2}(-\nabla X)^{i_3} \cdots \partial_X^{i_{r-1}}(-\nabla X)^{i_r}.$$

A detailed calculation shows

$$\|g_\ell(t)\|_{L^2} + \sigma(t)^{\frac{1-s_{\ell-1}}{2}} \|\nabla g_\ell(t)\|_{L^2} \le \mathcal{H}_{\ell-1}(t)\big(1 + \|\nabla \mathrm{X}^{\ell-1}(t)\|_{W^{1,p}}\big),$$
$$\|\partial_t g_\ell(t)\|_{L^2} \le \|v\|_{L^\infty} \|\nabla \mathrm{v}^\ell\|_{L^2} + \mathcal{H}_{\ell-1}(t) \sigma(t)^{-\left(1-\frac{s_{\ell-1}}{2}\right)}\big(1 + \|\nabla \mathrm{X}^{\ell-1}(t)\|_{W^{1,p}}\big),$$

and the application of Lemma 2.5 together with the estimates for $(F_\ell, \nabla \operatorname{div} g_\ell)$ implies

$$\|\mathrm{v}^\ell\|_{L_t^\infty(L^2) \cap L_t^2(\dot{H}^1)}^2 + \|\sigma^{\frac{1-s_\ell}{2}} \nabla \mathrm{v}^\ell\|_{L_t^\infty(L^2)}^2 + \big\|\sigma^{\frac{1-s_\ell}{2}}(\partial_t \mathrm{v}^\ell, \nabla^2 \mathrm{v}^\ell, \nabla \pi^\ell)\big\|_{L_t^2(L^2)}^2$$
$$\le C_0 \exp\big(C_0(1+t^2)\big) + \mathcal{H}_{\ell-1}(t)\Big(1 + \int_0^t \sigma^{-\left(1-\frac{\epsilon}{2k}\right)} \|\nabla \mathrm{X}^{\ell-1}(t')\|_{W^{1,p}}^2 \, dt'\Big). \tag{vell:H1}$$

We can also get by inductive arguments that $\operatorname{div} D_t \mathrm{v}^\ell = \operatorname{div} \mathfrak{a}_\ell$, $\operatorname{div} D_t^2 \mathrm{v}^\ell = \operatorname{div} \mathfrak{b}_\ell$ with

$$\mathfrak{a}_\ell = D_t \mathrm{v}^{\ell-1} \cdot \nabla X + \mathcal{E}_X \mathfrak{a}_{\ell-1}, \quad \mathfrak{b}_\ell = D_t^2 \mathrm{v}^{\ell-1} \cdot \nabla X + \mathcal{E}_X \mathfrak{b}_{\ell-1}.$$

Thanks to the inductive assumption $(S_{\ell-1})$ and also (vell:H1) we arrive at a similar estimate for $\mathfrak{a}_\ell, \mathfrak{b}_\ell$ as (a1b1), and then Lemma 2.4 yields a similar estimate for $D_t\mathrm{v}^\ell$ as (Dv1:H1). Finally $W^{2,p}$-estimate for $\mathrm{X}^{\ell-1}$ follows as for (X:W2p). This completes (I).

Acknowledgments. We would like to thank Professor Jean-Yves Chemin and Professor Raphaël Danchin for profitable discussions of this topic. Part of this work was done when we were visiting Morningside Center of Mathematics (MCM), Chinese Academy of Sciences. We appreciate the hospitality and the financial support from MCM. X. Liao is supported by SFB 1060 in University Bonn. P. Zhang is partially supported by NSF of China under Grants 11371347 and 11688101, and innovation grant from National Center for Mathematics and Interdisciplinary Sciences.

References

[1] H. Abidi, Équations de Navier-Stokes avec densité et viscosité variables dans l'espace critique, *Rev. Mat. Iberoam.*, **23** (2007), 537–586.

[2] H. Abidi and M. Paicu, Existence globale pour un fluide inhomogène, *Ann. Inst. Fourier*, **57** (2007), 883–917.

[3] H. Abidi, G. Gui and P. Zhang, On the wellposedness of 3−D inhomogeneous Navier-Stokes equations in the critical spaces, *Arch. Ration. Mech. Anal.*, **204** (2012), 189–230.

[4] H. Abidi, G. Gui and P. Zhang, Wellposedness of 3−D inhomogeneous Navier-Stokes equations with highly oscillating initial velocity field, *J. Math. Pures Appl.*, **100** (2013), 166–203.

[5] H. Bahouri, J.-Y. Chemin and R. Danchin, *Fourier Analysis and Nonlinear Partial Differential Equations*, Grundlehren der mathematischen Wissenschaften, Springer, 2010.

[6] A.-L. Bertozzi and P. Constantin, Global regularity for vortex patches, *Comm. Math. Phys.*, **152** (1993), 19-28.

[7] J.-Y. Chemin, Calcul paradifférentiel précisé et applications à des équations aux dérivées partielles non semilinéaires, *Duke Math. J.*, **56** (1988), 431–469.

[8] J.-Y. Chemin, Sur le mouvement des particules d'un fluide parfait incompressible bidimensionnel, *Invent. Math.*, **103** (1991), 599–629.

[9] J.-Y. Chemin, Persistance de structures géomšŠtriques dans les fluides incompressibles bidimensionnels, *Ann. Sci. École Norm. Sup.*, **26** (1993), 517-542.

[10] J.-Y. Chemin, *Perfect incompressible fluids*, Oxford Lecture Series in Mathematics and its Applications, **14**. The Clarendon Press, Oxford University Press, New York, 1998.

[11] W. Craig, X. Huang and Y. Wang, Global well-posedness for the 3D inhomogeneous incompressible Navier-Stokes equations, *J. Math. Fluid Mech.*, **15**(4), (2013), 747-758.

[12] R. Danchin, Poches de tourbillon visqueuses, *J. Math. Pures Appl. (9)*, **76**

(1997), 609-647.
[13] R. Danchin, Density-dependent incompressible viscous fluids in critical spaces, *Proc. Roy. Soc. Edinburgh Sect. A*, **133** (2003), 1311–1334.
[14] R. Danchin and P.B. Mucha, A Lagrangian approach for the incompressible Navier-Stokes equations with variable density, *Comm. Pure Appl. Math.*, **65** (2012), 1458-1480.
[15] R. Danchin and P.B. Mucha, Incompressible flows with piecewise constant density, *Arch. Ration. Mech. Anal.*, **207** (2013), 991–1023.
[16] R. Danchin and P.B. Mucha, The incompressible Navier-Stokes equations in vacuum, *arXiv:1705.06061.*
[17] R. Danchin and P. Zhang, Inhomogeneous Navier-Stokes equations in the half-space, with only bounded density, *J. Funct. Anal.*, **267** (2014), 2371-2436.
[18] R. Danchin and X. Zhang, On the persistence of Hölder regular patches of density for the inhomogeneous Navier-Stokes equations, *arXiv:1612.00282.*
[19] P. Gamblin and X. Saint-Raymond, On three-dimensional vortex patches, *Bull. Soc. Math. France*, **123** (1995), 375-424.
[20] F. Gancedo and E. Garcia-Juarez, Global regularity of 2D density patches for inhomogeneous Navier-Stokes, *arXiv:1612.08665.*
[21] T. Hmidi, Régularité Höldérienne des poches de tourbillon visqueuses, *J. Math. Pures Appl. (9)*, **84** (2005), 1455–1495.
[22] T. Hmidi, Poches de tourbillon singuliéres dans un fluide faiblement visqueux, *Rev. Mat. Iberoam.*, **22** (2006), 489-543.
[23] D. Hoff, Global solutions of the Navier-Stokes equations for multidimensional compressible flow with discontinuous initial data, *J. Differential Equations* , **120** (1995), 215-254.
[24] J. Huang, M. Paicu and P. Zhang, Global well-posedness of incompressible inhomogeneous fluid systems with bounded density or non-Lipschitz velocity, *Arch. Ration. Mech. Anal.*, **209** (2013), 631–682.
[25] A.V. Kazhikhov, Solvability of the initial-boundary value problem for the equations of the motion of an inhomogeneous viscous incompressible fluid, *Dokl. Akad. Nauk SSSR*, **216**, (1974), 1008-1010.
[26] O.A. Ladyv zenskaja and V.A. Solonnikov, The unique solvability of an initial-boundary value problem for viscous incompressible inhomogeneous fluids. (Russian) Boundary value problems of mathematical physics, and related questions of the theory of functions, 8, *Zap. Nauv cn. Sem. Leningrad. Otdel. Mat. Inst. Steklov. (LOMI)*, **52** (1975), 52-109, 218-219.
[27] J. Leray, Sur le mouvement d'un liquide visqueux remplissant l'espace, *Acta Mathematica*, **63** (1934), 193–248.
[28] X. Liao and Y. Liu, Global regularity of three dimensional density patches for inhomogeneous incompressible viscous flow, *Accepted in Sci. China Math.*.
[29] X. Liao and P. Zhang, On the global regularity of 2-D density patch for inhomogeneous incompressible viscous flow, *Arch. Ration. Mech. Anal.*, **220** (2016), 937-981.
[30] X. Liao and P. Zhang, Global regularity of 2-D density patches for viscous

inhomogeneous incompressible flow with general density: low regularity case, *Accepted in Comm. Pure Appl. Math.*.

[31] X. Liao and P. Zhang, Global regularity of 2-D density patches for viscous inhomogeneous incompressible flow with general density: high regularity case, *arXiv:1604.07922*.

[32] P.-L. Lions, *Mathematical topics in fluid mechanics. Vol. 1. Incompressible models.* Oxford Lecture Series in Mathematics and its Applications, **3**. Oxford Science Publications. The Clarendon Press, Oxford University Press, New York, 1996.

[33] M. Paicu and P. Zhang, Global solutions to the 3-D incompressible inhomogeneous Navier-Stokes system, *J. Funct. Anal.*, **262** (2012), 3556–3584.

[34] M. Paicu, P. Zhang and Z. Zhang, Global unique solvability of inhomogeneous Navier-Stokes equations with bounded density, *Comm. Partial Differential Equations*, **38** (2013), 1208-1234.

[35] P. Zhang and Q. Qiu, Propagation of higher-order regularities of the boundaries of 3-D vortex patches, *Chinese Ann. Math. Ser. A*, **18** (1997), 381–390.

[36] J. Simon, Nonhomogeneous viscous incompressible fluids: existence of velocity, density, and pressure, *SIAM J. Math. Anal.*, **21** (1990), 1093–1117.

Mathematical Institute, University Bonn
Endenicher Allee 60, 53115 Bonn, Germany
E-mail address: xianliao@math.uni-bonn.de

Academy of Mathematics & Systems Science and Hua Loo-Keng Key Laboratory of Mathematics, Chinese Academy of Sciences
and School of Mathematical Sciences, University of Chinese Academy of Sciences
Beijing 100049, China
E-mail address: zp@amss.ac.cn

Proceedings of the 7th International
Congress of Chinese Mathematicians
ALM 43, pp. 395–405

From Mariño-Vafa Formula to Witten Conjecture

Kefeng Liu and Shengmao Zhu

Abstract

Mariño-Vafa formula, which is a Hodge integral identity with triple λ-classes, has very powerful applications in the intersection theory of moduli space of stable curves. In particular, Mariño-Vafa formula implies the celebrated Witten conjecture, ELSV formula and λ_g-conjecture. These results are scattered in various literatures. In this short survey, we intend to describe the derivations of these results from Mariño-Vafa formula in a concise way.

Mathematics Subject Classification: 14H10, 14N35, 14N10.

Keywords and Phrases: Hodge integrals, Witten conjecture, Mariño-Vafa formula.

1 Introduction

The large N duality between the quantum Chern-Simons theory and topological string theory which was systemically developed by Gopakumar, Ooguri and Vafa et al [11, 26, 20] based on the early works of Witten [28, 30], provides new insights both in physics and geometry. In 2001, Mariño and Vafa [22] posted a beautiful and important paper on arXiv. Based on their careful computations with the large N duality between the Chern-Simons partition function of a framed unknot and the corresponding topological strings theory on the resolved conifold, they proposed a remarkable formula, which is a closed formula for certain Hodge integrals with triple λ-classes.

The first proof of the Mariño-Vafa formula was given in [18]. See also [25] for another proof. A key step in Liu-Liu-Zhou's proof [18] is to show that Mariño-Vafa formula satisfies the cut-and-join equation by using the functorial localization technique developed in [17]. Here the cut-and-join equation plays the essential role, which dominates both the geometric and algebraic structures of the Mariño-Vafa formula. Indeed, the derivations of the Witten conjecture [29], ELSV formula [5] and λ_g-conjecture [8] are based on this cut-and-join equation of the Mariño-Vafa formula.

In this survey, we mainly focus on the derivation of the Witten conjecture [29], which is one of the most fundamental result in the intersection theory of moduli spaces of stable curves. After the first proof given by Kontsevich [12], Witten conjecture has been reproved by many people [24, 14, 15, 21, 2, 1, 23] by now. We will follow the method in [1, 32] to describe the derivation of Witten conjecture from Mariño-Vafa formula in the rest of this article. ELSV formula, λ_g-conjecture and other Hodge integral identities from Mariño-Vafa formula will also be mentioned.

Acknowledgements. The authors would like to express their deep appreciations to Professor C. Vafa, who suggested them to write an article for non-experts to explain the derivation of the Witten conjecture from the Mariño-Vafa formula. Kefeng Liu is supported by NSFC (Grant No. 11531012) and NSF.

2 Mariño-Vafa formula

Let $\overline{\mathcal{M}}_{g,n}$ be the Deligne-Mumford moduli stack of stable curves of genus g with n marked points. Denoted by $\mathbb{E}$ the Hodge bundle and $\mathbb{L}_i, i = 1, .., n$ the n tautological line bundles over $\overline{\mathcal{M}}_{g,n}$. We define the ψ-classes and λ-classes by $\psi_i = c_1(\mathbb{L}_i)$ and $\lambda_j = c_j(\mathbb{E})$ respectively. Hodge integral is an integral of the form

$$\int_{\overline{\mathcal{M}}_{g,n}} \psi_1^{j_1} \cdots \psi_n^{j_n} \lambda_1^{k_1} \cdots \lambda_g^{k_g}$$

which is defined to be zero if the dimension condition

$$\sum_{i=1}^{n} j_i + \sum_{i=1}^{g} ik_i = 3g - 3 + n \tag{2.1}$$

is not satisfied. Hodge integrals arise in the calculations of Gromov-Witten invariants by localization techniques [13, 10], and also in the intersection theory of moduli space of curves [7].

Mariño-Vafa formula proposed in [22] is a closed formula of Hodge integral identity with triple λ-classes. Let

$$\Lambda_g^\vee(u) = u^g - \lambda_1 u^{g-1} + \cdots + (-1)^g \lambda_g$$

be the Chern polynomial of the dual Hodge bundle $\mathbb{E}^\vee$. For a partition μ given by

$$\mu_1 \geq \mu_2 \geq \cdots \geq \mu_{l(\mu)} > 0,$$

we will use the standard notation $|\mu| = \sum_{i=1}^{l(\mu)} \mu_i$. We let

$$\Gamma_g(\tau) = \Lambda_g^\vee(1)\Lambda_g^\vee(-\tau - 1)\Lambda_g^\vee(\tau)$$

and define

$$\mathcal{C}_{g,\mu}(\tau) = -\frac{\sqrt{-1}^{|\mu|+l(\mu)}}{|Aut(\mu)|}[\tau(\tau+1)]^{l(\mu)-1} \prod_{i=1}^{l(\mu)} \frac{\prod_{a=1}^{\mu_i-1}(\mu_i\tau + a)}{(\mu_i - 1)!} \int_{\overline{\mathcal{M}}_{g,l(\mu)}} \frac{\Gamma_g(\tau)}{\prod_{i=1}^{l(\mu)}(1 - \mu_i\psi_i)}$$

where the Deligne-Mumford stack $\overline{\mathcal{M}}_{g,l}$ is the moduli space of stable curves satisfying the stability condition $2g - 2 + l > 0$. For the unstable cases $(g, l) = (0, 1)$

and $(0,2)$, we define

$$\int_{\overline{\mathcal{M}}_{0,1}} \frac{1}{1-\mu\psi} = \frac{1}{\mu^2}, \quad \int_{\overline{\mathcal{M}}_{0,2}} \frac{1}{(1-\mu_1\psi_1)(1-\mu_2\psi_2)} = \frac{1}{\mu_1+\mu_2}.$$

For the formal variable $p = (p_1, p_2, ...)$, we define $p_\mu = p_{\mu_1} \cdots p_{\mu_{l(\mu)}}$, and $s_\mu = \sum_{|\nu|=|\mu|} \frac{\chi_\mu(C_\nu)}{\mathfrak{z}_\nu} p_\nu$.

We introduce the following two generating functions

$$\mathcal{C}(\lambda;\tau;p) = \sum_{g\geq 0, |\mu|\geq 1} \mathcal{C}_{g,\mu}(\tau)\lambda^{2g-2+l(\mu)} p_\mu,$$

$$\mathcal{R}(\lambda;\tau;p) = \log\left(\sum_{|\mu|\geq 0} (-1)^{\tau|\mu|} R_\mu(\lambda;\tau) s_\mu\right),$$

where $R_\mu(\lambda;\tau)$ is a combinatoric expression related to the colored HOMFLY-PT invariants of the framed unknot.

Mariño-Vafa formula states that

$$\mathcal{C}(\lambda;\tau;p) = \mathcal{R}(\lambda;\tau;p). \tag{2.2}$$

The first author, together with C.-C. Liu and J. Zhou, proved Mariño-Vafa formula by showing that

Theorem 2.1 ([18]). *Both $\mathcal{C}(\lambda;\tau;p)$ and $\mathcal{R}(\lambda;\tau;p)$ satisfy the following cut-and-join equation*

$$\frac{\partial \mathcal{F}}{\partial \tau} = \frac{\sqrt{-1}}{2}\lambda \sum_{i,j\geq 1}\left((i+j)p_i p_j \frac{\partial \mathcal{F}}{\partial p_{i+j}} + ijp_{i+j}\frac{\partial^2 \mathcal{F}}{\partial p_i \partial p_j} + ijp_{i+j}\frac{\partial \mathcal{F}}{\partial p_i}\frac{\partial \mathcal{F}}{\partial p_j}\right),$$

and have the common initial value $\mathcal{C}(\lambda;0;p) = \mathcal{R}(\lambda;0;p)$.

The ELSV formula [5] related to the Hurwitz numbers and Hodge integrals with one λ-class can be immediately proved by the same method used in the proof of Theorem 2.1. Furthermore, ELSV formula can also be derived by taking the large τ limit of the Mariño-Vafa formula (2.2). We refer to [19] for details.

In the following, we describe an effective method to convert the cut-and-join equation of Mariño-Vafa formula into a polynomial identity. This method was first introduced by Goulden, Jackson and Vakil in [9] to convert the cut-and-join equation of Hurwitz numbers into a polynomial identity, from which they derived the λ_g-integrals. Later, also by using this method, the first author together with L. Chen and Y. Li [2], derived the DVV recursion which is an equivalent form of the Witten conjecture. The second author derived a λ_{g-1}-Hodge integral identity [31]. Furthermore, an analogous method called Laplace transform was used in [6, 23]. Finally, the above techniques were generalized by L. Chen [1] and the second author [32], to deal with the cut-and-join equation of the Mariño-Vafa formula.

3 Polynomial identities from the Mariño-Vafa formula

We follow the notations in [32]. Let $\mathcal{C}_g(\mu;\tau) = |Aut(\mu)|\mathcal{C}_{g,\mu}(\tau)$, we define the Laplace operator "Lap" acting on $\mathcal{C}_g(\mu;\tau)$ as

$$\mathrm{Lap}(\mathcal{C}_g(\mu;\tau)) = \sum_{\mu\in\mathbb{N}^l} \frac{1}{\sqrt{-1}^{l+|\mu|}}\mathcal{C}_g(\mu;\tau)e^{-(\mu_1 w_1+\cdots+\mu_l w_l)}$$

where we have denoted $l = l(\mu)$. We introduce the new variable $t = \frac{1}{1-(\tau+1)y}$, where $y = e^{-w} + \sum_{k\geq 2} \frac{\prod_{a=0}^{k-2}(k\tau+a)e^{-kw}}{k!}$. By some combinatoric computations, we obtain

$$\mathrm{Lap}(\mathcal{C}_g(\mu;\tau)) = -(\tau(\tau+1))^{l-1} \sum_{\substack{b_i\geq 0\\ i=1,..,l}} \langle\prod_{i=1}^{l}\tau_{b_i}\Gamma_g(\tau)\rangle_g \prod_{i=1}^{l}\widehat{\Psi}_{b_i}(t_i;\tau)$$

where $\widehat{\Psi}_n(t;\tau)$ is a polynomial of t defined by

$$\widehat{\Psi}_n(t;\tau) = \left(\frac{t(t-1)(t\tau+1)}{\tau+1}\frac{d}{dt}\right)^n \frac{t-1}{\tau+1} \tag{3.1}$$

for $n\geq 0$.

By Theorem 2.1, $\mathcal{C}(\lambda;\tau;p)$ satisfies the cut-and-join equation:

$$\frac{\partial\mathcal{C}}{\partial\tau} = \frac{\sqrt{-1}}{2}\lambda\sum_{i,j\geq 1}\left((i+j)p_ip_j\frac{\partial\mathcal{C}}{\partial p_{i+j}} + ijp_{i+j}\frac{\partial^2\mathcal{C}}{\partial p_i\partial p_j} + ijp_{i+j}\frac{\partial\mathcal{C}}{\partial p_i}\frac{\partial\mathcal{C}}{\partial p_j}\right). \tag{3.2}$$

For every choice of $g\geq 1$ and a partition μ, taking the coefficient of $p_\mu\lambda^{2g-2+l(\mu)}$ in (3.2), we get

$$\begin{aligned}\frac{\partial\mathcal{C}_g(\mu;\tau)}{\partial\tau} &= \sqrt{-1}\sum_{i<j}(\mu_i+\mu_j)\mathcal{C}_g((\mu(\widehat{i},\widehat{j}),\mu_i+\mu_j);\tau)\\ &+\frac{\sqrt{-1}}{2}\sum_{i=1}^{l}\sum_{\alpha+\beta=\mu_i}\alpha\beta\left(\mathcal{C}_{g-1}(\mu(\widehat{i});\tau)) + \sum_{\substack{g_1+g_2=g\\ \nu_1\coprod\nu_2=\mu(\alpha,\widehat{i})}}\mathcal{C}_{g_1}(\nu_1;\tau)\mathcal{C}_{g_2}(\nu_2;\tau)\right).\end{aligned}$$

where $\mu(\widehat{i},\widehat{j}) = (\mu_1,\mu_2,..,\widehat{\mu}_i,..,\widehat{\mu}_j,..,\mu_{l(\mu)})$, $\mu(\widehat{i}) = (\mu_1,\mu_2,..,\widehat{\mu}_i,...,\mu_{l(\mu)})$ and $\mu(\alpha,\widehat{i}) = (\mu_1,\mu_2,...,\widehat{\mu}_i,..,\mu_{l(\mu)},\alpha,\mu_i-\alpha)$.

Applying the Lapalace operator Lap to the above formula, we obtain

Theorem 3.1 (cf. Theorem 1.1 in [32]). *For $g \geq 1$ and $l \geq 1$,*

$$\begin{aligned}
&-(\tau^2+\tau)^{l-2}\sum_{b_L\geq 0}\Big((l-1)(2\tau+1)\langle\tau_{b_L}\Gamma_g(\tau)\rangle_g \\
&+(\tau^2+\tau)\langle\tau_{b_L}\frac{d}{d\tau}\Gamma_g(\tau)\rangle_g\Big)\widehat{\Psi}_{b_L}(t_L;\tau)-(\tau^2+\tau)^{l-1}\sum_{b_L\geq 0}\langle\tau_{b_L}\Gamma_g(\tau)\rangle_g \\
&\cdot\sum_{i=1}^{l}\left(\frac{\partial}{\partial\tau}\widehat{\Psi}_{b_i}(t_i;\tau)+\frac{1}{t_i\tau+1}\widehat{\Psi}_{b_i+1}(t_i;\tau)\right)\widehat{\Psi}_{b_{L\setminus\{i\}}}(t_{L\setminus\{i\}};\tau) \\
&=T_1+T_2+T_3
\end{aligned} \tag{3.3}$$

where

$$\begin{aligned}
T_1 :=& -\frac{(\tau^2+\tau)^{l-2}}{\tau+1}\sum_{1\leq i<j\leq l}\sum_{\substack{a\geq 0\\ b_{L\setminus\{i,j\}}\geq 0}}\langle\tau_a\tau_{b_{L\setminus\{i,j\}}}\Gamma_g(\tau)\rangle_g\widehat{\Psi}_{b_{L\setminus\{i,j\}}}(t_{L\setminus\{i,j\}};\tau) \\
&\cdot\frac{(t_j-1)(t_i^2\tau+t_i)\widehat{\Psi}_{a+1}(t_i;\tau)-(t_i-1)(t_j^2\tau+t_j)\widehat{\Psi}_{a+1}(t_j;\tau)}{t_i-t_j},
\end{aligned}$$

$$\begin{aligned}
T_2 :=& \frac{(\tau^2+\tau)^l}{2}\sum_{i=1}^{l}\sum_{\substack{a_1\geq 0\\ a_2\geq 0\\ b_{L\setminus\{i\}}\geq 0}}\langle\tau_{a_1}\tau_{a_2}\tau_{b_{L\setminus\{i\}}}\Gamma_{g-1}(\tau)\rangle_{g-1} \\
&\cdot\prod_{n=1}^{2}\widehat{\Psi}_{a_n+1}(t_i;\tau)\widehat{\Psi}_{b_{L\setminus\{i\}}}(t_{L\setminus\{i\}};\tau),
\end{aligned}$$

$$\begin{aligned}
T_3 :=& -\frac{(\tau^2+\tau)^{l-1}}{2}\sum_{i=1}^{l}\sum_{\substack{a_1\geq 0\\ a_2\geq 0\\ b_{L\setminus\{i\}}}}\sum_{\substack{g_1+g_2=g\\ \mathcal{I}\coprod\mathcal{J}=L\setminus\{i\}\\ 2g_1-1+|\mathcal{I}|>0\\ 2g_2-1+|\mathcal{J}|>0}}\langle\tau_{a_1}\tau_{b_\mathcal{I}}\Gamma_{g_1}(\tau)\rangle_{g_1}\langle\tau_{a_2}\tau_{b_\mathcal{J}}\Gamma_{g_2}(\tau)\rangle_{g_2} \\
&\cdot\prod_{n=1}^{2}\widehat{\Psi}_{a_n+1}(t_i;\tau)\widehat{\Psi}_{b_{L\setminus\{i\}}}(t_{L\setminus\{i\}};\tau),
\end{aligned}$$

with $L=\{1,2,..,l\}$ an index set, and for any subset $I\subset L$,

$$t_I=(t_i)_{i\in I},\ b_I=(b_i)_{i\in I},\ \tau_{b_I}=\prod_{i\in I}\tau_{b_i},\ \widehat{\Psi}_{b_I}(t_I,\tau)=\prod_{i\in I}\widehat{\Psi}_{b_i}(t_i,\tau).$$

Considering the following expansions of τ

$$\widehat{\Psi}_b(t;\tau)=\sum_{k=0}^{b}\frac{\tau^k}{(\tau+1)^{b+1}}\Psi_b^k(t) \tag{3.4}$$

and

$$\Gamma_g(\tau) = \sum_{m=0}^{2g} \Lambda_g^\vee(1) a_m(\lambda)\tau^m. \tag{3.5}$$

Let $\widetilde{\tau} = \tau + 1$, substituting (3.4) and (3.5) into the formula (3.3), we will obtain a Laurent polynomial of $\widetilde{\tau}$ with highest degree $2g+l-3$, for fixed $g \geq 1$ and $l \geq 1$ due to the dimension condition (2.1). Collecting the coefficients of $\widetilde{\tau}^{2g+l-3}$, we obtain

Corollary 3.2 (cf. Corollary 1.2 in [32]). *For $g \geq 1$ and $l \geq 1$,*

$$\begin{aligned} L := \sum_{b_L \geq 0} \langle \tau_{b_L} \Lambda_g^\vee(1)\rangle_g [(2g-2+l)\psi_{b_L}(t_L) \\ + \sum_{i=1}^{l} (t_i^2 - t_i)\frac{\partial}{\partial t_i}\psi_{b_i}(t_i)\psi_{b_{L\backslash\{i\}}}(t_{L\backslash\{i\}})] = R_1 + R_2 + R_3. \end{aligned} \tag{3.6}$$

where

$$\begin{aligned} R_1 := \sum_{1\leq i<j\leq l} \sum_{\substack{a\geq 0 \\ b_{L\backslash\{i,j\}}\geq 0}} \langle \tau_a \tau_{b_{L\backslash\{i,j\}}} \Lambda_g^\vee(1)\rangle_g \psi_{b_{L\backslash\{i,j\}}}(t_{L\backslash\{i,j\}}) \\ \cdot \frac{(t_j-1)t_i^2\psi_{a+1}(t_i) - (t_i-1)t_j^2\psi_{a+1}(t_j)}{t_i - t_j} \end{aligned}$$

$$R_2 := \frac{1}{2}\sum_{i=1}^{l} \sum_{\substack{a_1\geq 0 \\ a_2 \geq 0 \\ b_{L\backslash\{i\}}\geq 0}} \langle \tau_{a_1}\tau_{a_2}\tau_{b_{L\backslash\{i\}}}\Lambda_{g-1}^\vee(1)\rangle_{g-1}\psi_{a_1+1}(t_i)\psi_{a_2+1}(t_i)\psi_{b_{L\backslash\{i\}}}(t_{L\backslash\{i\}})$$

$$\begin{aligned} R_3 := \frac{1}{2}\sum_{i=1}^{l} \sum_{\substack{a_1\geq 0 \\ a_2 \geq 0 \\ b_{L\backslash\{i\}}\geq 0}} \sum_{\substack{g_1+g_2=g \\ \mathcal{I}\cup\mathcal{J}=L\backslash\{i\} \\ 2g_1-1+|\mathcal{I}|>0 \\ 2g_2-1+|\mathcal{J}|>0}} \langle \tau_{a_1}\tau_{b_{\mathcal{I}}}\Lambda_{g_1}^\vee(1)\rangle_{g_1} \cdot \langle \tau_{a_2}\tau_{b_{\mathcal{J}}}\Lambda_{g_2}^\vee(1)\rangle_{g_2} \\ \cdot \psi_{a_1+1}(t_i)\psi_{a_2+1}(t_i)\psi_{b_{L\backslash\{i\}}}(t_{L\backslash\{i\}}). \end{aligned}$$

where $\psi_b(t) := \Psi_b^b(t)$. By the definitions (3.1) and (3.4), we have

$$\psi_b(t) = \left((t^3 - t^2)\frac{d}{dt}\right)^b (t-1), \ \ \textit{for } b \geq 0. \tag{3.7}$$

Similarly, one can also collect the coefficients of $\widetilde{\tau}$ of other degrees in the formula (3.3) to get more Hodge integral identities like the form of Corollary 3.2. The interested reader is referred to [32] for more results obtained in this way. In fact, an equivalent form of Corollary 3.2 was first proved by Goulden-Jackson-Vakil [9]. Corollary 3.2 can also be viewed as the Laplace transform [6] of the cut-and-join equation for ELSV formula. This also shows that ELSV formula is a special case of the Mariño-Vafa formula.

4 Proof of the Witten conjecture

Now we can prove the Witten conjecture by using Corollary 3.2. We introduce two operators acting on the ring $\mathbb{Q}[t_1, ..., t_l]$ of polynomials with rational coefficients. Suppose $g(t_1, ..., t_l) = \sum_{i_1,..,i_l \geq 0} c_{i_1 i_2 \cdots i_l} t_1^{i_1} \cdots t_l^{i_l} \in \mathbb{Q}[t_1, .., t_l]$, we define

$$\mathbf{F}_d(g(t_1, \cdots, t_l)) = \sum_{\sum_{k=1}^l i_k = d} c_{i_1 i_2 \cdots i_l} t_1^{i_1} \cdots t_l^{i_l} \quad \text{for} \quad d \geq 0.$$

and

$$[t_1^{j_1} \cdots t_l^{j_l}] g(t_1, \cdots, t_l) = c_{j_1 j_2 \cdots j_l}.$$

By formula (3.7), we write

$$\psi_b(t) = \left((t^3 - t^2) \frac{d}{dt} \right)^b (t-1) = \sum_i f(b,i) t^i.$$

It is clear that $f(b,i) = 0$, if $i \leq b$ or $i \geq 2b+2$, and for $b \geq 0$

$$f(b+1, b+2) = (-1)^{b+1}(b+1)!, \quad f(b+1, 2b+3) = (2b+1)!!$$

Then we have

$$\psi_b(t) = (-1)^b b! t^{b+1} + \cdots + (2b-1)!! t^{2b+1}. \tag{4.1}$$

Given fixed $g, l \geq 1$. For $b_i \geq 0$ and $\sum_{i=1}^l b_i = 3g - 3 + l$, we apply the operator

$$\mathbf{H} := [t_1^{2(b_1+1)} t_2^{2b_2+1} \cdots t_l^{2b_l+1}] \mathbf{F}_{6g+3l-5}$$

to (3.6). By formula (4.1), we have

$$\mathbf{H}(L) = \langle \tau_{b_L} \rangle_g (2b_1+1)!! \prod_{i=2}^l (2b_i - 1)!!.$$

$$\mathbf{H}(R_1) = \sum_{j=2}^l \langle \tau_{b_1+b_j-1} \tau_{b_{L \setminus \{1,j\}}} \rangle_g (2(b_1+b_i-1)+1)!! \prod_{k \neq 1,j}^l (2b_k - 1)!!$$

$$\mathbf{H}(R_2) = \frac{1}{2} \sum_{a_1+a_2=b_1-2} \langle \tau_{a_1} \tau_{a_2} \tau_{b_{L \setminus \{1\}}} \rangle_{g-1} (2a_1+1)!!(2a_2+1)!! \prod_{i=2}^l (2b_i - 1)!!$$

$$\mathbf{H}(R_3)=\frac{1}{2}\sum_{a_1+a_2=b_1-2}\sum_{\substack{g_1+g_2=g\,\mathcal{I}\cup\mathcal{J}=L\backslash\{1\}\\ 2g_1-1+|\mathcal{I}|>0\,2g_2-1+|\mathcal{J}|>0}}\langle\tau_{a_1}\tau_{b_{\mathcal{I}}}\rangle_{g_1}\langle\tau_{a_2}\tau_{b_{\mathcal{J}}}\rangle_{g_2}$$

$$\cdot(2a_1+1)!!(2a_2+1)!!\prod_{i=2}^{l}(2b_i-1)!!$$

Multiplying the constant $\prod_{i=2}^{l}(2b_i+1)$, we finally obtain the following recursion formula

$$\langle\widetilde{\tau}_{b_1}\widetilde{\tau}_{b_{L\backslash\{1\}}}\rangle_g=\sum_{j=2}^{l}(2b_j+1)\langle\widetilde{\tau}_{b_1+b_j-1}\widetilde{\tau}_{b_{L\backslash\{1,j\}}}\rangle_g \tag{4.2}$$

$$+\frac{1}{2}\sum_{a_1+a_2=b_1-2}[\langle\widetilde{\tau}_{a_1}\widetilde{\tau}_{a_2}\widetilde{\tau}_{b_{L\backslash\{1\}}}\rangle_{g-1}+\sum_{\substack{g_1+g_2=g\\ \mathcal{I}\cup\mathcal{J}=L\backslash\{1\}\\ 2g_1-1+|\mathcal{I}|>0\\ 2g_2-1+|\mathcal{J}|>0}}\langle\widetilde{\tau}_{a_1}\widetilde{\tau}_{b_{\mathcal{I}}}\rangle_{g_1}\langle\widetilde{\tau}_{a_2}\widetilde{\tau}_{b_{\mathcal{J}}}\rangle_{g_2}],$$

where $\widetilde{\tau}_b=(2b+1)!!\tau_b$. Identity (4.2) is called the Dijkgraaf-Verlinde-Verlinde formula [3, 27] which is equivalent to the Witten conjecture.

Remark 4.1. Similarly, we can derive the λ_g-conjecture from formula (3.6). For $b_i\geq 0$ and $\sum_{i=1}^{l}b_i=2g-3+l$, we define the operator

$$\mathbf{L}:=[t_1^{b_1+1}t_2^{b_2+1}\cdots t_l^{b_l+1}]\mathbf{F}_{2g-3+2l}.$$

Applying the operator $\mathbf{L}$ to (3.6), we obtain

$$\langle\tau_{b_L}\lambda_g\rangle_g=\frac{1}{l-1}\sum_{1\leq i<j\leq l}\frac{(b_i+b_j)!}{b_i!b_j!}\langle\tau_{b_i+b_j-1}\tau_{b_{L\backslash\{i,j\}}}\lambda_g\rangle_g.$$

Therefore

$$\langle\tau_{b_L}\lambda_g\rangle_g=\binom{2g+l-3}{b_1,..,b_l}c_g$$

where $c_g=\langle\tau_{2g-2}\lambda_g\rangle_g$.

Another derivation of the λ_g-conjecture from Mariño-Vafa formula (2.2) was given in the first author's joint work with C.-C. Liu and J. Zhou [19], where they derived the following formula

$$\langle\tau_{b_L}\lambda_g\rangle_g=\binom{2g+l-3}{b_1,...,b_l}\frac{2^{2g-1}-1}{2^{2g-1}}\frac{|B_{2g}|}{(2g)!} \tag{4.3}$$

directly from Mariño-Vafa formula (2.2). Formula (4.3) implies

$$c_g=\frac{2^{2g-1}-1}{2^{2g-1}}\frac{|B_{2g}|}{(2g)!}.$$

5 Conclusions

Mariño-Vafa formula is a closed formula including the Hodge integrals with triple λ-classes. It is natural to consider the question how to compute the Hodge integrals included in this formula. This question have been studied in several works [19, 16, 2, 1, 32]. Here we summarize several interesting results which have been derived from the Mariño-Vafa formula (2.2) and its cut-and-join form (3.3).

Besides the ELSV formula, Witten conjecture and λ_g-conjecture described previously in this article, the following two Hodge integral identities were extracted directly from the Mariño-Vafa formula in [19]

$$\int_{\overline{\mathcal{M}}_g} \lambda_{g-2}\lambda_{g-1}\lambda_g = \frac{1}{2(2g-2)!}\frac{|B_{2g-2}|}{2g-2}\frac{|B_{2g}|}{2g},$$

$$\int_{\overline{\mathcal{M}}_{g,1}} \frac{\lambda_{g-1}}{1-\psi_1} = c_g \sum_{i=1}^{2g-1}\frac{1}{i} - \frac{1}{2}\sum_{\substack{g_1+g_2=g\\ g_1,g_2>0}} \frac{(2g_1-1)!(2g_2-1)!}{(2g-1)!} c_{g_1}c_{g_2}.$$

In [16], Y. Li found the following Hodge integral identity from the Mariño-Vafa formula. For $g \geq 2$,

$$\int_{\overline{\mathcal{M}}_{g,1}} \psi_1^{2g-3}\lambda_g\lambda_1\psi_1^{2g-3} = \frac{1}{12}[g(2g-3)c_g + c_1c_{g-1}].$$

Furthermore, the Lapalace transform (3.3) of the cut-and-join equation for Mariño-Vafa formula provides us a general way to compute the Hodge integrals included in the Mariño-Vafa formula. As illustrated in previous sections, at first, we expand (3.3) in variable $\widetilde{\tau} = \tau + 1$. Then by considering certain level of $\widetilde{\tau}$, we obtain a polynomial identity in variable t_L. Taking the coefficients of a degree of t_L, we get a Hodge integral identity. This algorithm can be implemented on a computer. It is interesting to compare this algorithm with the one in [4] to compute similar triple Hodge integrals.

References

[1] L. Chen, *Symmetrized cut-join equation of the Mariño-Vafa formula*, Pacific Journal of Math. **235**, No.2, 201-212 (2008).

[2] L. Chen, Y. Li and K. Liu, *Localization, Hurwitz numbers and the Witten conjecture*, Asian J. Math. **12** No.4, 511-518 (2008).

[3] R. Dijkgraaf, *Intersection Theory, Integrable Hierarchies and Topological Field Theory,* New symmetry principles in quantum field theory (Cargse, 1991), 95-158, NATO Adv. Sci. Inst. Ser. B Phys., 295, Plenum, New York, 1992.

[4] Dubrovin B., Liu S.-Q., Yang D., Zhang Y.: *Hodge integrals and tau symmetric integrable hierarchies of Hamiltonian evolutionary PDEs.* Adv. Math. 293, 382-435 (2016).

[5] T. Ekedahl, S. Lando, M. Shapiro and A. Vainshtein, *Hurwitz numbers and intersections on moduli spaces of curves*, Invent. Math. **146** No. 2, 297-327 (2001).

[6] B. Eynard, M. Mulase and B. Safnuk, *The Laplace transform of the cut-and-join equation and the Bouchard-Mariño conjecture on Hurwitz numbers*, Publ. RIMS Kyoto Univ. 47 (2011), 629-670.

[7] C. Faber, *A conjectural description of the tautological ring of the moduli space of curves.* In Moduli of curves and abelian varieties, Aspects Math., E33, Vieweg, Braunschweig, Germany, 1999. 109-129.

[8] C. Faber and R. Pandharipande, *Hodge integrals, partition matrices, and the λ_g conjecture*, Ann. Math.(2) **157** No.1, 97-124 (2003).

[9] I.P. Goulden, D.M. Jackson and R. Vakil, *A short prooof of the λ_g-Conjecture without Gromov-Witten theory: Hurwitz theory and the moduli of curves*, math.AG/0604297.

[10] T. Graber and R. Pandharipande, *Localization of virtual classes*, Invent. Math. **135** (1999), no. 2, 487-518.

[11] R. Gopakumar and C. Vafa, *On the gauge theory/geometry correspondence*, Adv. Theor. Math. Phys.3(5) (1999) 1415-1443.

[12] M. Kontsevich, *Intersection theory on the moduli spaces of curves and the matrix. Airy function*, Comm. Math. Phys.,vol.147(1992),1-23.

[13] M. Kontsevich, *Enumeration of rational curves via torus actions*, in The moduli space of curves, R. Dijkgraaf, C. Faber, and G. van der Geer, eds., Birkhauser, 1995, pp 335-368.

[14] Y.-S. Kim, K. Liu, *A simple proof of Witten conjecture through localization*, preprint: math.AG/0508384.

[15] M.E. Kazarian, S. K. Lando, *An algebro-geometric proof of Witten's conjecture*, J. Amer. Math. Soc. 20 (2007), no. 4, 1079-1089.

[16] Y. Li, *Some results of the Mariño-Vafa Formula*, Math. Res. Lett. **13**, 10001-10018 (2006).

[17] B. Lian, K. Liu and S.-T. Yau, *Mirror principle I*, Asian J. Math. 4 (1997), 729-763.

[18] C.-C. Liu, K. Liu and J. Zhou, *A proof of a conjecture of Mariño-Vafa on Hodge integrals*, J. Differential Geom. **65** (2003).

[19] C.-C. Liu, K. Liu and J. Zhou, *Mariño-Vafa formula and Hodge integral identities*, J. Algebric Geom. **15**, 379-398 (2006).

[20] J.M.F. Labastida, M. Mariño and C. Vafa, *Knots, links and branes at large N*, J. High Energy Phys. 2000, no. **11**, Paper 7.

[21] M. Mirzakhani, *Weil-Petersson volumes and intersection theory on the moduli space of curves*, J. Amer. Math. Soc. 20 (2007), no. 1, 1-23.

[22] M. Mariño and C. Vafa, *Framed knots at large N*, Orbifolds in mathematics and physics (Madison, WI, 2001), 185-204, Contemp. Math., **310**, Amer. Math. Soc., Providence, RI, 2002.

[23] M. Mulase and N. Zhang, *Polynomial recursion formula for linear Hodge integrals*, Communications in Number Theory and Physics **4** No.2, 267-294.

[24] A. Okounkov, R. Pandharipande, *Gromov-Witten theory, Hurwitz numbers, and matrix models*, Algebraic geometry¡ªSeattle 2005. Part 1, 325-414, Proc.

Sympos. Pure Math., 80, Part 1, Amer. Math. Soc., Providence, RI, 2009.
[25] A. Okounkov, R. Pandharipande, *Hodge integrals and invariants of the unknot*, Geom. Topol. 8, 675-699 (2004).
[26] H. Ooguri, C. Vafa, *Knot invariants and topological strings*. Nucl. Phys. B 577(3), 419-438 (2000).
[27] E. Verlinde and H. Verlinde, *A solution of two-dimensional topological quantum gravity*, Nucl. Phys. B348 (1991) 457.
[28] E. Witten, *Quantum field theory and the Jones polynomial*, Commun. Math. Phys. **121** (1989) , 351.
[29] E. Witten, *Two-dimensional gravity and intersection theory on moduli space*, Surv. Differ. Geom. **1**, (1991) 243-310.
[30] E. Witten, *Chern-Simons Gauge Theory As A String Theory*, Prog. Math. **133**, 637 (1995).
[31] S. Zhu, *Hodge integral with one λ-class*, Sci. China Math. **55** No.1 (2012), 119-130.
[32] S. Zhu, *Hodge integral recursion from cut-and-join equation of Mariño-Vafa formula*, Pure Appl Math Q. 8 No.4 (2012), 1147-1177.

School of Mathematics, Capital Normal University
Beijing, 100048, China
and Department of Mathematics, University of California at Los Angeles
California 90095, USA
E-mail address: liu@math.ucla.edu

Center of Mathematical Sciences, Zhejiang University
Hangzhou, 310027, China
E-mail address: szhu@zju.edu.cn

Proceedings of the 7th International
Congress of Chinese Mathematicians
ALM 43, pp. 407–438

A Survey on Optimal L^2 Extension Theorems and Vanishing Theorems

Xiankui Meng, Xiangyu Zhou and Langfeng Zhu

Abstract

In this paper, we give a survey on optimal L^2 extension theorems and vanishing theorems, and we also explain some key ideas in the proofs.

Mathematics Subject Classification: 32D15, 32L20, 32Q15, 32L10, 14F18.

Keywords and Phrases: Optimal L^2 extension, vanishing theorem, plurisubharmonic function, multiplier ideal sheaf.

1 Introduction

L^2 extension theorems have become a powerful tool in several complex variables and complex geometry. The original L^2 extension theorem was established by Ohsawa and Takegoshi in [63] as follows.

Theorem 1.1 ([63]). *Let X be a bounded pseudoconvex domain contained in $\{z = (z^1, \cdots, z^n) \in \mathbb{C}^n : |z^n| < 1\}$, φ be a plurisubharmonic function on X and $H = \{z \in \mathbb{C}^n : z^n = 0\}$. Then there exists a uniform constant C independent of X, φ and H such that, for any holomorphic function f on $X \cap H$ satisfying*

$$\int_{X\cap H} |f|^2 e^{-\varphi}(\sqrt{-1})^{n-1} dz^1 \wedge d\bar{z}^1 \wedge \cdots \wedge dz^{n-1} \wedge d\bar{z}^{n-1} < \infty,$$

there exists a holomorphic function F on X satisfying $F = f$ on $X \cap H$ and

$$\begin{aligned}&\int_X |F|^2 e^{-\varphi}(\sqrt{-1})^n dz^1 \wedge d\bar{z}^1 \wedge \cdots \wedge dz^n \wedge d\bar{z}^n \\ &\leq C \int_{X\cap H} |f|^2 e^{-\varphi}(\sqrt{-1})^{n-1} dz^1 \wedge d\bar{z}^1 \wedge \cdots \wedge dz^{n-1} \wedge d\bar{z}^{n-1}.\end{aligned}$$

In [60], Ohsawa established the following generalized L^2 extension theorem with negligible weights, which can imply Theorem 1.1 by taking $\psi = 0$.

Meng was partially supported by NSFC (grant No. 11771231). Zhou was partially supported by NSFC (grant No. 11688101 and No. 11431013). Zhu was partially supported by NSFC (No. 11201347 and No. 11671306). This work was partially supported by the China Scholarship Council.

Theorem 1.2 ([60]). *Let X be a pseudoconvex domain in $\mathbb{C}^n$, φ be a plurisubharmonic function on X and $H := \{z \in \mathbb{C}^n : z^n = 0\}$. Then, for any plurisubharmonic function ψ on X such that $\sup_{z\in X}(\psi(z)+2\log|z^n|) \leq 0$, there exists a uniform constant $\mathbf{C}$ independent of X, φ, ψ and H such that, for any holomorphic function f on $X \cap H$ satisfying*

$$\int_{X\cap H} |f|^2 e^{-\varphi-\psi}(\sqrt{-1})^{n-1} dz^1 \wedge d\bar{z}^1 \wedge \cdots \wedge dz^{n-1} \wedge d\bar{z}^{n-1} < +\infty,$$

there exists a holomorphic function F on X satisfying $F = f$ on $X \cap H$ and

$$\begin{aligned}&\int_X |F|^2 e^{-\varphi}(\sqrt{-1})^n dz^1 \wedge d\bar{z}^1 \wedge \cdots \wedge dz^n \wedge d\bar{z}^n \\ &\leq 2\mathbf{C}\pi \int_{X\cap H} |f|^2 e^{-\varphi-\psi}(\sqrt{-1})^{n-1} dz^1 \wedge d\bar{z}^1 \wedge \cdots \wedge dz^{n-1} \wedge d\bar{z}^{n-1}.\end{aligned}$$

It is quite natural and useful to optimize the uniform constants in the final L^2 estimates of Theorem 1.2 and its various generalizations. For example, Guan and Zhou established an optimal L^2 extension theorem in a general setting and found some unexpected applications of their theorem in [35].

The constant $\mathbf{C}$ was calculated by many people. However, new method was needed to establish the optimal constant estimate. In [37] and [82], the authors initiated a method of undetermined function with ODE and reduced the constant $\mathbf{C}$ to 1.954 in the case of Stein manifolds with an arbitrary negligible weight (i.e., the setting of [60]). Later, in [6] Błocki gave a similar proof of the L^2 extension theorem by Guan-Zhou-Zhu for the special case of pseudoconvex domains in $\mathbb{C}^n$ with a special negligible weight and got the same constant 1.954 by using the same ODE in [82]. After that, in [7] Błocki got the optimal constant $\mathbf{C} = 1$ in the case of pseudoconvex domains in $\mathbb{C}^n$ with the same negligible weight by improving the ODE in [82]. In [31], Guan and Zhou got the optimal constant $\mathbf{C} = 1$ in the case of Stein manifolds with an arbitrary negligible weight by solving a new system of ODEs.

For the higher codimensional case in more general frameworks than [60], Guan and Zhou ([32], [34], [33] and [35]) proved several L^2 extension theorems with optimal constants in the framework of the generalized L^2 extension theorem by Ohsawa in [61] and more general framework than [61] respectively. The manifolds in their theorems are assumed to be quasi-Stein manifolds (containing Stein manifolds as Zariski open subsets).

In [81], Zhou and Zhu proved an optimal L^2 extension theorem of holomorphic sections of holomorphic line bundles equipped with singular metrics on weakly pseudoconvex Kähler manifolds (the setting discussed by Manivel and Demailly in [48], [16]). Moreover, they obtained the optimal constant estimate corresponding to variable factors in their L^2 estimate. In this setting, some difficulties arose because the manifolds considered may not necessarily contain a Stein manifolds as Zariski open subsets.

An important problem related to extension theorems is the vanishing of the cohomology groups valued in holomorphic vector bundles or coherent analytic sheaves. Cohomology vanishing theorems valued in homomorphic vector bundles and coherent analytic sheaves play an important role in several complex variables and complex algebraic geometry. It began with Kodaira vanishing theorem in 1953 and lots of generalizations appeared after then. Vanishing theorems are based on various concepts of positivity, which appear in algebraic geometry and several complex variables and are closely related. In this survey, we will discuss briefly some of the vanishing theorems in the literature.

The rest sections of this paper are organized as follows. At first, we discuss optimal L^2 extension theorems in Section 2, Section 3, Section 4 and Section 5. Then, we discuss vanishing theorems in Section 6, Section 7 and Section 8.

2 Initiation of the method related to solving ODEs

In [37] and [82], the authors obtained the following result by introducing the undetermined function and initiating the ODE method in the study of the optimal constant problems in L^2 extension theorems.

Theorem 2.1 ([37], [82]). *Let X be a Stein manifold of dimension n. Let φ and ψ be two plurisubharmonic functions on X. Assume that w is a holomorphic function on X such that $\sup_X(\psi + 2\log|w|) \leq 0$ and dw does not vanish identically on any branch of $w^{-1}(0)$. Put $Y = w^{-1}(0)$ and $Y_0 = \{x \in Y : dw(x) \neq 0\}$. Then there exists a uniform constant $\mathbf{C} < 1.954$ independent of X, φ, ψ and w such that, for any holomorphic $(n-1)$-form f on Y_0 satisfying*

$$c_{n-1}\int_{Y_0} e^{-\varphi-\psi} f \wedge \bar{f} < \infty,$$

where $c_k = i^{k^2}$ for $k \in \mathbb{Z}$, there exists a holomorphic n-form F on X satisfying $F = dw \wedge \tilde{f}$ on Y_0 with $\theta^* \tilde{f} = f$ and

$$c_n \int_X e^{-\varphi} F \wedge \bar{F} \leq 2\mathbf{C}\pi c_{n-1} \int_{Y_0} e^{-\varphi-\psi} f \wedge \bar{f},$$

where $\theta : Y_0 \longrightarrow X$ is the inclusion map.

We will sketch the proof of Theorem 2.1. The key idea in the proof is to introduce an undetermined function χ and to transform the optimal constant problem to the problem of solving an appropriate ODE of χ.

The following two lemmas are used in the proof. One is an inequality deduced from the twisted Bochner-Kodaira identity, and the other is a proposition about solving the $\bar{\partial}$ equations with estimates.

Lemma 2.2 ([63], [4], [51], [71]). *Let (X,ω) be a Kähler manifold of dimension n with a Kähler metric ω, $\Omega \subset\subset X$ be a pseudoconvex domain with C^∞ boundary $b\Omega$ and $\phi \in C^\infty(\overline{\Omega})$ be plurisubharmonic in Ω. Assume that $\tau, A \in C^\infty(\overline{\Omega})$ are positive functions. Then for any $(n,q+1)$-form*

$$\alpha = \sum_{|I|=n,|J|=q+1}{}' \alpha_{I\bar{J}} dz^I \wedge d\bar{z}^J \in Dom_\Omega(\bar{\partial}^*) \cap C^\infty_{(n,q+1)}(\overline{\Omega}),$$

we have

$$\begin{aligned}&\int_\Omega (\tau + A)|\bar{\partial}^*_\phi \alpha|^2 e^{-\phi} dV_X + \int_\Omega \tau |\bar{\partial}\alpha|^2 e^{-\phi} dV_X \\ &\geq \sum_{|I|=n,|K|=q}{}' \sum_{i,j=1}^n \int_\Omega \Big(-\partial_i \bar{\partial}_j \tau - \frac{\partial_i \tau \bar{\partial}_j \tau}{A} \Big) \alpha_{I}{}^{i}{}_{\bar{K}} \overline{\alpha_{\bar{I}j K}} e^{-\phi} dV_X. \qquad (2.1)\end{aligned}$$

Lemma 2.3 ([4]). *Let (X,ω) be a Kähler manifold of dimension n with a Kähler metric ω, $\Omega \subset\subset X$ be a strictly pseudoconvex domain in X with C^∞ boundary $b\Omega$ and $\phi \in C^\infty(\overline{\Omega})$. Let w be a holomorphic function on X which doesn't vanish identically and λ be the current $\bar{\partial}\frac{1}{w} \wedge \tilde{F}$, where $\tilde{F}$ is a holomorphic n-form on X. Assume the inequality*

$$|(\lambda,\alpha)_{\Omega,\phi}|^2 \leq C \int_\Omega |\bar{\partial}^*_\phi \alpha|^2 \frac{e^{-\phi}}{\mu} dV_X,$$

where $\frac{1}{\mu}$ is an integrable positive function on Ω and C is a constant, holds for all $(n,1)$-form $\alpha \in Dom_\Omega(\bar{\partial}^*) \cap Ker(\bar{\partial}) \cap C^\infty_{(n,1)}(\overline{\Omega})$. Then there is a solution u to the equation $\bar{\partial} u = \lambda$ such that

$$\int_\Omega |u|^2 \mu e^{-\phi} dV_X \leq C.$$

Without loss of generality, we can assume that $Y = Y_0$ and that both φ and ψ are smooth on X. Since X is Stein, there is a holomorphic n-form $\tilde{F}$ on X such that $\tilde{F} = dw \wedge \tilde{f}$ on Y with $\theta^* \tilde{f} = f$. Since X is Stein, we can find a sequence of strictly pseudoconvex domains $\{D_v\}_{v=1}^\infty$ with smooth boundaries satisfying $D_v \subset\subset D_{v+1}$ for all v and $\bigcup\limits_{v=1}^\infty D_v = X$.

Let $\mathcal{M}$ be the function class defined by

$$\Big\{ \chi \in C^\infty(-\infty,0] : \chi(t) \geq 0,\ \chi'(t) \leq -1 \text{ and } \chi''(t) < 0 \text{ for all } t \in (-\infty,0] \Big\}.$$

Let $a, \gamma \in (0,1)$ be two positive numbers and put $\sigma_\varepsilon = \gamma\big(\psi + \log(|w|^2 + \varepsilon^2) - a\big)$.

Let $\phi = \psi + \varphi$ and $\mu = e^{\gamma(\psi + \log|w|^2)}$. For any $\chi \in \mathcal{M}$, define $\tau_\varepsilon = \chi(\sigma_\varepsilon)$ and $A_\varepsilon = -\frac{(\chi'(\sigma_\varepsilon))^2}{\chi''(\sigma_\varepsilon)}$.

Since X is Stein, there is a holomorphic n-form $\tilde{F}$ on X such that $\tilde{F} = dw \wedge \tilde{f}$ on Y with $\theta^* \tilde{f} = f$. Define $\lambda = \bar{\partial}\frac{1}{w} \wedge \tilde{F}$. Then by the lemmas above and some careful calculation, we can solve the equations $\bar{\partial} u_{\gamma,v} = \lambda$ and obtain uniform L^2 estimates. Define $F_{\gamma,v} = w u_{\gamma,v}$. The weak limit of some weakly convergent subsequence of $\{F_{\gamma,v}\}_{\gamma,v}$ gives us the desired extension F with the constant $\mathbf{C}$ defined by

$$\inf_{\chi \in \mathcal{M}} \sup_{t \leq 0} \left\{ \left(\chi(t) - \frac{(\chi'(t))^2}{\chi''(t)} \right) e^t \right\}.$$

Then the constant $\mathbf{C}$ can be estimated by solving the ODE

$$\left(\chi(t) - \frac{(\chi'(t))^2}{\chi''(t)} \right) e^t = \mathbf{C} \tag{2.2}$$

with $\chi'(0)$ and $\mathbf{C}$ adjusted so that $\lim_{t \to -\infty} \chi'(t) = -1$. A standard Runge-Kutta integration with a computer then shows that one can obtain $\mathbf{C} < 1.954$ when $\chi(0) = 0$ and $\chi'(0) = -4.331$.

3 Improvement of the method related to solving ODEs

In [31], Guan and Zhou improved the method in [82] by introducing an additional weight to obtain a system of ODEs, and then they obtained the optimal constant $\mathbf{C} = 1$ in Theorem 2.1 by solving the system of ODEs. We will sketch their improvement to the method mentioned in the last section.

In their proof, (2.1) is replaced by

$$\begin{aligned} &\int_\Omega (\tau + A) |\bar{\partial}^*_\phi \alpha|^2 e^{-\phi} dV_X + \int_\Omega \tau |\bar{\partial} \alpha|^2 e^{-\phi} dV_X \\ \geq &\sum_{|I|=n, |K|=q}' \sum_{i,j=1}^n \int_\Omega \left(\tau \partial_i \bar{\partial}_j \phi - \partial_i \bar{\partial}_j \tau - \frac{\partial_i \tau \bar{\partial}_j \tau}{A} \right) \alpha_{I\ \bar{K}}^{\ i} \overline{\alpha^{\bar{I} j K}} e^{-\phi} dV_X. \end{aligned} \tag{3.1}$$

The key improvement in their proof is to introduce another undetermined function $\zeta : (-\infty, 0) \to \mathbb{R}$. Then they added the additional weight $\zeta(\sigma)$ to the weight function $\phi := \psi + \varphi$, where $\sigma := \psi + \gamma_v \log |w|^2$, $\gamma_v \in (0,1)$ and $\lim_{v \to +\infty} \gamma_v = 1$. Then they replaced $\tau_\varepsilon = \chi(\sigma_\varepsilon)$ and $A_\varepsilon = -\frac{(\chi'(\sigma_\varepsilon))^2}{\chi''(\sigma_\varepsilon)}$ in Section 2 with $\tau = \chi(\sigma)$ and $A = \frac{(\chi'(\sigma))^2}{\chi(\sigma)\zeta''(\sigma) - \chi''(\sigma)}$ respectively ($\chi \in \mathcal{M}$ is not assumed here).

By using modified Lemma 2.2, Lemma 2.3 and some similar arguments, they obtained the desired extension F with the constant $\mathbf{C}$ determined by the existence of a solution to the following system of ODEs defined on $(-\infty, 0)$:

$$\left(\chi(t) + \frac{(\chi'(t))^2}{\chi(t)\zeta''(t) - \chi''(t)} \right) e^{\zeta(t) + t} = \mathbf{C}, \tag{3.2}$$

$$\chi(t)\zeta'(t) - \chi'(t) = 1, \tag{3.3}$$

where $\chi > 0$, $\chi' < 0$, $\zeta > 0$ and $\zeta' > 0$. (3.2) is just (2.2) when no additional weight is introduced (take $\zeta = 0$).

Then they obtained the optimal constant $\mathbf{C} = 1$ by solving it (see [31]). The solution is

$$\chi = \frac{-t - 1 + e^t}{1 - e^t},$$
$$\zeta = -\log(1 - e^t).$$

4 Optimal L^2 extensions in more general frameworks

At first, let us recall some notation and symbols in [61], [33] and [35].

Definition 4.1. Let X be a complex manifold with a continuous volume form dV_X, and Y be a closed complex subvariety of X. We say (X, Y) satisfies condition (ab) if X and Y satisfy the following conditions:

- (a) Σ is locally negligible with respect to L^2 holomorphic functions, i.e., for any local coordinate neighborhood $U \subset X$ and for any L^2 holomorphic function f on $U \backslash \Sigma$, there exists an L^2 holomorphic function $\tilde{f}$ on U such that $\tilde{f}|_{U\backslash\Sigma} = f$ with the same L^2 norm;
- (b) $X \setminus \Sigma$ is a Stein manifold which intersects with every component of Y, such that $Y_{\text{sing}} \subset \Sigma$.

Let X be a complex n-dimensional manifold, and Y be a closed complex subvariety of X. Let dV_X be a continuous volume form on X. We consider a class of the upper-semicontinuous function Ψ from X to the interval $[-\infty, a)$, where $a \in (-\infty, +\infty]$, such that

1. $\Psi^{-1}(-\infty) \supset Y$, and $\Psi^{-1}(-\infty)$ is a closed subset of X;
2. if Y is l-dimensional around a point $x \in Y_{\text{reg}}$ (Y_{reg} is the regular part of Y), there exists a local coordinate $(z_1, \cdots, z_n)$ on a neighborhood U of x such that $z_{l+1} = \cdots = z_n = 0$ on $Y \cap U$ and

$$\sup_{U\backslash Y} \left|\Psi(z) - (n-l)\log \sum_{j=l+1}^{n} |z_j|^2\right| < +\infty.$$

The set of such polar function Ψ will be denoted by $\#_a(Y)$.

For each $\Psi \in \#_a(Y)$, one can associate a positive measure $dV_X[\Psi]$ on Y_{reg} as the minimum element of the partial ordered set of positive measures $d\mu$ satisfying

$$\int_{Y_l} f\, d\mu \geq \limsup_{t\to+\infty} \frac{2(n-l)}{\sigma_{2n-2l-1}} \int_M f\, e^{-\Psi} \mathbb{I}_{\{-t-1<\Psi<-t\}} dV_X$$

for any nonnegative continuous function f with $\operatorname{supp} f \subset\subset X$, where Y_l denotes the l-dimensional component of Y_{reg}, σ_m denotes the volume of the unit sphere in $\mathbb{R}^{m+1}$ and $\mathbb{I}_P$ is the characteristic function associated to a set P.

Let h be a Hermitian metric on a holomorphic vector bundle Q. Let $\Delta_{a,h,\delta}(Y)$ be the subset of functions Ψ in $\#_a(Y)$ which satisfies that both $he^{-\Psi}$ and $he^{-(1+\delta)\Psi}$

are semipositive in the sense of Nakano on $X\setminus(\Sigma\cup Y)$. Let $\Delta_a(Y)$ be the subset of plurisubharmonic functions Ψ in $\#_a(Y)$. The symbols $\#(Y)$, $\Delta_{h,\delta}(Y)$ and $\Delta(Y)$ in Theorem 4.1 below are $\#_0(Y)$, $\Delta_{0,h,\delta}(Y)$ and $\Delta_0(Y)$ respectively.

In [61], Ohsawa obtained the following generalization of Theorem 1.2, which is the higher codimensional case in a more general framework.

Theorem 4.1 ([61]). *Let (X,Y) satisfy condition (ab), h be a smooth metric on a holomorphic vector bundle Q on X with rank r. Then, for any function Ψ on X such that $\Psi\in\Delta_{h,\delta}(Y)\cap C^\infty(X\setminus Y)$, there exists a uniform constant $\mathbf{C}=\max\limits_{1\le k\le n}\frac{2^8\pi}{\frac{\pi^k}{k!}}$ such that, for any holomorphic section f of $K_X\otimes Q\big|_Y$ on Y satisfying*

$$\sum_{k=1}^{n}\frac{\pi^k}{k!}\int_{Y_{n-k}}|f|_h^2dV_X[\Psi]<+\infty,$$

there exists a holomorphic section F of $K_X\otimes Q$ on X satisfying $F=f$ on Y and

$$\int_X|F|_h^2dV_X\le\mathbf{C}(1+\delta^{-\frac{3}{2}})\sum_{k=1}^{n}\frac{\pi^k}{k!}\int_{Y_{n-k}}|f|_h^2dV_X[\Psi].$$

Especially, if $\Psi\in\Delta(Y)\cap\Delta_{h,\delta}(Y)\cap C^\infty(X\setminus Y)$, then there exists a holomorphic section F of $K_X\otimes Q$ on X satisfying $F=f$ on Y and

$$\int_X|F|_h^2dV_X\le\mathbf{C}\sum_{k=1}^{n}\frac{\pi^k}{k!}\int_{Y_{n-k}}|f|_h^2dV_X[\Psi].$$

In [33], Guan and Zhou optimized Theorem 4.1 by proving the following theorem.

Theorem 4.2 (Theorem 1.21 and Corollary 1.22 in [33]). *Let (X,Y) satisfy condition (ab), h be a smooth metric on a holomorphic vector bundle Q on X with rank r. Then, for function Ψ on X such that $\Psi\in\Delta_{h,\delta}(Y)\cap C^\infty(X\setminus Y)$, there exists a uniform constant $\mathbf{C}=1$, such that, for any holomorphic section f of $K_X\otimes Q|_Y$ on Y satisfying*

$$\sum_{k=1}^{n}\frac{\pi^k}{k!}\int_{Y_{n-k}}|f|_h^2dV_X[\Psi]<+\infty,$$

there exists a holomorphic section F of $K_X\otimes Q$ on X satisfying $F=f$ on Y and

$$\int_X|F|_h^2dV_X\le\mathbf{C}(1+\delta^{-1})\sum_{k=1}^{n}\frac{\pi^k}{k!}\int_{Y_{n-k}}|f|_h^2dV_X[\Psi].$$

Especially, if we assume that $\Psi\in\Delta(Y)$ further, the last inequality above can be replaced by

$$\int_X|F|_h^2dV_X\le\mathbf{C}\sum_{k=1}^{n}\frac{\pi^k}{k!}\int_{Y_{n-k}}|f|_h^2dV_X[\Psi].$$

Remark 4.1. Both the constants $\mathbf{C} = 1$ and $1 + \delta^{-1}$ in Theorem 4.2 are optimal (see Remark 1.8 in [33]).

Remark 4.2. Theorem 4.2 holds for a singular metric h when Q is a holomorphic line bundle (see Corollary 1.9 and Corollary 1.13 in [33]).

In [52], McNeal and Varolin generalized Theorem 1.2 such that some variable factors are allowed to appear in the final L^2 estimate.

At first, McNeal and Varolin defined a function class $\mathfrak{D}$.

Definition 4.2. The class $\mathfrak{D}$ consists of nonnegative functions with the following three properties.

(1) Each $R \in \mathfrak{D}$ is continuous and increasing.

(2) For each $R \in \mathfrak{D}$ the improper integral

$$C_R := \int_1^{\infty} \frac{dt}{R(t)} < +\infty.$$

(3) For each $R \in \mathfrak{D}$ there exists a constant $\delta > 0$ such that

$$K_\delta(R) := \sup_{x \geq 1} \frac{x + R_\delta(x)}{R(x)} < +\infty,$$

where $\delta \in (0, +\infty)$,

$$R_\delta(x) := \int_1^x \frac{1 - H_\delta(y)}{H_\delta(y)} dy$$

and

$$H_\delta(y) := \frac{1}{1+\delta}\left(1 + \frac{\delta}{C_R} \int_1^y \frac{dt}{R(t)}\right).$$

The L^2 extension theorem proved by McNeal and Varolin is stated below.

Theorem 4.3 ([52]). *Let X be a Kähler manifold of complex dimension n. Assume there exists a holomorphic function w on X, such that $\sup_X |w| = 1$ and dw is never zero on the set $Y = \{w = 0\}$. Assume there exists an analytic subvariety $V \subset X$ such that Y is not contained in V and $X \backslash V$ is Stein. Let L be a holomorphic line bundle over X together with a singular Hermitian metric. Let $\psi : X \longrightarrow [-\infty, +\infty]$ be a locally integrable function such that for any local representative $e^{-\varphi}$ of the metric of L over an open set U, the function $\psi + \varphi$ is not identically $+\infty$ or $-\infty$ on $Y \cap U$. Let R be a function in $\mathfrak{D}$. Assume that ψ satisfies that for all $\gamma > 1$ and $\varepsilon > 0$ sufficiently small (depending on $\gamma - 1$),*

$$\begin{aligned}
&\sqrt{-1}\partial\bar{\partial}(\varphi + \psi + \log |w|^2) \geq 0, \\
&R^{-1}\big(e^{-\psi} R(1 - \log |w|^2)\big) \geq 1 \textit{ and} \\
&\alpha - R^{-1}\big(e^{-\psi} R(\alpha)\big) \textit{ is plurisubharmonic},
\end{aligned}$$

where $\alpha = \gamma - \log(|w|^2 + \varepsilon^2)$. Then for any holomorphic $(n-1)$-form $f \in C^\infty(Y, \wedge^{n-1} T_Y^* \otimes L)$ on Y with values in L such that

$$\int_Y |f|^2 e^{-\varphi - \psi} \, dV_Y < +\infty,$$

there is a holomorphic n-form $F \in C^\infty(X, \wedge^n T_X^* \otimes L)$ with values in L such that $F|_Y = f \wedge dw$ and

$$\int_X \frac{|F|^2 e^{-\varphi}}{|w|^2 R\big(\log \frac{e}{|w|^2}\big)} dV_X \le 8\pi\big(K_\delta(R) + \frac{1+\delta}{\delta} C_R\big) \int_Y |f|^2 e^{-\varphi-\psi}\, dV_Y.$$

In [35], Guan and Zhou proved several optimal L^2 extension theorems, which are more general than Theorem 4.2 and Theorem 4.3. Moreover, they obtained optimal constants with respect to variable factors involved in the final L^2 estimates.

Given $\delta > 0$, let $c_a(t)$ be a positive function on $(-a, +\infty)$ $(a \in (-\infty, +\infty))$, which is in $C^\infty((-a, +\infty))$ and satisfies both

$$\int_{-a}^{+\infty} c_a(t) e^{-t} dt < +\infty$$

and

$$\Big(\frac{1}{\delta} c_a(-a) e^a + \int_{-a}^{t} c_a(t_1) e^{-t_1} dt_1\Big)^2$$
$$> c_a(t) e^{-t}\Big(\int_{-a}^{t}\Big(\frac{1}{\delta} c_a(-a) e^a + \int_{-a}^{t_2} c_a(t_1) e^{-t_1} dt_1\Big) dt_2 + \frac{1}{\delta^2} c_a(-a) e^a\Big)$$

for any $t \in (-a, +\infty)$. If $c_a(t)e^{-t}$ is decreasing with respect to t, then the last inequality above holds.

In [35], Guan and Zhou established the following optimal L^2 extension theorem with respect to the function $c_a(t)$.

Theorem 4.4 (Main Theorem 1 in [35]). *Let (X, Y) satisfy conditions (ab), and let h be a smooth metric on a holomorphic vector bundle Q on X with rank r. Let $\Psi \in \#_a(Y) \cap C^\infty(X \setminus Y)$, which satisfies*

(1) *$he^{-\Psi}$ is semipositive in the sense of Nakano on X on $X \setminus (Y \cup \Sigma)$ (Σ is as in the definition of condition (ab));*
(2) *there exists a continuous function $a(t)$ on $(-a, +\infty]$, such that $0 < a(t) \le \tilde{s}(t)$ and*

$$a(-\Psi)\sqrt{-1}\Theta_{he^{-\Psi}} + \sqrt{-1}\partial\bar\partial\Psi$$

is semipositive in the sense of Nakano on $X \setminus (Y \cup \Sigma)$, where

$$\tilde{s}(t) = \frac{\int_{-a}^{t}\big(\frac{1}{\delta} c_a(-a)e^a + \int_{-a}^{t_2} c_a(t_1)e^{-t_1} dt_1\big) dt_2 + \frac{1}{\delta^2} c_a(-a) e^a}{\frac{1}{\delta} c_a(-a) e^a + \int_{-a}^{t} c_a(t_1) e^{-t_1} dt_1}.$$

Then there exists a uniform constant $\mathbf{C} = 1$, which is optimal, such that, for any holomorphic section f of $K_X \otimes Q|_Y$ on Y satisfying

$$\sum_{k=1}^{n} \frac{\pi^k}{k!} \int_{Y_{n-k}} |f|_h^2 dV_X[\Psi] < +\infty, \tag{4.1}$$

there exists a holomorphic section F of $K_X \otimes Q$ on X satisfying $F = f$ on Y and

$$\int_X c_a(-\Psi)|F|_h^2 dV_X$$
$$\leq \mathbf{C}\left(\frac{1}{\delta}c_a(-a)e^a + \int_{-a}^{+\infty} c_a(t)e^{-t}dt\right)\sum_{k=1}^{n}\frac{\pi^k}{k!}\int_{Y_{n-k}} |f|_h^2 dV_X[\Psi], \qquad (4.2)$$

where $c_a(t)$ satisfies $c_a(-a)e^a := \lim_{t\to -a^+} c_a(t)e^{-t} < +\infty$ and $c_a(-a)e^a \neq 0$.

In [35], Guan and Zhou also proved a singular metric version of Theorem 4.4, which is stated below.

Theorem 4.5 (Theorem 5.2 in [35]). *Let (X, Y) satisfy condition (ab), and let h be a singular metric on a holomorphic line bundle L on X, which is locally integrable on X. Then, for any function Ψ on X such that $\Psi \in \Delta_{a,h,\delta}(Y)$, there exists a uniform constant $\mathbf{C} = 1$, which is optimal, such that, for any holomorphic section f of $K_X \otimes L\big|_Y$ on Y satisfying the L^2 integrable condition* (4.1), *there exists a holomorphic section F of $K_X \otimes L$ on X satisfying $F = f$ on Y and the optimal estimate* (4.2).

Now we consider a useful and simpler class of functions. Let $c_a(t)$ be a positive function in $C^\infty((-a, +\infty))$ ($a \in (-\infty, +\infty]$), satisfying

$$\int_{-a}^{+\infty} c_a(t)e^{-t}dt < +\infty$$

and

$$\left(\int_{-a}^{t} c_a(t_1)e^{-t_1}dt_1\right)^2 > c_a(t)e^{-t}\int_{-a}^{t}\int_{-a}^{t_2} c_a(t_1)e^{-t_1}dt_1dt_2$$

for any $t \in (-a, +\infty)$. When $c_a(t)e^{-t}$ is decreasing with respect to t and a is finite, the last inequality above holds.

For such a simpler and sufficiently useful class of functions, Guan and Zhou established the following L^2 extension theorem with an optimal estimate.

Theorem 4.6 (Main Theorem 2 in [35]). *Let (X, Y) satisfy condition (ab), and let Ψ be a plurisubharmonic function in $\Delta_a(Y) \cap C^\infty(X \setminus (Y \cup \Sigma))$ (Σ is as in the definition of condition (ab)). Let h be a smooth metric on a holomorphic vector bundle Q on X with rank r, such that $he^{-\Psi}$ is semipositive in the sense of Nakano on $X\setminus(Y \cup \Sigma)$ (when Q is a line bundle, h can be chosen as a semipositive singular metric). Then there exists a uniform constant $\mathbf{C} = 1$, which is optimal, such that, for any holomorphic section f of $K_X \otimes Q\big|_Y$ on Y satisfying*

$$\sum_{k=1}^{n}\frac{\pi^k}{k!}\int_{Y_{n-k}} |f|_h^2 dV_X[\Psi] < +\infty,$$

there exists a holomorphic section F of $K_X \otimes Q$ on X satisfying $F = f$ on Y and

$$\int_X c_a(-\Psi)|F|_h^2 dV_X \leq \mathbf{C}\int_{-a}^{+\infty} c_a(t)e^{-t}dt\sum_{k=1}^{n}\frac{\pi^k}{k!}\int_{Y_{n-k}} |f|_h^2 dV_X[\Psi].$$

Remark 4.3. Theorem 4.6 optimizes the uniform constant in the final L^2 estimate in Theorem 4.3 (see Corollary 3.20 in [35]).

Now we sketch the key idea in the proof of Theorem 4.4 (Theorem 4.5 and Theorem 4.6 were proved similarly).

In the proof in [35], Lemma 2.2 and Lemma 2.3 are replaced by the following two lemmas respectively.

Lemma 4.7 ([60], [63]). *Let Q be a Hermitian vector bundle on a Kähler manifold X of dimension n with a Kähler metric ω. Assume that $\tau, A > 0$ are smooth functions on X. Then for every form $v \in \mathcal{D}(X, \wedge^{n,q}T_X^* \otimes Q)$, which is the space of smooth differential forms with values in Q with compact support, we have*

$$\begin{aligned}&\int_X (\tau + A)|\mathrm{D}''^* v|_Q^2 dV_X + \int_X \tau |\mathrm{D}'' v|_Q^2 dV_X \\ \geq &\int_X \left\langle \left[\tau\sqrt{-1}\Theta_Q - \sqrt{-1}\partial\bar{\partial}\tau - \sqrt{-1}\frac{\partial\tau \wedge \bar{\partial}\tau}{A}, \Lambda\right] v, v\right\rangle_Q dV_X.\end{aligned}$$

Lemma 4.8 ([16], [18]). *Let (X, ω) be a complete Kähler manifold equipped with a (non necessarily complete) Kähler metric ω, and let Q be a Hermitian vector bundle over X. Assume that τ and A are smooth and bounded positive functions on X and let*

$$\mathrm{B} := [\tau\sqrt{-1}\Theta_Q - \sqrt{-1}\partial\bar{\partial}\tau - \sqrt{-1}A^{-1}\partial\tau \wedge \bar{\partial}\tau, \Lambda].$$

Assume that $\delta \geq 0$ is a number such that $\mathrm{B} + \delta\mathrm{I}$ is semipositive definite everywhere on $\wedge^{n,q}T_X^* \otimes Q$ for some $q \geq 1$. Then given a form $g \in L^2(X, \wedge^{n,q}T_X^* \otimes Q)$ such that $\mathrm{D}''g = 0$ and

$$\int_X \langle (\mathrm{B} + \delta\mathrm{I})^{-1} g, g\rangle_Q dV_X < +\infty,$$

there exists an approximate solution $u \in L^2(X, \wedge^{n,q-1}T_X^* \otimes Q)$ and a correcting term $h \in L^2(X, \wedge^{n,q}T_X^* \otimes Q)$ such that $\mathrm{D}''u + \sqrt{\delta}h = g$ and

$$\int_X \frac{|u|_Q^2}{\tau + A} dV_X + \int_X |h|_Q^2 dV_X \leq \int_X \langle (\mathrm{B} + \delta\mathrm{I})^{-1} g, g\rangle_Q dV_X.$$

In their proof of Theorem 4.4, Guan and Zhou used convolution with mollifiers to obtain a family of smooth functions approximating $\mathbb{I}_{\{-t-1<\Psi<-t\}}$. Then, instead of solving $\bar{\partial}$ equations of currents as in Section 2 and Section 3, they used Lemma 4.8 to solve $\bar{\partial}$ equations of smooth forms.

As in Section 3, Guan and Zhou introduced two undetermined function $s : (-a, +\infty) \to \mathbb{R}$ and $u : (-a, +\infty) \to \mathbb{R}$, where s was used to construct twisted factors, and u was used to construct additional weights. By careful calculations and some arguments, they transformed the optimal constant problem to the existence of a solution to the following system of ODEs defined on $(-a, +\infty)$:

$$\left(s(t) + \frac{(s'(t))^2}{s(t)u''(t) - s''(t)}\right) e^{u(t)-t} = \frac{\mathbf{C}}{c_a(t)},$$

$$s'(t) - s(t)u'(t) = 1,$$

where $s \geq \frac{1}{\delta}$, $su'' - s'' > 0$ and

$$\lim_{t\to+\infty} u(t) = -\log\left(\frac{1}{\delta}c_a(-a)e^a + \int_{-a}^{+\infty} c_a(t)e^{-t}dt\right).$$

The above ODEs are just (3.2) and (3.3) when $c_a(t) = 1$.

Then they obtained the optimal constant $\mathbf{C} = 1$ by solving it (see [35]). The solution is

$$s = \frac{\int_{-a}^{t}\left(\frac{1}{\delta}c_a(-a)e^a + \int_{-a}^{t_2} c_a(t_1)e^{-t_1}dt_1\right)dt_2 + \frac{1}{\delta^2}c_a(-a)e^a}{\frac{1}{\delta}c_a(-a)e^a + \int_{-a}^{t} c_a(t_1)e^{-t_1}dt_1},$$

$$u = -\log\left(\frac{1}{\delta}c_a(-a)e^a + \int_{-a}^{t} c_a(t_1)e^{-t_1}dt_1\right).$$

5 Optimal L^2 extension on weakly pseudoconvex Kähler manifolds

In [48], Manivel generalized Theorem 1.1 from a more geometric viewpoint. In [16], Demailly simplified Manivel's approach and proved the following L^2 extension theorem on weakly pseudoconvex Kähler manifolds.

Theorem 5.1 ([16]). *Let (X, ω) be a weakly pseudoconvex complex n-dimensional manifold possessing a Kähler metric ω, and let L (resp. E) be a Hermitian holomorphic line bundle over X (resp. a Hermitian holomorphic vector bundle of rank m over X $(1 \leq m \leq n)$) equipped with a smooth Hermitian metric, and s a global holomorphic section of E. Assume that s is generically transverse to the zero section, and let*

$$Y = \{x \in X : s(x) = 0, \wedge^m ds(x) \neq 0\}.$$

Moreover, assume that the $(1,1)$-form $\sqrt{-1}\Theta_L + m\sqrt{-1}\partial\bar{\partial}\log|s|_E^2$ is semipositive and that there is a continuous function $\alpha \geq 1$ such that the following two inequalities hold everywhere on X:

(a) $\sqrt{-1}\Theta_L + m\sqrt{-1}\partial\bar{\partial}\log|s|_E^2 \geq \dfrac{\{\sqrt{-1}\Theta_E s, s\}_E}{\alpha|s|_E^2}$,

(b) $|s|_E \leq e^{-\alpha}$.

Then for every holomorphic section f of $K_X \otimes L$ over Y, such that

$$\int_Y \frac{|f|_L^2}{|\wedge^m (ds)|_E^2} dV_Y < +\infty,$$

there exists a holomorphic extension F to X such that $F\big|_Y = f$ and

$$\int_X \frac{|F|_L^2}{|s|_E^{2m}(-\log|s|_E)^2} dV_X \leq C \int_Y \frac{|f|_L^2}{|\wedge^m (ds)|_E^2} dV_Y,$$

where C is a numerical constant depending only on m.

In [81], we proved the following L^2 extension theorem, which optimizes the constant C in Theorem 5.1 and allows the smooth metric of the holomorphic line bundle L to be a singular metric.

Let $\mathfrak{R}$ be the class of functions defined by

$$\Big\{R \in C^\infty(-\infty, 0] : R > 0,\ R' \le 0,\ \int_{-\infty}^{0} \frac{1}{R(t)} dt < +\infty$$
$$\text{and } e^t R(t) \text{ is bounded above on } (-\infty, 0]\Big\}.$$

We will denote $\int_{-\infty}^{0} \frac{1}{R(t)} dt$ by C_R. The function $R(t)$ is equal to the function $\frac{1}{c_a(-t)e^t}$ defined in Section 4 when $a = 0$.

Theorem 5.2 ([81]). *Let (X, ω) be a weakly pseudoconvex complex n-dimensional manifold possessing a Kähler metric ω, ψ be a plurisubharmonic function on X, E be a holomorphic vector bundle of rank m over X equipped with a smooth Hermitian metric $(1 \le m \le n)$, and s be a global holomorphic section of E. Assume that s is transverse to the zero section, and let*

$$Y := \{x \in X : s(x) = 0\}.$$

Let L be a holomorphic line bundle over X equipped with a singular Hermitian metric h_L, which is written locally as $e^{-\varphi_L}$ for some function $\varphi_L \in L^1_{\text{loc}}$ with respect to a local holomorphic frame of L. Assume that $\varphi_L + \psi$ is quasi-plurisubharmonic and φ_L is locally bounded above. Moreover, assume that

(i) $\sqrt{-1}\Theta_L + \sqrt{-1}\partial\bar\partial\psi + m\sqrt{-1}\partial\bar\partial \log |s|_E^2 \ge 0$ *holds on* $X \setminus Y$,

and that there is a continuous function $\alpha > 0$ on X such that the following two inequalities hold on $X \setminus Y$:

(ii) $\sqrt{-1}\Theta_L + \sqrt{-1}\partial\bar\partial\psi + m\sqrt{-1}\partial\bar\partial \log |s|_E^2 \ge \dfrac{\{\sqrt{-1}\Theta_E s, s\}_E}{\alpha |s|_E^2}$,

(iii) $\psi + m \log |s|_E^2 \le -2m\alpha$.

Then for every holomorphic section f on Y with values in the line bundle $K_X \otimes L$ (restricted to Y), such that

$$\int_Y \frac{|f|_L^2 e^{-\psi}}{|\wedge^m (ds)|_E^2} dV_Y < +\infty,$$

there exists a holomorphic section F on X with values in $K_X \otimes L$, such that $F = f$ on Y and

$$\int_X \frac{|F|_L^2}{e^{\psi + m \log |s|_E^2} R(\psi + m \log |s|_E^2)} dV_X \le C_R \frac{(2\pi)^m}{m!} \int_Y \frac{|f|_L^2 e^{-\psi}}{|\wedge^m (ds)|_E^2} dV_Y. \tag{5.1}$$

Remark 5.1. By slight modifications of our proof, we can in fact replace the curvature assumption (ii) in Theorem 5.2 with the following weaker one: *assume that there exists a nonnegative number* α_0 *such that*

$$(ii)' \quad \sqrt{-1}\Theta_L + \sqrt{-1}\partial\bar{\partial}\psi + m\sqrt{-1}\partial\bar{\partial}\log|s|_E^2 \geq \frac{m\{\sqrt{-1}\Theta_E s, s\}_E}{\widetilde{\chi}(-2m\alpha)|s|_E^2}$$

holds on $X \setminus Y$, *where*

$$\widetilde{\chi}(t) := \alpha_0 + \frac{\alpha_0 \int_t^0 \big(\frac{1}{R(0)} - \frac{1}{R(t_1)}\big)dt_1 + \int_t^0 \big(\int_{t_2}^0 \frac{1}{R(t_1)}dt_1\big)dt_2}{\frac{\alpha_0}{R(0)} + \int_t^0 \frac{1}{R(t_1)}dt_1}.$$

Then the constant $C_R\frac{(2\pi)^m}{m!}$ in (5.1) should be replaced by $\big(\frac{\alpha_0}{R(0)} + C_R\big)\frac{(2\pi)^m}{m!}$.

It is not hard to see that $\widetilde{\chi}$ is a smooth strictly decreasing function from $(-\infty, 0)$ to $(\alpha_0, +\infty)$. It is also not hard to verify that $\widetilde{\chi}(t) \geq -\frac{t}{2}$ when $\alpha_0 = 0$. Hence $(ii)'$ is weaker than (ii). Furthermore, we can prove that $\big(\frac{\alpha_0}{R(0)} + C_R\big)\frac{(2\pi)^m}{m!}$ is the optimal constant corresponding to the assumption $(ii)'$.

Now we explain some key ideas in the proof of Theorem 5.2.

By using a method similar to the one in Section 4, we obtained the optimal constant $C_R\frac{(2\pi)^m}{m!}$ in (5.1). The main problem which needs to be fixed is how to treat the singular metric of L on the weakly pseudoconvex Kähler manifold X. In this case, a difficulty arose, because, unlike the case of Stein or projective manifolds, there is a loss of positivity in the regularization process of singular weights on such manifolds.

In order to overcome the difficulty in dealing with singular weights on weakly pseudoconvex Kähler manifolds, not only a regularization theorem (see Theorem 5.3 below) and the error term method of solving $\bar{\partial}$ equations (see Lemma 4.8) are needed, but also a limit problem about L^2 integrals with singular weights needs to be solved. In [81], by replacing a fixed holomorphic function with a family of holomorphic functions, we solved the limit problem (see Proposition 5.5 below). Then by using Proposition 5.5 and the strong openness property of multiplier ideal sheaves (see Theorem 5.4 below) as the key tools, we constructed a family of special smooth extensions and overcame the difficulty in dealing with singular weights.

The following theorem is a regularization result concerning the approximation of quasi-plurisubharmonic functions.

Theorem 5.3 (Theorem 6.1 in [15]). *Let* (X, ω) *be a complex manifold equipped with a Hermitian metric* ω, *and* $\Omega \subset\subset X$ *be an open subset. Assume that* $T = \widetilde{T} + \frac{\sqrt{-1}}{\pi}\partial\bar{\partial}\varphi$ *is a closed* $(1,1)$*-current on* X, *where* $\widetilde{T}$ *is a smooth real* $(1,1)$*-form and* φ *is a quasi-plurisubharmonic function. Let* γ *be a continuous real* $(1,1)$*-form such that* $T \geq \gamma$. *Suppose that the Chern curvature tensor of* T_X *satisfies*

$$(\sqrt{-1}\Theta_{T_X} + \varpi \otimes \mathrm{Id}_{T_X})(\kappa_1 \otimes \kappa_2, \kappa_1 \otimes \kappa_2) \geq 0,$$

$$\forall \kappa_1, \kappa_2 \in T_X \text{ with } \langle \kappa_1, \kappa_2 \rangle = 0,$$

for some continuous nonnegative $(1,1)$*-form* ϖ *on* X*. Then there is a family of closed* $(1,1)$*-currents* $T_{\varsigma,\rho} = \widetilde{T} + \frac{\sqrt{-1}}{\pi}\partial\bar{\partial}\varphi_{\varsigma,\rho}$ *on* X ($\varsigma \in (0,+\infty)$ *and* $\rho \in (0,\rho_1)$ *for some positive number* ρ_1) *independent of* γ*, such that*

(*i*) $\varphi_{\varsigma,\rho}$ *is quasi-plurisubharmonic on a neighborhood of* $\overline{\Omega}$*, smooth on* $X\backslash E_\varsigma(T)$*, increasing with respect to* ς *and* ρ *on* Ω*, and converges to* φ *on* Ω *as* $\rho \to 0$*,*

(*ii*) $T_{\varsigma,\rho} \geq \gamma - \varsigma\varpi - \delta_\rho\omega$ *on* Ω*,*

where $E_\varsigma(T) := \{x \in X : \nu(T,x) \geq \varsigma\}$ ($\varsigma > 0$) *is the* ς*-upperlevel set of Lelong numbers, and* $\{\delta_\rho\}$ *is an increasing family of positive numbers such that* $\lim\limits_{\rho\to 0}\delta_\rho = 0$.

The following theorem is the strong openness property of multiplier ideal sheaves for plurisubharmonic functions, which was conjectured by Demailly in [17] and [18], and proved by Guan and Zhou in [36].

Theorem 5.4 ([36]). *Let* φ *be a negative plurisubharmonic function on the unit polydisk* $\Delta^n \subset \mathbb{C}^n$*. Assume that* F *is a holomorphic function on* Δ^n *satisfying*

$$\int_{\Delta^n} |F|^2 e^{-\varphi} dV_n < +\infty,$$

where dV_n *is the* $2n$*-dimensional Lebesgue measure on* $\mathbb{C}^n$*. Then there exists* $r \in (0,1)$ *and* $\beta \in (0,+\infty)$ *such that*

$$\int_{\Delta_r^n} |F|^2 e^{-(1+\beta)\varphi} dV_n < +\infty,$$

where $\Delta_r^n := \{(z^1,\cdots,z^n) \in \mathbb{C}^n : |z^k| < r, 1 \leq k \leq n\}$.

The following proposition is a convergence property for L^2 integrals with plurisubharmonic weights.

Proposition 5.5 (Proposition 5.2 in [81]). *Let* $\phi(z',z'')$ *be a plurisubharmonic function on* $\mathbb{B}_r^m \times \mathbb{B}_r^{n-m}$ ($1 \leq m \leq n, r > 0$) *such that* $\sup\limits_{\mathbb{B}_r^m \times \mathbb{B}_r^{n-m}} \phi < +\infty$*, where* $\mathbb{B}_r^k$ *is the open ball in* $\mathbb{C}^k$ *centered at* 0 *with radius* r*. Let* $f(z'')$ *be a holomorphic function on* $\mathbb{B}_r^{n-m}$ *and* $h(z',z'')$ *be a nonnegative continuous function on* $\mathbb{B}_r^m \times \mathbb{B}_r^{n-m}$*. Assume that* C*,* β*,* β_1*,* c_1*,* c_2*,* r'*,* r'' *and* $\varepsilon_{r'}$ *are positive numbers such that* $\beta_1 < 1$*,* $c_1 < c_2$*,* $r' < r'' < r$ *and* $\varepsilon_{r'} < \frac{r'}{4c_2}$*. Suppose that* $f_\varepsilon \in \mathcal{O}(\mathbb{B}_r^m \times \mathbb{B}_r^{n-m})$ ($\varepsilon \in (0,\varepsilon_{r'})$) *are a family of holomorphic functions satisfying* $f_\varepsilon(0,z'') = f(z'')$ ($\forall z'' \in \mathbb{B}_r^{n-m}$*,* $\forall \varepsilon \in (0,\varepsilon_{r'})$)*,*

$$\sup_{\mathbb{B}_{r''}^m \times \mathbb{B}_{r''}^{n-m}} |f_\varepsilon| \leq C\varepsilon^{-\beta_1}, \ \forall \varepsilon \in (0,\varepsilon_{r'}), \tag{5.2}$$

and

$$\frac{1}{\varepsilon^{2m}} \int_{(z',z'')\in \mathbb{B}_{2c_2\varepsilon}^m \times \mathbb{B}_r^{n-m}} |f_\varepsilon(z',z'')|^2 e^{-(1+\beta)\phi(z',z'')} dV_n \leq C, \ \forall \varepsilon \in (0,\varepsilon_{r'}). \tag{5.3}$$

Then

$$\lim_{\varepsilon\to 0}\frac{1}{\mu(\mathbb{B}^m)}\int_{(\mathbb{B}^m_{c_2\varepsilon}\setminus\mathbb{B}^m_{c_1\varepsilon})\times\mathbb{B}^{n-m}_{r'}}\frac{\varepsilon^2 h(z',z'')|f_\varepsilon(z',z'')|^2 e^{-\phi(z',z'')}}{m(|z'|^2+\varepsilon^2)^2|z'|^{2m-2}}dV_n$$
$$=\left(\frac{1}{c_1^2+1}-\frac{1}{c_2^2+1}\right)\int_{z''\in\mathbb{B}^{n-m}_{r'}} h(0,z'')|f(z'')|^2 e^{-\phi(0,z'')}dV_{n-m},$$

where $\mu(\mathbb{B}^m)$ *is the* $2m$*-dimensional Lebesgue measure of* $\mathbb{B}^m_1$*, and* dV_k *is the* $2k$*-dimensional Lebesgue measure on* $\mathbb{C}^k$*.*

6 Vanishing theorems for line bundles

6.1 Kodaira vanishing theorem

In this subsection, we will discuss the vanishing theorems of holomorphic line bundles over compact complex manifolds along the line of differential geometric techniques. All these results listed here are based on the Bochner-Kodaira-Nakano identity and the Hodge theory for compact Kähler manifolds.

Following Kodaira [40], a holomorphic line bundle $L\to X$ is positive (resp. semipositive) if there is a metric h on L such that its Chern curvature form

$$\sqrt{-1}\Theta_{L,h}=\sqrt{-1}\sum_{1\le j,k\le n} c_{jk}(z)dz_j\wedge d\bar{z}_k$$

is positive definite (resp. semi-definite) at every point $z\in X$.

Theorem 6.1 (Kodaira [40])**.** *Let* X *be a compact complex manifold of dimension* n *and let* L *be a positive line bundle over* X*, then*

$$H^q(X,K_X\otimes L)=0 \quad \text{for} \quad q>0.$$

Equivalently,

$$H^q(X,L^{-1})=0 \quad \text{for} \quad q<n.$$

Theorem 6.2 (Akizuki-Nakano [2])**.** *Under the assumptions made in Theorem 6.1, let* Ω^p_X *denote the sheaf of germs of holomorphic form of degree* p *on* X*, then*

$$H^q(X,\Omega^p_X\otimes L)=0 \quad \text{for} \quad p+q>n.$$

As an illustration of the vanishing theorems, we give the cohomology groups of all line bundles over $\mathbb{P}^n$, which is a very special case of Bott's general results and will be needed for the proofs of vanishing theorems for ample vector bundles.

Theorem 6.3 (Bott [9])**.**

$$h^q(\mathbb{P}^n,\Omega^p_{\mathbb{P}^n}\otimes\mathcal{O}_{\mathbb{P}^n}(k))=\begin{cases}\binom{k+n-p}{k}\binom{k-1}{p} & \text{for } q=0,\ 0\le p\le n,\ k>p\\ \binom{-k+p}{-k}\binom{-k-1}{n-p} & \text{for } q=n,\ 0\le p\le n,\ k<p-n\\ 1 & \text{for } k=0,\ 0\le p=q\le n\\ 0 & \text{otherwise.}\end{cases}$$

It is important for applications to obtain vanishing theorems under weaker assumptions of the positivity of the bundles.

Theorem 6.4 (Girbau [26]). *Let X be a compact Kähler manifold of dimension n and let L be a line bundle over X. If there is a metric h on L such that the curvature $i\Theta_{L,h}$ is semipositive and has at least $n-s+1$ positive eigenvalues at every point, then*

$$H^q(X, \Omega_X^p \otimes L) = 0 \quad for \quad p+q \geq n+s.$$

It should be noted that Girbau's result does not hold for $p < n$ if the curvature $i\Theta_{L,h}$ is only assumed to have $n-s+1$ positive eigenvalues on a dense open set. The following example is due to Ramanujam [65]. Let $\sigma : X \to \mathbb{P}^n$ be the blow-up of $\mathbb{P}^n$ in one point and let $L = \sigma^*(\mathcal{O}_{\mathbb{P}^n}(1))$. However, $H^p(X, \Omega_X^p \otimes L) \neq 0$ for $0 \leq p \leq n-1$.

Theorem 6.5 (Grauert-Riemenschneider [28]). *Let X be a compact connected Kähler manifold, and let L be an line bundle over X. If there is a metric h on L such that $i\Theta_{L,h}$ is semipositive on X and positive in at least one point of X, then*

$$H^q(X, K_X \otimes L) = 0 \quad for \quad q > 0.$$

Geometrically, the assumptions in this theorem mean that the Grauert tube $\{v \in L^*; |v|_{h_{L^*}} < 1\}$ have smooth pseudoconvex boundary which is strictly pseudoconvex at some point. The proof of this theorem uses again the harmonic theory.

The Kähler assumption in this theorem is in fact unnecessary: this is the so-called "Grauert-Riemenschneider conjecture" solved by Siu [70] and Demailly [12]. The Grauert-Riemenschneider vanishing theorem holds also for torsion free coherent sheaves and on reduced singular complex spaces, with suitably choice of the canonical sheaf.

The relative version of Grauert-Riemenschneider vanishing theorem says that the higher direct images of the canonical sheaf vanish. This was generalized by Takegoshi to

Theorem 6.6 ([76]). *Let $f : X \to S$ be a proper surjective morphism of complex spaces with X smooth. Suppose X is bimeromorphic to a Kähler manifold. Then $R^q f_*(K_X) = 0$ for $q > \dim_{\mathbb{C}} X - \dim_{\mathbb{C}} S$.*

Theorem 6.6 has the following important generalization.

Theorem 6.7 (Kollár [41]). *Let X be a smooth projective variety, Y an arbitrary (reduced) projective variety, $f : X \to Y$ a surjective map. Then*

(i) $R^q f_*(K_X)$ is torsion-free for $q \geqslant 0$.
(ii) $R^q f_*(K_X) = 0$ if $q >$ generic fiber dimension.
(iii) If L is ample on Y, then

$$H^p(Y, L \otimes R^q f_*(K_X)) = 0 \quad for \quad p > 0$$

The last statement of the theorem is due to Ohsawa [58] for $q = 0$. For further developments, one may refer to the articles of Enoki, Ohsawa, Fujino and Matsumura ([22], [62], [24], [50], etc.).

6.2 Algebraic generalizations

In this subsection, we firstly give some algebraic notions of positivity and then discuss the algebraic generalizations of Kodaira vanishing theorems. One may refer to [23] for a more comprehensive exposition.

If L is a holomorphic line bundle over a compact complex manifold X, then one can define a meromorphic map

$$\Phi_L : X \longrightarrow \mathbb{P}(H^0(X, L)).$$

It is holomorphic when restricted $X \backslash B_L$, where B_L is the base locus.

The line bundle L is said to be very ample if the map Φ_L is a regular embedding (By this we mean in particular the base locus B_L is empty). L is called ample if some multiple L^m, $m > 0$, is very ample. By Kodaira's embedding theorem, the notion of ampleness is equivalent to the notion of positivity for homomorphic line bundles.

The Kodaira-Iitaka dimension $\kappa(L)$ is the supremum of the rank of the holomorphic map $\Phi_{L^m} : X \backslash B_{L^m} \longrightarrow \mathbb{P}(H^0(X, L^m))$. In case that $H^0(X, L^m) = 0$ for all $m \geq 1$, we set $\kappa(L) = -\infty$. The line bundle L is said to be big if $\kappa(L) = \dim X$.

A holomorphic line bundle L over a projective manifold X is said to be numerically effective, nef for short, if $L \cdot C \geq 0$ for every curve $C \subset X$. The definition of an nef bundle can be generalized to arbitrary compact complex manifolds (cf. [18]): A holomorphic line bundle L over a compact complex manifold X is said to be nef if for every $\varepsilon > 0$, there is a smooth Hermitian metric h_ε on L such that $i\Theta_{L,h_\varepsilon} \geq -\varepsilon\omega$, where ω is a fixed Hermitian metric on X.

If L is a nef line bundle on a compact Kähler manifold X, then the numerical dimension of L is defined to be

$$\mathrm{nd}(L) = \max\{k | 0 \leq k \leq \dim X, \quad c_1(L)^k \neq 0 \quad \text{in} \quad H^{2k}(X, \mathbb{R})\}.$$

If L is an arbitrary holomorphic line bundle over a compact Kähler manifold, the numerical dimension of L can also be defined. For the details, one may refer to [10] and [18].

Now, we are ready to give some striking algebraic generalization of Kodaira vanishing theorems. We start with the following vanishing theorem due to Ramanujam.

Theorem 6.8 (Ramanujam [65]). *Let L be a holomorphic line bundle over a projective manifold X. Suppose that $\dim \Phi_{L^m}(X) = \kappa$ for some positive integer m, then*

$$H^q(X, L^{-1}) = 0$$

for $q < \min\{\kappa, \dim B_{L^m}\}$, where $B_{L^m} = \bigcap_{s \in H^0(X, L^m)} s^{-1}(0)$ is the base locus of $H^0(X, L^m)$.

The notion of k-ampleness (Sommese [74]) is very useful for particular purposes. Let X be a compact complex space. A holomorphic line bundle L over X

is said to be k-ample if there is some $m \in \mathbb{N}_+$ such that L^m is generated by global sections and such that the induced holomorphic map

$$\Phi_{L^m} : X \longrightarrow \mathbb{P}(H^0(X, L))$$

has at most k-dimensional fibers. A holomorphic vector bundle E over X is called k-ample if the Serre line bundle $\mathcal{O}_{\mathbb{P}(E)}(1)$ is k-ample.

The "algebraic version" of Girbau's vanishing theorem is the following result of Sommese which is similar to Ramanujam's vanishing theorem.

Theorem 6.9 (see [74], [66]). *Let L be a holomorphic line bundle over the projective manifold X of dimension n. Let $m > 0$ and $k \geq 0$ be integers such that the map $\Phi_{L^m} : X \backslash B_{L^m} \longrightarrow \mathbb{P}(H^0(X, L^m))$ has at most k-dimensional fibers, where B_{L^m} is the base locus of $H^0(X, L^m)$. Then*

$$H^q(X, \Omega_X^p \otimes L) = 0$$

for $p + q > n + \max\{k, \dim B_{L^m}\}$. In particular, if L is k-ample, then

$$H^q(X, \Omega_X^p \otimes L) = 0 \quad for \quad p + q > n + k.$$

A related result is the Bogomolov vanishing theorem which is one of the principal inputs of Viehweg's (cf. [77]) proof of Kawamata-Viehweg vanishing. A generalization is obtained by using the numerical dimension of line bundles (Boucksom [10]).

Theorem 6.10 (Bogomolov [8]). *Let D be a normal crossing divisor on a projective manifold X, and let L be a holomorphic line bundle over X. Then*

$$H^0(X, \Omega_X^p(\log D) \otimes L^{-1}) = 0$$

for $p < \kappa(L)$, where $\Omega_X^p(\log D)$ is the sheaf of p-forms with logarithmic poles along D.

An important vanishing theorem in algebraic geometry is the following generalization of Kodaira vanishing theorem discovered by Kawamata [39] and Viehweg [77]. For the statement, we need the following definition: An effective divisor $D = \sum D_i$ on projective manifold X has simple normal crossings if D is reduced, each component D_i is smooth, and if D is defined in a neighborhood of any point by an equation in local coordinate of the type $z_1 \cdots z_k = 0$ for some $k \leq \dim X$. A divisor $E = \sum a_i D_i$ has simple normal crossing support if the underlying reduced divisor $\sum D_i$ has simple normal crossings.

Theorem 6.11 (Kawamata-Viehweg). *Let X be a projective manifold of dimension n, and let L be a line bundle over X. Assume that some multiple mL can be written as $mL = F + D$, where F is a nef line bundle and $D = \sum a_i D_i$ is an integral divisor with simple normal crossing support and coefficients:*

$$0 \leq a_i < m \quad for\ all \quad i.$$

Then

$$H^q(X, K_X \otimes L) = 0 \quad for \quad q > n - \mathrm{nd}(F).$$

An effective $\mathbb{Q}$-divisor $D = \Sigma \alpha_j D_j$ is said to be integrable at a point $x_0 \in X$ if the function $\prod |g_j|^{-2\alpha_j}$ which is equal to the product of local generators g_j of $\mathcal{O}(-D_j)$ at x_0 is integrable on a neighborhood of x_0.

Based on the L^2-estimates for $\bar{\partial}$ deduced from the Bochner-Kodaira-Nakano inequality, Demailly proved the following improved version of Kawamata-Viehweg vanishing theorem.

Theorem 6.12 (Demailly [14]). *Let L be a line bundle over a projective manifold X. Assume that some positive power L^k can be written $L^k = F \otimes \mathcal{O}(D)$ with F nef and with an effective divisor D such that $k^{-1}D$ is integrable on $X \setminus B$.*

Then

$$H^q(X, L^{-1}) = 0 \quad \textit{for} \quad q < \min\{\max\{\mathrm{nd}(F), \kappa(L)\}, \mathrm{codim} B\}$$

In the setting of compact Kähler manifolds, Kawamata-Viehweg vanishing theorem is still valid.

Theorem 6.13 (Enoki [22]). *Let L be a nef line bundle over a compact n-dimensional Kähler manifold X. Then*

$$H^q(X, K_X \otimes L) = 0 \quad \textit{for} \quad q > n - \mathrm{nd}(L).$$

For the proof, the Calabi-Yau theorem ([78]) on solutions of Monge-Ampère equations is used in an essential way. One can also refer to [20] for related results.

6.3 Vanishing theorems on weakly pseudoconvex manifolds

A complex manifold X is said to be weakly pseudoconvex (or weakly 1-complete) if it possesses a smooth plurisubharmonic exhaustion functions. Using the L^2 method for $\bar{\partial}$ equations, almost all vanishing theorems for positive vector bundles over compact complex manifolds are also valid for weakly pseudoconvex manifolds. Now we list some of results for examples.

Theorem 6.14 (Nakano [56]). *If L is a positive line bundle over a weakly pseudoconvex manifold X of dimension n, then*

$$H^q(X, \Omega_X^p \otimes L) = 0 \quad \textit{for} \quad p + q > n.$$

Theorem 6.15 (Takegoshi [75]). *Let L be a line bundle over a connected weakly pseudoconvex Kähler manifold X. If there exists a metric h on L such that its curvature $\sqrt{-1}\Theta_{L,h}$ is semipositive on X and has at least $n-s+1$ positive eigenvalues on $X \backslash K$, where K is a proper compact subset of X. Then*

$$H^q(X, K_X \otimes L) = 0 \quad \textit{for} \quad q \geq s.$$

Girbau's vanishing theorem can also be generalized to weakly pseudoconvex manifolds.

Theorem 6.16 (Abdelkader [1], Ohsawa [57]). *Let X be a weakly pseudoconvex Kähler manifold of dimension n, and L be a line bundle over X. If there is a metric h on L such that the curvature $\sqrt{-1}\Theta_{L,h}$ is semipositive and has at least $n - s + 1$ positive eigenvalues at every point, then*

$$H^q(X, \Omega_X^p \otimes L) = 0 \quad \textit{for} \quad p + q \geq n + s.$$

6.4 Nadel vanishing theorem

Now, we are going to talk about vanishing theorems with multiplier ideal sheaves. We need to introduce some concepts. A singular Hermitian metric h on a holomorphic line bundle L over a complex manifolds X is a metric which is given in any local trivialization $\theta : L \restriction \Omega \longrightarrow \Omega \times \mathbb{C}$ by

$$\|\xi\|_h^2 = |\theta(\xi)|^2 e^{-\varphi(x)}, \quad x \in \Omega, \quad \xi \in L_x,$$

where $\varphi \in L^1_{\text{loc}}(\Omega)$ is an arbitrary function, called the weight of the metric with respect to the trivialization θ. We note that the curvature $\Theta_{L,h} = \partial\bar{\partial}\varphi$ is a well-defined closed $(1,1)$-current on X.

The line bundle L is called pseudoeffective if there is a singular metric h such that $i\Theta_{L,h} \geq 0$ in the sense of current (equivalently, the local weight φ is plurisubharmonic).

If φ is a plurisubharmonic function on an open subset $\Omega \subset \mathbb{C}^n$, the multiplier ideal sheaf $\mathcal{I}(\varphi)$ is the ideal sheaf of $\mathcal{O}_\Omega$ defined by

$$\mathcal{I}(\varphi)_x = \{f \in \mathcal{O}_{\Omega,x} : \exists\, U \ni x \ \text{ such that } \int_U |f|^2 e^{-\varphi} d\lambda < +\infty\},$$

where U is an open neighborhood of x, and $d\lambda$ is the Lebesgue measure on $\mathbb{C}^n$. By Nadel [54], the sheaf $\mathcal{I}(\varphi)$ is a coherent sheaf of ideals over Ω.

Let L be a holomorphic line bundle over X with a singular metric h. If φ is a local weight of h, then the ideal sheaf $\mathcal{I}(\varphi)$ is independent of the choice of the trivialization and we can obtain a global coherent ideal sheaf $\mathcal{I}(h)$ on X.

If $D = \Sigma\alpha_j D_j$ is an effective $\mathbb{Q}$-divisor, we define the multiplier ideal sheaf $\mathcal{I}(D)$ to be equal to $\mathcal{I}(\varphi)$ where $\varphi = \Sigma\alpha_j \log|g_j|^2$ is the corresponding plurisubharmonic function defined by generators g_j of $\mathcal{O}(-D_j)$. In terms of multiplier ideal sheaves, Kawamata-Viehweg vanishing theorem can be stated in the following form.

Theorem 6.17 ([18]). *Let X be a projective manifold of dimension n, and let L be a line bundle over X. Assume that some multiple mL can be written as $mL = F + D$, where F is a nef line bundle and D an effective divisor. Then*

$$H^q(X, K_X \otimes L \otimes \mathcal{I}(m^{-1}D)) = 0 \quad \textit{for} \quad q > n - \text{nd}(F).$$

We now come to the Nadel vanishing theorem for multiplier ideals which is one of the most useful tools in complex analytic and algebraic geometry. It can be seen as the analytic generalization of the Kawamata-Viehweg vanishing theorems. For the approach of Kawamata-Viehweg vanishing theorem (in the form of Theorem 6.17) via Nadel vanishing, one may refer to [18]. The version of Nadel vanishing theorem stated here is contained in [18].

Theorem 6.18 (Nadel vanishing theorem). *Let (X, ω) be a weakly pseudoconvex Kähler manifold, and L be a holomorphic line bundle over X with a singular*

Hermitian metric h. Assume that $\sqrt{-1}\Theta_{L,h} \geq \varepsilon\omega$ in the sense of current for some continuous positive function ε on X, then

$$H^q(X, K_X \otimes L \otimes \mathcal{I}(h)) = 0$$

for all $q \geq 1$.

The proof follows from the L^2-estimates for $\bar{\partial}$-equations with singular weights. As in the case of smooth metrics, it is important to relax the strict positivity assumptions of the singular metric. Let (L, h) be a pseudoeffective line bundle over a compact Kähler manifold X, denote by $\mathrm{nd}(L, h)$ the numerical dimension of (L, h) as defined in [11].

By Guan-Zhou's proof of Demailly's strong openness conjecture in [36] (combining with the vanishing result in Cao [11]), Nadel's vanishing theorem can be generalized as follows.

Theorem 6.19 ([11], [36]). *Let (L, h) be a pseudoeffective line bundle over a compact n-dimensional Kähler manifold X. Then*

$$H^q(X, K_X \otimes L \otimes \mathcal{I}(h)) = 0 \quad \text{for all} \quad q > n - \mathrm{nd}(L, h).$$

Recently, Meng and Zhou removed the Kähler condition in Nadel vanishing theorem for holomorphically convex manifolds.

Theorem 6.20 ([53]). *Let (X, ω) be a Hermitian holomorphically convex manifold, and let L be a holomorphic line bundle over X equipped with a singular Hermitian metric h. Assume that $i\Theta_{L,h} \geq \varepsilon\omega$ for some continuous positive function ε on X. Then*

$$H^q(X, K_X \otimes L \otimes \mathcal{I}(h)) = 0 \quad \text{for} \quad q \geq 1.$$

7 Vanishing theorems for vector bundles

In the case of line bundles, positivity and ampleness are equivalent by Kodaira's embedding theorem. However, for vector bundles of higher rank, the notions of positivity and ampleness are not always equivalent.

Following Hartshorne [38], a vector bundle E over a projective variety X is ample if the Serre bundle $\mathcal{O}_{\mathbb{P}(E)}(1)$ is an ample line bundle over the projectivized bundle $\mathbb{P}(E)$.

In [27], Grauert defined the positivity in the following way: let E be a vector bundle over a compact complex space X, E is called negative if the zero section of E is exceptional; a vector bundle F over a compact complex space is called positive if its dual F^* is negative.

Equivalently, E is negative if there is a tube W around the zero section of E with a strictly pseudoconvex (1-convex) differential boundary. If L is a positive line bundle in the sense of Kodaira, then the Grauert tube $\{v \in L^*; |v|_{h_{L^*}} < 1\}$ is a strictly pseudoconvex domain with smooth boundary, and hence the zero section of the dual bundle L^* is exceptional. Therefore, the positivity notions of Kodaira and Grauert coincide for line bundles over compact complex manifolds. It follows that the positivity notions of Hartshorne and Grauert are the same for vector bundles.

We note that the definition of positivity is still valid for coherent analytic sheaves. If $\mathscr{E}$ be a coherent analytic sheaf over a compact complex space X, then $\mathscr{E}$ is said to be positive if the zero section of the linear space $V(\mathscr{E})$ (see [27]) associated to $\mathscr{E}$ is exceptional. For more propositions of positive sheaves and ample vector bundles, one may refer to [27] and [38].

Next we give some differential-geometric notions of positivity. Let E be a holomorphic Hermitian vector bundle of rank r over an n-dimensional complex manifold X. In the local coordinates $(z_1, \cdots, z_n)$ of X and a C^∞ local orthonormal frame $(e_1, \cdots, e_r)$ of E, the curvature can be written as

$$\Theta_{E,h} = \sum_{1\le j,k\le n, 1\le\lambda,\mu\le r} c_{jk\lambda\mu} dz_j \wedge d\bar{z}_k \otimes e_\lambda^* \otimes e_\mu, \quad c_{jk\lambda\mu} = \bar{c}_{kj\mu\lambda}.$$

This leads in a natural way to the following positivity notions for vector bundles.

(1) E is called positive in the sense of Griffiths if there is a Hermitian metric h on E such that its curvature satisfies the following condition (in local coordinates):

$$\sqrt{-1} \sum_{j,k,\lambda,\mu} c_{jk\lambda\mu}(x)\xi_j\bar{\xi}_k v_\lambda \bar{v}_\mu > 0$$

for all $x \in X$ and $0 \ne \sum \xi_j \frac{\partial}{\partial z_j} \in T_{X,x}$ and all $0 \ne v = \sum v_\lambda e_\lambda \in E_x$.

(2) E is said to be positive in the sense of Nakano if there is a Hermitian metric h on E such that

$$\sqrt{-1} \sum_{j,k,\lambda,\mu} c_{jk\lambda\mu}(x) v_{j,\lambda}\bar{v}_{k,\mu} > 0$$

for all $x \in X$ and all $0 \ne v = v_{j,\lambda}\frac{\partial}{\partial z_j} \otimes e_\lambda \in T_{X,x} \otimes E_x$.

Various concepts of positivity (dual Nakano positive, m-positive ([19]), Nakano q-positive ([69]), etc.) can be defined in a similar way, and these notations of positivity yield corresponding vanishing theorems. However, we will not talk about them here.

In the case of line bundles, the notions of positivity, Griffiths positivity and Nakano positivity coincide. For vector bundles, it is clear that Nakano positivity is stronger than Griffiths positivity. However, the converse is not true in general. Actually, the holomorphic tangent bundle of $\mathbb{CP}^n$, $n \ge 2$, is Griffiths positive, but it is not Nakano positive. Demailly-Skoda [21] gives a "converse" result in the following sense: if E is Griffiths positive, then $E \otimes \det E$ is Nakano positive.

It is not hard to see that Griffiths positive vector bundles are ample. The converse problem is the so called Griffiths conjecture: if E is an ample vector bundle of rank $r > 1$ over a compact complex manifold X, then E is positive in the sense of Griffiths. This problem has been solved in the affirmative when X is a curve, but the general case is still open. Berndtsson [5] shows that if E is ample, then $E \otimes \det E$ is Nakano positive.

As in the case of line bundles, the Bochner-Kodaira-Nakano identity and the Hodge theory imply the following results.

Theorem 7.1 (Nakano vanishing theorem). *Let X be a compact complex manifold of dimension n, and E be a holomorphic vector bundle over X. If E is Nakano positive, then*

$$H^q(X, K_X \otimes E) = 0$$

for all $q > 0$.

As in the Grauert-Riemenschneider vanishing theorem, the Nakano positivity can be weakened as follows: E is Nakano semipositive on X and Nakano positive in at least one point.

Next we discuss vanishing theorems for ample vector bundles. The proofs of these results for vector bundles are deduced from vanishing theorems for line bundles over associated projectivized bundles or flag bundles, using the associated spectral sequences.

Theorem 7.2 (Griffiths [30]). *If E is a ample vector bundle over compact Kähler manifold X, then*

$$H^q(X, K_X \otimes S^m E \otimes \det E) = 0$$

for all $q > 0$ and $m \geq 0$.

The example in [64] shows that the factor $\det E$ can not be omitted in Griffiths (and hence Demailly) vanishing theorem.

Theorem 7.3 (Le Potier vanishing theorem). *Let X be a compact manifold of dimension n. If E is an ample vector bundle of rank $r \geq 1$ over X. then*

$$H^q(X, K_X \otimes \Lambda^m E) = 0 \quad \textit{for} \quad m > 0 \textit{ and } \quad q \geq r - m,$$

$$H^q(X, \Omega_X^p \otimes E) = 0 \quad \textit{for} \quad p + q \geq n + r.$$

It is not true in general that $H^q(X, \Omega_X^p \otimes \Lambda^m E) = 0$ for E ample and $p + q > n + r - m$. Counter-examples are constructed in [13].

Theorem 7.4 (Sommese [74]). *Let $\{E_i | 1 \leq i \leq m\}$ be a set of vector bundles over a projective manifold X. Assume that E_i are k_i-ample and of rank r_i, then*

$$H^q(X, \Omega_X^p \otimes \Lambda^{s_1} E_1 \otimes \cdots \otimes \Lambda^{s_m} E_m) = 0,$$

where $p + q > \dim X + \sum_i s_i(r_i - s_i) + \min_i\{k_i\}$ if either one of the k_i is zero or for each i one of the following holds:

(1) $s_i = 1$,
(2) *E_i is generated by global sections.*

Theorem 7.5 (Demailly [13]). *Let X be a compact complex n-dimensional manifold, E be a vector bundle of rank r, and L be a line bundle over X. Assume that E is ample and L is semi-ample (i.e., some power of L is globally generated), or that E is semi-ample and L is ample. Then*

$$H^q(X, K_X \otimes \Gamma^a E \otimes (\det E)^h \otimes L) = 0$$

for $q \geq 1$, where $\Gamma^a E$ is the bundle associated to E by the irreducible power representation of $\mathrm{Gl}(E)$ of highest weight $a \in \mathbb{Z}^r$ and $h \in \{1, \cdots, r-1\}$ such that $a_1 \geq a_2 \geq \cdots a_h > a_{h+1} = \cdots = a_r = 0$. Moreover,

$$H^q(X, \Omega_X^p \otimes E^k \otimes (\det E)^l) = 0$$

for all $p + q \geq n + 1$, $k \geq 1$, $l \geq n - p + r - 1$.

For the definition of $\Gamma^a E$, one may refer to Demailly's paper [13]. In particular, $\Gamma^a E = S^k E$ for $a = (k, 0, \cdots, 0)$ and $\Gamma^a E = \Lambda^k E$ for $a = (1, \cdots, 1, 0, \cdots, 0)$ with k numbers "1". In case of $\Gamma^a E = S^k E$, it is exactly the Griffiths vanishing theorem.

Theorem 7.6 (Manivel [49]). *Let E be a holomorphic vector bundle of rank r, and L be a line bundle over a projective manifold X of dimension n. Suppose that E is ample and L is nef, or that E is nef and L is ample. Then, for any sequences of integers $k_1, \cdots, k_l$ and $j_1, \cdots, j_m$,*

$$H^q(X, \Omega_X^p \otimes S^{k_1} E \otimes \cdots \otimes S^{k_l} E \otimes \Lambda^{j_1} E \otimes \cdots \otimes \Lambda^{j_m} E \otimes (\det E)^{l+n-p} \otimes L) = 0$$

as soon as $p + q > n + \sum_{s=1}^{m} (r - j_s)$.

Let us mention that there are many vanishing theorems for vector bundles over smooth projective manifolds. It is impossible to collect all of them here, we refer to some references: [42], [43], [44], [45], [46], etc.

Since these results are obtained from the vanishing theorems of line bundles over compact complex manifolds, it is easy to see that most of the vanishing theorems (Griffiths, Le Potier, Demailly, Manivel, $\cdots$) are also valid for weakly pseudoconvex manifolds. For example, the Griffiths vanishing theorem can be stated as follows.

Theorem 7.7. *Let E be a Griffiths positive vector bundle over a weakly pseudoconvex manifold X. Then*

$$H^q(X, K_X \otimes S^m E \otimes \det E) = 0$$

for all $q > 0$ and $m \geq 0$.

8 Coarse vanishing theorems

Theorem 8.1 (Serre [67]). *Let E be a ample vector bundle on the projective scheme X. For any coherent sheaf $\mathscr{F}$, there is a positive integer $m_0 = m_0(\mathscr{F})$ such that*

$$H^q(X, S^m E \otimes \mathscr{F}) = 0$$

for $q > 0$ and $m \geq m_0$.

Actually, Serre's vanishing theorem characterizes ample vector bundles. More explicitly, E is ample if and only if for any given coherent sheaf $\mathscr{F}$ over X, there is a positive integer $m_0(\mathscr{F})$ such that $H^1(X, S^m E \otimes \mathscr{F}) = 0$ for all $m \geq m_0$.

The following result of Fujita shows that the Serre vanishing theorem is uniform with respect to twists by nef divisors. It is very useful in applications.

Theorem 8.2 (Fujita [25]). *Let X be a projective manifold and L be an ample line bundle over X. Given any coherent sheaf $\mathscr{F}$ over X, there exists an integer $m_0 = m_0(\mathscr{F}, L)$ such that*

$$H^q(X, L^m \otimes F \otimes \mathscr{F}) = 0$$

for all $q > 0$, $m \geq m_0$ and any nef line bundle F over X.

We next consider the analytic versions of Serre type vanishing theorems and generalizations. The following result is due to Grauert.

Theorem 8.3 ([27],[29]). *Let X be a compact complex space, and $\mathscr{E}$ be a positive sheaf on X. For any coherent sheaf $\mathscr{F}$, there is a positive integer $m_0 = m_0(\mathscr{F})$ such that*

$$H^q(X, S^m \mathscr{E} \otimes \mathscr{F}) = 0$$

for $q > 0$ and $m \geq m_0$.

Similar characterizations are also valid for positive coherent sheaves. If the coherent sheaves $\mathscr{F}$ in Grauert's vanishing theorem are of the form m_x, m_{xy} and m_x^2, then we can get an embedding of X into the Grassmann manifolds if $\mathscr{E}$ is locally free. If $\mathscr{E}$ is merely a positive coherent sheaf of rank $r > 0$, then we can construct an almost defined map to a Grassmann manifold and conclude that X is Moishezon. It follows that a compact complex space is projective if and only if it carries a positive line bundle and an irreducible reduced compact complex space is a Moishezon space if and only if it carries a positive torsion free coherent sheaf $\mathscr{L}$ with $\mathrm{supp}(\mathscr{L}) = X$.

In the following we recall the notions of q-convexity, which generalize plurisubharmonicity and pseudoconvexity. Assume that $G \subset \mathbb{C}^n$ is a domain, $z \in G$ is an arbitrary point and ψ is a smooth real function in G. The function ψ is called strictly q-convex at z if the Levi form $i\partial\bar{\partial}\psi$ at z has at least $n - q + 1$ positive eigenvalues (with $q = 1, \cdots, n + 1$). The function ψ is called strictly q-convex in G if ψ is strictly q-convex everywhere in G.

Let X be an arbitrary complex space. If ϕ is a continuous real function in X, and $x_0 \in X$ is a point, then ϕ is said to be strictly q-convex in x_0 if there is a neighborhood U of x_0 and a biholomorphic embedding $\rho : U \to G$ of U in a domain $G \subset \mathbb{C}^n$ and a strictly q-convex function ψ in G such that $\phi|_U = \psi \circ \rho$. The notion of q-convexity does not depend on the choice of the embedding ρ.

A complex space X is said to be q-convex if X has a smooth exhaustion function ϕ which is strictly q-convex outside of a compact subset $K \subset X$. X is called q-complete if ϕ can be chosen so that $K = \emptyset$. If X is compact, then X is called 0-convex.

The following finiteness theorem is a generalization of the Cartan-Serre finiteness theorem for compact complex spaces.

Theorem 8.4 (Andreotti and Grauert [3]). *Let $\mathscr{F}$ be a coherent analytic sheaf over a q-convex complex space X. Then the cohomology group $H^k(X, \mathscr{F})$ is finite dimensional for $k \geq q$. Moreover, if X is q-complete, then we have $H^k(X, \mathscr{F}) = 0$ for all $k \geq q$.*

By Ohsawa [59], an n-dimensional complex space X is n-complete if it has no compact irreducible component of dimension n. Then one can get the following theorem of Siu. The special case when X is smooth and $\mathscr{F}$ is locally free is due to Malgrange [47].

Theorem 8.5 (Siu [68]). *If no irreducible component of an n-dimensional complex space X is compact, then $H^n(X, \mathscr{F}) = 0$ for all coherent analytic sheaf $\mathscr{F}$.*

From Andreotti-Grauert finiteness theorem, they derived the following two vanishing theorems. They are generalizations of Theorem 8.3.

Theorem 8.6 ([29]). *Assume that X is a q-convex complex space, that $\mathscr{F}$ is a coherent sheaf on X, and that V is a positive vector bundle over X. Then there is an integer m_0 such that for all $m \geq m_0$ the cohomology $H^k(X, \mathscr{F} \otimes S^m V)$ vanishes for $k \geq q$.*

Another interesting case is that X is compact complex space (0-convex) and E is a holomorphic vector bundle over X. E is called q-negative if there is a q-convex tube $W \Subset E$ around the zero section. The dual of E is called q-positive. A vector bundle is q-positive if and only if it is positive.

Theorem 8.7 ([29]). *Assume that X is a compact complex space, that $\mathscr{F}$ is a coherent sheaf on X and that V is a q-positive vector bundle over X. Then there is a integer m_0 such that for all $m \geq m_0$ the cohomology $H^k(X, \mathscr{F} \otimes S^m V)$ vanishes for $k \geq q$.*

The basic ideas in the proofs are as follows: The coherent sheaf $\mathscr{F}$ can be lifted to a coherent sheaf $\mathscr{H}$ over the complex space E and the expanding of power series gives an injection of the products of cohomology classes of $\mathscr{F}$ and $\mathscr{F} \otimes S^m E$ ($m \geq 1$) into the cohomology classes of $\mathscr{H}$ over a q-convex neighborhood of the zero section of E. Then the desired vanishing theorems are obtained from the Andreotti-Grauert finiteness theorem.

It's natural to discuss the vanishing theorem valued in a psef line bundle over a strongly 1-convex manifold. Using the strong openness property of the multiplier ideal sheaves by Guan and Zhou [36], Meng and Zhou obtain the following vanishing theorem

Theorem 8.8 ([53]). *Let X be a strongly 1-convex manifold and (L, h) be a psef line bundle over X ($i\Theta_{L,h} \geq 0$ in the sense of current). Then*

$$H^q(X, K_X \otimes L \otimes \mathcal{I}(h)) = 0 \quad \text{for} \quad q \geq 1.$$

References

[1] O. Abdelkader, *Annulation de la cohomologie d'une variété kählérienne faiblement 1-complète à valeur dans un fibré vectoriel holomorphe semi-positif*, C. R. Acad. Sci. Paris 290 (1980) 75–78.

[2] Y. Akizuki, S. Nakano, *Note on Kodaira-Spencer's proof of Lefschetz theorems*, Proc. Japan Acad. 30, (1954), 266–272.

[3] A. Andreotti, H. Grauert, *Théorème de finitude pour la cohomologie des espaces complexes*, Bull. Soc. Math. France 90 (1962), 193–259.

[4] B. Berndtsson, *The extension theorem of Ohsawa-Takegoshi and the theorem of Donnelly-Fefferman*, Ann. Inst. Fourier (Grenoble) 46 (1996), no. 4, 1083–1094.

[5] B. Berndtsson, *Curvature of vector bundles associated to holomorphic fibrations*, Ann. of Math. (2) 169 (2009), no. 2, 531–560.

[6] Z. Błocki, *On the Ohsawa-Takegoshi extension theorem*, Univ. Iagel. Acta Math. No. 50, [2012 on articles] (2013), 53–61.

[7] Z. Błocki, *Suita conjecture and the Ohsawa-Takegoshi extension theorem*, Invent. Math. 193 (2013), no. 1, 149–158.

[8] F. Bogomolov, *Unstable vector bundles and curves on surfaces*, Proceedings of the International Congress of Mathematicians (Helsinki, 1978), 517–524.

[9] R. Bott, *Homogeneous vector bundles*, Ann. of Math. (2) 66 (1957), 203–248.

[10] S. Boucksom, *Cônes positifs des variétés complexes compactes*, Thesis, Grenoble, 2002.

[11] J.Y. Cao, *Numerical dimension and a Kawamata-Viehweg-Nadel type vanishing theorem on compact Kähler manifolds*, Compositio Math. 150 (2014), 1869–1902.

[12] J.-P. Demailly, *Champs magnétiques et inégalités de Morse pour la d''-cohomologie*, Ann. Inst. Fourier 35 (1985), 189–229.

[13] J.-P. Demailly, *Vanishing theorems for tensor powers of an ample vector bundle*, Invent. Math. 91 (1988), no. 1, 203–220.

[14] J.-P. Demailly, *Transcendental proof of a generalized Kawamata-Viehweg vanishing theorem*, Geometrical and algebraical aspects in several complex variables (Cetraro, 1989), 81-94, Sem. Conf., 8, EditEl, Rende, 1991.

[15] J.-P. Demailly, *Regularization of closed positive currents of type* $(1,1)$ *by the flow of a Chern connection*, Actes du Colloque en l'honneur de P. Dolbeault (Juin 1992), édité par H. Skoda et J.-M. Trépreau, Aspects of Mathematics, Vol. E 26, Vieweg, 1994, 105–126.

[16] J.-P. Demailly, *On the Ohsawa-Takegoshi-Manivel L^2 extension theorem*, Complex analysis and geometry (Paris, 1997), Progr. Math., 188, Birkhäuser, Basel, 2000, 47–82.

[17] J-P. Demailly, *Multiplier ideal sheaves and analytic methods in algebraic geometry*, School on Vanishing Theorems and Effective Results in Algebraic Geometry (Trieste, 2000), ICTP Lect. Notes 6, Abdus Salam Int. Cent. Theoret. Phys., Trieste, 2001, 1–148.

[18] J.-P. Demailly, *Analytic methods in algebraic geometry*, Surveys of Modern Mathematics 1, Higher Education Press, Beijing, 2010.

[19] J-P. Demailly, *Complex analytic and differential geometry*, available at https://www-fourier.ujf-grenoble.fr/~demailly/manuscripts/agbook.pdf.
[20] J-P. Demailly, Th. Peternell, *A Kawamata-Viehweg vanishing theorem on compact Kähler manifolds*, J. Differential Geom. 63 (2003), no. 2, 231-277.
[21] J.-P. Demailly, H. Skoda, *Relations entre les notions de positivités de P. A. Griffiths et de S. Nakano pour les fibrés vectoriels*, Séminaire Pierre Lelong-Henri Skoda (Analyse), Années 1978/79, 304–309, Lecture Notes in Math., 822, Springer, Berlin, 1980.
[22] I. Enoki, *Kawamata-Viehweg vanishing theorem for compact Kähler manifolds*, Einstein metrics and Yang-Mills connections (ed. T. Mabuchi, S. Mukai), Marcel Dekker, (1993), 59–68.
[23] H. Esnault, E. Viehweg, *Lectures on vanishing theorems*, DMV Seminar, 20. Birkhäuser Verlag, Basel, 1992.
[24] O. Fujino, *A transcendental approach to Kollár's injectivity theorem II*, J. Reine Angew. Math. 681 (2013), 149–174.
[25] T. Fujita, *Vanishing theorems for semipositive line bundles*, Algebraic geometry (Tokyo/Kyoto, 1982), 519–528, Lecture Notes in Math., 1016, Springer, Berlin, 1983.
[26] J. Girbau, *Sur le théorème de Le Potier d'annulation de la cohomologie*, C. R. Acad. Sci. Paris 283 (1976), 355–358.
[27] H. Grauert, *Über Modifikationen und exzeptionelle analytische Mengen*, Math. Ann. 146 (1962) 331–368.
[28] H. Grauert, O. Riemenschneider, *Verschwindungssätze für analytische Kohomologiegruppen auf komplexen Räumen*, Invent. Math. 11 (1970), 263–292.
[29] H. Grauert, Th. Peternell, R. Remmert, editors. *Several complex variables*, VII. volume 74 of Encyclopaedia of Mathematical Sciences. Springer-Verlag, Berlin, 1994. Sheaf-theoretical methods in complex analysis.
[30] P. Griffiths, *Hermitian differential geometry, Chern classes, and positive vector bundles*, Global Analysis (Papers in Honor of K. Kodaira) Univ. Tokyo Press, Tokyo 1969, 185–251.
[31] Q.A. Guan, X.Y. Zhou, *Optimal constant problem in the L^2 extension theorem*, C. R. Acad. Sci. Paris, Ser. I 350 15-16 (2012) 753–756.
[32] Q.A. Guan, X.Y. Zhou, *Generalized L^2 extension theorem and a conjecture of Ohsawa*, C. R. Acad. Sci. Paris, Ser. I 351 (2013) 111–114.
[33] Q.A. Guan and X.Y. Zhou, *Optimal constant in an L^2 extension problem and a proof of a conjecture of Ohsawa*, Sci. China Math. 58 (2015), no. 1, 35–59.
[34] Q.A. Guan, X.Y. Zhou, *An L^2 extension theorem with optimal estimate*, C. R. Math. Acad. Sci. Paris 352 (2014), no. 2, 137–141.
[35] Q.A. Guan and X.Y. Zhou, *A solution of an L^2 extension problem with an optimal estimate and applications*, Ann. of Math. (2) 181 (2015), no. 3, 1139–1208.
[36] Q.A. Guan and X.Y. Zhou, *A proof of Demailly's strong openness conjecture*, Ann. of Math. (2) 182 (2015), no. 2, 605–616.
[37] Q.A. Guan, X.Y. Zhou, L.F. Zhu, *On the Ohsawa-Takegoshi L^2 extension theorem and the twisted Bochner-Kodaira identity*, C. R. Acad. Sci. Paris

349 (2011) 797–800.

[38] R. Hartshorne, *Ample vector bundles*, Inst. Hautes Études Sci. Publ. Math. No. 29 (1966) 63–94.

[39] Y. Kawamata, *A generalization of Kodaira-Ramanujam's vanishing theorem*, Math. Ann. 261 (1982), no. 1, 43–46.

[40] K. Kodaira, *On a differential-geometric method in the theory of analytic stacks*, Proc. Nat. Acad. Sci. U. S. A. 39, (1953), 1268–1273.

[41] J. Kollár, *Higher direct images of dualizing sheaves*, I. Ann. of Math. (2) 123 (1986), no. 1, 11-42.

[42] F. Laytimi, W. Nahm, *A generalization of Le Potier's vanishing theorem*, Manuscripta Math. 113 (2004), 165-189.

[43] F. Laytimi, W. Nahm, *A vanishing theorem*, Nagoya Math. J. 180 (2005), 35-43.

[44] F. Laytimi, W. Nahm, *On a vanishing problem of Demailly*, Int. Math. Res. Not. (2005), 2877-2889.

[45] K. Liu, X. Yang, *Effective vanishing theorems for ample and globally generated vector bundles*, Comm. Anal. Geom. 23 (2015), 797-818.

[46] K. Liu, X. Sun, X. Yang, *Positivity and vanishing theorems for ample vector bundles*, J. Algebraic Geom. 22 (2013), 303-331.

[47] B. Malgrange, *Existence et approximation des solutions des équations aux dérivées partielles et des équations de convolution*, Ann. Inst. Fourier, Grenoble 6 (1955-1956), 271–355.

[48] L. Manivel, *Un théorème de prolongement L^2 de sections holomorphes d'un fibré vectoriel*, Math. Zeitschrift, 212 (1993), 107–122.

[49] L. Manivel, *Vanishing theorems for ample vector bundles*, Invent. Math. 127 (1997), no. 2, 401–416.

[50] S. Matsumura, *A vanishing theorem of Kollár-Ohsawa type*, Math. Ann. 366 (2016), 1451-1465.

[51] J. McNeal, *On large values of L^2 holomorphic functions*, Math. Res. Let. 3 (1996), 247–259.

[52] J. McNeal and D. Varolin, *Analytic inversion of adjunction: L^2 extension theorems with gain*, Ann. Inst. Fourier (Grenoble) 57 (2007), no. 3, 703–718.

[53] X.K. Meng, X.Y. Zhou, *Pseudo-effective line bundle over holomorphically convex manifold*, to appear in JAG.

[54] A. Nadel, *Multiplier ideal sheaves and existence of Kähler-Einstein metrics of positive scalar curvature*, Ann. of Math. (2) 132 (1990), no. 3, 549–596.

[55] S. Nakano, *On complex analytic vector bundles*, J. Math. Soc. Japan 7 (1955), 1–12.

[56] S. Nakano, *Vanishing theorems for weakly 1 -complete manifolds, II*, Publ. Res. Inst. Math. Sci. 10 (1974/75), no. 1, 101–110.

[57] T. Ohsawa, *On $H^{p,q}(X, B)$ of weakly 1-complete manifolds*, Publ. Res. Inst. Math. Sci. 17 (1981), no. 1, 113–126.

[58] T. Ohsawa, *Vanishing Theorems on complete Kähler manifolds*, Publ. RIMS, Kyoto Univ. 20 (1984), 21-38.

[59] T. Ohsawa, *Completeness of noncompact analytic spaces*, Publ. Res. Inst. Math. Sci. 20 (1984), no. 3, 683–692.

[60] T. Ohsawa, *On the extension of L^2 holomorphic functions, III: Negligible weights*, Math. Z. 219 (1995), no. 2, 215–225.
[61] T. Ohsawa, *On the extension of L^2 holomorphic functions, V: effects of generalization*, Nagoya Math. J. 161 (2001), 1–21. *Erratum to "On the extension of L^2 holomorphic functions, V: Effects of generalization"*, Nagoya Math. J. 163 (2001), 229.
[62] T. Ohsawa, *On a curvature condition that implies a cohomology injectivity theorem of Kollár-Skoda type*, Publ. Res. Inst. Math. Sci. 41 (2005), no. 3, 565–577.
[63] T. Ohsawa and K. Takegoshi, *On the extension of L^2 holomorphic functions*, Math. Z. 195 (1987), 197–204.
[64] T. Peternell, J. Le Potier, M. Schneider, *Vanishing theorems, linear and quadratic normality*, Invent. Math. 87 (1987), no. 3, 573–586.
[65] C. Ramanujam, *Remarks on the Kodaira vanishing theorem*, J. Indian Math. Soc. 36 (1972), 41–51.
[66] B. Schiffman, A. Sommese, *Vanishing theorems on complex manifolds*, Progress in Mathematics 56, Birkhäuser 1985.
[67] J.-P. Serre, *Faisceaux algébriques cohérents*, Ann. of Math. (2) 61 (1955). 197–278.
[68] Y.-T. Siu, *Analytic sheaf cohomology groups of dimension n of n-dimensional noncompact complex manifolds*, Pacific J. Math. 28 (1969) 407–411.
[69] Y.-T. Siu, *Complex-analyticity of harmonic maps, vanishing and Lefschetz theorems*, J. Differential Geom. 17 (1982), 55-138.
[70] Y.-T. Siu, *A vanishing theorem for semi-positive line bundles over non-Kähler manifolds*, J. Differential Geom. 19 (1984), 431–452.
[71] Y.-T. Siu, *The Fujita conjecture and the extension theorem of Ohsawa-Takegoshi*, Geometric complex analysis (Hayama, 1995), World Sci. Publ., River Edge, NJ, 1996, 577–592.
[72] Y.-T. Siu, *Extension of twisted pluricanonical sections with plurisubharmonic weight and invariance of semipositively twisted plurigenera for manifolds not necessarily of general type*, Complex geometry, Collection of papers dedicated to Hans Grauert, Springer-Verlag, Berlin, 2002, 223–277.
[73] Y.-T. Siu, *Multiplier ideal sheaves in complex and algebraic geometry*, Sci. China Ser. A 48 (2005), suppl., 1–31.
[74] A. Sommese, *Submanifolds of Abelian varieties*, Math. Ann. 233 (1978), 229–256.
[75] K. Takegoshi, *A generalization of vanishing theorems for weakly 1-complete manifolds*, Publ. Res. Inst. Math. Sci. 17 (1981), no. 1, 311–330.
[76] K. Takegoshi, *Relative vanishing theorems in analytic spaces*, Duke Math. J. 52 (1985), 273-279.
[77] E. Viehweg, *Vanishing theorems*, J. Reine Angew. Math. 335 (1982), 1–8.
[78] S.-T. Yau, *On the Ricci curvature of a compact Kähler manifold and the complex Monge-Ampère equation*, I. Comm. Pure Appl. Math. 31 (1978), 339-411.
[79] X.Y. Zhou, *A Survey on L^2 extension problem*, Complex geometry and dynamics, Abel Symp., 10, Springer, Cham, 2015, 291–309.

[80] X.Y. Zhou and L.F. Zhu, *Ohsawa-Takegoshi L^2 extension theorem: revisited*, Fifth International Congress of Chinese Mathematicians, Part 1, 2, AMS/IP Stud. Adv. Math., 51, pt. 1, 2, Amer. Math. Soc., Providence, RI, 2012, 475–490.

[81] X.Y. Zhou and L.F. Zhu, *An optimal L^2 extension theorem on weakly pseudoconvex Kähler manifolds*, to appear in J. Differential Geom..

[82] L.F. Zhu, Q.A. Guan, X.Y. Zhou, *On the Ohsawa-Takegoshi L^2 extension theorem and the Bochner-Kodaira identity with non-smooth twist factor*, J. Math. Pures Appl. 97 (2012) 579–601.

Yau Mathematical Sciences Center, Tsinghua University
Beijing 100084, China
E-mail address: mengxiankui@amss.ac.cn

Institute of Mathematics, AMSS, and Hua Loo-Keng Key Laboratory of Mathematics, Chinese Academy of Sciences, Beijing 100190, China and School of Mathematical Sciences, University of Chinese Academy of Sciences, Beijing 100049, China
E-mail address: xyzhou@math.ac.cn

School of Mathematics and Statistics, Wuhan University
Wuhan 430072, China
E-mail address: zhulangfeng@amss.ac.cn

Proceedings of the 7th International Congress of Chinese Mathematicians
ALM 43, pp. 439–455

On the Geometry of Orbifold Gromov-Witten Invariants

Hsian-Hua Tseng

Abstract

We consider the question of how geometric structures of a Deligne-Mumford stack affect its Gromov-Witten invariants. The two geometric structures studied here are *gerbes* and *root constructions*. In both cases, we explain conjectures on Gromov-Witten theory for these stacks and survey some recent progress on these conjectures.

Mathematics Subject Classification: 14N35, 53D45, 14D23.

Keywords and Phrases: Gromov-Witten theory, Deligne-Mumford stacks, gerbes, root constructions.

1 Introduction

1.1 Overview

Gromov-Witten theory of Deligne-Mumford stacks was introduced in [4, 5] after the work [15] in symplectic setting. It has been an active area of research in the past decade, in particular in connection with the *crepant resolution conjecture* [18], [19], [13] and with mirror symmetry (see e.g. [16]).

In [42], the author gave a survey on some aspects of orbifold Gromov-Witten theory. This paper can be considered as a sequel to [42]: here we further elaborate on some questions discussed in [42] and give an update on recent progress. In Section 1.2 we explain some natural geometric structures of Deligne-Mumford stacks. In Section 1.3, we indicate that these structures lead to the following two questions:

1. the structure of Gromov-Witten theory of gerbes, which we discuss in Section 3.
2. the determination of the Gromov-Witten theory of a root stack, which we discuss in Section 4.

Basic definitions and constructions of orbifold Gromov-Witten theory are given in many places, for example [4, 5], [15], and [42], and are reviewed in Section 2 below. Throughout this paper, we work over $\mathbb{C}$.

1.2 Geometry of Deligne-Mumford stacks

We discuss the geometric structure of a smooth Deligne-Mumford stack in relation to its coarse moduli space.

Let $\mathcal{X}$ be a smooth Deligne-Mumford stack. By the main result of [28], $\mathcal{X}$ admits a coarse moduli space X which we assume to be a scheme. The canonical morphism

$$\mathcal{X} \to X$$

is universal in the sense that any morphism $\mathcal{X} \to Z$ with Z a scheme must factor through X, i.e. there exists a morphism $X \to Z$ such that the following diagram is commutative:

$$\begin{array}{ccc} \mathcal{X} & & \\ \downarrow & \searrow & \\ X & \longrightarrow & Z. \end{array}$$

Let G be the generic stabilizer group of $\mathcal{X}$. The construction of *rigidification* in [2] allows one to "remove" the group G from stabilizers. More precisely, this yields a stack $\mathcal{X}_1$ together with a morphism[1]

$$\mathcal{X} \to \mathcal{X}_1$$

such that the generic stabilizer group of $\mathcal{X}_1$ is G/G, i.e. the trivial group, and $\mathcal{X}_1$ also has X as its coarse moduli space. We thus have a factorization

$$\begin{array}{ccc} \mathcal{X} & & \\ \downarrow & \searrow & \\ \mathcal{X}_1 & \longrightarrow & X. \end{array}$$

By construction, the morphism $\mathcal{X} \to \mathcal{X}_1$ has the structure of a *gerbe*. The stacky locus (i.e. the locus where the stabilizer groups are not trivial) of $\mathcal{X}_1$ is of codimension at least 1.

By assumption, the scheme X has only quotient singularities. Étale locally X is a quotient of a smooth scheme by a finite group. These local presentations of X assemble to an étale groupoid. Let

$$\mathcal{X}_{can}$$

denote the stack quotient of this groupoid. The coarse moduli space of $\mathcal{X}_{can}$ is also X, and we have a canonical morphism

$$\mathcal{X}_{can} \to X.$$

$\mathcal{X}_{can}$ is the *canonical stack* associated to X, in the sense of [20, Section 4.1]. By [20, Theorem 4.6], for any Deligne-Mumford stack $\mathcal{Y}$ with trivial generic stabilizer

[1]This morphism is also constructed in [12].

and a dominant codimension preserving morphism $\mathcal{Y} \to X$, there exists a unique factorization

$$\begin{array}{ccc} \mathcal{Y} & \longrightarrow & \mathcal{X}_{can} \\ \downarrow & \swarrow & \\ X. & & \end{array}$$

In particular, there is a morphism

$$\mathcal{X}_1 \to \mathcal{X}_{can}.$$

Therefore we have a factorization of $\mathcal{X} \to X$:

$$\mathcal{X} \to \mathcal{X}_1 \to \mathcal{X}_{can} \to X. \tag{1.1}$$

By [21], under some mild hypothesis, the map $\mathcal{X}_1 \to \mathcal{X}_{can}$ is a composition of the *root construction* of [5] and [14]. A review of root constructions is given in Section 4.1.

1.3 Structures of Gromov-Witten theory

In this paper, we take the following perspective: the factorization (1.1) leads to structural results of Gromov-Witten invariants of $\mathcal{X}$.

1.3.1 Stage 1

The first stage of the factorization, $\mathcal{X} \to \mathcal{X}_1$, is a gerbe. In Section 3 below we explain the conjectural description of Gromov-Witten theory of $\mathcal{X}$ obtained from this gerbe structure, see Conjecture 3.1. As evidence supporting the conjecture, we also explain in some details the state-of-the-art result on the structure of Gromov-Witten invariants of a large class of gerbes whose bands are trivial.

1.3.2 Stage 2

The second stage of the factorization, $\mathcal{X}_1 \to \mathcal{X}_{can}$, is a composition of root constructions. It is natural[2] to expect that the Gromov-Witten theory of a root construction can be explicitly determined by the input data, namely the base stack and the divisor along which the root is taken. However, such a result is not proven in full generality. In Section 4, we formulate this expectation more precisely as Conjecture 4.1. We also discuss some details of a proof of Conjecture 4.1 in a special case[3].

1.3.3 Stage 3

The third stage of this factorization, $\mathcal{X}_{can} \to X$, is canonical: as mentioned above, the stack $\mathcal{X}_{can}$ is canonically associated to the scheme X. By construction

[2]Apparently experts have believed this to be true.

[3]Despite the expectation, even this special case was proven very recently, in 2015.

the stacky locus of $\mathcal{X}_{can}$ is of codimension at least 2. Therefore, the morphism $\mathcal{X}_{can} \to X$ is birational and *crepant* (i.e. it identifies the canonical divisors $K_{\mathcal{X}_{can}}$ and K_X via pull-back). As advocated by M. Satriano and others, $\mathcal{X}_{can} \to X$ should be considered as a *stacky (crepant) resolution*. The so-called *crepant resolution conjecture*, originated from String Theory, predicts an equivalence between the Gromov-Witten theory of $\mathcal{X}_{can}$ and Gromov-Witten theory of other crepant resolutions of X (if existed). As noted above, crepant resolution conjecture has been the main driving force for much of the developments in orbifold Gromov-Witten theory in the past decade. In this paper we do not attempt to give a survey on the status of crepant resolution conjecture. The papers [19] and [17] contains nice discussions about this conjecture, to which we refer the readers.

1.4 Acknowledgment

The author is very grateful to Xiang Tang and Fenglong You for inspiring and fruitful collaborations on results surveyed in this paper. The author is supported in part by NSF grant DMS-1506551 and a Simons Foundation Collaboration Grant.

2 Gromov-Witten theory of Deligne-Mumford stacks

In this Section we give a review on the basic constructions of Gromov-Witten theory of Deligne-Mumford stacks. In addition to the original papers [15], [4], and [5] mentioned above, expository accounts of Gromov-Witten theory of Deligne-Mumford stacks can be found in [1] and [41].

Let $\mathcal{X}$ be a smooth proper Deligne-Mumford stack over $\mathbb{C}$ with projective coarse moduli space X. The Gromov-Witten theory of $\mathcal{X}$ is based on moduli stacks of stable maps[4] to $\mathcal{X}$. Let $g, n \in \mathbb{Z}_{\geq 0}$ and let $d \in H_2(X, \mathbb{Q})$. The moduli stack

$$\overline{\mathcal{M}}_{g,n}(\mathcal{X}, d)$$

of n-pointed genus g stable maps to $\mathcal{X}$ of degree d parametrizes objects of the following kind:

$$(f : \mathcal{C} \to \mathcal{X}, p_1, ..., p_n)$$

where

1. $\mathcal{C}$ is a nodal orbifold curve. This means that $\mathcal{C}$ is a connected nodal curve with isolated stacky points of the following two types:
   (a) If a stacky point is not nodal, then it has an étale neighborhood of the form

$$[\mathrm{Spec}\mathbb{C}[x]/\mu_r]$$

   for some μ_r where $\zeta \in \mu_r$ acts on the coordinate x by multiplication.

[4]These are often called *twisted stable maps* [6] or *orbifold stable maps* [15].

(b) If a stacky point is nodal, then it has an étale neighborhood of the form

$$[(\mathrm{Spec}\mathbb{C}[x,y]/(xy))/\mu_r]$$

for some μ_r where $\zeta \in \mu_r$ acts on the coordinates[5] x, y via

$$x \mapsto \zeta x, \quad y \mapsto \zeta^{-1} y.$$

2. $f : \mathcal{C} \to \mathcal{X}$ is a representable morphism of stacks. This means that for $x \in \mathcal{C}$, the group homomorphism between stabilizer groups induced by f,

$$Aut(x) \to Aut(f(x)),$$

is injective.
3. $p_1, ..., p_n \in \mathcal{C}$ are distinct (possibly stacky) points which are not nodal.
4. $f_*[\mathcal{C}] = d$. Here we use the fact that $H_2(\mathcal{X}, \mathbb{Q}) \simeq H_2(X, \mathbb{Q})$.
5. The data $(f : \mathcal{C} \to \mathcal{X}, p_1, ..., p_n)$ have only finitely many automorphisms. Combinatorially this means the following:
   (a) On an irreducible component of $\mathcal{C}$ of genus 0, the number of special points (i.e. marked points and nodes) is at least 3.
   (b) On an irreducible component of $\mathcal{C}$ of genus 1, the number of special points is at least 1.

Furthermore, we require a technical condition. For a scheme S, consider an S-family of stable maps to $\mathcal{X}$:

$$\begin{array}{ccc} (\mathcal{C}, p_1, ..., p_n) & \xrightarrow{f} & \mathcal{X} \\ \downarrow & & \\ S. & & \end{array} \tag{2.1}$$

Suppose the i-th marked point p_i is a stacky point with stabilizer group μ_{r_i}. Then the locus of the i-th marked point of $\mathcal{C}$ is a μ_{r_i}-gerbe over S. We require a technical condition that this gerbe has sections. This condition has several benefits. For instances, the universal family of stable maps,

$$\begin{array}{ccc} \mathcal{C} & \xrightarrow{f} & \mathcal{X} \\ {\scriptstyle\pi}\downarrow & & \\ \overline{\mathcal{M}}_{g,n}(\mathcal{X}, d), & & \end{array} \tag{2.2}$$

admits sections given by marked points. Such sections are useful in the calculations of [41].

[5]The action of μ_r on coordinates x, y considered here, which leaves xy invariant, is called *balanced*. Such a node is called a balanced node. The more general notion of stable maps to $\mathcal{X}$ introduced in [6] allows nodes which are not balanced. However, for Gromov-Witten theory we require the nodes to be balanced.

Also, for the stable map (2.1), since the gerbe of the i-th marked point has sections, this gerbe is trivial. Hence the locus of the i-th marked points in $\mathcal{C}$ is isomorphic to $S \times B\mu_{r_i}$. Restricting the map f to this locus yields a map from $S \times B\mu_{r_i}$ to $\mathcal{X}$. Following [5, Definition 3.1.5], this is an S-point of the cyclotomic inertia stack of $\mathcal{X}$. As explained in [4, Section 4.4], the cyclotomic inertia stack of $\mathcal{X}$ is the same as the *inertia stack* of $\mathcal{X}$,

$$I\mathcal{X} := \mathcal{X} \times_{\Delta, \mathcal{X}\times\mathcal{X}, \Delta} \mathcal{X},$$

where $\Delta : \mathcal{X} \to \mathcal{X} \times \mathcal{X}$ is the diagonal morphism. Thus, this restriction construction yields a morphism

$$ev_i : \overline{\mathcal{M}}_{g,n}(\mathcal{X}, d) \to I\mathcal{X},$$

called the i-th evaluation map. Without this technical condition, the evaluation map takes values in the *rigidified* inertia stack, see [4], [5] for details. For the purpose of this paper it is more convenient to work with inertia stacks, hence we impose this technical condition.

By the main results of [6], the moduli stack $\overline{\mathcal{M}}_{g,n}(\mathcal{X}, d)$ is a proper Deligne-Mumford stack with projective coarse moduli space. In general, the stack $\overline{\mathcal{M}}_{g,n}(\mathcal{X}, d)$ can be pretty bad: it can be arbitrarily singular, with many components of various dimensions, etc. To obtain intersection numbers with reasonable properties, we need to use virtual fundamental classes. As in [5], the object $R^*\pi_* f^* T_{\mathcal{X}}$ in the derived category of $\overline{\mathcal{M}}_{g,n}(\mathcal{X}, d)$ yields a perfect obstruction theory in the sense of [11]. By the recipe of [11], this yields a virtual fundamental class

$$[\overline{\mathcal{M}}_{g,n}(\mathcal{X}, d)]^w \in H_*(\overline{\mathcal{M}}_{g,n}(\mathcal{X}, d), \mathbb{Q}).$$

We should point out that, because of the technical condition on marked points mentioned above, this virtual fundamental class needs a minor modification (a suitable rescaling). See [4] and [41] for details.

The i-th tautological line bundle on $\overline{\mathcal{M}}_{g,n}(\mathcal{X}, d)$,

$$L_i \to \overline{\mathcal{M}}_{g,n}(\mathcal{X}, d),$$

has fibers over a stable map $(f : \mathcal{C} \to \mathcal{X}, p_1, ..., p_n)$ the cotangent line $T^*_{\bar{p}_i} C$ of the *coarse curve* C at the (image of) i-th marked point. The descendant classes on $\overline{\mathcal{M}}_{g,n}(\mathcal{X}, d)$ are first Chern classes of L_i,

$$\psi_i := c_1(L_i), \quad 1 \le i \le n.$$

We are now ready to define Gromov-Witten invariants of $\mathcal{X}$. Given integers $k_1, ..., k_n \ge 0$ and classes in the *Chen-Ruan orbifold cohomology of* $\mathcal{X}$, $a_1, ..., a_n \in H^*(I\mathcal{X})$, we define the following descendant Gromov-Witten invariant of $\mathcal{X}$,

$$\left\langle \prod_{i=1}^n a_i \psi_i^{k_i} \right\rangle_{g,n,\beta}^{\mathcal{X}} := \int_{[\overline{\mathcal{M}}_{g,n}(\mathcal{X},d)]^w} \prod_{i=1}^n ev_i^*(a_i)\psi_i^{k_i} \in \mathbb{Q}. \tag{2.3}$$

3 Codimension 0: gerbes

In this Section we discuss Gromov-Witten theory of gerbes. Our study is guided by a conjecture originated in Physics [26] and further developed in [37]. This is Conjecture 3.1 below.

3.1 The conjecture

Essentially a gerbe is a stack with non-trivial generic stabilizers. Let $\mathcal{Y}$ be a smooth Deligne-Mumford stack whose generic stabilizer is a finite group G. Then there is a map

$$\pi : \mathcal{Y} \to \mathcal{B}$$

where $\mathcal{B}$ is a smooth Deligne-Mumford stack with trivial generic stabilizer. The map π is known as a *G-gerbe* over $\mathcal{B}$. The only G-gerbe over $\mathcal{B} = \mathrm{pt}$ is

$$BG = [\mathrm{pt}/G].$$

Given an open cover $\{U_i\}$ of $\mathcal{B}$, one can construct $\mathcal{Y}$, the total space of a G-gerbe over $\mathcal{B}$, by gluing $U_i \times BG$ using the following data:

$$\begin{aligned} &\phi_{ij} \in Aut(G) \quad \text{for each double overlap } U_{ij} := U_i \cap U_j, \\ &g_{ijk} \in G \quad \text{for each triple overlap } U_{ijk} := U_i \cap U_j \cap U_k. \end{aligned} \tag{3.1}$$

The following requirements are imposed:

$$\begin{aligned} &\phi_{jk} \circ \phi_{ij} = \mathrm{Ad}_{g_{ijk}} \circ \phi_{ik} \quad \text{on } U_{ijk} \\ &g_{jkl} g_{ijl} = \phi_{kl}(g_{ijk}) g_{ikl} \quad \text{on } U_i \cap U_j \cap U_k \cap U_l. \end{aligned} \tag{3.2}$$

Let $Out(G) := Aut(G)/Inn(G)$ be the group of outer automorphisms of G. By (3.2), the images of ϕ_{ij} under the quotient map $Aut(G) \to Out(G)$ satisfy the usual compatibility requirement for principal bundles and can be used to construct a principal $Out(G)$-bundle over $\mathcal{B}$, which we denote by

$$\overline{\mathcal{Y}} \to \mathcal{B}.$$

This principal bundle $\overline{\mathcal{Y}} \to \mathcal{B}$ is called the *band* of the G-gerbe π. As in [26], for a G-gerbe $\pi : \mathcal{Y} \to \mathcal{B}$, one can construct a pair

$$(\widehat{\mathcal{Y}}, c),$$

where $\widehat{\mathcal{Y}}$ is a disconnected space with an étale map $\widehat{\pi} : \widehat{\mathcal{Y}} \to \mathcal{B}$ and c is a $U(1)$-valued 2-cocycle on $\widehat{\mathcal{Y}}$, see also [37]. More precisely,

$$\widehat{\mathcal{Y}} := [\overline{\mathcal{Y}} \times \widehat{G}/Out(G)].$$

Here $\widehat{G}$ is the set of isomorphism classes of irreducible complex representations of G, treated as a space of isolated points. The group $Out(G)$ acts on $\widehat{G}$ by precompositions and acts on $\overline{\mathcal{Y}} \times \widehat{G}$ diagonally. The $U(1)$-valued 2-cocycle c measures the failure of certain family of vector spaces to be a locally free sheaf on $\widehat{\mathcal{Y}}$, see [37] for more details. The Gromov-Witten theory of $\widehat{\mathcal{Y}}$ incorporating a twist by the cocycle c can be obtained from the general construction of [35].

Conjecture 3.1 (see [26] and [37]). *For a G-gerbe $\mathcal{Y} \to \mathcal{B}$, the Gromov-Witten theory of $\mathcal{Y}$ is equivalent to the Gromov-Witten theory of $\widehat{\mathcal{Y}}$ twisted by c.*

Upon first look, Conjecture 3.1 seems strange and unmotivated. It is thus surprising that Conjecture 3.1 turns out to be correct in all examples studied so far: it is proven for specific classes of examples in [27], [8, 9, 10], and [37, 38]. In [38] Conjecture 3.1 is proven for G-gerbes $\mathcal{Y} \to \mathcal{B}$ whose bands $\overline{\mathcal{Y}} \to \mathcal{B}$ are trivializable principal bundles. This is so far the largest class of examples. In the rest of this Section we give a survey on the results of [38].

3.2 The trivial band case

When the band $\overline{\mathcal{Y}} \to \mathcal{B}$ is trivializable, i.e. $\overline{\mathcal{Y}} \simeq \mathcal{B} \times Out(G)$, the construction implies that

$$\widehat{\mathcal{Y}} \simeq \mathcal{B} \times \widehat{G}.$$

In other words, $\widehat{\mathcal{Y}}$ is a disjoint union of $|\widehat{G}|$ copies of $\mathcal{B}$. Furthermore, on each copy of $\mathcal{B}$ in $\widehat{\mathcal{Y}}$, the map $\widehat{\pi}$ restricts to the identity map on $\mathcal{B}$. For $[\rho] \in \widehat{G}$, we write

$$\mathcal{B}_\rho \subset \widehat{\mathcal{Y}}$$

for the copy of $\mathcal{B}$ indexed by $[\rho]$. The restriction of the cocycle c onto $\mathcal{B}_\rho$, which we denote by c_ρ, defines a cohomology class

$$[c_\rho] \in H^2(\mathcal{B}, U(1)).$$

This class has a very simple description as follows. By the classification of gerbes with trivial bands (see e.g. [22]), the G-gerbe $\pi : \mathcal{Y} \to \mathcal{B}$ determines a class

$$[\pi] \in H^2(\mathcal{B}, Z(G)),$$

where $Z(G) \subset G$ is the center of G. For an irreducible representation $\rho : G \to GL(V_\rho)$, by Schur's lemma, the restriction $\rho|_{Z(G)} : Z(G) \to GL(V_\rho)$ determines a homomorphism $\rho|_{Z(G)} : Z(G) \to U(1)$. Using this to change coefficients, we obtain a map

$$H^2(\mathcal{B}, Z(G)) \to H^2(\mathcal{B}, U(1)).$$

The class $[c_\rho]$ is the image of $[\pi]$ under this map.

The Gromov-Witten theory of $\widehat{\mathcal{Y}}$ twisted by c is understood to be the Gromov-Witten theory of $\mathcal{B}_\rho$ twisted by c_ρ for all $[\rho] \in \widehat{G}$ put together. For the precise meaning of this, see Theorem 3.1 below. By the construction of [35], the Gromov-Witten theory of $\mathcal{B}_\rho$ twisted by c_ρ depends only on the class $[c_\rho]$, not the cocycle c_ρ.

The "state space" of the Gromov-Witten theory of $\mathcal{Y}$ is the Chen-Ruan orbifold cohomology $H^*_{CR}(\mathcal{Y})$. Additively,

$$H^*_{CR}(\mathcal{Y}) := H^*(I\mathcal{Y}, \mathbb{C})$$

is the cohomology of $I\mathcal{Y} := \mathcal{Y} \times_{\mathcal{Y}\times\mathcal{Y}} \mathcal{Y}$, the inertial stack of $\mathcal{Y}$. The state space of the Gromov-Witten theory of $\mathcal{B}$ twisted by c_ρ is the twisted orbifold cohomology $H^*_{orb}(\mathcal{B}, c_\rho)$ of $(\mathcal{B}, c_\rho)$ introduced in [36]. Additively,

$$H^*_{orb}(\mathcal{B}, c_\rho) := H^*(I\mathcal{B}, \mathcal{L}_{c_\rho})$$

is the cohomology of the inertial stack $I\mathcal{B}$ with coefficients in a line bundle $\mathcal{L}_{c_\rho} \to I\mathcal{B}$ known as inner local system. As in [36], the line bundle $\mathcal{L}_{c_\rho}$ can be explicitly constructed[6] using c_ρ as a cocycle.

By a result[7] of [37], there is an isomorphism of graded vector spaces,

$$I : H^*_{CR}(\mathcal{Y}) \to \oplus_{[\rho]\in\widehat{G}} H^*_{orb}(\mathcal{B}, c_\rho).$$

For $\delta \in H^*_{CR}(\mathcal{Y})$, we write $I(\delta) = \sum_{[\rho]\in\widehat{G}} I(\delta)_\rho$ where $I(\delta)_\rho \in H^*_{orb}(\mathcal{B}, c_\rho)$.

The main result of [38] is the following, which confirms Conjecture 3.1 for G-gerbes with trivial bands in a precise form.

Theorem 3.1 ([38], Theorem 1.1). *Let $\mathcal{Y} \to \mathcal{B}$ be a G-gerbe over $\mathcal{B}$ whose band is trivializable. Let $g \geq 0$, $n > 0$, and $a_1, ..., a_n \geq 0$ be integers. Let $\beta \in H_2(\mathcal{B}, \mathbb{Q})$ be a curve class. Let $\delta_1, ..., \delta_n \in H^*_{CR}(\mathcal{Y})$. Then the following equality of descendant Gromov-Witten invariants holds:*

$$\left\langle \prod_{j=1}^{n} \delta_j \psi_j^{a_j} \right\rangle_{g,n,\beta}^{\mathcal{Y}} = \sum_{[\rho]\in\widehat{G}} \left(\frac{dimV_\rho}{|G|}\right)^{2-2g} \left\langle \prod_{j=1}^{n} I(\delta_j)_\rho \psi_j^{a_j} \right\rangle_{g,n,\beta}^{\mathcal{B},c_\rho}. \tag{3.3}$$

In what follows we explain some ideas involved in the proof of Theorem 3.1. Gromov-Witten invariants of $\mathcal{Y}$ are intersection numbers on moduli spaces

$$\overline{\mathcal{M}}_{g,n}(\mathcal{Y}, \beta)$$

of stable maps to $\mathcal{Y}$. Gromov-Witten invariants of $\mathcal{B}$ twisted by c_ρ are intersection numbers on moduli spaces

$$\overline{\mathcal{M}}_{g,n}(\mathcal{B}, \beta)$$

of stable maps to $\mathcal{B}$. Given a stable map $\mathcal{C} \to \mathcal{Y}$, one can obtain a stable map to $\mathcal{B}$ by stabilizing the composition $\mathcal{C} \to \mathcal{Y} \xrightarrow{\pi} \mathcal{B}$. This construction defines a map

$$\pi_{g,n,\beta} : \overline{\mathcal{M}}_{g,n}(\mathcal{Y}, \beta) \to \overline{\mathcal{M}}_{g,n}(\mathcal{B}, \beta).$$

The first ingredient in the proof of Theorem 3.1 is the *virtual pushforward property* for $\pi_{g,n,\beta}$. More precisely, we prove that the pushforward

$$\pi_{g,n,\beta*}[\overline{\mathcal{M}}_{g,n}(\mathcal{Y}, \beta)]^w \in H_*(\overline{\mathcal{M}}_{g,n}(\mathcal{B}, \beta), \mathbb{Q})$$

[6]The isomorphism class of $\mathcal{L}_{c_\rho}$ admits a formal description. There is a natural map $H^2(\mathcal{B}, U(1)) \to H^1(L\mathcal{B}, U(1))$, where $L\mathcal{B}$ is the loop space of $\mathcal{B}$. The inertia stack $I\mathcal{B}$ is contained in $L\mathcal{B}$ as the locus of constant loops. The image of the class of c_ρ restricts to the class of $\mathcal{L}_{c_\rho}$ via the inclusion $I\mathcal{B} \subset L\mathcal{B}$.

[7]This isomorphism I is also constructed in [37] for G-gerbes whose bands are not trivial. The construction of I is rather complicated, as it involves Morita equivlanece between explicit groupoid algebras. However, an explicit formula of I was found in [37].

can be written as a linear combination of components of the virtual fundamental class $[\overline{\mathcal{M}}_{g,n}(\mathcal{B},\beta)]^w$. To compute coefficients of this linear combination, we need to study the degrees of $\pi_{g,n,\beta}$ over components of $\overline{\mathcal{M}}_{g,n}(\mathcal{B},\beta)$. More precisely, given a stable map $f:\mathcal{C}\to\mathcal{B}$ with $\mathcal{C}$ nonsingular, we need to classify all possible commutative diagrams

$$\begin{array}{ccc} \mathcal{C}' & \xrightarrow{f'} & \mathcal{Y} \\ \downarrow & & \downarrow{\scriptstyle \pi} \\ \mathcal{C} & \xrightarrow{f} & \mathcal{B} \end{array}$$

where $f':\mathcal{C}'\to\mathcal{Y}$ is a stable map and $\mathcal{C}'\to\mathcal{C}$ is an isomorphism over the non-stacky locus $\mathcal{C}^{ns}$. The class of the pull-back gerbe

$$f^*\mathcal{Y}\to\mathcal{C}$$

lies in $H^2(\mathcal{C},Z(G))$. Let

$$\mathcal{C}^0\subset\mathcal{C}$$

be obtained from $\mathcal{C}$ by removing small disks centered at marked points. Since $\mathcal{C}^0$ is homotopic to its 1-skeleton, we have

$$H^2(\mathcal{C}^0,Z(G))=0.$$

So the restriction of $f^*\mathcal{Y}\to\mathcal{C}$ to $\mathcal{C}^0$ is a trivializable G-gerbe. This simple observation leads to a characterization of all possible f' in terms of conditions on orbifold structures on $\mathcal{C}'$. Consequently, the count of all such f' can be obtained from the Frobenius-Mednyh formula [31]. Solution to this counting problem can be extended to nodal $\mathcal{C}$ by a simple splitting argument.

Now let $\varphi_1,\ldots,\varphi_n\in H^*_{orb}(\mathcal{B},c_\rho)$. By definition we have

$$\left\langle\prod_{j=1}^n\varphi_j\psi_j^{a_j}\right\rangle_{g,n,\beta}^{\mathcal{B},c_\rho}:=\int_{[\overline{\mathcal{M}}_{g,n}(\mathcal{B},\beta)]^w}\theta_*(\prod_{j=1}^n ev_j^*(\varphi_j))\cup\prod_{j=1}^n\psi_j^{a_j},$$

here $\theta:\otimes_{j=1}^n ev_j^*\mathcal{L}_{c_\rho}\to\underline{\mathbb{C}}$ is the trivialization appearing in the construction of [35]. As discussed above, there is an equality[8]

$$\pi_{g,n,\beta*}[\overline{\mathcal{M}}_{g,n}(\mathcal{Y},\beta)]^w=d_{g,n,\beta}[\overline{\mathcal{M}}_{g,n}(\mathcal{B},\beta)]^w$$

for a certain degree function $d_{g,n,\beta}$. Therefore we have

$$\left\langle\prod_{j=1}^n\varphi_j\psi_j^{a_j}\right\rangle_{g,n,\beta}^{\mathcal{B},c_\rho}=\frac{1}{d_{g,n,\beta}}\int_{[\overline{\mathcal{M}}_{g,n}(\mathcal{Y},\beta)]^w}\pi^*_{g,n,\beta}\theta_*(\prod_{j=1}^n ev_j^*(\varphi_j))\cup\prod_{j=1}^n\psi_j^{a_j},$$

[8] At the risk of not being modest enough, we want to point out that the determination of this degree function $d_{g,n,\beta}$ in [38] is quite difficult. Also, the degrees appears in previous applications of virtual pushforwards are very simple. This makes our application of virtual pushforwards technically very sophisticated.

(note that the descendant classes ψ_j are compatible under various pull-backs). We need to compare this with

$$\left\langle \prod_{j=1}^{n} I^{-1}(\varphi_j)\psi_j^{a_j} \right\rangle_{g,n,\beta}^{\mathcal{Y}} = \int_{[\overline{\mathcal{M}}_{g,n}(\mathcal{Y},\beta)]^w} \prod_{j=1}^{n} ev_j^*(I^{-1}(\varphi_j))\psi_j^{a_j}.$$

When G is abelian, this can be done easily from an explicit formula for the inverse map I^{-1}. For non-abelian G, the short exact sequence

$$1 \to Z(G) \to G \to G/Z(G) =: K \to 1$$

leads to a factorization of $\pi : \mathcal{Y} \to \mathcal{B}$ into a composition

$$\mathcal{Y} \to \mathcal{Y}' \to \mathcal{B}$$

where $\mathcal{Y} \to \mathcal{Y}'$ is a $Z(G)$-gerbe with trivial band and $\mathcal{Y} \to \mathcal{B}$ is a K-gerbe with trivial band. Theorem 3.1 for $\mathcal{Y} \to \mathcal{Y}'$ is proven first. Since the center of K is trivial, the K-gerbe $\mathcal{Y}' \to \mathcal{B}$ is necessarily trivial, and Theorem 3.1 for $\mathcal{Y}' \to \mathcal{B}$ follows from either the results on Gromov-Witten theory of trivial gerbes in [8], or by a direct calculation using the virtual pushforward property. Theorem 3.1 for $\mathcal{Y} \to \mathcal{B}$ then follows by combining these two cases.

3.3 Outlook

Naturally, the next goal is to prove Conjecture 3.1 for G-gerbes with non-trivial bands. It is hopeful that the approach used to prove Theorem 3.1 can be applied to this general setting as well. However, there are several difficulties:

1. If the band of $\mathcal{Y} \to \mathcal{B}$ is not trivializable, then components of the "dual" $\widehat{\mathcal{Y}}$ are not necessarily isomorphic to $\mathcal{B}$. This further complicates the study of Gromov-Witten theory of $\widehat{\mathcal{Y}}$, twisted by c or not.
2. In case of non-trivial bands, the class of c does not seem to have a simple description. This makes the study of Gromov-Witten theory of $\widehat{\mathcal{Y}}$ twisted by c much harder.
3. If the band is not trivial, then computation of the degree of the map $\pi_{g,n,\beta} : \overline{\mathcal{M}}_{g,n}(\mathcal{Y},\beta) \to \overline{\mathcal{M}}_{g,n}(\mathcal{B},\beta)$ must be more involved. While the general idea used in the trivial band case still seems hopeful, it is not clear how to, for example, take into account the information about the band.

4 Codimension 1: root stacks

In this Section, we discuss Gromov-Witten theory of root constructions.

4.1 Conjecture

A way to introduce stack structures in codimension 1 is the construction of root stacks, introduced in [14] and [5, Appendix B] (see also [29]). We review the construction. Let $\mathcal{X}$ be a Deligne-Mumford stack and

$$\mathcal{D} \subset \mathcal{X}$$

a smooth irreducible divisor. Given a positive integer r, the stack of r-th roots of $\mathcal{D}$, denoted by $\mathcal{X}_r$, can be defined to be the moduli stack whose objects over a scheme $f : S \to \mathcal{X}$ are

$$(M, \phi, \tau),$$

where

1. M is a line bundle on S,
2. $\phi : M^{\otimes r} \simeq f^*\mathcal{O}_{\mathcal{X}}(\mathcal{D})$ is an isomorphism,
3. τ is a section of M such that $\phi(\tau^r)$ is the tautological section of $f^*\mathcal{O}_{\mathcal{X}}(\mathcal{D})$.

The construction comes with a natural map $\mathcal{X}_r \to \mathcal{X}$, which is an isomorphism over $\mathcal{X} \setminus \mathcal{D}$. Let

$$\mathcal{D}_r \subset \mathcal{X}_r$$

be the divisor lying over $\mathcal{D} \subset \mathcal{X}$.

In [21], it is shown that the root construction above is essentially the only way stack structures can arise in codimension 1. Naturally one can make the following

Conjecture 4.1 (see [39], Conjecture 1.1)**.** *In the situation above, the Gromov-Witten theory of $\mathcal{X}_r$ is determined by the Gromov-Witten theory of $\mathcal{X}, \mathcal{D}$ and the restriction map $H^*_{CR}(\mathcal{X}) \to H^*_{CR}(\mathcal{D})$.*

However natural it may seem, Conjecture 4.1 is not proven in the generality stated. The evidence known so far is the following

Theorem 4.1 ([39], Theorem 1.3)**.** *Conjecture 4.1 holds true when $\mathcal{D}$ does not intersect the stacky locus of $\mathcal{X}$.*

The above condition on $\mathcal{D}$ implies that $\mathcal{D}$ is a scheme. This is a rather restrictive assumption. However there is one instance in which this assumption holds even in repeated applications of root constructions: an orbifold curve $\mathcal{C}$ is obtained from its coarse moduli space, which is a nonsingular curve C, by a sequence of root constructions (see e.g. [5]). Consequently, Theorem 4.1 implies that the Gromov-Witten theory of an orbifold curve $\mathcal{C}$ is determined by the Gromov-Witten theory of its coarse moduli space, the nonsingular curve C. By the trilogy [32], [33], [34], the Gromov-Witten theory of C is completely determined. Therefore, the Gromov-Witten theory of an orbifold curve $\mathcal{C}$ is completely determined[9].

4.2 Proof of a special case

We now give an outline of the arguments for Theorem 4.1. The first idea, which is quite standard in Gromov-Witten theory, is to use the deformation to the normal cone of $\mathcal{D}_r \subset \mathcal{X}_r$. This yields a degeneration of $\mathcal{X}_r$ to a union

$$\mathcal{X}_r \cup_{\mathcal{D}_r} \mathcal{Y}$$

[9] Unfortunately, not explicitly determined at this moment.

with

$$\mathcal{Y} := \mathbb{P}_{\mathcal{D}_r}(N_{\mathcal{D}_r/\mathcal{X}_r} \oplus \mathcal{O})$$

the projectivization of the normal bundle $N_{\mathcal{D}_r/\mathcal{X}_r}$ of $\mathcal{D}_r \subset \mathcal{X}_r$. The degeneration formula for orbifold Gromov-Witten theory, proven in [3], applies to express Gromov-Witten invariants of $\mathcal{X}_r$ in terms of relative Gromov-Witten invariants of the two pairs

$$(\mathcal{X}_r, \mathcal{D}_r), \quad (\mathcal{Y}, \mathcal{D}_0 \simeq \mathcal{D}_r),$$

where $\mathcal{D}_0 \subset \mathcal{Y}$ is the 0-section. Thus we need to determine Gromov-Witten invariants of the pairs $(\mathcal{X}_r, \mathcal{D}_r)$ and $(\mathcal{Y}, \mathcal{D}_r)$.

By [3, Proposition 4.9], Gromov-Witten invariants of $(\mathcal{X}_r, \mathcal{D}_r)$ are determined by Gromov-Witten invariants of $(\mathcal{X}, \mathcal{D})$. Because $\mathcal{D}$ is assumed to be a scheme, we can employ the proof of the "relative-in-terms-of-absolute" result in [30] to show that the relative Gromov-Witten theory of $(\mathcal{X}, \mathcal{D})$ is determined by the Gromov-Witten theory of $\mathcal{X}, \mathcal{D}$ and the restriction map $H^*_{CR}(\mathcal{X}) \to H^*_{CR}(\mathcal{D})$.

It remains to treat the Gromov-Witten theory of $(\mathcal{Y}, \mathcal{D}_r)$. The torus $\mathbb{C}^*$ acts on $\mathcal{Y}$ by scaling the fibers. The fixed loci are

$$\mathcal{D}_0, \mathcal{D}_\infty \subset \mathcal{Y},$$

the "0- and ∞-sections. Both divisors are isomorphic to $\mathcal{D}_r$. This $\mathbb{C}^*$ action on $\mathcal{Y}$ induces $\mathbb{C}^*$-actions on relevant moduli spaces, which allows us to apply virtual localization formula [24], [25] to express Gromov-Witten invariants of $(\mathcal{Y}, \mathcal{D}_r)$ as integrals over $\mathbb{C}^*$-fixed loci. Analyzing the explicit virtual localization formula for relativce Gromov-Witten theory given in [25], we find that the relative Gromov-Witten theory of $(\mathcal{Y}, \mathcal{D}_0)$ is determined by the Gromov-Witten theory of $\mathcal{Y}$ and the "rubber theory[10]" for $(\mathcal{Y}, \mathcal{D}_0, \mathcal{D}_\infty)$.

Virtual localization formula expresses Gromov-Witten invariants of $\mathcal{Y}$ in terms of integrals over the moduli space $\overline{\mathcal{M}}_{g,n}(\mathcal{D}_r, d)$ whose integrands involve the inverse $\mathbb{C}^*$-equivariant Euler class of certain K-theory class arising from the line bundle $N_{\mathcal{D}_r/\mathcal{X}_r}$. These quantities are examples of the so-called *twisted Gromov-Witten invariants* of $\mathcal{D}_r$, in the sense of [41]. The main theorem of [41], usually called the orbifold quantum Riemann-Roch theorem, expresses these twisted Gromov-Witten invariants in terms of the usual Gromov-Witten invariants of $\mathcal{D}_r$ in a somewhat explicit fashion.

We are thus led to studying Gromov-Witten theory of $\mathcal{D}_r$. Since $\mathcal{D}_r$ is a root gerbe over $\mathcal{D}$ and $\mathcal{D}$ is a scheme, we can apply the results of [9, 10] (which are special cases of [38] surveyed in Section 3) to explicitly write Gromov-Witten invariants of $\mathcal{D}_r$ in terms of Gromov-Witten invariants of $\mathcal{D}$.

It remains to study the rubber theory for $(\mathcal{Y}, \mathcal{D}_0, \mathcal{D}_\infty)$. Firstly, we apply the rigidification argument in [30] to write rubber invariants of $(\mathcal{Y}, \mathcal{D}_0, \mathcal{D}_\infty)$ as relativce Gromov-Witten invariants of $(\mathcal{Y}, \mathcal{D}_0, \mathcal{D}_\infty)$ with some "target descendant" insertions. We then apply an elaborated recursion called rubber calculus in [30] to remove these target descendants. This allows us to express rubber invariants of $(\mathcal{Y}, \mathcal{D}_0, \mathcal{D}_\infty)$ in terms of relative Gromov-Witten invariants of $(\mathcal{Y}, \mathcal{D}_0, \mathcal{D}_\infty)$.

[10]This is a variant of relative Gromov-Witten theory.

It can be seen from the construction that $\mathcal{Y}$ is obtained from a $\mathbb{P}^1$-bundle over $\mathcal{D}$, $\overline{\mathcal{Y}} \to \mathcal{D}$, by applying the r-th root constructions to its 0- and ∞-sections $\overline{\mathcal{D}_0}, \overline{\mathcal{D}_\infty} \subset \overline{\mathcal{Y}}$. Therefore by [3, Proposition 4.9], Gromov-Witten invariants of $(\mathcal{Y}, \mathcal{D}_0 \cup \mathcal{D}_\infty)$ are determined by .Gromov-Witten invariants of $(\overline{\mathcal{Y}}, \overline{\mathcal{D}_0} \cup \overline{\mathcal{D}_\infty})$. Because $\mathcal{D}$ is a scheme, the quantum Leray-Hirsch theorem in [30] is applicable to the $\mathbb{P}^1$-bundle $\overline{\mathcal{Y}} \to \mathcal{D}$, implies that the Gromov-Witten theory of $(\overline{\mathcal{Y}}, \overline{\mathcal{D}_0} \cup \overline{\mathcal{D}_\infty})$ is determined by the Gromov-Witten theory of $\mathcal{D}$ and the restriction map $H^*_{CR}(\mathcal{X}) \to H^*_{CR}(\mathcal{D})$.

The proof of Theorem 4.1 is thus complete.

4.3 Outlook

Naturally, we would like to prove Conjecture 4.1 in general. It is hopeful that the approach used in the proof of Theorem 4.1 can be modified to address the general case, with no condition on $\mathcal{D}$. Indeed, it is not hard to see that several steps in the above outline do not use the condition on $\mathcal{D}$. However the condition on $\mathcal{D}$ is used in an essential way in the study of Gromov-Witten theory of $(\mathcal{Y}, \mathcal{D}_r)$. Therefore another approach must be taken.

In the author's opinion, the main difficulty lies in proving a quantum Leray-Hirsch result for Deligne-Mumford stacks of the form $\mathbb{P}_{\mathcal{D}}(L \oplus \mathcal{O})$ where L is a line bundle over $\mathcal{D}$. The relative Gromov-Witten theory of $\mathbb{P}_{\mathcal{D}}(L \oplus \mathcal{O})$ relative to the 0- and ∞-sections when the curve classes lie in the fiber of $\mathbb{P}_{\mathcal{D}}(L \oplus \mathcal{O}) \to \mathcal{D}$ requires solving the rubber theory for 1-dimensional targets of the form

$$([\mathbb{P}^1/G], [0/G] \cup [\infty/G]).$$

This is recently worked out in [40]. However a quantum Leray-Hirsch result for $\mathbb{P}_{\mathcal{D}}(L \oplus \mathcal{O})$ relative to its 0- and ∞-sections that covers curve classes of mixed type (i.e. combinations of fiber classes and classes from $\mathcal{D}$) is not yet proven.

It is also very desirable to have an *effective* version of Conjecture 4.1. More precisely, we would like to have some explicit formulas for Gromov-Witten invariants of $\mathcal{X}_r$ in terms of Gromov-Witten invariants of $\mathcal{X}$ and $\mathcal{D}$. It is reasonable to start with genus 0. In various results on genus 0 Gromov-Witten theory of a Deligne-Mumford stack $\mathcal{X}$, an essential role is played by the *J-function* of $\mathcal{X}$. This is a generating function of 1-point genus 0 descendant Gromov-Witten invariants of $\mathcal{X}$ which are important in the structures of genus 0 Gromov-Witten theory. Thus it is very natural to ask the following

Question 4.2. *Is there an explicit formula for the J-function of $\mathcal{X}_r$, or a slice of the Lagrangian cones [23] defined by the genus 0 Gromov-Witten theory of $\mathcal{X}_r$, given in terms of genus 0 Gromov-Witten invariants of $\mathcal{X}$ and $\mathcal{D}$?*

To the author's knowledge, no answers to this Question is known in any examples. Since genus 0 Gromov-Witten theory of toric stacks are explicitly calculated [16], it makes sense to try to study this Question when $\mathcal{X}$ is a toric stack and $\mathcal{D}$ is a toric prime divisor. Experiments in this toric setting may suggest possible answers to this Question in general.

References

[1] D. Abramovich, *Lectures on Gromov-Witten invariants of orbifolds*, In: "Enumerative invariants in algebraic geometry and string theory", 1–48, Lecture Notes in Math., 1947, Springer, Berlin, 2008.

[2] D. Abramovich, A. Corti, A. Vistoli, *Twisted bundles and admissible covers*, Comm. Algebra 31 (2003), no. 8, 3547–3618.

[3] D. Abramovich, B. Fantechi, *Orbifold techniques in degeneration formulas*, Ann. Sc. Norm. Super. Pisa Cl. Sci. (5), vol. XVI (2016), 519–579.

[4] D. Abramovich, T. Graber, A. Vistoli, *Algebraic orbifold quantum products*, In: [7], 1–24.

[5] D. Abramovich, T. Graber, A. Vistoli, *Gromov-Witten theory of Deligne-Mumford stacks*, Amer. J. Math., 130 (2008), no. 5, 1337–1398.

[6] D Abramovich, A. Vistoli, *Compactifying the space of stable maps*, J. Amer. Math. Soc. 15 (2002), no. 1, 27–75

[7] A. Adem, J. Morava, Y. Ruan (Ed.), *Orbifolds in mathematics and physics (Madison, WI, 2001)*, Contemp. Math., 310, Amer. Math. Soc. Providence, RI, (2002).

[8] E. Andreini, Y. Jiang, H.-H. Tseng, *Gromov-Witten theory of product stacks*, Comm. Anal. Geom. 24 (2016), no. 2, 223–277.

[9] E. Andreini, Y. Jiang, H.-H. Tseng, *Gromov-Witten theory of root gerbes, I: structure of genus 0 moduli spaces*, J. Differential Geom. 99 (2015), no. 1, 1–45.

[10] E. Andreini, Y. Jiang, H.-H. Tseng, *Gromov-Witten theory of banded gerbes over schemes*, arXiv:1101.5996.

[11] K. Behrend, B. Fantechi, *The intrinsic normal cone*, Invent. Math. 128 (1997), no. 1, 45–88.

[12] K. Behrend, B. Noohi, *Uniformization of Deligne-Mumford curves*, J. Reine Angew. Math. 599 (2006), 111–153.

[13] J. Bryan, T. Graber, *The crepant resolution conjecture*, In: "Algebraic geometry–Seattle 2005", Part 1, 23–42, Proc. Sympos. Pure Math., 80, Part 1, Amer. Math. Soc., Providence, RI, 2009.

[14] C. Cadman, *Using stacks to impose tangency conditions on curves*, Amer. J. Math. 129 (2007), no. 2, 405–427.

[15] W. Chen, Y. Ruan, *Orbifold Gromov-Witten theory*, In: [7], 25–85.

[16] T. Coates, A. Corti, H. Iritani, H.-H. Tseng, *A Mirror Theorem for Toric Stacks*, Compos. Math. 151 (2015), no. 10, 1878–1912.

[17] T. Coates, H. Iritani, Y. Jiang, *The Crepant Transformation Conjecture for Toric Complete Intersections*, arXiv:1410.0024.

[18] T. Coates, H. Iritani, H.-H. Tseng, *Wall-crossings in toric Gromov-Witten theory. I. Crepant examples*, Geom. Topol. 13 (5):2675–2744, 2009.

[19] T. Coates, Y. Ruan, *Quantum cohomology and crepant resolutions: a conjecture*, Ann. Inst. Fourier (Grenoble), 63(2):431–478, 2013

[20] B. Fantechi, E. Mann, F. Nironi, *Smooth toric Deligne-Mumford stacks*, J. Reine Angew. Math. 648 (2010), 201–244.

[21] A. Geraschenko, M. Satriano, *A "bottom up" characterization of smooth*

Deligne-Mumford stacks, arXiv:1503.05478.

[22] J. Giraud, *Cohomologie non abélienne*, Die Grundlehren der mathematischen Wissenschaften 179, Springer-Verlag, (1971).

[23] A. Givental, *Symplectic geometry of Frobenius structures*, In: "Frobenius manifolds", 91–112, Aspects Math., E36, Friedr. Vieweg, Wiesbaden, 2004.

[24] T. Graber, R. Pandharipande, *Localization of virtual classes*, Invent. Math. 135 (1999), no. 2, 487–518.

[25] T. Graber, R. Vakil, *Relative virtual localization and vanishing of tautological classes on moduli spaces of curves*, Duke Math. J. 130 (2005), no. 1, 1–37.

[26] S. Hellerman, A. Henriques, T. Pantev, E. Sharpe, M. Ando, *Cluster decomposition, T-duality, and gerby CFTs*, Adv. Theor. Math. Phys. 11 (2007), no. 5, 751–818.

[27] P. Johnson, *Equivariant Gromov-Witten theory of one dimensional stacks*, Comm. Math. Phys. (2014), 327, no. 2, 333–386.

[28] S. Keel, S. Mori, *Quotients by groupoids*, Ann. of Math. (2) 145 (1997), no. 1, 193–213.

[29] K. Matsuki, M. Olsson, *Kawamata-Viehweg vanishing as Kodaira vanishing for stacks*, Math. Res. Letters 12 (2005), 207–217.

[30] D. Maulik, R. Pandharipande, *A topological view of Gromov-Witten theory*, Topology 45 (2006), no. 5, 887–918.

[31] A. D. Mednyh, *Determination of the number of nonequivalent coverings over a compact Riemann surface*, Dokl. Akad. Nauk SSSR, 239 (1978), 269–271 (Russian). Soviet Math. Dokl., 19 (1978), 318–320 (English transl.).

[32] A. Okounkov, R. Pandharipande, *Gromov-Witten theory, Hurwitz theory, and completed cycles* Ann. of Math. (2) 163 (2006), no. 2, 517–560.

[33] A. Okounkov, R. Pandharipande, *The equivariant Gromov-Witten theory of* $\mathbf{P}^1$, Ann. of Math. (2) 163 (2006), no. 2, 561–605.

[34] A. Okounkov, R. Pandharipande, *Virasoro constraints for target curves*, Invent. Math. 163 (2006), no. 1, 47–108.

[35] J. Pan, Y. Ruan, X. Yin, *Gerbes and twisted orbifold quantum cohomology*, Sci. China Ser. A 51 (2008), no. 6, 995–1016.

[36] Y. Ruan, *Discrete torsion and twisted orbifold cohomology*, J. Symplectic Geom. 2 (2003), no. 1, 1–24.

[37] X. Tang, H.-H. Tseng, *Duality theorems of étale gerbes on orbifolds*, Adv. Math. 250 (2014), 496–569.

[38] X. Tang, H.-H. Tseng, *A quantum Leray-Hirsch theorem for banded gerbes*, arXiv:1602.03564.

[39] H. -H. Tseng, F. You, *On orbifold Gromov-Witten theory in codimension one*, J. Pure Appl. Algebra 220 (2016), no. 10, 3567–3571.

[40] H. -H. Tseng, F. You, *Double ramification cycles on the moduli spaces of admissible covers*, arXiv:1606.03770.

[41] H.-H. Tseng, *Orbifold quantum Riemann-Roch, Lefschetz, and Serre*, Geom. Topol. 14 (2010) 1–81.

[42] H.-H. Tseng, *Notes on orbifold Gromov-Witten theory*, In: "Fifth International Congress of Chinese Mathematicians", Part 1, AMS/IP Stud. Adv. Math., vol. 51, Amer. Math. Soc., Providence, RI, 2012, pp. 327–340.

Department of Mathematics, Ohio State University
100 Math Tower, 231 West 18th Ave., Columbus, OH 43210, USA
E-mail address: hhtseng@math.ohio-state.edu

Proceedings of the 7th International Congress of Chinese Mathematicians
ALM 43, pp. 457–472

Quasi-local and Total Angular Momentum in General Relativity

Mu-Tao Wang

Abstract

The notion of angular momentum is of most fundamental importance in any branch of physics. However, there have been great difficulties in finding a physically acceptable definition of this concept in general relativity except for a few cases such as the axially symmetric case and the asymptotically flat case. I shall explain these difficulties and report recent progresses. In particular, new definitions of angular momentum at both the quasi-local and total level introduced in [10, 11] are discussed. It is shown that the definitions satisfy desirable properties such as finiteness, invariance, and conservation. At the end, a formula for the total angular momentum in the harmonic asymptotic case and a formula for the total angular momentum in the asymptotically hyperbolic case are reviewed. The paper is based on joint work [6, 7, 10, 11, 12] with Po-Ning Chen and Shing-Tung Yau.

Mathematics Subject Classification: 35Q76, 83C05, 83C30.

Keywords and Phrases: Angular momentum, conserved quantities, symmetry, quasi-local mass, energy-momentum.

1 Introduction

I shall discuss the notion of angular momentum in general relativity. In particular, I shall review a new definition of *quasilocal* angular momentums [10, 11] and the limit of this definition at the infinity of an isolated system as the definition of total angular momentum in both the asymptotically flat and asymptotically hyperbolic settings. The fundamental question of the definition of *quasilocal* angular momentum can be formulated as follows : given a spacelike region Ω in spacetime, define the angular momentum (per unit mass) in terms of data on the boundary 2-surface Σ of Ω. In 1982, Penrose [27] compiled a list of major unsolved problems in classical general relativity. The first problem is the definition of quasilocal mass and the second one is the definition of quasilocal angular momentum. In order to motivate such a quasilocal definition, we consider the notions of mass and conserved quantities in Newtonian gravity and special relativity first.

This work was partially supported by a grant from the Simons Foundation (#305519 to Mu-Tao Wang). The author was partially supported by the National Science Foundation under grant DMS-1405152. He would like to thank his collaborators Po-Ning Chen and Shing-Tung Yau, to whom he owes gratitude.

The fundamental equation in Newtonian gravity is

$$\Delta\Phi = 4\pi\rho,$$

where ρ is the mass density. For a domain $\Omega \subset \mathbb{R}^3$, the total mass is simply $\int_\Omega \rho$, which can be turned into a boundary integral $\frac{1}{4\pi}\int_{\partial\Omega}\frac{\partial\Phi}{\partial\nu}$.

In special relativity, one considers a matter field in $\mathbb{R}^{3,1}$, whose energy-momentum density tensor is a symmetric $(0,2)$ tensor $T_{\mu\nu}$. Given a spacelike bounded region (hypersurface) Ω, the energy as seen by an observer t^ν (future timelike unit) is the flux integral

$$\int_\Omega T_{\mu\nu}t^\mu u^\nu$$

where u^ν is the future timelike unit normal of the hypersurface Ω. The dominant energy condition implies $T_{\mu\nu}t^\mu u^\nu \geq 0$. $T_{\mu\nu}$ also satisfies the conservation law $\nabla^\mu T_{\mu\nu} = 0$.

Suppose t^μ is a constant future-directed timelike unit vector field (this is a translating Killing field) in $\mathbb{R}^{3,1}$. By the conservation law, $T_{\mu\nu}t^\mu$ is divergence free and thus,

$$\int_{\Omega_1} T_{\mu\nu}t^\mu u^\nu = \int_{\Omega_2} T_{\mu\nu}t^\mu u^\nu$$

if $\partial\Omega_1 = \partial\Omega_2$.

Minimizing among all such unit Killing observers t^μ gives the quasi-local mass. Moreover, the dual $\int_\Omega T_{\mu\nu}u^\nu$ defines the quasi-local energy-momentum 4 vector. Replacing t^μ by other Killing fields of $\mathbb{R}^{3,1}$ gives the notions of quasi-local angular momentum (rotation Killing field) and quasi-local center of mass (boost Killing field).

Two major difficulties in generalizing these notions to general relativity are the lack of density for gravitation and the lack of Killing fields in a general spacetime. There have been great efforts to search for such quasi-local descriptions of mass and angular momentum in the past decades, in particular along the Hamilton-Jacobi approach, see the survey paper [34] for more details.

1.1 Review of quasi-local mass

The new definition of quasilocal angular momentum is built on the theory of quasilocal mass developed by Wang-Yau in [36, 37], which is reviewed in the following. Given a two-dimensional spacelike surface Σ (of the topological type of a sphere) in spacetime, the quasilocal mass and quasilocal angular momentum are defined in terms of the physical data $(\sigma, |H|, \alpha_H)$ of Σ, where σ is the induced Riemannian metric on Σ and H is the mean curvature vector field, the unique normal vector field along Σ that satisfies

$$\delta_V|\Sigma| = -\int_\Sigma \langle H, V\rangle$$

for any variation field V. Here $|\Sigma|$ is the surface area and $\langle\cdot,\cdot\rangle$ denotes the Lorentzian product in spacetime. We assume that H is spacelike and extract from it a positive function $|H|$ and a connection one-form α_H (see Section 2 of [10] for the precise definition of α_H).

Consider an isometric embedding of σ into the Minkowski spacetime $\mathbb{R}^{3,1}$, i.e. $X : S^2 \to \mathbb{R}^{3,1}$ such that $\langle dX, dX\rangle = \sigma$. Such an embedding can be obtained in the following way. Take any smooth function τ on S^2 and isometrically embed $\sigma + d\tau \otimes d\tau$ as a surface $\hat{\Sigma}$ in $\mathbb{R}^3$ (as the orthogonal complement of ∂_t, or the time axis, in $\mathbb{R}^{3,1}$) by the theorem of Nirenberg [25] and Pogorelov [26] (suppose the Gauss curvature of $\sigma + d\tau \otimes d\tau$ is positive). Take the graph of τ over $\hat{\Sigma}$ and denote the embedding of the graph by X. It is not hard to see that the induced metric on the image surface of X is exactly σ and $\tau = -\langle X, \partial_t\rangle$. We can replace ∂_t by any future timelike unit Killing field T_0 in $\mathbb{R}^{3,1}$. Let $\widehat{X}$ denote the embedding by projecting X onto $\mathbb{R}^3$ and denote by $\widehat{H}$ the mean curvature of $\widehat{X}$ as a surface in $\mathbb{R}^3$. It is clear that such isometric embedding is parametrized by the "time function" τ. The question is, among all these isometric embeddings, is there a "best match" of the physical surface Σ? We adopt a variational approach to answer this question.

Consider the quasilocal energy $E(\Sigma, X, T_0)$ with respect to the pair (X, T_0)

$$\frac{1}{8\pi}\int_{\hat{\Sigma}} \widehat{H} - \frac{1}{8\pi}\int_{\Sigma}[\sqrt{|H|^2(1+|\nabla\tau|^2)+(\Delta\tau)^2} + \theta\Delta\tau - \alpha_H(\nabla\tau)] \tag{1.1}$$

where θ is defined by $\sinh\theta = \frac{-\Delta\tau}{|H|\sqrt{1+|\nabla\tau|^2}}$, and ∇ and Δ are defined with respect to the metric σ.

The expression comes from the Hamilton-Jacobi analysis of gravitational action [4, 21]. In particular, $\frac{1}{8\pi}\int_{\hat{\Sigma}} \widehat{H}$ is the reference term. The general expression depends on a gauge choice. Brown-York [4] chose the gauge that corresponds to an isometric embedding into a totally geodesic $\mathbb{R}^3$ in $\mathbb{R}^{3,1}$, or $\tau = 0$. The quasilocal energy (1.1) is defined with respect to a canonical gauge choice that matches the expansions along the surfaces [36, 37].

The choice of canonical gauge was first justified by the positivity of the quasi-local energy [36, 37]. It is closely related to a gravitational conservation law which later played an instrumental role in several other applications, see Section 5 of [10]. To find a "best match" of the physical data $(\sigma, |H|, \alpha_H)$, we minimize $E(\Sigma, X, T_0)$ over the pairs (X, T_0), which are considered as various observers [34]. By Lorentz invariance, it suffices to consider variation with respect to $\tau = -\langle X, T_0\rangle$, or the time function. The Euler-Lagrange equation is:

$$-(\widehat{H}\hat{\sigma}^{ab} - \hat{\sigma}^{ac}\hat{\sigma}^{bd}\hat{h}_{cd})\frac{\nabla_b\nabla_a\tau}{\sqrt{1+|\nabla\tau|^2}} + div_\sigma\left(\frac{\nabla\tau}{\sqrt{1+|\nabla\tau|^2}}\cosh\theta|H| - \nabla\theta - \alpha_H\right) = 0. \tag{1.2}$$

In order to better understand the energy and the equation, we rewrite them in terms of data on the image surface of X. Let H_0 be the mean curvature vector and again we extract $|H_0|$ and α_{H_0} of the image surface of X. Consider the following function ρ and one-form j_a on Σ defined by:

$$\rho = \frac{\sqrt{|H_0|^2 + \frac{(\Delta\tau)^2}{1+|\nabla\tau|^2}} - \sqrt{|H|^2 + \frac{(\Delta\tau)^2}{1+|\nabla\tau|^2}}}{\sqrt{1+|\nabla\tau|^2}}, \tag{1.3}$$

and

$$j_a = \rho\nabla_a\tau - \nabla_a[\sinh^{-1}(\frac{\rho\Delta\tau}{|H_0||H|})] - (\alpha_{H_0})_a + (\alpha_H)_a. \tag{1.4}$$

It can be checked [8, 10] that the quasi-local energy can be written as

$$E(\Sigma,\tau) = \frac{1}{8\pi}\int_\Sigma (\rho + j_a\nabla^a\tau).$$

Minimizing among τ, the Euler-Lagrange equation becomes $\nabla^a j_a = 0$. The quasi-local mass is defined to be

$$\frac{1}{8\pi}\int_\Sigma \rho$$

and ρ is called the quasi-local mass density.

The solution of optimal isometric embedding system for $(\sigma, |H|, \alpha_H)$ seeks for (X, T_0) with $\tau = -\langle X, T_0\rangle$ such that the following system is satisfied

$$\begin{cases} \langle dX, dX\rangle & = \sigma \\ \nabla^a j_a = 0. \end{cases} \tag{1.5}$$

Though it is not known how to obtain the general solution of (1.5), we established following theorems concerning the linearized equation and the minimizing property of critical points with positive quasilocal mass density.

Theorem 1 *[9, 7] Let $(\sigma, |H|, \alpha_H)$ be the data of a spacelike surface Σ in the Minkowski spacetime. Suppose the projection of Σ onto the orthogonal complement of T_0 is convex. Then the kernel of the linearized optimal isometric embedding system* (1.5) *consists precisely of Lorentz transformations.*

This follows from Theorem 3 in [7]. Since the optimal isometric embedding system is the Euler-Lagrange equation of the quasilocal energy, the linearized optimal isometric embedding system correspond to the second variation of the quasilocal energy. The kernel was pinpointed by tracing the proof of the positivity of quasi-local energy and identifying the equality case in each step.

Theorem 2 *[9, 7] Let $(\sigma, |H|, \alpha_H)$ be the data of a spacelike surface Σ in a general spacetime. Suppose that τ_0 is a critical point of the quasi-local energy $E(\Sigma,\tau)$ with convex projection as in Theorem 1, and that the corresponding quasilocal mass density ρ is positive, then τ_0 is a local minimum for $E(\Sigma,\tau)$.*

Note that $\rho > 0$ implies $|H_0| > |H|$ by (1.3). The above statement then follows from Theorem 2 in [9]. Under the assumption of Theorem 2, the local solvability of (1.5) near τ_0 is proved in Gimre's thesis [20].

Chen-Wang-Yau [10] defined the following quasi-local conserved quantities with respect to an optimal isometric embedding:

Definition 3 *For each optimal isometric embedding $X : \Sigma \to \mathbb{R}^{3,1}$ satisfying* (1.5) *and any Killing field K in $\mathbb{R}^{3,1}$, the corresponding quasi-local conserved quantity is defined to be:*

$$-\frac{1}{8\pi}\int_{\Sigma}\left[\langle K, T_0\rangle\rho + (K^{\top})^a j_a\right],$$

where $K^{\top}$ is the component of K that is tangential to the image surface of X.

Each rotation Killing field $K = x^i\partial_j - x^j\partial_i, i < j$ defines an angular momentum with respect to ∂_t. When K is a boost Killing field, the definition gives the center of the mass, see [35] for a survey of the corresponding results.

2 Total angular momentum of an asymptotically flat initial data set

Recall that (M, g, k) is an asymptotically flat initial data set if outside a compact subset, there exists an asymptotically flat coordinate system (x^1, x^2, x^3) on each end, such that

$$g = \delta + O_2(r^{-q}) \text{ and } k = O_1(r^{-p}) \text{ for } q > \frac{1}{2} \text{ and } p > \frac{3}{2} \tag{2.1}$$

where $r = \sqrt{\sum_{i=1}^{3}(x^i)^2}$ is the radial function. The ADM energy momentum [1] of an end of M is the four-vector (E, P_1, P_2, P_3) in which

$$E = \lim_{r\to\infty}\frac{1}{16\pi}\int_{S_r}\sum_{i,j=1}^{3}(\partial_j g_{ij} - \partial_i g_{jj})\nu^i$$

is the total energy and

$$P_k = \lim_{r\to\infty}\frac{1}{16\pi}\int_{S_r}\sum_{i=1}^{3}2(p_{ik} - \delta_{ik}p_{jj})\nu^i, k = 1, 2, 3 \tag{2.2}$$

is the total linear momentum. Here S_r is a coordinate sphere of radius r on the end and ν is the outward unit normal of S_r.

The positive mass theorem (Schoen-Yau [30], Witten [40]) asserts that under the dominant energy condition, the four-vector (E, P_1, P_2, P_3) is future-directed time-like, i.e.

$$E \geq 0 \text{ and } -E^2 + P_1^2 + P_2^2 + P_3^2 \leq 0.$$

Given an asymptotically flat initial data set (M, g, k), consider the coordinate sphere S_r. Suppose the ADM mass of (M, g, k) is positive, then under suitable assumptions it is proved in [9] that there is a unique, locally energy-minimizing, optimal isometric embedding of S_r whose image approaches a large round sphere in $\mathbb{R}^3$. Taking the limit as $r \to \infty$ of the quasi-local conserved quantities on S_r, we obtain a full set of total conserved quantities (E, P_i, J^i, C^i) where (E, P_i) is the same as the ADM energy-momentum [38].

There have been several proposals of the quasilocal angular momentum. In the axi-symmetric case, there is the Komar type angular momentum, see for example Section 6 of [10]. For a general spacetime, there are several definitions proposed by, Penrose [28] etc, which are based on twistor or spinor constructions, see [33] for more details.

A very natural question is what justifies a good definition of quasilocal angular momentum. For the definition of quasilocal mass, there are several obvious criteria [14, 36, 37]. For example, the mass should be positive under suitable energy condition and should be zero for surfaces in the Minkowski space. Especially, the mass should approach the well-defined ADM mass at spatial infinity. However, there are some difficulties even for the total angular momentum of an asymptotically flat initial data set.

We recall that, in addition to the ADM mass (energy-momentum), there is also a companion definition of angular momentum that is attributed to ADM (see also [2, 13]), which is defined by, for each $(i,j), i<j$,

$$J_{(i,j)} = \lim_{r\to\infty} \frac{1}{8\pi} \int_{S_r} \pi(x^i\partial_j - x^j\partial_i, \nu), \tag{2.3}$$

where $\pi = k - (tr_g k)g$ is the conjugate momentum. Here $x^i\partial_j - x^j\partial_i$ is considered to be an asymptotically Killing field that depends on the asymptotically flat coordinates. In particular, the formula (2.3) applying to the maximal slice of Kerr spacetime recovers the angular momentum parameter a. However, the general case is more subtle, as the defining expression (2.3), unlike the one for linear momentum (2.2) is apparently divergent under the asymptotically flat condition (2.1). (Note that an alternative form of P_k is $P_k = \lim_{r\to\infty} \frac{1}{8\pi}\int_{S_r} \pi(\partial_k, \nu)$.)

Regge-Teitelboim [29] proposed a parity condition on (g,k) to assure finiteness of the expression (2.3) and there are important gluing constructions and density theorems by Corvino-Schoen [19], Chruściel-Delay [16], Huang-Schoen-Wang [22], etc. under this parity condition.

In [5], it is shown that there exist spacelike hypersurfaces (M, g, k) in $\mathbb{R}^{3,1}$ or the Schwarzschild spacetime with finite, nonzero ADM angular momentum (2.3), with $g = \delta + O(r^{-\frac{4}{3}})$ and $k = O(r^{-\frac{5}{2}})$. Note that this is still within the range of asymptotic condition (2.1) in which the ADM energy-momentum definition is valid. On the other hand, Chru sciel showed in [15] that the ADM expression (2.3) is finite if $p+q>3$

It is natural to ask to what extent the ADM angular momentum definition (2.3) is a valid one. All previous known definitions of quasilocal angular momentum satisfy the covariant properties with respect to the Poincare group. But there does not seem be an effective criterion that is relevant to general relativity. We thus propose the following criterion: a quasilocal definition of angular momentum is good if it gives good limits in an asymptotically flat spacetime.

In this regards, we prove an invariance theorem in the Kerr spacetime for the newly defined total angular momentum.

Theorem 4 *Any strictly spacelike hypersurface in the Kerr spacetime has the same total angular momentum defined in [9, 10].*

A strictly spacelike hypersurface in the Kerr spacetime is a hypersurface defined by $t = O(cr)$ for $|c| < 1$ in the Boyer-Linguist coordinates. Note that this invariance theorem in particular implies the new angular momentum vanishes on the spacelike hypersurfaces considered by [5] in $\mathbb{R}^{3,1}$ or the Schwarzschild spacetime.

3 Total conserved quantities for order one data

In this section, we review the definition of total conserved quantities using the limits of quasi-local conserved quantities (see Section 6 of [10]). First we recall the definition of asymptotically flat initial data set of order one.

Definition 5 *(M, g, k) is said to be asymptotically flat of order one if there is a compact subset K of M such that $M \backslash K$ is diffeomorphic to a union of ends and such that in the asymptotically flat coordinate system on each end, we have the following decay condition for g and k:*

$$
\begin{aligned}
g_{ij} &= \delta_{ij} + \frac{g_{ij}^{(-1)}}{r} + \frac{g_{ij}^{(-2)}}{r^2} + o(r^{-2}) \\
k_{ij} &= \frac{k_{ij}^{(-2)}}{r^2} + \frac{k_{ij}^{(-3)}}{r^3} + o(r^{-3}).
\end{aligned}
$$

On each level set of r, Σ_r, we can use $\{u^a\}_{a=1,2}$ as coordinate system to express the geometric data we need in order to define the quasi-local conserved quantities:

$$
\begin{aligned}
\sigma_{ab} &= r^2\tilde{\sigma}_{ab} + r\sigma_{ab}^{(1)} + \sigma_{ab}^{(0)} + o(1) \\
|H| &= \frac{2}{r} + \frac{h^{(-2)}}{r^2} + \frac{h^{(-3)}}{r^3} + o(r^{-3}) \\
\alpha_H &= \frac{\alpha_H^{(-1)}}{r} + \frac{\alpha_H^{(-2)}}{r^2} + o(r^{-2})
\end{aligned}
$$

where $\tilde{\sigma}_{ab}$ is the standard metric on a unit round sphere S^2. $\sigma_{ab}^{(1)}$ and $\sigma_{ab}^{(0)}$ are considered as symmetric 2-tensors on S^2, $h^{(-2)}$ and $h^{(-3)}$ are functions on S^2 and $\alpha_H^{(-1)}$ and $\alpha_H^{(-2)}$ are 1-forms on S^2. We shall see that all total conserved quantities on (M, g, k) can be determined from these data.

Using the result of [8], there is a unique family of isometric embeddings $X(r)$ and observers $T_0(r)$ which minimizes the quasi-local energy locally. The embedding $X(r) = (X^0(r), X^i(r))$ has the following expansion

$$
\begin{aligned}
X^0(r) &= (X^0)^{(0)} + \frac{(X^0)^{(-1)}}{r} + o(r^{-1}) \\
X^i(r) &= r\tilde{X}^i + (X^i)^{(0)} + \frac{(X^i)^{(-1)}}{r} + o(r^{-1}) \\
T_0(r) &= (a^0, a^i) + \frac{1}{r}(0, (a_i)^{(-1)}) + o(r^{-1}).
\end{aligned}
\tag{3.1}
$$

Using these, we compute

$$|H_0| = \frac{2}{r} + \frac{h_0^{(-2)}}{r^2} + \frac{h_0^{(-3)}}{r^3} + o(r^{-3})$$

$$\alpha_{H_0} = \frac{\alpha_{H_0}^{(-1)}}{r} + \frac{\alpha_{H_0}^{(-2)}}{r^2} + o(r^{-2})$$

and the expansion for ρ and j_a

$$\rho = \frac{\rho^{(-2)}}{r^2} + \frac{\rho^{(-3)}}{r^3} + o(r^{-3})$$

$$j_a = \frac{j_a^{(-1)}}{r} + \frac{j_a^{(-2)}}{r^2} + o(r^{-2}).$$

The isometric embedding equation gives

$$-\partial_a X^0 \partial_b X^0 + \sum_i \partial_a X^i \partial_b X^i = \sigma_{ab}.$$

From the expansions of X^0 and X^i in equation (3.1), we obtain the following two equations:

$$\partial_a \tilde{X}^i \partial_b (X^i)^{(0)} + \partial_a (X^i)^{(0)} \partial_b \tilde{X}^i = (\sigma_{ab})^{(1)} \tag{3.2}$$

$$\partial_a \tilde{X}^i \partial_b (X^i)^{(-1)} + \partial_a (X^i)^{(-1)} \partial_b \tilde{X}^i + \sum_i \partial_a (X^i)^{(0)} \partial_b (X^i)^{(0)} \tag{3.3}$$

$$= (\sigma_{ab})^{(0)} + \partial_a (X^0)^{(0)} \partial_b (X^0)^{(0)}.$$

On the other hand, the optimal embedding equation reads $div_\sigma j = 0$. From the expansion of metric and j_a, we have $div_{\tilde{\sigma}} j^{(-1)} = 0$ which determines $(X^0)^{(0)}$ and (a^0, a^i). The equation reads

$$div_{\tilde{\sigma}}(\rho^{(-2)} \tilde{\nabla} \tau^{(1)}) - \frac{1}{4} \tilde{\Delta}(\rho^{(-2)} \tilde{\Delta} \tau^{(1)}) - \frac{1}{2} \tilde{\Delta}(\tilde{\Delta} + 2)(X^0)^{(0)} + div_{\tilde{\sigma}}(\alpha_H^{(-1)}) = 0 \tag{3.4}$$

where $\tau^{(1)} = -\sum_i a^i \tilde{X}^i$.

We recall the definitions in [10].

Definition 6 *Suppose $T_0(r) = A(r)(\frac{\partial}{\partial x^0})$ for a family of Lorentz transformation $A(r)$. The CWY total center of mass integral is defined to be*

$$C^i = \lim_{r \to \infty} E(\Sigma_r, X(r), T_0(r), A(r)(x^i \frac{\partial}{\partial x^0} + x^0 \frac{\partial}{\partial x^i})),$$

and the CWY total angular momentum integral is defined to be

$$J_i = \lim_{r \to \infty} \epsilon_{ijk} E(\Sigma_r, X(r), T_0(r), A(r)(x^j \frac{\partial}{\partial x^k} - x^k \frac{\partial}{\partial x^j}))$$

where Σ_r are the coordinate spheres and $(X(r), T_0(r))$ is the unique family of optimal isometric embeddings of Σ_r such that $X(r)$ converges to a round sphere of radius r with the center at the origin in $\mathbb{R}^3$ when $r \to \infty$.

In the following, we summarize the properties of the CWY total conserved quantities.

1. The definition depends only on the geometric data (g, k) and the foliation of surfaces at infinity. In particular, it does not depend on an asymptotically flat coordinate system or the existence of an asymptotically Killing field.
2. All CWY total conserved quantities vanish on any spacelike hypersurface in the Minkowski spacetime, regardless of the asymptotic behavior. In fact, all the quasi-local conserved quantities vanishes for surfaces in the Minkowski spacetime since the definition uses comparison of physical Hamiltonian with reference Hamiltonian. As pointed out in [5], this is not true for the ADM angular momentum and BORT center of mass.
3. For the ADM angular momentum and BORT center of mass, there are issues of finiteness since the rotational and boost Killing fields are of higher order. By comparison, the CWY total angular momentum and total center of mass are always finite on any vacuum asymptotically flat initial data set of order one. See [10, Theorem 7.4].
4. The CWY total angular momentum satisfies the conservation law. In particular, the CWY total angular momentum on any strictly spacelike hypersurface of the Kerr spacetime is the same. See Theorem 4 and [10, Theorem 8.4].
5. Under the vacuum Einstein evolution of initial data sets, the CWY total angular momentum is conserved, $\partial_t J_i(t) = 0$, and the CWY total center of mass obeys the dynamical formula $\partial_t C^i(t) = \frac{P^i}{E}$ where (E, P^i) is the ADM energy-momentum four vector. The precise statement in [10] is the following:

   Theorem 7 *Let (M, g, k) be a vacuum asymptotically flat initial data set of order one. Let $(M, g(t), k(t))$ be the solution to the initial value problem $g(0) = g$ and $k(0) = k$ for the vacuum Einstein equation with lapse function $N = 1 + O(r^{-1})$ and shift vector $\gamma = \gamma^{(-1)}r^{-1} + O(r^{-2})$. Let $C^i(t)$ and $J_i(t)$ be the the total center of mass and total angular momentum of $(M, g(t), k(t))$ respectively. We have*

   $$\begin{aligned}\partial_t C^i(t) =& \frac{P^i}{E},\\ \partial_t J_i(t) =& 0\end{aligned}$$

   for $i = 1, 2, 3$ where (E, P^i) is the ADM energy momentum of (M, g, k).

   In fact, the dynamical formula was proved for initial data sets with much weaker asymptotics. See [10, Theorem 10.2].

4 The total angular momentum of a harmonic asymptotic initial data set

In this section, we consider the total conserved quantities for harmonic asymptotic initial data sets proposed by Corvino-Schoen [19].

We recall that an initial data consists of (M, g, k) where g is the induced metric and k is the second fundamental form of a spacelike hypersurface M in a given spacetime. In particular, an initial data set in a vacuum spacetime satisfies the vacuum constraint equation.

Definition 8 *An initial data set (M, g, k) is said to satisfy the vacuum constraint equation if*

$$\begin{aligned} R(g) + (tr_g k)^2 - |k|_g^2 =& 0 \\ div_g(k - (tr_g k)g) =& 0. \end{aligned}$$

Here $R(g)$ is the scalar curvature of the metric g.

The following notion of harmonic asymptotic initial data set was proposed by Corvino-Schoen [19] in their gluing construction of solutions of the vacuum constraint equation.

Definition 9 *An initial data set (M, g, k) is said to be of harmonic asymptotics if it is asymptotically flat, and on each end there is a function u and a vector field Y such that*

$$\begin{aligned} g_{ij} =& u^4 \delta_{ij} \\ k_{ij} =& u^2 (Y_{i,j} + Y_{j,i} - \frac{Y_{k,k}}{2} \delta_{ij}) \end{aligned} \tag{4.1}$$

where u tends to 1 and Y tends to 0 at infinity.

For simplicity, we denote

$$(\mathfrak{L}Y)_{ij} = Y_{i,j} + Y_{j,i} - (Y_{k,k})\delta_{ij}.$$

For a harmonic asymptotic initial data, the vacuum constraint equation is reduced to the following elliptic system for u and Y_i:

$$\begin{aligned} 8\Delta_\delta u =& u(-|\mathfrak{L}Y|^2 + \frac{1}{2}(Tr_\delta(\mathfrak{L}Y))^2) \\ \Delta_\delta Y_i =& 2u^{-1}u_i Tr_\delta(\mathfrak{L}Y) - 4u^{-1}u_j(\mathfrak{L}Y)_i^j, \end{aligned} \tag{4.2}$$

where Δ_δ is the Laplacian with respect to the flat metric δ_{ij} on $\mathbb{R}^3$.

Suppose u and Y_i admit expansions of the following form:

$$\begin{aligned} u =& 1 + \frac{A}{r} + \frac{u^{(-2)}}{r^2} + o(r^{-2}) \\ Y_i =& \frac{B_i}{r} + \frac{Y_i^{(-2)}}{r^2} + o(r^{-2}). \end{aligned} \tag{4.3}$$

The expansions follows naturally from the ansatz of the equation (4.1), the vacuum constraint equation (4.2), the asymptotically flatness and expansions of integral order. Here $u^{(-2)}$ and $Y_i^{(-2)}$ are functions on S^2 independent of r. Let $\tilde{\sigma}$ be the standard unit round metric on S^2, and $\tilde{\nabla}$ and $\tilde{\Delta}$ be the covariant derivative and Laplacian with respect to $\tilde{\sigma}$, respectively.

We show that $u^{(-2)}$ and $Y_i^{(-2)}$ can be determined explicitly by (4.2) in the following lemma.

Lemma 10 *Let (M,g,k) be a harmonic asymptotic initial data satisfying the vacuum constraint equation where u and Y have expansion as in equation (4.3). Then*

$$\begin{aligned} u^{(-2)} =& c_i\tilde{X}^i - \frac{9}{64}\sum_i (B_i)^2 + \frac{1}{64}\sum_{i,j} B_iB_j\tilde{X}^i\tilde{X}^j \\ Y_i^{(-2)} =& d_{ij}\tilde{X}^j - \frac{5}{2}B_iA + \frac{A}{2}(\sum_k B_k\tilde{X}^k)\tilde{X}^i. \end{aligned}$$

where c_i and d_{ij} are constants.

The CWY total conserved quantities for harmonic asymptotic initial data sets were evaluated in [6]:

Lemma 11 *For a harmonic asymptotic initial data set satisfying (4.3), the solution of the optimal isometric embedding equation (1.2) has the following expansion:*

$$\begin{aligned} X^i =& r\tilde{X}^i + 2A\tilde{X}^i + r^{-1}(2u^{(-2)} + A^2)\tilde{X}^i + o(r^{-1}) \\ X^0 =& r^{-1}(X^0)^{(-1)} + o(r^{-1}) \\ T_0 =& (a^0, a^i) + r^{-1}(a_0^{(-1)}, a_i^{(-1)}) + o(r^{-1}) \end{aligned}$$

where

$$\frac{a^k}{a^0} = \frac{B_k}{4A}.$$

Remark 12 *The terms $(X^0)^{(-1)}$ and $a_0^{(-1)}, a_i^{(-1)}, i = 1,2,3$ can be solved from the second order terms of the optimal isometric embedding equation. However, the total conserved quantities do not depend on them.*

Theorem 13 *For a harmonic asymptotic initial data set, let $X(r), T_0(r)$ be the solution of optimal embedding equation on Σ_r obtained in Lemma 11. Let $T_0(r) = A(r)\frac{\partial}{\partial x^0}$ where*

$$A(r) = A_{\alpha\beta} + O(r^{-1}).$$

The CWY total angular momentum and center of mass are

$$\begin{aligned} C_i =& \frac{2c_i}{a^0} + \frac{(c_iB_j - c_jB_i)A_{0j}}{4A} + \frac{(d_{ij} - d_{ji})A_{0j}}{4} \\ J_i =& \sum_{j,k,l} \frac{\epsilon_{ijk}}{4}[\frac{A_{kl}(c_jB_l - c_lB_j)}{A} + A_{kl}(d_{jl} - d_{lj})], \end{aligned}$$

where A, B_i, c_i, d_{ij} are the coefficients in the expansions of u and Y as in Lemma .

5 Total angular momentum of an asymptotically hyperbolic initial data set

The CWY quasi-local angular momentum and center of mass can also be used to define total conserved quantities for asymptotically hyperbolic initial data sets

[11]. They are defined using the limit of quasi-local conserved quantities in a similar fashion. The main difference is that the total conserved quantities in the asymptotically hyperbolic case is finite only for the foliation with vanishing linear momentum. Assuming the total energy-momentum is timelike, there exists a foliation with vanishing linear momentum by applying a coordinate boost. The limit of quasi-local conserved quantities is evaluated for such a foliation in terms of the expansion of the metric and second fundamental form. We discuss this application to an asymptotically hyperbolic initial data set (M, g, k) in this section. For such initial data sets, the definition of mass was considered by X. Wang [39], Chruściel-Herzlich [17], Sakovich[32], Chruściel-Jezierski-Łęski, [18], Zhang [41] etc. The model case is an umbilical slice in the Schwarzschild spacetime where $k = g$. More specifically, we consider the Schwarzschild spacetime with the metric

$$-(1-\frac{2m}{r})dt^2+(1-\frac{2m}{r})^{-1}dr^2+r^2(d\theta^2+\sin^2\theta d\phi^2).$$

Take a hypersurface M_1 defined by $t = f_1(r)$ such that

$$f_1'(r)=\frac{r}{(1-\frac{2m}{r})\sqrt{1-\frac{2m}{r}+r^2}}.$$

The induced metric is

$$\frac{1}{1-\frac{2m}{r}+r^2}dr^2+r^2(d\theta^2+\sin^2\theta d\phi^2)$$

and the second fundamental is the same as the induced metric, see Section 3.2 of [3] for the calculation therein.

There is also a spacelike hypersurface M_2 that is isometric to the hyperbolic 3-space. In fact, take $t = f_2(r)$ such that

$$f_2'(r)=\frac{\sqrt{r^2+\frac{2m}{r}}}{(1-\frac{2m}{r})\sqrt{1+r^2})},$$

and it is not hard to see that the induced metric is

$$\frac{1}{1+r^2}dr^2+r^2(d\theta^2+\sin^2\theta d\phi^2), \tag{5.1}$$

or the metric of the hyperbolic 3-space.

In most previous work on this subject, the second fundamental form k of (M, g, k) is assumed to be the same as the induced metric g and thus there is no linear momentum or angular momentum. It is nevertheless important to take into account of k, even for the definition of mass, in view of the above example M_2: there is no way to read off the mass from the induced metric (5.1) of M_2.

Suppose $g_{\mathbb{H}^3} = \frac{1}{r^2+1}dr^2 + r^2\tilde{\sigma}_{ab}du^a du^b$ is the hyperbolic metric in coordinates $r, u^a, a = 1, 2$, where $\tilde{\sigma}_{ab}du^a du^b = d\theta^2 + \sin^2\theta d\phi^2$ is the standard metric on a round 2-sphere.

Definition 14 *(M, g, k) is said to be an asymptotically hyperbolic initial data set if there exists a compact subset K of M such that $M\backslash K$ is diffeomorphic to a finite union of ends $\cup_i(\mathbb{H}^3\backslash B_i)$ where each B_i is a geodesic ball in $\mathbb{H}^3$. Under the diffeomorphism on each end, we have*

$$g = g_{rr}dr^2 + 2g_{ra}drdu^a + g_{ab}du^a du^b \text{ and } k = g + p,$$

where

$$\begin{aligned} g_{rr} =& \frac{1}{r^2} - \frac{1}{r^4} + \frac{g_{rr}^{(-5)}}{r^5} + \frac{g_{rr}^{(-6)}}{r^6} + O(r^{-7}), \qquad g_{ra} = \frac{g_{ra}^{(-3)}}{r^3} + O(r^{-4}), \\ g_{ab} =& r^2\tilde{\sigma}_{ab} + g_{ab}^{(0)} + \frac{g_{ab}^{(-1)}}{r} + \frac{g_{ab}^{(-2)}}{r^2} + O(r^{-3}), \end{aligned}$$

and

$$\begin{aligned} p_{rr} =& \frac{p_{rr}^{(-4)}}{r^4} + O(r^{-5}) \qquad p_{ra} = \frac{p_{ra}^{(-3)}}{r^3} + O(r^{-4}), \\ p_{ab} =& p_{ab}^{(0)} + \frac{p_{ab}^{(-1)}}{r} + \frac{p_{ab}^{(-2)}}{r^2} + O(r^{-3}). \end{aligned}$$

Each term in the expansion is considered to be the corresponding tensor on a standard round 2-sphere that is independent of r. Here $\tilde{\sigma}_{ab}$ is the standard round metric on the unit 2-sphere S^2. $g_{rr}^{(-5)}, g_{rr}^{(-6)}, p_{rr}^{(-4)}$ are considered to be functions on S^2 that do not depend on r, $g_{ra}^{(-3)}, p_{ra}^{(-3)}$ are considered to be one-forms on S^2 that do not depend on r, and $g_{ab}^{(0)}, g_{ab}^{(-1)}, g_{ab}^{(-2)}, p_{ab}^{(0)}, p_{ab}^{(-1)}, p_{ab}^{(-2)}$ are considered to be symmetric two-tensors on S^2 that do not depend on r. Furthermore, we assume that $p_{ab}^{(0)}$ and $g_{ab}^{(0)}$ are traceless with respect to $\tilde{\sigma}_{ab}$.

By solving the optimal isometric equation equation on each coordinate sphere and taking the limit of quasilocal conserved quantities, it is shown in [12] that the total energy-momentum is

$$\begin{aligned} E =& \frac{1}{8\pi}\int_{S^2} m\, dS^2 \\ P^i =& \frac{1}{8\pi}\int_{S^2} \tilde{X}^i m\, dS^2, i = 1, 2, 3 \end{aligned}$$

where $\tilde{X}^i, i = 1, 2, 3$ are the three standard coordinate functions on S^2 and the mass aspect function m is defined to be

$$m = \frac{3}{2} tr_{S^2} g_{ab}^{(-1)} + tr_{S^2} p_{ab}^{(-1)} + g_{rr}^{(-5)}. \tag{5.2}$$

In particular, (E, P^i) is non-spacelike by the positivity of quasi-local mass if the dominant energy condition is satisfied. Set $\mu = \sqrt{E^2 - \sum_{i=1}^3 (P^i)^2}$, then along the vacuum Einstein equation with $(M, g(t), k(t))$.

$$\partial_t \mu(t) = \frac{-1}{8\pi}\int_{S^2} |p_{ab}^{(0)} + g_{ab}^{(0)}|^2 dS^2.$$

For example, the case X. Wang [39] considered for asymptotically hyperbolic Riemannian manifolds corresponds to $g_{ra} = 0$ and $g_{rr} = \frac{1}{r^2+1}$. The BMS group acts on (M, g, k) by changing the foliations and this induces an $SO(3,1)$ action on (E, P^i). The angular momentum is finite only for the foliation with vanishing P^i, or aligning with the energy-momentum vector. In this case, the angular momentum is given by:

$$J^i = \frac{1}{8\pi} \int_{S^2} \tilde{X}^i \tilde{\epsilon}^{ab} \tilde{\nabla}_b (g_{ra}^{(-3)} + p_{ra}^{(-3)}) dS^2.$$

We refer to [12] for more details.

References

[1] R. Arnowitt, S. Deser and C. W. Misner, The dynamics of general relativity, in *Gravitation: An introduction to current research*, 227–265, Wiley, New York.

[2] A. Ashtekar and R. O. Hansen, *A unified treatment of null and spatial infinity in general relativity. I. Universal structure, asymptotic symmetries, and conserved quantities at spatial infinity,* J. Math. Phys. 19 (1978), no. 7, 1542–1566.

[3] S. Brendle and M.-T. Wang, *A Gibbons-Penrose inequality for surfaces in Schwarzschild spacetime*, Comm. Math. Phys. 330 (2014), no. 1, 33–43. arXiv:1303.1863

[4] J. D. Brown and J. W. York, Jr., *Quasilocal energy and conserved charges derived from the gravitational action, Phys.* Rev. D (3) **47** (1993), no. 4, 1407–1419.

[5] P.-N. Chen, L.-H. Huang, M.-T. Wang, and S.-T. Yau, *On the validity of the definition of angular momentum in general relativity*, Ann. Henri Poincaré 17 (2016), no. 2, 253–270. arXiv:1401.0597

[6] P.-N. Chen and M.-T. Wang, *Conserved Quantities of harmonic asymptotic initial data sets*, Surveys in differential geometry 2015. One hundred years of general relativity, 227–248, Surv. Differ. Geom., 20, Int. Press, Boston, MA, 2015, arXiv:1409.5150

[7] P.-N. Chen and M.-T. Wang, *Rigidity and minimizing properties of quasi-local mass*, Surveys in differential geometry 2014. Regularity and evolution of nonlinear equations, 49–61, Surv. Differ. Geom., 19, Int. Press, Somerville, MA, 2015, arXiv:1411.6251

[8] P.-N. Chen, M.-T. Wang, and S.-T. Yau, *Evaluating quasilocal energy and solving optimal embedding equation at null infinity*, Comm. Math. Phys. **308** (2011), no.3, 845–863

[9] P.-N. Chen, M.-T. Wang, and S.-T. Yau, *Minimizing properties of critical points of quasi-local energy*, Comm. Math. Phys. **329** (2014), no. 3 919–935

[10] P.-N. Chen, M.-T. Wang, and S.-T. Yau, *Conserved quantities in general relativity: from the quasi-local level to spatial infinity*, Comm. Math. Phys. 338 (2015), no. 1, 31–80. arXiv:1312.0985

[11] P.-N. Chen, M.-T. Wang, and S.-T. Yau, *Quasilocal angular momentum and center of mass in general relativity*, Adv. Theor. Math. Phys. 20, no. 4 (2016), 671–682, arXiv:1312.0990

[12] P.-N. Chen, M.-T. Wang, and S.-T. Yau, *Conserved quantities on asymptotically hyperbolic initial data set*, to appear in Adv. Theor. Math. Phys., arXiv:1409.1812

[13] D. Christodoulou, *Mathematical problems of general relativity. I,* Zürich Lectures in Advanced Mathematics. European Mathematical Society (EMS), Zürich, 2008.

[14] D. Christodoulou and S.-T. Yau, S.-T. *Some remarks on the quasi-local mass,* Mathematics and general relativity (Santa Cruz, CA, 1986), 9–14, Contemp. Math., 71, Amer. Math. Soc., Providence, RI, 1988.

[15] P. -T. Chruściel, *On angular momentum at spatial infinity,* Classical Quantum Grav. **4** (1987), no. 6, L205–L210.

[16] P.-T. Chruściel and E. Delay, *On mapping properties of the general relativistic constraints operator in weighted function spaces, with applications*, Mém. Soc. Math. Fr. (N.S.) 94 (2003).

[17] P. T. Chruściel and M. Herzlich, *The mass of asymptotically hyperbolic Riemannian manifolds*, Pacific J. Math. **212** (2003), no. 2, 231–264.

[18] P. T. Chruściel, J. Jezierski and S. Łęski, *The Trautman-Bondi mass of hyperboloidal initial data sets*, Adv. Theor. Math. Phys. **8** (2004), no. 1, 83–139.

[19] J. Corvino and R. Schoen, *On the asymptotics for the vacuum Einstein constraint equations,* J. Diff. Geom. 73(2), 185–217 (2006).

[20] K. Gimre, *Quasi-local energy and isometric embedding* Ph. D. thesis, Columbia University 2016.

[21] S. W. Hawking and G. T. Horowitz, *The gravitational Hamiltonian, action, entropy and surface terms*, Classical Quantum Gravity **13** (1996), no. 6, 1487–1498.

[22] L.-H. Huang, R. Schoen and M.-T. Wang, *Specifying angular momentum and center of mass for vacuum initial data sets*, Comm. Math. Phys. **306** (2011), no. 3, 785–803.

[23] C.-C. M. Liu and S.-T. Yau, *Positivity of quasi-local mass. II*, J. Amer. Math. Soc. **19** (2006), no. 1, 181–204

[24] Ludvigsen, M. and Vickers, J.A.G., *Momentum, angular momentum and their quasi-local null surface extensions*, J. Phys. A: Math. Gen., **16**, 1155–1168 (1983).

[25] L. Nirenberg, *The Weyl and Minkowski problems in differential geometry in the large,* Comm. Pure Appl. Math. **6**, (1953). 337–394.

[26] A. V. Pogorelov, *Regularity of a convex surface with given Gaussian curvature,* (Russian) Mat. Sbornik N.S. **31**(73), (1952). 88–103.

[27] R. Penrose, *Some unsolved problems in classical general relativity*, in *Seminar on Differential Geometry*, 631–668, Ann. of Math. Stud., 102, Princeton Univ. Press, Princeton, NJ.

[28] R. Penrose, *Quasilocal mass and angular momentum in general relativity*, Proc. Roy. Soc. London Ser. A **381** (1982), no. 1780,

[29] T. Regge and C. Teitelboim, *Role of Surface Integrals in the Hamiltonian*

Formulation of General Relativity, Ann. Phys. 88, 286–318 (1974).

[30] R. Schoen and S.-T. Yau, *On the proof of the positive mass conjecture in general relativity.* Comm. Math. Phys. 65 (1979), no. 1, 45–76.

[31] Y. Shi and L.-F. Tam, *Positive mass theorem and the boundary behaviors of compact manifolds with nonnegative scalar curvature*, J. Differential Geom. **62** (2002), no. 1, 79–125.

[32] A. Sakovich, *A study of asymptotically hyperbolic manifolds in mathematical relativity*, Ph. D. Thesis, (2012), KTH, Stockholm, Sweden.

[33] L. B. Szabados, *Quasi-Local Energy-Momentum and Angular Momentum in General Relativity*, Living Rev. Relativity **12**, (2009).

[34] M.-T. Wang, *Quasilocal mass and surface Hamiltonian in spacetime*, XVIIth International Congress on Mathematical Physics, 229–238, World Sci. Publ., Hackensack, NJ, 2014.

[35] M.-T. Wang, *Energy, momentum, and center of mass in general relativity,* Surveys in differential geometry 2016. Advances in geometry and mathematical physics, 343–352, Surv. Differ. Geom., 21, Int. Press, Boston, MA, 2015,arXiv: 1605.04968

[36] M.-T. Wang and S.-T. Yau, *Quasilocal mass in general relativity,* Phys. Rev. Lett. **102** (2009), no. 2, no. 021101, 4 pp.

[37] M.-T. Wang and S.-T. Yau, *Isometric embeddings into the Minkowski space and new quasi-local mass.* Comm. Math. Phys. **288** (2009), no. 3, 919-942.

[38] M.-T. Wang and S.-T. Yau, *Limit of quasilocal mass at spatial infinity.* Comm. Math. Phys. **296** (2010), no.1, 271–283.

[39] X. Wang, *The mass of asymptotically hyperbolic manifolds*, J. Differential Geom. **57** (2001), no. 2, 273–299.

[40] E. Witten, *A new proof of the positive energy theorem*, Comm. Math. Phys. **80** (1981), no. 3, 381–402.

[41] X. Zhang, *A definition of total energy-momenta and the positive mass theorem on asymptotically hyperbolic 3-manifolds. I*, Comm. Math. Phys. **249** (2004), no. 3, 529–548.

Department of Mathematics, Columbia University
2990 Broadway, New York, NY 10027, USA
E-mail address:

Proceedings of the 7th International Congress of Chinese Mathematicians
ALM 43, pp. 473–488

On Global Dynamics of the Maxwell-Klein-Gordon Equations

Shiwu Yang and Pin Yu

Abstract

On the three dimensional Euclidean space, for data with finite energy, it is well-known that the Maxwell-Klein-Gordon equations admit global solutions. However, the asymptotic behaviours of the solutions for the data with non-vanishing charge and arbitrary large size are unknown. It is conjectured that the solutions disperse as linear waves and enjoy the so-called peeling properties for pointwise estimates. We propose a gauge independent proof of the conjecture.

1 A quick introduction

The Maxwell-Klein-Gordon (MKG) equations are a nonlinear system modeling the motion of a charged particle moving in an electric-magnetic field. From the physics point of view, the electric-magnetic field fades away and the final state of the particle must be static. The mathematical interpretation of this conjecture is that the MKG equations admit global *decaying* solutions. The global solvability of this system has been well understood since the work of Eardley-Moncrief in 1980's. But very little is known on the aysmptotic behaviours of the solutions for the general large initial data. The aim of our current study is to show that the solutions enjoy the peeling estimates for the massless case.

We begin with a quick introduction to the MKG equations. Let $A = A_\mu dx^\mu$ be a $\mathbb{R}$-valued connection 1-form for a given complex line bundle $\mathbf{L}$ over the Minkowski spacetime $\mathbb{R}^{3+1}$. The curvature 2-form $F = \big(F_{\mu\nu}\big)$ associated to A is simply dA, i.e., $F_{\mu\nu} = \partial_\mu A_\nu - \partial_\nu A_\mu$. In particular, F is a closed 2-form or equivalently $\nabla_{[\alpha}F_{\beta\gamma]} = 0$. The pair of square brackets denote the anti-symmetrization of the three indices α, β and γ. We will use Einstein summation convention throughout the paper. The above equation is also called the Bianchi equation of F. It is also equivalent to

$$\nabla^\mu {}^*F_{\mu\nu} = 0,$$

where ${}^*F_{\mu\nu}$ is the Hodge dual of $F_{\mu\nu}$. We recall that the Hodge dual ${}^*G_{\alpha\beta}$ of a 2-form $G_{\alpha\beta}$ is $\frac{1}{2}\mathscr{E}_{\alpha\beta\gamma\delta}G^{\gamma\delta}$, where $\mathscr{E}_{\alpha\beta\gamma\delta}$ is the volume form of the Minkowski metric $m_{\alpha\beta}$.

A section of the bundle **L** can be represented by a $\mathbb{C}$-valued function ϕ. The covariant derivative of ϕ with respect to A is

$$D_\mu\phi = \partial_\mu\phi + \sqrt{-1}A_\mu \cdot \phi.$$

The curvature form measures the non-commutativity of the covariant derivatives

$$[D_\mu, D_\nu]\phi = \sqrt{-1}F_{\mu\nu} \cdot \phi.$$

The massless MKG equations is a system of equations for a connection A on **L** and a section ϕ of **L**:

$$\begin{cases} \nabla^\mu F_{\mu\nu} = -J_\nu, \\ \Box_A\phi = 0, \end{cases} \tag{1.1}$$

where $J_\mu = \Im(\phi \cdot \overline{D_\mu\phi})$ is called the *current* and $\Box_A = D^\mu D_\mu$. It can be derived as the Euler-Lagrange equations for the action

$$\mathcal{L}(A,\phi) = \frac{1}{2}\int_{\mathbb{R}^{3+1}} D^\mu\phi \cdot \overline{D_\mu\phi} + \frac{1}{4}\int_{\mathbb{R}^{3+1}} F^{\mu\nu}F_{\mu\nu}.$$

We use the volume form of the Minkowski metric in the action. The system is a $\mathbf{U}(1)$-gauge theory, namely, if (A,ϕ) is a solution of (1.1), then $(A - d\chi, e^{i\chi}\phi)$ is also a solution for any smooth function χ.

The total charge of the system is given by

$$q_0 = \frac{1}{4\pi}\int_{\mathbb{R}^3} \mathbf{div}E dx = \frac{1}{4\pi}\int_{\mathbb{R}^3} \Im(\overline{D_t\phi} \cdot \phi) = \frac{1}{4\pi}\lim_{r\to\infty}\int_{|\omega|=1} r^2 E_i \frac{x_i}{r} d\omega,$$

where $E_i = F_{0i}$ is the electric field. It is easy to check that the total charge is conserved. This in particular implies that the electric field E has the nontrivial tail $q_0 r^{-3}x$ at any fixed time.

The pioneering works [6] and [7] of Eardley-Moncrief established the celebrated global existence result to the general Yang-Mills-Higgs equations with sufficiently smooth initial data. Around ten years later, by introducing the weighted Sobolev spaces, Klainerman-Machedon systematically studied the bilinear estimates of the null forms. As a consequence, they derived the notable global existence result for data merely bounded in the energy space. The idea of proving bilinear estimates of null form introduced in [10] leads to an revolution on the global well-posedness of PDEs of classical field theory, such as MKG equations, Yang-Mills equations, wave maps, etc., aiming at studying low regularity initial data in order to construct global solutions, see [11] and references therein. For a more recent and comprehensive summary of the progresses along this line, we refer to the work of Oh-Tataru [17]. The common feature of all these works is to construct a local solution with rough data so that the global well-posedness follows from conserved energy quantities. However regarding the global dynamics of the solutions, very little can be obtained through this approach.

The long time dynamics of solutions of MKG equations have only been well understood for sufficiently small initial data or data which are essentially compactly supported. The robust vector field method introduced by Klainerman in [9] has been successfully applied to derive the decay estimates for linear fields in [4] or nonlinear spin fields in [20] with small initial data. If the data are compactly supported, one can also use the conformal compactification method (see e.g. [2]) and this approach requires strong decay of the initial data, which in particular forces the total charge to be vanishing. To tackle the general case with nonzero charge, Shu in [21] proposed a frame work but without details. A complete proof towards this direction was contributed by Lindblad-Sterbenz in [15], also see a recent work [1] of Bieri-Miao-Shahshahani. However all these works are restricted to the small data regime or can be viewed as global stability problems of trivial solutions.

As for the large data problem, by using the conformal compactification method together with Eardley-Moncrief's results, Petrescu in [18] obtained the asymptotic decay properties of solutions to MKG equations with essentially compactly supported data, i.e., the scalar field has compact support and the Maxwell field is electrostatic outside the support. A similar result for Yang-Mills equation on $\mathbb{R}^{3+1}$ was obtained by Georgiev-Schirmer in [8] but with spherically symmetric data bounded in the conformal energy space (in particular charges must be vanishing!). For general initial data, the global asymptotic behaviour is only partially known. A Morawetz type of integrated local energy estimate was obtained by Psarelli in [19] for solutions of massive Maxwell-Klein-Gordon equations with data bounded in the energy space. For massless MKG equations, the first author Yang in [23] derived the stronger inverse polynomial decay of the energy flux through outgoing null cones with data bounded in weighted energy space. Both results allow the existence of nonzero charges. However the decay estimates in [19] are too weak to see the effect of the charges while Yang's work in addition affirmatively answered a conjecture of Shu in [20] that the nonzero charge can only affect the asymptotic behaviours of the solutions outside a forward light cone. Another consequence of the method used in [23] allowed the author to improve the small data results to data merely small on the scalar field while the Maxwell field can be arbitrarily large, see details in [22]. This result can be interpreted as the global nonlinear stability of linear Maxwell fields.

It is of great interests to remove the restriction of essentially compactly supported data without any smallness assumption. This is the final state conjecture of charged scalar fields: the solution should eventually decay as long as it decays suitably initially. In this work, we propose a sequence of new ideas to handle the long range effect of the large charge and this will lead to the resolution of the conjecture for rapidly decaying data.

2 The main result

2.1 The statement of the main theorem

Throughout the paper, we use the following conventions:

- The Greek letters $\alpha, \beta, \cdots$ denote indices from 0 to 3. The capital Latin letters $A, B, \cdots$ denote indices from 1 to 2. The little Latin letters $i, j, k, \cdots$ denote indices from 1 to 3.
- (ϕ, F) is a *given* finite energy smooth solution of the MKG equations. It exists globally and remains smooth according to the classical result of Klainerman-Machedon [10].
- The letter f denotes an *arbitrary* section of the bundle $\mathbf{L}$ (it may not be ϕ). The letter G denotes an *arbitrary* 2-form $G_{\mu\nu}$ (it may not be F).
- We define $\psi = r\phi$.

We use two coordinate systems on the Minkowski spacetime $\mathbb{R}^{3+1}$: the Cartesian coordinates $(x^0 = t, x^1, x^2, x^3)$ and the polar coordinates (t, r, ϑ). The optical functions u and v are defined as

$$u = \frac{1}{2}(t-r), \quad v = \frac{1}{2}(t+r), \quad u_+ = 1 + |u|, \quad v_+ = 1 + |v|.$$

A *null frame* is defined by $(e_1, e_2, e_3 = \underline{L}, e_4 = L)$, where $L = \partial_t + \partial_r$, $\underline{L} = \partial_t - \partial_r$ and e_1, e_2 is an orthonormal complement of L and $\underline{L}$.

The level sets of u and v define (locally) null foliations of the Minkowski spacetime. Given $r_2 > r_1 > 0$, we define the outgoing (or incoming) null hypersurfaces $\mathcal{H}_{r_1}^{r_2}$ (or $\underline{\mathcal{H}}_{r_2}^{r_1}$) as

$$\mathcal{H}_{r_1}^{r_2} := \big\{(t, r, \vartheta) \,\big|\, t \geq 0, u = -\frac{1}{2}r_1, r_1 \leq r \leq r_2\big\} \quad \text{or}$$

$$\underline{\mathcal{H}}_{r_2}^{r_1} := \big\{(t, r, \vartheta) \,\big|\, t \geq 0, v = \frac{1}{2}r_2, r_1 \leq r \leq r_2\big\}$$

respectively. On the initial time slice $\big\{t = 0\big\}$ where the Cauchy datum is given, we define

$$\mathcal{B}_{r_1}^{r_2} := \big\{(t, r, \vartheta) \,\big|\, t = 0, r_1 \leq r \leq r_2\big\}.$$

In the limiting case where $r_2 = \infty$, we write $\mathcal{H}_{r_1} = \mathcal{H}_{r_1}^{\infty}$, $\underline{\mathcal{H}}_{r_1} = \underline{\mathcal{H}}_{r_1}^{\infty}$ and $\mathcal{B}_{r_1} = \mathcal{B}_{r_1}^{\infty}$. Three hypersurfaces $\mathcal{H}_{r_1}^{r_2}$, $\underline{\mathcal{H}}_{r_2}^{r_1}$ and $\mathcal{B}_{r_1}^{r_2}$ bound a spacetime region and it is denoted by $\mathcal{D}_{r_1}^{r_2}$. In the following picture, the gray region is $\mathcal{D}_{r_1}^{r_2}$. The truncated light cones $\mathcal{H}_{r_1}^{r_2}$ and $\underline{\mathcal{H}}_{r_2}^{r_1}$ are denoted by the dashed line segments. Their intersection is a 2-sphere of radius $\frac{r_1+r_2}{2}$ and it is the tip of $\mathcal{D}_{r_1}^{r_2}$ in the picture. We denote this sphere by $\mathcal{S}_{r_1}^{r_2}$. The dashed-dotted line segment on the bottom is $\mathcal{B}_{r_1}^{r_2}$.

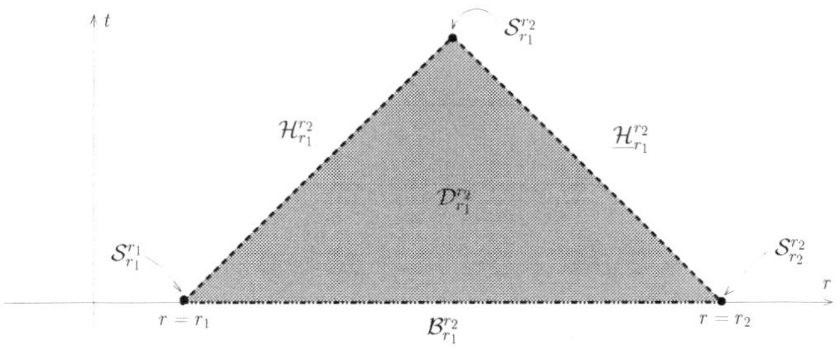

In the null frame, we have $\nabla_L L = \nabla_L \underline{L} = \nabla_{\underline{L}} L = \nabla_{\underline{L}} \underline{L} = 0$. Moreover, we have

$$\nabla_{e_A} L = \frac{1}{r} e_A, \ \nabla_{e_A} \underline{L} = -\frac{1}{r} e_A, \ \ \nabla_{e_A} e_B = \nabla\!\!\!\!/_{e_A} e_B + \frac{1}{2r} g\!\!\!/_{AB} (\underline{L} - L),$$

where $\nabla\!\!\!\!/_{e_A} e_B$ is the projection of $\nabla_{e_A} e_B$ to a 2-sphere $\mathcal{S}_{r_1}^{r_2}$ (or to the span of e_1 and e_2) and $g\!\!\!/_{AB}$ is the restriction of the Minkowski metric to $\mathcal{S}_{r_1}^{r_2}$.

We can decompose $G_{\mu\nu}$ with respect to the null frame:

$$\alpha(G)_A := G(L, e_A), \ \underline{\alpha}(G)_A := G(\underline{L}, e_A), \ \rho(G) := \frac{1}{2} G(\underline{L}, L), \ \sigma(G)_{AB} := G_{AB}.$$

For the special case $G_{\mu\nu} = F_{\mu\nu}$, we write

$$\alpha_A = F(L, e_A), \ \underline{\alpha}_A = F(\underline{L}, e_A), \ \rho := \frac{1}{2} F(\underline{L}, L), \ \sigma_{AB} := F_{AB}.$$

Since σ_{AB} is a 2-form on $\mathcal{S}_{r_1}^{r_2}$, there exists a function σ so that $\sigma_{AB} = \sigma \epsilon\!\!/_{AB}$ where $\epsilon\!\!/_{AB}$ is the volume form on $\mathcal{S}_{r_1}^{r_2}$. For the Hodge dual ${}^*F$ of F, if we denote ${}^*\alpha_A = -\epsilon\!\!/_A{}^B \alpha_B$ (the Hodge dual of α on $\mathcal{S}_{r_1}^{r_2}$), we have

$$\alpha_A({}^*F) = {}^*\alpha_A, \ \underline{\alpha}_A({}^*F) = -{}^*\underline{\alpha}_A, \ \rho({}^*F) = \sigma, \ \sigma({}^*F)_{AB} = -\rho \epsilon\!\!/_{AB}.$$

We consider Cauchy problem to (1.1) with initial data given by

$$\phi_0(x) = \phi(0, x), \ \phi_1(x) = \partial_t \phi(0, x), \ E_i^{(\text{ini})}(x) = E_i(0, x), \ B_i^{(\text{ini})}(x) = B_i(0, x).$$

The initial data set $(\phi_0, \phi_1, \ E^{(\text{ini})}, \ B^{(\text{ini})})$ is said to be *admissible* if it satisfies the compatibility condition

$$\mathbf{div}(E^{(\text{ini})}) = \Im(\phi_0 \cdot \overline{\phi_1}), \quad \mathbf{div}(B^{(\text{ini})}) = 0, \tag{2.1}$$

To impose precise assumptions on the initial data, split the electric field $E^{(\text{ini})}$ into the divergence free part E^{df} and the curl free part E^{cf}, that is,

$$\mathbf{div}(E^{df}) = 0, \quad \mathbf{curl}(E^{cf}) = 0, \quad E^{(\text{ini})} = E^{df} + E^{cf}.$$

From the above constraint equation, E^{cf} is uniquely determined by $\Im(\phi_0 \cdot \overline{\phi_1})$. In particular we can freely assign $(\phi_0, \phi_0, E^{df}, B^{(\text{ini})})$ as long as E^{cf}, $B^{(\text{ini})}$ are divergence free on the initial hypersurface $\{t = 0\}$. We require this part of data decay rapidly and belong to certain weighted Sobolev space. However, since E^{cf} satisfies an elliptic equation on $\mathbb{R}^3$, it has a nontrivial tail $\frac{q_0 x}{r^3}$ even with (ϕ_0, ϕ_1) compactly supported. To describe the asymptotic behaviour of the solutions, we need to precisely capture the asymptotic behaviour of the solution contributed by the charge. By formally expanding the Green's function for Laplacian:

$$|x-y|^{-1} = |x|^{-1} + |x|^{-3}x\cdot y + \sum_{i,j=1}^{3} \frac{1}{2}|x|^{-3}(3|x|^{-2}x_i x_j - \delta_{ij})y_i y_j + o(|y|^2),$$

we can define a potential function $V(x)$ as

$$V(x) = |x|^{-1}\frac{1}{4\pi}\int_{\mathbb{R}^3}(1+|x|^{-2}x\cdot y+\frac{1}{2}|x|^{-2}(3|x|^{-2}(x\cdot y)^2-|y|^2))\Im(\phi_0\cdot\bar{\phi}_1)dy, \quad |x|>0.$$

The potential is well defined if the initial data (ϕ_0, ϕ_1) of the scalar field decay rapidly. With the potential $V(x)$, we can define the general charge 2-form $F[q_0]$ with components

$$F[q_0]_{0i} = E_i[q_0] = \partial_i V(x), \quad F[q_0]_{ij} = 0.$$

It is straightforward to check that $F[q_0]$ satisfies the linear Maxwell equation on the region away from the axis $\{x=0\}$. Moreover, there is a constant C, depending only on ϕ_0 and ϕ_1, so that

$$|\rho(F[q_0])| \le Cr^{-2}, \quad |\underline{\alpha}(F[q_0])| = |\alpha(F[q_0])| \le Cr^{-3}, \quad |\sigma(F[q_0])| = 0. \tag{2.2}$$

We remark that most commonly one uses $F[q_0] = \frac{q_0}{r^2}dt\wedge dr$ to denote the charge part near special infinity and it is a special case of the above construction.

Let ε_0 be a small positive constant (say 10^{-2}). We assume that the initial data is bounded in the following gauge invariant weighted Sobolev norm

$$C_0 := \sum_{k\le 2}\int_{\mathbb{R}^3}\Big[(1+r)^{2k+6+8\varepsilon_0}\big(|D^{k+1}\phi_0|^2 + |D^k\phi_1|^2 + |\nabla^k(E^{(\text{ini})} - E[q_0]\mathbf{1}_{|x|\ge 1})|^2 \\ + |\nabla^k B^{(\text{ini})}|^2\big) + r^{4+8\varepsilon_0}|\phi_0|^2\Big]dx. \tag{2.3}$$

The main theorem of the paper is as follows:

Theorem 2.1 (Main result). *Consider the Cauchy problem to the massless MKG equation (1.1) with admissible initial data $(\phi_0, \phi_1, E_i^{(ini)}, B_i^{(ini)})$ bounded in the above weighted norm (2.3). Then there is a global in time solution (ϕ, F) satisfying the following pointwise peeling estimates*

$$|\phi| \le Cu_+^{-1}v_+^{-1}, \quad |D_L(r\phi)| \le Cu_+^{-1}v_+^{-2}, \quad |\alpha(\mathring{F})| \le Cu_+^{-1}v_+^{-3}, \\ |\rho(\mathring{F})| + |\sigma(\mathring{F})| + |\not{D}\phi| \le Cu_+^{-2}v_+^{-2}, \quad |\underline{\alpha}(\mathring{F})| + |D_{\underline{L}}\phi| \le Cu_+^{-3}v_+^{-1}. \tag{2.4}$$

for some constant C depending only on C_0, where $\mathring{F} = F - F[q_0]\mathbf{1}_{\{1+t\le|x|\}}$ with $\mathbf{1}_{\{1+t\le|x|\}}$ the characteristic function of the exterior region $\{(t,x)|t+1\le|x|\}$.

We give several remarks.

Remark 2.2. *There is no restriction on the size or on the support of the data. In particular, the charge q_0 can be large. Besides the above pointwise estimates, uniform energy estimates as well as weighted energy estimates can also be derived in the course of the proof.*

Remark 2.3. *The peeling estimates (2.4) for the chargeless part of the solution together with the trivial bound (2.2) of the charge part describe the asymptotic behaviour of the full solution in the exterior region. Moreover the estimate implies that the nontrivial charge can only affect the asymptotic behaviour of the solution in the exterior region. This confirms the conjecture of Shu in [20].*

Remark 2.4. *There is a heuristic explanation of the construction of the charge part $F[q_0]$ from the dipole expansion perspective: if we expand the Maxwell field F in a Taylor series near spatial infinity $r = \infty$ as*

$$F = F_2 + F_3 + F_4 + F_5 + \cdots ,$$

where $F_k = O(r^{-k})$. The formal expansion of the Green function gives the $F[q_0] = F_2 + F_3 + F_4$. In this work we require that the perturbation starts from F_5. Indeed, the main reason for doing this is to make $F - F[q_0]$ decay sufficiently fast initially so that the chargeless part is bounded in the weighted Sobolev norm defined in (2.3).

Remark 2.5. *Regarding the dependence of the constant C on the size of the initial data, our proof can easily imply that C depends exponentially on the zeroth order weighted energy (without derivative of the initial data) but polynomially on the higher order weighted energies. Simply from the charge part, it seems that exponential dependence on the zeroth order energy can not be improved. However from the point of view of the bilinear estimates in [10], we conjecture that the dependence on higher order energy should be linear.*

2.2 An outline of the proof: difficulties, ideas and novelties

The proof uses almost all the existing techniques and results for Maxwell-Klein-Gordon equations: the vector field method, the conformal compactification, the conformal analogues in the vector field method and the low regularity existence results of Klainerman-Machedon. Besides these, we will also introduce new commutation vector fields, new null forms and study some new structure of the nonlinearities. In the rest of the section, we will first sketch the proof in three steps. Then, we will present the difficulties in each step and provide heuristic ideas to handle these difficulties. Finally, we will summarize some new aspects of the proof.

2.2.1 The structure of the proof

The proof consists of three steps:

Step 1 We take a positive number R_* and it determines the so-called exterior region $\mathcal{D}_{R_*}$ (grey part).

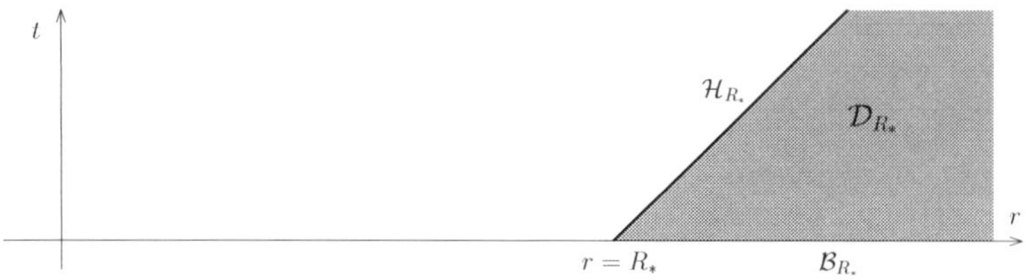

For large R_*, by restricting data on the region where $r \geq R_*$, i.e., $\mathcal{B}_{R_*}$ (as the bottom of the grey region), we can assume the chargeless part of the restricted data is small. Since the grey region is the domain of dependence of $\mathcal{B}_{R_*}$, the solution in $\mathcal{D}_{R_*}$ is completely determined by the restricted data on $\mathcal{B}_{R_*}$. We therefore study the long time behaviour of solutions of MKG equations in the grey region $\mathcal{D}_{R_*}$ with data small in the chargeless part. We emphasize that this is not a small data problem as the charge part of the solution is large and is independent of the radius R_*.

Step 2 This step connects the first step to the third. First of all, we will carefully choose a hyperboloid in $\mathcal{D}_{R_*}$ (on which we have precise control on the solution from the previous step). This hypersurface is denoted by Σ_+ in the next picture.

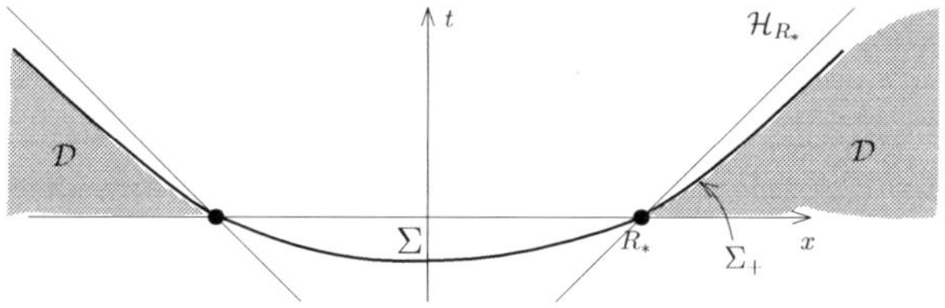

The solution restricted on this hyperboloid can be viewed as initial datum for the solution in the interior region which is unknown so far. This step is devoted to showing that the solution obtained from the previous step is sufficient regular on Σ_+ so that we can conduct the next step.

Step 3 In this last step, we will study the asymptotics of the solution in the causal future $\mathcal{J}^+(\Sigma)$ which is the grey region in the left figure (this is the white region in the previous picture).

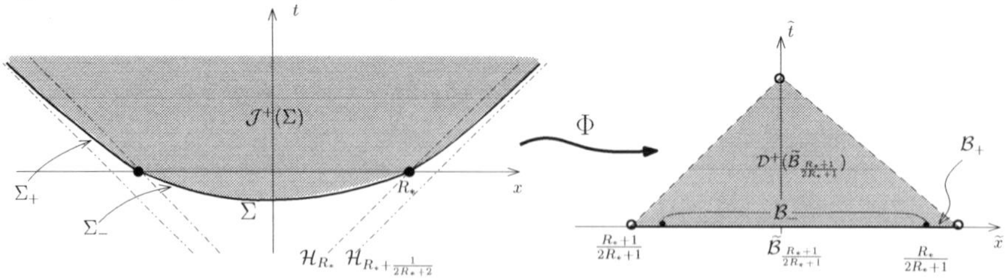

The hypersurface Σ consists of two parts: Σ_- and Σ_+. Since Σ_- is finite, the solution on Σ_- can be well controlled by the data on the compact region $\{t = 0, |x| \leq R_*\}$. This indeed follows from the classical result of Eardley-Moncrief or the result of Klainerman-Machedon. Together with Step 2, the restriction of the solution on Σ will be well-understood in terms of the initial data.

Then we will perform a conformal transformation Ψ to map $\mathcal{J}^+(\Sigma)$ to a backward finite light cone (the grey cone on the right hand side of the picture). The hypersurface Σ will be mapped to the bottom of the cone. By multiplying conformal factors appropriately, the global dynamics of solutions to MKG equations defined on the left of the picture is then reduced to understanding the solution to MKG equations defined on the right of the picture. The estimates from Step 2 provide a bound of the H^2-norm of the solution on the bottom of the cone on the right hand side of the picture. This allows us to use the classical theory of Klainerman-Machedon to bound the solution on the cone up to two derivatives hence the L^∞ norm of the solution. Finally, we undo the conformal transformation by rewriting the solution on the left hand side in terms of the solutions on the right hand side. The conformal factors then give the decay estimates of the solution in $\mathcal{J}^+(\Sigma)$. Together with the decay estimates from Step 1, we can derive the peeling estimates in the main theorem.

2.2.2 Difficulties in the proof

We list several difficulties which did not appear in previous works on MKG equations. We would like to emphasize that the first difficulty (the largeness of charge) listed below is related to all the rest. The remaining difficulties arise in course of the resolution of the first one. We also want to point out that the most difficult part of the proof is Step 1.

1. The large nonzero charge.

   Although the energy norm of the chargeless part of the data in Step 1 is small, the charge q_0 can be large. First of all, the traditional conformal compactification method used in [3] by Christodoulou-Bruhat requires strong decay of the data which forces the charge to be vanishing. Secondly, the presence of nonzero charge may cause a logarithmic divergence in the energy estimates, see a more thorough discussion in the work [15] of Lindblad-Sterbenz for the purely small data case. The error term caused by the charge can in fact be absorbed if the charge is sufficiently small. We overcome this large charge difficulty by using the method developed by Yang in [23].

   We would also like to compare this charge difficulty with the massive case of recent work [14] of Klainerman-Wang-Yang, in which they studied the massive MKG equations with small initial data. Their method also applies to the case with arbitrary large charge. However due to the existence of mass which gives control of the scalar field itself, the effect of nonzero charge can be easily controlled (see more detailed discussion in the next subsection). The main difficulty there, however, lies in the inconsistent asymptotic behaviours of Maxwell fields and solutions of Klein-Gordon equations.

2. The sharp peeling estimates.

   Since in Step 3 we have to compactify $\mathcal{J}^+(\Sigma)$, the estimates for the solution obtained from Step 1 on Σ must be sufficiently regular so that the solution on its conformal compactification are bounded in the right Sobolev spaces. In particular it requires to obtain the sharp decay estimates such as $D_L(r\phi) = O(r^{-2})$ and $\alpha = O(r^{-3})$ along outgoing light cones. So far as we know, even for the small data regime (with small charge), these estimates are unknown.

3. New commutators to prove the necessary sharp peeling estimates.

   The idea to obtain the above sharp peeling estimates is straightforward: we need to put more r-weights in the usual energy estimates so that the r-weights will be converted into extra decay via Sobolev inequality. We will use the conformal Morawetz vector field K as commutators. This vector field is of order 2 in terms of weights r and is used traditionally only as multipliers. In such a way, the structure of the nonlinear terms after commutation becomes the primary concern and we will show that it has some new null structure.

4. The hidden null structure of the MKG equations related to commutators.

   This is related to point (2) above. When one commutes vector fields with the MKG equations, although it may generate many error terms, one needs to at least make sure some of the fundamental structures remain unchanged. Very often, these structures are important in the analytic perspective and more precisely they should be phrased in such a way that they fit into the energy estimates. We will show that there is a new null structure of the nonlinear terms which is invariant after commuting correct vector fields. Also, there is another important type structure, which we will call it *reduced structure*, also remains unchanged after commutations.

5. The choice of conformal compactifications.

   The presence of nonzero charge prevents us to use the usual Penrose type compactification for the entire spacetime (see [3]): the ρ-component of the Maxwell field behaves as $\frac{q_0}{r^2}$ which cannot be compactified near the spatial infinity. However this effect of charge does not propagate from the spatial infinity to the future null infinity so that we can indeed perform a conformal transformation inside a null cone to avoid spatial infinity.

6. Precise energy estimates on the hypersurface Σ in Step 2.

   Since Σ is a hyperboloid in Minkowski spacetime, the energy estimates on Σ, especially those needed in the Klainerman-Machedon theory after the compactification, are not straightforward. Nevertheless, this part is less serious compared to all the previous ones and can be derived by using the classical energy estimates in a geometric way.

2.2.3 Key ideas and novelties of the proof

In the subsection, we will list all the ideas and new features of the proof in order to deal with the difficulties mentioned in the previous subsection.

1. The reduced structure and converting spatial decay against the logarithmic growth.

   We first explain the reduced structure of the nonlinearity. Let $F = dA$. We may think of A as ϕ. Thus, the Maxwell equations are reduced to the form

$$\Box A = \phi \cdot D\phi.$$

While most commonly, a nonlinear wave equation with quadratic interaction looks like

$$\Box \phi = \nabla \phi \cdot \nabla \phi.$$

The MKG equations is one derivative less in the nonlinearities. This is the reduced structure.

In terms of energy estimates, the reduced structure will be reflected in the following formula:

$$\int_{\mathcal{H}_{r_1}^{\infty}} |D_L \phi|^2 \leq C_1 \mathring{\varepsilon} r_1^{-\gamma_0} + C_2 \int_{r_1}^{\infty} \int_{\mathcal{H}_s^{\infty}} \frac{|q_0|}{r^2} |\phi||D_L\phi|.$$

The left hand side is a classical energy flux term through outgoing null cones $\{u = r_1\}$. The first term on the right hand side is coming from the data and the exponent $-\gamma_0$ reflects the decay of the data near spatial infinity. The second term on the right hand side contains a ϕ without any derivative acting on it. We remark that the $\frac{q_0}{r^2}$ factor is arising from the charge. Heuristically for waves, a $\frac{1}{r}$ factor can be regard as D_L-derivative so that we should think of the second term as $\frac{1}{r}|D_L\phi|^2$ thus we see that there is a logarithmic growth when we integrate. We remark here that for the massive case in [14], since solution of massive Klein-Gordon equation decays as quickly as its derivatives, i.e., one can regard ϕ as $D_L\phi$, the above error term can be easily absorbed by using Gronwall's inequality.

We use an idea introduced by Yang in [23] to handle this logarithmic loss. Morally speaking, to obtain the estimates for the energy flux, we can afford a loss in r instead of in time:

$$\int_{\mathcal{H}_{r_1}^{\infty}} |D_L \phi|^2 \leq C \mathring{\varepsilon} \cdot r_1^{-\gamma_0 + \varepsilon_0}.$$

In other words, the decay rate near null infinity changes from γ_0 to $\gamma_0 - \varepsilon_0$.

2. The ε_0-reductive argument for higher order energy estimates.

The argument is designed to make a better use of the reduced structure of the nonlinearity when we do higher order energy estimates. Although the charge vanishes when taking derivatives, the above type error term arises from the connection field A and has the same structure as described previously. We design an ansatz which allows higher order derivative to lose more decay. To be more precise, we will lose $2(k+1)\varepsilon_0$-decay for the k-th order derivatives. We will use the following example to illustrate how the argument works. In the course of deriving energy estimates for the first order derivatives, schematically, the nonlinear terms look like $\int |\nabla\phi||\nabla^2\phi|$. On the other hand, $|\nabla\phi|$ is already controlled when one derives estimates for the solution itself without commuting vector fields with equations, thus the estimates on $\nabla\phi$

only lose $2\varepsilon_0$ decay. Compared to the $4\varepsilon_0$ loss in the first order derivative case, we indeed have a gain in decay for the nonlinear terms. This gain will play an essential role in closing the estimates.

3. Morawetz vector field as commutator and new commutation formulas.

   Traditionally, the Morawetz vector field K is only used as multipliers in the energy estimates. In this work, we will commute K with the equation. The extra weights compared to the classical commutators such as rotations and scaling provide an extra decay factor for the solutions near null infinity. This extra decay factor is indispensable when we perform the conformal compactification. We would also like to remark that, since K is the image of ∂_t under the inversion map, commuting K with the equation can be regarded as the usual commutation of ∂_t after the conformal transformation. Thus, this idea should be viewed as a vector field method version of conformal transformations.

   More precisely, for $Z \in \mathcal{Z} = \{T, \Omega_{12}, \Omega_{23}, \Omega_{31}, S, K\}$, where T is the time translation, Ω_{ij} are rotations and S is scaling, for **Div** (the principle part of the Maxwell equations) and $\Box_A$, we have the following two formulas

$$[r^2\mathbf{Div}\,, \mathcal{L}_Z]G = 0, \quad [r^2\Box_A, D_Z + \frac{Z(r)}{r}]\phi = r^2 Q(\phi, F; Z) \qquad (2.5)$$

   for any closed 2-form G and complex scalar field ϕ. We emphasize that the formula holds for K and $Q(\phi, F; Z)$ is quadratic in ϕ and $F = dA$. We also remark that to our knowledge these commutator formulas are new.

4. A new null form.

   The quadratic form $Q(\phi, F; Z)$ is indeed a null form. Take $Z = S$ for example. It can be shown that

$$\begin{aligned}|Q(\phi, F; Z)| \lesssim &\big(\frac{r}{|u|}|\rho| + |\underline{\alpha}|\big)|D_L(r\phi)| + \big(\frac{r}{|u|}|\alpha| + \frac{|u|}{r}|\underline{\alpha}| + |\sigma|\big)|\not{D}(r\phi)| \\ &+ \big(|\alpha| + \frac{|u|}{r}|\rho|\big)|D_{\underline{L}}(r\phi)| + \big(|\rho| + |\sigma|\big)|\phi| + \text{cubic terms}.\end{aligned}$$

   Similar estimates hold for other vector fields in $\mathcal{Z}$. We remark that rather than ϕ itself, the derivatives of $r\phi$ appear naturally in the above null structure estimate.

   The most remarkable property of $Q(\phi, F; Z)$ is that it has an iterative structure. This is crucial when we commutate multiple derivatives with equations. More precisely, if we define $\widehat{D}_Z = D_Z + \frac{Z(r)}{r}$, we can show that

$$\big[\widehat{D}_Y, [\widehat{D}_X, r^2\Box_A]\big]\phi = -r^2 Q(\phi, F; [Y, X]) - r^2 Q(\phi, \mathcal{L}_Y F; X) + 2r^2 F_{Y\mu} F_X{}^{\mu}\phi.$$

   The right hand side after commutating two derivatives can still be expressed in terms of Q and it still satisfies the null structure. This is one of the keys in the proof.

   We remark that to our knowledge this null structure is also new.

5. The algebraic structure of J.

   We have seen that $r\phi$ appears naturally in the null form estimates. We would like to point out another perspective. We mentioned previously that $D_L(r\phi) = O(\frac{1}{r^2})$. We can also show that the best decay estimates for $D_L\phi$ is still $O(\frac{1}{r^2})$ instead of $O(\frac{1}{r^3})$. From this point of view, we may consider $r\phi$ to be "better" than ϕ itself. On the other hand, for the Maxwell equation, instead of commuting with the operator **Div**, we commute with r^2**Div**. It thus requires to analyze $r^2 \cdot J$, where the charge density J has components $J_\mu = \Im(\phi \cdot \overline{D_\mu \phi})$. The special algebraic form implies

$$r^2 \cdot J_\mu[\phi] = \Im\big((r\phi) \cdot \overline{D_\mu(r\phi)}\big) = J_\mu[r\phi].$$

   Therefore, we only have to deal with the "better" field $r\phi$ rather than ϕ itself. This special cancellation from the algebraic structure is crucial to obtain the sharp peeling estimates and to close the energy estimates.

6. The conformal compactification.

   Since the trace of the energy momentum tensor for MKG equations are not zero, this field theory is not conformal. However, for special conformal transformations, it can still be conformally invariant, e.g., if $\Box\Lambda = 0$ where Λ is the conformal factor. The inversion map restricted in the forward light cone is such a conformal map in $\mathbb{R}^{3+1}$ (not in other dimensions).

   On the other hand, there is another important observation: although the presence of a nonzero charge does not allow compactification around the spatial infinity, this effect indeed does not appear on the null infinity. This was first pointed out by Shu in [20]. The following computation for $F[q_0]$ justifies this observation: on a outgoing light cone $\mathcal{H}_u$ defined by $r - t = 2u$, the conformal energy flux passing through this light cone (this is the basic energy quantity needed after the conformal transformation) is given by

$$\mathcal{E}\big[F[q_0]\big] \approx \int_{\mathcal{H}_u} |u|^4 |\rho|^2.$$

   Since $|\rho| \approx \frac{q_0}{r^2}$ (as $F[q_0]$ has the leading term $q_0 dt \wedge dr$) and u is a constant on $\mathcal{H}_u$, the above energy flux is finite. On the other hand, if one considers conformal energy on a constant time slice, the factor u^4 would be replace by r^2u^2 (near spatial infinity) so that the contribution of the charge part of the field would be divergent. This is why we choose inversions as the conformal mappings.

7. r^p-weighted energy estimates.

   We use the r^p-weighted energy estimates which was first introduced by Dafermos-Rodnianski in [5] for the study of decay of linear waves on black hole spacetimes. The method has also been used in the first author's works on MKG equations, see [23, 22], where $p < 2$. The new point in the current work is that we have to deal with the end point case $p = 2$ to get the sharp peeling estimates.

2.3 Further discussions

It is instructive to make a comparison to the works [13], [12] of Klainerman-Nicolò to prove higher peeling estimates near Minkowski spacetime in an exterior region and the work [16] of Luk-Oh for proving global nonlinear stability of dispersive solutions to Einstein equations. Indeed, for a given initial datum of the vacuum Einstein field equations, one can work in the region $r \geq R_*$ and can assume that the datum is small provided R_* is sufficiently large. The mass m for the Einstein equations plays a similar role as the charge q_0 for the Maxwell-Klein-Gordon equations: they all represent a slow decay tail representing a static solution at spatial infinity (which is the Schwarzschild solutions in the Einstein equations' case). The proof of Klainerman and Nicolò indeed does not use the smallness of the mass m and this is similar to our case where we do not assume q_0 is small. For vacuum Einstein field equations, the mass m comes in through the ρ-component of the curvature:

$$\rho = \frac{m}{r^3} + \mathring{\rho},$$

where $\mathring{\rho}$ decays as $\frac{1}{r^4}$. However, for MKG equations, the charge q_0 comes in through the ρ-component of the Maxwell field:

$$\rho = \frac{q_0}{r^2} + \mathring{\rho}.$$

The r^{-3} decay is sufficient to apply Gronwall's inequality in the Einstein equations' case while for MKG equations we have to find a new way to compensate the logarithmic loss as we mentioned before.

Alternatively, for this large mass issue for Einstein equations, Luk-Oh in [16] choose a special gauge condition so that such mass problem does not appear. Since our approach in this paper is gauge invariant and the charge is inherited in the connection field A, the charge difficulty is essentially different from the mass problem for Einstein field equations.

For Einstein field equations coupled with other fields, say a scalar field, the coupling field may bring a slower decay tail. We believe that our method in the exterior region can also be applied to these cases to derive sharp peeling estimates.

References

[1] L. Bieri, S. Miao, and S. Shahshahani. Asymptotic properties of solutions of the Maxwell Klein Gordon equation with small data. *Comm. Anal. Geom.*, 25(1):25–96, 2017.

[2] Y. Choquet-Bruhat and D. Christodoulou. Existence of global solutions of the Yang-Mills, Higgs and spinor field equations in 3 + 1 dimensions. *Ann. Sci. École Norm. Sup. (4)*, 14(4):481–506 (1982), 1981.

[3] Y. Choquet-Bruhat and D. Christodoulou. Existence of global solutions of the Yang-Mills, Higgs and spinor field equations in 3 + 1 dimensions. *Ann. Sci. École Norm. Sup. (4)*, 14(4):481–506 (1982), 1981.

[4] D. Christodoulou and S. Klainerman. Asymptotic properties of linear field equations in Minkowski space. *Comm. Pure Appl. Math.*, 43(2):137–199, 1990.

[5] M. Dafermos and I. Rodnianski. A new physical-space approach to decay for the wave equation with applications to black hole spacetimes. In *XVIth International Congress on Mathematical Physics*, pages 421–432. World Sci. Publ., Hackensack, NJ, 2010.

[6] D. Eardley and V. Moncrief. The global existence of Yang-Mills-Higgs fields in 4-dimensional Minkowski space. I. Local existence and smoothness properties. *Comm. Math. Phys.*, 83(2):171–191, 1982.

[7] D. Eardley and V. Moncrief. The global existence of Yang-Mills-Higgs fields in 4-dimensional Minkowski space. II. Completion of proof. *Comm. Math. Phys.*, 83(2):193–212, 1982.

[8] V. Georgiev and P. Schirmer. The asymptotic behavior of Yang-Mills fields in the large. *Comm. Math. Phys.*, 148(3):425–444, 1992.

[9] S. Klainerman. Uniform decay estimates and the lorentz invariance of the classical wave equation. *Comm. Pure Appl. Math.*, 38(3):321–332, 1985.

[10] S. Klainerman and M. Machedon. On the Maxwell-Klein-Gordon equation with finite energy. *Duke Math. J.*, 74(1):19–44, 1994.

[11] S. Klainerman and M. Machedon. Finite energy solutions of the Yang-Mills equations in $\mathbb{R}^{3+1}$. *Ann. of Math. (2)*, 142(1):39–119, 1995.

[12] S. Klainerman and F. Nicolò. *The evolution problem in general relativity*, volume 25 of *Progress in Mathematical Physics*. Birkhäuser Boston, Inc., Boston, MA, 2003.

[13] S. Klainerman and F. Nicolò. Peeling properties of asymptotically flat solutions to the Einstein vacuum equations. *Classical Quantum Gravity*, 20(14):3215–3257, 2003.

[14] S. Klainerman, Q. Wang, and S. Yang. Global solution for massive Maxwell-Klein-Gordon equations. 2018. arXiv:1801.10380.

[15] H. Lindblad and J. Sterbenz. Global stability for charged-scalar fields on Minkowski space. *IMRP Int. Math. Res. Pap.*, pages Art. ID 52976, 109, 2006.

[16] J. Luk and S. Oh. Global nonlinear stability of large dispersive solutions to the Einstein equations. preprint.

[17] S. Oh and D. Tataru. Global well-posedness and scattering of the $(4+1)$-dimensional Maxwell-Klein-Gordon equation. *Invent. Math.*, 205(3):781–877, 2016.

[18] D. Petrescu. Time decay of solutions of coupled Maxwell-Klein-Gordon equations. *Comm. Math. Phys.*, 179(1):11–23, 1996.

[19] M. Psarelli. Time decay of Maxwell-Klein-Gordon equations in 4-dimensional Minkowski space. *Comm. Partial Differential Equations*, 24(1-2):273–282, 1999.

[20] W. Shu. Asymptotic properties of the solutions of linear and nonlinear spin field equations in Minkowski space. *Comm. Math. Phys.*, 140(3):449–480, 1991.

[21] W. Shu. Global existence of Maxwell-Higgs fields. In *Nonlinear hyperbolic equations and field theory (Lake Como, 1990)*, volume 253 of *Pitman Res. Notes Math. Ser.*, pages 214–227. Longman Sci. Tech., Harlow, 1992.

[22] S. Yang. Decay of solutions of Maxwell-Klein-Gordon equations with arbitrary

Maxwell field. *Anal. PDE*, 9(8):1829–1902, 2016.
[23] S. Yang. On the global behavior of solutions of the Maxwell–Klein–Gordon equations. *Adv. Math.*, 326:490–520, 2018.

Beijing International Center for Mathematical Research
Peking University, Beijing, China
E-mail address: shiwuyang@math.pku.edu.cn

Department of Mathematics and Yau Mathematical Sciences Center
Tsinghua University, Beijing, China
E-mail address: yupin@mail.tsinghua.edu.cn

Proceedings of the 7th International
Congress of Chinese Mathematicians
ALM 43, pp. 489–505

Some Recent Topics on Random Matrix Theory

Jun Yin

Abstract

In recent years, random matrix theory (RMT) has become a very active field of research. Here we introduce some recent topics in this field. We will focus on the universality phenomenon on the eigenvalues and eigenvectors of Wigner type random matrices. Some other topics on random band matrices and random non-Hermitian matrices will also be included.

Keywords and Phrases: Random matrix, universality.

1 Introduction

Random matrix theory (RMT) is a very active field of research. One of the main reasons is that interesting properties of physical systems or estimation problems encountered in real data can be represented mathematically as matrix problems: for example, the spectra of heavy atoms in physics [105], statistical estimation of covariance matrices [106], Erdős-Rényi graphs [39, 38], capacity of quantum channels [59], speech recognition [30, 63], mathematical finance [84, 71, 77], wireless communication [102], physics of mixture [90], and data analysis and statistical learning [61], etc. But there is a more fundamental reason that originates with Wigner and is widely believed now that the random matrix statistics and Poisson statistics represent two paradigms of many-body systems: Poisson for independent systems and random matrix for highly correlated systems. (It means systems are universal in the sense that they depend only on the symmetry class of the physical system but not on other detailed structures). This belief is strongly supported by many important results and conjectures in a variety of diverse fields. (1): The length of the longest increasing subsequence of random permutations [19] (2): The fluctuations of the solutions of KPZ equations [23, 85, 33]. (3): The Bohiga-Giannoni-Schmit conjecture [21] asserts that spectral statistics of quantum systems whose classical counterparts exhibit chaotic behavior are described by random matrix theory. (4): The zeros of the Riemann zeta function (and other L-functions) is modeled by the distribution of eigenvalues of certain random matrices [66] Montgomery and Dyson.

Partially supported by NSF Career Grant DMS-1552192 and Sloan fellowship.

As the first result of this type, Wigner proved a type of law of large numbers for the density of eigenvalues for Wigner matrices, which we will explain in the next section. On the scale of individual eigenvalues, Wigner predicted that the fluctuations of the gaps are universal and their distribution is given by a new law. This might be viewed as the random matrix analogue of the central limit theorem. After Wigner's discovery, Dyson, Gaudin and Mehta achieved several fundamental mathematical results, in particular they were able to compute the gap distribution and the local correlation functions of the eigenvalues for Gaussian ensembles. They are called Gaussian Orthogonal Ensemble (GOE) and Gaussian Unitary Ensemble (GUE) corresponding to the two most important symmetry classes, the real symmetric and complex Hermitian matrices. Universality problem for the limiting distribution of individual eigenvalues is nowadays known as the Wigner-Dyson-Mehta (WDM) universality conjecture. As presented in Mehta [79], this conjecture asserts that the local eigenvalue statistics for large random matrices with independent entries are universal, i.e., they do not depend on the particular distribution of the matrix elements. In particular, they coincide with those in the Gaussian case that were computed explicitly. This universality phenomenon also extended to the limiting distributions of eigenvectors and many more general random matrices, as Wigner predicted that for large complicated quantum systems are universal in the sense that they depend only on the symmetry class of the physical system but not on other detailed structures. In the next two sections, we will focus on the recent process on the study of universality phenomenon of Wigner type random matrices.

For most Wigner type $N \times N$ random matrices H, the main properties are

$$H = H^{\dagger}, \quad Var(H_{ij}) = O(N^{-1})$$

and H_{ij} $(i \leq j)$ are independent to each other. In this work, besides Wigner type random matrices, we will also introduce some recent works on random band matrix (in section 4) and non-Hermitian matrix (in section 5). More precisely, for random band matrix H, with some $a < 1$, we have

$$Var(H_{ij}) \sim N^{-a}\mathbf{1}(|i-j| \leq N^{a}).$$

The delocalization - localization problem of random band matrix is related to the metal-insulator transition is the Anderson model [7]. For non-Hermitian matrix case, it is well known that the eigenvalue density of *i.i.d.* random matrices satisfies the circular law [16, 98, 57, 99, 28, 29, 108]. In section 5, we will introduce some new results on the eigenvalue density of more general non-Hermitian random matrices, including some cases in which the matrix entries are not independent to each other.

2 Universality of Wigner matrices

2.1 Bulk universality

We define a *Wigner ensemble* [42, 95] $H_N = \{(H_N)_{ij}\}_{i,j=1}^{N}$ as Hermitian or symmetric random matrices such that

$$\mathbb{E}(H_N)_{ij} = 0, \quad \mathbb{E}|(H_N)_{ij}|^2 = N^{-1} \tag{2.1}$$

and H_{ij}'s are independent entries up to symmetry. If $\sqrt{N}(H_N)_{ij}$ are complex (real) Gaussians, this is called the Gaussian unitary (orthogonal) ensemble: GUE (GOE). Usually one also needs some decay properties: for example for large enough $s > 0$, we have

$$\sup_{i,j,N} \mathbb{E}(\sqrt{N}|(H_N)_{ij}|)^s < \infty \tag{2.2}$$

or for some $s > 0$,

$$\sup_{i,j,N} \mathbb{E}\left(s^{\sqrt{N}|(H_N)_{ij}|}\right) < \infty \tag{2.3}$$

For any $k = 1, 2, \ldots, N$, let $p^k(\lambda_1, \ldots, \lambda_k)$ be the k-point correlation function of the eigenvalues of Wigner matrix H. The case $k = 1$ point function $p^1(\lambda_1)$ gives the density $\varrho(\lambda)$ of the eigenvalues. In a macroscopic energy window the celebrated Wigner's semicircle law shows:

$$\varrho(\lambda) \overset{weak}{\longrightarrow} \varrho_{sc}(\lambda) = (2\pi)^{-1}\sqrt{4-\lambda^2}\,\mathbf{1}(|\lambda| \le 2)\,. \tag{2.4}$$

In a microscopic energy window (window scale of $1/N$), the asymptotics of local statistics in the bulk of the spectrum was first proved in the GUE case: Fix $\varepsilon > 0$, for any $|E| \le 2 - \varepsilon$ (bulk condition), the rescaled k-point correlation function around E, $p_E^k(a_1, a_2, \cdots, a_k)$

$$:= \frac{1}{[\varrho_{sc}(E)]^k} p^k\left(E + \frac{a_1}{\varrho_{sc}(E)N}, E + \frac{a_2}{\varrho_{sc}(E)N}, \cdots, E + \frac{a_k}{\varrho_{sc}(E)N}\right) \tag{2.5}$$

converges in a weak sense to

$$\det|f(a_i - a_j)|_{i,j=1}^k, \quad f(u) = \frac{\sin \pi u}{\pi u}.$$

i.e.

$$p_E^k(a_1, a_2, \cdots, a_k) \quad \to \quad \det|f(a_i - a_j)|_{i,j=1}^k, \quad |E| < 2 - \epsilon \tag{2.6}$$

It is widely believed that Wigner matrices have the same asymptotic correlation function as GUE (or GOE). This phenomenon is called universality. Finding and proving the conditions for the universality phenomenon has been being a main topic in this field for decades.

For Hermitian Wigner matrix, the basic idea is first to establish a local version of the semicircle law and use it as an input to control the correlation function asymptotics in the Brezin-Hikami formula (which is related to Harish - Chandra/Itzykson - Zuber formula). This provides universality for the so-called Gaussian divisible models with a very small Gaussian component, or "noise" (previously, this universality was established by Johansson [62] when the noise is of order 1). The last step is an approximation of a general Wigner ensemble by Gaussian divisible ones. By using the local semicircle law [41] as an input, Tao-Vu [95] proved a comparison theorem that provides an approximation result for Hermitian Wigner matrices. Finally, the conditions for tail distributions of the matrix elements were greatly relaxed in Erdős, et al, and Tao, et al.[38, 46, 101].

For real symmetric matrices, no algebraic formula in the spirit of Brezin-Hikami is known. A completely new method based on relaxation of the Dyson Brownian motion (DBM) to local equilibrium was developed in a series of papers of Erdős, Schlein, Yau and Yin [43, 45, 49]. This approach is very robust and applies to all symmetry classes of random matrices, including also sample covariance matrices and sparse matrices [38, 49], but it yields only the average energy universality. Although the energy averaging is on a very small scale, i.e.

$$\int_{|E-E_0|\le N^{-1}} \frac{1}{2N^{-1}} p_E^k(a_1, a_2, \cdots, a_k) \quad \to \quad \det |f(a_i - a_j)|_{i,j=1}^k \tag{2.7}$$

It so far cannot be completely removed with this relaxation method. The fixed energy universality of symmetric Wigner matrices was finally proved by Bourgade, Erdős, Yau and Yin recently in [25] with a coupling method on Dyson Brownian motion. (This method also works for Hermitian case). Furthermore the result in [25] also holds for *generalized Wigner matrix*, which is defined as Wigner matrix except that (2.1) is replaced with

$$\mathbb{E}(H_N)_{ij} = 0, \quad \mathbb{E}|(H_N)_{ij}|^2 = O(N^{-1}), \quad \sum_j \mathbb{E}|(H_N)_{ij}|^2 = 1 \tag{2.8}$$

2.2 Edge universality

Similarly, for the eigenvalues at the edge of the spectrum, we define the rescaled k-point correlation function near the edge of spectrum as

$$p_{edge}^k(a_1, \cdots, a_k) := p^k\left(2 + a_1 N^{-2/3}, \cdots, 2 + a_k N^{-2/3}\right) \tag{2.9}$$

It was first proved by Tracy Widom [103, 104] and Fokas [50] that the limiting distribution of p_{edge}^k is Tracy-Widom distribution for GOE and GUE (i.e., Wigner matrix with Gaussian entries). The edge universality for Wigner matrices was first proved by Soshinikov [93] under the assumption that the distribution of the matrix elements is symmetric and has finite moments for all orders. Soshnikov used an elaborated moment matching method and the conditions on the moments are not easy to improve. Later, a completely different method based on the Green function comparison theorem was initiated in Erdős, Yau and Yin's work [47]. (A different comparison method was used in Tao and Vu [96].) This method is more flexible in many ways and it removes essentially all restrictions in previous works. In particular, in [76] with some improvement, Lee and Yin proved the necessary and sufficient condition on the distribution of the matrix elements for edge universality of Wigner matrices as follows:

$$p_{edge}^k \to TW \quad \Longleftrightarrow \quad \lim_{s\to+\infty} \mathbb{P}(N^{1/2}|(H_N)_{ij}| > s)s^4 = 0$$

For generalized Wigner matrices, as defined in (2.8), in [47] with resolvent comparison method it was proved that if two generalized Wigner matrices share the same matrix entry variance, then they have the same p_{edge}^k, i.e.

$$\mathbb{E}|(H_N)_{ij}|^2 = \mathbb{E}|(\widetilde{H}_N)_{ij}|^2 \quad \Longrightarrow p_{edge}^k - \widetilde{p}_{edge}^k \to 0$$

But it was not shown that the statistics are independent of the variances themselves. In [24] Bourgade, Erdős and Yau provided this missing step, by studying the local equilibrium of beta random matrix ensembles. Thus it proves the edge universality for the generalized Wigner matrices.

2.3 Universality of eigenvectors

Universality questions have been traditionally formulated for eigenvalues, but they naturally extend to eigenvectors as well. Note that by a complete $U(N)$ or $O(N)$ symmetry, the eigenfunctions of a GUE or GOE matrix are distributed according to the Haar measure on the N-dimensional unit sphere, moreover the distribution of a particular coordinate $\mathbf{u}_\alpha(i)$ of an fixed eigenvector $\mathbf{u}_\alpha$ is asymptotically Gaussian. The same result holds for generalized Wigner matrices by Theorem 1.2 of Bourgade and Yau's [27] (a similar result was also obtained in Knowles and Yin's [67] for eigenvectors whose eigenvalues are near the edge of spectrum, and in [97] under the condition that the first four moments match with those of the Gaussian). In [27] the main method is moment flow. More precisely, they showed that under the Dyson Brownian motion, the moments of eigenvector components satisfy a flow equation, which only depends on the eigenvalues and these moments themselves. On the other hand, in [67, 97] the main method is Lindebuger swapping method. In these work, a key input was the entry wise local semi law, which was proved in Erdő, Yau and Yin's [47]. Furthermore, with the isotropic local semicircle law proved in [68, 13] (Erdős et al.), which will be introduced in the next subsection, these results can be extended to limiting distribution of the general component of the eigenvectors, i.e., the inner product of $\mathbf{u}_\alpha$ and any deterministic unit vector $\mathbf{w} \in \mathbb{C}^N$:

$$N^{1/2}\,(\mathbf{u}_\alpha,\ \mathbf{w}) \to \mathcal{N}(0,1).$$

2.4 (Isotropic) local semicircle law

In these recent work on universality phenomenon, the (isotropic) local semicircle law plays a key role for understanding the local behaviors of eigenvalues and eigenvectors of Wigner - type random matrices. The (isotropic) local semicircle laws are estimates of the Green's function of the random matrices. In the last decade, the development of estimates is always accompanied with the proof of important results on Wigner - type random matrices.

For a fixed spectral parameter, $z = E + i\eta$ with $E \in \mathbb{R}$, $\eta > 0$, the Green's function of Wigner matrix H is defined as

$$G(z) := (H - z)^{-1}$$

Let $m_{sc}(z)$ be the Stieltjes transform of the semicircle law (see ϱ_{sc} in (2.4)) $m_{sc}(z) = \int_{\mathbb{R}} \frac{\varrho_{sc}(x)}{x-z}dx$. In early works (see book [17] for details) , it was proved that for $\eta = \operatorname{Im} z \sim 1$, i.e., $C^{-1} \le \eta \le C$ for some positive constant C,

$$N^{-1}\mathrm{Tr}G(z) - m_{sc}(z) = o(1).$$

This estimate immediately implies the Wigner's semicircle law, i.e., (2.4). Later it was extended by Erdős, Schlein and Yau, to the case $|E| < 2$ and $\eta \gg 1/N$ [44]. This result provides a key step towards establishing universality for Wigner matrices in two famous breakthrough papers [42, 95](Erdős et al, Tao et al). Later, in Erdős, Yau and Yin's work [47], the error estimate was improved to (almost) optimal bound in the optimal scale: let H be a generalized Wigner matrix, for any $\delta > 0$, with high probability,

$$N^{-1}\mathrm{Tr}G(z) - m_{sc}(z) \leq N^{\delta}(N\eta)^{-1}, \quad \mathrm{Im}\, z \geq N^{-1+\eta}.$$

Based on this result and the Helffer-Sjőstrand functional calculus, the following rigidity properties of the eigenvalues was proved [47]: for generalized Wigner matrix, for any $1 \leq k \leq N$, define γ_k as the classical location of the kth smallest eigenvalue:

$$\gamma_k : N \int_{-\infty}^{\gamma_k} \varrho_{sc}(x) = k$$

then let λ_k be the k-th eigenvalue of H, for any $\delta > 0$, with high probability,

$$|\gamma_k - \lambda_k| \leq N^{\delta}|\gamma_k - \gamma_{k\pm 1}|.$$

Furthermore, instead of studying its trace, Green's function itself was also well studied. In [48, 49, 47], with self consistent equation method, Erdős, Yau and Yin showed that for generalized Wigner matrix and any $\delta > 0$, with high probability,

$$G_{ij}(z) - \delta_{ij}m_{sc}(z) \leq N^{\delta}\Psi(z), \quad \Psi(z) := \sqrt{\frac{\mathrm{Im}\, m_{sc}}{N\eta}} + \frac{1}{N\eta}$$

Furthermore, with Green's function comparison method and/or polynomialization method in [68, 13], the above entry wise bounds were improved to the isotropic law: for any deterministic unit vectors $\mathbf{v}$ and $\mathbf{w}$, for any $\delta > 0$, with high probability,

$$\Big(\mathbf{w}, [G(z) - m_{sc}(z)I]\mathbf{v}\Big) \leq N^{\delta}\Psi(z), \quad \Psi(z) := \sqrt{\frac{\mathrm{Im}\, m_{sc}}{N\eta}} + \frac{1}{N\eta} \tag{2.10}$$

Up to N^{δ} factor, this bound is almost optimal. It worths to note that these bounds played a key role in proofs of bulk universality, edge universality and eigenvector universality.

3 Deformed Wigner matrices

In this section, we review some recent work on the local statistics of more general Wigner type matrices.

3.1 Non-zero-mean Wigner matrices

First case is none-zero mean Wigner matrix, here deformed Wigner matrix $\widetilde{H}$ can be written as

$$\widetilde{H} := H + A, \quad A = \mathbb{E}\widetilde{H}$$

where H is regular Wigner matrix and A is a deterministic $N \times N$ Hermitian (symmetric) matrix. Usually we assume that $\|A\| = O(1)$, which is comparable with the spectrum of H.

The early research was focused on the case that A is a rank $O(1)$ deformation. For example, Capitaine, Donati-Martin, Féral Füredi, Komlós. Benaych-Georges, Guionnet, Maïda, Nadakuditi 's work [52, 10, 11, 12, 31, 32]. The spectrum of the deformed matrix undergoes a transition, associated with the creation or annihilation of an outlier, when an eigenvalue d_i of the deformation crosses a critical value ± 1. This transition happens on the scale $|d_i| - 1 \sim N^{-1/3}$. In a seminal work [9, 81], Baik, Ben Arous, and Péché investigated the spectrum of deformed (spiked) complex GUE (the Gaussian Unitary Ensemble). They established a phase transition, sometimes referred to as the BBP transition. Subsequently, the results of [9, 81] were extended to the other Gaussian ensembles, such as GOE (the Gaussian Orthogonal Ensemble), by Bloemendal and Virág [15]. In [68, 69], Knowles and Yin studied the deformed general Wigner matrices. With the isotropic law in (2.10), they provided an exact formula for the distribution of the outliers. Furthermore, they showed that if $||d_i| - 1| \gg N^{-1/3}$, then the non-outliers close to the spectral edge have Tracy Widom distribution.

For general full rank A, recently in [73], with Dyson Brownian motion flow method, Lee and Schnelli proved that for a large class of diagonal matrices A, the rescaled distribution of the extremal eigenvalues is again given by the Tracy-Widom distribution. (Note: it is a very large class of diagonal matrices, actually most A are in this class). Later, Knowles and Yin established anisotropic law of deformed Wigner matrix in [70], and use this law to extend the above result to non-diagonal (general) A case. More precisely they proved taht the rescaled distribution of the extremal eigenvalues is invariant with respect to

$$A \to UAU^{\dagger}.$$

Similarly for bulk universality, the universality of averaged correlation function i.e., (2.7) was first proved in Lee, Schnelli, Stetler and Yau's work [75] for diagonal case A. In this proof, the key step is analyzing the convergence of Dyson Brownian motion to local equilibrium. More precisely, it was proved that the convergence in (2.7) holds if H contains a small Gaussian component. Later Landon and Yau improved this result by removing the average in [72], i.e. the convergence in (2.6) holds if H contains a small Gaussian component. Another key input of the proof of bulk universality is establish a local law for the Green's function of $\widetilde{H} := H + A$. It was first proved in [70] with comparison method under the condition that $\mathbb{E}H_{ij}^3 = 0$. Recently this condition was removed in [60](by He, Knowles and Rosenthal) with cumulant expansion method.

3.2 Wigner matrix with various variations

The second type of variation of Wigner matrix can be called various variance Wigner type random matrices. More precisely in this case, we only assume that

$$\mathbb{E}H_{ij}^2 = O(N^{-1})$$

instead of $\mathbb{E}H_{ij}^2 = O(1/N)$. Note: usually it is called generalized Wigner matrix (in most papers) if (2.8) holds. In above discussion, we have seen the universality phenomenon for the generalized Wigner matrix. In a recent work, Ajanki, Erdős, Torben Krüger [1] proved that some universality results also hold for various variance Wigner matrices. It also relies on an estimate on the local law of the Green's functions of these matrices. To see that, in their proof they provided a detailed analysis on quadratic vector equations on complex upper half-plane, which was in their early work [2]. In their other work [3] and [4], they also studied some random matrices in which there exist some weak correlations between the matrix entries.

3.3 Random covariance matrices

The last type of variation of Wigner matrix is random covariance matrix. The covariance matrix is defined as

$$XX^\dagger, \quad X_{ij}\text{'s are i.i.d. random variables},$$

where X is a $N \times M$ matrix and

$$\mathbb{E}X_{ij} = 0, \quad \mathbb{E}X_{ij}^2 = 1/N \tag{3.1}$$

It can be considered as a variation of Wigner matrix, since the eigenvalues and eigenvectors of covariance matrix can be studied via the following Winger type random matrix:

$$\begin{pmatrix} 0 & X \\ X^\dagger & 0 \end{pmatrix}$$

Covariance matrices, and especially their deformations with finite rank deterministic matrices are fundamental objects in modern multivariate statis- tics where the advance of technology has led to high-dimensional data. They have manifold applications in various applied fields; see [35, 63, 64, 65] (by Dieng, Tracy and Johnstone) for an extensive account on statistical applications, [58] (by Harding) for applications in economics and [83] (by Patterson, etc) in population genetics, to name a few.

The empirical density of eigenvalues λ_i of $XX^\dagger$ converges to the Marchenko-Pastur law [78] in the limit when $M,$, $N \to +\infty$ such that $d = N/M$ is fixed. For local statistics, bulk universality were proved in Erdős etc,[45] and improved in [82] by Pillai and Yin. The edge universality is also well studied in Péché [81], Feldheim and Sodin [51], Tao and Vu [100], Pillai and Yin [82].

For applications in principal component analysis, via studying the $O(1)$ rank perturbed covariance matrices, the main focus is on the extreme eigenvalues and the outliers whose detailed analysis was given in [14] by Bloemendal, Knowles, Yau and Yin. Furthermore, for general case i.e., full rank perturbation, in [70] Knowles and Yin studied the following random matrices,

$$TXX^\dagger T^\dagger, \quad \|T\| = O(1)$$

where T is $N' \times N$ deterministic matrix and X is *i.i.d.* random matrices. They showed that edge universality phenomenon also holds for this type of random matrices. Similar result was first proved for diagonal T case in [74] by Lee and Schnelli.

In a different general way, in [5], Alt, Erdős and Krüger showed the local statistics of the random Gram matrices. In Gram matrices, the assumption (3.1) can be reduced to

$$\mathbb{E}X_{ij} = 0, \quad \mathbb{E}X_{ij}^2 = O(1/N) \tag{3.2}$$

As above various variance Wigner matrices, the proof in [5] also relies on a detailed analysis on quadratic vector equations on complex upper half-plane.

4 Random Band matrices

First we define random band matrices. Let $\Lambda = [[1, L]]^d$ have periodic boundary conditions, and $H = (H_{ij})_{ij\in\Lambda}$ have independent entries (up to the symmetry $(ij) \leftrightarrow (ji)$). We consider a *band matrix* with band width $W < L/2$, i.e.

$$\mathbb{E}|H_{ij}|^2 \neq 0 \text{ only if } |i-j| < W, \tag{4.1}$$

where $|\cdot|$ is the periodic L^1 distance on $\mathbb{Z}^d$ and all non-trivial entries H_{ij} have a variance σ_{ij}^2 with the same order of magnitude, normalized so that $\sum_j \sigma_{ij}^2 = 1$ for any $i \in \Lambda$. General Wigner matrices correspond to $W = L/2$

As a quantum mechanical Hamiltonian, a Wigner random matrix H is a model of *mean-fild type*: the entries H_{ij}, which represent the quantum transition rates between two quantum states with labels i and j, are all of comparable size. More realistic models involve quantum states with more geometric structure, as typical quantum transitions only occur between nearby states. As one can see from the above definition, the random band matrices were introduced for approaching some spectral problems on (more realistic) disordered quantum system.

For disordered quantum systems represented by a random Hamiltonian H, the fundamental *metal-insulator* phase transition holds. More precisely, eigenstates are delocalized in the metallic phase, and localized in the insulator one. Together with delocalization, eigenvalues statistics supposedly coincide with GOE/GUE statistics, while localization is thought to occur together with Poisson spectral statistics. For the corresponding infinite volume operators, delocalization corresponds to an absolutely continuous spectrum while localization happens together with pure-point.

One main model for the metal-insulator transition is the Anderson model [7], a Hamiltonian of type $-\Delta + V$ on $\mathbb{Z}^d$ where Δ is the discrete Laplacian and V a random potential such that the real random variables $V(x)$, $x \in \mathbb{Z}^d$, are i.i.d and centered with variance σ^2. Amongst the many mathematical contributions to the analysis of the Anderson model, localization was proved by Fröhlich and Spencer [51] in the high disorder regime. An alternative proof was given by Aizenman and Molchanov [8], and Minami [80] proved the Poisson spectral statistics.

As we can see that the random band matrices [94] defined in (4.1) is a prominent model interpolating between the Wigner matrices and the random Schrödinger operator: $-\Delta + V$. A fundamental conjecture [53] states that the local spectral statistics of a band matrix H are governed by random matrix statistics for large W and by Poisson statistics for small W. The transition is conjectured to be sharp in [53] (Fyodorov and Mirlin) and [94] (Spencer) for the band matrices in one spatial dimension around the critical value $W = N^{1/2}$. In other word, for $W \gg N^{1/2}$, we expect the universality results of Wigner matrices will also hold. Furthermore, the eigenvectors of H are expected to be completely delocalized in this range. (For the case $W \sim N$, the delocalization was proved by Erdős, Yau and Yin in [49]). On the other hand, for $W \ll N^{1/2}$, one expects that the eigenvectors are exponentially localized in this range. The only rigorous work indicating the $\sqrt{N}$ threshold concerns the second mixed moments of the characteristic polynomial for a special class of Gaussian band matrices in Shcherbina's work [88, 89].

The localization length for band matrices in one spatial dimension was recently investigated in numerous works. For general distribution of the matrix entries, eigenstates were proved to be localized by Schenker in [87] for $W \ll N^{1/8}$, and weakly delocalization of most eigenvectors in a certain averaged sense holds for $W \gg N^{6/7}$ [37] by Erdős and Knowles, improved to $W \gg N^{4/5}$ [40] by Erdős, Knowles, Yau and Yin. (Here *weakly* delocalization means most e.vectors are not strongly exponentially localized.) The GreenÕs function $(H - z)^{-1}$ was controlled down to the scale $Imz \gg W^{-1}$ in [49], implying a lower bound of order W for the localization length of all eigenvectors. When the entries are Gaussian with some specific covariance profiles, supersymmetry techniques are applicable to obtain stronger results. This approach has first been developed by physicists (see Efetov's [36] for an overview); the rigorous analysis was initiated by Spencer (see [94] for an overview), with an accurate estimate on the expected density of states on arbitrarily short scales for a three-dimensional band matrix ensemble in [34] (Disertori, Pinson and Spencer). More recent works include universality for $W = \Omega(N)$ in Shcherbina's [89], and the control of the GreenÕs function down to the optimal scale $Imz \gg N^{-1}$, hence delocalization in a strong sense for all eigenvectors, when $W \gg N^{6/7}$ (see [18] of Bao and Erdős) with first four moments matching the Gaussian ones (both results require a block structure and hold in part of the bulk spectrum). These rigorous results based on the supersymmetric method so far assumed the complex hermitian condition.

For universality at the spectral edge, Sodin proved that [91]: extreme eigenvalues follow the Tracy-Widom law for $W \gg N^{5/6}$, an essentially optimal condition. On the other hand, the local statistics in the bulk of the spectrum was much less known. Even for the case that $W \sim N$, the universality was not proved until very recently [26](Bourgade, Erdős, Yau and Yin). More precisely, in [49] bulk universality has been proved for generalized Wigner matrix, but at there one needs

$$\max_{ij} \mathbb{E}|H_{ij}|^2 \geq cN^{-1}$$

for some constant $c > 0$. In [26], the main idea is to use the mean-field reduction to convert the problem of universality of the band matrices H to a matrix ensemble of the form $H + A$ with H a Wigner ensemble of the size of the band, independent of A.

5 General Circular law

In this section we discuss some recent results on non-Hermitian matrix. For an $N \times N$ matrix X with complex entries and eigenvalues $\lambda_1, \lambda_2, \cdots, \lambda_N$ (counted with multiplicity and labeled in some arbitrary fashion), the empirical spectral distribution (ESD) is given by

$$\mu_X := \frac{1}{N}\sum_{i=1}^{N}\delta_{\lambda_i}$$

A seminal result in non-Hermitian random matrix theory is the circular law, which describes the asymptotic global distribution of the spectrum for matrices with *i.i.d.* entries of finite variance. More precisely, if X_{ij}'s $(1 \le i, j \le N)$ are i.i.d. random variables with zero mean and unit variance, then the ESDs of $\mu_{\frac{1}{\sqrt{N}}X}$ converge weakly to the circular measure, i.e.,

$$\mu_{\frac{1}{\sqrt{N}}X} \Longrightarrow \frac{1}{\pi}\mathbf{1}_{|z|\le 1}dzd\bar{z}$$

It was established by Tao and Vu [99], and is the culmination of the work of many authors: Ginibre, Girko, Bai, Silverstein, Gőtze, Tikhomirov , Pan, Zhou, Tao and Vu, see the survey [22] for a more detailed account of these works. Clearly, it is another instance of the universality phenomenon in random matrix theory. The universal circular law has found a wide range of applications. In neuroscience, it was used to derive a so-called "transition to chaos" for a mean field model of dynamical neural networks [92]. Here a matrix with i.i.d. entries was used to model the synaptic matrix governing the interactions between neurons. The key idea of proving the circular law is so called Hermitization technique first used by Girko [56]. More precisely for any smooth function $F : \mathbb{C} \to \mathbb{C}$,

$$\mu_X(F) = \frac{1}{4\pi}\int_{\mathbb{C}} \Delta F(z)\sum_j \log\lambda_j(z)dzd\bar{z}$$

where $\lambda_j(z)$ is the j-th eigenvalue of the *Hermitian* matrix $(X - z)(X - z)^*$. Therefore, one can study the non-Hermitian matrix via the *Hermitian* matrix $(X - z)(X - z)^*$. This $(X - z)(X - z)^*$ matrix is similar to the covariance matrix, which we have discussed above.

In [28, 29, 108], Bourgade, Yau and Yin showed that the circular law holds on the optimal scale of $N^{-1/2+\epsilon}$, which is slightly above the typical eigenvalue spacing. These results were extended in [107] by Xi, Yang and Yin to the more general non-Hermitian matrices with the form TX, where T is a deterministic matrix such that all of the non-zero singular values of T is bounded from below and above by some constants, and X is an *i.i.d.* random matrix. It worth to note that in this case, the matrix entries of TX are highly related to the others. In [107], they established the local law for TX (up to the almost optimal scale) and showed that the limiting measure is axial symmetric, and only depends on the spectrum of $TT^\dagger$.

For another type of general circular law, in [20] (Cook, Hachem, Najim, Renfrewand) and [6] (Alt, Erdő s and Krüger), the authors studied the non-Hermitian matrices X with a general matrix of variances $S = (s_{ij})$, i.e.,

$$\mathbb{E}X_{ij} = 0, \quad \mathbb{E}|X_{ij}|^2 = s_{ij} = O(1)$$

with some different technique assumptions in both papers. In these work, they established a local inhomogeneous circular law in the bulk spectrum on the optimal scale for X.

Compared with the eigenvalues, the local statistics of the eigenvectors of the $i.i.d$ random matrix is much less known. In [86] Rudelson and Vershynin proved the delocalization of eigenvectors of random matrices with independent entries. It seems to be the strongest result we know about the eigenvectors of non-Hermitian matrices. The limiting distribution of the eigenvector is still a open question, though it is widely believed that the limiting distributionshould be $U(N)$ invariant.

References

[1] O. Ajanki, L. Erdő s, and T. Krüger, *Quadratic vector equations on complex upper half-plane* Preprint arXiv:1506.05095. ,

[2] O. Ajanki, L. Erdő s, and T. Krüger, *Universality for general Wigner-type matrices*, Preprint arXiv:1506.05098.

[3] O. Ajanki, L. Erdő s, and T. Krüger, *Local eigenvalue statistics for random matrices with general short range correlations*, Preprint arXiv:1604.08188.

[4] O. Ajanki, L. Erdő s, and T. Krüger, *Local spectral statistics of Gaussian matrices with correlated entries*, J. Stat. Phys. 163 (2016), 280–302.

[5] J. Alt, L. Erdő s, and T. Krüger, *Local law for random Gram matrices*

[6] J. Alt, L. Erdő s, and T. Krüger, *Local inhomogeneous circular law*

[7] P.W. Anderson, *Absence of diffusion in certain random lattices*, Phys. Rev. 109 (1958), 1492

[8] M. Aizenman and S. Molchanov, *Localization at large disorder and at extreme energies: An elementary derivation*, Comm. Math. Phys. 157 (1993), 245-278.

[9] J. Baik, G. Ben Arous, S. Peche, *Phase transition of the largest eigenvalue for nonnull complex sample covariance matrices*, Ann. Prob. 33 (2005), 1643–1697.

[10] F. Benaych-Georges, A. Guionnet, and M. Maïda, *Fluctuations of the extreme eigenvalues of finite rank deformations of random matrices*, Electron. J. Probab. Volume 16 (2011), paper no. 60, 1621-1662.

[11] F. Benaych-Georges, A. Guionnet, and M. Maïda, *Large deviations of the extreme eigenvalues of random deformations of matrices*, Prob. Theor. Rel. Fields (2010), 1–49.

[12] F. Benaych-Georges and R.R. Nadakuditi, *The eigenvalues and eigenvectors of finite, low rank perturbations of large random matrices,* Adv. Math. 227 (2011), 494-521.

[13] A. Blomendal, L. Erdoes, A. Knowles, H.T. Yau, J. Yin. *Isotropic local laws for sample covariance and generalized Wigner matrices.* Elect. J. Prob. Vol 19 (2014), no. 33, 1-53.

[14] A. Bloemendal, A. Knowles, H.T. Yau, J. Yin. *On the principal components of sample covariance matrices* To appear in Prob. Theor. Rel. Fields.
[15] A. Bloemendal and B. Virág, *Limits of spiked random matrices I*, Prob. Theor. Rel. Fields, Vol 156, Issue 3, pp 795-825 (2013).
[16] Z.D. Bai, *Circular law.* Ann. Probab. 25 (1): 494–529. (1997).
[17] Z. D. Bai, J. Silverstein, *Spectral Analysis Of Large Dimensional Random Matrices* published by Science Press, 2006 2nd ed. 2010, XVI, 552p.
[18] Z. G. Bao and L. Erdos *Delocalization for a class of random block band matrices* arXiv: 1503.07510.
[19] Baik, J.; Deift, P.; Johansson, K. *On the distribution of the length of the longest increasing subsequence of random permutations*, Journal of the American Mathematical Society 12 (4): 1119–1178
[20] Nicholas Cook, Walid Hachem, Jamal Najim, David Renfrew *Limiting spectral distribution for non-Hermitian random matrices with a variance profile*
[21] O. Bohigas, M. J. Giannoni, and C. Schmidt, *Characterization of Chaotic Quantum Spectra and Universality of Level Fluctuation Laws* Phys. Rev. Lett. 52, **1** (1984)
[22] C. Bordenave and D. Chafaï. *Around the circular law.* Probab. Surv., 9:1-89, 2012.
[23] A. Borodin, I. Corwin, P. Ferrari *Free energy fluctuations for directed polymers in random media in 1+1 dimension* Comm. Pure Appl. Math. Vol 67, Issue 7, pp 1129-1214, (2014).
[24] P. Bourgade, L. Erdős, H.T. Yau, *Universality of General β-Ensembles* Volume 163, Number 6 (2014), 1127-1190.
[25] P. Bourgade, L. Erdoes, H.T. Yau, J. Yin. *Fixed energy universality for generalized Wigner matrices* To appear in Comm. Pure Appl. Math.
[26] P. Bourgade, L. Erdoes, H.T. Yau, J. Yin. *Universality for a class of random band matrices*
[27] P. Bourgade, H.T. Yau *The Eigenvector Moment Flow and local Quantum Unique Ergodicty* Comm. Math. Phys. Vol 350, Issue 1, pp 231-278 (2017).
[28] P. Bourgade, H.T. Yau, J. Yin, *Local Circular Law for random matrices* Prob. Theor. Rel. Fields. Vol 159, no. 3-4 (2014) 545-595.
[29] P. Bourgade, H.T. Yau, J. Yin, *The local circular law II: the edge case* Prob. Theor. Rel. Fields. Vol 159, Issue 3-4, (2014) 619-660.
[30] A. Buja, T. Hastie, and R. Tibshirani. *Penalized discriminant analysis.* Ann. Statist., 23:73–102, 1995
[31] Capitaine M., Donati-Martin C., and Féral D., *The largest eigenvalue of finite rank deformation of large Wigner matrices: convergence and non universality of the fuctuations*, Ann. Probab., 37, (1), 1-47 (2009).
[32] Capitaine M., Donati-Martin C., and Féral D.,*Central limit theorems for eigenvalues of deformations of Wigner matrices*, Ann. Inst. H. Poincaré Probab. Statist. Volume 48, Number 1 (2012), 107-133.
[33] I. Corwin *The Kardar-Parisi-Zhang equation and universality class* arxiv: 1106.1596 (Workshop Summary).
[34] M. Disertori, H. Pinson, and T. Spencer, *Density of states for random band matrices*, Comm. Math. Phys. 232 (2002), 83-124.

[35] Dieng, M. and Tracy, C. A. *Application of random matrix theory to multivariate statistics. In Random Matrices, Random Processes and Integrable Systems* 443–507. Springer, New York. MR2858441

[36] K. Efetov. *Supersymmetry in disorder and chaos.* Cambridge University Press, Cambridge, 1997.

[37] L. Erdős, A. Knowles *Quantum Diffusion and Delocalization for Band Matrices with General Distribution* Ann. Henri Poincaré 12 (2011), 1227-1319

[38] L. Erdős, A. Knowles, H. T. Yau, J. Yin, *Spectral Statistics of Erdős-Rényi Graphs II: Eigenvalue Spacing and the Extreme Eigenvalues.* Comm. Math. Phys. Vol 314, Issue, pp 587-640. (2012)

[39] L. Erdős, A. Knowles, H. T. Yau, J. Yin, *Spectral Statistics of Erdős-Rényi Graphs I: Local Semicircle Law.* Ann. Prob. Vol 41, No. 3B (2013), 2279-2375.

[40] L. Erdős, A. Knowles, H. T. Yau, J. Yin, *Delocalization and Diffusion Profile for Random Band Matrices* To appear in Comm. Math. Phys.

[41] L. Erdős, B. Schlein, H.T.Yau, *Local semicircle law and complete delocalization for Wigner random matrices*, Comm. Math. Phys. 287 (2009), no. 2, 641–655. doi:10.1007/s00220-008- 0636-9

[42] L. Erdős, B. Schlein, H. T. Yau, *Bulk Universality for Wigner Matrices*, Commun. Pure Applied Math. 63, 895-925, (2010).

[43] L. Erdős, B. Schlein, H. T. Yau, *Universality of Random Matrices and Local Relaxation Flow,* Inventiones Mathematicae DOI: 10.1007/s00222-010-0302-7.

[44] L. Erdős, B. Schlein, H. T. Yau, *Semicircle law on short scales and delocalization of eigenvectors for Wigner random matrices*

[45] L. Erdős, B. Schlein, H.-T. Yau , J. Yin, *The local relaxation flow approach to universality of the local statistics for random matrices.* Annales Inst. H. Poincare (B), Probability and Statistic. Vol 48, No. 1, pp 1-46. (2012).

[46] L. Erdős, H. T. Yau, *A comment on the Wigner-Dyson-Mehta bulk universality conjecture for Wigner matrices.* Electron. J. Probab. 17 (2012), no. 28, 5 pp. doi:10.1214/EJP.v17-1779

[47] L. Erdős, H. T. Yau, J. Yin, *Rigidity of Eigenvalues of Generalized Wigner Matrices.* Adv. Math. Vol 229, Issue 3, pp 1435 - 1515. (2012).

[48] L. Erdős, H. T. Yau, J. Yin, *Universality for generalized Wigner matrices with Bernoulli distribution.* J. of Combinatorics. Vol 2, No. 1 pp 13-82 (2011)

[49] L. Erdős, H. T. Yau, J. Yin, *Bulk Universality for Generalized Wigner Matrices*, Prob. Theor. Rel. Fields. Vol 154, Issue 1-2, pp 340-407. (2012)

matrix models in 2D quantum gravity. *Comm. Math. Phys.* **147** (1992), 395–430.

[50] Forrester, P. *The Spectral Edge of Random Matrix Ensembles*, Nucl. Phys. B 402, 709-728 (1993).

[51] J. Fröhlich, T. Spencer *Absence of diffusion in the Anderson tight binding model for large disorder or low energy*, Commun. Math. Phys. 88 (1983), 151–184.

[52] Z. Füredi and J. Komlós, *The eigenvalues of random symmetric matrices*, Combinatorica 1 (1981), 233-241.

[53] Y.V. Fyodorov, A.D. Mirlin, *Scaling properties of localization in random band matrices: a σ-model approach*, Phys. Rev. Lett. 67 (1991), 2405-2409.

[54] Gaudin, M.: Sur la loi limit de l'espacement des valeurs propres d'une matrice aléatoire. *Nucl. Phys.* **25**, 447-458.
[55] R. A. Janik, W. Norenberg, M. A. Nowak, G. Papp, I. Zahed, *Correlations of eigenvectors for non-Hermitian random-matrix models*, Physical Review E, 60(3):2699–2705, 1999.
[56] V. L. Girko, *The circular law*, Teor. Veroyatnost. i Primenen. 29 (1984), no. 4, 669–679 (Russian).
[57] Gotze, F.; Tikhomirov, A. *The circular law for random matrices.* Ann. Probab. 38 (4): 1444–1491 (2010).
[58] Harding, M., Explaining the single factor bias of arbitrage pricing models in finite samples, *Economic Letters*, **99**, 85-88 (2008).
[59] M. B. Hastings, *Superadditivity of communication capacity using entangled inputs*, Nature Phys. 5, 255–257 (2008)
[60] Y. He, A. Knowles, R. Rosenthal, *Isotropic self-consistent equations for mean-field random matrices*
[61] D. Hoyle and M. Rattray. *Limiting form of the sample covariance eigenspectrum in PCA and kernel PCA.* Proceedings of Neural Information Processing Systems 2003.
[62] Johansson, K.: Universality of the local spacing distribution in certain ensembles of Hermitian Wigner matrices. *Comm. Math. Phys.* **215** (2001), no.3. 683–705.
[63] Johnstone, I. M., On the distribution of the largest eigenvalue in principal components analysis. *Ann. Statist.* **29**, No. 2, 295-327 (2001).
[64] Johnstone, I.M., High Dimensional Statistical Inference and Random Matrices, *International Congress of Mathematicans*, Vol. **1**, 307-333, Eur. Math. Soc. Zurich (2007).
[65] Johnstone, I.M., Multivariate analysis and Jacobi ensembles: largest eigenvalue, Tracy-Widom limits and rates of convergence, *Ann. of Stat.*, **36**, No. 6, 2638-2716 (2008).
[66] J. Keating *The Riemann zeta-function and quantum chaology* Proc. Internat. School of Phys. Enrico Fermi CXIX: 145–185.(1993)
[67] A. Knowles, J. Yin, *Eigenvector Distribution of Wigner Matrices.* Prob. Theor. Rel. Fields. Vol 155, Issue 3-4, pp 543-582. (2013)
[68] A. Knowles, J. Yin, *The Isotropic Semicircle Law and Deformation of Wigner Matrices* Comm. Pure Appl. Math. Vol 66, No. 11(2013), 1663-1750.
[69] A. Knowles, J. Yin, *The Outliers of a Deformed Wigner Matrix.* Vol 42, no. 5, (2014) 1980-2031.
[70] A. Knowles, J. Yin, *Anisotropic local law of random matrices* PTRF.
[71] L. Laloux, P. Cizeau, M. Potters, and J. Bouchaud. *Random matrix theory and financial correlations.* Intern. J. Theor. Appl. Finanace, 3(3):391–397, 2000.
[72] B. Landon, H.T Yau, *Convergence of local statistics of Dyson Brownian motion*
[73] J. O. Lee, K. Schnelli *Edge Universality for Deformed Wigner Matrices* Reviews in Mathematical Physics 27 (2008), 1550018
[74] J. O. Lee, K. Schnelli *Tracy-Widom Distribution for the Largest Eigenvalue*

of Real Sample Covariance Matrices with General Population

[75] J. O. Lee, K. Schnelli, B. Stetler, H. T. Yau *Bulk universality for deformed Wigner matrices*

[76] J. O. Lee, J. Yin *A Necessary and Sufficient Condition for Edge Universality of Wigner matrices.* Duke math. J. Vol 163, No. 1, 117-173 (2014).

[77] Y. Malevergne and D. Sornette. Collective origin of the coexistance of apparent RMT noise and factors in large sample correlation matrices. http://xxx.lanl.gov/abs/cond-mat/0210115

[78] Marcenko, V.A. and Pastur, L.A. *Distribution of eigenvalues for some sets of random matrices.* Sbornik: Mathematics 1, 457–483. (1967).

[79] Mehta, M. L.: *Random matrices.* Academic Press, Inc., Boston, MA, second edition, 1991.

[80] N. Minami, *Local fluctuation of the spectrum of a multidimensional Anderson tight binding model*, Commun. Math. Phys. 177 (1996), 709-725.

[81] S. Péché, *The largest eigenvalue of small rank perturbations of Hermitian random matrices*, Prob. Theor. Rel. Fields 134 (2006), 127-173.

[82] N. Pillai, J. Yin *Universality for Covariance matrices* Ann. App. Prob. 24, no. 3 (2014): 935-1001.

[83] Patterson, N., Price, A.L., Reich, D., Population structure and eigen analysis, *PLoS Genetics*, No. **2 (12)** (2006):e190.

[84] V. Plerous, P. Gopikrishnan, B. Rosenow, L. Amaral, T. Guhr, and H. Stanley. *Random matrix approach to cross correlations in financial data.* Phys. Rev. E, 65(6):066126, 2002.

[85] J. Quastel *Introduction to KPZ*

[86] M. Rudelson, R. Vershynin *Delocalization of eigenvectors of random matrices with independent entries* arXiv:1306.2887

[87] J. Schenker, *Eigenvector localization for random band matrices with power law band width*, Comm. Math. Phys. 290 (2009), 1065-1097.

[88] T. Shcherbina, *On the second mixed moment of the characteristic polynomials of 1D band matrices.* Comm. Math. Phys., 328(1):45–82, 2014.

[89] T. Shcherbina, *Universality of the local regime for the block band matrices with a finite number of blocks*, Preprint. arXiv:1309.2120 (2013).

[90] R. Sear and J. Cuesta. *Instabilities in complex mixtures with a large number of components.* Phys. Rev. Lett., 91(24):245701, 2003.

[91] S. Sodin *The spectral edge of some random band matrices* arXiv: 0906.4047

[92] H. Sompolinsky, A. Crisanti, and H.-J. Sommers. *Chaos in random neural networks.* Phys. Rev. Lett., 61(3):259–262, 1988.

[93] Soshnikov, A.: Universality at the edge of the spectrum in Wigner random matrices. *Comm. Math. Phys.* **207** (1999), no.3. 697-733.

[94] T. Spencer, *Random banded and sparse matrices (Chappter 23) in Oxford Handbook of Random Matrix Theory* edited by G. Akemann, J. Baik, and P. Di Francesco (2010).

[95] T. Tao and V. Vu, *Random matrices: universality of local eigenvalue statistics* Acta Math. 206 (2011), no. 1, 127-204. doi:10.1007/s11511-011-0061-3

[96] T. Tao and V. Vu, *Random matrices: Universality of local eigenvalue statistics up to the edge* arXiv:0908.1982v1

[97] Tao, T. and Vu, V., *Random matrices: universal properties of eigenvectors.* Random Matrices Theory Appl., 1(1):1150001, 27, 2012.

[98] Tao, T.; Vu, V.H. *Random matrices: the circular law* Commun. Contemp. Math. 10 (2): 261–307. (2008).

[99] Tao, T.; Vu, V. *Random matrices: Universality of ESD and the Circular Law* (with appendix by M. Krishnapur)", Annals of Probability 38 (5): 2023–2065, (2010).

[100] Tao, T. and Vu, V. *Random matrices: The distribution of the smallest singular values.* Geom. Funct. Anal. 20 260–297. 2010.

[101] Tao, T. and Vu, V. *The Wigner-Dyson-Mehta bulk universality conjecture for Wigner matrices.* Electronic J. Probab. 16, (2011), 2104 - 2121.

[102] E. Telatar. *Capacity of milti-antenna Gaussian channels. European transactions on Telecommunications,* 10(6):585–595, 1999.

[103] Tracy, C., Widom, H. *Level-Spacing Distributions and the Airy Kernel,* Comm. Math. Phys. 159, 151-174 (1994).

[104] Tracy, C., Widom, H. *On Orthogonal and Symplectic Matrix Ensembles,* Comm. Math. Phys. 177, 727-754 (1996).

[105] Wigner, E. *Characteristic vectors of bordered matrices with infinite dimensions.* Ann. of Math. **62** (1955), 548-564.

[106] Wishart, J. *The generalized product moment distribution in samples from a Normal multivariate population.* Biometrika **20A**, 32-52 (1928)

[107] H. Xi, F. Yang, and J. Yin. *Local circular law for the product of a deterministic matrix with a random matrix.* arXiv preprint arXiv:1603.04066, 2016.

[108] J. Yin *The Local Circular Law III: General Case* Prob. Theor. Rel. Fields. Vol 160, Issue 3-4, (2014) 679-732.

Department of Mathematics
University of Wisconsin at Madison
E-mail address: jyin@math.uwisc.edu

Proceedings of the 7th International
Congress of Chinese Mathematicians
ALM 43, pp. 507–520

On Linear Independence of Generalized Logarithms

Jing Yu

Abstract

This short article is aimed at homogeneous forms in generalized logarithms of algebraic "numbers". Fix (K_v, k, v), with global field k, place v of k, and K_v an algebraically closed field containing k which is complete relative to a fixed extension of v. We introduce LA-sets and graded LA-sets for our generalized logarithms. Elements of a LA-set which are linearly independent over k are required to be linearly independent over the algebraic closure of k inside K_v. Graded LA-sets are modeled after homogeneous forms in logarithms of algebraic numbers. There are many interesting sets in classical number theory which are conjectured to be LA-sets. We report that relating to the arithmetic of global function fields, there are already many confirmed LA-sets and graded LA-sets. Associated with given graded LA-set, there is natural graded weighted k-algebra. The structure of this algebra is interwoven with the collections of homogeneous linear relations among generalized logarithms of these algebraic "numbers".

Mathematics Subject Classification: 11J81, 11J85, 11J86, 11J89, 11J93.

Keywords and Phrases: Global fields, logarithms, Baker's Theorem, v-adic logarithms, polylogarithms, multiple polylogarithms, Carlitz module, Drinfeld module, t-Modules, Anderson-Thakur polynomials, multiple zeta values, weight, log-algebraicity.

1 Introduction

In mathematics, we are interested in special values arising from various mathematical contexts and structures, like π, $\log(2)$, or the Riemann zeta values at integers $n > 1$, denoted by $\zeta(n)$. These special values are believed to be transcendental numbers, meaning that they are not algebraic numbers, substituting such a number into any non-zero polynomial in one variable with integral coefficients, you will never get 0. Since C. Hermite in late 19-th century, the transcendence nature of many such special values has been confirmed. However there are many more classical special values for which proofs of transcendence are still not in sight.

When we are confronted with several such special values, a natural question is asking for their algebraic relationships, if they are algebraically related or not. This is a even more subtle challenge than proving individual transcendence. Ideally, we would like to know whether, say, $\log 3$ and $\log 2$ are algebraically dependent, i.e. whether there are two variables polynomials $P(X, Y)$ with coefficients from $\mathbb{Z}$ not all zero so that $P(\log 2, \log 3) = 0$. One sees no reason that these two values should be algebraically dependent. Whenever a set of (would be transcendental) special values is given to us, we want to find and explain all the possible algebraic relations among them. Only after that, can we say that we really have understood these special values.

The question of algebraic relations between $\log 3$ and $\log 2$ originated from Hilbert 7-th problem which asks for the transcendence of the ratio of these two logarithms. Gelfond-Schneider (1929) solved this problem of Hilbert affirmatively. However, till today no one knows whether $\log 2$ and $\log 3$ are algebraically independent.

Natural logarithms of algebraic numbers have been studied extensively in the last 100 years or so. The great theorem of Alan Baker (1966) extends Gelfond-Schneider's investigation from 2 logarithms to any n logarithms of algebraic numbers, with delicate precise estimates of the lower bound on linear logarithmic forms of algebraic numbers. This has many far-reaching consequences in arithmetic, cf. [B2].

In late 1980's, we noticed that interesting phenomena analogous to the qualitative Baker's theorem can also be found over global function fields. This leads us to the following considerations. Let (K_v, k, v) be a triple, where k is a global field, v is a fixed place of k, and K_v is an algebraically closed field extension of k completed with respect to a selected extension of the valuation at v. We also fix an embedding of the algebraic closure of k inside K_v, to be denoted by $\bar{k}$.

Given a subset $\mathcal{P} \subset K_v$ consisting of elements in K_v transcendental over k, set $\mathcal{L} := \mathrm{Span}_{\bar{k}}\, \mathcal{P}$, and $\mathcal{L}_0 := \mathrm{Span}_k\, \mathcal{P}$ be vector subspaces inside K_v. We say that $\mathcal{P}$ is a logarithmic set (abbreviated LA-set) with respect to the triple (K_v, k, v) if $\mathcal{L} \cap \bar{k} = \{0\}$ and the following property holds:

$$\mathcal{L} \cong \bar{k} \otimes_k \mathcal{L}_0.$$

As a consequence, any linear combination of elements from $\mathcal{P}$ with coefficients in $\bar{k}$ is either zero or trancendental over k, and the vector space generated over $\bar{k}$ by $\mathcal{P}$ has a natural k-structure.

Given a k-linear subspace $\mathcal{L}_0 \subset K_v$ which contains a k-basis that is algebraically independent over k, then any k-basis of $\mathcal{L}_0$ is algebraically independent over k. Let $\mathcal{P}'$ be a subset of the LA-set $\mathcal{P}$, and let $\mathcal{P}'_0$ be a maximally k-linearly independent subset of $\mathcal{P}'$, then $\mathcal{P}'_0$ must be algebraically independent over k; hence $\mathcal{P}'$ is also LA-set w.r.t. (K_v, k, v).

Developments of arithmetic geometry in the last century yield interesting LA-sets, both in the number field case and also in the function field (positive characteristic) case. In this report, we contend and conjecture that there are many more such sets in mathematics. This contention is deeply interwoven with some of the very important wide open problems in Mathematics.

2 Classical logarithmic type families and Graded LA-sets

The best-known LA-set is $\mathcal{P}_{\mathbb{G}_m}$ w.r.t. $(\mathbb{C}, \mathbb{Q}, \infty)$, consisting of the logarithms of all non-zero algebraic numbers. That these numbers have the desired property is the qualitative form of Alan Baker's great theorem (1966) [B2], asserting that logarithms of algebraic numbers linearly independent over $\mathbb{Q}$ must in fact be linearly independent over the field of all algebraic numbers. Here logarithms means natural logarithms taking at any branch.

Staying with $\mathbb{G}_m$, there is also the p-adic version of Baker's theorem (Coates, Vinogradov-Sprindzuk cf.[Sprin]), p any prime number. One considers the triple $(\mathbb{C}_p, \mathbb{Q}, p)$, where $\mathbb{C}_p$ is the completion of a fixed algebraic closure of the field of p-adic numbers $\mathbb{Q}_p$. To get a LA-set, one takes the p-adic logarithms of p-adic algebraic numbers α whose (normalized) p-adic absolute values satisfy $\mid \alpha - 1 \mid_p < p^{\frac{-1}{p-1}}$. This is applied to the study of the value taken at 1 of the p-adic zeta function of Kubota-Leopoldt, cf. [Washington] Chap.5. These p-adic special values can be expressed as linear combinations of p-adic logarithms of algebraic numbers with algebraic coefficients.

Replace (after classical works of Siegel-Schneider) $\mathbb{G}_m$ by an elliptic curve E defined over $\overline{\mathbb{Q}}$ with Weierstrass elliptic function $\wp(z)$. Let $\mathcal{P}_E$ be the set of complex numbers u with $\wp(u) \in \overline{\mathbb{Q}} \cup \{\infty\}$. If E does not have complex multiplications, then Masser-Bertrand (1974, 1980) prove that $\mathcal{P}_E$ is a LA-set w.r.t. the triple $(\mathbb{C}, \mathbb{Q}, \infty)$. If on the other hand E has complex multiplications from imaginary quadratic field k, then $\mathcal{P}_E$ is a LA-set w.r.t. $(\mathbb{C}, k, \infty)$, cf. [Wald].

LA-sets usually come in as families. We therefore introduce **graded** LA-set with respect to the triple (K_v, k, v). This is a LA-set $\mathbb{P}$ which is a disjoint union of subsets $\mathbb{P} = \dot{\bigcup}_{n=1}^{\infty} \mathcal{P}_n$, satisfying

$$\mathcal{P}_n \mathcal{P}_m \subset (\mathcal{L}_{n+m})_0 := \mathrm{Span}_k\, \mathcal{P}_{n+m}.$$

Given a graded LA-set $\mathbb{P}$, we have a graded k-subalgebra $\mathcal{A}$ and $\bar{k}$-subalgebra $\overline{\mathcal{A}}$ inside K_v, both generated by $\mathbb{P}$. We then have :

$$\mathcal{A} = k \oplus \bigoplus_{n=1}^{\infty} (\mathcal{L}_n)_0, \quad \bar{\mathcal{A}} \cong \mathcal{A} \otimes_k \bar{k}.$$

For a graded LA-set $\mathbb{P}$, we note that if $n \neq m$, the ratio of two non-zero elements, one from $\mathcal{P}_n$ and one from $\mathcal{P}_m$, must be transcendental over k. The most interesting question for a given graded LA-set is to determine the structure of its associated k-algebra $\mathcal{A}$.

Given a subset $\mathbb{P}'$ of the graded LA-set $\mathbb{P} = \dot{\bigcup}_{n=1}^{\infty} \mathcal{P}_n$. We let $\mathcal{P}'_n := \mathcal{P}_n \cap \mathcal{P}'$. Suppose $\mathcal{P}'_n \cdot \mathcal{P}'_m \subset \mathrm{Span}_k\, \mathcal{P}'_{n+m}$ holds, then $\mathbb{P}'$ is also graded LA-set.

Given an (ungraded) LA-set $\mathcal{P}$, a natural question is whether monomials in its elements form a graded LA-set, with $\mathcal{P}_n$ consisting of products of any n elements from $\mathcal{P}$. For the classical LA-sets $\mathcal{P}_{\mathbb{G}_m}$, one conjectures that their monomials do form a graded LA-set. Furthermore the associated algebra $\mathcal{A}_{\mathbb{G}_m}$ is precisely the polynomial algebra $\mathbb{Q}[\mathcal{P}']$, where $\mathcal{P}'$ can be any maximal subset of $\mathcal{P}$ which is linearly independent over $\mathbb{Q}$. In other words, any set of logarithms of algebraic numbers which are linearly independent over $\mathbb{Q}$ must be algebraically independent over $\mathbb{Q}$. Similar phenomenon should occur also for the set $\mathcal{P}_E$ associated to an elliptic curve E, with or without complex multiplications.

Take $\mathcal{P} = \{\log 2, \log 3\}$. Under the hypothesis that $\log 2, \log 3$ are algebraically independent over $\mathbb{Q}$, the monomials in $\log 2$, and $\log 3$ form a graded LA-set whose associated algebra $\mathcal{A}$ is just the polynomial algebra $\mathbb{Q}[\log 2, \log 3]$.

Naturally graded LA-sets w.r.t. $(\mathbb{C}, \mathbb{Q}, \infty)$ come also from the multiple zeta values introduced by Euler. Let $\mathfrak{s} := (s_1, \ldots, s_r) \in \mathbb{N}^r$,

$$\zeta(\mathfrak{s}) := \zeta(s_1, \ldots, s_r) := \sum_{n_1 > \cdots > n_r \geq 1} \frac{1}{n_1^{s_1} \cdots n_r^{s_r}},$$

where $s_1, \ldots, s_r$ are positive integers with $s_1 \geq 2$, $w(\mathfrak{s}) := s_1 + \cdots + s_r$ is the weight and r is the depth. When depth r is 1, these are the Riemann zeta values taken at integers $s = w \geq 2$. We take $\mathcal{P}_n$, $n \geq 2$, to be the set of all weight n multiple zeta values, and $\mathcal{P}_1 := \emptyset$. The well-known shuffle relations for these values give $\mathcal{P}_n \mathcal{P}_m \subset (\mathcal{L}_{n+m})_0$. One conjectures that all these multiple zeta values are transcendental, also a conjecture of Goncharov and Zagier go further to declare that the algebra $\mathcal{A}$ generated by these multiple zeta values should be graded by weight. This leads to the problem of asking whether $\cup_{n=1}^{\infty} \mathcal{P}_n$ forms graded LA-set, and determining the structure of the associated algebra $\mathcal{A}$. This is the focus of current research on that subject.

Still with triple $(\mathbb{C}, \mathbb{Q}, \infty)$, another generalization of logarithms is to go to polylogarithms, cf. [Zhao]. For an integer $l > 1$, a degree l polylogarithms is defined by:

$$\mathrm{Li}_l(z) := \sum_{m=1}^{\infty} \frac{z^m}{m^l}, \quad |z| < 1.$$

Then by analytic continuation it can be extended to the whole z-plane as multiple-valued function having singularity at $z = 1$. For $l = 1$ also take

$$\mathrm{Li}_1(z) := -\log(1 - z), \quad |z| < 1.$$

Consider degree l **polylogarithms** of algebraic numbers:

$$\mathcal{P}_l := \{\mathrm{Li}_l(\alpha) \mid \alpha \in \bar{\mathbb{Q}}, \ |\alpha| < 1\}.$$

An interesting "naive"question to raise here is, whether $\mathcal{P}_l$ is LA-set for each $l \geq 1$. When $l = 1$, this reduces back to Baker's theorem. Note $\zeta(l) = \mathrm{Li}_l(1)$ for $l \geq 2$, therefore this question includes the assertion that $\zeta(l)$ for $l \geq 2$ are transcendental numbers. It is still unknown whether $\zeta(3) = \sum_{m=1}^{\infty} \frac{1}{m^3}$ is transcendental. On the other hand, Zagier [Zagier] has conjectured that for any algebraic number field F, the values of its Dedekind zeta function $\zeta_F(m)$ at integer $m > 1$ can be expressed in terms of the polylogarithmic functions at suitable algebraic points. Thus our naive question is related to the problem of transcendence of Dedekind zeta values at positive integers.

More generally, introducing multipolylogarithms: $\mathfrak{s} := (s_1, \ldots, s_r) \in \mathbb{N}^r$,

$$\mathrm{Li}_{\mathfrak{s}}(z_1, \ldots, z_r) := \sum_{n_1 > \cdots > n_r \geq 1} \frac{z_1^{n_1} \cdots z_r^{n_r}}{n_1^{s_1} \cdots n_r^{s_r}},$$

where $w(\mathfrak{s}) := s_1 + \cdots + s_r$ is the weight, r is the depth. The multiple zeta values are given by $\zeta(s_1, \ldots, s_r) = \mathrm{Li}_{\mathfrak{s}}(1, \ldots, 1)$, for $s_1 \geq 2$. When $w = 1$, multiple polylogarithms are just the usual logarithms.

Given w, let $\mathcal{P}_w$ be the set of weight w multipolylogarithms taken at algebraic points, $\mathrm{Li}_{\mathfrak{s}}(a_1, \ldots, a_r)$, $a_i \in \bar{\mathbb{Q}}$ and for which these values are defined. A natural question is whether $\mathcal{P}_w$ is still an LA-set if $w \geq 2$, and furthermore whether this gives a graded LA-set w.r.t. $(\mathbb{C}, \mathbb{Q}, \infty)$. The classical "stuffle "relation shows already: $\mathcal{P}_w \mathcal{P}_{w'} \subset \mathrm{Span}_{\mathbb{Q}} \mathcal{P}_{w+w'}$. Replacing the archimedean place ∞ by any prime number p, it also make sense to ask p-adic versions of the above questions on p-adic multiple polylogarithms at algebraic points w.r.t. $(\mathbb{C}_p, \mathbb{Q}, p)$.

In the next Section, we report the recent progress on studies of LA-sets and graded LA-sets in the positive characteristic world, where we can go much further answering the analogues of all the above open questions.

3 Background for positive characteristic Arithmetic

Notations:

$\mathbb{F}_q$:= the finite field of q elements, with q a power of prime p.
$k := \mathbb{F}_q(\theta)$:= the rational function field in the variable θ over $\mathbb{F}_q$.
$\bar{k}$:= fixed algebraic closure of k.

t := another variable distinct from the variable θ.

$k_\infty := \mathbb{F}_q((\frac{1}{\theta}))$, the "completion"of k with respect to the infinite "place".

$\overline{k_\infty}$:= a fixed algebraic closure of k_∞ containing $\bar{k}$.
$\bar{k}$:= the algebraic closure of k inside $\overline{k_\infty}$.
$|\cdot|_\infty$:= the non-archimedean absolute value with $|\theta|_\infty = q$ on $\overline{k_\infty}$.
$\mathbb{C}_\infty$:= fixed completion of $\overline{k_\infty}$ with respect to $|\cdot|_\infty$.

Interesting LA-sets for positive characteristic triple $(\mathbb{C}_\infty, k, \infty)$ come from the arithmetic objects known as Drinfeld modules (1974). The first Drinfeld module was introduced earlier by L. Carlitz, the Carlitz module (1930).

We consider the $\mathbb{F}_q$-linear $\mathbb{F}_q[t]$-action on the additive group $\mathbb{G}_a$ given by:

$$\phi_C(t) : x \longmapsto \theta x + x^q.$$

There is the Carlitz exponential $\exp_C$ making the following diagram commuts:

$$\begin{array}{ccc} \overline{k_\infty} & \xrightarrow{\exp_C} & \mathbb{G}_a(\overline{k_\infty}) = \overline{k_\infty} \\ \theta(\cdot)\downarrow & & \downarrow \phi_C(t) \\ \overline{k_\infty} & \xrightarrow{\exp_C} & \mathbb{G}_a(\overline{k_\infty}) = \overline{k_\infty} \end{array}$$

The kernel of the Carlitz exponential is described by the short exact sequence:

$$0 \to \mathbb{F}_q[\theta]\,\tilde{\pi} \to \overline{k_\infty} \xrightarrow{\exp_C} \overline{k_\infty} \to 0.$$

Here $\tilde{\pi}$ is the fundamental Carlitz period transcendental over the rational field k (Wade 1946):

$$\tilde{\pi} = \tilde{\pi}(q) := (-\theta)^{\frac{q}{q-1}} \prod_{i=1}^{\infty} \left(1 - \theta^{1-q^i}\right)^{-1}.$$

This period is determined up to a sign in $\mathbb{F}_q[\theta]$, i.e. a constant in $\mathbb{F}_q^\times$. Note that the power $\tilde{\pi}^n$ is in field k_∞ if and only if n is divisible by $q-1$.

The Carlitz exponential is written as an "entire" function :

$$\exp_C(z) := \sum_{h=0}^{\infty} \frac{z^{q^h}}{D_h} = z \prod_{a \neq 0 \in \mathbb{F}_q[\theta]} (1 - \frac{z}{a\tilde{\pi}}),$$

where $D_0 := 1$, and $D_i := (\theta^{q^i} - \theta^{q^{i-1}}) \cdots (\theta^{q^i} - \theta)$, for $i \geq 1$. Note D_i is exactly the product of all the monic polynomials in $\mathbb{F}_q[\theta]$ of degree equal to i. This exponential function is additive, $\mathbb{F}_q$-linear, and $\tilde{\pi}\mathbb{F}_q[\theta]$-periodic. The Carlitz logarithm is the compositional inverse to the Carlitz exponential which is given by:

$$\log_1(z) := \log_C(z) := \sum_{i=0}^{\infty} \frac{z^{q^i}}{L_i}, \quad |z|_\infty < q^{\frac{q}{q-1}}.$$

Here $L_0 := 1$ and $L_i := \prod_{j=1}^{i} (\theta - \theta^{q^j})$ for $i \geq 1$. Up to $(-1)^i$, L_i is exactly the least common multiple of all monic polynomials in $\mathbb{F}_q[\theta]$ of degree i. Given $\alpha \in \bar{k}$, any element in the inverse image $\exp_c^{-1}(\alpha)$ is called a Carlitz logarithm of the algebraic α.

Let X be a smooth, projective, geometrically connected curve over $\mathbb{F}_q$. We fix a closed point ∞ on X and denote by A the ring of functions on X regular away from ∞. Let k be the function field of X over $\mathbb{F}_q$, and let k_∞ be the completion of k at ∞. Let $\mathbb{C}_\infty$ be the completion of a fixed algebraic closure of k_∞ at ∞. Let $\Lambda \subset \mathbb{C}_\infty$ be an A-lattice of projective rank r, i.e., a finitely generated discrete A-module in $\mathbb{C}_\infty$, and then form the exponential function

$$e_\Lambda(z) := z \prod_{0 \neq \lambda \in \Lambda} (1 - \frac{z}{\lambda}).$$

The Drinfeld A-module ρ^Λ of rank r associated to Λ is the $\mathbb{F}_q$-linear ring homomorphism denoted by ρ^Λ from A into the endomorphism ring of the additive group $\mathbb{G}_{a/\mathbb{C}_\infty}$ that satisfies for all $a \in A$:

$$\begin{array}{ccc} \mathbb{C}_\infty & \xrightarrow{e_\Lambda} & \mathbb{G}_a(\mathbb{C}_\infty) = \mathbb{C}_\infty \\ {\scriptstyle a(\cdot)}\downarrow & & \downarrow{\scriptstyle \rho^\Lambda(a)} \\ \mathbb{C}_\infty & \xrightarrow{e_\Lambda} & \mathbb{G}_a(\mathbb{C}_\infty) = \mathbb{C}_\infty \end{array} .$$

Thus ρ^Λ induces an A-module structure on $\mathbb{C}_\infty$, which is isomorphic to $\mathbb{C}_\infty/\Lambda$ via $e_\Lambda(z)$. The field of multiplications of Drinfeld module ρ_Λ is the fraction field of the integral domain $\{b \in \bar{k} \mid b\lambda \subset \Lambda\}$. This is always a finite extension of k of degree $\leq$ the rank. We say that this Drinfeld module ρ^Λ is defined over $\bar{k}$ if all the endomorphisms in $\rho^\Lambda(A)$ have coefficients in $\bar{k}$ when written as $\mathbb{F}_q$-linear polynomials. The multi-valued inverse of the Drinfeld exponential $e_\Lambda(z)$ is called the Drinfeld logarithm associated to the Drinfeld A-module ρ^Λ.

Carlitz (1935) also introduced one analogue of Riemann zeta values in positive characteristic. For $m \geq 1$ (here m can be 1),

$$\zeta_A(m) := \sum_{a \in \mathbb{F}_q[\theta]_+} \frac{1}{a^m} \in \mathbb{F}_q\big((\frac{1}{\theta})\big) = k_\infty,$$

where $\mathbb{F}_q[\theta]_+$ consists of monic polynomials in $A := \mathbb{F}_q[\theta]$. One amusing fact is: $\zeta_A(1) = \log_C(1)$. Another is the fact that when $q = 2$, there is no "sign"in that world, and $\zeta_A(n) \in k^\times \tilde{\pi}^n$ does indeed hold.

Call a positive integer m even (w.r.t. q), if $m \equiv 0 \pmod{q-1}$. Then there are Euler-Carlitz relations for zeta values at these even m :

$$\zeta_A(m) = \frac{\tilde{\pi}^m \widetilde{B_m}}{\Gamma_{m+1}},$$

The $\widetilde{B_m}$ are the Bernoulli-Carlitz rationals in the rational field $k = \mathbb{F}_q(\theta)$ coming from coefficients of the Carlitz exponential series. The Γ_n are (Carlitz) factorials of the integer ring A.

More precisely, take the q-adic expansion $\sum_{i=0}^{\infty} n_i q^i$ of n, and let

$$\Gamma_{n+1} := \prod_{i=0}^{\infty} D_i^{n_i} .$$

The $\widetilde{B_m} \in \mathbb{F}_q(\theta)$ are the Bernoulli-Carlitz "numbers" given by

$$\frac{z}{\exp_C(z)} = \sum_{m=0}^{\infty} \widetilde{B_m} \frac{z^m}{\Gamma_{m+1}}.$$

In the arithmetic of polynomials over given finite field, there is a very fascinating family of integral polynomials in two variables introduced by Anderson-Thakur [AT90] which is somewhat like the Bernoulli polynomials, but contain precious information about $\zeta_A(n)$ for all $n \geq 1$. Put $G_0 = 1$. For $i \in \mathbb{N}$, set

$$G_i = \prod_{j=1}^{i} (t^{q^i} - \theta^{q^j}) \in \mathbb{F}_q[t, \theta] := A[t].$$

For $n = 0, 1, 2, \ldots$, the Anderson-Thakur polynomial $H_n \in A[t]$ is defined by the generating series identity

$$\left(1 - \sum_{i=0}^{\infty} \frac{G_i}{D_i|_{\theta=t}} x^{q^i}\right)^{-1} = \sum_{n=0}^{\infty} \frac{H_n}{\Gamma_{n+1}|_{\theta=t}} x^n.$$

For $n \leq q - 1$, one simply has $H_n \equiv 1$. These polynomials play key roles in the constructions to be given in the next Section.

Fixed a finite place, i.e. a monic irreducible in $\mathbb{F}_q[\theta]$, and let k_v be the v-adic completion of $\mathbb{F}_q(\theta)$. Following D. Goss, one also introduces v-adic zeta values $\zeta_v(n) \in k_v$, for positive integers n. Goss's approach is through "analytic continuation"; he defines $\zeta_C(m) \in \mathbb{F}_q(\theta)$ for integers $m \leq 0$, then puts

$$\zeta_v(m) := (1 - v^{-m})\zeta(m), \quad m \leq 0,$$

$$\zeta_v(n) := \lim_{i \to \infty} \zeta_v(n - (q^{\deg v} - 1)p^i, \quad n \geq 1.$$

Finally D. Thakur (2004) [T04] initiated the study of the depth r **multiple zeta values** (MZVs) over given finite field $\mathbb{F}_q$:

$$\zeta_A(s_1, \ldots, s_r) := \zeta_A(\mathfrak{s}) := \sum \frac{1}{a_1^{s_1} \cdots a_r^{s_r}} \in k_\infty,$$

where the sum is over $(a_1, \ldots, a_r) \in \mathbb{F}_q[\theta]_+^r$ with $\deg_\theta a_i$ strictly decreasing. He showed that these MZVs are always non-vanishing, and gives very interesting special values in positive characteristic. When depth $r = 1$, these are just the Carlitz zeta values.

The transcendence over $\mathbb{F}_q(\theta)$ of all these special zeta values, in particular $\zeta_A(n)$ for all $n \geq 1$, can be verified by expressing these special values as rational linear combinations of "generalized" logarithms at algebraic points ([AT90], [C14]) to be explained in the following Section.

4 Generalized Logarithms in Positive Characteristic

Let $\mathcal{P}_C$ be the set of all non-zero Carlitz logarithms of algebraic elements in $\bar{k}$. Yu showed this constitutes an LA-set w.r.t. $(\mathbb{C}_\infty, k, \infty)$. This construction can be generalized from the Carlitz module to any Drinfeld module over given global function field, with Drinfeld logarithm playing the role of Carlitz logarithm. In 2008 [Papa], Papanikolas successfully developed Tannakian duality Theory proving that monomials of elements from $\mathcal{P}_C$ forms a graded LA-set, thereby proves that linear independence of any set in $\mathcal{P}_C$ implies algebraic independence of that very set. This theorem is generalized in [CP] to all Drinfeld modules defined over the algebraic closure of any given global field of positive characteristic.

Replacing the infinite place ∞ of $k := \mathbb{F}_q(\theta)$ by a finite place v corresponding to monic irreducible $v \in \mathbb{F}_q[\theta]$, we also have the v-adic analogue of Baker's theory. For the triple $(\mathbb{C}_v, k, v)$, the Carlitz logarithm series converges v-adically in the disc consisting of $\alpha \in \mathbb{C}_v$ satisfying $\mathrm{Ord}_v(\alpha) > \frac{1}{q^{\deg v}-1}$. We ([Yu91]) take as LA-set the v-adic non-zero Carlitz logarithm of elements inside that disc which are algebraic over k. Again this can be generalized to all Drinfeld modules defined over the algebraic closure of any given global field of positive characteristic.

To accommodate special zeta values, we introduce further generalizations of logarithms. Following Anderson [A86], consider t-modules of higher dimension: Put $\tau(x) = x^q$ and denote by $\mathbb{C}_\infty\{\tau\}$ the twisted polynomial ring in τ over $\mathbb{C}_\infty$ with the relation $\tau\beta = \beta^q\tau$ for all $\beta \in \mathbb{C}_\infty$. For $d \in \mathbb{N}$, a d-dimensional t-module is a pair (E, ρ), where E is the d-dimensional algebraic group $\mathbb{G}_a^d$ and ρ is an $\mathbb{F}_q$-linear ring homomorphism

$$\rho : \mathbb{F}_q[t] \to \mathrm{Mat}_d(\mathbb{C}_\infty[\tau]) = \mathrm{End}\, \mathbb{G}^d_{a/\mathbb{F}_q}.$$

such that $\rho_t = \alpha_0 + \sum_i \alpha_i \tau^i$ where $\alpha_i \in \mathrm{Mat}_d(\mathbb{C}_\infty)$, and $\alpha_0 - \theta I_d$ is a nilpotent matrix. This action of $\mathbb{F}_q[t]$ on $\mathbb{G}_a^d$ is required to satisfy:

$$\begin{array}{ccc} \mathbb{C}^d_\infty & \xrightarrow{\exp_\rho} & \mathbb{G}^d_a(\mathbb{C}_\infty) = \mathbb{C}^d_\infty \\ {\scriptstyle d\rho_t}\downarrow & & \downarrow{\scriptstyle \rho_t} \\ \mathbb{C}^d_\infty & \xrightarrow{\exp_\rho} & \mathbb{G}^d_a(\mathbb{C}_\infty) = \mathbb{C}^d_\infty \end{array},$$

with an entire exponential map $\exp_\rho$. This exponential map is normalized so that $\exp_\rho(\mathbf{z}) \equiv \mathbf{z}$ mod higher degree terms for vector $\mathbf{z} \in \mathbb{C}^d_\infty$. The compositional inverse of $\exp_\rho$ is the logarithmic map $\log_\rho := (\log_{\rho 1}, \dots, \log_{\rho d})$ for the t-module ρ. We call $\log_{\rho d}$ which is defined on small disc around $0 \in \mathbb{C}^d_\infty$, the last coordinate logarithm of the t-module ρ.

A typical example of t-module of dimension n is the n-th tensor power of the Carlitz module $C^{\otimes n} := (\mathbb{G}_a^n, [\cdot]_n)$ given by

$$[t]_n := \begin{pmatrix} \theta & 1 & \cdots & 0 \\ & \ddots & \ddots & \vdots \\ & & \ddots & 1 \\ & & & \theta \end{pmatrix} + \begin{pmatrix} 0 & 0 & \cdots & 0 \\ 0 & 0 & \vdots & 0 \\ \vdots & \vdots & \vdots & \vdots \\ 1 & 0 & \cdots & 0 \end{pmatrix} \tau.$$

Anderson-Thakur [AT90] shows that $\ker(\exp_{\mathbf{C}^{\otimes n}})$ is a rank one $\mathbb{F}_q[t]$-module (via the $d[a]_n$-action) with a generator λ_n:

$$\lambda_n = \begin{pmatrix} * \\ \vdots \\ \tilde{\pi}^n \end{pmatrix} \in \mathbb{C}_\infty^n,$$

where $\tilde{\pi}$ is the Carlitz period. On the other hand,

$$\log_{[n]} \begin{pmatrix} 0 \\ \vdots \\ z \end{pmatrix} = \begin{pmatrix} * \\ \vdots \\ \sum_{i=0}^{\infty} \frac{z^{q^i}}{L_i^n} \end{pmatrix}.$$

Furthermore for each $n \geq 1$, [AT90] found "logarithmic"vector $\mathbf{z}_n \in \mathbb{C}_\infty^n$ satisfying the following crucial formula:

$$\mathbf{Z}_n := \exp_{[n]}(\mathbf{z}_n) = \exp_{[n]} \begin{pmatrix} * \\ \vdots \\ * \\ \Gamma_n \zeta_A(n) \end{pmatrix} \in \mathbf{C}^{\otimes n}(\mathbb{F}_q[\theta]).$$

In other words $\Gamma_n \zeta_A(n)$ is a last coordinate logarithm of an rational integral point on the t-module $\mathbf{C}^{\otimes n}$. Let $\mathcal{P}^{\otimes n}$ be the set of all last coordinate logarithms of algebraic points $\mathbf{Z}$ on $\mathbf{C}^{\otimes n}$, with $\mathbf{Z} = (z_1, \ldots, z_n)$ satisfying $| z_i |< q^{i-n+\frac{nq}{q-1}}$. Then Yu [Yu91] showed $\mathcal{P}^{\otimes n}$ is an LA-set w.r.t. $(\mathbb{C}_\infty, k, \infty)$. Particularly transcendence of the last coordinate logarithms taking values at algebraic points of the Carlitz tensor power (or any other sufficiently nice t-module defined over $\bar{k}$) is obtained. The transcendence of $\zeta_A(n)$ and linear independence of $\zeta_A(n)$ with $\tilde{\pi}$ over $\bar{k}$ when n is not divisible by $q-1$, follows immediately.

Chang-Yu (2007) [CY07] discovered that natural LA-sets also come from higher degree Carlitz polylogarithms, with degree $s \geq 1$:

$$\log_s(z) := \sum_{i=0}^{\infty} \frac{z^{q^i}}{L_i^s}, \ |z|_\infty < q^{\frac{sq}{q-1}}.$$

Here we take $\mathcal{P}_s$ as the set of degree s non-zero Carlitz polylogarithms taking values at algebraic $\alpha \in \bar{k}$ satisfying $|\alpha|_\infty < q^{\frac{sq}{q-1}}$. Chang-Yu also went further showing that any set of degree s polylogarithms at algebraic elements must be algebraically independent if they are linearly independent over k.

The story about $\zeta_A(n)$ for **all** n was unfolded in Anderson-Thakur (1990) [AT90] and Yu (1991) [Yu91]. For any $n \geq 1$, $\zeta_A(n)$ is also shown to be a $\mathbb{F}_q(\theta)$-linear combination of degree n (Carlitz) polylogarithms taken at the elements $\theta^i, 0 \leq i \leq l_n \leq \frac{nq}{q-1}$. The non-trivial recipe for writing up these linear combinations comes

$$\Gamma_n \zeta_A(n) = \sum_{i=0}^{l_n} h_{n,i} \operatorname{Log}_n(\theta^i).$$

The story about v-adic Carlitz zeta values also come out in that two papers. For any positive integers n, the last coordinate logarithm of $\mathbf{C}^{\otimes n}$ viewed as v-adic series $\log_{n,v}$ converges on small enough discs and the set of those non-zero v-adic last coordinate logarithms of small algebraic points on $\mathbf{C}^{\otimes n}$ also forms an LA-set w.r.t. $(\mathbb{C}_v, k, v)$. This set contains $\log_{n,v}([v^n - 1]\mathbf{Z}_n) = v^n\Gamma_n\zeta_v(n)$. It follows that the v-adic Carlitz zeta values $\zeta_v(n)$ is either transcendental over k or is zero. It is transcendental if n is not divisible by $q-1$, and it is zero if n is divisible by $q-1$. In other words, the ∞-adic Carlitz zeta value $\zeta_A(n)$ is a rational multiple of $\tilde{\pi}^n$ if and only if the corresponding v-adic zeta value vanishes.

To go one step further, we also introduce Carlitz multipolylogarithms over $\mathbb{F}_q[\theta]$ with $\mathfrak{s} = (s_1, \ldots, s_r) \in \mathbb{N}^r$:

$$\mathrm{Li}_{\mathfrak{s}}(z_1, \ldots, z_r) := \sum_{i_1 > \cdots > i_r \geq 0} \frac{z_1^{q^{i_1}} \cdots z_r^{q^{i_r}}}{L_{i_1}^{s_1} \cdots L_{i_r}^{s_r}}, \quad |z_i|_\infty < q^{\frac{s_i q}{q-1}}.$$

where $\mathfrak{w} = \mathfrak{w}(\mathfrak{s}) := s_1 + \cdots + s_r$ is the weight, r is the depth. When depth $r = 1$, it reduces to polylogarithms of degree $s = s_1$. Let $\mathcal{P}_{\mathfrak{w}}$ be the set of weight $\mathfrak{w}$ non-zero Carlitz multipolylogarithms (abbreviated CMPLs) at algebraic points $(\alpha_1, \ldots, \alpha_s) \in \bar{k}^r$ where the above multiple series converge. Then Chang 2014 [C14] showed that this gives an LA-set w.r.t. $(\mathbb{C}_\infty, \mathbb{F}_q(\theta), \infty)$ for any weight $\mathfrak{s}$. Moreover as the stuffle relation goes through here also, Chang was able to establish the fact that the CMPLs at algebraic points form a graded LA-set. A particularly interesting consequence here is what we call the Euler dichotomy phenomenon: any CMPL of weight $\mathfrak{w}$ at algebraic point is either a rational multiple of the Carlitz period $\tilde{\pi}^{\mathfrak{w}}$ or must be algebraically independent from $\tilde{\pi}$. A v-adic theory of Carlitz multipolylogarithms is developed by Chang-Mishiba 2016 [CM]. The v-adic CMPLs at algebraic points where they are defined also forms a graded LA-set w.r.t. $(\mathbb{C}_v, k, v)$. Again there is the phenomenon that ∞-adic Eulerian corresponds to v-adic vanishing.

D. Thakur (2004) [T04] initiated the study of the depth r **multiple zeta values** (MZVs) over given finite field $\mathbb{F}_q$:

$$\zeta_A(s_1, \ldots, s_r) := \zeta_A(\mathfrak{s}) := \sum \frac{1}{a_1^{s_1} \cdots a_r^{s_r}} \in k_\infty,$$

where the sum is over $(a_1, \ldots, a_r) \in \mathbb{F}_q[\theta]_+^r$ with $\deg_\theta a_i$ strictly decreasing. Thakur also showed that these MZVs are always non-vanishing. Afterwards Chang (2014) established the fact that $\zeta_A(\mathfrak{s})$ of weight $\mathfrak{w}$ can be written as rational linear combination of weight $\mathfrak{w}$ CMPLs (at algebraic points).

For $\mathfrak{w} \geq 1$, let $\mathcal{PZ}_{\mathfrak{w}}$ be the set of all weight $\mathfrak{w}$ MZVs. This is a subset of $\mathcal{P}_{\mathfrak{w}}$ and also turns out to be an LA-set w.r.t. $(\mathbb{C}_\infty, \mathbb{F}_q(\theta), \infty)$. It follows in particular that all the MZVs in positive characteristic are transcendental over $\mathbb{F}_q(\theta)$. As Thakur has already observed from doing "shuffle"with these MZVs, $\mathcal{PZ}_w \cdot \mathcal{PZ}_{w'} \subset \mathrm{Span}_{\mathbb{F}_p} \mathcal{PZ}_{w+w'}$ holds, where p is the characteristic. Chang is able to establish that the union $\mathcal{PZ}$ of all these $\mathcal{PZ}_{\mathfrak{w}}$ form a wonderful graded LA-set.

Just as with the classical characteristic zero multiple zeta values, there is a rich source of beautiful rational linear relations among these zeta values. These linear relations are coded in the structure of the associated infinite dimensional k-algebra $\mathcal{A}$, and the analogue of the conjectures of Goncharov, Zagier indeed hold.

We have more precisely

$$\mathcal{A} = k \oplus \bigoplus_{\mathfrak{w}=1}^{\infty} (\mathcal{PZ}_{\mathfrak{w}})_0, \quad \bar{\mathcal{A}} \cong \mathcal{A} \otimes_k \bar{k}.$$

This k-algebra $\mathcal{A}$ is graded by weights $\mathfrak{w} \in \mathbb{N}$. For each $\mathfrak{w}$, we take a maximal k-linear independent subset of $\mathcal{PZ}_{\mathfrak{w}}$, then take the union $\mathcal{B}$ of all these subsets as weight varies. It follows that $\mathcal{B}$ gives a transcendence basis for $\mathcal{A}$ over k. All the k-linear relations among MZVs generate a homogeneous ideal inside the infinite and graded polynomial ring $k[\mathcal{B}]$. Thus this algebra $\mathcal{A}$ corresponds to a very interesting "infinite"weighted projective variety. The collection of linear relations inside each $\mathcal{PZ}_{\mathfrak{w}}$ reflects the geometry of this weighted projective variety coming from the MZVs. This points to another story.

All the proofs about LA-sets described in this Section are essentially based on a "Motivic Transcendence Theory"in positive characteristic, as developed in the last 3 decades by : Anderson, Brownawell, Chang, Papanikolas, Thakur, and Yu.

5 On Special Values in Arithmetic Geometry

We introduce the terminology: LA-sets and graded LA-sets with respect to triple (K_v, k, v), where k is a global field with v a selected place, and K_v is a universal domain containing k. From the given arithmetico-geometric structure defined over k, one usually carries out natural v-adic analytic processes, such as taking "periods", or forming L-series, which results in "special"values inside the domain K_v. These special values are supposed to tell about arithmetic geometry of the structure in question. Particularly we want to know the non-vanishing, the transcendence, and the algebraic independence among the special values confronting us. We even try to determine all the algebraic relations between a set of chosen special values.

For special L-values at 1 or Riemann zeta values at positive even integers, it is the old story, involving only the usual logarithms or π. In recent years, great efforts and progress have been made on understanding special L-values or zeta values taken at integers greater than 1. In particular, various conjectures have been successfully spelled out in expressing the zeta or L-values at integers as linear combinations of **polylogarithms** taking on specific algebraic points (Characteristic 0: works of Zagier, Goncharov... Positive characteristic: works of Anderson, Papanikolas, Pellarin, Taelman, ...). Once sufficiently nice formulas are at hands, e.g. from "Log-Algebraicity"in positive characteristic, the LA-sets and graded LA-sets in question will reveal the nature of these special values.

Acknowledgment

We would like to thank D. Brownawell for carefully reading a preliminary version of this paper and gives valuable comments.

References

[A86] Anderson, G. W., t-motives, Duke Math. J. 53 (1986), 457-502.

[A94] Anderson, G. W., Rank one elliptic A-modules and A-harmonic series, Duke Math. J. 73(1994), No. 3, 491-542.

[A96] Anderson, G. W., Log-algebraicity of twisted A-harmonic series, and special values of L-series in characteristic p, JNT 60(1996, No.1, 165-209.

[ABP] Anderson, G.W., Brownawell, W.D. and Papanikolas, M.A., Determination of the algebraic relations among special Γ-values in positive characteristic, Ann. of Math. (2) 160 (2004), 237-313..

[AT90] Anderson, G.W. and Thakur, D. S., Tensor powers of the Carlitz module and zeta values, Ann. of Math. (2) 132 (1990), No.1, 159-191.

[AT09] Anderson, G.W. and Thakur, D. S., Multizeta values for $\mathbf{F}_q[t]$, their period interpretation, and relations between them, IMRN (2009), No.11, 2038-2055.

[Andre] André, Y., *Une Introduction aux Motifs*, Société Mathématique de France 2004.

[B1] Baker, A., *Transcendental Number Theory*, Cambridge University Press 1975.

[Ba-Wu] Baker, A. and Wüstholz, G., *Logarithmic Forms and Diophantine Geometry*, Cambridge University Press 2007.

[B2] Baker, A., Linear forms in the logarithms of algebraic numbers I, Mathematika 13 (1966), 204-216.

[Car] Carlitz, L., On certain functions connected with polynomials in a Galois field, Duke Math. J., 1(1935), No.2, 137-168.

[C14] Chang, C.-Y., Linear independence of monomials of multizeta values in positive characteristic, Compositio Math., 150(2014), 1789-1808.

[CEP] Chang, C. Y., El-Guindy, A., and Papanikolas, M., Log-algebraic identities on Drinfeld modules and special L-values, preprint 2017.

[CM] Chang, C.-Y. and Mishiba,Y., On multiple polylogarithms in characteristic p : v-adic vanishing versus ∞-adic Eulerianness, to appear in IMRN 2017.

[CP] Chang, C.-Y. and Papanikolas, M.A. Algebraic Independence of Periods and Logarithms of Drinfeld Modules, Jour. Amer. Math. Soc., 25(2012), No.1, p.123-150.

[CPY] Chang, C.-Y., Papanikolas, M.A. and Yu, J., An effective criterion for Eulerian multizeta values in positive characteristic, to appear in Jour. European Math. Soc. 2017.

[CY07] Chang, C.-Y. and Yu, J., Determination of algebraic relations among special zeta values in positive characteristic, Adv. Math. 216(2007), No.1, 321-345.

[Goss] Goss, D. *Basic Structures of Function Field Arithmetic*, Springer-Verlag 1996.

[Papa] Papanikolas, M., Tannakian duality for Anderson-Drinfeld motives and algebraic independence of Carlitz logarithms, Invent. Math., 171(2008), No.1, 123-174.

[Papa2] Papanikolas, M., *Log-Algebraicity of the tensor powers of the Carlitz Module and special values of Goss L-Functions*, preprint 2016.

[Pell1] Pellarin, F. Aspects de l'indépendance algébrique en caractéristiques non nulle, Sém. Bourbaki, vol 2006/2007, Astérisque 317(2008), No. 973, viii, 205-242.

[Pell2] Pellarin, F., Values of certain L-series in positive characteristic, Ann. of Math. (2) 176(2012), No. 3, 2055-2093.

[Sprin] Sprinžugk, V. G., Concerning Baker's theorem on linear forms in logarithms, Dpk;. Akad. Nauk 11(1967), 767-769.

[Ta12] Taelman, L., Special L-values of Drinfeld modules, Ann. of Math. (2) 175(2012), No. 1, 369-391.

[T04] Thakur, D.S., *Function Field Arithmetic*, World Scientific Publishing, River Edge NJ, 2004.

[Wald] Waldschmidt, M., Elliptic Functions and Transcendence, *Surveys in Number Theory*, Dev. Math. 17 (2008), 143-188, Springer.

[Washington] Washington, L., *Introduction to Cyclotomic Fields*, 1997 Springer GTM.

[WW] Wolfart, J. and Wüstholz, G., Der Überlagerungsradius gewisser algebraischer Kurven und die Werte der Betafunktion an rationalen Stellen, Math. Ann., 273(1985), 1-15.

[Yu91] Yu, J., Transcendence and special zeta values in characteristic p, Ann. of Math. (2) 134 (1991), No.1, 1-23.

[Yu97] Yu, J., Analytic homomorphisms into Drinfeld modules, Ann. of Math. (2) 145 (1997), No.2, 215-233.

[Zagier] Zagier, D., Polylogarithms, Dedekind Zeta Functions, and the Algebraic K-Theory of Fields, *Arithmetic algebraic geometry* (Texel, 1989), 391-430, Birkhäuser, 1991.

[Zhao] Zhao, J., Multiple Zeta Functions, Multiple Polylogarithms and Their Special Values, World Scientific Publishing, 2015.

Department of Mathematics, Taiwan University
E-mail address: yu@math.ntu.edu.tw

Proceedings of the 7th International
Congress of Chinese Mathematicians
ALM 43, pp. 521–536

On Faltings Heights of Abelian Varieties with Complex Multiplication

Xinyi Yuan

Abstract

This expository article introduces some conjectures and theorems related to the Faltings heights of abelian varieties with complex multiplication. The topics include the Colmez conjecture, the averaged Colmez conjecture, and the André–Oort conjecture.

Mathematics Subject Classification: 11G15, 11G18, 14G10, 14G40.

Keywords and Phrases: Faltings height, abelian variety with complex multiplication, CM point, Shimura curve, Hodge bundle, Colmez conjecture, André–Oort conjecture, logarithmic derivative, Artin L-function.

1 Introduction

The celebrated Gross–Zagier formula is an equality between the central derivative of certain Rankin-Selberg L-function of modular form of weight two and the Néron–Tate height of certain Heegner cycle on an modular curve. It is an astonishing identity of type "*derivative* = *height*."

In this survey, we introduce Colmez's conjectural equality between the Faltings height of a CM abelian variety and the logarithmic derivative of the relevant Artin L-function, which is another identity of type "*derivative* = *height*." We will also introduce some known results related to this conjecture. The topics of this survey include the following:

(1) the definition of Faltings height by Faltings [Fa],
(2) the Colmez conjecture by Colmez [Co],
(3) the averaged Colmez conjecture proved by Yuan–Zhang [YZ] and Andreatta–Goren–Howard–Madapusi-Pera [AGHM] independently.
(4) the proof of the André–Oort conjecture assuming the averaged Colmez conjecture by Tsimerman [Ts].

Acknowledgements. This article is written for the International Congress of Chinese Mathematicians 2016 held in Beijing. The author would like to thank the hospitality of the Morningside Center of Mathematics. The author is supported by the NSF grant DMS-1601943.

2 Faltings heights

To define the Faltings height of an abelian variety over a number field, we start with the arithmetic degree of a hermitian line bundle in the setting of Arakelov geometry.

2.1 Arithmetic degrees

Let F be a number field, and denote by O_F its ring of integers. Recall that a line bundle $\mathcal{L}$ (i.e. invertible sheaf) over $\mathrm{Spec}(O_F)$ corresponds to an O_F-module, locally free of rank one. By abuse of notations, we also write $\mathcal{L}$ for this O_F-module.

By a *hermitian line bundle* over $\mathrm{Spec}(O_F)$, we mean a pair $\overline{\mathcal{L}} = (\mathcal{L}, \|\cdot\|)$, consisting of a line bundle $\mathcal{L}$ over $\mathrm{Spec}(O_F)$ and a collection $\|\cdot\| = \{\|\cdot\|_\sigma\}_\sigma$, where $\|\cdot\|_\sigma$ is a metric of the complex line $\mathcal{L}_\sigma = \mathcal{L} \otimes_\sigma \mathbb{C}$ for each embedding $\sigma : F \to \mathbb{C}$. The metrics are required to be invariant under the action of the complex conjugate.

The *arithmetic degree* of the hermitian line bundle $\overline{\mathcal{L}} = (\mathcal{L}, \|\cdot\|)$ is defined to be

$$\widehat{\deg}(\overline{\mathcal{L}}) = \log \#(\mathcal{L}/sO_F) - \sum_{\sigma:F\to\mathbb{C}} \log \|s\|_\sigma.$$

Here $s \in \mathcal{L}$ is a nonzero element, and we will see that the definition is independent of the choice of s. In fact, for any (nonzero) prime ideal $\wp \subset O_F$, define a $F_\wp$-metric on $\mathcal{L} \otimes_{O_F} O_{F_\wp}$ by setting $\|s\|_\wp = N(\wp)^{-\mathrm{ord}_\wp(s)}$. Then the definition is equivalent to the more symmetric formula

$$\widehat{\deg}(\overline{\mathcal{L}}) = -\sum_{\wp \subset O_F} \log \|s\|_\wp - \sum_{\sigma:F\to\mathbb{C}} \log \|s\|_\sigma.$$

Then we see that the definition is independent of the choice of $s \in \mathcal{L}$ since a different choice is of the form as for some $a \in F^\times$, which does not change the arithmetic degree by the product formula of absolute values of F.

One can take inverse and tensor products of hermitian line bundles, and the arithmetic degree is additive under these operations.

2.2 Faltings heights

Let A be an abelian variety of dimension g over a number field F. Denote by O_F the integer ring of F as before. Let $\mathcal{A}$ be the Néron model of A over O_F. The *Hodge bundle* of A is defined to be

$$\omega_{\mathcal{A}} = \epsilon^* \Omega^g_{\mathcal{A}/O_F} = \pi_* \Omega^g_{\mathcal{A}/O_F},$$

where $\pi : \mathcal{A} \to \mathrm{Spec}(O_F)$ denotes the structure morphism and $\epsilon : \mathrm{Spec}(O_F) \to \mathcal{A}$ denotes the identity section. Then $\omega_{\mathcal{A}}$ is a line bundle over $\mathrm{Spec}(O_F)$.

There is a *canonical hermitian metric* $\|\cdot\| = \{\|\cdot\|_\sigma\}_\sigma$ on $\omega_{\mathcal{A}}$ given by

$$\|\alpha\|^2_\sigma = \frac{1}{(2\pi)^g} \left| \int_{A_\sigma(\mathbb{C})} \alpha \wedge \overline{\alpha} \right|,$$

where $\sigma : F \to \mathbb{C}$ is any embedding and

$$\alpha \in \omega_{\mathcal{A}} \otimes_\sigma \mathbb{C} = \Gamma(A_\sigma(\mathbb{C}), \Omega^g_{A_\sigma(\mathbb{C})/\mathbb{C}})$$

is any global holomorphic g-form on the complex torus $A_\sigma(\mathbb{C})$. Note that we have used the normalizing factor $1/(2\pi)^g$, while the normalizing factor $1/2^g$ was used by Faltings [Fa, §3].

The pair $\overline{\omega}_{\mathcal{A}} = (\omega_{\mathcal{A}}, \|\cdot\|)$ forms a hermitian line bundle over $\mathrm{Spec}(O_F)$. Define the *unstable Faltings height* of A to be

$$h'(A) = \frac{1}{[F:\mathbb{Q}]} \widehat{\deg}(\overline{\omega}_{\mathcal{A}}).$$

We put the word "unstable" in the name because the height may decrease under base change unless A has a semistable reduction over F. Therefore, define the *stable Faltings height* of A to be

$$h(A) = h'(A_{F'})$$

where F' is a finite extension of F such that $A_{F'}$ has everywhere semistable reduction over $O_{F'}$. The definition does not depend on the choice of F', and is invariant under base change.

The Colmez conjecture is about the stable Faltings height. For simplicity, when we say "Faltings height" in the following, we always mean "stable Faltings height."

Remark 2.1. Let A be an abelian variety of dimension g over a number field F. By replacing F by a finite extension if necessary, we can assume that A has semistable reduction over F and that the class of $\omega_{\mathcal{A}}$ in $\mathrm{Cl}(O_F)$ is trivial. Thus we can write $\omega_{\mathcal{A}} = O_F\alpha$ for some $\alpha \in \Gamma(\mathrm{Spec}(O_F), \omega_{\mathcal{A}}) = \Gamma(\mathcal{A}, \Omega^g_{\mathcal{A}/O_F})$. If $g = 1$, α is the usual Néron differential over the elliptic curve. In terms of α, the Faltings height is just

$$h(A) = -\frac{1}{2[F:\mathbb{Q}]} \sum_{\sigma: F \to \mathbb{C}} \log \left(\frac{1}{(2\pi)^g} \left| \int_{A_\sigma(\mathbb{C})} \alpha \wedge \overline{\alpha} \right| \right).$$

It is a rational linear combination of the logarithms of periods in the sense of Kontsevich–Zagier [KZ].

3 The Colmez conjecture

The goal of this section is to introduce the Colmez conjecture. We have already defined Faltings height, so the major work here is to describe the Artin L-function appeared in the conjecture. We will also recall the classical Chowla–Selberg formula, which can be considered as a special case of the Colmez conjecture.

3.1 Abelian varieties with complex multiplication

We first recall the basics of abelian varieties with complex multiplications. Most results below can be found in [Mi, §10]. For a complete treatment, we refer to Shimura [Sh].

Let E/F be an CM extension; i.e., F is a totally real number field, and E is a totally imaginary quadratic field extension of F. Denote $g = [F : \mathbb{Q}]$ and $c : E \to E$ the nontrivial element of $\mathrm{Gal}(E/F)$.

By a *CM type* of E, we mean a subset Φ of $\mathrm{Hom}(E, \mathbb{C})$ such that

$$\Phi \cap \Phi^c = \emptyset, \quad \Phi \cup \Phi^c = \mathrm{Hom}(E, \mathbb{C}).$$

In other words, Φ is a subset of $\mathrm{Hom}(E, \mathbb{C})$ which picks exactly one element in each conjugate pair of elements of $\mathrm{Hom}(E, \mathbb{C})$. We also say that (E, Φ) is a *CM type*.

By *an abelian variety with complex multiplication by* E, we mean an abelian variety A of dimension g over $\mathbb{C}$, together with an injection $i : E \to \mathrm{End}_{\mathbb{C}}(A) \otimes_{\mathbb{Z}} \mathbb{Q}$ of rings. One can verify that there is a unique CM type Φ of E such that

$$\mathrm{tr}(i(\alpha)|_{\mathrm{Lie}(A)}) = \sum_{\sigma \in \Phi} \sigma(\alpha), \quad \forall \alpha \in E.$$

In this case, we also say that A is *of CM type* (E, Φ). If moreover the maximal order O_E acts on A, i.e., $i(O_E) \subset \mathrm{End}_{\mathbb{C}}(A)$, then we say that A is *of CM type* (O_E, Φ).

For convenience, for any subfield $k \subset \mathbb{C}$, we say that an abelian variety A over k has complex multiplication by E (resp. is of CM type (E, Φ), of CM type (O_E, Φ)) if the base change $A_{\mathbb{C}}$ over $\mathbb{C}$ satisfies this property.

The following are some basic results of the theory of abelian varieties with complex multiplication:

(1) Let (E, Φ) be a CM type with $[E : \mathbb{Q}] = 2g$. Let $O_E \to \mathbb{C}^g$ be the injection given by the (distinct) elements of Φ. Then O_E is a lattice of $\mathbb{C}^g$ under this injection, and the quotient $\mathbb{C}^g/O_E$ is an abelian variety with CM type (O_E, Φ).
(2) Any two abelian varieties of the same CM type (E, Φ) are isogenous over $\mathbb{C}$.
(3) Any abelian variety with complex multiplication is defined over a number field.
(4) Any abelian variety with complex multiplication over a number field has potentially good reduction everywhere.

We refer to [Mi, §10] for the proofs.

Let A be an abelian variety of CM type (O_E, Φ) over $\mathbb{C}$. Then A is defined over a number field H, so it makes sense to consider the Faltings height $h(A) \in \mathbb{R}$, which is independent of the choice of H. By [Co, Theorem 0.3(ii)], the height $h(A)$ depends only on the CM type Φ. We denote it by $h(A_\Phi)$ to emphasize the dependence on Φ.

3.2 Colmez conjecture

Colmez [Co] associates an Artin character to the CM type (E,Φ) as follows. Denote by $\mathbb{Q}^{\mathrm{CM}}$ the compositum of all CM fields over $\mathbb{Q}$, By [Sh, Lem. 18.2], the compositum of two CM fields is also a CM field, and the Galois closure of a CM field is also a CM field. Therefore, $\mathbb{Q}^{\mathrm{CM}}$ is Galois over $\mathbb{Q}$. Denote $G=\mathrm{Gal}(\mathbb{Q}^{\mathrm{CM}}/\mathbb{Q})$. The complex conjugate c is an element in the center of G. Fix inclusions $\mathbb{Q}^{\mathrm{CM}}\subset\overline{\mathbb{Q}}\subset\mathbb{C}$. Then G acts on $\mathrm{Hom}(E,\mathbb{C})=\mathrm{Hom}(E,\mathbb{Q}^{\mathrm{CM}})$ by its action on $\mathbb{Q}^{\mathrm{CM}}$. Define a function $A^0_{E,\Phi}:G\to\mathbb{C}$ by

$$A^0_{E,\Phi}(\sigma)=\int_G|\Phi\cap(\tau\sigma\tau^{-1}\Phi)|d\tau.$$

Here the integral uses the Haar measure on G with total volume 1, and the integrand is the order of the intersection of the two subsets of $\mathrm{Hom}(E,\mathbb{Q}^{\mathrm{CM}})$. The integral is essentially a finite sum by compactness, and $A^0_{E,\Phi}$ takes only finitely many values. See [Co, §0.4] for an equivalent definition of $A^0_{E,\Phi}$.

By definition, we have

$$A^0_{E,\Phi}(\sigma)+A^0_{E,\Phi}(c\sigma)=g,\quad \sigma\in G.$$

Define a function $f_\Phi:G\to\mathbb{C}$ by

$$f_\Phi(\sigma)=A^0_{E,\Phi}(\sigma)-\frac{1}{2}g.$$

Then we have

$$f_\Phi(\sigma)+f_\Phi(c\sigma)=0,\quad \sigma\in G.$$

By definition, f_Φ is a class function on G, i.e., it is a function on the conjugacy classes of G. One can write f_Φ uniquely as a linear combination of characters associated to finite-dimensional irreducible representations π of G over $\mathbb{C}$, and then use these irreducible representations to define L-functions. Alternatively, for the sake of analytic continuation, by Brauer's induction theorem, we can write

$$f_\Phi=\sum_{(M,\chi)}a_{(M,\chi)}\mathrm{Ind}^G_{\mathrm{Gal}(\mathbb{Q}^{\mathrm{CM}}/M)}(\chi),\quad a_{(M,\chi)}\in\mathbb{C}.$$

Here the pair (M,χ) consists of a finite extension M of $\mathbb{Q}$ in $\mathbb{Q}^{\mathrm{CM}}$ and a continuous homomorphism $\chi:\mathrm{Gal}(\mathbb{Q}^{\mathrm{CM}}/M)\to\mathbb{C}^\times$. The summation has only finitely many nonzero terms. By abuse of notation, $\mathrm{Ind}^G_{\mathrm{Gal}(\mathbb{Q}^{\mathrm{CM}}/M)}(\chi)$ means the character of the induced representation in the summation. Define

$$Z(f_\Phi,s)=\sum_{(M,\chi)}a_{(M,\chi)}\frac{L'(\chi,s)}{L(\chi,s)},\qquad \log\mu(f_\Phi)=\sum_{(M,\chi)}a_{(M,\chi)}\log\mu_\chi.$$

Here μ_χ is the conductor of χ for the functional equation. The definition is independent of the choice of the decomposition of f_Φ.

Finally, the Colmez conjecture ([Co, Thm. 0.3, Conj. 0.4]) is as follows:

Conjecture 3.1 (Colmez conjecture). *For any CM type* (E,Φ),

$$h(A_\Phi) = -Z(f_\Phi, 0) - \frac{1}{2}\log \mu(f_\Phi).$$

The above equation is equivalent to the conjectural formula of [Co, Thm. 0.3, Conj. 0.4] on the Faltings height. In fact, to convert our equation to the one in the loc. cit., use

$$Z(f_\Phi, 0) = Z(A^0_{E,\Phi}, 0) - \frac{1}{2}g \cdot \frac{\zeta'(0)}{\zeta(0)} = Z(A^0_{E,\Phi}, 0) - \frac{1}{2}g\log(2\pi).$$

This gives an extra term $\frac{1}{2}g\log(2\pi)$ on the right-hand side, which is also equal to the difference of the Faltings heights due to the different normalizing factors of Hermitian metrics.

Now we check that the function $Z(f_\Phi, s)$ is holomorphic at $s = 0$, so that $Z(f_\Phi, 0)$ is well-defined. Go back to the decomposition

$$f_\Phi = \sum_{(M,\chi)} a_{(M,\chi)} \mathrm{Ind}^G_{\mathrm{Gal}(\mathbb{Q}^{\mathrm{CM}}/M)}(\chi).$$

Note that M is either a CM field or a totally real field by [Sh, Lem. 18.2]. We first claim that we can choose a decomposition such that all M are totally real. In fact, if M is a CM field, denote by M_+ its maximal totally real subfield. Then we have an isomorphism

$$\mathrm{Gal}(\mathbb{Q}^{\mathrm{CM}}/M_+) = \mathrm{Gal}(\mathbb{Q}^{\mathrm{CM}}/M) \times \{1, c\}.$$

As a consequence, the 2-dimensional representation $\mathrm{Ind}^{\mathrm{Gal}(\mathbb{Q}^{\mathrm{CM}}/M_+)}_{\mathrm{Gal}(\mathbb{Q}^{\mathrm{CM}}/M)}(\chi)$ is a direct sum of two 1-dimensional characters. Thus we can replace M by M_+ by using these two characters. Now we assume that every M in the decomposition is totally real. By $f_\Phi(c\sigma) = -f_\Phi(\sigma)$ and $\chi(c\sigma) = \chi(c)\chi(\sigma)$, we can assume that $\chi(c) = -1$ for every χ by removing the ones with $\chi(c) = 1$. It is reduced to the classical statement that *if N is a totally imaginary abelian extension of a totally real field M and $\chi : \mathrm{Gal}(N/M) \to \mathbb{C}^\times$ is a character with $\chi(c) = -1$, then $L(\chi, s)$ is holomorphic and nonzero at $s = 0$.* Note that $L(\chi, s)$ is holomorphic and nonzero at $s = 1$ by decomposing $\zeta_M(s)$ into the product of $L(\omega, s)$ for all characters ω of $\mathrm{Gal}(N/M)$. This gives the result at $s = 0$ by the functional equation, since the gamma factor is holomorphic and nonzero at $s = 0, 1$.

The conjecture is known in the the following cases. If $F = \mathbb{Q}$, the conjecture is equivalent to the classical Chowla–Selberg formula [CS] proved in 1949, which will be discussed in the next subsection. If $E/\mathbb{Q}$ is abelian, the conjecture was proved by Colmez [Co] up to rational multiples of $\log 2$, and the rational multiples were eliminated by Obus [Ob] later. If $[E : \mathbb{Q}] = 4$, the conjecture was proved by Yang [Ya1, Ya2].

In history, there are a lot of works to prove, reformulate or generalize the Chowla–Selberg formula. A very interesting geometric proof of the Chowla–Selberg formula was discovered by Gross [Gr]. He also made a conjecture with

Deligne for the periods of motives with complex multiplication by an abelian field. Anderson [An] reformulated the conjecture of Deligne and Gross in terms of the logarithmic derivatives of odd Dirichlet L-functions at $s = 0$. All these treatments were only up to algebraic numbers. Finally, motivated by a far-reaching conjectural product formula from the p-adic Hodge theory, Colmez [Co] used the Faltings height, instead of just the archimedean periods, to formulate the precise conjecture.

3.3 Chowla–Selberg formula

Here we review the classical Chowla–Selberg formula following the treatment of Weil [We]. Let E be an imaginary quadratic field of discriminant $-d < 0$, class number h, and $w = |O_E^\times|$. Fix an embedding $E \subset \mathbb{C}$. For any fractional ideal $\mathfrak{a}$ of O_E, denote

$$F(\mathfrak{a}) = \Delta(\mathfrak{a})\Delta(\mathfrak{a}^{-1}) = N(\mathfrak{a})^{12}|\Delta(\mathfrak{a})|^2,$$

where

$$\Delta(\mathfrak{a}) = g_2(\mathfrak{a})^3 - 27g_3(\mathfrak{a})^2$$

is the modular discriminant of $\mathfrak{a}$. By definition, $F(\mathfrak{a})$ depends only on the ideal class of $\mathfrak{a}$. In the case $\mathfrak{a} = \omega_1\mathbb{Z} + \omega_2\mathbb{Z}$ with $\tau = \omega_1/\omega_2 \in \mathcal{H}$, one has

$$F(\mathfrak{a}) = d^{-6}(2\mathrm{Im}(\tau))^{12}|\Delta(\tau)|^2,$$

where

$$\Delta(\tau) = (2\pi)^{12} q \prod_{n=1}^{\infty}(1-q^n)^{24}, \quad q = e^{2\pi i\tau}.$$

The Chowla–Selberg formula takes many classical forms. One form in [We, p. 92] is

$$\prod_{\mathfrak{a}\in \mathrm{Cl}(O_E)} F(\mathfrak{a}) = \left(\frac{2\pi}{d}\right)^{12h} \prod_{i=1}^{d-1} \Gamma\left(\frac{i}{d}\right)^{6w\eta(i)}.$$

In the last line of [We, p. 91], the formula takes the form

$$\frac{\zeta_E'(0)}{\zeta_E(0)} = \frac{1}{12h}\sum_{\mathfrak{a}\in \mathrm{cl}} \log F(\mathfrak{a}).$$

This is close to the setting of the current paper. In fact, by $\zeta'(0)/\zeta(0) = \log(2\pi)$, the left-hand side has the main term $L'(0,\eta)/L(0,\eta)$. Here $\eta : (\mathbb{Z}/d\mathbb{Z})^\times \to \{\pm 1\}$ is the quadratic character associated to E. By the Kodaira spencer map on the modular curve $X_0(1)$, the right-hand side is the sum of $-2h(A)$ and some minor terms, where A is an elliptic curve with CM by O_E. Therefore, the Chowla–Selberg formula is equivalent to

$$h(A) = -\frac{1}{2}\frac{L'(\eta,0)}{L(\eta,0)} - \frac{1}{4}\log(d).$$

Here $L(\eta, s)$ is the usual Dirichlet L-function (without the gamma factors).

Example 3.2. If $E = \mathbb{Q}(i)$, then the Chowla–Selberg formula is equivalent to

$$\Delta(i) = \pi^{-6}\,\Gamma(\frac{1}{4})^{24}.$$

Remark 3.3. Recall that the class number formula gives

$$L(0,\eta) = 2h/w.$$

The Chowla–Selberg formula can be viewed as an arithmetic version of this formula. The relation of these two formulas is like the relation between the Waldspurger formula and the Gross–Zagier formula stated in [YZZ, §1.3-1.4].

4 The averaged Colmez conjecture

The Colmez conjecture is far-reaching due to the generality of the CM type. For example, the Artin L-function in the conjecture is very mysterious. However, after we average the equality in the conjecture over all CM types Φ of a fixed CM extension E/F, the right-hand side is essentially given by the logarithmic derivative of the L-function of the quadratic character associated to the quadratic extension E/F. More precisely, we have the following result.

Theorem 4.1 (averaged Colmez conjecture). *Let E/F be a CM extension, η be the corresponding quadratic character of $\mathbb{A}_F^\times/F^\times$ associated to the extension E/F, and d_F (resp. $d_{E/F}$) be the absolute discriminant of F (resp. the norm of the relative discriminant of E/F). Then*

$$\frac{1}{2^g}\sum_{\Phi} h(A_\Phi) = -\frac{1}{2}\frac{L'(\eta,0)}{L(\eta,0)} - \frac{1}{4}\log(d_{E/F}d_F),$$

where Φ runs through the set of all CM types of E/F. Here $L(\eta,s)$ is the L-function without the Gamma factors.

The averaged formula was explicitly conjectured in [Co, p. 634]. It is proved independently by Yuan–Zhang [YZ] and Andreatta–Goren–Howard–Madapusi-Pera in [AGHM]. These two proofs are very different. The proof of [AGHM] is based on the idea of Yang [Ya1, Ya2] (on the Colmez conjecture for $g = 2$) and computes arithmetic intersection numbers over high-dimensional Shimura varieties of orthogonal type. The proof of [YZ] is inspired by the work of Yuan–Zhang–Zhang [YZZ] (on the Gross–Zagier formula) and computes arithmetic intersection numbers over Shimura curves.

In the following, we will sketch the main ideas of the proof of [YZ]. The proof will take two separate steps. The first step expresses the averaged Faltings height in terms of the height of a single CM point on a quaternionic Shimura curve. The second step proves that the height of the single CM point is given by the expected logarithmic derivative.

4.1 Step 1: averaged Faltings height vs. quaternionic height

Let E/F be the CM extension as before. Take a quaternion algebra B over F with an embedding $E \to B$ over F. Assume that B is indefinite at exactly one archimedean place of F. Let $U \subset B^\times(\mathbb{A}_f)$ be a maximal compact subgroup containing the image of $\widehat{O}_E^\times$ in $B^\times(\mathbb{A}_f)$. Consider the *Shimura curve*

$$\mathrm{Sh}(B^\times, U) = B^\times \backslash \mathcal{H}^\pm \times B^\times(\mathbb{A}_f)/U.$$

By the theory of canonical models (cf. [De]), $\mathrm{Sh}(B^\times, U)$ descends to a smooth curve X over F. Assume that B is not a matrix algebra over F so that X is projective over F. Based on [BC, Ca], we have a canonical integral model $\mathcal{X}$ of X over O_F. Note that X is the the coarse moduli space of the corresponding Deligne–Mumford stack, and our treatment can be also written in terms of the stack.

Let $\overline{\mathcal{L}} = (\mathcal{L}, \|\cdot\|)$ be the arithmetic Hodge bundle of $\mathcal{X}$, which is a hermitian $\mathbb{Q}$-line bundle in the setting of Arakelov geometry. Here $\mathcal{L}$ is the relative dualizing sheaf of $\mathcal{X}$ over O_F modified by some "ramification divisor," and the hermitian metric at an archimedean place $\sigma : F \to \mathbb{C}$ is the Petersson metric

$$\|d\tau\|_\sigma = 2\,\mathrm{Im}(\tau)$$

in terms of the complex uniformization of the Shimura curve.

Let $P \in X(E^{\mathrm{ab}})$ be the algebraic point of X represented by the pair $[\tau_0, 1]$ in terms of the complex uniformization, where $\tau_0 \in \mathcal{H}$ is the unique fixed point of $E^\times$. The *height* of P is defined by

$$h_{\overline{\mathcal{L}}}(P) = \frac{1}{[F(P):F]}\widehat{\deg}(\overline{\mathcal{L}}|_{\overline{P}}),$$

where $F(P)$ denotes a field of definition of P and $\overline{P} : \mathrm{Spec}(O_{F(P)}) \to \mathcal{X}$ denotes the extension of P to $\mathcal{X}$. The arithmetic degree is defined as in §2.1. The first step is to prove the following result.

Theorem 4.2. *Let d_B be the norm of the product of finite primes of O_F over which B is ramified. Assume that there is no finite place of F ramified simultaneously in B and E. Then*

$$\frac{1}{2^g}\sum_\Phi h(A_\Phi) = \frac{1}{2}h_{\overline{\mathcal{L}}}(P) - \frac{1}{4}\log(d_B d_F).$$

This theorem is proved by several manipulations of heights in Part I of [YZ]. Note that X is not of PEL type, so it does not parametrize abelian variety naturally.

Let A/H be an abelian variety of CM type (O_E, Φ) over a number field H. Assume that A has semistable reduction over H, which can be achieved by enlarging H if necessary. Then it actually has everywhere good reduction by the CM theory. As a consequence, the Néron model $\mathcal{A}$ of A is an abelian scheme over

O_H. Recall that the Faltings height $h(A)$ is defined to be the arithmetic degree of the Hodge bundle

$$\omega_{\mathcal{A}} = \det \Omega_{\mathcal{A}}.$$

Here

$$\Omega_{\mathcal{A}} = \epsilon^* \Omega^1_{\mathcal{A}/O_H} = \pi_* \Omega^1_{\mathcal{A}/O_H}$$

is a vector bundle on $\mathrm{Spec}(O_H)$ of rank $g = \dim A$, where $\pi : \mathcal{A} \to \mathrm{Spec}(O_H)$ denotes the structure morphism and $\epsilon : \mathrm{Spec}(O_H) \to \mathcal{A}$ denotes the identity section.

We further assume that H contains the Galois closure of E. Considering the action of O_E on $\Omega_{\mathcal{A}}$, we can decompose it into a direct sum of line bundles on $\mathrm{Spec}(O_H)$. Accordingly, we have a decomposition

$$h(A_\Phi) = \sum_{\tau \in \Phi} h(\Phi, \tau)$$

up to some error term coming from ramifications of the number fields involved.

Let (Φ_1, Φ_2) be a nearby pair of CM types in the sense that $|\Phi_1 \cap \Phi_2| = g-1$. Let τ_i be the complement of $\Phi_1 \cap \Phi_2$ in Φ_i for $i = 1, 2$. Define

$$h(\Phi_1, \Phi_2) = \frac{1}{2}(h(\Phi_1, \tau_1) + h(\Phi_2, \tau_2)).$$

It suffices to compute the average of $h(\Phi_1, \Phi_2)$ over all nearby pairs (Φ_1, Φ_2). In [YZ], we have actually proved

$$g \cdot h(\Phi_1, \Phi_2) = \frac{1}{2} h_{\overline{\mathcal{L}}}(P) - \frac{1}{4} \log(d_B).$$

This is surprising since the right-hand side is independent of the choice of (Φ_1, Φ_2). To prove the last formula, $h(\Phi_1, \Phi_2)$ is further reduced to the height of a CM point on the Shimura curve of certain unitary group $U(1,1)$. By some delicate local comparison of the unitary Shimura curve with the quaternionic Shimura curve, we eventually relate the heights on these two Shimura curves.

4.2 Step 2: Compute the quaternionic height

The second step is to prove the following result.

Theorem 4.3. *In the setting of Theorem 4.1 and Theorem 4.2,*

$$h_{\overline{\mathcal{L}}}(P) = -\frac{L'(\eta, 0)}{L(\eta, 0)} + \frac{1}{2} \log \frac{d_B}{d_{E/F}}.$$

This theorem is proved by extending the method of the proof of the Gross–Zagier formula in [YZZ]. Recall that the Gross–Zagier formula is an identity between the derivative of L-series of a Hilbert modular form of parallel weight two and the height of a CM point on a modular abelian variety of GL(2)-type. This formula is proved by a comparison of the analytic kernel $\Pr I'(0, g, \phi)$ and the geometric kernel $2Z(g,(1,1),\phi)$ parametrized by certain modified Schwartz functions $\phi \in \overline{\mathcal{S}}(\mathbb{B} \times \mathbb{A}^\times)$. Here $\mathbb{B}$ is obtained from $B(\mathbb{A})$ by changing every archimedean component to the Hamiltonian algebra. More precisely, we have proved that the difference

$$\mathcal{D}(g,\phi) = \Pr I'(0,g,\phi) - 2Z(g,(1,1),\phi), \qquad g \in \mathrm{GL}_2(\mathbb{A}_F)$$

is perpendicular to cusp forms. The matching of the "main terms" of $\mathcal{D}(g,\phi)$ eventually implies the Gross–Zagier formula; On the other hand, the matching of the "degenerate terms" implies Theorem 4.3.

On the analytic side, the function $I(s,g,\phi)$ is a mixed Eisenstein–theta series, i.e. a finite linear combination of automorphic forms of the form $\theta(g)E(s,g)$. Here the theta series $\theta(g)$ and the Eisenstein series $E(s,g)$ each has weight one. The main contribution of $I'(0,g,\phi)$ to Theorem 4.3 is given by the intertwining part of the constant term of $E'(0,g)$. Via analytic continuation, the intertwining part of $E(s,g)$ is equal to a simple multiple of $L(s,\eta)/L(s+1,\eta)$. Hence, its contribution in $I'(0,g,\phi)$ gives the logarithmic derivative $L'(0,\eta)/L(0,\eta)$. Here the functional equation can be used to convert $L'(1,\eta)/L(1,\eta)$ to $L'(0,\eta)/L(0,\eta)$.

On the geometric side, the series

$$Z(g,(1,1),\phi) = \langle Z(g,\phi)P^\circ, P^\circ\rangle_{\mathrm{NT}}$$

is a Néron-Tate height pairing. Here $P^\circ = P - \xi$, where ξ is the normalized Hodge bundle, i.e., a rational multiple of the Hodge bundle $\mathcal{L}_F$ which has degree one on the geometrically connected component of X containing P. Here $Z(g,\phi)$ is Kudla's generating function, which is an automorphic form with coefficients in $\mathrm{Pic}(X^2)\otimes_{\mathbb{Z}}\mathbb{C}$. A primitive version of the generating function is the classical series $\sum_{N\geq 0} T_N q^N$ of Hecke correspondences on the square of the usual modular curve. In terms of Arakelov geometry, we can compute $\langle Z(g,\phi)P^\circ, P^\circ\rangle_{\mathrm{NT}}$ in terms of arithmetic intersection numbers. The divisor $Z(g,\phi)P$ is literally a sum of CM points of X, and it contains a multiple of P, which is our "degenerate term." The contribution of this degenerate term in $Z(g,(1,1),\phi)$ is a multiple of

$$\langle P, P\rangle = -\widehat{\deg}(\overline{\mathcal{L}}|_{\overline{P}}) = -[F(P):P]\cdot h_{\overline{\mathcal{L}}}(P).$$

Here the left-hand side is certain arithmetic intersection number on $\mathcal{X}$, and the first equality is a version of the arithmetic adjunction formula.

Finally, the matching of these two "degenerate terms" in $\mathcal{D}(g,\phi)$ gives the desired height formula in Theorem 4.3. To get this matching, we have the matching of the "main terms" of $\mathcal{D}(g,\phi)$ by explicit local computations, which is mostly done in [YZZ]. By the modularity of $\mathcal{D}(g,\phi)$ and the theory of pseudo-theta functions in [YZ], the matching of the "main terms" implies the matching of the "degenerate terms."

5 The André–Oort conjecture

By the recent work of Tsimerman [Ts], which is based on the previous works of Pila et al, the averaged Colmez conjecture implies the André–Oort conjecture for the Siegel modular variety. In this section, we give a sketch of this implication.

Conjecture 5.1 (André–Oort conjecture, [And, Oo]). *Let X be a Shimura variety over $\mathbb{C}$. Let $Y \subset X$ be a closed subvariety which contains a Zariski dense subset of special points of X. Then Y is a special subvariety.*

There are enormous progresses on the conjecture made by André, Edixhoven, Klingler, Ullmo, Yafaev, Pila, Zannier, Tsimerman et al. We describe some of them as follows, and refer to [Sc, PT] for more detailed descriptions. Klingler–Ullmo–Yafaev [UY, KY] proved the conjecture assuming the generalized Riemann hypothesis for CM fields. Pila [Pi] proved the conjecture when X is a power of modular curves. Tsimerman [Ts] proved the conjecture when X is a Siegel modular variety (and thus all Shimura varieties of abelian type) assuming the averaged Colmez conjecture (cf. Theorem 4.1). Consequently, the André–Oort conjecture is proved for all Shimura varieties of abelian type.

In the following, we sketch the proof of Tsimerman [Ts]. From averaged Colmez conjecture to the André–Oort conjecture (for the Siegel modular variety), there is an intermediate conjecture proposed by Edixhoven, which we will start with.

5.1 Edixhoven's conjecture implies André–Oort conjecture

Let $\mathcal{S}_g$ be the Siegel modular variety over $\mathbb{Q}$, i.e. the coarse moduli space of principally polarized abelian varieties of dimension g. Let $x \in \mathcal{S}_g(\overline{\mathbb{Q}})$ be a special point. Then x corresponds to a principally polarized abelian variety A_x over $\overline{\mathbb{Q}}$, isogenous to a product of abelian varieties with complex multiplication as defined in §3.1. Denote by R_x the center of the endomorphism algebra $\mathrm{End}_{\overline{\mathbb{Q}}}(A_x)$, which is an order in a direct sum of CM fields. Then it makes sense to talk about the discriminant $\mathrm{disc}(R_x) \in \mathbb{Z}$ of R_x. In [EMO, Problem 14], the following statement is proposed as a question.

Conjecture 5.2 (Edixhoven's conjecture). *There is a positive real number c_g depending only on g such that*

$$[\mathbb{Q}(x) : \mathbb{Q}] \geq |\mathrm{disc}(R_x)|^{c_g}$$

for any special point $x \in \mathcal{S}_g(\overline{\mathbb{Q}})$.

If $g = 1$, then A_x is an elliptic curve with complex multiplication by R_x. In this case, $\mathbb{Q}(x)$ is almost the ring class field of R_x, whose degree is given by the class number of R_x. Then the conjecture follows from the classical Brauer–Siegel theorem (cf. [Br]). If $g > 1$, such an argument does not work since $\mathbb{Q}(x)$ is much more complicated than the ring class field.

The key result of our interest here is Pila–Tsimerman [PT, Theorem 7.1], which asserts that Edixhoven's conjecture implies the André-Oort conjecture for the case $X = \mathcal{S}_g$. The proof of this theorem is built up on the works of Bombieri, Pila, Zannier, Wilkie et al. We refer to [PT, Sc, Ts] for more details, while we sketch some of the ideas here. The setup is to consider the complex uniformization $\pi : \mathcal{H}_g \to \mathcal{S}_g(\mathbb{C})$. Take a "nice" fundamental domain $F_g \subset \mathcal{H}_g$ of $\mathcal{S}_g(\mathbb{C})$ and denote $G_g = F_g \cap \pi^{-1}(Y(\mathbb{C}))$. A special point $x \in Y(\overline{\mathbb{Q}})$ gives roughly $[\mathbb{Q}(x) : \mathbb{Q}]$ special points of $Y(\overline{\mathbb{Q}})$ by the Galois action, and these points lift to algebraic points of G_g of degree at most $2g$. By Conjecture 5.2, they give "many" algebraic points of G_g. This suggests that connected components of G_g "behave like" hermitian symmetric subdomains of $\mathcal{H}_g$. In fact, a fundamental theorem of Pila–Wilkie [PW] implies that G_g contains a semi-algebraic set of $\mathcal{H}_g$. Finally, the hyperbolic Ax-Lindemann theorem proved in [PT] finishes the argument.

Therefore, it remains to prove Edixhoven's conjecture, which is the main theorem of Tsimerman [Ts]. We sketch it in the following.

5.2 Averaged Colmez conjecture implies Edixhoven's conjecture

Tsimerman's proof is a combination of a few important theorems. We describe it in a few steps.

Step 1: reduce to simple case

Note that A_x is isogenous to a product of abelian varieties with complex multiplication by CM fields. It can actually be reduced to a (simple) abelian variety with complex multiplication by a maximal order. This is done in the proof of [Ts, Theorem 5.1]. Then the goal becomes to prove that if E/F is a CM extension and A is an abelian variety of primitive CM type (O_E, Φ), then

$$[\mathbb{Q}(A):\mathbb{Q}] \geq |d_E|^{c_g}$$

Here c_g depends only on $g = [F:\mathbb{Q}]$. Here $\mathbb{Q}(A)$ is the intersection of all subfields of $\mathbb{C}$ to which A can be descended to.

Step 2: bound Faltings height

The goal here is to prove

$$h(A) \leq |d_E|^{o_g(1)}.$$

Assuming the averaged Colmez conjecture in Theorem 4.1, the bound is a consequences of the following bounds:

(1) $0 < L(1,\eta) \leq |d_E|^{o_g(1)}$,
(2) $L'(1,\eta) \leq |d_E|^{o_g(1)}$,
(3) $h(A') \geq O_g(1)$.

Note we can switch $-L'(0,\eta)/L(0,\eta)$ to $L'(1,\eta)/L(1,\eta)$ by the functional equation. Part (1) follows from the classical Brauer–Siegel theorem (cf. [Br]). Part (2) follows form a standard subconvexity bound in analytic number theory. Part (3) holds for any abelian variety A' over $\overline{\mathbb{Q}}$ of dimension g, which is a result of Bost [Bo].

Step 3: isogeny of large degree

We claim that there is an abelian variety B over $\mathbb{C}$ of CM type (O_E, Φ) such that

$$\deg(A,B) \geq |d_E|^{1/4-o_g(1)}.$$

Here $\deg(A,B)$ denotes the smallest value among the degrees of all isogenies between A and B over $\mathbb{C}$. Note that any isogeny between A and B over $\mathbb{C}$ can be descended to a number field. To prove the claim, denote by $S(O_E, \Phi)$ the set of the isomorphism classes of abelian varieties over $\mathbb{C}$ with CM type (O_E, Φ). There is a bijection $\mathrm{Cl}(O_E) \to S(O_E, \Phi)$ sending a fractional ideal $\mathcal{I}$ of O_E to $\mathbb{C}^g/\mathcal{I}$, so

$$|S(O_E,\Phi)| = h_E \geq h_E/h_F \geq |d_E|^{1/4-o_g(1)}.$$

Here the last equality follows from the Brauer–Siegel theorem again. With the lower bound of $|S(O_E, \Phi)|$, the claim can be obtained by an easy argument.

Step 4: isogeny theorem

The last step is of an application of the powerful isogeny theorem of Masser-Wüstholz [MW] from diophantine geometry. We are need the version of the theorem proved in in [GR, Theorem 1.4], which asserts that for any isogenous abelian varieties A and B of dimension g over a number field H, we have

$$\deg(A,B) \leq \max\{h(A), [H:\mathbb{Q}]\}^{O_g(1)}.$$

Combine it with the inequality in Step 3, and note the upper bound of $h(A)$ in Step 2. This finishes the proof.

References

[AGHM] F. Andréatta, E. Goren, B. Howard, and K. Madapusi-Pera, *Faltings heights of abelian varieties with complex multiplication.* arXiv:1508.00178.

[An] G. Anderson, *Logarithmic derivatives of Dirichlet L-functions and the periods of abelian varieties.* Compositio Math. 45 (1982), no. 3, 315-332.

[And] Y. André, *G-functions and Geometry.* Aspects of Mathematics E13, Vieweg, Braunschweig,1989.

[BC] J.-F. Boutot; H. Carayol, *Uniformisation p-adique des courbes de Shimura: les théorèmes de Čerednik et de Drinfel'd.* Courbes modulaires et courbes de Shimura (Orsay, 1987/1988). Astérisque No. 196-197 (1991), 7, 45-158 (1992).

[Bo] J.B. Bost, *Arakelov geometry of abelian varieties.* Proceedings of a conference on arithmetic geometry, technical report, Max-Planck Institut für Mat., Bonn, Vol. 9651 (1996).

[Br] R. Brauer, *On the Zeta-Functions of Algebraic Number Fields.* American Journal of Mathematics, Vol. 69, No. 2, 243–250.

[Ca] H. Carayol, *Sur la mauvaise réduction des courbes de Shimura.* Compositio Mathematica 59.2 (1986): 151-230.

[Co] P. Colmez, *Périodes des variétés abéliennes à multiplicaton complexe*, Annals of Mathematics, 138 (1993), 625-683.

[CS] S. Chowla and A. Selberg, *On Epstein's zeta-function.* J. Reine Angew. Math. 227 1967 86-110.

[De] P. Deligne, *Travaux de Shimura,* Séminaire Bourbaki, 23ème année (1970/71), Exp. No. 389, pp. 123–165. Lecture Notes in Math., Vol. 244, Springer, Berlin, 1971.

[EMO] S.J. Edixhoven, B.J.J. Moonen, F. Oort, editors, Open problems in algebraic geometry, Bull. Sci. Math. 125 (2001), 1–22.

[Fa] G. Faltings, *Endlichkeitssätze für abelsche Varietäten über Zahlkörpern.* Invent. Math. 73 (1983), no. 3, 349–366.

[GR] E. Gaudron, G. Rémond, *Polarisations et isogénies.* Duke Math. J. 163 (2014), 2057–2108.

[Gr] B. Gross, *On the periods of abelian integrals and a formula of Chowla and Selberg.* With an appendix by David E. Rohrlich. Invent. Math. 45 (1978), no. 2, 193–211.

[KY] B. Klingler, A. Yafaev, *The André–Oort conjecture.* Annals of Mathematics 180 (2014), Issue 3, 867–925.

[KZ] M. Kontsevich, D. Zagier, *Periods.* In Mathematics unlimited–2001 and beyond, pages 771–808. Springer, Berlin, 2001.

[Mi] J. S. Milne, *Introduction to Shimura varieties.* Harmonic analysis, the trace formula, and Shimura varieties, 265–378, Clay Math. Proc., 4, Amer. Math. Soc., Providence, RI, 2005.

[MW] D. W. Masser, G. Wüstholz, *Factorization estimates for abelian varieties.* Publications Mathématiques de l'IHÉS, 1995, Vol. 81, 5–24

[Ob] A. Obus, *On Colmez product formula for periods of CM-abelian varieties,* Mathematische Annalen, Volume 356, Issue 2, pp 401–418.

[Oo] F. Oort, *Canonical lifts and dense sets of CM points.* Arithmetic Geometry, Cortona, 1994, 228–234, F. Catanese, editor, Symposia. Math., XXXVII, CUP, 1997.

[Pi] J. Pila, *O-minimality and the André–Oort conjecture for* $\mathbb{C}^n$. Ann. of Math. 173, (2011), 1779–1840.

[PT] J. Pila, J. Tsimerman, *Ax–Lindemann for* A_g. Annals of Mathematics 179 (2014), Issue 2, 659–681.

[PW] J. Pila and A. J. Wilkie, *The rational points of a definable set.* Duke Math. J. 133 (2006), 591–616.

[Sc] T. Scanlon, *A proof of the André–Oort conjecture via mathematical logic [after Pila, Wilkie, and Zanner].* S'eminaire Bourbaki, 1037, 63ème ann'ee 2010–2011.

[Sh] G. Shimura, Abelian varieties with complex multiplication and modular functions, Princeton University Press, 1998.

[Ts] J. Tsimerman, *A proof of the André-Oort conjecture for* A_g. arXiv:1506.01466 [math.NT].

[UY] E. Ullmo and A. Yafaev, *Galois orbits and equidistribution of special subvarieties: towards the André–Oort conjecture.* Annals of Mathematics 180 (2014), Issue 3, 823–865.

[We] A. Weil, *Elliptic functions according to Eisenstein and Kronecker.* Ergebnisse der Math. und ihrer Grenzgebiete 88, Springer-Verlag, 1976.

[Ya1] T. Yang, *An arithmetic intersection formula on Hilbert modular surfaces,* Amer. J. Math., 132 (2010), 1275–1309.

[Ya2] T. Yang, *Arithmetic intersection and Faltings' height.* Asian J. Math., 17 (2013), 335–382.

[YZ] X. Yuan, S. Zhang, *The averaged Colmez conjecture.* arXiv:1507.06903 [math.NT].

[YZZ] X. Yuan, S. Zhang, W. Zhang, *The Gross–Zagier Formula on Shimura Curves.* Annals of Mathematics Studies, No. 184, Princeton University Press, 2012.

Department of Mathematics, University of California, Berkeley CA 94720, USA

E-mail address: yxy@math.berkeley.edu

Advanced Lectures in Mathematics (ALM)

Executive editors: Shing-Tung Yau, Kefeng Liu, and Lizhen Ji

www.intlpress.com/alm

ALM 46: Open Problems in Arithmetic Algebraic Geometry

ALM 45: Tsinghua Lectures in Mathematics

ALM 44: Proceedings of the Seventh International Congress of Chinese Mathematicians, Volume II

ALM 43: Proceedings of the Seventh International Congress of Chinese Mathematicians, Volume I

ALM 42: Uniformization, Riemann-Hilbert Correspondence, Calabi-Yau Manifolds, and Picard-Fuchs Equations

ALM 41: Handbook of Group Actions, Volume IV

ALM 40: Handbook of Group Actions, Volume III

ALM 39: Hodge Theory and L^2-analysis

ALM 38: Complex Geometry from Riemann to Kähler–Einstein and Calabi–Yau

ALM 37: Proceedings of the Sixth International Congress of Chinese Mathematicians, Volume II

ALM 36: Proceedings of the Sixth International Congress of Chinese Mathematicians, Volume I

ALM 35: The Legacy of Bernhard Riemann After One Hundred and Fifty Years, Volumes I and II

ALM 34: Some Topics in Harmonic Analysis and Applications

ALM 33: Introduction to Modern Mathematics

ALM 32: Handbook of Group Actions, Volume II

ALM 31: Handbook of Group Actions, Volume I

ALM 30: Automorphic Forms and L-functions

ALM 29: Selected Expository Works of Shing-Tung Yau with Commentary, Volume II

ALM 28: Selected Expository Works of Shing-Tung Yau with Commentary, Volume I

ALM 27: Number Theory and Related Areas

ALM 26: Handbook of Moduli, Volume III

ALM 25: Handbook of Moduli, Volume II

ALM 24: Handbook of Moduli, Volume I

ALM 23: Recent Developments in Geometry and Analysis

ALM 22: Differential Geometry: Under the Influence of S.-S. Chern

ALM 21: Advances in Geometric Analysis

ALM 20: Surveys in Geometric Analysis and Relativity

ALM 19: Arithmetic Geometry and Automorphic Forms

ALM 18: Geometry and Analysis, No. 2

ALM 17: Geometry and Analysis, No. 1

ALM 16: Transformation Groups and Moduli Spaces of Curves

ALM 15: An Introduction to Groups and Lattices: Finite Groups and Positive Definite Rational Lattices

ALM 14: Handbook of Geometric Analysis, No. 3

ALM 13: Handbook of Geometric Analysis, No. 2

ALM 12: Cohomology of Groups and Algebraic K-theory

ALM 11: Recent Advances in Geometric Analysis

ALM 10: Trends in Partial Differential Equations

ALM 9: Automorphic Forms and the Langlands Program

ALM 8: Recent Developments in Algebra and Related Areas

ALM 7: Handbook of Geometric Analysis, No. 1

ALM 6: Geometry, Analysis and Topology of Discrete Groups

ALM 4: Variational Principles for Discrete Surfaces

ALM 3: Computational Conformal Geometry

ALM 2: Asymptotic Theory in Probability and Statistics with Applications

ALM 1: Superstring Theory

Made in the USA
Monee, IL
12 May 2020

30739803R20306